SOLID WASTES:
Origin, Collection, Processing, and Disposal

SOLID WASTES:
origin, collection, processing, and disposal

C. L. MANTELL, Ph.D.
consulting engineer

Fellow, American Institute of Chemical Engineers
Senior Member, American Institute of Mining and Metallurgical Engineers

A Wiley-Interscience Publication
JOHN WILEY & SONS

New York • London • Sydney • Toronto

Library of Congress Cataloging in Publication Data:

Mantell, Charles Letnam, 1897–
Solid wastes.

"A Wiley-Interscience publication."
Includes bibliographical references.
1. Factory and trade waste. 2. Refuse and refuse disposal. I. Title.

TD897.M36 628′.445 74-26930
ISBN 0-471-56777-9

Printed in the United States of America

10 9 8 7 6 5 4 3 2 1

CONTRIBUTORS

L. K. Arnold
Department of Chemical Engineering
Iowa State University
Ames, Iowa

J. C. Barber
Office of Agricultural and Chemical Development
Tennessee Valley Authority
Muscle Shoals, Alabama

E. W. Bauman
Managing Director
National Slag Association
Silver Spring, Maryland

Thomas W. Bendixen
Environmental Protection Agency
Cincinnati, Ohio

James A. Buckham, Manager
Division of Nuclear and Chemical Technology
Idaho Nuclear Corporation
Idaho Falls, Idaho

Leo J. Cohan
Combustion Engineering, Inc.
Windsor, Connecticut

U. K. Custred
American Cyanamid Company
Bradley, Florida

Mr. J. R. Edwards
Vice President
The American Paper Institute
New York, New York

Robert N. Farvolden, Chairman
Department of Earth Science
University of Waterloo
Ontario, Canada

John H. Fernandes
Senior Project Engineer
Combustion Engineering, Inc.
Windsor, Connecticut

Richard H. Forbes
Eastman Kodak Corporation
Rochester, New York

Glenn H. Glover
Cotton Producers Association
Atlanta, Georgia

James D. Hainley
Ackenheil & Associated
Baltimore Maryland, Inc.
Baltimore, Maryland

Leslie C. Hardison
Vice President
Air Resources, Inc.
Palatine, Illinois

John T. Hays
Technical Information Division
Hercules Inc.
Hercules Research Center
Wilmington, Delaware

Raiford L. Holmes
P.O. Box 908
Bogalusa, Louisiana

J. R. Hoskins, Head
Department of Mining Engineering and Metallurgy
University of Idaho
Moscow, Idaho

Harold B. Jensen
Manager of Purchases
Lukens Steel Company
Coatesville, Pennsylvania

John S. Jeris, Director
Department of Environmental Engineering and Sciences
Manhattan College
Bronx, New York

James Kirkpatrick
244 Columbia Avenue
Park Ridge, Illinois

Alan C. Klein
Regional Manager
Commercial Building Systems
Aluminum Company of America
Lincoln Park, New Jersey

R. J. Logan
Commercial Building Systems
Aluminum Company of America
Lincoln Park, New Jersey

Philip Meikle
Assistant Director
National Ash Association, Inc.
Washington, D. C.

Dwight L. Miller
Northern Regional Research Center
U.S. Department of Agriculture
Agricultural Research Service
Peoria, Illinois

R. D. Mobley
Cotton Producers Association
Talanta, Georgia

Arthur J. Muehling
Cooperative Extension Service
College of Agriculture
University of Illinois
Urbana, Illinois

W. J. Nellen
Product Manager
AMF Microflake Tobacco Products
East Windsor Hill, Connecticut

Rudy G. Novak
Superintendent of Waste-Control Operations
Midland Division
Dow Chemical Company
Midland, Michigan

Bernard O'Reilly
Plant Metallurgist
Nassau Smelting & Refining
Tottenville, Staten Island, New York

Charles Ostrander
Agricultural Extension
New York College of Agriculture
Cornell University
Ithaca, New York

Newton Penny, Head
Department of Agricultural Economics
Georgia Experimental Station
Griffin, Georgia

Richard F. Peters, Chief
Bureau of Vector Control
Division of Environmental Sanitation
California State Department of Public Health
Sacramento, California

Richard A. Pfluger
Maxwell House Division
General Foods Corporation
Hoboken, New Jersey

Norman D. Phillips
Group Manager
Air Pollution Research & Development
Fuller Company
Catasauqua, Pennsylvania

George F. Potter*
American Tung Oil Association
Poplarville, Mississippi

E. B. Pugsley
6902 South Jackson Way
Littleton, Colorado

Eric Quartly
Public Works Director
City of San Diego
San Diego, California

Raymond W. Regan
Department of Environmental Engineering and Sciences
Manhattan College
Bronx, New York

Michael Soderquist
Department of Food Science and Technology
Oregon State University
Corvallis, Oregon

* Current address: 1417 West 10th St., Bogalusa, Louisiana.

Thomas J. Sorg
Environmental Protection Agency
Cincinnati, Ohio

Ralph Stone, President
Ralph Stone & Company, Engineers
Los Angeles, California

Edwin E. Slover
Environmental Protection Department
Union Carbide Corporation
Charleston, West Virginia

P. E. Wehrspann
Projects Engineer
Papio Watershed Board
Omaha, Nebraska

Gary E. Welch
Senior Research Engineer
Zinc Smelting Division
St. Joseph Lead Company
Monaca, Pennsylvania

K. J. Williamson
Department of Food Science and Technology
Oregon State University
Corvallis, Oregon

Ivan A. Wolff
Eastern Regional Research Center
Agricultural Research Service
U.S. Department of Agriculture
Philadelphia, Pennsylvania

PREFACE

The origin, collection, processing, and disposal of solid wastes are problems of modern society, its people, organization, and political structure; these problems were met in earlier civilizations in much the same way as they are met now: by burning or by dumping and burying in the earth or at sea.

The rag picker, the chimney sweep, the scavenger at the dump, the handcart junkman, and the old-clothes man now exist only in the romantic literature of Dickens' era. Today our affluent society turns its eyes away from waste with an "out-of-sight, out-of-mind" attitude; we say "get rid of it but don't let it cost too much."

Recently, we have begun to talk of waste management, devoting our engineering and political acumen to the disposal of waste tonnages equal in volume to thousands of Cheops pyramids or Empire State buildings. Now we refine with engineering skill the methods of dumping, burning, processing, biologically destroying, or burying in the land or sea our mountains of waste, created by an affluent society where each one of us has in power, machinery, services, and comforts, the equivalent of 500 slaves of an ancient pharaoh.

This book contains many statistics. Solid waste from animal husbandry and agriculture, today operated in the manner of productions factories, is many times the trash of our cities. The statistics are astounding; we are overwhelmed by the complexity of the problems—mostly social—of our industrial society. Litter has always been associated with civilizations, and it is a peoples' housekeeping job to eliminate this aesthetic insult to our senses with its accompanying damage to our health and environment.

This book tries to bring the problems of solid wastes into focus, often by case studies of what has been done. Solid wastes can be managed, but we must realize that the dignity of man has eternal vigilance as its price. We cannot revert to the pace and standards of our forefathers when coping with problems of a modern society; and we cannot survive without righting the environmental insults we and those before us have perpetuated.

In preparing this book, we were fortunate to have the help of a large number of authorities in specialized areas, as well as the data of the city, municipal, state, and federal organizations whose work was studied. There were so many such authorities that we have used abbreviations (defined in Appendix 1) where possible; reference should be made to the list of contributors and the list of abbreviations for more complete information.

My heartfelt thanks go to my many friends in industry who cooperated and helped in the publication of this book, and particularly to my wife, Adelaide, for her help and encouragement, without which this book would never have been completed.

A saying of Solomon from the Bible is applicable to waste management: "A little sleep, a little slumber, a little folding of the hands to rest, and so shall thy poverty come."

C. L. MANTELL

Manhasset, New York
January, 1975

CONTENTS

SOLID WASTES:
Origin, Collection, Processing, and Disposal

PART I
origin of solid wastes

INTRODUCTION

Solid wastes have been a result of man's activities from his earliest civilization, made more prominent during the Industrial Revolution, intensified by technological developments of the twentieth century, and given political and economic exposure by the passage of the Solid Waste Disposal Act of 1965, which is presented in Chapter I.1. Chapter I.2 presents some of the major statistical data and discusses urban wastes and their management. Packaging is treated in Chapter I.3 as a waste problem associated with an affluent society.

Chapters 1.4 and 1.5 present the statistical aspects of industrial, agricultural, and mineral waste. Manpower and engineering talent, present and future, are treated statistically in Chapter I.6. Legislation and cooperation of people, agencies, waste producers, and disposers are presented through a case study of Erie County, Pa., in Chapter I.7, as an approach to the waste problems. These will be examined in greater detail later in this volume, to which Part I is really an introduction. References are at the end of Part I.

CHAPTER I.1
the national solid wastes act

Solid wastes were given national attention with the passage of the Solid Waste Disposal Act by the 89th Congress as Public Law 89-272, reprinted here.* (It has since been amended.)† Now all phases, particularly the engineering, planning, sanitation, and health aspects, have funds provided for development and experimental approaches.

Public Law 89-272
89th Congress, S. 306
October 20, 1965

AN ACT

. . . to authorize a research and development program with respect to solid-waste disposal, and for other purposes.

Be it enacted by the Senate and House of Representatives of the United States of America in Congress assembled.

TITLE II—SOLID WASTE DISPOSAL

SHORT TITLE

SEC. 201. This title (hereinafter referred to as "this Act") may be cited as the "Solid Waste Disposal Act".

FINDINGS AND PURPOSES

SEC. 202. (a) The Congress finds—

(1) that the continuing technological progress and improvement in methods of manufacture, packaging, and marketing of consumer products has resulted in an ever-

* Public Law 89-272 includes two separate Acts, "Title I—Amendments to Clean Air Act," and "Title II—Solid Waste Disposal." Title I has been omitted here because it does not relate directly to the program of the Office of Solid Wastes.

† Legislative History: House Report No. 899 (Comm. On Interstate & Foreign Commerce).
Senate Report No. 192 (Comm. on Public Works).
Congressional Record, Vol. 111 (1965): May 18: Considered and passed Senate. Sept. 23: Considered in House. Sept. 24: Considered and passed House, amended. Oct. 1: Senate concurred in House amendments.

mounting increase, and in a change in the characteristics, of the mass of material discarded by the purchaser of such products;

(2) that the economic and population growth of our Nation, and the improvements in the standard of living enjoyed by our population, have required increased industrial production to meet our needs, and have made necessary the demolition of old buildings, the construction of new buildings, and the provision of highways and other avenues of transportation, which, together with related industrial, commercial, and agricultural operations, have resulted in a rising tide of scrap, discarded, and waste materials;

(3) that the continuing concentration of our population in expanding metropolitan and other urban areas has presented these communities with serious financial, management, intergovernmental, and technical problems in the disposal of solid wastes resulting from the industrial, commercial, domestic, and other activities carried out in such areas;

(4) that inefficient and improper methods of disposal of solid wastes result in scenic blights, create serious hazards to the public health, including pollution of air and water resources, accident hazards, and increase in rodent and insect vectors of disease, have an adverse effect on land values, create public nuisances, otherwise interfere with community life and development;

(5) that the failure or inability to salvage and reuse such materials economically results in the unnecessary waste and depletion of our natural resources; and

(6) that while the collection and disposal of solid wastes should continue to be primarily the function of State, regional, and local agencies, the problems of waste disposal as set forth above have become a matter national in scope and in concern and necessitate Federal action through financial and technical assistance and leadership in the development, demonstration, and application of new and improved methods and processes to reduce the amount of waste and unsalvageable materials and to provide for proper and economical solid-waste disposal practices.

(b) The purposes of this Act therefore are—

(1) to initiate and accelerate a national research and development program for new and improved methods of proper and economic solid-waste disposal, including studies directed toward the conservation of natural resources by reducing the amount of waste and unsalvageable materials and by recovery and utilization of potential resources in solid wastes; and

(2) to provide technical and financial assistance to State and local governments and interstate agencies in the planning, development, and conduct of solid-waste disposal programs.

DEFINITIONS

SEC. 203. When used in this Act—

(1) The term "Secretary" means the Secretary of Health, Education, and Welfare; except that such term means the Secretary of the Interior with respect to problems of solid waste resulting from the extraction, processing, or utilization of minerals or fossil fuels where the generation, production, or reuse of such waste is or may be controlled within the extraction, processing, or utilization facility or facilities and where such control is a feature of the technology or economy of the operation of such facility or facilities.

(2) The term "State" means a State, the District of Columbia, the Commonwealth of Puerto Rico, the Virgin Islands, Guam, and American Samoa.

(3) The term "interstate agency" means an agency of two or more municipalities in different States, or an agency established by two or more States, with authority to provide for the disposal of solid wastes and serving two or more municipalities located in different States.

(4) The term "solid waste" means garbage, refuse, and other discarded solid materials, including solid-waste materials resulting from industrial, commercial, and agricultural operations, and from community activities, but does not include solids or dissolved material in domestic sewage or other significant pollutants in water resources, such as silt, dissolved or suspended solids in industrial waste water effluents, dissolved materials in irrigation return flows or other common water pollutants.

(5) The term "solid-waste disposal" means the collection, storage, treatment, utilization, processing, or final disposal of solid waste.

(6) The term "construction", with respect to any project of construction under this Act, means (A) the erection or building of new structures and acquisition of lands or interests therein, or the acquisition, replacement, expansion, remodeling, alteration, modernization, or extension of existing structures, and (B) the acquisition and installation of initial equipment of, or required in connection with, new or newly acquired structures or the expanded, remodeled, altered, modernized or extended part of existing structures (including trucks and other motor vehicles, and tractors, cranes, and other machinery) necessary for the proper utilization and operation of the facility after completion of the project; and includes preliminary planning to determine the economic and engineering feasibility and the public health and safety aspects of the project, the engineering, architectural, legal, fiscal, and economic investigations and studies, and any surveys, designs, plans, working drawings, specifications, and other action necessary for the carrying out of the project, and (C) the inspection and supervision of the process of carrying out the project to completion.

RESEARCH, DEMONSTRATIONS, TRAINING, AND OTHER ACTIVITIES

SEC. 204. (a) The Secretary shall conduct, and encourage, cooperate with, and render financial and other assistance to appropriate public (whether Federal, State, interstate, or local) authorities, agencies, and institutions, private agencies and institutions, and individuals in the conduct of, and promote the coordination of, research, investigations, experiments, training, demonstrations, surveys, and studies relating to the operation and financing of solid-waste disposal programs, the development and application of new and improved methods of solid-waste disposal (including devices and facilities therefor), and the reduction of the amount of such waste and unsalvageable waste materials.

(b) In carrying out the provisions of the preceding subsection, the Secretary is authorized to—

(1) collect and make available, through publications and other appropriate means, the results of, and other information pertaining to, such research and other activities, including appropropriate recommendations in connection therewith;

(2) cooperate with public and private agencies, institutions, and organizations, and with any industries involved, in the preparation and the conduct of such research and other activities; and

(3) make grants-in-aid to public or private agencies and institutions and to individuals for research, training projects, surveys, and demonstrations (including construction of facilities), and provide for the conduct of research, training, surveys, and demonstrations by contract with public or private agencies and institutions and with individuals; and such contracts for research or demonstrations or both (including contracts for construction) may be made in accordance with and subject to the limitations provided with respect to research contracts of the military departments in title 10, United States Code, section 2353, except that the determination, approval, and certification required thereby shall be made by the Secretary.

(c) Any grant, agreement, or contract made or entered into under this section shall contain provisions effective to insure that all information, uses, processes, patents and other developments resulting from any activity undertaken pursuant to such grant, agreement, or contract will be made readily available on fair and equitable terms to industries utilizing methods of solid-waste disposal and industries engaging in furnishing devices, facilities, equipment, and supplies to be used in connection with solid-waste disposal. In carrying out the provisions of this section, the Secretary and each department, agency, and officer of the Federal Government having functions or duties under this Act shall make use of and adhere to the Statement of Government Patent Policy which was promulgated by the President in his memorandum of October 10, 1963. (3 CFR, 1963 Supp., p. 238.)

(d) Notwithstanding any other provision of this Act, the United States shall not make any grant to pay more than two-thirds of the cost of construction of any facility under this Act.

INTERSTATE AND INTERLOCAL COOPERATION

Sec. 205. The Secretary shall encourage cooperative activities by the States and local governments in connection with solid-waste disposal programs; encourage, where practicable, interstate, interlocal, and regional planning for, and the conduct of, interstate, interlocal, and regional solid-waste disposal programs; and encourage the enactment of improved and, so far as practicable, uniform State and local laws governing solid-waste disposal.

GRANTS FOR STATE AND INTERSTATE PLANNING

Sec. 206. (a) The Secretary may from time to time, upon such terms and conditions consistent with this section as he finds appropriate to carry out the purposes of this Act, make grants to State and interstate agencies of not to exceed 50 per centum of the cost of making surveys of solid-waste disposal practices and problems within the jurisdictional areas of such States or agencies, and of developing solid-waste disposal plans for such areas.

(b) In order to be eligible for a grant under this section the State, or the interstate agency, must submit an application therefor which—

(1) designates or establishes a single State agency (which may be an interdepartmental agency) or, in the case of an interstate agency, such interstate agency, as the sole agency for carrying out the purposes of this section;

(2) indicates the manner in which provision will be made to assure full consideration of all aspects of planning essential to statewide planning (or in the case of an interstate agency jurisdictionwide planning) for proper and effective solid-waste disposal consistent with the protection of the public health, including such factors as population growth, urban and metropolitan development, land use planning, water pollution control, air pollution control, and the feasibility of regional disposal programs;

(3) sets forth its plans for expenditure of such grant, which plans provide reasonable assurance of carrying out the purposes of this section;

(4) provides for submission of a final report of the activities of the State or interstate agency in carrying out the purposes of this section, and for the submission of such other reports, in such form and containing such information, as the Secretary may from time to time find necessary for carrying out the purposes of this section and for keeping such records and affording such access thereto as he may find necessary to assure the correctness and verification of such reports; and

(5) provides for such fiscal-control and fund-accounting procedures as may be necessary to assure proper disbursement of and accounting for funds paid to the State or interstate agency under this section.

(c) The Secretary shall make a grant under this section only if he finds that there is satisfactory assurance that the planning of solid-waste disposal will be coordinated, so far as practicable, with other related State, interstate, regional, and local planning activities, including those financed in part with funds pursuant to section 701 of the Housing Act of 1954.

LABOR STANDARDS

Sec. 207. No grant for a project of construction under this Act shall be made unless the Secretary finds that the application contains or is supported by reasonable assurance that all laborers and mechanics employed by contractors or subcontractors on projects of the type covered by the Davis-Bacon Act, as amended (40 U.S.C. 276A—276a-5), will be paid wages at rates not less than those prevailing on similar work in the locality as determined by the Secretary of Labor in accordance with that Act; and the Secretary of

Labor shall have with respect to the labor standards specified in this section the authority and functions set forth in Reorganization Plan Numbered 14 of 1950 (15 F.R. 3176; 5 U.S.C. 133z-15) and section 2 of the Act of June 13, 1934, as amended (40 U.S.C. 276c).

OTHER AUTHORITY NOT AFFECTED

SEC. 208. This Act shall not be construed as superseding or limiting the authorities and responsibilities, under any other provisions of law, of the Secretary of Health, Education, and Welfare, the Secretary of the Interior, or any other Federal officer, department, or agency.

PAYMENTS

SEC. 209. Payments of grants under this Act may be made (after necessary adjustment on account of previously made underpayments or overpayments) in advance or by way of reimbursement, and in such installments and on such conditions as the Secretary may determine.

APPROPRIATIONS

SEC. 210. (a) There is hereby authorized to be appropriated to the Secretary of Health, Education, and Welfare, to carry out this Act, not to exceed $7,000,000 for the fiscal year ending June 30, 1966, not to exceed $14,000,000 for the fiscal year ending June 30, 1967, not to exceed $19,200,000 for the fiscal year ending June 30, 1968, and not to exceed $20,000,000 for the fiscal year ending June 30, 1969.

(b) There is hereby authorized to be appropriated to the Secretary of the Interior, to carry out this Act, not to exceed $3,000,000 for the fiscal year ending June 30,1966, not to exceed $6,000,000 for the fiscal year ending June 30, 1967, not to exceed $10,800,000 for the fiscal year ending June 30, 1968, and not to exceed $12,500,000 for the fiscal year ending June 30, 1969.

Approved October 20, 1965, 9:10 a.m.

CHAPTER I.2

statistics of solid waste

Solid waste collection and disposal has long been a concern of public health agencies and those dealing with disease prevention and control, as evidenced by bibliographies of Weaver,[1] on refuse collection and disposal, 1954–55, Williams,[2] 1956–57, and Williams and Black,[3] 1958–59.

However, statistics were needed. Combustion Engineering, Inc., Windsor, Connecticut, completed contract PH 86-66-163 in the form of an extensive report, "Technico-Economic Study of Solid Waste Disposal Needs and Practices,"[4] embracing municipalities, industries of wide variety, development of an information system, and a technical economic overview.

During the fall of 1966 and spring of 1967, approximately 600 cities were surveyed and municipal officials in 50 of them were interviewed. Waste surveys of the industries statistically recorded by the USDC census of manufacturers were investigated and the data coordinated.

The conclusions reached in the report were as follows.

Statistical Data

1. Approximately 1,380,000,000 lb of residential, commercial, and industrial wastes are generated each day in the United States. Approximately one billion pounds of this solid waste must be disposed of in either municipal or private contractor facilities; the remainder is disposed of in industrial sites.

2. It is estimated that a typical urban area must dispose of approximately 5.5 lb/cap/day of solid waste from all sources, and that the "national" figure based on a population of 200 million people is 5.1 lb/cap/day.

3. Approximately 20% of the communities of over 25,000 population use incineration to dispose of their solid wastes; the remaining communities use sanitary landfill, open dumping, open burning, or composting.

4. Approximately 9% of municipal refuse is incinerated.

5. As of December 31, 1966, there were 74,600 tons/day of installed incinerator capacity operating in the United States.

6. As of June 1967, there were approximately 730 tons/day of installed composting capacity in operation in the United States.

7. Many cities are not faced with long hauling distances to their current disposal site, with approximately one-half of the cities reporting a hauling distance of less than five miles.

8. Approximately 50% of the cities had over 25,000 population currently using sanitary landfill less than 6 yr of life left in their existing facility, and many of these do not know at this time where the next facility will be located.

Planning

1. Little or no data are kept by the typical municipality on the physical make up of refuse.

2. Significant improvements can be made in the data gathering and record keeping of most municipalities.

3. There are apparent differences in solid waste problem areas from one population stratum to another, and these should be examined separately.

4. If present trends continue and further long-range planning is not expanded, facilities for handling solid wastes will be lacking by 1975.

5. Regionalization is recognized to be a practical solution to the solid waste disposal problem in many areas, but there are political and emotional objections to its implementation.

6. A possible short-term solution to the facilities gap is expansion of the private contractor's role because of his ability to cross municipal and political boundaries. For example, he could use a private disposal site for the refuse of several communities.

R. D. Vaughn, the director of the Bureau of Solid Waste Management, in a preface to "Technico-Economic Study of Solid Waste Disposal Needs and Practices"[4-6] stated that

> Rising population in the United States, increasing urbanization of this population, industrial growth, and the unparalleled affluence of American society have resulted in an ever-increasing volume of wastes in the solid state that must be regularly collected, transported, and ultimately disposed of. Per capita generation of solid wastes has risen from 2.75 pounds in 1920 to 5.3 pounds in 1968, and this figure may rise to 8 pounds by 1980. At the very time that these larger amounts of solid wastes must be managed, cities are faced with shortages of suitable disposal sites, and with

present solid waste management practices that are inadequate to protect the environment.

Upon assuming responsibilities under this Act, the Federal solid wastes program was confronted with a lack of comprehensive information to define the solid waste problems of municipalities and industries in specific terms, and to assess the existing state of solid waste technology.

Since submission of this report, the Bureau of Solid Waste Management has completed a National Survey of Community Solid Waste Practices that provides a statistically reliable estimate of the prevailing costs, modes of collection, processing and disposal, and the quality of solid waste management in the United States. Persons interested in obtaining basic information in the solid waste field are referred to the National Survey.[5,6] In cases of statistical discrepancy between the publications, the National Survey should be considered authoritative.

By 1968 there was a comprehensive assessment of solid waste problems and needs in the Eliassen[7] report, bringing together the Office of Science and Technology, the Solid Wastes Program, various segments of HEW, the USDI, the USBM, the USDD, the USAF, the USDA, the USBB, and representatives of the offices in which the report was written:

> The various public statements in the past, whether expressed in tons of solid wastes or in dollars of expenditure, have usually been limited to the quantities of wastes for which municipalities have been responsible or represent an estimate of gross municipal costs. It is estimated that almost one and one-half billion pounds of urban solid wastes are generated in the nation each day and that costs of community collection and disposal approximate $4 billion per year.

The traditional view of solid wastes, including urban garbage and rubbish, augmented by quantities of abandoned vehicles, demolition-construction wastes, and various other items, requires some reexamination. There is more to the problem than just handling and disposing of community or urban solid wastes, despite the fact that even this one aspect of the problem is far from solution. Congress directed attention to the overall solid waste problem in the Solid Waste Disposal Act of 1965, which defined solid wastes as "garbage, refuse, and other discarded materials, including solid-waste materials resulting from *industrial, commercial,* and *agricultural operations,* and from community activities" (italics supplied). Disposal was also broadly defined to include "the collection, storage, treatment, utilization, processing or final disposal of solid waste."

The term "solid waste" describes that material which is normally solid, which arises from animal or human life and activities, and which is discarded as useless or unwanted. Solid waste also includes deposited waste particulates, even when temporarily suspended in air or water. It refers to the heterogeneous mass of throwaways from the urban community as well as the more homogeneous accumulations of agricultural, industrial, and mineral wastes.

The Eliassen report also states:

> Nature has demonstrated its capacity to disperse, degrade, absorb or otherwise dispose of unwanted residues in the natural sinks of the atmosphere, inland waterways, oceans, and the soil. But the major concern are those residues that may poison, damage, or otherwise effect one or more species in the biosphere, with a resultant ecological shift.
>
> Practical problems in solid waste operations have often been treated without full use of available information. Short-range remedies have held sway over long-range solutions. Local views have prevailed over state or regional planning. Standards set may have met single objectives but have failed to meet other objectives. In some instances, decisions may be made on the basis of dollar cost-benefit analysis alone. Yet, in most situations overriding benefits to health, conservation, recreation, and aesthetics must enter and dictate solutions.
>
> It is impractical to think of impeding urban affluence, progress of industry, and advances of agriculture by strict control, but the trend for the future must include the development of solid waste management systems that permit progress without harmful side effects. The preservation of public health and aesthetics of the environment should be of paramount consideration in management decisions involving solid wastes.
>
> A prime solution to the resource depletion problem created by profligate use of metals and other elements would be to improve methods for recycling and reuse. To some extent, the salvage industry has traditionally served this purpose, but the concept might well be more widely applied. It is not inconceivable that present-day tailings dumps, landfills, and automobile graveyards may be looked upon in the future as "mines" for minerals whose natural ores have been depleted, or that remain only in deposits which can be mined only at greater cost than required for recycling of wastes.

Furthermore, the report sets forth a method of attack on the problem, as to funding, operation, development, and organization. Some tables from the report have been updated and are presented in this book.

Black et al.[8] presented an interim report on the national solid wastes survey, covering all the incorporated communities of 5000 or more population in the 1960 census. The data were collected on a prescribed basis[9]; Muhich et al. give a data analysis.[5] In the report Hickman[8] states:

> This country has over 12,000 land disposal sites being utilized by collection services, control indicates that 94 percent are unacceptable and represent disease potential, threat of pollution, and land blight. By no stretch of the imagination do these sites resemble a sanitary landfill. The waste management field must face the challenge of studying and evaluating these sites to determine their ability for conversion to sanitary landfills. We must develop the necessary plans, finances, and action programs to convert those sites that can function as a sanitary landfill. In many instances it will be necessary to close and abandon many of these sites. Local government then must locate and develop new sites now for immediate use and to provide necessary capacities for the increase of the future.
>
> There are approximately 300 incinerators in this country, 70 percent of which are without adequate air pollution control devices. Many produce residues and quench waters that are released into the environment without adequate treatment and control. These incinerators must be studied, evaluated, and upgraded to meet the more stringent pollution control regulations now being promulgated in this country. In many instances, complete abandonment of these facilities will be necessary because of their antiquated designs and capabilities. Because many communities are without adequate space for sanitary landfill, new incinerator plans will be needed. The current level of the solid waste load being treated by incineration (8%) may have to be increased to help do the job.

However, even at the present time the most convenient means for disposal of solid waste, garbage, and trash, is an open burning dump. This out-of-sight, out-of-mind, method is cheap. However, what may have been acceptable at an earlier time can no longer be tolerated in a country of over 200 million persons, 70% of whom live in urban and suburban areas.

With our increasing efficiency in agricultural and animal husbandry, it is no longer unusual to have a million penned chickens in an egg-laying and collecting establishment. The digestive excreta of such installations produce enormous amounts of concentrated putrescible waste. In a similar fashion there has been increasing acceptance of feedlots of raising livestock. Animals, while confined to feedlots to fatten more rapidly for the market, produce enormous and concentrated quantities of manure that cannot be readily and safely assimilated in the soil. It has little value when contrasted to chemical fertilizers, and it is not dispersed as would occur under conventional open grazing practices.

Mineral and coal mining and related processing industries, such as sand, gravel, and slate, have generated more than 20-billion tons of mineral waste in the last 30 years. These are now great slag heaps, burning culm piles (now being extinguished and covered, with state bond issues carrying the cost), mounds of mill tailings, overburden, and spoil, which have accumulated in the area of the extraction or processing operation. We have learned how to reclaim the land in areas affected by these wastes.

It is estimated that total solid waste generated from municipal, commercial, and industrial sources in the United States amounts to more than 360 million tons annually. Agricultural waste including crop residues, animal manures, and food processing residues are well over 2 billion tons, while mineral solid waste generation is almost 1.1–1.5 billion tons with an estimated rise to 2 billion tons in 1980. Summing these up, there are more than 3.5 billion tons of solid wastes generated in the United States each year. The safe and economic disposal of such quantities of solids imposes enormous problems. This is particularly true because of the insistence on doing the job in the cheapest possible fashion. It would appear, however, that this economic phase of the problem must give way to realities. We cannot afford to bury ourselves in our own waste, nor live on it in layers as the Trojans did.

Americans have developed a feeling that when they are through with something and deposit it in a refuse can or a waste disposal container, and a garbage truck picks it up, or when they dispose of their household waste through a grinder or disposal, the product passes to the sewer system and the job is therefore done.

Three-quarters of the 4–5 billion dollars spent annually for solid waste disposal is accounted for by collection and transportation costs. The remainder is for processing and disposal; consequently, and despite public objections, the use of the open dump is the predominant method of disposal.

Technological and marketing considerations have had a major influence on the amount of waste and its disposal. Planned obsolescence is a fact of the marketplace; we are urged to buy the new and discard the old, and the consumer has become accustomed to one-way packaging without regard to its potential reuse or recycling. Thus junked cars and electric appliances, as well as empty cans and bottles, and foil and wrappings of all kinds, constitute a large part of modern urban waste.

Cities all over the world have merely transported wastes beyond their confines and discarded them. Such practices were reasonably satisfactory so long as municipal jurisdictions were separated by relatively uninhabited space, but urban and suburban sprawl are now erasing this space.

Regional planning is necessary for effective collection, processing, and disposal of solid waste; such planning is lacking in many communities. Planning requires a long-range view of a total area or region, but many communities rely on short-term solutions, neglecting population growth and density, industrial and commercial zoning, and technological development.

Traditionally, urban refuse disposal has been concerned with household waste, that is, garbage. Disposal areas have become unavailable as cities and suburbs have expanded beyond their former confines. Around some cities specialty enterprises, such as dairy, poultry and egg production, and animal feeding in feedlots, have changed the character of former rural areas.

In many suburbs, waste is collected by private or public agencies on a periodic basis in compactor trucks and transported to local incinerators, at a cost to the householder that is high but acceptable. In addition to garbage, gardening wastes such as grass cuttings, leaves, branches, and trimmings are included in this waste.

Urban waste includes food waste (garbage), paper and paper products, wood, cans, glass, dirt, ashes, dead animals, abandoned cars and trucks; demolition residues and residues from new construction; occasionally radioactive materials, explosives, pathological wastes from hospitals; and this list is not exhaustive. Commercial organizations contribute their share of similar waste products.

THE VISIBILITY OF WASTE

Municipal solid waste includes street litter, sweepings, discarded automobile bodies, power plant and incinerator ashes and residues, and sewage sludge. Old cars are abandoned in city streets, highways, and vacant lots, since owners find this cheaper than to have them towed to a collection point. In New York City approximately 14,000 cars were abandoned in 1964; only 2500 were abandoned in 1960. By 1966 that figure almost doubled again to 25,000 vehicles, and by 1970 it was closer to 30,000. Nationwide, more than half a million cars and trucks were abandoned annually from 1967 on.

Annually electric power generating plants consume well over 300 million tons of coal and lignite and generate about 10% of this as fly ash and cinders, the ratio being about 80% fly ash to 20% bottom ash. These wastes, their handling and disposal, are treated in a separate section of this book.

When collected, urban wastes are handled by public and private contractors and transported to disposal facilities. However, substantial amounts do not go through channels of collection but may be burned in apartment houses, home incinerators, or backyards, whereas in some commercial establishments, or on demolition sites, some portions are disposed of in garbage grinders connected to sewers. A number of estimates made of urban and industrial waste generated in the United States averaged 10 lb/cap, whereas collected waste averaged about 5.3 lb/cap.

The composition of the municipal refuse varies widely over the country, but the composite is reported as follows: paper, 42–57%; metals, 1.5–8%; glass, 2–15%; rags, 0.6–2%; garden debris, 10–12%; and ashes, 5–19%.

An analysis of municipal wastes is provided in the Quad-City Solid Wastes Project,[10] covering waste collection in the cities of Clifton, Passaic, and Paterson and the town of Wayne, New Jersey. The quantities, analysis, compositions, and population are given in Tables 1 and 2.

The official 1960 census was used as a basis for estimating per capita data. Three assays conducted in October 1966 and February and May 1967 provided data on municipally collected solid wastes. Plant visits to 32% of the industrial establishments in the four communities provided information for projection of industrial waste data. Field visits provided information on construction/demolition production. Community public works departments and contract scavengers furnished data on periodic cleanup and street cleaning production.

The three-assay data indicate that the municipal waste production is 2.74 lb/cap/day. The industrial survey shows a Quad-City production of 3.31 lb/cap/day. Total waste production in the four communities amounts to an estimated 9.39 lb/cap/day. Production in the Quad-City

Table 1 *Quad-City Solid Wastes Project First Annual Report Fact Sheet*

Item	Paterson	Clifton	Passaic	Wayne	Quad-City
1. Total wastes production					
Municipal					
lb/cap/day	2.46	2.67	2.47	4.18	2.74
$/ton	9.89	7.32	9.70	10.21	9.27
tons/day	185.0	117.7	64.0	91.0	457.7
Industrial					
lb/cap/day	2.89	3.78	3.06	1.59	3.31
tons/day	205.98	154.66	82.32	23.33	508.51
Construction					
lb/cap/day	0.20	0.59	0.21	1.57	0.44
tons/day	14.42	24.30	5.59	23.01	67.33
Demolition					
lb/cap/day	2.20	1.32	2.77	1.87	2.03
tons/day	156.68	53.97	74.52	27.40	312.57
Cleanup week					
lb/cap/day	0.22	0.28	0.11	1.23	0.31
tons/day	15.65	11.45	2.96	18.05	47.62
Street sweepings					
lb/cap/day	0.14	1.00	0.45	1.57	0.56
tons/day	9.8	41.0	12.2	23.00	86.0
Total waste					
lb/cap/day	8.11	9.64	9.07	12.01	9.39
tons/day	586.55	403.08	241.59	205.79	1479.73

Table 1 *(Continued)*

Item	Paterson	Clifton	Passaic	Wayne	Quad-City
2. % Composition of municipal waste					
Paper	39.80	46.80	44.41	53.66	45.63
Wood	3.38	3.18	3.45	1.69	3.00
Plastic	2.97	2.34	2.33	2.39	2.52
Glass	9.96	3.73	5.24	5.38	6.22
Metal	12.91	6.95	8.68	6.96	9.08
Stone, sand	7.04	8.53	8.65	6.16	7.65
Organics	20.22	26.76	21.43	22.51	22.62
Rags	5.07	2.20	7.76	1.69	4.45
3. Chemical characteristics					
As received lab. sample					
% Moisture	41.7	39.1	45.3	47.2	43.3
% Carbon	26.2	29.3	24.6	25.8	26.5
% Nitrogen	0.68	0.83	0.90	0.58	0.75
Btu/lb	5,199	4,635	4,976	4,332	4,786
% Volatiles	45.9	49.9	43.1	44.9	46.0
% Ash	9.7	10.7	11.4	8.0	10.0
4. Adjusted estimate chemical characteristics					
As collected					
% Moisture	27.5	28.1	30.2	34.8	30.1
Btu/lb	3,426	3,281	3,598	3,145	3,364
% Volatiles	34.6	37.3	32.4	34.4	34.7
% Ash	37.8	34.6	37.3	31.0	35.2
Dry basis					
Btu/lb	5,917	5,236	6,563	6,361	6,016
% Volatiles	58.4	61.5	58.0	65.2	60.7
% Ash	41.6	38.5	42.1	34.7	39.3

5. % Composition in total industrial waste					
Paper	45.50	38.33	35.96	63.51	42.71
Wood	13.03	2.95	11.23	1.05	7.34
Plastic	7.00	10.81	16.95	25.39	12.61
Glass	6.51	1.73	2.08	0.36	3.03
Metals	5.45	6.06	2.03	0.75	4.48
Stone sand	5.59	6.48	4.38	—	5.09
Organic chemical	0.14	5.48	5.26	0.21	3.32
Textiles, rags	6.97	1.17	2.56	2.71	3.28
Ceramics	—	0.51	0.02	—	1.57
Inorganic chemical	—	0.53	—	0.18	0.19
Petro-chem.	0.92	4.98	0.81	0.38	0.23
Rubber	—	1.22	5.59	0.28	2.45
Mixed commercial	0.15	7.45	5.62	0.01	4.15
Food	4.22	8.28	—	4.49	5.00
Miscellaneous	4.57	4.02	7.51	0.68	4.53
Total	100.00%	100.00%	100.00%	100.00%	100.00%
6. Miscellaneous data					
Population (1960)	142,301	81,831	53,770	29,353	307,255
Area (sq. mi.)	8.3	11.2	3.2	25.7	48.4
No. of industries	562	233	208	40	1043
No. of industries surveyed	125	106	71	22	324

Table 2 *Quad-City Solid Wastes Project: Industrial Waste Study. Composition of Industrial Waste Among Industries Surveyed*

Type of Material	Amount Produced[a] (tons/wk)	(tons/day)	(%)
Paper	540.63	108.13	42.73
Wood	92.89	18.58	7.34
Plastic	159.48	31.90	12.61
Glass	38.31	7.66	3.03
Metal	56.65	11.33	4.48
Stone, sand	64.32	12.86	5.09
Organic chemicals	42.08	8.42	3.32
Textiles	41.51	8.30	3.28
Rubber	19.93	3.99	1.57
Ceramics	2.43	0.49	0.19
Inorganic chemicals	2.92	0.58	0.23
Petro-chemicals	31.04	6.21	2.45
Mixed commercial	52.70	10.54	4.15
Food	63.23	12.65	5.00
Miscellaneous	57.32	11.46	4.53
Total	1265.44	253.10	100.00

[a] Based on a five-day week.

community currently amounts to 460 tons of municipal waste, 500 tons of industrial waste, and a total of 1500 tons/day.

Btu of municipal waste will run about 3400 in the field with about 30% moisture content and about 35% by weight inert or noncombustible materials based on laboratory assay.

Municipal waste includes approximately 46% paper and paper products and 23% organic garbage. Glass, metal, and plastics account for about 18% of the production. By comparison, industrial waste composition includes about 43% paper and 5% organic food wastes. Glass, metal, and plastics amount to about 20% of the total. Industrial waste composition is given in Table 1.

Dr. Merril Eisenbud, Professor of Environmental Medicine at New York University Medical Center, the first administrator of the Environmental Protection Administration of the Environmental Protection Administration of the City of New York, wrote[11]:

> Dust and soot are the most annoying form of air pollution in many cities. The sources of the particulate emissions in New York City are shown in Table [3], which indicates that space heating,

municipal incineration, apartment house incineration and power generation account for about 80 percent of the 69,100 tons (1 ton = 907 kilograms) emitted per year to the atmosphere as of November 1969.

During 1969, three of the city's eleven municipal incinerators were shut down, and another is scheduled to be closed as soon as alternate means of handling refuse can be arranged in the next year to two. The remaining seven incinerators are sufficiently modern so that air-cleaning equipment can be installed at a cost of about $12 million. However, because no equipment manufacturer would offer performance warranties, and in the absence of experience, an experimental program was designed to obtain the information needed to make the required engineering decisions. In addition to pilot plant tests of various air-cleaning techniques, full-scale installations have been made of two electrostatic precipitators and one Venturi scrubber. The early experience at these installations has been encouraging, but, because of the corrosive nature of the effluents and the generally arduous service to which equipment of this kind must be put, many months of testing will be necessary. These are the first installations of this type in the United States, and the information being obtained will be generally useful to communities throughout the country.

The largest single source of particulate emissions to the air of New York City is space heating from about 30,000 apartment houses that burn No. 6 residual fuel oil. The black smoke that one sees curling up from apartment house rooftops during the heating season is usually the result of improper operation of residual fuel

Table 3 *Sources of particulate emissions to the atmosphere of New York City in November 1969 [from Ref. 11]*

Source	Amount (tons/year)	Percent
Space heating	22,300	32.3
Municipal incineration	13,330	19.3
On-site incineration	12,690	18.4
Mobile sources	9,900	14.3
Power generation	6,400	9.2
Industrial sources	4,500	6.5
Total	69,120	100.0

oil boilers. Local Law 14 mandates installation of equipment modifications that will result in increased combustion efficiency and less particulate emission, and these are working well in about 1,500 furnaces where the change has been made.

The second largest source of particulates is apartment house incinerators, about 17,000 of which were constructed between 1947 and 1967.

The improvements required for incinerators and residual oil burners proved practical, but the apartment owners nevertheless brought suit against the city, charging that the law was unconstitutional and imposed unreasonable hardships on the landlords. This suit has stalled compliance with the provisions of Local Law 14 that pertain to apartment house oil burners and incinerators.

SANITARY LANDFILLS

One method of solid waste disposal is designated as sanitary landfill: burying the waste and covering it with soil. Problems associated with this method of disposal include groundwater pollution, generation of methane or other noxious gases and odors, and breeding or harboring of rodents, flies, and other pests.

Construction waste and demolition debris have been used as landfill for centuries. Free from putrescibles, it is clean fill.

In 1683 the British on Manhattan Island feared a French attack and dumped dirt on the southern tip of the island, extending the shore line to mount a cannon. Since then the many inlets, drainage brooks, and swamps have been filled with debris and excavation residue, and the land contours have been smoothed out. Crumbling piers, old sailing vessels sunk in their slips, and abandoned scows and lighters have been entombed under tons of fill that pushed the shoreline further into the rivers. From 1683, 3600 acres of new land have been added to Manhattan, so that about one-fourth of the island is man-made (see Figure 1). Moreover, as the result of the lack of availability of disposal areas in Manhattan at that time, many thousands of tons of stone from the old Pennsylvania Railroad Station were buried in the Jersey meadows.

Building in navigable waterways requires permission from Congress and approval by the Army Corps of Engineers, which makes tests to make sure that currents will not be adversely affected by landfill.

Landfills are in operation at Virginia Beach, Virginia, and one is planned for the Chicago area. The latter will rise above the contour of

Figure 1. Changes in Manhattan Island from the 17th to the 20th century.

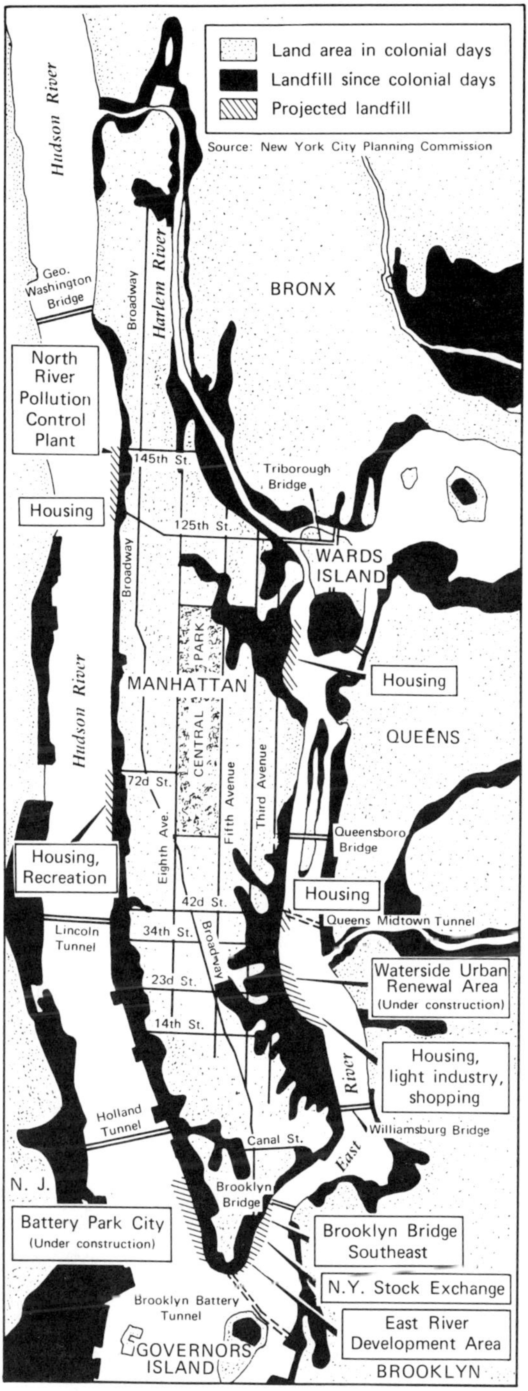

Land area in colonial days
Landfill since colonial days
Projected landfill
Source: New York City Planning Commission
Hudson River
Geo. Washington Bridge
Broadway
Harlem River
BRONX
North River Pollution Control Plant
145th St.
Triborough Bridge
Housing
125th St.
WARDS ISLAND
Broadway
CENTRAL PARK
MANHATTAN
Housing
QUEENS
Hudson River
72d St.
Eighth Ave.
Fifth Avenue
Third Avenue
Queensboro Bridge
Housing, Recreation
Housing
42d St.
Queens Midtown Tunnel
Lincoln Tunnel
34th St.
Broadway
Waterside Urban Renewal Area (Under construction)
23d St.
14th St.
Housing, light industry, shopping
East River
Holland Tunnel
Williamsburg Bridge
Canal St.
N. J.
Brooklyn Bridge
Battery Park City (Under construction)
Brooklyn Bridge Southeast
N.Y. Stock Exchange
Brooklyn Battery Tunnel
East River Development Area
GOVERNORS ISLAND
BROOKLYN

the surrounding countryside to form a huge man-made hill, designated by columnists as "Mount Trashmore." Upon completion these sanitary landfills will be developed as recreation areas or parks.

Use of landfill moves the emphasis from disposal to reduction of volume, either by incineration or pyrolysis for generation of power.

Trash, rubbish, waste, and littering, are problems of all of our cites, but in general New York handles them as well as any city in the nation by American standards, and in a manner far superior to Asian continental European, or South and Central American cities. New York also spends a far greater amount of money per capita than any other city in the world on waste disposal.

New York City's eight million people live on 6000 miles of streets. These people are joined each workday by an influx of more than two million people, approximately the population of the nation's second largest city, who come from outlying suburbs to earn their living. The rate of solid waste generation is increasing by 2–4%/yr. Depending on the part of the city, the cost of collecting refuse varies from $15 to $30/ton, and has increased steadily. The sanitation industry is one of the few in which wages have increased during the past decades without a commensurate increase in the productivity of labor. It is frequently said that the only change in the technology of garbage collection is that the internal combustion engine has replaced the horse.

The garbage can is the principal impediment to higher efficiency and is long overdue for replacement. Numerous options are available as alternatives that will make the job easier for the sanitation man, increasing his productivity and making it possible to provide cleaner streets at less cost. Experiments conducted during 1969 demonstrated that plastic or paper bags are an efficient and sanitary alternative and that their use should be encouraged. The main advantage is that the sanitation employee is no longer required to pick up a heavy can and laboriously shake the refuse from it. Bags have been found to be popular with both the householder and the sanitation employee, and their use is increasing. The cost to the householder is approximately 8¢ per day, and this will decrease as the bags are made available in mass distribution.

New high-rise apartment houses are still being built with little provision for refuse handling other than the garbage can. One large housing complex was planning to use 400 cans per day. Although plastic or paper bags offer a suitable alternative for private homes or small multidwelling buildings, a whole spectrum of more efficient alternatives is available for the larger buildings. These range from containers of 1 yd^3 capacity, which can be handled manually, to large 10-yd^3 containers, which are handled mechanically by trucks.

Another significant problem is the scrap automobile. Until a few years ago the market for scrap steel was such that a scrap car could be disposed of by its owner at a price that offered incentive for him to arrange for its removal from the city streets. Changes in economics of the steel industry have altered this situation to the extent that in most parts of the city it costs more to remove a car than it is worth. This has resulted in automobiles being abandoned on the streets of New York at an increasing rate. The city has franchised scrap dealers to collect these cars from various parts of the city. In some cases the scrap dealer is subsidized by the city, and in others he pays the city a small price for the car. It is illegal to abandon a car in the city streets, but when the last owner has removed the license plates and has filed off the engine number, it becomes prohibitively costly to trace him.

The nonreturnable bottle and the aluminum and steel can are other forms of litter that have their origins in the economics of our times. The beverage distributors insist that until recently a deposit bottle made as many as 30 round trips between the distributor and the customer, but that because of the indifference of the consumers to even a 5¢ deposit, the number of round trips in many communities gradually diminished to as few as four or five before the bottle was discarded. This is given as the reason for the shift from deposit bottles. The consumer prefers the nonreturnable container, as does the supermarket, some of which will not handle deposit bottles.

Ever since colonial days the city has followed the common practice of disposing of its solid wastes by filling its lowlands, and at present about 11% of the land area of the city has been created in this way, including some of the most valuable commercial and recreational areas. The largest landfill in the world, for example, is at Fresh Kills on Staten Island, but this will be filled by 1975. Smaller landfills exist in other parts of the city, and they too will be exhausted by that time.

The city presently routes about 7000 tons/day of refuse through municipal incinerators before sending it to the landfills. A basic strategy, therefore, must be to increase the municipal incineration capacity, to reduce the volume of waste, and to convert the refuse to a more manageable form. A $200-million capital program was begun; and construction of four giant incinerators is now contemplated; these will reduce the weight of refuse by about 75%, leaving a relatively innocuous ash that will occupy about 10% of the original volume. The present generation of municipal incinerators is one of the principal sources of atmospheric particulates. These incinerators are being equipped with air cleaning equipment, and all new units will be provided with modern stack cleaning equipment. Since contemporary refuse has a heat value of about 5000 Btu/lb, every effort should be made to dispose of the heat either for power production or for generation of steam. It is estimated that the city can in this way recover about $2/ton of refuse burned or as much as $12 million/yr. Owing to financial stress, the $200,000,000 program was cut back in 1971 and the construction of the four giant incinerators was abandoned.

COLLECTION AND HANDLING

Moderate in-plant compacting, say to 25% of original volume, could cut handling and transportation costs by as much as one-half. If waste is relatively incompressible, because of high liquid content, larger transfer trucks might be investigated. Transfer trailers of 60–80 yd^3 capacity with no compaction equipment cost between $20,000 and $25,000. Tractors cost about $18,000. Typical municipal-type refuse trucks with rear hydraulic compactor bodies and a capacity of 20 yd^3 sell for about $20,000. Municipal refuse is compacted about 60% with such equipment.

Waste is brought to a central point in average-size trucks. Here, its volume is often reduced for a trip in much larger vehicles to distant landfill sites. Typically, a 20-yd^3 truck with rear compactor and a three-man crew costs $150/day to operate when fuel, amortization, and maintenance are included. The average truck is driven about 50 miles/day, disposing of two 4-ton loads. Thus, operating costs are about $19/ton.

If a new dump site 10 miles further away is needed, the increased travel time is more than 1.5 hr at 25 miles/hr. So only 1.5 loads per/day, or 6 tons, could be collected, raising the collection cost close to $26/ton. The $7 difference could be applied to a secondary transportation system, which generally costs $6 to $10/ton for compaction, transportation, and landfilling.

Pneumatic solid waste systems, developed in Europe, are slowly finding applications here in high-rise apartments and hospitals. Water transport of food wastes, ground in kitchen garbage disposals, has been a fact for 30 yr; and, of course, solid human wastes, amounting to about one-fourth of the urban waste load, travel by water. The possibility of using sewage systems to transport organic solid wastes such as ground paper is being explored.

Techniques are available to control landfill leaching caused by local conditions such as waste with high moisture content or groundwater seepage. For landfills with unbaled trash, the lining for the entire area to be filled is a membrane barrier. The trapped water flows from the membrane to collection points where dissolved metals and biological materials can be removed to protect ground water supplies or nearby rivers. A 46-acre landfill using the system is in operation near Tullytown, Pennsylvania.

Baled waste is easier to contain. The individual bales may be wrapped in plastic. Depending on the waste mix, compacted bales may be nearly impervious to water, even without wrapping. A Japanese technique for coating compressed blocks of refuse with asphalt was hailed at one time, but proved to be a failure.

Some municipalities, including New York City, have been asking residents to place their refuse in plastic or kraft bags, making collection easier. Although the plastic bags impede degradation in landfills, they do seal odors that would attract rodents to trash waiting to be picked up at curbside.

Incineration is the second most popular disposal method. There are 300 municipal and many thousands of industrial incinerators operating in the United States, but only one-fourth of the municipal systems meet current air codes because they are overloaded. Strict emission control, which will be necessary to meet ambient air quality standards set for 1975, will probably retard the use of incinerators in large cities. Industrial incinerator systems continue to be popular, particularly models in the 200–500 lb/hr range, often with two-stage combustion.

Small to medium-sized industrial incinerators are comparatively low in cost and can handle mixed waste loads, including fluid wastes. Batch-loading and continuous-loading systems are available from dozens of manufacturers. Volume reductions of 95% are common, with reductions of 98% possible under some conditions.

Energy recovery, either to heat water for steam or as burnable gases, becomes economical in large units. Chicago dedicated a 1600-tons/day water-wall incinerator in 1971. Some of the 440,000 lb/hr of steam generated will be used to operate the plant, with the remainder available for sale to surrounding industries. About a dozen large steam-generating incinerators are operating in Europe.

A similar system operating in North America is a 1200 tons/day plant in Montreal, which also began operating in 1972. The Montreal plant's four boilers generate a total of 100,000 lb/hr of steam. Montreal calculates incineration costs at about $7/ton, and only half that once markets are found for all the steam. That figure includes amortization of the incinerator's $15 million cost.

Incineration in a rotary kiln is often considered where large volumes of mixed waste must be disposed of. A rotary kiln built at Bayer's Leverkusen works in West Germany (1968) for \$4.5 million disposes about 45,000 tons/yr of mixed wastes, including plastics. Even though the kiln provides up to 30 tons of steam per hour for process heat, disposal costs are \$20/ton.

Since it is difficult to provide excess air in a rotary kiln for complete combustion, volatile products are produced that often must be treated in a second combustion chamber, or afterburner. Many of today's low-pollution incinerator designs for industrial and small-town use also use an air-starved primary chamber to partially pyrolyze waste, followed by an afterburner to clean the exhaust gases.

Sludge from wastewater treatment can be disposed of in many ways. In the past, the two most common methods have been ocean dumping and drying in open beds followed by landfill. Odor problems are causing a reevaluation of open drying, even where land is available. Scientists at Penn State have pioneered the spraying of dilute sludge on poor soils to increase fertility. The idea is getting a large-scale tryout with the disposal of sewage sludge from Chicago.

Incineration is one of the cheapest ways to reduce sludge volume, after it is concentrated to 25–40% solids by other means. The most common incinerator for sludge disposal is the multihearth furnace, a refractory-lined steel shell with refractory hearths located one above the other. Sludge is fed through the top. A rotating air-cooled central shaft with air-cooled rabble arms and teeth plows the sludge across the open hearth to drop holes. It falls to the next hearth and then the next until ash is discharged at the bottom. Total cost per ton of dry solids removed is between \$8 and \$14, of which \$0.50 to \$5 is operating expense.

Thinner sludges can be handled by fluidized bed incinerators. The sludge is fed onto a bed, usually of sand, that has been heated. The bed is "fluidized" by an updraft of warm air or combustion gases. This allows the sludge and the sand to come into intimate contact for efficient incineration. Total cost is about \$20/ton of solids removed, with actual operating costs between \$3 and \$5.

The thinnest of sludges can be handled by the Zimpro wet air oxidation process. Combustible matter is oxidized under heat and pressure in the process, which is self-sustaining when there is sufficient heat value in the feed. About 45 units are in operation around the world. Units operating at lower temperatures and pressures are available for sludge conditioning.

Sludge can be concentrated to about one-third of its original volume.

It is pumped from a sedimentation basin or digestor through a heat exchanger to the reactor. There, steam enters through a steam-jet circulator that ensures intimate mixing of sludge and steam at 230–390°F and 180 to 230 psi. After 30–45 min conditioned sludge passes back through the heat exchanger, giving up heat to incoming sludge and the decanting vessel. Volume is usually reduced further by final dewatering with vacuum filters, filter presses, or other mechanical equipment. The product contains as little as 35% moisture and is an easily handled sterile cake suitable for compost or landfill.

The shortest route to recycling wastes may be through pyrolytic techniques, in which wastes are baked in an airfree chamber at temperatures as high as 3000°F. About 800 such units are operating now worldwide, all but a few of them in industrial applications. Pyrolytic systems have been generally more expensive than incinerators and have been resorted to mainly with wastes that are difficult to burn. Pyrolytic techniques may replace incineration for waste disposal in large cities.

A few wastes may lend themselves to direct reuse after pyrolysis. Pyrolyzing waste plastics into hard and soft waxes, adhesives, tars, and greases is a continuous process.

A pyrolytic system for the joint handling of sludge and municipal refuse was put into operation in South Houston, Texas in 1970. The $600,000 system is capable of handling 100 tons/wk. The sludge is pumped from a tertiary treatment plant to a concentrator for dewatering. It is then discharged into the pyrolysis plant. A small tractor is used to push the trash and sludge alternately into the loading devices.

Montsanto Enviro-Chem Systems built a 35-ton/day pilot version of its Landgard system for municipal wastes. The company proposed to build a 1000-ton/day unit for New York City, but financing was lacking. By-product recovery could have been added to the system. Operating costs without resource recovery would have been $8/ton, about the same as incineration.

Waste can be compressed to 5–10% of its original volume in a hydraulic press, compared to 20% on mechanical compactors. Completely nondegradable items, such as old refrigerators, can be handled with available compaction equipment, which comes in a wide range of sizes and capacities.

The main decomposition product of baled municipal refuse is carbon dioxide, with no methane to pose a fire hazard in dumps. This suggests that bale decomposition is aerobic.

About one-half of all urban wastes and some industrial waste can be degraded biochemically to a sanitary humus-like material with little or no odor. Costs typically are in the $8/ton range before sale of the

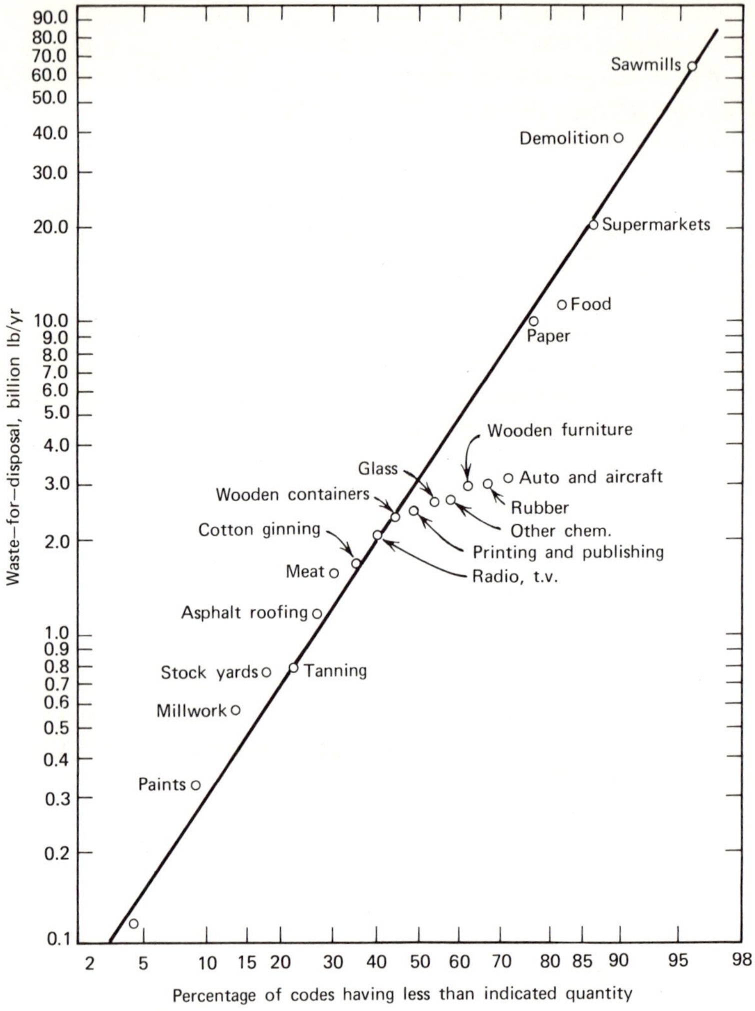

Figure 2. Distribution among 21 code groups of waste. *Source:* Technical-economic study of solid waste disposal needs and practices, report by Combustion Engineering, Inc. to Solid Wastes Program, 1967.

product compost. The greatest drawback of composting operations is that the potential market for compost is far less than the potential supply.

Figure 2 shows the distribution of 21 code groups of waste for disposal.

CHAPTER I.3

packaging and waste

Packaging and the solid waste problem are closely linked in the public view. Discarded packages are visible all around us. It is impossible to stroll through city streets, drive in the country, walk in a park, or ride in a boat on our rivers and lakes without encountering the telltale signatures of our affluence: discarded cans and bottles, cigarette packs and paper sacks, and similar objects. It is not surprising, therefore, that packaging materials are placed high on the list of items that create blight.

Packaging materials are only one part—and a relatively small one—of the total solid waste problem. Food wastes (garbage), lawn trimmings and leaves, ashes, newspapers and magazines, discarded toys and tools, rags, furniture, and a variety of other items such as Christmas trees, old tires, aged swing sets, and an occasional washer or dryer, account for about 80% of all residential wastes. Packaging accounts for about 20%. Materials other than packaging wastes represent 92% of total industrial and commercial discards; packaging represents the remainder.

The materials flowing into the waste processing system are the products of sophisticated industries—fiber processing, chemicals refining, metal shaping, and glass making. In short, the kinds of waste are changing. The availability of goods at throwaway prices, the rapidly rising economic well-being of a large segment of the population, and the growth in population are increasing the quantity of wastes.

The sheer quantity of packaging wastes imposes a load on the disposal facilities of the nation. Collection of packaging materials is costly if containers are thrown away carelessly, easy if they are discarded in waste cans. Processing of packaging wastes also may be easy or difficult, depending on whether they are simply dumped or whether they are incinerated or landfilled. Different materials cause different degrees of difficulty. Reuse and recycling of packaging wastes present yet another aspect, distinct from collection or disposal.

Immediately after World War II, 35% of all U.S. paper fiber requirements were met by waste paper or paperstock, as it is called. By 1956, paperstock accounted for slightly more than 26%; in 1966, for 21%; and by 1980 paperstock should supply only 17.5% of total paper required. The decline in percentage figures does not mean that tonnage of paperstock used has been declining; on the contrary, it rose from 7.3 million tons in 1946 to 10.2 million in 1966.

The declining popularity of waste paper as a raw material may be traced to cost. Pulptree harvesting and processing have been improved by new technology, driving down virgin pulp prices, while costs of handling secondary paper have been rising, partly because labor costs are increasing, but, more significantly, because the waste papers are increasingly contaminated. Papers are coated with plastics, printed with inks that resist de-inking, and are laminated to metals. To separate the useful and useless components of waste paper takes hand labor, and the more sorting that is required, the higher the price. Not surprisingly, paperstock dealers prefer to buy from sources such as printing shops and binderies, where the waste is already fairly clean, rather than from office buildings or residential sources, where paper wastes are frequently mixed with organic materials.

Much the same situation characterizes salvage of other materials used in packaging. To give one last example, to make a ton of cullet (broken glass of uniform color), a dealer has to handle between 1900 and 4600 bottles; and for a ton, he can obtain $15.00 at best. In some cities, sorting costs exceed $30/ton, and consequently cullet is not handled at all.

Darnay and Franklin[12] reported that 52 million tons of packaging materials were used in the United States in 1966 and greater quantities have been used each succeeding year.

R. D. Vaughn, then director of the Bureau of Solid Waste Management, in a foreword to PHS 1855,[12] states that packaging is increasing much more rapidly than population. Estimated national per capita consumption was 404 lb (1958), 528 lb (1966), and expected to be 661 lb by 1976. The factors influencing this rise are self-service merchandizing, advancing technology of production, public desire for convenience, personal affluence, and the persuasive nature of packaging.

Vaughn concludes that the aims of packaging and waste disposal are mutually exclusive. The packager wants, and his consumer buys, a container that will not burn, break, crush, degrade, or dissolve in water. The waste processor is equipped to handle combustible, breakable, and compactible or degradable forms.

Statistics are obtainable from Business and Defense Services Administration; Containers and Packaging and its Current Industrial Report Series; trade associations such as the American Paper Institute, the Can Manufacturers Institute, and the Glass Container Manufacturers Institute; and trade publications such as *Modern Packaging, Modern Plastics,* and *Paperboard Packaging.*

Packaging materials are used to convey goods from manufacturer, processor, and marketer to consumer, who destroys and discards the package after a single trip. One exception is steel drums and pails, which are often relaundered and conditioned for reuse. The package protects against impact, abrasion, scratching, oxidation, heat, cold, and destruction or change by light, bacteria, or contamination, often allowing visibility of contents or a base for an identifying label or product identification. The manufacturer joins paper, metals, foils, glass, wood, fibers, film and foamed plastics, textiles, adhesives, waxes, sealants, desiccants, and minerals, in unnumbered varieties of mixtures of materials, making economic recovery after use virtually impossible. Increasing application of large, well-engineered plastic bags for trash, waste garbage, and discards is a rapidly growing packaging application.

Table 1 shows the applications of packaging, Table 2 the applications in food, beverages, and chemicals.

Paper and paperboard is a highly integrated activity of paper companies as well as container makers, capable of forest ownership and management through lumber, by-product utilization, paper manufacture, to packages. The buyer needs only to fill, seal, and label the container. The exception is set up box production serving limited geographical areas.

About 10% of the steel production such as sheet, plate, tinplate, and related items ends up as containers, primarily metal cans. These are produced by can-making companies, and by consumers who have captive production.

Aluminum packages, foil, containers, tubes, and composites, are integrated in some cases but there are also independent fabricators.

The glass manufacturer is the container producer. He ships to the packager for filling, sealing, and shipment and furnishes closures.

The plastic resin producers (chemical and petroleum producers) have integrated into resin conversion and make films, bottles, containers, and tubes. Package manufacturers have integrated into plastic resins since packaging consumes about one-quarter of the tonnage of the plastics industry. The Plastics Institute projects that the

Table 1 Applications of Packaging

End Use	Percentage
1960–1965	
Food	44
Beverages	13
Chemicals, allied products	12
Paper, printed and allied	4
Textiles and clothing	4
Hardware	2
Petroleum and oils	2
Tobacco, etc.	1
Toys, jewelry	1
Other	19
Beverages	
Beer and malt liquor	8
Soft drinks	2.8
Others	3.2
Total beverages	14
Food	
Meat	3.3
Dairy products	7.7
Canned foods	
Seafoods	0.6
Specialties	2.8
Fruits and vegetables	6.7
Dehydrated	0.5
Pickles, sauces, etc.	1.4
Frozen foods	
Fish	0.4
Fruits and vegetables	2.6
Other foods: TV dinners, pies, canned and frozen foods	2.3
Total foods	28
Grain and flour mills	2.1
Baked products	4.2
Sugar and confectionary	6.3
Fats and oils	2.0
Salt, pepper, spices, etc.	1.0
Miscellaneous	2.6
Total	18.2
Process industries products	
Medicinals and pharmaceuticals	2.1
Cosmetics, etc.	7.2
Paints and varnishes	1.3
Miscellaneous	1.2
Total process industry products	12.8
All other industries	26

Source: U. S. Dept. of Commerce, Business and Defense Administration, Containers and Packaging.

Table 2 Packaging Materials Consumption, 1966 and 1976 (In Millions of Tons)

	1966		1976	
Material	Tonnage	Percent of Total	Tonnage	Percent of Total
Paper and paperboard	25.2	55	36.9	57
Glass	8.2	18	11.9	19
Metals	7.1	16	8.4	13
Wood	4.1	9	4.4	7
Plastics	1.0	2	2.5	4
Total	45.6	100	64.1	100

growth of plastics in packaging will exceed the rate of consumption in engineering. The packaging plastic materials are not recycled but join the solid waste stream to be disposed of.

Darnay and Franklin[12] conclude that the outlook for the salvage of paper-based packaging is not encouraging. Three-fourths of the recycled waste paper is from wastes other than packaging. Other than newspaper, none of the recycled paper comes from residences, except for the recycling of 2,500,000–3,000,000 tons of old corrugated boxes.

Salvage operators can profitably pack and grade only those materials separated at the source, and with known contaminants from converters, printers, paper mills, and processors. Once paper has entered a garbage compactor truck, contamination has rendered it worthless. Plastic-coated items, milk cartons, and laminated materials are not acceptable for recycling; waste paper dealers simply say "Burn them."

Collection of packaging wastes presents no special problems if the householder places rubbish in waste containers. Increase in packaging, however, will put an added burden on waste collecting agencies by 1976. Just to haul the increase in packaging volume, waste collection trucks will have to make four million trips in 1976 that they did not have to make in 1966. If new equipment has to be installed to transport the increase, expenditures exceeding $100 million will have to be made in the 10-yr period.

A waste collection problem is presented by packaging materials carelessly flung from cars or left in parks and along rivers and lakes. It is not possible to ascertain just how much of the packaging material tonnage turns into litter, but indications are that well over half the items in litter are packages or package components. In 1976, the "litterbug" will have considerably more ammunition at his disposal; particularly noteworthy will be the greatly expanded use of beverage containers (Table 3): disposal bottles and cans are expected to replace the traditional deposit bottle.

In 1966, eight beverage containers out of every 100 were returnable. Ten years later, only three out of 100 will be returnable. In the same period, per capita consumption of beverage containers will have increased from 155 units in 1966 to 283 units in 1976.

In the two processes, sanitary landfilling and incineration, packaging materials appear to present relatively few difficulties to the waste processor. One reason for this is that more than half of all packaging is paper, which is the most readily disposable of all packaging materials. The other major materials, in an ascending order of difficulty, are wood, plastics, metal, and glass; metals and glass are roughly equal in terms of disposal difficulty.

In landfill operations, the greatest difficulty presented by packaging materials is the resistance of certain materials to degradation in the soil. Aluminum, glass, and plastic containers are virtually indestructible and persist for decades. Even paper tends to have a long life underground.

In incinerators, metals and glass show up as inert residue that must be hauled away; plastics can cause trouble in high concentrations. This can occur when industrial discards, high in plastics content, must be

Table 3 *Beverage Container Consumption, 1966 and 1976 (In Billions of Units)*[a]

	1966	1976
Total	30.5	62.9
Returnable glass	2.5	1.7
Nonreturnable glass	9.5	25.2
Metal cans	18.5	36.0
Beer containers	18.5	28.1
Glass	5.6	9.1
Metal	12.9	19.0
Soft drink containers	9.5	31.7
Glass	3.9	14.7
Metal	5.6	17.0
Liquor and wine (glass)	2.5	3.0

[a] Source: Glass Manufacturers Institute.

handled. Packaging materials do not yet contain sufficiently high percentages of these materials to disrupt operations consistently.

By 1976, the relative difficulties of processing packaging wastes will increase somewhat, but the relative dominance of materials will remain about the same. Paper will still be in the lead and will, in fact, have enlarged its share of total packaging. Plastics will have doubled in tonnage, but in 1976 they will still amount to only 7% that of paper.

In an ideal system, packaging materials would never be discarded—they would be reprocessed by industry and made into new packages or other products. Of the 46 million tons of packaging materials produced in 1966, only about 4.5 million tons were recovered. Recovered packaging materials consist of 2.5 million tons of old corrugated boxes, 1.1 million tons of returnable glass containers, and 900,000 tons of all other materials.

When packaging materials have passed through the trash barrel and the compactor truck, they are thoroughly contaminated with organic matter, dirt, and moisture and intimately mixed with other wastes. It would be virtually impossible to separate them economically from the trash.

This situation might be improved slightly if packaging wastes could be separated before disposal. But the American consumer has a marked distaste for separating rubbish, and is, at the same time, an eager buyer of "disposable" packaging. Monumental efforts at public education would be required to bring about a change in the consumer's attitude in a period of relative national security and prosperity. Even if the consumer could be persuaded to separate rubbish into bins labeled steel, aluminum, plastics, paper, glass, and so forth, these efforts would only ease the problem.

The aims of the package manufacturer and the conservationist not only fail to coincide, but conflict. The purposes of conservation would be best served by packages that are homogeneous in composition and relatively uniform in size and shape. The evolution of packaging, on the other hand, is in the direction of diversity in regard to types of packages and combinations of dissimilar materials.

Relatively few packages are homogeneous in composition: Steel cans contain tin, lead, and organic adhesives, and are overwrapped with paper. Some glass bottles come with twist-off caps that leave a ring of aluminum adhering to the bottle neck. Steel bottle caps are commonly gasketed with plastic or cork. Paper is laminated to metal; coated with clay, plastics, or wax; fastened with organic glues or steel staples; printed with a variety of inks. Plastics are combined with paper or laminated with other plastics and sealed with steel or aluminum closures.

Waste collection remains the last residential utility that is not automated. In part, this may always remain so. A high proportion of residential wastes, however, including most of the packaging materials that reach the home, could be conveyed to central facilities by pipe. Automation could reduce the costs of collection: Gathering and transport of wastes absorb about 90% of all expenditures on waste handling today, and the bulk of collection outlays is in the form of labor costs. Separation of wastes by type, however, cannot be accomplished without manual labor. Innovative techniques of sensing materials and separating them by materials-handling equipment are needed. Furthermore, technology needs be developed to refine heterogeneous waste masses that can be ground to a uniform consistency and conveyed to a "waste refinery" by pipe. Such a technology might encompass chemical refining and extraction, mechanical separation, magnetic removal, mechanical sorting, and so on. Our ingenuity in the processing of virgin materials would indicate that such a technology is within reach.

CHAPTER I.4
industrial waste

Mechanical industry wastes are primarily scrap metal, chips, sand and grit, oily sludges from stamping and assembly, scrap plastic, rubber, and rejects. Here again the industry is solving its own problems. A similar situation exists in household appliance manufacturing.

It is the mechanized "industrial" agriculture and the technologically advanced animal husbandry with other agricultural and forest wastes that are the largest problem and the greatest producers of solid wastes. These are classified in Tables 1–4; they include animal manures, vineyard and orchard prunings, crop harvesting residues, animal carcasses, greenhouse waste, and pesticide containers.

The wastes from chemical industries are a special problem and will be treated separately in this volume. It is estimated that the chemical industry generates about 150 million lb of solid waste each year, 20% being process waste and 80% plant trash and shipping waste. The industry is, however, solving its own problems of disposal without added technology.

Drug industry wastes are related basically to chemical industry waste, and the material is processed in a similar fashion.

Rubber industry wastes are just beginning to receive concentrated attention and this industry will probably solve its problems during the next five years.

Various estimates place mineral waste, spoil, processing residues, mill tailings, slags, and flotation wastes at more than 1.25 billion tons a year. Some estimates are given in Table 5. These wastes will be individually discussed, as will be the slags and smelter wastes from the metallurgical process industries, such as iron and steel, copper, lead, and zinc foundries, mill scales from rolling mills, gypsum wastes, sand and gravel spoil, coal and waste banks from the processing of coal, and related fuels.

Before 1965 it was estimated that 5 million acres, or 700 square miles, were covered with metal and fossil fuel mine spoil and

Table 1 *Sources and Types of Industrial Wastes*[a]

Code	SIC Group Classification	Waste Generating Processes	Expected Specific Wastes
17	Plumbing, heating, air conditioning Special trade contractors	Manufacturing and installation in homes, buildings, factories	Scrap metal from piping and duct work; rubber, paper, insulating materials, misc. construction, demolition debris
19	Ordnance and accessories	Manufacturing and assembling	Metals, plastic, rubber, paper, wood, cloth, chemical residues
20	Food and kindred products	Processing, packaging, shipping	Meats, fats, oils, bones, offal vegetables, fruits, nuts and shells, cereals
22	Textile mill products	Weaving, processing, dyeing, and shipping	Cloth and fiber residues
23	Apparel and other finished products	Cutting, sewing, sizing, pressing	Cloth, fibers, metals, plastics, rubber
24	Lumber and wood products	Sawmills, mill work plants, wooden container, misc. wood products, manufacturing	Scrap wood, shavings, sawdust; in some instances metals, plastics, fibers, glues, sealers, paints, solvents
25	Furniture, wood	Manufacture of household and office furniture, partitions, office and store fixtures, mattreses	Those listed under Code 24; in addition, cloth and padding residues
25	Furniture, metal	Manufacture of household and office furniture, lockers, bedsprings, frames	Metals, plastics, resins, glass, wood, rubber, adhesives, cloth, paper
26	Paper and allied products	Paper manufacture, conversion of paper and paperboard, manufacture of paperborad boxes and containers	Paper and fiber residues, chemicals, paper coatings and fillers, inks, glues, fasteners
27	Printing and publishing	Newspaper publishing, printing, lithography, engraving, and bookbinding	Paper, newsprint, cardboard, metals, chemicals, cloth, inks, glues

28	Chemicals and related products	Manufacture and preparation of inorganic chemicals (ranges from drugs and soups to paints and varnishes, and explosives)	Organic and inorganic chemicals, metals, plastics, rubber, glass, oils, paints, solvents, pigments
29	Petroleum refining and related industries	Manufacture of paving and roofing materials	Asphalt and tars, felts, asbestos, paper, cloth, fiber
30	Rubber and miscellaneous plastic products	Manufacture of fabricated rubber and plastic products	Scrap rubber and plastics, lampblack, curing compounds, dyes
31	Leather and leather products	Leather tanning and finishing; manufacture of leather belting and packing	Scrap leather, thread, dyes, oils, processing and curing compounds
32	Stone, clay, and glass products	Manufacture of flat glass, fabrication or forming of glass; manufacture of concrete, gypsum, and plaster products; forming and processing of stone and stone products, abrasives, asbestos, and misc. nonmineral products	Glass, cement, clay, ceramics, gypsum, asbestos, stone, paper, abrasives
33	Primary metal industries	Melting, casting, forging, drawing, rolling, forming, extruding operations	Ferrous and nonferrous metals scrap, slag, sand, cores, patterns, bonding agents
34	Fabricated metal products	Manufacture of metal cans, hand tools, general hardware, nonelectric heating apparatus, plumbing fixtures, fabricated structural products, wire, farm machinery and equipment, coating and engraving of metal	Metals, ceramics, sand, slag, scale, coatings, solvents, lubricants, pickling liquors

Table 1 *(Continued)*

Code	SIC Group Classification	Waste Generating Processes	Expected Specific Wastes
35	Machinery (except electrical)	Manufacture of equipment for construction, mining, elevators, moving stairways, conveyors, industrial trucks, trailers, stackers, machine tools, etc.	Slag, sand, cores, metal scrap, wood, plastics, resins, rubber, cloth, paints, solvents, petroleum products
36	Electrical	Manufacture of electric equipment, appliances, and communication apparatus, machining, drawing, forming, welding, stamping, winding, painting, plating, baking, firing operations	Metal scrap, carbon, glass, exotic metals, rubber, plastics, resins, fibers, cloth residues
37	Transportation equipment	Manufacture of motor vehicles, truck and bus bodies, motor vehicle parts and accessories, aircraft and parts, ship and boat building and repairing motorcycles and bicycles and parts, etc.	Metal scrap, glass, fiber, wood, rubber, plastics, cloth, paints, solvents, petroleum products
38	Professional, scientific controlling instruments	Manufacture of engineering, laboratory, and research instruments and associated equipment	Metals, plastics, resins, glass, wood, rubber, fibers, abrasives
39	Miscellaneous manufacturing	Manufacture of jewelry, silverware, plated ware, toys, amusement, sporting and athletic goods, costume novelties, buttons, brooms, brushes, signs, advertising displays	Metals, glass, plastics, resins, leather, rubber, composition, bone, cloth, straw, adhesives, paints, solvents

[a] Source: Standard industrial classification manual, 1967.

Table 2 *Solid Wastes from Agriculture (USDA)*

Source	Waste	Composition
Farms, ranches Greenhouses	Crop waste	Cornstalks, prunings, sugarcane bagasse, drops and culls from fruit and vegetables washer waste, stubble and straw, hulls, fertilizer bags, etc.
	Forest slash	Trees, stumps, limbs, debris
Animal husbandry	Animal manure Pouch manure Poultry manure Swine manure	Lignin, organic fibers, nitrogen, phosphorus, potassium, proteins, fats, carbohydrates
	Animal carcasses	Ammonia and nitrates, proteins, etc., flesh, blood, fat, oil, grease
	Pesticides, insecticides, herbicides, fungicides, vermicides and microbiocides	Chlorinated organics, phosphorus, complex organic and inorganic salts

processing waste. Mining operations of about 20,000 active service and strip mines carve out about 153,000 acres annually as waste piles.

Incinerators, refuse dumps, sanitary landfills, automobile graveyards, compost plants, biologic oxidation units, and solids from water treatment plants are common and visible manifestations of solid wastes. Industrial wastes in warehouses and back lots of industrial properties, the piles and acres of sludges, slags, wastes, plastics, scrap metal, bales of rags and paper, and drums of off-grade products are not so readily visible; nor are the disposal problems of service stations, transportation industries, and numerous mechanical and electrical manufacturing agencies visible. These wastes are as diversified as industry itself, with wide variations from industry to industry, perhaps with the greatest variation in the chemical process operations.

In the latter category, wastes at one stage are often the raw material for a second, and thus by-products are the conventional designation and utilization. However, even this chain reaches its end when it is not feasible to incorporate the waste into new processes, and the cost of storing it must be evaluated against forms of disposal. Individual companies specialize in waste disposal largely through sophisticated types of incinerators, markedly different from those employed for urban wastes.

Table 3 Major Agricultural Crops (USDA)

Crop	Acres Harvested (10^6)	Field Waste (tons/cap/yr)	Field Waste (10^6 tons/yr)	Nature
Corn, grain	57	4.5	256	Leaves, stalks
Wheat	50	1.3	65	Stubble, straw
Oats	18	1.8	33	Stubble, straw
Barley	11	1.8	19	Stubble, straw
Rye	1.5	1.3	1.8	Stubble, straw
Mixed grains	1	1.3	1.3	Stubble, straw
Rice	2	3.	5.9	Stubble, straw
Flax	2.7	0.8	1.1	Leaves, stalks, straw
Alfalfa, clover seed	1.4	0.8	1.1	Leaves, stalks, straw
Sorghum for grain	13.0	3.0	39.0	Leaves, stalks, straw
Cotton	10.0	2.0	20.0	Leaves, stalks, straw
Beans, dry	2.0	2.0	3.0	Leaves, stalks, straw
Peas, dry	0.4	2.0	0.7	Leaves, stalks, straw
Soybeans for beans	37.0	2.0	74.0	Leaves, stalks, straw
Peanuts for nuts	1.6	3.0	4.4	Leaves, stalks, straw
Potatoes	1.7	3.0	4.4	Leaves, vines, culls
Sweet potatoes	0.2	3.0	0.6	Leaves, vines, culls
Tobacco	1.0	0.5	0.5	Leaves, vines, culls
Sugar cane	0.7	0.5	0.4	Leaves, vines, culls
Sugar beets	1.2	3.0	3.5	Leaves, vines, culls
Vegetables	4.0	3.0	11.0	Leaves, stalks, culls
Fruits and nuts	5.0	2.0	10.0	

With addition of omitted specialty crops, tung nut, castor oil bean, guar, etc., this waste is of the order of 3 tons/cap/yr or 16.5 lb/cap/day.

Table 4 Waste of Major Farm Animals[a]

Animal	Population ($\times 10^6$)	Manure (tons/unit/yr)	(10^6 tons)
Cattle	110	10	1100
Hogs	48	8	400
Sheep	22	3	66
Horses, mules	1	10	10
Poultry			
Broilers	2600	0.005	13
Turkeys	116	0.025	3
Layers	350	0.05	16
Ducks, pigeons, rock hens, etc.	10	0.05	0.5
Pets			
Cats	21	0.06	1.25
Dogs	25	0.10	2.5

Approximately 8 tons/cap/yr or 45 lb/cap/day

[a] Calculated from data furnished by Prof. Tainganides, University of Urbana, Illinois.

Table 5 Wastes from Mineral Extraction[a]

Material	Mine Wastes (10^6 tons)	Mill Tailings (10^6 tons)	Washing Plant (10^6 tons)	Slag	Processing Plant (10^6 tons)	Total (10^6 tons)
Copper	300	200		6		506
Iron and steel	120	110		15	2	247
Coal, bituminous	13		90			103
Coal, anthracite	1		3			4
Phosphates	1		60	4	10	75
Lead–zinc	3	20	2			25
Aluminum					6	6
Coal ash					26	26
Others	10	10	30	15	235	300
Total	448	340	185	40	279	1292

[a] Source: U.S. Bureau of Mines.

The typical industrial worker contributes some eight times as much solid waste as the average community resident, while the same typical employee contributes as plant trash, office papers, scrap coffee cups, or other containers, paper towels, sanitary items, and newspapers about as much solid waste per day as he generates at home. Table 1 gives the distribution among 21 code groups of waste for disposal, in a survey by Combustion Engineering, 1967.[4]

There are two basic options for handling industrial solids; they reduce to one plan of action: Turn whatever you can into something useful, then reduce whatever is left into the smallest practical volume and put it into the most innocuous place you can find. Both options are subject to economic constraints. There is not much "useful" about a fly ash cinderblock if it costs more to make and market than to throw the ash away in the first place. There is not much sense in reducing waste volume if you can dump all you want in the back yard of the factory. Furthermore, reclamation costs, which are 70% labor, are generally rising faster than the prices at which recycled material can be sold.

However, technology is evolving especially for sorting wastes and turning them into energy. Grinders, hammermills, and wet pulpers are being tried for sorting, and water-wall incinerators and pyrolytic techniques are becoming increasingly common in the U.S. for recovering energy. Paper and lumber needs, for instance, will outpace natural forest growth by about 1980. Air pollution controls for an incinerator may cost more than the incinerator itself. Widespread environmental concern has reduced the stigma of recycled materials in items such as textile fibers and plastics.

CHAPTER I.5
agricultural waste

The growing, harvesting, processing, packaging, preservation, and preparation for market of food crops generate large tonnages of solid wastes. Agricultural experts have estimated that of the weight of corn grown for canning, about 50% is field waste, about 30% is process waste, and less than 20% is actual corn in the can. Mechanical harvesting has caused a shift in solid waste location from the rural to the urban setting because all the hand-harvesting methods were more selective in sorting out the usable crop portion. Culls, plant parts, and other wastes were left in the field, where they were piled up or burned. With mechanical harvesting, more of the crop may be waste that is transported from the usable crop portion to a central location, where previous waste handling techniques cannot be used. Production of processed foods has more than doubled in 25 years and is on the increase, with concomitant increases in the generation of solid waste. Rice generates a stubble problem, and the milling of rice results in 20% of the crop being disposed of as hulls, containing 18% silica, not readily burnable, difficult to plow under, and with no application.

ANIMAL WASTE

Domestic animals produce about one billion tons of fecal waste a year and over 400 million tons of liquid wastes. Used bedding, pouch manures from abattoirs, and dead carcasses result in an annual animal production of solid wastes of more than 2 billion tons. Over 50% of the animal wastes are generated in concentrated growing operations close to urban areas. Large-scale animal raising operations have developed rapidly in the last twenty years. One feedlot with 10,000 head of cattle can produce 260 tons of manure a day, all in the wrong place and not competitive with chemical fertilizer. A dairy with 100 milk cows produces 14 tons of solid waste a day. A poultry enterprise, with a million eggs per day, produces 50 tons of waste a day. Tables 2 and 3 in Chapter II.4 give an estimate of solid waste generation from major agricultural crops, but the figures do not include processing wastes. These materials will be discussed individually. Table 4 in Chapter II.4 gives an estimate of the solid waste generated by major farm animals.

CHAPTER I.6

manpower of waste handling

The number of people engaged in solid waste disposal in the United States is small compared to the number engaged in government, teaching, and the professions. Tables 1–4 list some of the relevant statistics. Each individual generates about a ton of waste a year, with the result that millions of tons are produced; agricultural and food processing wastes add billions of tons annually.

The conclusion from these and other statistics is that as a people we are more interested in vocational and professional services than we are in intelligently disposing of the waste that inevitably results from those services; and it is only when some unusual crisis, like the garbagemen's strike in New York City, forces the matter on our attention that we even become aware of the problem.

We have all the technology required to solve our waste problem, esthetically and technically, if we are willing to pay the price, as we have been with our schools, our drugs, our national parks, our housing, and our poor.

In the chemical process industries, wastes generally become useful by-products. Even here, however, the chemical engineer has not been accorded his rightful place in helping to solve the problems.

The National Solid Wastes Survey,[13] referring to the use of labor in contrast to engineers, made the following statement:

> All of these needed and necessary improvements and increased efforts are going to require large increases in manpower, capital investments, and operating expenses. To operate existing disposal sites as sanitary landfills may require well over 10,000 new sanitary landfill operators. Further, efforts must be made on an intensive capital development basis to provide the maximum amount of automation and control for incinerators and the minimum amount of manpower. The current concept of the manpower required in incineration operation must be completely changed. We can no longer afford to overstaff our incinerators in

Table 1 *Origin and Collection of Solid Wastes. Sample: 46% of Population; 92,500,000 Persons*

	Urban	Rural	Industry
Population (%)	73–75	25–27	
Operating collection systems (%)	64	22–24	12
Collection (%)	Household waste	Commercial	Industrial
Public	56	25	13
Private	32	62	57
Individual	12	13	30
Separation (%)	Garbage and waste	Combined systems	Simultaneous
	33	56	12
Collection frequency	Weekly	Twice a week	More often
(%)	48	32	20

Table 2 *Manpower and Machinery*

	Public Sector	Private Sector	Total
Manpower, 340,000 people			
1 for every 600 inhabitants	53%	47%	
Operations			
Collecting household and commercial waste	75%	65%	
Industrial waste	5%	27%	
Street cleaning	20%	8%	
Vehicles, compactor trucks, units	44,650	50,350	95,000
1 for every 2100 inhabitants			
Collecting household and commercial waste	94%	71%	
Industrial	4%	23%	
Street cleaning	2%	6%	
Vehicles other than compactor trucks	36,000	144,000	180,000
1 for every 1100 inhabitants	20%	80%	
Collecting household and commercial wastes	47%	30%	
Industrial	5%	68%	
Street cleaning	48%	2%	

Table 3 *Type of Worker and Number of People Served*

Type of Worker	No. of People Served
State governmental worker	21
Federal governmental worker	75
Truck drivers	99
Elementary and secondary school teachers	100
Waiters, bartenders and countermen	165
Carpenters	217
Bankers	227
Charwomen and janitors	274
College professors	315
Engineers, all varieties	375
Nurses, professional	376
Mechanics and repairmen	377
Machinists and job setters	377
Painters and paper hangers	478
Attendants, auto parking	560
Electricians	586
Telephone operators	581
Solid waste operators	600
Plumbers and pipe fitters	625
Insurance agents and brokers	625
Physicians	678
Laundry and dry cleaning operators	692
Policemen and sheriffs	747
Linemen, telephone, telegraph, and power	758
Hairdressers and cosmetologists	761
Midwives	942
Cooks	958
Masons	981
Lawyers and judges	1004
Clergymen	1050
Toolmakers	1120
Barbers	1182
Firemen	1489
Dentists	1882

an effort to operate antiquated and poorly designed facilities beyond their capabilities. To attract and retain the manpower with the qualifications needed, salaries and benefits for solid waste personnel must be increased several fold. The solid waste management field, both public and private, must make efforts to train personnel and provide the opportunity for a feeling of pride in their work.

Table 4 *Activities of Engineers in Environmental Control and Human Factors*

Water supply	5100
Sewage and sewage treatment	3500
Safety	2300
Soil mechanics	2200
Hydrology	2100
Traffic	2100
Water pollution	2000
Surveying and mapping	1900
Air pollution	1800
Environmental control	1800
Drainage	1600
Bioengineering	1300
Illumination	1300
Irrigation	800
Reclamation	800
Industrial waste	500
Biomechanics	200
Geodesy	200
Photogrammetry	200
Meteorology	200
Industrial hygiene	100
Total	32,000

One solid waste collector or driver is employed in the public or private sector for every 590 persons. Extrapolating to the nation, this indicates that more than 337,000 persons are involved in collecting and transporting solid waste from various sources. About 53 percent of these people are employed in the public sector with 47 percent being used in the private sector. For the public sector, about 75 percent are involved in collecting household and commercial wastes, 5 percent are used for industrial wastes and 20 percent for street cleaning. In the private sector only 65 percent are employed in collecting household and commercial wastes while 27 percent are employed in the industrial area with only 8 percent for street cleaning.

CHAPTER I.7

example of legislation and cooperation: erie county, pennsylvania

Solid waste is a people's housekeeping problem in which many private, municipal, county, and state governmental agencies must be involved and cooperate. In Erie County, Pennsylvania, the solid waste problem is more serious than water pollution. For water pollution each new approved septic system and every foot of sanitary sewer installed, each industrial violation eliminated, reduces the water pollution problem. However, in the area of solid waste, the continually expanding population generates increasing amounts of wastes; this contributes to environmental problems, since there is no comprehensive program for solid waste control. The first organized effort came from the recommendations made in an environmental health survey report titled ERIELAND 2000. Erie County was selected as the first area by the Pennsylvania State Department of Health and the USPHS to conduct a demonstration on Urban Planning for Environmental Health. The study began in September 1965 and was presented in the spring of 1967. An advisory committee composed of representatives from a college, the Pennsylvania Department of Health, the Erie County Department of Health, City and County Planning Departments, a planning consultant, local governments, medical association, the USPHS, and lay people provided the guidance. The ERIELAND 2000 study proposed the following:

1. The Commissioners of Erie County, with the support of the local officials, should establish a "Solid Waste Disposal Authority" which would provide the entire county with controlled and suitable disposal facilities and sites.
2. Application should be made to the State of Pennsylvania and to the Federal Government for the funds to make a study of solid waste problems in Erie County and to develop a master plan.

3. The Solid Waste Disposal Authority should coordinate a master solid waste plan for Erie County, as soon as it has been developed, with future land-use plans and implement its operation.

4. Necessary legislation should be established to eliminate open dumping and open burning in Erie County.

5. The Solid Waste Disposal Authority should assume the responsibility, and develop a plan with industry in Erie County, for the handling and disposal of industrial wastes in an acceptable manner.

6. The Solid Waste Disposal Authority should assume the responsibility for the control of junked autos and other bulky waste and should establish regulations to ensure storage, handling, processing, and disposal of these wastes.

Implementation of the recommendations began before the final ERIELAND 2000 study was presented in 1967. In May 1966, application was made for a federal grant to conduct an engineering study of present solid waste disposal sites and practices and to devise a plan for disposal of solid wastes. On August 16, 1966, an engineering firm began the county-wide study under Federal Demonstration Grant DO1-SW-0035. The study was presented in March 1968. The engineering report pointed out the need for a comprehensive waste management plan, and recommended an interim plan to serve until the county-wide program could be implemented.

The interim plan suggested the establishment of landfills to fill the immediate critical need, but not as the total solution of the problem for an indefinite period; there were few suitable disposal sites because of soil conditions in Erie County. Following the engineering report, the County Department of Health recommended that the County Commissioner appoint an Advisory Board to study the report. The Advisory Board was appointed in 1968. Upon completing the study, the Advisory Board recommended that the County Commissioners establish a Solid Waste Management Authority with the responsibility of establishing an interim Solid Waste Plan and a detailed comprehensive plan.

The authority was formed in May 1969. The County Solid Waste Management Authority so created has the power to acquire, hold, construct, maintain, operate, own, improve, and lease in capacity either of lessor or lessee, projects, facilities, and equipment for the collection, removal, or disposal of ashes, garbage, rubbish, and other refuse and waste materials by incineration, landfill, or other methods. The authority engaged in preparing the specifications was to initiate the interim Solid Waste Disposal Plan so that the project could be placed out for bids.

The authority negotiated with a Solid Waste Management engineer to prepare a master plan for storage, collection, and disposal of waste for the entire county. The original engineering study covered only the aspect of disposal.

The County Solid Waste Authority has been studying all current and experimental methods of waste management. The authority considered standard and high-temperature incineration, composting, transporting of wastes by rail, transfer stations, fuel cells, sanitary landfills, and combinations of methods. The county made an application for a grant for a study of the extraction and recycling (reusing or reclaiming) of solid wastes. This study and investigation grant will be necessary if a plan is utilized to produce fuel cells to remove deleterious substances and could ensure the economic feasibility for extraction–recycling. Also, since many of the disposal practices under study require the use of landfills, and such landfill sites are limited in Erie County, the information relative to extraction and recycling is necessary for the authority to formulate a comprehensive plan.

Realizing that the success of any solid waste management program would depend on public acceptance of the program, an attitude survey was conducted in June 1969. The attitude survey has been included as part of the Federal Demonstration Grant DO1-SW-0035. The survey indicated that the public's knowledge relative to solid waste practices was fragmented, disorganized, and often erroneous. Therefore, the county has applied for a grant to modify favorably attitudes concerning solid waste management.

In the summer of 1967, the Erie County Department of Health conducted a survey in all 40 municipalities in the county to determine the storage practices, volume of refuse collected, and the methods of disposal. The survey also indicated that few communities had satisfactory solid waste laws. Upon completion, a survey was made of all industries in Erie County in 1968 to determine the volume of refuse generated and the disposal practices. The study was in line with the recommendations made in the ERIELAND 2000 study. The information was also necessary for the comprehensive plan.

Also, in 1969 the State Solid Waste Management Act became effective. Under provisions of this act, state-wide regulations and standards have been established for solid waste management. The act required that disposal sites that do not meet the standards must be closed by 1970. The act also required counties to provide comprehensive solid waste management plans by January 1, 1971. Erie County has demonstrated its determination to provide a comprehensive solid waste management plan by the formation of one of the three Solid Waste Authorities in the country. Each step in the county program has been logical and deliberate, so that the comprehensive plan upon completion can be implemented.

Extending the concept that solid wastes are a people's housekeeping project, the Pennsylvania Solid Waste Management Act of August 1968, is included here. Note the citizens' involvement in Section 4 and the political and interagency relationships and their complications. The act is introduced only to show these relationships and coordination. (Amended acts and the latest forms should be consulted for further information.)

NO. 241; AN ACT

SB 514

Providing for the planning and regulation of solid waste storage, collection, transportation. processing and disposal systems; requiring municipalities to submit plans for solid waste management systems in their jurisdiction; authorizing grants to municipalities; requiring permits for operating processing or disposal systems; imposing duties on and granting powers to municipalities; authorizing the Department of Health to adopt rules, regulations, standards and procedures; creating an advisory committee; providing remedies, prescribing penalties, and making an appropriation.

The General Assembly of the Commonwealth of Pennsylvania hereby enacts as follows:

Section 1. Short Title.—This act shall be known and may be Cited as the "Pennsylvania Solid Waste Management Act."

Section 2. Legislative finding; declaration of policy.—It is hereby determined and declared as a matter of legislative finding that, since improper and inadequate solid waste practices create public health hazards, environmental pollution and economic loss, it is the purpose of this act to:

(1) Establish and maintain a cooperative state and local program of planning and technical and financial assistance for comprehensive solid waste management;

(2) Utilize, wherever feasible and desirable, the capabilities of private enterprise in accomplishing the desired objectives of an effective solid waste management program; and

(3) Require permits for the operation of processing and disposal systems.

Section 3. Definitions.—The following words and phrases shall have the meaning ascribed to them in this section unless the context clearly indicates otherwise:

(1) "Department" means the Department of Health of the Commonwealth of Pennsylvania and its authorized representatives.

(2) "Secretary" means the Secretary of Health of the Commonwealth of Pennsylvania

(3) "Solid waste" means garbage, refuse and other discarded materials including, but not limited to, solid and liquid waste materials resulting from industrial, commercial, agriculture and residential activities.

(4) "Solid waste management system" means the entire process of storage, collection, transportation, processing and disposal of solid wastes by any person engaging in such process as a business or any municipality, authority, county or any combination thereof.

(5) "Municipality" means a city, incorporated town, township and borough.

(6) "Person" means individual, partnership, corporation, association, institution, cooperative enterprise, or legal entity.

Section 4. Advisory Committee.—(a) An Advisory Committee consisting of twenty-two members shall be appointed by the governor, membership of which shall include one representative of the Department of Agriculture, Department of Commerce, Department of Community Affairs, State Planning Board, Pennsylvania State Association of Township Supervisors, Pennsylvania State Association of Township Commissioners, Pennsylvania State Association of Boroughs, Pennsylvania League of Cities, Pennsylvania Association of County Commissioners, Pennsylvania Municipal Authorities Association, Pennsylvania State University, Drexel Institute of Technology, Pennsylvania State Grange, Pennsylvania Farmers' Association, Pennsylvania Home Builders' Association, Pennsylvania Canners and Food Processors Association and Pennsylvania Society of Professional Engineers.

(b) The Advisory Committee may recommend to the secretary the adoption, amendment or repeal of such rules, regulations, standards and procedures as it deems necessary and advisable for the implementation of the act.

(c) The members of the committee shall not receive any compensation for their services but shall be reimbursed for their actual and necessary expenses incurred in the performance of their duties.

Section 5. Submission of Plans.—(a) Each municipality with a population density of three hundred or more inhabitants per square mile shall submit to the department an officially adopted plan for a solid waste management system or systems serving areas within its jurisdiction, within two years of the effective date of this section, and shall, from time to time, submit such revisions of said plan as it deems necessary or as the department may require.

(b) When more than one municipality has authority over an existing or proposed solid waste management system or systems, or any part thereof, the required plan or any revision thereof may be submitted jointly by the municipalities concerned or by an authority or county or by one or more of the municipalities with the concurrence of the others.

(c) Every plan, and any revision thereof, shall delineate areas where solid waste management systems are in existence and areas where the solid waste management systems are planned to be available within a ten-year period.

(d) Every plan shall:

(1) Provide for the orderly extension of solid waste management systems in a manner consistent with the needs and plans of the whole area, and in a manner which will not create pollution of the waters or air of the Commonwealth, nor constitute a public nuisance and shall otherwise provide for the safe and sanitary disposal of solid waste.

(2) Take into consideration all aspects of planning, zoning, population estimates, engineering and economics so as to delineate with all practicable precision those portions of the area which may reasonably be expected to be served by a solid waste management system within the next ten years as well as those areas where it is not reasonably foreseeable that a solid waste management system will be needed after ten years;

(3) Take into consideration any existing State plan affecting the development, use and protection of air, water or land resources;

(4) Set forth a time schedule and proposed methods of financing the development, construction and operation of the planned solid waste management systems, together with the estimated cost thereof;

(5) Include a provision for periodic revision of the plan;

(6) Include such other information as the department shall require.

(e) The plan shall be reviewed by appropriate official planning agencies within a municipality including a planning agency with areawide jurisdiction, if one exists, and the county planning commission, for consistency with programs of planning for the area, and all such reviews shall be transmitted to the department with the proposed plan.

(f) The department is hereby authorized to approve or disapprove plans for solid waste management systems submitted in accordance with this act. Any plan which has not been disapproved within one year of the date of its submission shall be deemed an approved plan. In case any plan is disapproved, a hearing shall be held thereon before the department within fifteen days after request therefore is made by the municipality, municipalities, county or authority whose plan is disapproved. Within seven days following the date of such hearing, the department shall notify all parties in writing of the determination of said hearing and the reasons therefor. Any party aggrieved by this determination shall have the right of appeal in accordance with the provisions of the act of June 4, 1945 (P.L.1388), known as the "Administrative Agency Law."

(g) The department is authorized to provide technical assistance to counties, municipalities and authorities in coordinating plans for solid waste management systems required by this act, including revisions of such plans.

(h) The department may establish priorities for the time within which plans shall be submitted and may, in appropriate cases, recommend the submission of joint plans.

(i) The department may institute an action in mandamus in the court of common pleas of the county in which the municipality is located to compel municipalities to submit plans in accordance with this act and the rules, regulations and procedures of the department.

Section 6. Powers and Duties of the Department.—The department shall have the power and its duty shall be to:

(1) Administer the solid waste management program pursuant to the provisions of this act.

(2) Cooperate with appropriate Federal, State, interstate and local units of government and with appropriate private organizations in carrying out its duties under this act.

(3) Adopt such rules, regulations, standards and procedures as shall be necessary to conserve the air, water and land resources of the Commonwealth protect the public health, prevent public nuisances, and enable it to carry out the purposes and provisions of this act.

(4) Develop a State-wide solid waste management plan in cooperation with local governments, the Department of Community Affairs and the State Planning Board. When feasible, emphasis shall be given to area-wide planning.

(5) Provide technical assistance to municipalities, counties and authorities including the training of personnel.

(6) Report to the legislature from time to time on further assistance that will be needed to administer the solid waste management program.

(7) Initiate, conduct and support research, demonstration projects,

and investigations and coordinate all State agency research programs pertaining to solid waste management systems.

(8) Establish policies for effective solid waste management systems.

(9) Issue such permits and orders and conduct such inspections as may be necessary to implement the provisions of this act and the rules, regulations and standards adopted pursuant to the act.

Section 7. Applications and Permits.—(a) It shall be unlawful for any person, municipality, county or authority to use or continue to use their land or the land of any other person, municipality, county or authority as a solid waste processing or disposal area of a solid waste management system without first obtaining a permit from the department: Provided, however, that this section shall not apply to farmers and they shall not be required to obtain a permit for normal farming operations: And, provided further, that this section shall not apply to the storage of by-products which are utilized in the processing or manufacturing of other products.

(b) Application for a permit shall be in writing and shall be made on a form prescribed, prepared and furnished by the department and shall set forth such information and be accompanied by such data as the department may require.

(c) Upon approval of the application, the department shall issue a permit for the operation of each solid waste processing or disposal facility or area set forth in the application.

(d) Plans, designs and relevant data for the construction or alteration of solid waste processing and disposal facilities and the location of solid waste processing and disposal areas shall be prepared by a registered professional engineer and shall be submitted to the department for approval prior to the construction, alteration or operation of such facility or area except when food process wastes are used for agricultural purposes in a manner which will not create a public health hazard or pollution of the air or water.

(e) Any permit granted by the department, as provided in this act, shall be revocable or subject to suspension at any time the department shall determine that the solid waste processing or disposal facility or area is, or has been conducted in violation of this act or the rules, regulations, or standards adopted pursuant to the act, or is creating a public nuisance.

(f) In any case where a permit is required by this section for the disposal of solid wastes produced by a public utility or a municipally owned facility producing a public utility service, the department shall not refuse an application, or revoke or suspend a permit previously granted, unless it first obtains from Pennsylvania Public Utility Commission a certification that such refusal, revocation or suspension will not adversely affect utility service to the public.

(g) In case any permit is denied, suspended or revoked, a hearing shall be held thereon before the department within fifteen days after request therefor is made by the person, municipality, county or authority whose permit is denied, suspended or revoked. Within seven days following the date of such hearing the department shall notify all parties in writing of the determination of said hearing and the reasons therefor. Any party aggrieved by this determination shall have the right of appeal in accordance with the provisions of the act of June 4, 1945 (P.L.1388), known as the "Administrative Agency Law."

Section 8. State Agencies.—All State institutions and agencies, including the General State Authority and the State Public School Buildings Authority, shall obtain a permit from the department under the provisions of section 7 of this act and shall also comply with all other provisions of this act.

Section 9. Prohibited Acts.—It shall be unlawful for any person, municipality, county, or authority to:

(1) Dump or deposit, or permit the dumping or depositing of any solid wastes onto the surface of the ground or into the waters of the Commonwealth without having obtained a permit as required by section 7: Provided, that this provision shall not prohibit the use of solid wastes in normal farming operations or in the processing or manufacturing of other products in a manner that will not create a public nuisance or adversely affect the public health: And, provided further, that this provision shall not prohibit individuals from dumping or depositing solid wastes resulting from their own residential activities onto the surface of ground owned or leased by them when such wastes do not thereby create a public nuisance or adversely affect the public health.

(2) Construct, alter or operate a solid waste processing or disposal facility or area of a solid waste management system without a permit or other approval from the department or in violation of the rules, regulations, standards, or orders of the department.

(3) Burn solid wastes except in a manner approved by the Air Pollution Commission or the department.

(4) Store, collect, transport, process or dispose of solid waste contrary to the rules, regulations, standards or orders of the department or in such a manner as to create a public nuisance.

(5) Refuse or hinder entry and inspection by an agent or employe of the department after such agent or employe identifies himself and gives notice of his purpose.

No person shall be held responsible under the provisions of this section for the dumping or depositing of any solid waste on ground owned or leased by him without his expressed or implied consent, permission or knowledge.

Section 10. Powers and Duties of Municipalities.—

(a) Each municipality with a population density of three hundred or more inhabitants per square mile shall be responsible for the collection, transportation, processing and disposal of solid wastes within its boundaries.

(b) In carrying out its responsibilities, any such municipality may adopt ordinances, regulations and standards for the storage and collection of solid wastes which shall be in conformity with the rules, regulations, standards and procedures adopted by the department for the storage, collection, transportation, processing and disposal of solid waste.

(c) Municipalities may contract with any person, other municipality, county or authority to carry out their responsibilities for the collection, transportation, processing and disposal of solid wastes.

Section 11. Orders to Municipalities.—(a) If the department finds that the storage, collection, transportation, processing or disposal of solid waste from a municipality subject to the provisions of section 10 (a) is causing pollution of the land, air or waters of the Commonwealth or is creating a public nuisance, the department may order the municipality to alter its storage, collection or transportation systems or provide such storage, collection or transportation systems as will prevent pollution and public nuisances. Such order shall specify the length of time, after receipt of the order, within which the facility or area shall be repaired, altered, constructed or reconstructed. Any party aggrieved by an order under this section shall have the right of appeal in accordance with the provisions of the act of June 4, 1945 (P.L. 1388), known as the "Administrative Agency Law."

(b) Any municipality ordered by the department to repair, alter, construct or reconstruct a solid waste facility or area shall take such steps for the repair, alteration, construction or reconstruction of the facility or area as may be necessary for the processing and disposal of its solid waste in compliance with this act and the rules, regulations, standards, and orders of the department.

(c) The department may institute an action in mandamus in the court of common pleas of the county in which the municipality is located to compel compliance with an order issued under subsection (a) of this section.

Section 12. Grants Authorized; Appropriation.—

(a) The department is authorized to assist counties, municipalities, and authorities by administering grants to pay up to fifty per cent of the costs of preparing official plans for solid waste management systems in accordance with the requirements of this act and the rules, regulations and standards adopted pursuant to this act, and for carrying out related studies, surveys, investigations, inquiries, research and analyses.

(b) All grants shall be made from funds appropriated for this purpose by the General Assembly.

(c) Any municipality with a population density of less than three hundred inhabitants per square mile may elect to be governed by the provisions of this act or to establish within such municipality waste disposal districts subject to the provisions of this act, and such municipality shall thereby become eligible for grants under this section.

(d) The sum of fifty thousand dollars ($50,000), or as much thereof as may be necessary, is hereby specifically appropriated to the department for the fiscal year July 1, 1968 to June 30, 1969.

Section 13. Restraining Violations.—In addition to any other remedies provided in this act, the secretary may institute a suit in equity in the name of the Commonwealth in the court of common pleas of the county where the violation or nuisance exists for an injunction to restrain a violation of this act or the rules, regulations or standards adopted thereunder and to restrain the maintenance of a public nuisance.

Section 14. Penalties.—(a) Any person violating this act or the rules, regulations or standards thereunder shall, upon conviction thereof in a summary proceeding, be sentenced to pay a fine of not more than three hundred dollars ($300) and costs and, in default of the payment of such fine and costs, shall undergo imprisonment for not more than thirty days.

(b) Violations on separate days shall be considered separate and distinct offenses under subsection (a) of this section.

(c) All fines and penalties imposed under the provisions of this section shall be paid into the General Fund of the Commonwealth.

Section 15. Severability Clause.—The provisions of this act are severable and if any provision or part thereof shall be held invalid or unconstitutional or inapplicable to any person or circumstances, such invalidity, unconstitutionality or inapplicability shall not affect or impair the remaining provisions of the act.

Section 16. Saving Clause.—Nothing in this act shall be deemed to affect, modify, amend or repeal any provision of any act administered by the Department of Health, Sanitary Water Board, Air Pollution Commission, Department of Mines and Mineral Industries or any other department, board, commission or agency of the Commonwealth.

Section 17. Effective date.—Section 4 shall take effect immediately. Section 5 and subsection (a) of section 12 shall take effect January 1, 1969. The remainder of the act shall take effect January 1, 1970.

APPROVED—July 31, 1968.

RAYMOND P. SHAFER

In this volume, after presenting the problem, we examine, often in the form of case studies, the origin, collection, processing, and disposal of solid waste. This is primarily the problem of engineers: sanitation has been the area of the civil engineer, who now often addes "environmental" to his title. The chemical engineer is expert in corrosion and processing, and his industries have often been leaders in solving difficult waste problems. The mechanical engineer is involved in incinerator design; geologists and mining engineers contribute their expertise in the mineral sector. Metallurgists have been engaged in recycling for generations. Electrical engineers provide the necessary power and controls, and the practitioners of the life sciences are required to study the problems of pest control and other disease vectors that may be generated by solid wastes.

All are united in a common cause with lawmakers, government agencies, and the people themselves, in attempting to reconcile our seemingly boundless appetites for the goods provided by industry with our—up to now—unwillingness to pay the price of disposing of the effluent of that same industry.

REFERENCES

1. Leo Weaver, Refuse Collection and Control Bibliography 1954–1955, PHS91B, USHEW, PHS, BDPEC, NCUIH, 1963, 32 pp.
2. Edward R. Williams, PHS 91C, USHEW, BDPC, NCUIH, 1958, 44 p.
3. Edward R. Williams and Ralph J. Black, PHS91, USHEW, PHS, BDPC, NCUIH, 1961, 73 p.
4. PHS 1886, USHEW, ECA, BSWM, Rockville, Maryland, Clearing House for Federal Scientific and Technical Information, USDC, Washington, D.C., PB 187712, 1969.

5. A. J. Mulich, A. J. Klee, and P. W. Britton, "Preliminary Data Analysis, 1968, National Survey of Community Solid Waste Practice," PHS 1867, USGPO, Washington, D.C., 1968, 483 p.

6. A. J. Muhich, A. J. Klee, and C. R. Hampel, "National Survey of Community Wastes," PHS 1866, USGPO, Washington, D.C.

7. R. Eliassen, "Solid Waste Management," prepared by the Ad Hoc Group for Office of Science and Technology, Executive Office of the President, May 1969, USGPO, Washington, D.C.

8. R. J. Black, A. J. Muhich, A. J. Klee, and H. L. Hickman, Jr., "The National Solid Wastes Survey—An Interim Report," USHEW, PHS, CPEHS, ECA.

9. Manual of instructions and sample problem for use in conducting the national survey of community solid waste practices, SWP, July 1967, Cincinnati, Ohio, 65 p.

10. "Quad City Solid Wastes Project—An Interim Report," USHEW, PHS, NCUIH, Cincinnati, Ohio, 1968, under Grant D01-U1-00026.

11. Merril Eisenbud, *Science*, AAAS, **170**, 706 ff, Nov. 1970.

12. Arsen Darney and Walter J. Franklin, "The Role of Packaging in Solid Waste Management 1966, 1976," Contract PHS6-67-114, USHEW, PHS, ECA, BSWM, Rockville, Md.

13. R. J. Black, A. J. Muhich, A. J. Klee, H. L. Hickman, Jr., and R. D. Vaughn, USHEW, PHS, ECA, Presented APWA, Miami Beach, Fla., Oct. 1968.

14. The Resource Recovery Act, Public Law 91-512, 91st Congress HR 11833 Oct. 26, 1970, USGPO, Washington, D.C., 9 pp.

15. C. E. Spongle, Summaries; Solid Waste Demonstration Grants Projects 1969, PHS 1821, USGPO, Washington, D. C., 1969, 175 pp.

16. C. A. Clemons and R. J. Black, Summaries of Solid Wastes Program Contracts 6/1/65 to 6/30/1968, PHS 1897, USGPO, 1969, 46 pp.

17. L. W. Lefke, Summaries of solid waste research and training grants, PHS 1596, USGPO, 1968, 48 pp.

18. A. J. Klee and D. Carruth, "Sample weights in solid waste composition studies," *J. Sanitary Eng. Div.*, **ASCE 96(SA4)**, 945–954, August 1970.

19. W. C. Achinger and L. E. Daniels, "An evaluation of seven incinerators," Proc. 1970 Natl. Incinerator Conference, Cincinnati, Ohio, May 17–20, 1970; ASME, New York, pp. 32–64.

20. R. A. Perkins, "Satellite vehicles for solid waste collection; evaluation and application." OSWMP, National Technical Information Service, Washington, D.C., 1971.

21. R. L. Cumins, Study of recreation solid wastes for USDA Forest Service, PHS 1991, USGPO, 1969, 134 pp.

22. J. DeMarco, D. J. Keller, J. Leckman, and J. L. Newton, "Incinerator Guidelines 1969," PHS 2012, USGPO, 1969, 98 pp.

23. D. R. Brunner, S. J. Hubbard, D. J. Keller, and J. L. Newton, "Closing Open Dumps," USGPO, 1971, 19 pp.

24. "Ad Hoc Conference on Solid Waste Training," Robert A. Taft Sanitary Engineering Center, USHEW, PHS, 1964, Cincinnati, Ohio, 72 pp.

25. "Applying Technology to Unmet Needs," USHEW, PHS, February 1966, 12 pp.

26. "Can Engineering Cope with the Debris of Affluence?" *Product Engineering,* Oct. 9, 1967, 9 pp.

27. "Cities' Rubbish Woes Grow as Volume Rises, Dumping Sites Fill Up," Mitchell Gordon, *Wall St. Journal,* Oct. 18, 1961, 2 pp.

28. "Collection and Disposal of Solid Wastes for the Des Moines Metropolitan Area. A Systems Engineering Approach to the Overall Problems of Solid Waste Management—An Interim Report," Henningson, Durham and Richardson, Inc. under Grant No. DOI-UI-00060, USHEW, PHS, CPEHS, ECA, SWP, 1968, 324 pp.

29. Peter L. Davis and Ralph J. Black, "Effects of Garbage Grinding on Sewage Systems and Environmental Sanitation," *APWA Reporter,* Dec. 1962, 4 pp.

30. Richard D. Vaughn and Ralph J. Black, The Federal Solid Wastes Program—A Progress Report Presented at the 1968 National Incinerator Conference, ASME, New York City, May 8, 1968, 16 pp.

31. "Grant Programs Under the Solid Waste Disposal Act," USHEW, PHS, NCUIH, SWP, February 1968, 7 pp.

32. "Grants for Solid Waste Disposal Projects," USHEW, PHS, *Federal Register,* Mar. 30, 1966, 4 pp.

33. "A National Solid Wastes Program," USHEW, PHS, NCUIH, SWP, February 1968, 9 pp.

34. "Private Contractor Opportunities in the Solid Wastes Program," Ralph J. Black, The National Refuse Equipment Show and Congress, San Francisco, Calif. June 1967, USHEW, PHS, CPEHS, ECA, SWP, 13 pp.

35. "Proceedings—The Surgeon General's Conference on Solid Waste Management for Metropolitan Washington," July 19–20, 1967, PHS No. 1729, Leo Weaver (Ed.), USHEW, NCUIH, SWP, 1967, 194 pp.

36. Ralph J. Black, "Progress Under the Federal Solid Wastes Program," 94th Annual Meeting APHA, San Francisco, Calif. Nov. 2, 1966, USHEW, PHS, SWP 10 pp.

37. Robert J. Anderson, M.D., "The Public Health Aspects of Solid Waste Disposal," Public Health Reports, February 1964, 4 pp.

38. Leo Weaver, "Refuse and Litter Control in Recreation Areas," *Public Works Mag.,* April 1967, 4 pp.

39. Thrift G. Hanks, M.D., "Solid Waste/Disease Relationships—A Literature Survey," PHS No. 999-U1H-6. Final report of Aerojet-General Corp. Life Systems Division to the Solid Wastes Program under PHS Contract No. PH 86-66-151, USHEW, PHS, NCUIH, SWP, 1967, 179 pp.

40. Ralph J. Black, "Solid Wastes Handling in Hospitals," *Environmental Aspects of the Hospital, Vol. II, Hospital and Medical Facilities Series,* The Hill-Burton Program, March, 1967, PHS No. 930-C016, USHEW, PHS, CPEHS, ECA, SWP, 1968, 8 pp.

41. "Solid Waste Handling in Metropolitan Areas," PHS No. 1554, USHEW, PHS, BDPEC, NCUIH, February 1964, 41 pp.

42. Samuel A. Hart, Ph.D., "Solid Wastes Management in Germany—Report of the U.S. Solid Wastes Study Team Visit," June 25–July 8, 1967, PHS No. 1812, USHEW, PHS, NCUIH, SWP, 1968, 18 pp.

43. "Solid Waste Reduction/Salvage Plant—An Interim Report," City of Madison Pilot Plant, Demonstration Project, June 14 to Dec. 31, 1967. Prepared by City of Madison Engineering Staff, John J. Reinhardt and Gerard Rohlich, Grant No. D01-UI-00004, USHEW, PHS, NCUIH, SWP, 1968, 25 pp.

44. Wendell C. McElwee, M.P.H., and Maxwell J. Wilcomb, Ph.D, "Some Effects of Disposable Plastic Liners on Refuse Handling Efficiency," *J. Environ. Health,* March–April, 1968, 9 pp.

45. "Abstracts, Selected Patents on Refuse Handling Facilities for Buildings," PHS No. 1793, John A. Connolly, (Ed.), USHEW, PHS, NCUIH, SWP, 1968, 320 pp.

46. "Demonstration Project Abstracts—Solid Wastes Program," USHEW, PHS, BDPEC, NCUIH, SWP, March 1967, 28 pp.

47. "Demonstration Project Abstracts—Solid Wastes Program, Suppl. A," USHEW, PHS, BDPEC, NCUIH, SWP, August 1967, 28 pp.

PART II
disposal of waste

Part II deals with disposal methods for solid wastes. Burial, termed sanitary landfill, has a number of aspects. This section is organized as follows: The constructional aspects are treated by Sorg and Bendixen in Chapter II.1. Chapter II.2 by Quartly discusses the effect of increasing the density of waste and of disposal of more per unit volume of the disposal area. A case study involving the solid waste disposal problem for Fresno California is given by Peters in Chapter II.3; the use of gullies as sanitary landfill and reclamation of land is given by Wehrspann in Chapter II.4. Hydrogeologic aspects of landfills and preservation of the environment are presented by Farvolden in Chapter II.5; aerobic stabilization of landfills (which can blow up because of methane gas generation) are presented by Stone in Chapter II.6. In Chapter II.7 we discuss the handling of construction waste, which is highly heterogenous in character, biologic oxidation, useful for cannery wastes, animal wastes, and similar sludges, is introduced with a definition of technical terms in Chapter II.8; the nitrogen cycle and disposal of agricultural market and animal wastes as mechanisms are discussed in Chapter II.9; and a treatment of sludges is given in Chapter II.10. Composting, widely practiced in Europe and highly thought of by the "organic gardening" advocates, often financed for profit but seldom successful, but useful as a version of biologic oxidation, is discussed in Chapter II.11, which also includes a portion of Jeris and Regan on optimum conditions for composting. In Chapter II.12 we discuss pyrolysis, often thought of in the process industries as low-temperature, destructive distillation but extended to high temperatures for disposal of waste rubber tires, plastic insulation of reclaimed metal, cables and conductors. Cohan and Fernandes of Combustion Engineering, a major builder of large boilers, grates, incinerators, and fuel burning devices, discuss incineration in its engineering and economic aspects in Chapter II.13. Chapter II.14 treats rail transportation of wastes, sometimes thought of as a solution of the problems of cities, but to date of little value. Chapter II.15 is devoted to a discussion of burial and disposal at sea by coastal cities; disposal at sea is employed for debris, sludge, and residues from dredging, harbor improvement, and the like.

Since each chapter treats a specific theme, rather than a discussion on a general topic as in Part I, references are at the end of chapters rather than at the end of the part, as in Part I.

CHAPTER II.1
sanitary landfill

THOMAS J. SORG and THOMAS W. BENDIXEN

When the earth was sparsely populated and open land was abundant, people dumped their wastes in any uninhabited area. When people began to form communities, however, centralized dumping sites became necessary. These areas were usually isolated locations out of *sight* from the community.

Between 1900 and 1910 several U.S. and European cities began burying their wastes. This led to the use of heavy equipment in the 1930s to compact waste in order to conserve space and to dispose of the wastes in a sanitary manner. These techniques evolved into the present method of disposal called *sanitary landfilling.* Through the 1930s and 1940s, the term was generally applied to operations in which waste materials were placed in trenches and covered. In the 1950s, the Sanitary Engineering Division of the American Society of Civil Engineers formed a committee to prepare a manual on sanitary landfilling. The committee's definition of sanitary landfilling is the one now most commonly used:

> A method of disposing of refuse on land without creating nuisances or hazards to public health or safety, by utilizing the principles of engineering to confine the refuse to the smallest practical area, to reduce it to the smallest practical volume, and to cover it with a layer of earth at the conclusion of each day's operation, or at such more frequent intervals as may be necessary.[1]

Although sanitary landfilling is an acceptable method of solid waste disposal, it has not received wide public acceptance, principally because so many communities have called their open dump a sanitary landfill. The public's misconception of a sanitary landfill results from the fact that the vast majority of land disposal sites in the United States are not sanitary but open and burning dumps. Between 1966 and 1970 a national survey of land disposal practices was conducted by the solid waste agencies of state governments in cooperation with the

Table 1 *National Survey of Land Disposal Site Problems*

Problem(s)	Numbers	Percent
No cover	1,046	8.9
Burning	304	2.6
Water pollution	320	2.7
Burning and no cover	5,393	45.8
No cover and water pollution	764	6.5
Burning and water pollution	199	1.7
No cover, burning, water pollution	3,755	31.8
Total	11,781	100.0

BSWM of the USPHS.* This survey revealed that only 6% of over 17, 000 land disposal sites surveyed met the very minimum requirement for a sanitary landfill. The remainder were unacceptable because of open burning, water pollution, lack of daily cover, or various combinations thereof. An analysis of the data on the first 11,781 unacceptable disposal sites revealed that 93% had insufficient or no daily cover, 84% burned some waste, and 45% had a water pollution problem or a potential one (Table 1).

Between 1966 and 1970, many states passed legislation approving sanitary landfills and prohibiting open dumps. It is apparent, however, that old rules and regulations were either not adequate or not enforced, because a large number of dumps still exist. State programs are becoming more effective, and a definite trend is developing to eliminate open dumps and operate only sanitary landfills. One important factor is a common state requirement that an engineering plan be developed and submitted to the appropriate state agency before a new sanitary landfill can be approved. To do this effectively, both planners and engineers must have a knowledge and understanding of the engineering principles and practices involved in sanitary landfilling.

PLANNING

Planning is an important element of all engineering projects. It is a most essential and critical step for a sanitary landfill project because without proper planning, the operation will probably be unacceptable or not even get under way.

* Now the Office of Solid Waste Management Programs, USEPA.

At a minimum, the planning phase encompasses the development of local sanitary landfill rules, regulations, or codes, data collection and analysis, selection of potential site, analysis and selection of various financing methods, public education, and use of the completed site. Although some planners may have design capabilities, it is more likely that they will only help select the designer.

If at all possible, planners should develop a regional system whereby one or more sanitary landfills serve several communities. The major advantage is economic; unit costs are lower because of the large volume of solid waste handled, and hence funds for proper supervision, operation, and equipment can be more easily provided; furthermore, the elimination of duplicative efforts saves each community money. In addition to these economic advantages, the centralized operational control inherent in a regional arrangement gives each participating community more assurance that its wastes are being disposed of properly. Finally, more sites can be considered from which to select a good one, which is clearly a benefit to communities that do not have suitable landfill sites.

Data Development

Data on the quantity and characteristics of the solid waste produced in the communities served by the sanitary landfill must be available for both planning and design. If good data are not available from existing records, the necessary information should be collected by conducting a survey during the initial stage of the planning phase.

Waste Generation

Quantitative data on the amounts of solid waste that will be collected from all residential, commercial, industrial, municipal, and institutional sources in the area to be serviced must be obtained. Data should also be secured on such special types of waste as bulky waste, demolition material, sludges, and junked automobiles that might be handled. In addition, demographic, industrial, and commercial information should be obtained. This type of information is needed to accurately estimate the land area, personnel, and equipment requirements.

The report on the 1968 National Solid Waste Survey contains estimated per capita amounts of the solid waste collected in the United States during 1967 and 1968 (Table 2).[7] It should be noted that the rate fluctuated considerably from community to community. The estimated averages may be adequate, therefore, for rough planning, but final plans and designs should be based upon accurate data obtained from the communities participating in the project.

Table 2 *Average Amount of Solid Waste Collected*[2] *(lb/cap/day)*

Solid wastes	Urban	Rural	National
Residential	1.26	0.72	1.14
Commercial	0.46	0.11	0.38
Combined	2.63	2.60	2.63
Industrial	0.65	0.37	0.59
Demolition, construction	0.23	0.02	0.18
Street and alley	0.11	0.03	0.09
Miscellaneous	0.38	0.08	0.31
Totals	5.72	3.93	5.32

Seasonal variation in solid waste produced should be clearly identified. This information is needed to plan and design for peak load periods that require additional personnel and equipment and for slack months when vacations should be taken and machinery overhauled.

Because the seasonal patterns are influenced by climate, type of industry, tourism, construction work, and other factors, these factors should also be identified from data obtained from the communities participating in the project. An example of seasonal variation for solid waste collected in New Orleans, Louisiana is shown in Table 3.

Table 3 *Monthly Variation in Municipal Waste Collection, New Orleans, La.*[3] *(Percent of Average)*

Month	1965	1966	1967	Avg. 3 yr
January	93	94	94	94
February	86	90	89.5	88
March	100	102	103	101
April	101.5	103	98	101
May	99	105	107	103
June	104	107.5	106	106
July	108.5	109	109	108
August	109.5	111.5	112	110
September	104	106	99	103
October	95	96	96	95
November	94	94	88	91
December	102.5	101	98	100

Waste Characteristics

Various studies have been conducted to identify the physical and chemical characteristics of solid waste. However, not all of these studies have been conducted for the same purpose, and, consequently, various results have been obtained. The OSWMP of the EPA used a nine-category reporting scheme to characterize solid waste disposed at several incinerators, and this method is being widely adopted (Table 4).[4] The separation methods consisted of manually separating 200-lb samples into the nine classes. All studies that have been conducted by various agencies indicate that the paper content is increasing and the amount of food is decreasing.

Public Education Program

Getting public acceptance is one of the greatest problems in establishing a new sanitary landfill, primarily because most people equate this disposal method with a dump. In 1970, a psychometric study of attitude factors and problems of solid waste pointed out that only 7.5% of over 1600 people interviewed in ten cities could name one distinguishing characteristic of a sanitary landfill, and that more than half could not make a clear, accurate distinction between a sanitary landfill and an open dump.[5] The survey also illustrated that an educational approach can be beneficial. After being shown pictures of a sanitary landfill and having the operation explained, approximately 88% of the people thought it was an acceptable method of waste disposal.

Table 4 *Solid Waste Composition*[4] *(Percent by Weight)*

	Incinerator Location in the United States					
Component	Western	Eastern	Midwest	Southern	Southern	Southern
Food waste	7.4	6.1	8.5	12.2	18.3	11.0
Garden waste	3.4	8.4	0.5	1.6	0.6	9.8
Paper products	62.5	58.0	60.4	58.7	60.6	44.9
Plastic, rubber, leather	2.8	3.3	5.4	3.0	2.1	3.5
Textiles	2.4	3.1	2.4	1.8	1.8	3.2
Wood	2.4	1.4	5.4	0.4	2.3	3.1
Metal	9.0	8.2	9.0	8.6	8.5	8.1
Glass, ceramics	4.2	8.1	3.5	10.3	5.4	9.5
Ash, rock, dirt	5.9	3.4	4.9	3.4	0.4	6.9
Total	100.0	100.0	100.0	100.0	100.0	100.0

A planning program must, therefore, include a public education program. Every effort should be made to develop public support, particularly through the news media. Presentations outlining the operation and benefits of a sanitary landfill should be given at civic and neighborhood meetings. If the completed site will be a community asset—a golf course or playground, for example,—this use should be stressed. The use of visual aids, such as scale models and architectual renderings, is also recommended.

One of the best ways by which community officials can overcome public opposition is by converting open dumps into sanitary landfills to prove that their agency can run a good operation. The obvious improvement in appearance and operation should generate support for the establishment of completely new sanitary landfills.

Financing

No matter how well a sanitary landfill is designed, it has no future without adequate funding, and ways to secure the money must be determined during the planning phase. Various financing techniques should be examined; because of differences in ordinances, regulations, and statutes, the one or ones selected will probably vary from community to community. As the financing plan is developed, it is important that expert financial, legal, and engineering advice be sought.

The financing alternatives normally available to a community are general obligation and revenue bonds and a pay-as-you-go system. General obligation bonds can be issued only by legal entities with taxing powers. Tax limitation, tax rate, and the necessity of voter approval will generally influence the selection of this method. The present national trend indicates that any bond limitation submitted to a voter's referendum has little chance of passing. At the present time, few citizens are willing to support solid waste undertakings by increasing their property taxes.

Revenue bonds may be issued by a state, territory, or possession, or any of their political subdivisions, and voter approval is generally not required. The bonds are payable from revenues derived from charges levied on users of the facility. It is for this reason and because revenue bonds are not ordinarily subject to debt limits that they are often issued rather than general obligation bonds.

The pay-as-you-go system uses current revenues to finance active projects, and money must, therefore, be put aside if plans call for constructing a sanitary landfill. This form of financing does not have to be approved by referendum. Moreover, since no debt is incurred, the community's credit rating should improve, a development that would be beneficial in the event of a financial emergency. On the other hand, revenues may not be available for needed capital improvement if the system is carried to an extreme. A question of equity also arises, because current taxpayers finance an operation that will benefit future users. Since no one financing method appears to be ideal, various combinations may be the most practical for sanitary landfills.

SITE SELECTION

Social, political, economic, and engineering factors play important roles in selecting a site for a sanitary landfill, and it is essential, therefore, that a well-qualified individual or a team of specialists make this decision. Because community acceptance is critical, several alternative sites should be carefully evaluated and selected.

Land Requirement

The amount of land needed depends on the character and the quantity of the solid waste received, the density to which it is compacted, the ratio of the compacted waste to the cover material, the depth of the fill, and the desired life of the site. The data for making these estimates should be obtained during the planning phase and should cover all sources of waste. Waste collection information is usually reported in pounds per person per day; total waste projections are normally based on estimated population growth and increased per capita collection. It has been estimated that an average amount of solid waste collected per person in 1967 was 5.32 lb (Table 2), and an increase of 10% a year per person is suggested as a planning factor.

The site selected should provide enough capacity for at least a 20-yr period, particularly if the sanitary landfill will be the predominant method of disposal. If no 20-yr site is available, several smaller sites should be considered instead of reducing the design period.

The composition of the solid waste will also influence the land area required. Some types of waste, such as cardboard boxes, wooden crates, and glass bottles, can be substantially reduced in volume, but others, such as bed springs, mattresses, stoves, and tires, cannot be reduced. Composition is, therefore, one of the factors that determine the density achieved by compaction. The moisture content of the waste, the weight of the compaction equipment, the number of passes the equipment makes over the waste, and the depth of the fill also influence the density realized. For design purposes, 1000–1200 lb/yd^3 is suggested as a realistic range.

The land area required to hold a certain amount of solid waste is primarily a function of the depth of the fill, which determines the volume available. The higher the elevation of the completed fill, the larger the volume of waste per unit area. Although most sanitary landfills are not over 50–100 ft high, in some areas of the country, particularly California, depths of 700 ft and greater have been reached.

The ratio of the solid waste to the cover material must also be considered in the calculations for land area required and the general rule is 1 part earth cover to 4 parts solid waste. The density of the cover material will vary according to compaction procedures, but generally will be around 100 lb/ft^3.

The volume requirements may be calculated as follows:

$$V = 365 \frac{W \cdot P}{D} + C_v$$

where V is the volume required for waste disposal of per year in cubic yards, W is the amount of waste in pounds per capita per day handled at the site, P is the population of community using sanitary landfill, D is the density of landfill waste in pounds per cubic yard, and C_v is the volume of cover material in cubic yards. Using the general ratio of 1 part earth cover to 4 parts waste, the formula becomes

$$V = 365 \cdot \frac{W \cdot P}{D} \cdot (1.25) \qquad \text{or} \qquad \frac{456 \cdot W \cdot P}{D}$$

Example. A community of 50,000 people produces an average of 5.0 lb of solid waste per day to be disposed of at a sanitary landfill that will have a waste compaction density of 1000 lb/yd^3. The volume required is

$$V = \frac{(365 \times 5.0 \times 50{,}000)}{1000} (1.25) = 114{,}062 \text{ yd}^3$$

If the average depth of the fill is 25 ft, the number of acres of land required each year is thus

$$A = \frac{114{,}062 \times 27}{25} \times \frac{1}{43{,}560} = 2.8 \text{ acres}$$

Accessibility

Ideally, a disposal site should be located in the center of the waste generation area, but this is usually impractical because it would be in a residential area. Therefore, sites are generally near the fringe of the waste generation area. The distance that trucks must travel to a site is an important economic factor to consider when selecting the location, but travel time and truck capacity are often more crucial. A short haul time and large collection vehicles will lower the transportation cost. Since hauling time depends on the type of roads and speed limits, site locations near major highways should be sought. Moreover, the disposal site should be near major highways so that trucks do not have to travel through too many residential areas. It is also important to have several access routes to the site in case one of them becomes temporarily unusable. Because solid waste collection trucks are 16–25 yd^3 in size and carry 4–6 tons per load, the main highways and artery streets should be wide and of heavy construction to handle these trucks adequately when fully loaded during all weather conditions. Bridges, underpasses, and steep grades on all routes must be investigated to determine their capacity to accommodate the trucks.

Hydrology

Many of the problems that can arise with regard to ground and surface water pollution can be avoided if they are taken into consideration when a candidate site is being examined. The amount of leachate produced from the sanitary landfill depends on the moisture content of the solid waste and the quantity of water entering the fill. Once the solid waste has reached its capacity to hold water, the leachate that is formed can create a serious water pollution problem. It is important, therefore, that precipitation, surface run-off characteristics, evapotransformation, and the location and movement of groundwaters in relation to the solid waste be considered during site selection and the design phase.

A site should have topographic and soil characteristics that will keep the amount of surface water infiltrating the landfill to a minimum. It is wise not to select low-lying sites that might be drainage basins for surrounding areas or that could be flooded by rivers, streams, or lakes. Such sites can be used, but they require extensive engineering.

Detailed investigations on the quantity of water that may enter a landfill through the soil cover and on the quantity and quality of the leachate that may leave the fill and enter an aquifer or stream are extremely limited. One laboratory investigation has led to the claim that knowing the available water storage capacity, quantity and frequency of water infiltration, and the rates of evaporation and transformation for a cover material, it is possible to predict the quantity of surface water that will enter the underlying solid waste.[6] Because this claim was made from the results of laboratory tests under controlled conditions, it seems unlikely that accurate predictions for field conditions can be made. However, rough estimates might be determined to help evaluate a potential site.

Geological investigations should be made to determine the location and history of the water table, the direction and rate of flow of groundwater, and the quality of aquifer. A site should be rejected if it has an extremely high water table or contains highly permeable formations, such as gravel. If this is the only site available, then it must be so engineered that leachate cannot enter the groundwater. This is an extremely expensive project and many problems may arise while the landfill is operational and after it is completed.

Some state regulations require a minimum distance between the landfill and the water table, 5 ft being the most common requirement. Some states have special regulations requiring larger distances for permeable soils, such as gravel. In some cases, the regulations prohibit the construction of a sanitary landfill in such soils or require that a special barrier be placed between the landfill and the water table.

Cover Material

Since it is a prerequisite that the wastes at a sanitary landfill be covered with at least 6 in. of compacted soil by the close of each operating day, the site should have an adequate supply of soil, because earth is expensive to buy. Even if a large amount of soil is available, it might not be suitable for use as cover material. Hence a soil investigation must be made to determine its suitability and the quantity available. If the amount and type are not adequate, then another source must be found and the purchase and haul cost calculated to determine the additional cost of the potential site.

The soil cover is intended to serve many purposes and obviously cannot perform all of them to the degree desired (Table 5). For example, it should drain well so that loaded trucks do not bog down, but at the same time it should have low permeability if water is to be kept out of the fill. It is also advisable to have a permeable soil that will allow the landfill gases to vent into the atmosphere, but this type of cover also permits surface water to enter the fill.

Although many soils can be used as cover material, their characteristics greatly influence how they are placed and compacted. It is essential, therefore, that a soil analysis be made to determine the characteristics and the operational procedures that must be followed. Well-graded soils are easily handled, but poorly graded sands and hard clay soils are difficult to compact.

Table 5 Suitability of General Soil Types as Cover Material [7][a]

Function	Clean gravel	Clayey-silty gravel	Clean sand	Clayey-silty sand	Silt	Clay
Prevent rodents from burrowing or tunneling	G	F–G	G	P	P	P
Keep flies from emerging	P	F	P	G	G	E[b]
Minimize moisture entering fill	P	F–G	P	G–E	G–E	E[b]
Minimize landfill gas venting through cover	P	F–G	P	G–E	G–E	E[b]
Provide pleasing appearance and control blowing paper	E	E	E	E	E	E
Grow vegetation	P	G	P–F	E	G–E	F–G
Be permeable for venting decomposition gas[c]	E	P	G	P	P	P

[a] E, excellent; G, good; F, fair; P, poor.
[b] Except when cracks extend through the entire cover.
[c] Only if well drained.

Climate

Climate may dictate the type of operation, soil cover, and hence the location of the site; for example, low-lying sites that might frequently flood or become muddy during rainy weather should not be selected in high rainfall areas. Sites that require a trench operation should be avoided in areas with long and cold winters.

Political and Social Considerations

If it is apparent that the public will strongly oppose a potential landfill site, it should be avoided because public opposition is most difficult to overcome. An information program should be always carried out to help identify and reduce potential problems. It is also wise to select sites in areas that are zoned for landfill-type operations. Attempting to rezone an area after the site has been selected may be a long and expensive project, particularly if the public is opposed to the landfill.

DESIGN AND OPERATION

The designer of a sanitary landfill is responsible for (1) developing a detailed plan on how daily functions are to be carried out; (2) establishing a sequence of areas to be filled; (3) determining the number of lifts required; and (4) delineating the contours of the completed fill. He also draws up a written set of operating procedures, requirements, and specifications, which are used to estimate investment and operating costs, to obtain bids on equipment and facilities, to maintain operational control and efficiency, and, if required, to develop a contract for the operation of the sanitary landfill. Good plans and specifications are essential for the operator to run an acceptable and efficient operation.

Site Preparation

The plans and specifications should require that certain steps be carried out before operations begin. These steps include cleaning up and pregrading the site area and constructing of fences, signs, operating facilities, access roads, and utilities. In most cases, a site will have to be pregraded and cleaned up, and these requirements should be clearly spelled out in the plans and specifications. The grading may entail the excavation of drainage ditches, and it may be necessary to construct and install temporary and permanent culverts. A fence with lockable gates should be installed around the site, and signs should be placed at strategic points along the access roads showing the hours and days of operation, speed limits, fees (if any), and any other pertinent information. Directional signs and markers may also be necessary on nearby highways to direct the public to the sanitary landfill.

On-site access roads should be of all-weather construction and wide enough to permit two-way truck travel. Entrance roads should be so graded that large, fully loaded trucks can travel over them at a reasonable speed. Steeper grades can be tolerated on exit roads because the vehicles using them will be empty. Parking areas should also be provided.

Facilities

A number of facilities will be required, their number and size depending on the size of operation. Scales are recommended to obtain data on the amount of solid waste deposited. Such data provide information on the amount of waste collected in the community and disposed of at the site. Scale operators should be provided with a shelter that is comfortable in any weather. At a small site, the scale house may also serve as an office and contain sanitation facilities for the landfill personnel. A trailer or prefabricated building might be used, because they can be easily moved, if necessary.

At a large site, one or more buildings, of temporary or permanent construction, may be necessary for office and landfill personnel, drivers of collection trucks, and equipment maintenance. The landfill personnel building should contain lockers, sanitary facilities, showers, and a drinking fountain.

Water, electricity, sanitation lines, and telephone service should be provided at all sites; the extent of the services depends on the size and the life of the landfill. At small sites, these services may not be necessary if employees can use nearby ones.

Operational Methods

Although various titles are used to describe the operating methods employed at sanitary landfills, only two basic techniques—area and trench—are involved; the rest are merely modifications. At many sites, both methods are used, either simultaneously or sequentially.

Area Method

In this method, the solid waste is deposited on the surface, compacted, and then covered with a layer of compacted soil at the end of the working day (Figure 1). Use of the area method is seldom restricted by topography; flat or rolling terrain, canyons, gravel pits, and other types of depressions are all acceptable.

The cover material may come from on or off the site. If the ground is flat or rolling, the soil may be scraped up near the working face or taken from a *borrow* area on the site. If a canyon, gravel pit, or other type of depression is involved, the soil may be taken from its sides or from a borrow pit. If the soil is not suitable or available at the site, it must be hauled in from another area.

The width and the length of the operating face depend on the nature of the terrain, the volume of the solid waste delivered daily, and the number of trucks that will unload at one time. It should not be so narrow that trucks are held up for any extended time.

Trench Method

The trench method is mostly used on level or gently slopping land where the water table is not near the surface. In this method, a trench is excavated; the solid waste is placed in it; and the spoil from the trench is then laid on the waste and compacted.

The trench method can be carried out in several ways. One way is to excavate a long trench with a scrapper or dragline and place the excavated earth along the berm of the trench. After the solid waste is deposited and compacted, it is covered with the soil from the berm. It is

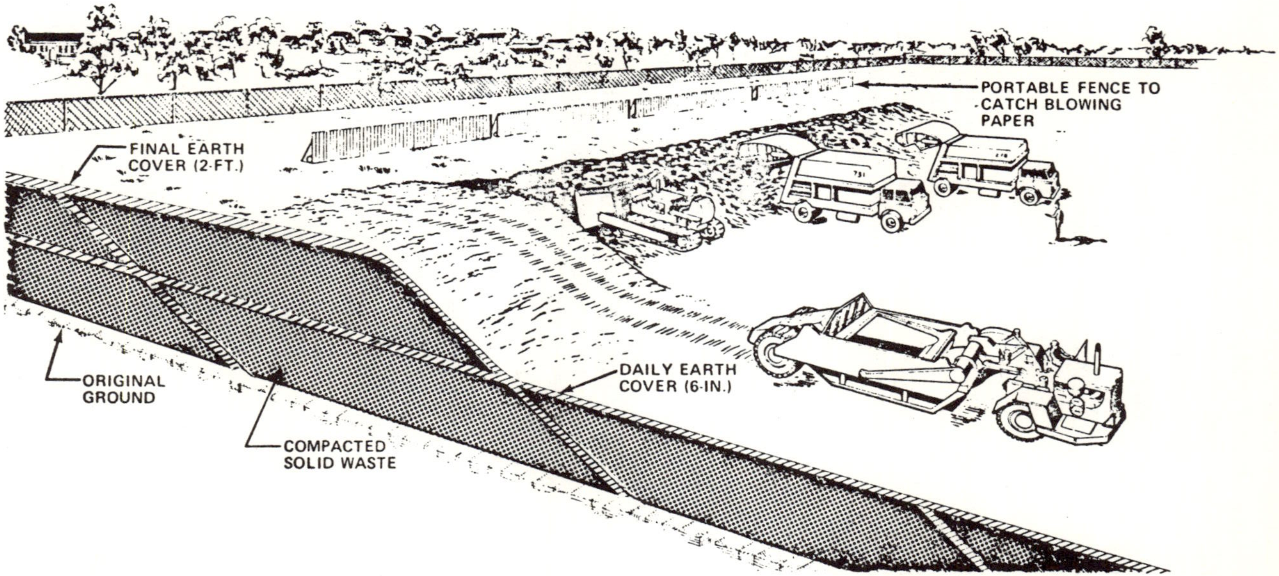

Figure 1. Area method. The bulldozer spreads and compacts solid wastes. The scraper (foreground) is used to haul the cover material at the end of the day's operation. Note the portable fence that catches any blowing debris.

important that the trenches be dug to be self-draining or that a pump be provided to remove any water.

Another way to operate is to excavate only a small section of the trench to handle 1–3 days of solid waste (Figure 2). At the end of the day, the bulldozer or other equipment is used to extend the trench and apply the excavated soil as the cover material. A third method is to excavate a trench parallel to the existing one and use the soil excavated from this trench as cover material. In this method, the new trench is excavated at the end of the day and applied directly to the waste with a dragline, scrapper, or other heavy equipment.

The advantage of the trench method is that cover material is readily available as a result of trench excavation. Stockpiles can be created by excavating long trenches, or the material can be dug up daily. The depth depends on the location of the groundwater or the character of the soil. Trenches should be at least twice as wide as the compacting equipment so that its treads or wheels can compact all material on the surface of the working face. There is no maximum width; trenches have been constructed as wide as 25–50 ft.

Spreading and Compacting

After the wastes are deposited at the base of the working area (toe), they should be spread in 2-ft layers. Spreading may be on a flat surface or slope; the use of a slope or ramp with the area method is commonly referred to as the ramp method. This technique, however, can also be used in a trench. Although the wastes can be spread down the slope it has been found that better compaction and control are achieved when they are spread upward.

Following the spreading step, the machinery makes several passes over the waste to compact it. The density achieved depends on the number of passes made, the character of the waste, and the weight and type of machine used. On an average, a cubic yard of waste that has been compacted by a crawler-type machine weighs 1000 lb; a landfill compactor-type vehicle can probably increase that figure by 10–30%. Adding moisture to the wastes improves compaction and, therefore, density, but the additional water also speeds bacteriological decomposition, which leads to increased gas generation.

Depth of Cells

Compacted waste that is completely covered by a continuous layer of compacted soil constitutes a cell (Figure 3). A sanitary landfill, therefore, is made up of a series of adjoining cells, and those that are

Figure 2. Trench method. The waste collection truck deposits its load into the trench where the bulldozer spreads and compacts it.

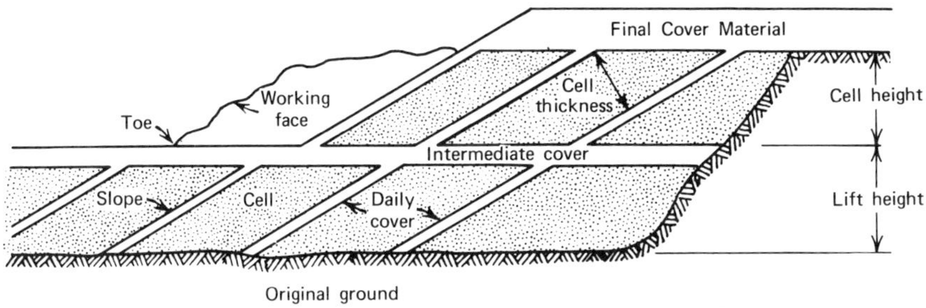

Figure 3. Sanitary landfill construction.

constructed to the same grade make a "lift." A completed sanitary landfill consists of one or more lifts.

Cell height varies from landfill to landfill, but 8 ft is commonly recommended as the maximum height. This maximum height has been selected because some designers think it will not cause severe settlement problems. It is not uncommon, however, to find cell heights of 15 ft or greater, and if all the cells are built to the same height, total settlement should not differ significantly.

Cover

The principal difference between an open dump and a sanitary landfill is the daily use of soil to cover wastes at the landfill. The daily cover provides vector, litter, fire, and moisture control functions. Intermediate and final soil covers are also used (Table 6).

Daily Cover

It is recommended that the compacted daily cover by 6 in. thick. It must be applied at least at the end of each operating day, but if necessary, more frequently. All of the waste, including the top and sides of the constructed cell, must be covered. The cover should be

Table 6 *Application of Cover Material*

Cover Material	Minimum Thickness	Exposure Time
Daily	6 in.	0–7 days
Intermediate	1 ft	7–365 days
Final	2 ft	>365 days

graded and compacted to prevent erosion and to keep water from ponding.

Intermediate Cover

If the daily cover will be exposed for more than a week but less than a year, intermediate cover should be applied. The recommended minimum compacted depth is 1 ft. Because an intermediate cover frequently serves as a road base, it may be necessary to grade and compact it periodically. If cracks or depressions develop, as a result of moisture loss and settlement, additional cover material should be applied.

Final Cover

If the surface of the completed fill will be exposed for more than a year, the cover material should be applied and compacted in 6 in. layers until it is 2 ft thick. Final cover serves basically the same functions as daily and intermediate covers, but it normally supports vegetation. The final cover may have to be more than 2 ft thick because of the soil type or the anticipated use of the completed landfill. It is extremely important that the final cover be graded in accordance with specifications developed during the design phase. The contours should be shown on all maps and drawings of the completed fill. The ponding of water must be avoided, and grades should not exceed 2–4%, to prevent erosion by water. Side slopes should be less than one vertical to three horizontal.

Maintenance

The clean appearance that is one of the hallmarks of a sanitary landfill and helps to distinguish it from an open dump will deteriorate quickly if the entire site is not maintained conscientiously every working day. Emphasis should be placed on litter control procedures—especially those involving paper, which constitutes about 60% of the solid waste load. The working face should be kept as small as possible and covered as quickly as is feasible. Temporary fences should be positioned around the working area and should be cleaned daily of any debris they have collected. If extremely high winds blow litter over these fences, the area beyond them will have to be policed.

Dust should not be allowed to blow about because it can cause excessive wear on equipment, can be a health hazard to operating personnel, and can be a public nuisance. Dust roads and unloading areas should be wetted down periodically with water or a deliquescent chemical, such as calcium chloride. They may also be soaked with waste oils. To keep dust from becoming a daily problem at large sites, asphalt or concrete roads should be constructed.

Salvage Operations

Salvage operations of any form are not recommended at sanitary landfills because they frequently interfere with the daily operation and lead to sloppy site conditions. As the need to recycle waste materials begins to play a major role in waste management, however, some type of reclamation activity will probably take place at sanitary landfills. Companies rather than individuals will most likely be involved, and their operations should be strictly controlled. Under no circumstances should salvage activity be permitted at the working face; on-site buildings are recommended to house the operations.

Cold- and Wet-Weather Operations

If a site is in an area where cold weather or rain is common, this factor should be considered when the operational plans are drawn up. When the trench method of landfilling is used and freezing weather can be expected, the trenches should be excavated before winter sets in. The stockpiled soil should be covered with straw, leaves, or other heat retaining material. If the ground freezes more than 6 in., dozers or loaders equipped with hydraulic rippers will be needed to loosen the soil or surface of the working area. All ice and snow should be removed from trenches before using them, and snow fences should be constructed to protect the access road and the working area. The trenches should be sloped so that rain collects at one end and can be pumped out before it freezes. It is essential that all machines have heated cabs and that personnel facilities are warm.

Rain often makes the soil too soft or slippery for the equipment to operate and may cause trucks to bog down. Wet weather can also seriously interfere with trenching and covering operations and impede the flow of traffic to and from the working face. If rain is commonplace, all-weather roads should be provided. It is good practice to stockpile concrete rubble, broken asphalt pavement, gravel, or crushed stone for use in making roads passable. At large sites that will be in existence for some time, it may be more economical to run asphalt or concrete roads into the center of the working area and lay gravel or crushed stone on the roads that lead to the working area.

If heavy rains are frequent and the normal working area becomes inaccessible, it is advisable to build a temporary wet-weather operational area adjacent to an all-weather road. Such areas usually have a gravel or stone surface and enough cover material to last for several days is available; they should be ready for use at all times.

Care should be taken that trucks do not carry mud onto the public road system, because this will draw complaints from residents and businessmen.

Public Health Aspects

Problems involving vectors and water and air pollution are alien to a properly operated and maintained sanitary landfill, but the potential problem remains and must be constantly guarded against.

Vector Control

The use of 6 in. of compacted cover should keep house flies inside the fill[8] and discourage rats from burrowing in to reach food sources. In some areas, flies may be a problem during the summer, and the use of an insecticide may be necessary (Table 7).

Rats are sometimes brought to the landfill along with the solid waste. If they are not buried and seek harborage nearby, they may become a nuisance; in this case, an extermination program under the supervision of an expert is necessary. Such programs usually last 2–3 wk, and signs should be erected to warn the public of the use of poisons. (Detailed information on rat control programs is included in the section "Dump Closing".)

Birds are sometimes attracted to landfills and may become a nuisance. To minimize the problem, the working face should be kept small and all wastes covered as soon as possible. Various devices such as tape

Table 7 *Insecticides for Fly Control*[9] *(Outdoor Space Sprays)*

Insecticide	Approximate lb/acre Dosage Required for Effective Kills (up to 200 ft)
Malathion	0.6
Dichloruos	0.3
Naled	0.1–0.2
Dimethoate	0.1–0.2
Diazinon	0.3
Fenthion	0.4
Ronnel	0.4

recordings of distress calls and blank cannon fire have also been used with some success to scare birds away.

Air Pollution

Burning is never permitted at a sanitary landfill, but if an accidental fire occurs, it should be extinguished immediately using earth, water, or chemicals.

Odors

Odors, which usually result from such special types of wastes as septic tank solid or anaerobically digested putrescible materials, can be controlled by covering the waste immediately and quickly sealing any surface cracks that develop.

Water Pollution

Ground and surface waters, in contact with solid waste, can become grossly polluted and unusable for domestic or irrigational purposes. It is essential, therefore, that the geological data developed during the planning phase be studied carefully and the landfill be designed to prevent chemical and bacteriological pollution.

Solid wastes in a landfill decompose chemically and biologically to produce solid, liquid, and gaseous products. The factors affecting the rate and products of decomposition include the physical, chemical, and biological properties of the wastes, the availability of oxygen and moisture in the fill, temperature, and microbial population.

Initially, the wastes decompose aerobically and carbon dioxide, water, and nitrate are produced. As the oxygen supply is exhausted, facultative and anaerobic microorganisms take over and produce methane, carbon dioxide, water, organic acids, nitrogen, ammonia, and sulfides of iron, manganese, and hydrogen.

Ground and surface water moving through a landfill will pick up dissolved and suspended solid matter and microbial waste products to form a solution termed *leachate.* The composition of leachate depends on the solid waste composition and the decomposition activities in the fill. Identification of leachate composition has been the object of several studies.[10-14] The results have exhibited a significant range of values (Table 8); for example, the chloride content varied from 96 to 2350 mg/liter. The leachate composition also varies with time; for example, the sulfate concentrations ranged from 730 mg/liter at the start of one test to 84 mg/liter near its conclusion.

Few studies have attempted to determine the long-term leaching effect on solid waste or the amount of leachate produced from a land-

Table 8 *Composition of Initial Leachate from Municipal Solid Waste*[10]

Components	Composition, mg/liter[a]	
	Low	High
pH	6.0	6.5
Hardness, $CaCO_3$	890	7,600
Alkalinity, $CaCO_3$	730	9,500
Ca	240	2,330
Mg	64	410
Na	85	1,700
K	28	1,700
Fe, total	6.5	220
Ferrous iron	8.7	8.7
Chloride	96	2,350
Sulfate	84	730
Phosphate	0.3	29
Organic-N	2.4	465
NH_4-N	0.22	480
BOD	21,700	30,300

[a] Average composition of first 1.3 liters of leachate per cubic foot of a compacted, representative municipal solid waste.

fill.[10, 12] Limited data indicate that the quantity of contaminants in leachate from a completed fill decrease with time, but more research is needed to adequately describe the phenomena.

Studies on the movement of leachate through soils indicate so far that the distance the contaminants travel depends on the composition of the soil, its permeability, and the type of contaminant.[13-15] Organic materials that are biodegradable do not travel far, but inorganic ions and refractory organics can. Furthermore, the rate of movement through some soils may be so slow that the full impact of contaminant travel may not be realized for years. Since natural purification processes are limited, treatment of leachate should be considered when large quantities are expected and a potential water pollution problem exists.

Gas Production

The principal gaseous products of decomposition are methane, nitrogen, carbon dioxide, hydrogen, and hydrogen sulfide. The quantity depends on the type of solid waste decomposing and the rate on the microbial composition.

Studies on the quantity, rate, and composition of gas production have been extremely limited. One study conducted over a 907-day period found that approximately 40 ft^3 of gas per cubic yard of solid waste was produced.[16] The study also indicated that the composition of the gas changes with age of the landfill; during the early years of the fill, the predominating gas is carbon monoxide while during the later years the gas is composed almost equally of carbon dioxide and methane (Table 9). Because methane is explosive, its movement must be controlled to eliminate any problems, particularly when building will be located on or adjacent to the landfill. (Additional discussion is contained in the section "Complete Sanitary Landfill".)

EQUIPMENT

A landfill operator must have the proper machinery to operate a sanitary landfill effectively and efficiently. The goal of providing the landfill operator with the best possible machinery is influenced not only by the operational aspects but also by the economic considerations, because equipment purchases can often represent up to 30% of the total cost. The decision-making process involves an analysis of the site characteristics, method of operation, and the performance capabilities of existing equipment. The objective is to select the best combination of equipment size, type, accessories, and number to perform

Table 9 *Landfill Gas Composition*[16]

Time Interval since Start of Cell Completion (months)	Average Percent by Volume		
	N_2	CO_2	CH_4
0–3	5.2	88	5
3–6	3.8	76	21
6–12	0.4	65	29
12–18	1.1	52	40
18–24	0.4	53	47
24–30	0.2	52	48
30–36	1.3	46	51
36–42	0.9	50	47
42–48	0.4	51	48

all of the operations successfully and economically. The person responsible for selecting the machines must, therefore, be thoroughly familiar with the sizes and types of equipment available and with the accessories that can enhance their performance. Furthermore, he must be thoroughly familiar with the proposed method of operation and the characteristics of the sanitary landfill site. The decision-making process is more difficult if a small rather than a large site is involved, because only one or two pieces of equipment will probably be purchased.

Equipment Functions

Sanitary landfill equipment for small landfill sites must perform many functions, whereas specialized equipment can be purchased for each special function for large landfill sites. The functions required of the equipment fall into three general categories: (1) those directly involved in waste handling, (2) those involved in handling cover material, and (3) those performing supportive functions. Special attention must also be given to the site characteristics such as the type of soil, weather conditions, and degree of compaction desired.

Waste Handling

Waste handling involves the spreading and compacting of all solid wastes. The quantity of solid wastes to be handled determines the type and number of machines needed to spread and compact the material. The type of solid waste, such as demolition wastes, junked cars, or other bulky material, and the desired density of the landfill also influence the type and size of equipment selected.

Cover Material Handling

The handling of cover material is similar to other earth-moving operations such as highway construction. In landfill operations, the equipment must excavate, haul, spread, and compact the cover material daily. The type and size of machinery to carry out these operations should be based upon the type and quantity of soil to be excavated, the haul distance, and the required in-place density of the cover material.

Support Functions

At most sanitary landfills, equipment must perform support functions, such as construction and maintenance of roads, dust control, fire protection, and assistance in unloading collection trucks and removing stuck vehicles. The necessary support functions should be defined for the specific site and the proper type and size of equipment selected to carry out these activities. Although special equipment may be necessary at large landfills, the equipment used for spreading and compacting can normally handle these operations.

Weather Conditions

Weather conditions may be determining factors in the type of equipment that can operate at a site. If cold or wet weather conditions are prevalent, this may lead to the exclusion of the types of wheeled vehicles that cannot work efficiently and economically in these conditions.

Soil Characteristics

The type of soil present is extremely important, because it affects the size and types of machines required. For example, wheeled equipment may be adequate for sand, gravel, and loamy clays and silts but may not be able to handle the tougher natural soils. For these types of soils, the use of tracked equipment may be necessary. Moreover, the tougher natural solid may also require special excavation equipment such as a crawler type of tractor equipped with a rock ripper. A scraper may also require an additional support crawler to excavate these types of soils.

Equipment Types and Characteristics

Tractor Machines

There are two basic types of tractors—the crawler machine and the rubber-tired machine (Figure 4). Each may use a dozer blade, landfill blade, front-end loader, or bull clam in combination with it. A crawler machine with front-end loader is commonly called a loader, track loader, or front-end loader. Basically, however, it is a crawler machine with a front-end bucket. Consequently, the primary difference in tractor machines is that one is a track-type machine and the other a rubber-tire machine. Each has advantages and disadvantages.

Crawler machines have good flotation and traction because of the self-laying tracks that provide a large ground contact area. The design is based on a low unit pressure under the treads since the mass weight is distributed over the large area of the tracks. Crawler tractors have the following characteristics: (1) They can be used to develop sites, maintain access roads, remove vehicles, and carry out other general operations; (2) have good stability on steep slopes; (3) have good excavation capability on all types of soils; (4) work well in wet and cold weather; (5) are suited for use in both the area and trench methods of

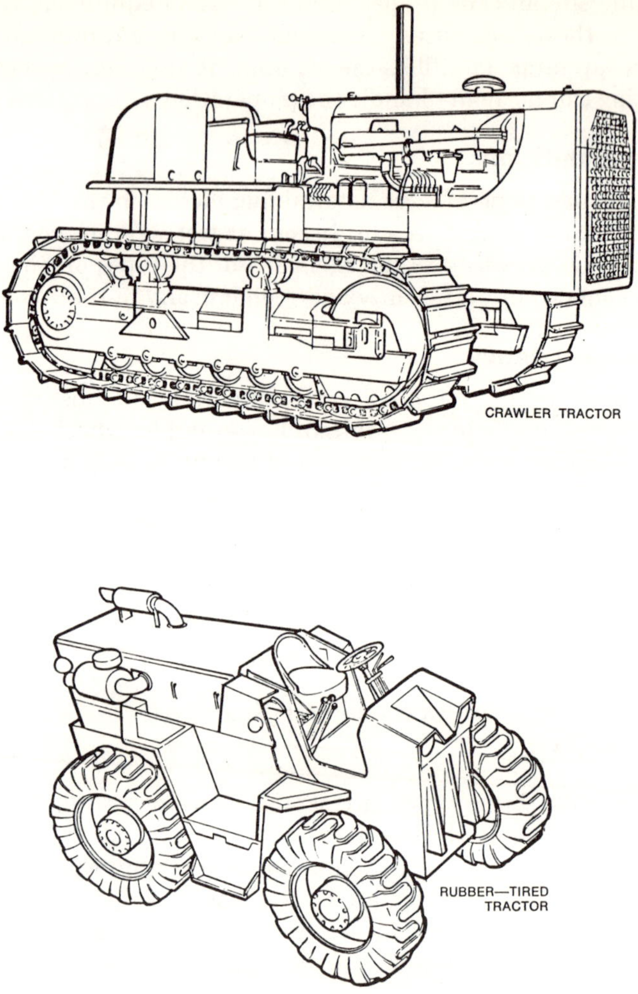

Figure 4. Standard landfill tractors.

operation; (6) can pull a scraper to transport cover material; and (7) can push cover material effectively and economically up to 300 ft. On the other hand, they cannot achieve high speeds (their maximum is 8 mph) and must be transported on a trailer.

A rubber-tire tractor is mostly used with a front-end bucket but can be equipped with a dozer blade or multipurpose bucket. When employed at a sanitary landfill, the machine is generally provided with steel-guarded tires called "rock tires" or "landfill tires." The rubber-tire tractor (1) moves faster than a crawler machine (maximum speed forward or reverse, about 29 mph); (2) does not need special equipment to transport it; (3) can be used for other jobs, such as snow removal; (4) probably has better compacting ability because its load is exerted on four points; and (5) can be used to transport cover material economically for distances up to 600 ft because of speed advantage. It has, however, only limited traction on snow, ice, and mud, its stability on steep slopes is restricted, and its excavating and grading abilities are low, particularly in certain hard natural soils.

Front-End Accessories

A crawler tractor is generally equipped with a dozer blade, trash blade, or a front-end loader, but it may also be fitted with a multipurpose bucket (Figure 5). A rubber-tire machine usually has a front-end

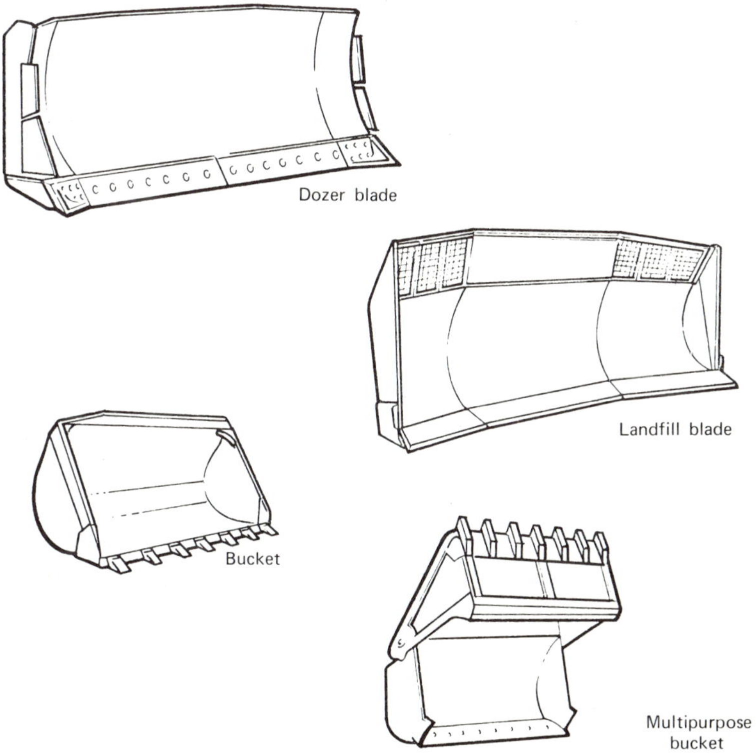

Figure 5. Front-end accessories.

loader but can accommodate a blade or a multipurpose bucket. The type of front-end accessory is determined by assessing operational requirements, and this decision may be as important as the type of tractor chosen (Table 10).

Dozer blades and trash blades are used to spread the waste material evenly over the working face and to move earth; they are similar in shape, but trash blades are much larger and can handle more solid waste. Tractor machines were designed to push earth having densities of 2500–3000 lb/yd^3. Solid waste, on the other hand, weighs about 350–600 lb/yd^3. Since a tractor has the capability to push more solid waste than soil, trash blades were designed to handle the larger quantities. Trash blades can be purchased directly from a manufacturer or a dozer blade can be modified by extending its sides and top. When used for pushing earth, the trash blade does not have any advantage over the dozer blade.

A front-end loader can be fitted on both the crawler and rubber-tire tractor to haul cover material or to load it from a borrow area into a transfer vehicle. It can be used in excavating operations, but it cannot

Table 10 *Recommended and Optional Accessories for Landfill Equipment*[7]

Accessory	Dozers		Loaders		Landfill compactor
	Crawler	Wheel	Track	Wheel	
Dozer blade	O[a]	O	—	—	O
U-blade	O	O	—	—	O
Landfill blade	R[b]	R	O	O	R
Hydraulic controls	R	R	R	R	R
Rippers	O	—	O	—	—
Engine screens	R	R	R	R	R
Radiator guards-hinged	R	R	R	R	R
Cab or helmet air conditioning	O	O	O	O	O
Ballast weights	O	O	R	R	R
Multipurpose bucket	—	—	R	R	—
General-purpose bucket	—	—	O	O	—
Reversible fan	R	R	R	R	R
Steel-guarded tires	—	R	—	R	—
Lift-arm extensions	—	—	O	O	—
Cleaner bars	—	—	—	—	R
Roll bars	R	R	R	R	R
Backing warning system	R	R	R	R	R

[a] O, optional.
[b] R, recommended.

push as much material as a dozer blade. It does not spread wastes over the working area smoothly.

The multipurpose bucket (bull clam) resembles the front-end loader, but it is able to pick up and, in some cases, to crush bulky items such as stoves, refrigerators, and car bodies. Like the front-end loader, it cannot handle larger quantities of wastes or spread them evenly on the working face.

Safety and Comfort Accessories

Because the heterogeneous nature of solid waste creates operational problems, the machines used should be provided with various accessories to protect them and their operators and to increase the effectiveness of both (Table 10). Overheating of landfill equipment is a frequent problem because paper and plastic film often become embedded in the radiator and restrict the flow of air. Landfill equipment engines, therefore, should be protected with side screens and a perforated hood. A reversible fan will also alleviate the problem by allowing the direction of air flow to be changed in just a few minutes. Although radiators may be protected and even have a reversible fan, they still must be cleaned periodically. It is advisable, therefore, that the radiator guard be hinged for easy access (Figure 6).

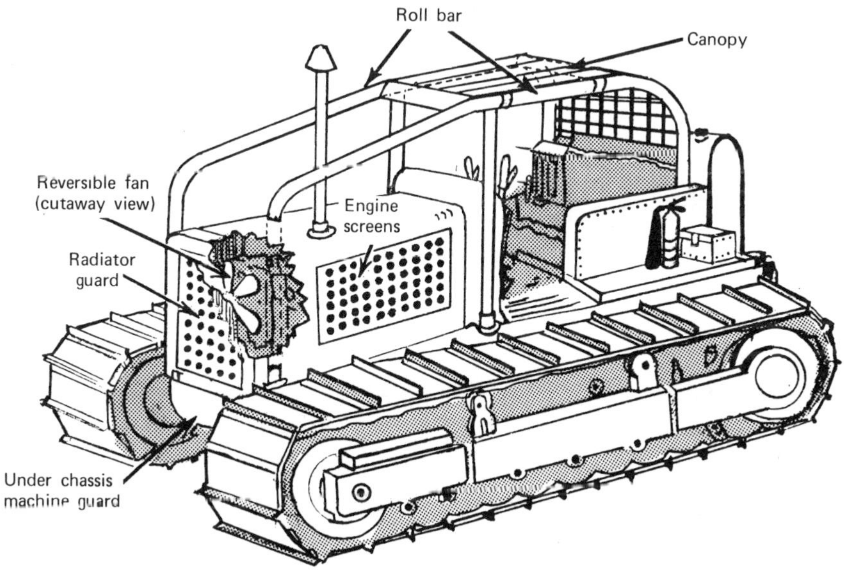

Figure 6. Machine guards.

Another critical area of most machines is the underside which contains vital engine and drive line components; to protect them, underside guarding should be provided. Furthermore, hydraulic lines and other essential items should be protected if they are susceptible to damage.

Operator comfort and safety should not be overlooked; both should increase efficiency. Roll bars, canopy, or cab may prevent a serious accident and are advisable for all landfill equipment. Enclosed cabs with air conditioning and heating are also recommended for machines operating in dusty and hot weather or in cold climates.

Different kinds of power trains are generally available for use on most large machines. Power shift drives and torque convertors are preferable to the dry clutch, director drive models because they achieve greater speed and put less strain on both engine and operator.

Size of Machine

The size of machine needed depends on the amount of waste material that must be handled and the degree of compaction desired. Larger machines, of course, can handle more wastes and can achieve greater compaction. If space is critical, or if the settlement of the completed landfill is important, a large machine should be selected to achieve maximum compaction. The approximate size of machinery necessary to handle various quantities of solid wastes has not been firmly established but general estimates have been made (Table 11).

Specialized Equipment

At large landfill sites, several pieces of equipment will normally be needed and specialized equipment may be warranted to improve efficiency. Specialized equipment, such as the landfill compactor, dragline, and scrapper, can perform specific functions such as spreading, compacting, excavation, and hauling of cover material (Fig. 7). General performance characteristics of this equipment and the tractor machines are provided in Table 12.

Compactor

Since about 1965, several equipment manufacturers have marketed special tractor vehicles that have rubber or steel wheels and are equipped with large trash blades to handle mass quantities of solid wastes. Most manufacturers claim that steel wheel machines impart greater crushing and compactive effort to the solid wastes than the rubber-tire or crawler machines but agree that the equipment should not be used for excavation or pushing cover material. In one of the few studies

Table 11 *Landfill Equipment Needs*[7]

Solid Waste Handled (tons/8 hr)	Crawler Loader		Crawler Dozer		Wheel Loader	
	Flywheel Horsepower	Weight[b] (lb)	Flywheel Horsepower	Weight[b] (lb)	Flywheel Horsepower	Weight[b] (lb)
0–20	<70	<20,000	<80	<15,000	<100	<20,000
20–50	70 to 100	20,000 to 25,000	80 to 110	15,000 to 20,000	100 to 120	20,000 to 22,500
50–130	100 to 130	25,000 to 32,500	110 to 130	20,000 to 25,000	120 to 150	22,500 to 27,500
130–250	150 to 190	32,500 to 45,000	150 to 180	30,000 to 35,000	150 to 190	27,500 to 35,000
150–500	Combination of machines		250 to 280	47,500 to 52,000	Combination of machines	
500–plus	Combination of machines					

[a] Compiled from assorted promotional material from equipment manufacturers and based on ability of one machine in stated class to spread, compact, and cover within 300 ft of working face.
[b] Basic weight without bucket, blade, or other accessories.

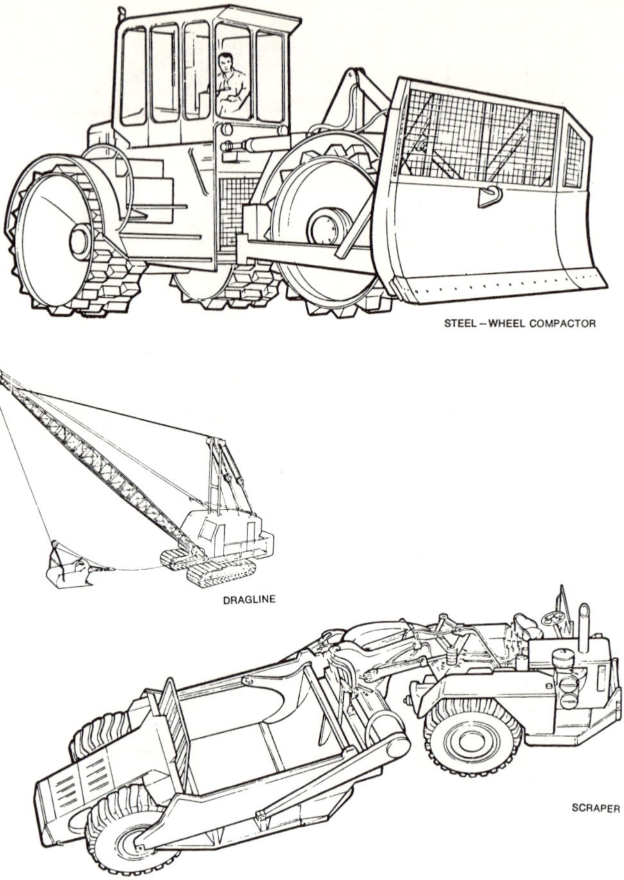

Figure 7. Specialized landfill equipment.

conducted, the in-place dry density of solid wastes compacted by a 4700-lb compactor was 13% greater, under the same conditions, than that obtained by the same machine equipped with rubber tires and by a 6200-lb crawler dozer.[17]

Scrapers

Self-propelled and towed scrapers are available in a wide range of capacities to excavate, haul, and spread cover material. The first type can economically haul material more than 1000 ft and the latter can do so at distances of 300–1000 ft. These machines can be used only at

Table 12 Performance Characteristics of Landfill Equipment [a,b] [7]

	Solid Waste		Cover Material			
Equipment	Spreading	Compacting	Excavating	Spreading	Compacting	Hauling
Crawler dozer	E	G	E	E	G	NA
Crawler loader	G	G	E	G	G	NA
Rubber-tired dozer	E	G	F	G	G	NA
Rubber-tired loader	G	G	F	G	G	NA
Landfill compactor	E	E	P	G	E	NA
Scraper	NA	NA	G	E	NA	E
Dragline	NA	NA	E	F	NA	NA

[a] Basis of evaluation: Easily workable soil, and cover material haul distance greater than 1000 ft.

[b] Rating key: E, excellent; G, good; F, fair; P, poor; NA, not applicable.

large fills where a tractor or other type of machine is available to spread and compact the solid waste.

Dragline

The dragline is generally found only at large landfills where the trench method is employed or where cover material is obtained from a borrow pit. They are used to excavate cover material and, in some cases, to spread the cover material over the compacted solid waste. Their outstanding characteristic is their ability to excavate moderately hard soils and to cast or throw them out of the excavation.

Other Equipment

Other equipment may be necessary such as a road grader to maintain the dirt and gravel roads on the landfill site, a water truck to control blowing paper and dust and for fire fighting. This equipment is generally only used at large landfill sites where their use is considered economical.

Standby Equipment

A sanitary landfill must have equipment capable of carrying out each day's operation because solid waste will continue to be generated and brought to the site even though equipment is not in good operating condition. It is essential, therefore, that standby equipment be available in case of a breakdown and during maintenance periods. This is particularly important at small disposal sites where only one machine is used.

Arrangements should be made in advance for obtaining a replacment from an equipment dealer, a local contractor, or a public works department. The need for replacement machines cannot be overstressed, and planning to provide this equipment on short notice is imperative if the sanitary landfill is to be continually operated properly.

COSTS

The sanitary landfill is the most economical of existing, acceptable waste disposal methods. Although a wide range of capital investment and operating costs for sanitary landfills has been reported, few cost data exist from which accurate estimates for new operations can be developed.

Capital Investment

The magnitude of the capital investment depends on the size and sophistication of the landfill. Rule-of-thumb estimates for investment cost on a per-ton basis are generally available from manufacturers for processing and disposal facilities, but they have not been developed for sanitary landfills. Since a sanitary landfill is not a package type of facility, capital investment costs can be determined only by summing the estimated expenditures for its components, such as land, building facilities, and equipment.

Land and equipment constitute the two greatest investment costs. The price of land may vary from several hundred to several thousand dollars per acre, but the cost—on a volume basis—is dramatically affected by the expected depth of the fill. The deeper the fill, the less the cost on a per-ton or cubic-yard basis. Evaluation of the economics of the land should be based, therefore, on the total volume available, not the area.

The sanitary landfill equipment market is very competitive, for both new and used machines. Prices generally range, however, from $20,000 to $70,000 per piece of equipment, depending on the size and type. The major capital investment costs for establishing a new sanitary landfill are site development (clearing access roads, utilities, and fencing), building (administration, maintenance, sanitation, and scales), and equipment. If funds are not available for direct purchases of equipment, consideration should be given to leasing or leasing with an option to buy later. In general, most new equipment costs $1.00 to $1.05 per ton of machine. Accessories can increase the total cost by $3000 to $5000.

Operating Costs

Operating expenditures depend primarily on the cost of labor and equipment and the method and efficiency of operation. The principal items to consider in estimating operating costs are personnel, equipment, cover material, administration, overhead, and such miscellaneous items as utilities and insurance. Limited data indicate that in costs $0.75 to $5.00 to dispose of a ton of waste (Fig. 8). Smaller sites are less efficient, because they are usually operated on a part-time basis. Full-time personnel, full-time use of equipment, specialized machines, and proper management reduce the unit cost of disposal.

When estimating the investment and operating costs of various types of disposal operations, it is essential that any return on the initial investment and the collection hauling costs be considered, not just disposal expenses, because the disposal method used may influence the location, which in turn affects haul costs. Sanitary landfills are generally farther away from the waste-generating areas than incinerators, and collection hauling costs are, therefore, higher. On the other hand, sanitary landfill may increase the value of the land if the completed site is converted into a playground, golf course, or park, thereby obtaining a major investment cost advantage over incineration. In the final analysis, location may be the determining factor when a disposal method is being selected, because it affects the cost of collection, which is the most expensive part of any total solid waste management system.

COMPLETED SANITARY LANDFILL

A completed sanitary landfill has many uses and thus can be of benefit to the community. The projected use is generally one of the strong selling points in convincing the public that it should accept a sanitary landfill in a certain area.

Maintenance

A completed landfill normally requires routine maintenance because it generates gas and because its surface settles differentially as the wastes decompose. The decomposition process varies according to the types of materials in the landfill. For example, food waste decomposes readily while paper products, wood, metals, plastics, and other materials decompose very slowly or not at all. Moreover, the rate of decomposition depends on the moisture content and temperature and normally is very slow; it is not uncommon to find landfills unchanged after 15–25 yr.

The gases produced are principally methane, carbon dioxide, nitrogen, hydrogen, and hydrogen sulfide. Because methane is explosive, gas production is especially important if final use is as a building, paved parking lot, or other structure that may prevent the flow of gas to the surface. If gases are prevented from venting to the atmosphere, they could accumulate in the structure or soil, cause explosions, and stunt or kill vegetation. Therefore, gas control systems must be designed so that they are compatible with the use of the fill, and they must be maintained to prevent any explosions.

Decomposition of the waste is the principal cause of settlement, and it is greatly influenced by the amount of water in the fill. In Seattle, for example, where rainfall exceeds 30 in./yr, a 20-ft fill settled 4-ft (20%) in 1 yr. In Los Angeles, where less than 15 in. of rain fall per year, 3 yr after the fill had been completed, a 75-ft-high area had settled only 2.3 ft (3%), and another that had been 46 ft high had settled a mere 1.3 ft (2.7%).

The amount of settlement that occurs also depends on the types of waste disposed of, the volume of the cover used in relation to the volume of waste, and the density achieved by compaction. The larger the quantity of cover material placed and the greater the density obtained, the less settlement should take place. When settlement occurs, the surface must be resloped and graded to maintain good drainage and to keep surface-covered cracks from developing. If wide cracks develop, the waste is exposed to rats and flies, water infiltrates, and the gases escape. Settlement also causes problems to structures on landfills and utilities or structures within landfills. Potential settlement, therefore, must be considered in the design of any facility built on or within a landfill.

Uses

Completed sanitary landfills are used for a variety of purposes but are most frequently converted into parks, playgrounds, or golf courses.

If they are used for agricultural purposes or are turned into open green areas, the final earth cover must be at least 2 ft to support vegetation or crops. Pasture and hay grasses are shallow-rooted and need only that amount, but alfalfa and clover require more.

The type of soil used for final cover also influences the choice of vegetation. Tall meadow oats grass thrives well on light sandy or gravelly soils, while other grasses, such as timothy, are better suited to the heavier clays and loams. Generally speaking, at least 4 ft of cover will be needed if crops are to be cultivated, and a 1–2-ft layer of relatively impermeable soil, such as clay, should be placed on top of the soil waste to prevent the top soil from drying out.

Although gas venting is not as critical for open green areas or agricultural uses as for structures, proper venting devices should be considered and installed, if found necessary. The use of gravel trenches or vent pipes can exhaust the gases to the atmosphere through the cover layer.

Because of the problems presented by settlement and gas generation, buildings on or near a completed sanitary landfill must be specially designed. As a result, they cost more than those erected on earth fills. If a structure is planned, a foundation engineering expert should be consulted.

Light buildings are sometimes constructed directly on the landfill surface, but most heavy structures are erected on pilings driven through the landfill. It is, however, often very difficult to do this, because of the wastes' bulky items, such as broken concrete and white goods. Because the solid waste is very corrosive, pilings must be protected with corrosion-resistant coatings. The use of pilings can be avoided by meticulous planning and well-executed operations. The technique involves the creation of solid fill islands in the landfill. They can consist of undisturbed soils that are bypassed during excavation or they can be formed of compacted soils.

It is imperative that gases generated by the decomposing waste not be allowed to accumulate in any structure; even a building on an island must be specially designed to keep this from happening. Various methods have been used including pipe vents, gravel trenches, the use of sand, and a layer of polyvinyl chloride sheeting. Furthermore, all utility connections constructed through the landfill must be made gasproof if they enter a structure below grade.

ADMINISTRATION

Personnel

Competent employees are essential for the proper operation of a sanitary landfill. Each employee should have his duties defined and explained to him so that he has a clear understanding of his responsibilities. Training should be provided periodically and should emphasize safety and emergency procedures. Employees should also be kept advised of all current work rules and management programs.

Operational Control

To continually monitor and evaluate the method and cost of operation, accurate records on all aspects of the sanitary landfill operation should be maintained. At a minimum, operational records should cover the amounts and types of wastes landfilled, the general location of the landfill material, the quantity of cover soil used, and the machine and man hours expended. Cost records should be kept on wages and salaries, maintenance and fuel charges, utility bills, depreciation and interest on buildings and equipment, and overhead. To obtain such data, a standard record keeping and cost accounting system should be set up for the landfill. A cost accounting system recommended for use has been developed by the federal government's solid waste program.[18]

DUMP CLOSING

A dump is a land disposal site where solid wastes are deposited with little or no regard for pollution controls or aesthetics. A dump may be referred to as an "open dump" because the wastes are left uncovered or as an "open burning dump," which is set on fire to reduce the volume of wastes at the dump. These types of disposal operations can be health hazards as well as pollutants of the air and water.

The public's growing awareness of shortcomings and environmental effects of open dumping and burning is resulting in state regulations prohibiting these practices. However, before a dump can be eliminated, an alternative must be provided. Two alternatives exist; providing a new and proper method of waste disposal or converting the dump to a properly operated sanitary landfill. Regardless of which is selected, the dump must be closed in an acceptable manner.

The first step is to develop a plan to eliminate the dump and establish an acceptable substitute. The plan describes the dump closing procedures and the funding arrangements. Before closing, the public, industry, and municipal agencies should be told about the forthcoming activities including why and how the dump will be closed, the alternative method of disposal, and the method of financing.

Dump closing consists of three basic steps: (1) Exterminating the rats; (2) extinguishing the fires; and (3) covering the site with earth. These steps can take from 3 wk to several months, depending on the size of the dump. The first step, rat extermination, is vital if the dump is an active one that contains food for the rats. If they are denied food but are not killed, they may migrate in numbers to populated areas in search of subsistence and harborage. Since rodents and insects might also be present at an old dump, the extermination program should never be omitted.

Various rat extermination techniques have been reported to be effective against the rat population at open dumps. Regardless of the method employed, it is essential that only trained personnel be allowed to conduct the operation because dangerous poisons are used. Fifty to a hundred pounds or more of poisoned bait may be required to bait a single heavily infested dump. Some characteristics of common rodenticides are shown in Table 13. Good results have been obtained by using the following techniques:

1. Wait about a day after the last load of waste is deposited at the dump. Provide drinking water if none is available.
2. Using a single-dose poison, bait heavily for several nights. If rodent ectoparasites or cockroaches are a problem, powder dust with a recommended insecticide.
3. On the day following the last night of single-dose poisoning, establish a number of anticoagulant roden site bait stations and service them regularly until all signs of rat feeding cease.

Dump fires are usually difficult and expensive to extinguish, because applying large quantities of water does not always achieve fast or good results. In many cases, burning solid waste has to be excavated and spread out so that the fire can burn itself out or so that water can be applied directly to the smoldering remains. Operators should be ordered to work cautiously when digging up underground fires because of the hazards involved.

As the final step, the dump is covered with at least 2 ft of compacted earth. There are various methods to cover a dump and, in most cases, crawler dozers will be necessary to carry out the operations. One method is to excavate a nearby trench and spread, compact, and cover the waste in the trench with the excavated soil. It is essential that the bottom of the trench be kept above the level of high groundwater.

Another method is to consolidate the waste as much as possible over a small area, compact it in thin layers, and then cover it with 2 ft of soil. The surface areas of both methods must be graded to avoid ponding of surface waters.

If a dump is in a marsh land or an area where underground or surface waters have been contaminated, the waste should be removed from the water to prevent future degradation of the water. The solid waste and water can be separated by either diverting the flow of water or by removing the solid waste from the water course. Regardless of the method selected, the waste should be spread, compacted, and covered properly to eliminate any future water pollution problems.

Table 13 *Characteristics of Common Rodenticides*

Poisons	Lethal Dose (mg/kg)	Percent Used in Bait	Degree of Effectiveness	Acceptance	Tolerance Developed	Solubility		Types of Bait Mixtures			Action (Cause of Death)
						Water	Oil	Dry	Fresh	Water	
Anticoagulants Warfarin Fumarin Pival[c]	1[h]	0.025	Good	Good	No	Yes	Yes	Yes	No	Yes	Inhibits clotting of blood; causes internal hemorrhages.[e]
Anticoagulant Diphacinone[c]	0.5[h]	0.005	Good	Good	No	No	Yes	Yes	No	No	Inhibits clotting of blood; causes internal hemorrhages.[e]
Antu[a]	8[i]	1.5	Good	Good	Yes	No	No	Yes	Yes	No	Pleural effusion (overproduction of fluid in the lungs).[e]
Arsenic[b]	100[j]	3.0	Fair	Fair	Yes	Yes	No	No	Yes	Yes	Kidney destruction; gastroenteritis; central nervous system affected[e]
Fluoroacetamide (1081)[c]	15-Norway 51-Mice	2.0	Good	Good	No	Yes	No	No	Yes	Yes	Similar to sodium fluoroacetate (1030).[f]
Phosphorus, yellow[b]	1.7	0.05[k]	Good	Good	No	No	Yes	Yes	No	No	Heart paralysis; gastrointestinal and liver damage[f]

Red squill[a]	500[l]	10.0	Fair	Fair	No	Yes	Yes	Yes	Yes	Yes	Heart paralysis.[e]
Sodium fluoroacetate (1080)[c]	5-Norway 2-Roof rat 10-Mice	½ oz/gal 1 oz/28 lb	Good	Good	No	Yes	No	No	Yes	Yes	Paralysis of heart and the central nervous system[f]
Strychnine (Alkaloid)[j]	6	0.6	Fair	Fair	Yes	Yes	No	Yes	No	No	Convulsions due to super-stimulation of nervous system; exhaustion; asphyxia[g]
Strychnine (Sulfate)[d]	8	0.8	Fair	Fair	Yes	Yes	No	Yes	No	No	Convulsions due to super-stimulation of nervous system; exhaustion; asphyxia.[g]
Zinc phosphide[c]	40	1.0	Good	Good	No	No	Yes	Yes	Yes	No	Sams as phosphorus.[f]

[a] Effective against Norway rats only.
[b] Effective against Norway rats and roof rats.
[c] Effective against Norway rats, roof rats, and house mice.
[d] Effective against mice only.
[e] Slow acting.
[f] Fast acting.
[g] Very fast acting.
[h] More or less. Successive doses required for 5–10 days or more.
[i] Norway rats only, on first exposure.
[j] Particle sizes of USP grades vary widely; some coarse powders test as high as 600 mg, micronized at 25 mg; latter recommended, at 1.5%.
[k] Commercial preparations vary from 1–3% in paste form, use as label directs.
[l] Minimum acceptable level; more toxic squills give better results.

REFERENCES

1. Committee on Sanitary Landfill Practice of the Sanitary Engineering Division, "Sanitary Landfill." *ASCE—Manuals of Engineering Practice,* No. 39, New York, ASCE, 1959, 61 pp.
2. R. J. Black, A. J. Muhich, A. J. Klee, H. L. Hickman, and R. D. Vaughan, "The National Solid Waste Survey—An Interim Report." Cincinnati, USHEW, 1968, 53 pp.
3. Albert Switzer and Associates, Inc., and Greenleaf/Telesca. "Master Plan for Solid Waste Collection and Disposal," Tri-parish metropolitan area of New Orleans; final report on a solid waste management demonstration. USPHS Pub. No. 1932, Washington, USGPO 1969, 359 pp.
4. W. C. Achinger, and L. E. Daniels, "An Evaluation of Seven Incinerators." Cincinnati, EPA, 1971, 70 pp.
5. B. T. Jensen and J. M. Stormes, "A Psychometric Study of Attitude Factors and Problems of Solid Wastes," Contract No. CPE-69-107, EPA, OSWMP, 1971.
6. I. Remson, A. A. Fungaroli, and A. W. Lawrence, "Water Movement in an Unsaturated Sanitary Landfill." *Proc. ASCE,* **94(SA2):**307–317, April 1968.
7. D. R. Brunner and D. J. Keller, "Sanitary Landfill Design and Operation." Cincinnati, EPA, 1971.
8. R. J. Black and A. M. Barnes, "Effects of Earth Cover on Fly Emergence from Sanitary Landfills," *Public Works, Mag.,* **89**(2):91–94, February 1958.
9. "Public Health Pesticides," *Pest Control,* **38**(3):15–54, March 1970.
10. California State Water Pollution Control Board, "Report on the Investigation of Leaching of a Sanitary Landfill," Publication No. 10, Sacramento, 1954, 92 pp.
11. "Summary Report—Pollution of Subsurface Water by Sanitary Landfill," PHS Grant No. 5-R01-UI00516, Drexel Institute of Technology, February 1969.
12. S. R. Qasim, "Chemical Characteristic of Seepage Water from Simulated Landfills," Ph.D. Dissertation, W.V. University, Morgantown, W.Va., 1965, 145 pp.
13. J. R. Anderson and J. M. Dornbush, "Influence of Sanitary Landfill on Groundwater," *AWWA,* **59**(4):457–470 April 1970.
14. G. M. Hughes, R. A. Landon, and R. N. Farvolden, "Hydrogeology of Solid Waste Disposal Sites in Northeastern Illinois—An Interim Report of a Solid Waste Demonstration Grant Project." Cincinnati, USHEW, 1969, 137 pp.

15. County of Los Angeles, Department of County Engineers and Engineering Science, Inc, "Development of Construction and Use Criteria for Sanitary Landfills—An Interim Report." Cincinnati, USHEW, 1969, 267 pp.

16. R. C. Merz and R. Stone, "Special Studies of a Sanitary Landfill." Washington, USHEW, 1970, 240 pp. (Distributed by National Technical Information Service, Springfield, Va., as PB-196 148.)

17. R. Stone and E. T. Conrad, "Landfill Compaction Equipment Efficiency." *Public Works Mag.*, **100**(5):111–113, May 1969.

18. Eric R. Zausner, "An Accounting System for Sanitary Landfill Operations," USPHS Publication 2007, USGPO, Washington, 1969, 18pp.

CHAPTER II.2

baling municipal refuse

ERIC QUARTLY

Baling of municipal refuse is among the newer approaches to the problem of solid waste disposal. Widely used for many years in industry and agriculture, the baling process is perhaps the simplest and most economical form of packaging. The compaction inherent in the process contributes to the conservation of space and consequently to economies in storage and transportation. In most industrial and agricultural applications, baling is used for the compressing and packaging of relatively homogeneous materials—cotton, hay, paper stock, and rags, for example. With such commodities the process is relatively standardized and more or less trouble-free, because the materials are relatively uniform in composition and in the size and shape of their constituent elements. They present few surprises. Potential compressibility is known within reasonable limits.

A major complication in adapting the baling process to municipal refuse is that this material is inhomogeous and unpredictable. On the average, the characteristics of such refuse may be distinctive in different parts of the country, from city to city, and even from area to area within the same community. Any given batch fed into a baler hopper may differ markedly from the average, may present unusual problems, and in the completed bale may vary widely from the norm in density.

In 1968 the city of San Diego completed and reported on a one-year study and investigation project dealing with what was then a new and previously untried technique—the application of the baling process to solid waste disposal.[1] The project was jointly financed by city funds and a grant from the Solid Wastes Program of the Public Health Service's National Center for Urban and Industrial Health. As a result of that study, a pilot refuse baling plant was being constructed in San Diego, also under city and federal auspices, in which the process was demonstrated on a production basis.

The purposes of the first phase of the project were (1) to study the technical and economic feasibility of refuse baling as a means of conserving the life of sanitary landfill sites, (2) to explore the possibility of reducing the distances of direct haul to disposal sites in refuse collection vehicles by studying the feasibility of baling as an integral feature of possible transfer station operation, (3) to study the utilization of refuse bales as fill material in the reclamation of small canyons in close proximity to residential areas, and (4) to develop yardsticks, formulas, and techniques that might be useful to other communities in dealing with refuse collection and disposal problems.

The first of the above purposes defined the major thrust of the study. To reach this objective it was necessary to compare the compaction of refuse now being obtained by conventional landfill methods with that potentially obtainable by baling. The first of these was investigated by making a comprehensive field survey at one of the city's landfills to determine the weight and compacted volume of refuse now being deposited, and the second by actually baling refuse in test situations.

The field survey showed that by standard landfill methods San Diego is currently achieving compaction ranging between 1189 and 1383 lb/yd^3, that better results are obtained with a compactor than with a bulldozer, and that the application of larger amounts of water to the refuse improves the effectiveness of compaction efforts.

The feasibility of baling refuse and the degree of compaction obtainable by this method were investigated at two levels: (1) an extended local test was conducted in which unprocessed refuse was baled in a leased light-duty baler installed at a test facility especially constructed for the purpose, and (2) arrangements were made for a demonstration of refuse baling by an interested baler manufacturer, using modern, heavy-duty equipment. In the local baling test a total of 48.5 tons of typical raw, unshredded city-collected refuse was processed during a 5 wk period into 162 bales averaging 599 lb apiece in weight and 689 lb/yd^3 in density. As a second phase of this local program, 64 additional bales were produced and "nested" in a simulated landfill situation. At the factory demonstration four bales were produced from shredded municipal refuse. They ranged from 1500 to 2490 lb in weight, from 0.92 to 1.56 yd^3 in volume, and from 1466 to 1593 lb/yd^3 in density.

Findings in the local test and the factory demonstration tended to supplement and reinforce each other. The balers used in both tests were of the same basic horizontal design that lends itself to more continuous production than other types. Locally it was shown that ordinary city-collected refuse can be successfully processed, even without preshredding, into reasonably well-formed bales which generally maintain their integrity during transportation and disposal. Since this test extended over a period of time, it established the feasibility of baling refuse that varies considerably in composition and moisture content from day to day and from one collection area to another. The factory test provided persuasive evidence of the desirability of shredding the refuse before it is fed into the baler and of using baling equipment of high compression capability. The factory-produced bales were not only denser but better shaped, with sharp, well-defined corners and relatively smooth surfaces, and they maintained their integrity without appreciable distortion during handling.

The densities obtained in the local test were not competitive with the approximately 1380 lb/yd^3 obtainable by efficient landfill methods, but the compaction obtained at the factory exceeded that level. New balers of this type have subsequently been developed, which are specifically designed to handle municipal refuse and which are expected to yield densities of 1890 lb/yd^3 with preshredded refuse having a moisture content of 30%. On this basis, assuming 5% of volume for voids between bales in place in a landfill, the space required per ton for refuse in bales would be only 77% of that required with conventional landfill methods. It follows that the potential saving in landfill life that is realizable by baling could run as high as 23%.

It was felt, however, that the potential economic benefits of baling refuse could be fully realized only in the operational setting of a transfer station. Preliminary studies were made of the economic feasibility of a full-scale baling transfer station in circumstances where the exhaustion of an existing favorably located landfill would require longer direct haul by collection vehicles to a more remote landfill. Under the San Diego conditions, when the additional round-trip distance required per truck by such haul exceeded about 10 miles the operation of a baling transfer station would be more economical than direct hauling. In the station capacity ranges studied (300–350 tons a day), a net saving could be realized by baling despite the baling operating costs because of the reduction in the costs of land for the station site, of rehaul equipment, and of compaction equipment and labor at the landfill. These offsets would make a baling transfer station (fully equipped with balers, a hogger, and conveyors) more economical than a conventional station in which refuse is simply transferred from collection to rehaul vehicles. This conclusion, however, might not be valid if the comparison were made with a conventional transfer station of larger capacity and consequently greater efficiency.

The pilot baling transfer station tested some of these tentative conclusions under actual production conditions. It also permitted (1) refining of baling techniques and routines, (2) ascertaining the optimum moisture content of the refuse being baled, (3) comparing the compaction obtainable with preshredded versus unprocessed refuse, (4) determining and dealing with any possible nuisance factors or health hazards (such as noise, dust, odor, and vector breeding) that may be encountered in the operation, (5) exploring the feasibility of other means of bale disposal, such as in the reclamation of small canyons near residential areas, and (6) developing cost comparisons.

The pilot plant processed 150 tons of city-collected refuse in a 6-hr operating day. It was designed for minimum land area usage and minimum labor requirements. A dumping pit with a pan conveyor floor held four 25-yd packer truck loads of refuse at one time. The pan conveyor moved the refuse to a shredder of the hammer mill type that was installed in a subterranean room (1) to eliminate the necessity for lifting the refuse from the dumping pit to the shredder hopper on a long conveyor belt, which would increase the area requirements of the plant, and (2) to suppress most of the noise inherent in hammer mill operations.

The hammer mill deposited the refuse, shredded to 5-in.-maximum particle size, on a slider bed conveyor which fed a vertical continuous-bucket elevator. The discharge chute of the elevator served as a storage extension of the baler hopper. To increase compaction in the completed bale, water was introduced into the shreddded refuse as it descended through the elevator discharge chute. The baler itself was enclosed in a concrete block building which also housed the electrical power panel and all plant controls. The baler was of the heavy-duty continuous horizontal type capable of baling a minimum of 25 tons/hr at an average density of more than 65 lb/ft^3. Completed bales were moved automatically on roller conveyors from the baler to a loading dock and thence onto a rehaul truck which itself had roller conveyors built into its bed.

In the pilot operation, loads of four bales, each weighing close to 2 tons, were hauled to a disposal site in Balboa Park and rolled off the truck. A specially designed bale grab operated from a crane boom picked up the bale and deposited it in the blocklike disposal area. When this pilot bale disposal site is completed it will be used as an archery range. The site is a canyon that is too small for conventional sanitary landfill operations.

San Diego has many such canyons, some of them strategically located with reference to refuse collection routes. Many of them could be converted into attractive recreational areas and "green belts" by using baled refuse as fill. It is hoped that such projects, which will be much quieter, cleaner, and less odorous than conventional landfill operations, will be acceptable to the public and particularly to neighboring householders.

CHAPTER II.3

the fresno project

RICHARD F. PETERS

The California Integrated Solid Wastes Management Project was devised to demonstrate the value of a rigorous systems-oriented study on management of solid wastes in an area that faced typical urban–rural expansion problems. A portion of Fresno County was selected for the study. This area, with about 390,000 people, consisted of a core city surrounded by ten incorporated and 16 unincorporated communities. Two of the incorporated cities are large enough to constitute a metropolitan area. Surrounding and interwoven with the periphery of the core area is a high-density agricultural belt. Approximately 1200 square miles are included in the study. The principal sources of income are agriculture and agribusiness.

As elsewhere, the traditional method of handling solid wastes in most communities in the Fresno area has been to transport them beyond the community complex and discard them in the least expensive manner that the public would tolerate. Standards for landfills were generally overlooked or very lax, and the public appeared unconcerned. Very little planned consideration was given to the interface where urban developments and agriculture joined. However, these attitudes have changed markedly during the past five years. Scarcity and rising costs of land, recognition of need for conserving prime agricultural areas, increased public demands for more healthful and aesthetic environments, and the mounting volume of solid wastes, created a sense of urgency among public officials to improve management procedures. Furthermore, there is now an enlightened new emphasis on the development of methods and techniques for recycling or converting these wasted materials to useful products.

The Fresno study was a joint effort by the State Department of Public Health, city and county agencies, and private industry. About half the total effort was performed, through contract, by the Aerojet-General Corporation assisted by Engineering-Science, Inc. This group possessed the engineering, systems analysis, computer, and related technical capabilities required to develop a comprehensive approach to the planning and design aspects. Other responsibilities, such as general supervision, agency coordination, environmental criteria, and assistance in public education, were performed by the State Department of Public Health.

METHOD

The Fresno Project was planned to incorporate five specific tasks or efforts that were timed and interrelated as shown in Figure 1. These tasks included (1) education of the public, (2) an intensive factual systems-oriented study of solid wastes, (3) special studies to delineate entomological problems of solid wastes, (4) implementation of the program planned by the systems study, and (5) development of management criteria.

Task 1. The public education program was designed to develop within the people of the Fresno area a readiness to accept the new concepts and recommendations that were to be developed by the project.

Task 2. The systems-oriented study was expected to determine, through a comprehensive study, an optimum solution to the Fresno region's solid wastes management problems, and to develop a technology and methodology for regional solid wastes studies to solve solid waste management problems in other urbanizing regions.

Task 3. Special entomological studies were devised to conduct special field studies of solid wastes for which sufficient data were unavailable. The results of the study were used to augment existing information used to develop preliminary solid wastes management criteria for the Fresno area.

Task 4. Criteria were developed to guide the proper evaluation of solid wastes management systems under study in Task 2. These criteria will serve as the environmental specification that the proposed Fresno regional system will be required to meet. After field testing and refinement, the criteria are expected to become the standards for solid wastes management in the Fresno area.

Task 5. Designs to implement recommendations have been a continuing effort of the Project since its inception, and the likelihood of implementation was one of the important factors in selecting the Fresno area. This task provided informed advice and consultation on the new concepts and recommendations developed by the Project to public agencies and private concerns in the area.

SYSTEMS APPROACH

The systems approach for this study was based on the premise that solid waste management involves such complexities and interdependencies that a highly sophisticated and systematic approach is needed if a genuine solution is to be achieved. The use of a systems concept provides a new perspective in analyzing the problems of solid wastes management. Almost all current methods are concerned with the waste material itself. The systems concept considers waste materials as an "imput" to a system that then examines them critically in a manner designed to give one or more "outputs," which in turn have predetermined desirable characteristics and utility. Thus the system focuses attention on all the details of the kinds and amounts of wastes, the hardware, and practices for handling them, the interrelated legal and socioeconomic structures, and the criteria or standards that control how the wastes should be managed to achieve the health, aesthetic, and projected management goals to provide an optimum environment for urban, agricultural, and industrial activities.

The basic objectives of the Fresno Region Solid Waste Management Study were (1) to determine an optimum system for the management of solid wastes in the Fresno region and (2) to develop a technology for that determination that would be applicable to other similar regions throughout California and the nation. The development of a method for evaluating the adverse environmental effects of solid wastes encompasses all the adverse effects of solid wastes including human disease, animal disease, flies, rodents, insects, plant disease and crop damage, safety hazards, toxicity hazards, and the aesthetic factors of odor and unsightliness. It becomes readily apparent that no one person, however experienced, has at his fingertips sufficient knowledge of all the environmental fields to properly assess their effects given any combination of solid wastes in any condition. Hence, it was necessary to use a large number of environmental scientists and engineers with expertise in the above fields to provide value judgments that could then be used in a systems analysis approach for the development of an integrating device.

The study was undertaken on the hypothesis that environmental improvement can be expressed in terms of the degree to which any postulated management system decreases the environmental degradation or bad effects of solid wastes. The performance scoring procedure developed in this program provides quantitative bad effect scores for unit quantities for each type of waste when placed in any of the considered conditions. Without the innovative measuring methods

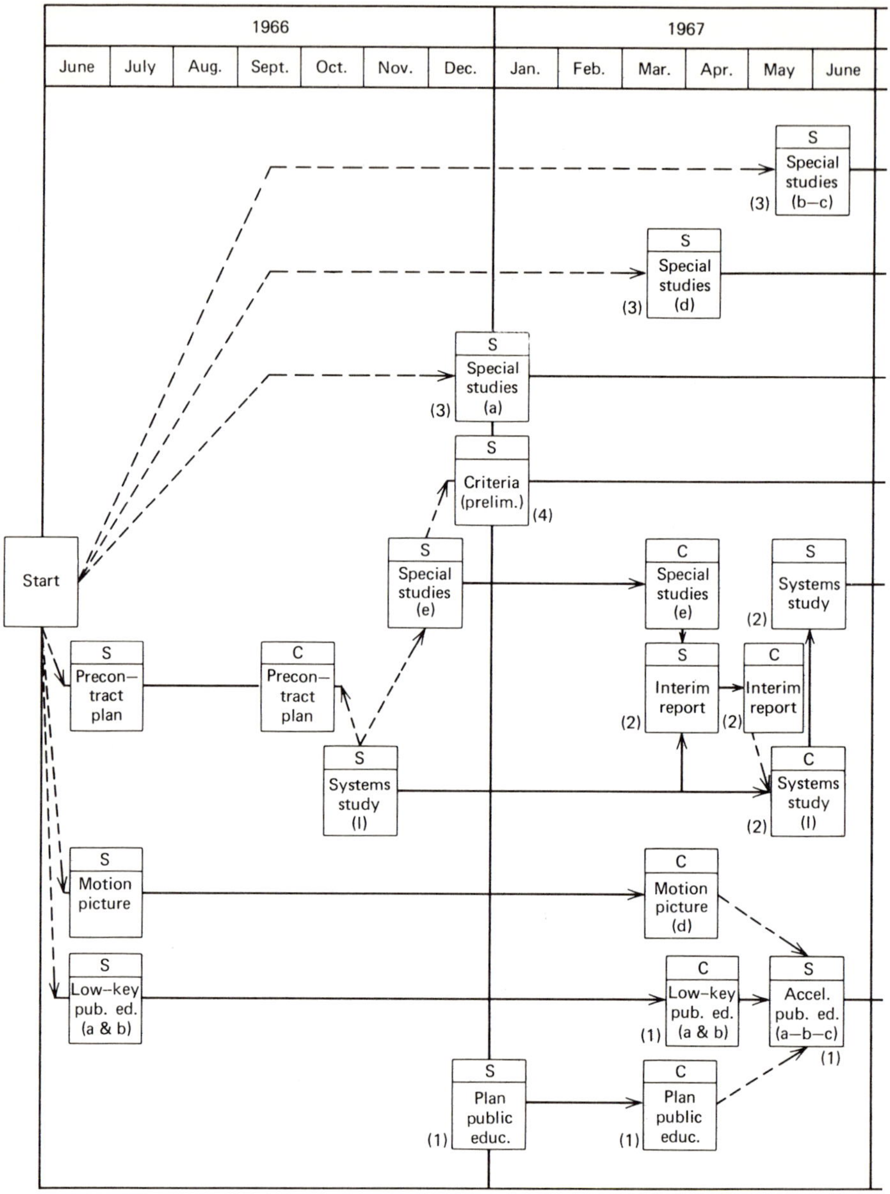

Figure 1. Diagram of the tasks in the Fresno project.

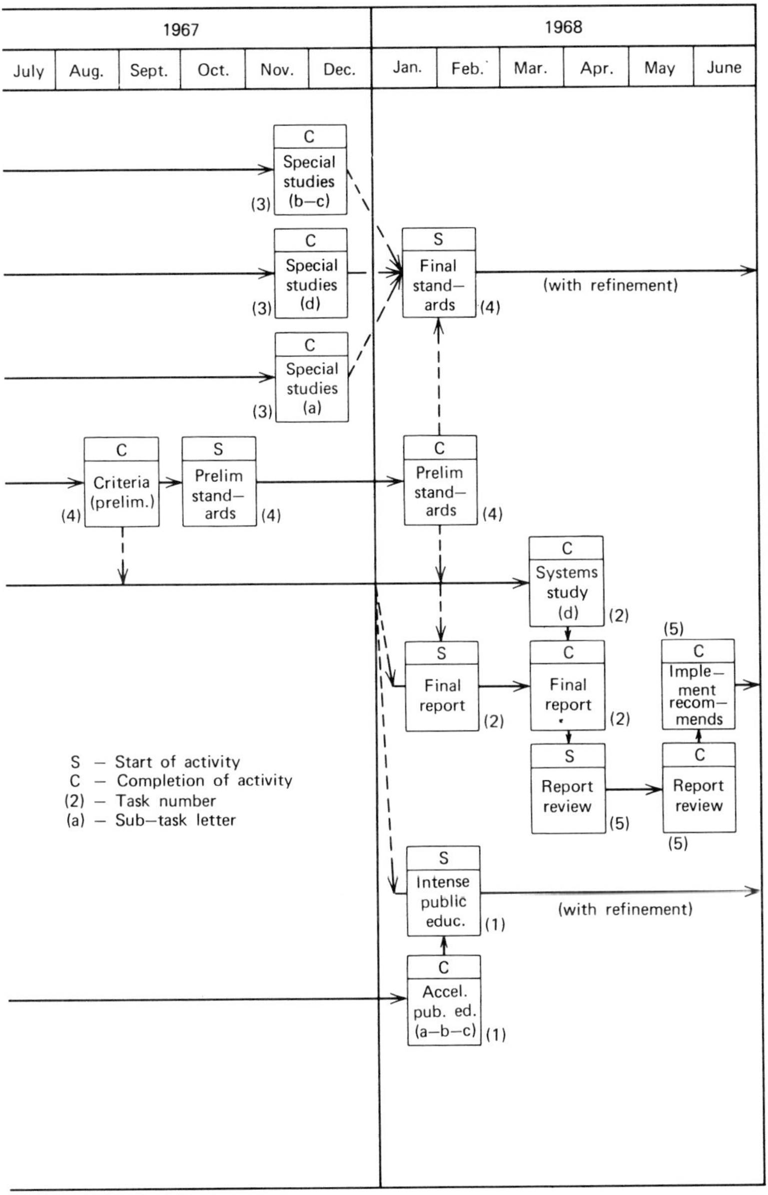

Figure 1. (Continued)

developed in this study, it is probable that several important factors would have been overlooked in determining a possible solution to the region's solid waste management problems. For example, more than two-thirds of the environmental effects in the Fresno region are caused by only three categories of solid wastes; almost one-third of the municipal region's environmental problems are created by on-site storage. Contrary to popular belief, sanitary landfills cause more problems in an agricultural region than in a municipal region.

The magnitude of the solid waste problem can more readily be envisioned when it is recognized that the Fresno Study region, with a population of 396,000, almost 2.5 million tons of solid waste are generated each year. By the year 200, when the region's population is expected to exceed 1 million, the rate of solid waste production is expected to increase to over 5.5 million tons/yr.

The three major categories of wastes mentioned above, namely, organic municipal refuse, organic industrial refuse, and animal manures, constitute almost two-thirds of the total amount of solid wastes generated within the Fresno region and, in addition, because of their highly putrescible nature, produce an even greater amount of the deleterious environmental effects. It was therefore particularly necessary to devise a solution that would provide major improvements in managing the three categories of gross environmental pollutants and to develop better storage techniques in the municipal subregions.

Any system for the study region would automatically delete open burning and open dumping because of their atmospheric and land polluting effects. Recommendation of a system that utilized sanitary landfilling totally, although quite effective, would require increasingly larger landfill sites to accommodate the increasing projected rate of waste production. Additionally, such a solution would pay insufficient attention to the long-range aims of conservation of our natural resources. Furthermore, it is highly desirable that any solution be amenable to salvage operations when they become economically feasible, and that as much as possible of the wastes be recycled as products useful to the region's economy. Another desirable objective was the built-in capability of allowing for an orderly transition from the existing system to that being proposed. Additional restraints on solution concepts included the region's projected economic capacity as well as legal, political, sociological, and practical factors.

SELECTED SYSTEM

The procedure for selecting a suitable management plan is based on the total systems cost and on the effectiveness with which the system reduces envirnmental damage. As an example, if one unit of odor from a constant source is cut in half when the source is loosely covered, the method would be 50% effective; if the same source were sealed in a container, the method would be nearly 100% effective.

Costs and performance scores were calculated for 22 different management systems that were conceived by professional engineers to fit the operational needs of the Fresno area. These were compared to the costs and performance scores of the present system, and acceptable levels of improvement and expenditures were determined.

When none of the subsystems for the municipal-industrial and the agricultural regions *combined* measured up to acceptable performance goals or fell within predetermined cost ranges, a final plan (Figure 2) was developed whereby large fractions of the total waste load would be handled individually on the basis of the severity of problems that the fractions caused. Thus, dead animals, garbage, manures, and other highly putrescible materials are handled differently from poorly degradable materials such as demolition wastes and others.

The Fresno area plan will reduce environmental damage due to solid wastes in the municipal-industrial regions by 84%. Reductions in the agricultural regions amount to 70%. The costs of the plan will be lower than the calculated maximum allowable amounts for the area by 17% and 1% for the same regions, respectively. The total estimated cost of operating the plan for the year 2000 is $78.6 million, or about $14 per ton of wastes that must be handled by that time. Many systems in municipal-industrial areas are operating currently at much higher costs.

Some of the present operations will be incorporated in the plan with only minor changes, while others such as collecting municipal refuse by vacuum system or conversion of manures to compost would change drastically in the direction of convenience and conservation of resources.

The study emphasizes that a high degree of flexibility within elements of the system has been incorporated deliberately so that new techniques, less expensive methods, assistance in financing, and other profitable changes can be incorporated readily in the future.

SIGNIFICANCE OF STUDY

The Fresno Project has produced an innovative conceptual design of an area-wide system that will provide a realistic solution to current and future solid waste problems in the Fresno area. The planned public information program has led to implementation of many program recommendations in Fresno County and additional progress is expected. It is anticipated that the regional program concepts of the project will be

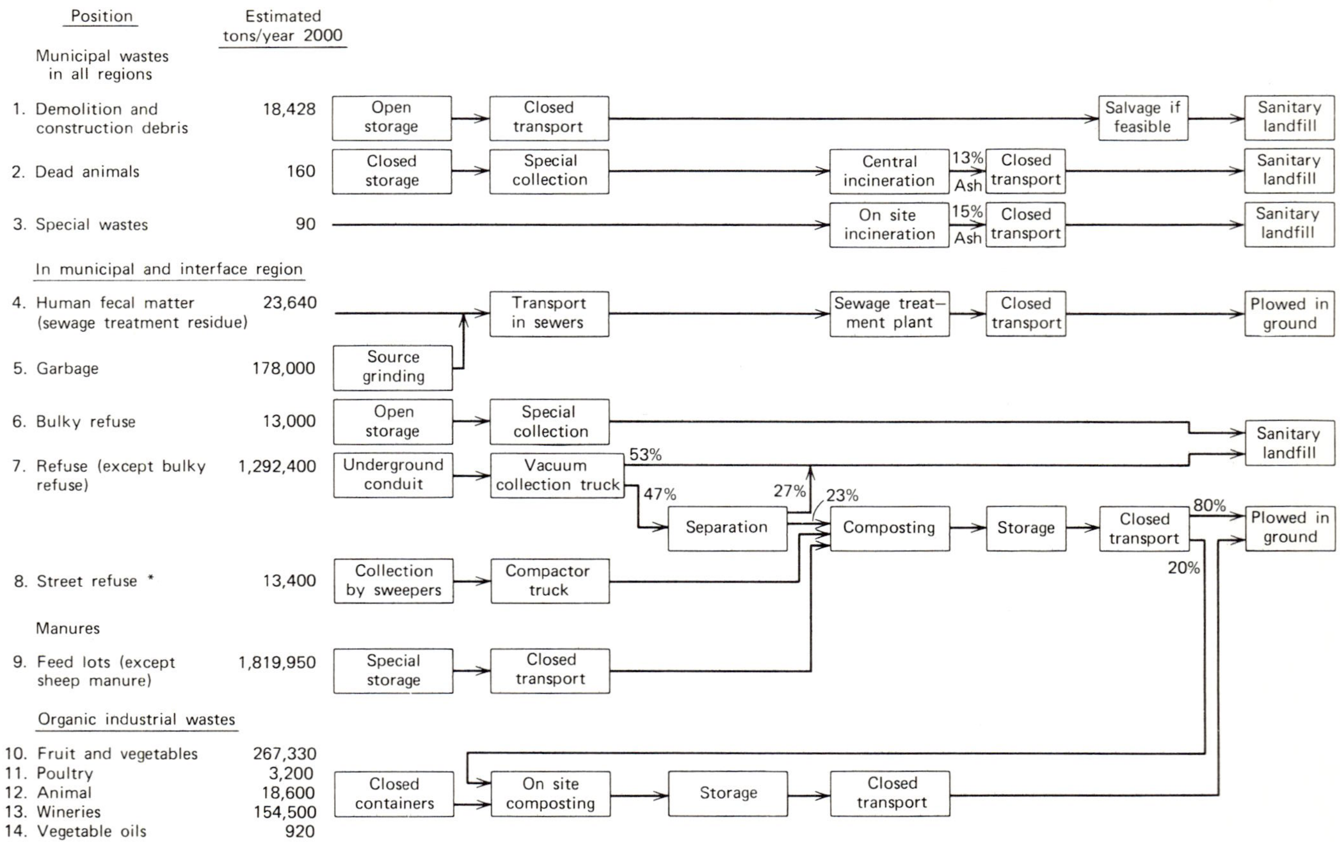
Position
Estimated tons/year 2000
Municipal wastes in all regions
1. Demolition and construction debris
18,428
Open storage
Closed transport
Salvage if feasible
Sanitary landfill
2. Dead animals
160
Closed storage
Special collection
Central incineration
13%
Ash
Closed transport
Sanitary landfill
3. Special wastes
90
On site incineration
15%
Ash
Closed transport
Sanitary landfill
In municipal and interface region
4. Human fecal matter (sewage treatment residue)
23,640
Transport in sewers
Sewage treat– ment plant
Closed transport
Plowed in ground
5. Garbage
178,000
Source grinding
6. Bulky refuse
13,000
Open storage
Special collection
Sanitary landfill
7. Refuse (except bulky refuse)
1,292,400
Underground conduit
Vacuum collection truck
53%
47%
Separation
27%
23%
Composting
Storage
Closed transport
80%
20%
Plowed in ground
8. Street refuse *
13,400
Collection by sweepers
Compactor truck
Manures
9. Feed lots (except sheep manure)
1,819,950
Special storage
Closed transport
Organic industrial wastes
10. Fruit and vegetables
267,330
11. Poultry
3,200
12. Animal
18,600
13. Wineries
154,500
14. Vegetable oils
920
Closed containers
On site composting
Storage
Closed transport

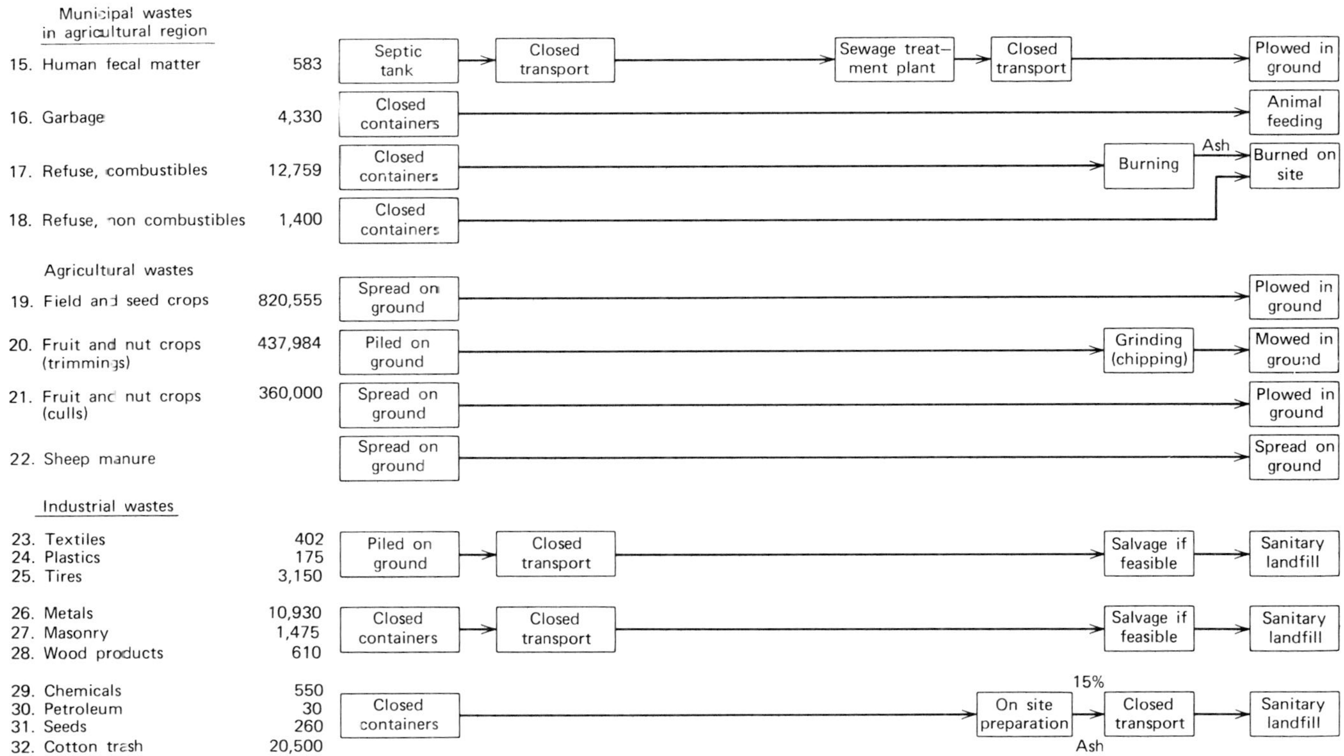

*Leaves (only) to composting; dirt and sand to landfill

Figure 2. Solid waste management system (Fresno region).

used in other localities of California and in other states, such as Kansas, where a similar intensive study is now in progress.

The study has special significance for the Fresno area. For the first time, the leadership of the area has access to detailed tabulations by census tract of the kinds and amounts of solid wastes produced by communities, agriculture, and industry. Realistic projections of the populations and waste loads in these tracts are tabulated by 10-yr periods for the next 30 yr.

Futhermore, the damaging effects on environment of mismanaged solid wastes are listed and documented. Any one of these damaging effects can become serious enough to cause local disasters unless controlled.

Finally, and perhaps of greatest importance to the area, the study has produced a suggested management plan that can serve all of the communities of the area for more than 30 yr at a cost that is geared specifically to the area's projected income and to the public services that will have to be provided as urbanization expands. The plan has flexibility to phase in new methods of handling wastes as they become available, and provides for tests of advanced concepts before they are installed.

Although the study concentrated on the rapidly urbanizing area of the county, it appears entirely feasible with a minimum of modification to extend the plan beyond its arbitrarily chosen boundaries.

CHAPTER II.4

sanitary landfills as a gully reclamation method

JERRY R. WEHRSPANN

A recent solid waste disposal facility survey of the Omaha metropolitan area by the Metropolitan Area Planning Agency indicates there are 21 disposal facilities open to the public. These serve an area of slightly more than 1500 square miles containing a present population of approximately 540,000. All of the facilities surveyed had some deficiencies, some of which could be corrected; however, many of the facilities must be classified as unregulated open burning dumps.

This fact and the need for gully reclamation were recognized by the Papio Watershed Advisory Board, who encouraged the Sarpy County Board of Commissioners to sponsor a gully reclamation sanitary landfill demonstration project in 1966. Financial assistance to carry out this demonstration project was received from the BSWM, USPHS, and USHEW.

The project, in operation from April 1967, to June 1968, is located in Sarpy County, Nebraska, on the southern fringe of the metropolitan Omaha area. Subsequent to completion of the initial landfill, a second site was opened by the county in the same vicinity. As a site is filled, other areas will be opened for public use.

The primary objective of the demonstration project was to convince public officials, community residents, and landowners that a properly planned and operated sanitary landfill can be permitted within reasonable distance from the source of refuse without annoyance to neighbors. It was also deemed necessary to determine if a sanitary landfill could be used to reclaim gullied land and provide a stable outlet for future runoff from the contributing watershed as well as being part of the community-wide plan for solid waste disposal. The project was also intended to serve as a practical guide to the development of future standards and regulations concerning sanitary landfills and to inventory and study management, operation and economic requirements necessary to operate a sanitary landfill at acceptable community standards.

The Sarpy County Soil and Water Conservation District expressed an interest in operation of the demonstration project because of its interest in gully and erosion control. At the outset, it was planned that the 4.5 acres of gullied land could be reclaimed for a higher use of land.

The voided and unusable land before project initiation totaled 4.5 acres. The natural gully volume after trees were removed was 32,100 yd^3. By trenching across the gully, the capacity was increased by 36, 800 yd^3. This resulted in a total gully capacity to a planned finish grade of 68,900 yd^3. Records were obtained to show that 7950 tons of solid waste or approximately 30,000 yd^3 were landfilled into the site during the 15 months of operation. The excavated 36,800 yd^3 plus an additional 2100 cubic yd^3 of soil were used for daily cover. The daily volume of solid waste received at the site averaged 85 yd^3/day, which indicated approximately the amount of solid waste one might expect from a city with a 20,000–25,000 population.

Figure 1 typifies the operation procedure followed at the landfill site. The north–south entrance road (1) is on the section line and located within an existing Sarpy County road right-of-way. The road was graded to a 24-ft width with a maximum 3% grade and rocked for all-weather access. Drainage structures were installed in the road embankment where required. A public disposal facility and turnaround was constructed at the north end of the entrance road (2). This area was graveled and developed as a 24-hr disposal point for solid waste hauled to the landfill site by private automobiles. A 20-ft by 8-ft metal building (3) that doubled as a scale house and office was located inside the main entrance gate at the north end of the access road. A 25-ton scale was located at this point to obtain gross and tare weights of incoming trucks. These trucks were then directed to the landfill site for dumping. A drop inlet erosion control structure (4) with a slotted flume outlet was constructed of compacted earth fill across the gully at the lower end of the landfill site. The principal spillway consisted of a 42-in. corrugated metal pipe riser, 30-in. corrugated metal pipe barrel, and slotted flume outlet. A 6-in.-diameter corrugated metal pipe intercepter drain was installed on the upstream side of the earth fill to intercept underground seepage water from the landfill area for later analysis. Diversion terraces (5) were located on each side of the landfill area to intercept surface runoff from the 30-acre drainage area above and around the landfill area. The diversions outletted into the 42-in. riser in the principal spillway of the dam.

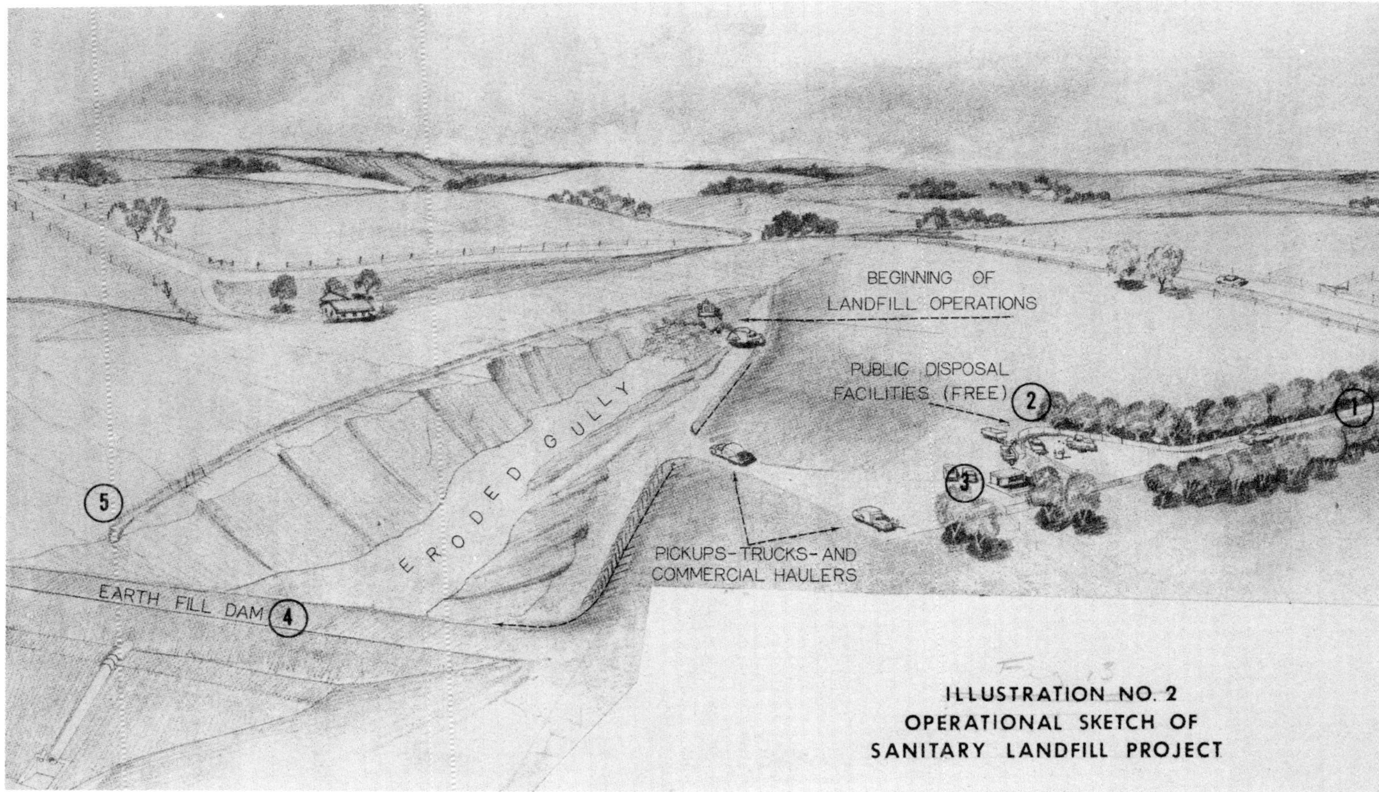

Figure 1. Operational sketch of sanitary landfill project.

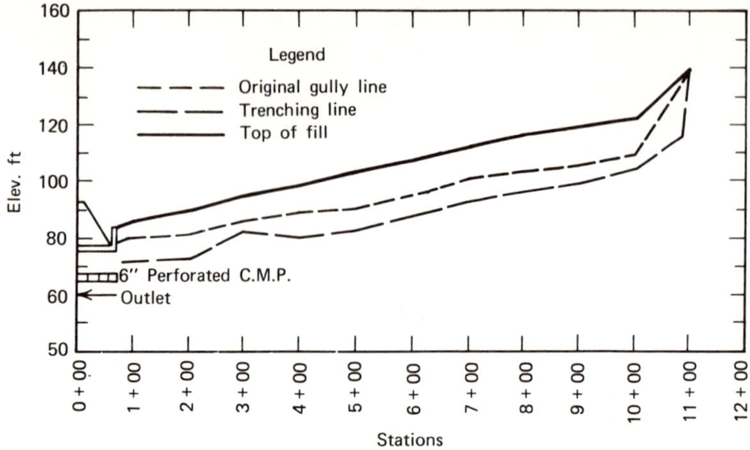

Figure 2. Sanitary landfill Papio watershed: - - · - - original gulley line; —X— trenching line; —·— top of fill.

Figure 2 illustrates the amount of excavation below the original gully flow line which was required as a part of the project. The additional trenching was required to provide cover material and also to provide additional volume for incoming refuse. Figure 3 shows the amount of total refuse received in each month during the 15-month period in

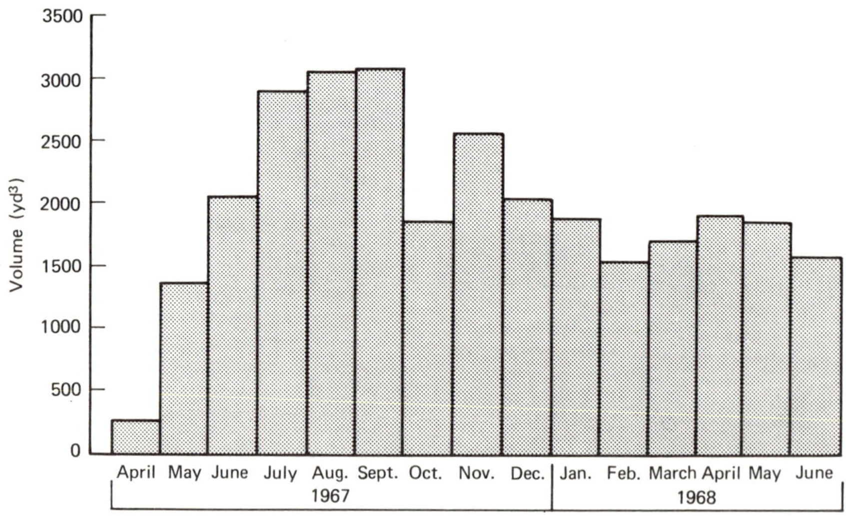

Figure 3. Sanitary landfill, monthly volume summary, Papio watershed.

which the demonstration project was in operation. The total amount of refuse received at the site was approximately 30,000 yd^3.

Samples of ground and seepage water below the dam were analyzed from February 1967, through September 1968, and no change in the chemical quality of the water was evident.

Settlement of the sanitary landfill during compaction and covering operations was not a serious problem; however, periodic visual inspections are planned to record the extent of settlement that might occur and become a problem over a long-range period.

Public relations and timely informational meetings with neighboring landowners near the proposed site are important as soon as preliminary site evaluations are completed. The two landowners at the demonstration site were contacted by the Project Manager to determine their willingness to participate in the proposed sanitary landfill demonstration. As soon as the application for a PHS demonstration grant assistance was approved, the neighboring landowners were contacted and the three-year demonstration project explained. Sufficient contact and explanation was completed so that no one appeared at the Sarpy County Planning Board hearing in opposition to the requested zoning change and special use permit.

Construction and operation costs for the project were not considered to be typical for a gully reclamation–sanitary landfill because of the demonstration aspect of the installation. Total equipment and initial construction costs were approximately \$40,000. Monthly operation costs averaged approximately \$2700 over the 14-month period of operation. Unit costs have been reduced on subsequent sites because of increased volume and increased operational efficiency which was gained from experience at the demonstration project site.

In this demonstration the primary advantage for landowner cooperation was the benefits they would receive from reclaiming gullied lands, control of active gullies, and more convenient field crossings for farming operations. Both property owners involved have stated their complete satisfaction with the demonstration project and have since signed easements allowing the county to operate a second site near the demonstration project.

With proper compaction and a final cover of 2 ft of clean soil, gully-type sanitary landfill sites can be considered for agriculture or park uses, thereby increasing the value and productivity of a plot of previously unusable land.

Originally the tree-lined gully occupied 4.5 acres of unfarmable land area. After the fill was completed the entire area was fertilized and seeded to bromegrass for erosion control. During the 1970 growing season, an additional 4.5 acres of corn was planted on the fill. Corn production in this area can average about $50 per acre net income, thereby offering the potential of $225 per year increase in net income to the landowner. There will also be some benefit received from improved farming operations since tractors and implements can now cross the previously gullied area for more efficient field travel.

Placement of the drop-inlet structure at the outlet of the gully controlled 22 vertical feet of drop through the corrugated metal pipe spillway. Two acres of land voiding by erosion that would have taken place during the next 25 yr has been prevented. Estimated average annual benefit to property is $25 per acre per year.

Construction of the grade control structure and filling the voided gully raised the waterway flow line an average of 20 ft. This provides a stable outlet for the terraces and grassed waterway applied to the 40 acres of cropland above the dam. The effect of this improvement is to triple the life of the terraces and waterways that are designed to hold in place an estimated 600 tons of soil that would have eroded off the land each year. Conservation practices applied are expected to reduce sediment yield from 20 tons to 5 tons/acre/yr. A frequently used value to estimate the benefits of reducing sediment from clogging waterways and reservoirs is $1 per ton. With that somewhat arbitrary value, the benefits to the property owner and the public for sediment reduction should exceed $600 per year.

Gullied land appears on the local property tax rolls at $5 per acre. Through proper gully landfill techniques, approximately 4.5 acres were restored to $500 per acre agriculture land. The demonstration site alone increases taxable land value after landfilling operations by over $1800, prevents an additional two acres of land voiding, and is now holding in place an estimated 600 tons of erosive soil each year. Multiple benefits for the landowner and the local taxpayer in addition to the value of solid waste disposal are calculated at $915 annually.

It was determined that gullies typical of those found in the loess-mantled hills of eastern Nebraska and western Iowa could be used satisfactorily for a sanitary landfill. Often this land will be made available for use without charge or could be leased or purchased for a fraction of the cost of productive land. A properly engineered drainage control structure should be required at the downstream end of the gully to protect against future erosion. A well-operated sanitary landfill need not be located in a remote section. Gully-type sanitary landfills can be planned to stop soil erosion and reclaim acres of voided land, provide a drainage outlet for the land above, and add to the beauty of the countryside.

REFERENCES

1. Project Report, "Sanitary Landfill Demonstration of a Gully Reclamation Method," Sarpy County Nebraska: USPHS Demonstration Project No. 5-DOI-UI-00011-01, 02, 03, November 1968.
2. Report No. 102, "Solid Waste Disposal," Omaha-Council Bluffs Metropolitan Area Planning Agency, 1969.

CHAPTER II.5

hydrogeologic aspects of solid waste disposal

R. N. FARVOLDEN

The common objections to "garbage dumps" can be avoided by converting all such waste disposal facilities to sanitary landfill operations. In a sanitary landfill there is little risk or evidence of either pollution or nuisance at the surface, but there is some risk of groundwater pollution. Uncertainties related to these risks led to a series of demonstration and research projects in the Chicago region (sponsored mainly by the USBSW, Grant No. D01-UI-00006) and in Southern Ontario (sponsored by the Ontario Waste Management Branch and the Grand River Conservation Authority). In all, some seven landfills have been investigated and instrumented to determine the extent of subsurface pollution, and to relate it to the hydrogeologic conditions at each site. The purpose of both projects is to develop criteria for site selection and design, based upon physical aspects of the site. This account draws heavily on the results of these projects, most of which have been published.[1, 2, 6]

BASIC CONSIDERATIONS

Certain hydrogeologic factors are basic to understanding the implications of burying solid wastes in the ground. In an arid climate it is possible to bury the waste in the unsaturated zone, between the surface and the water table, because the water table is deep over wide areas. Low precipitation and high evaporation together cause a soil moisture deficiency that is nearly permanent. There is seldom, if ever, sufficient water in the root zone to cause downward percolation, so water in the unsaturated zone is more or less fixed in place. Consequently there is no migration of fluid into, or out of, waste buried in the unsaturated zone. Under such ideal conditions there is little danger of groundwater pollution from the waste.

In humid climates the moisture content of the root zone is at or near field capacity much of the time. Any heavy rain is likely to cause groundwater recharge by downward movement of soil water to the water table. This water will seep through any waste buried in the unsaturated zone, and become highly charged with bacteria, viruses, and dissolved and suspended solids. Water polluted by solid waste is called leachate.

The migration or retention of contaminants in the ground depends on the nature of both the contaminant and the materials through which it flows. Groundwater is the vehicle that carries the contaminants, so the migration of contaminants depends on groundwater flow. Retention of contaminants depends on mineralogical and physical characteristics of the medium.

Water in the subsurface moves according to Darcy's law, which may be written

$$V = KI$$

where V is the velocity (cm/sec), K is the hydraulic conductivity (cm/sec), and I is the hydraulic gradient (dimensionless).

From the equation it is obvious that if there is a finite permeability and a gradient then there must be groundwater movement. The driving force is gravity, and movement is in the direction of the maximum gradient. Water levels in wells are a measure of the groundwater head and thus are significant in groundwater flow studies.

Groundwater moves freely through permeable rock units called aquifers but less freely through relatively impermeable aquitards. Consequently, the best aquifers are also the most susceptible to contamination, whereas the poor aquifers are relatively immune.

In a humid region groundwater heads are highest under hills and lowest in valleys because the water table follows the land surface rather closely. Thus groundwater moves from uplands zones of recharge to lowland zones of discharge. The dry-weather flow of streams consists mostly of groundwater discharge, which must be balanced by an equivalent amount of recharge in order to satisfy the requirements of continuity. The effects of varying basin geometry and geology on groundwater flow systems have been studied by Toth[9, 10] and Freeze and Witherspoon.[3-5] Depending on these controls, water may move only a short distance underground before returning to the surface, or it may be in part of the groundwater reservoir where regional flow occurs and may remain underground for hundreds of years. Surface manifestations of groundwater discharge are easy to recognize in an arid region usually as a well-watered location in an arid landscape. In a humid region the evidence is not so obvious since abundant soil moisture is the usual condition.

METHOD OF STUDY

Landfills were chosen for the investigations on the basis of age, accessibility, hydrogeologic environment, and size. Older sites were preferred in order to reduce the effect of time as a limiting factor in migration of pollutants. The problem of accessibility in urban areas is apparent. A variety of hydrogeologic environments included well-drained uplands, alluvial gravels of river bottom lands, flat and poorly drained uplands and lowlands, and several sites difficult to categorize. Some landfills seemed well suited for investigation but were rejected merely because they were too big to be properly tested.

The usual method of testing each site was to drill a series of holes through the glacial drift overburden and into bedrock. Continuous split spoon samples were collected for most of these initial test holes. Piezometer nests were then installed about the site in order that distribution of hydraulic head could be measured, mapped, and monitored. Core or drill cuttings were collected from these holes also to provide data for mapping the distribution of the subsurface units. Hydraulic properties were crudely measured in the field by various sorts of slug tests and pump tests, and laboratory measurements were made of permeability, porosity, and other pertinent properties.

Hydrograph and flow net analysis provided an estimate of recharge rates into the fills and flow rates out of the fills and in the immediate area. From this sort of information the pattern of groundwater flow, and hence the position of the site with respect to groundwater flow systems of the locality, was established. This in turn was used to decide where additional sample points were needed to test for pollutants in the groundwater.

Several innovations were used in order to make the necessary field measurements, but the most difficult problem encountered was in the sampling procedure for water quality from piezometers in clay. Before any samples were taken, each piezometer was cleaned by pumping. The chloride content of the discharged water was monitored and the well was considered clear of pollutants caused by the drilling operation when the chloride concentration stabilized.

Response time of piezometers in clay was improved by reducing the pipes to ¼-in. diameter. Continuous hydrographs were obtained from piezometers cased with 1¼-in. black plastic pipe by using Keck recorders coupled to a Leupold-Stevens type recorder.

RESULTS

At every site investigated in these studies a leachate is present. In localities with impermeable soils, the leachate has formed from water infiltrating into the waste, and only a small volume of soil and water immediately surrounding the waste is contaminated. But even at well-drained sites, where the waste is in the unsaturated zone, soil water is converted to leachate as it passes downward through the waste, and this leachate then reaches the water table and the groundwater reservoir.

Whether the waste is saturated or not, anaerobic conditions develop very quickly as decomposition progresses. Thus, in a humid region, efforts to ensure that the waste is buried within the unsaturated zone are useless, and regulations requiring such restrictions may in fact prohibit use of satisfactory sites.

A common practice is to carry the fill above the original grade with the intention of building an artificial hill for recreational purposes. Several unpleasant side effects of high landfills can be avoided by considering the hydrologic implications.

Unless the site is well drained, a groundwater mound forms under the landfill. The groundwater is converted to leachate by contact with the waste. The hydraulic gradient causes leachate to move away from the mound in all directions, causing the envelope of contamination to enlarge. Springs of leachate develop near the toe of the slope of the landfill and detract from the recreational value of the site. All this can be prevented by ensuring that the site is well drained, but this is not a complete solution because the ultimate disposal of the contaminants must still be considered.

According to classical groundwater theory, no downward percolation will reach the saturated zone and cause groundwater recharge until the materials of the unsaturated zone have reached field capacity. In this condition the water content of the soil is the maximum that can be retained against the pull of gravity, and consequently if any more water is added, the soil will release an equivalent amount to migrate downward. Remson et al.[8] calculate that in the climate of Pennsylvania, considerable time is required for the waste in a landfill to reach field capacity and therefore suggest that groundwater recharge through the waste cannot occur for an equivalent time. However, an observation well installed to the bottom of the new DuPage County landfill in northeastern Illinois shows that a groundwater mound is already forming, before the fill above is complete. Recharge has taken place while the bulk of the waste still has a water content below field capacity.

An apparent improvement in quality of the leachate with age of fill was observed, although there is not enough evidence from this investigation to reach firm conclusions. Sample points within the waste at the old DuPage County site in northeast Illinois produced a leachate varying from a total dissolved solids content of 11,250 ppm from a part of the fill about 4 yr old, to 2075 ppm from a part about 12 yr old. Data from the landfill of the city of Winnetka further illustrate this relationship:

Age of fill (yr)	*Total dissolved solids in leachate from fill (ppm)*
18	2941
15	3250
6	4750
2–3	5938

There is also a rather regular improvement in water quality with distance from the site. In most cases, the envelope of contamination has spread only 30 or 40 ft, or less. Even where contamination was detected farther from a site it was identified only by anomalously high values for chloride, and less often sulfate (as SO_4^{2-}) or hardness.

Other contaminants are removed from the leachate by migration through the granular materials of the near surface deposits. In no case could it be demonstrated that contaminants from these landfills had migrated more than 600 ft from the site.

However, it is obvious that in a humid region some envelope of contamination must form about the waste. For this reason, regulations that stipulate "no groundwater contamination" are unrealistic and should be changed, because it is not possible for an operator to comply with them. Regulations should require that an investigation of the site be conducted so that predictions can be made regarding the spread of the envelope of contamination and the feasibility of controlling the spread. Regulations should permit the location, design, and operation of a landfill only where one of the following objectives can be achieved:

1. Design to allow for predicted and acceptable migration of contaminants from the site.
2. Design to allow for predicted and acceptable migration, and then collection of contaminants to prevent further, unacceptable migration.
3. Design to retain and collect leachate within the landfill where no migration whatever is acceptable.

Social factors such as zoning and ultimate plans for the land use determine the design objective most appropriate for the site. Hydrogeology is the main factor in the cost of providing the necessary protection. Field examples can be used to illustrate the applications of design objectives.

1. *Design to allow for predicted and acceptable migration.* In many situations the migration of leachate into an aquifer can be considered acceptable. Contaminants are removed from groundwater as it migrates through granular materials. The main mechanisms for this removal are simple filtration and adsorption onto the fine-grained particles. Quality is also improved by dilution if lines of flow diverge and by biochemical reactions.

The old Dupage County landfill in northeast Illinois (Figures 1 and 2) was built up above the surrounding plains level to form a low mound of waste covered with a thin layer of soil. It is underlain by surficial silts, clays, and sands of glacial origin. A heavy clay till underlies these surface deposits, and an important regional acquifer ("the Silurian dolomite aquifer") is below the till at a depth of about 70 ft (Figure 1*b*). A groundwater mound has formed under the landfill, and springs of leachate occur along the south slope of the fill and to a lesser degree on the north slope (Figure 1*c*). Leachate has migrated into the surficial sand under the field to the south of the fill, but the only known consequences of this are a slight increase in the chloride content and a barely detectable increase in the sulfate content and hardness of the groundwater. This might be regarded as acceptable migration.

The hydrogeologic setting of the Elgin landfill is quite different. The site is an old gravel quarry, well drained, and separated from the Fox River by a narrow strip of land which is used as a marina and picnic area (Figure 3). The water table is below the fill, owing to the good drainage, and the gradient on the water table is toward the river. Groundwater immediately below the fill has been contaminated by leachate, and the envelope of contamination can be traced by chloride and sulfate content to the shallow wells near the river. These wells have a water level approximately equal to that of the river, but deeper wells have a higher head and flow, indicating an upward gradient typical of a groundwater discharge zone. Thus a leachate is being produced by decomposition of waste in the unsaturated zone. The contaminants reach the saturated zone and migrate as would have been predicted, and eventually reach the river. However, the dilution ratio is hundreds or thousands, so the contamination may be considered acceptable, even though shallow wells are contaminated, because of the availability of groundwater from the deeper aquifer. There is no risk of contamination of the deeper aquifer unless heavy pumping should lower the head sufficiently to reverse the gradient causing downward rather than upward movement.

2. *Design to allow for migration and subsequent collection of leachate.* A regulatory agency might well decide that, in the examples

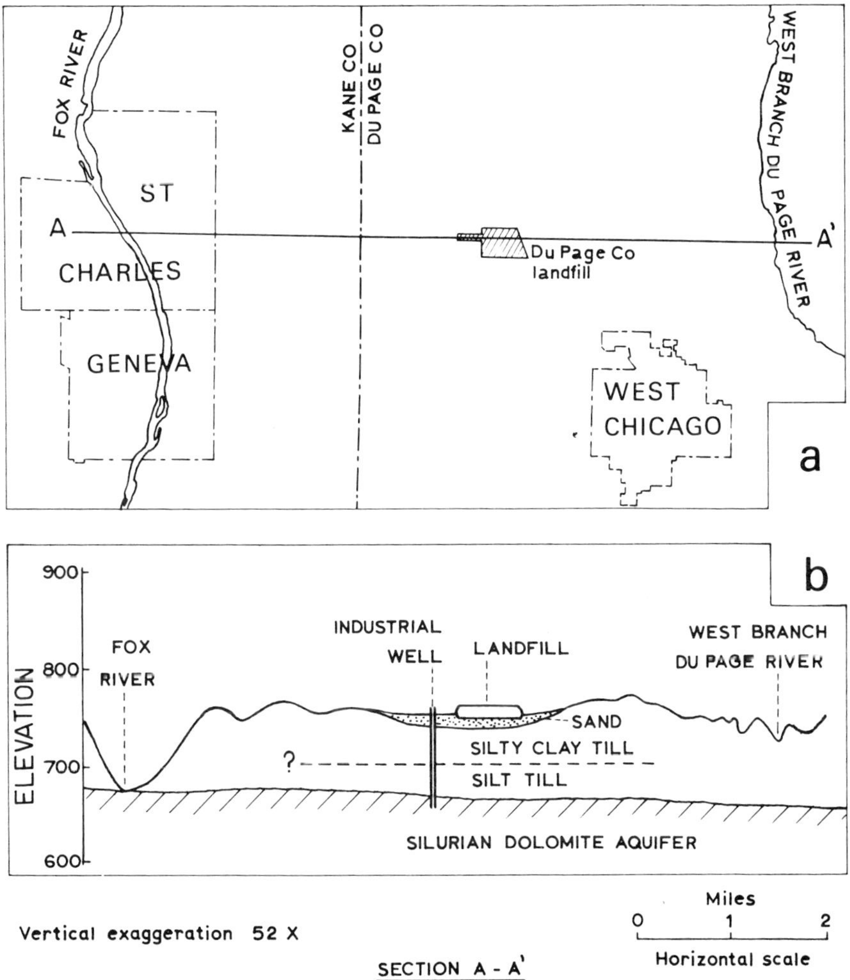

Figure 1. Old DuPage County, Illinois, landfill (after Hughes, Landon, and Farvolden). (*a*) Location; (*b*) generalized regional cross-section; (*c*) plan of site showing piezometer locations and groundwater contours.

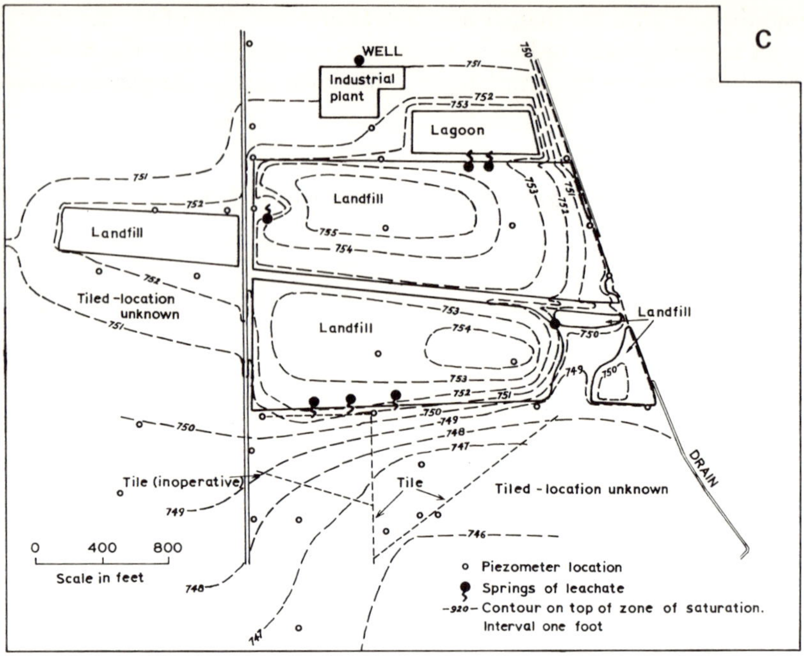

Figure 1. (Continued)

cited above, the migration of pollutants is not acceptable. In that case the spread of pollutants from the site could be controlled by strategically located wells, or drain tiles. Either scheme could be designed so that all leachate would be diverted and recovered, and then disposed of as a special sort of liquid waste. Important considerations would be the volume of fluid recovered and the degree of dilution, and once again hydrogeology is the most critical factor. In relatively impermeable materials a cone of influence might be created by pumping at a rate of only a few gallons per minute, but several wells might be needed in order to generate the groundwater gradient over a wide enough area. In relatively permeable materials, one well might provide a cone of influence of sufficient extent, but a high pumping rate would produce large volumes. The dilution factors would present an additional set of alternatives that would have to be evaluated on hydrogeologic considerations.

Near the old DuPage County landfill there is a high-capacity industrial well completed in the "Silurian dolomite aquifer" (Figures 3*a*

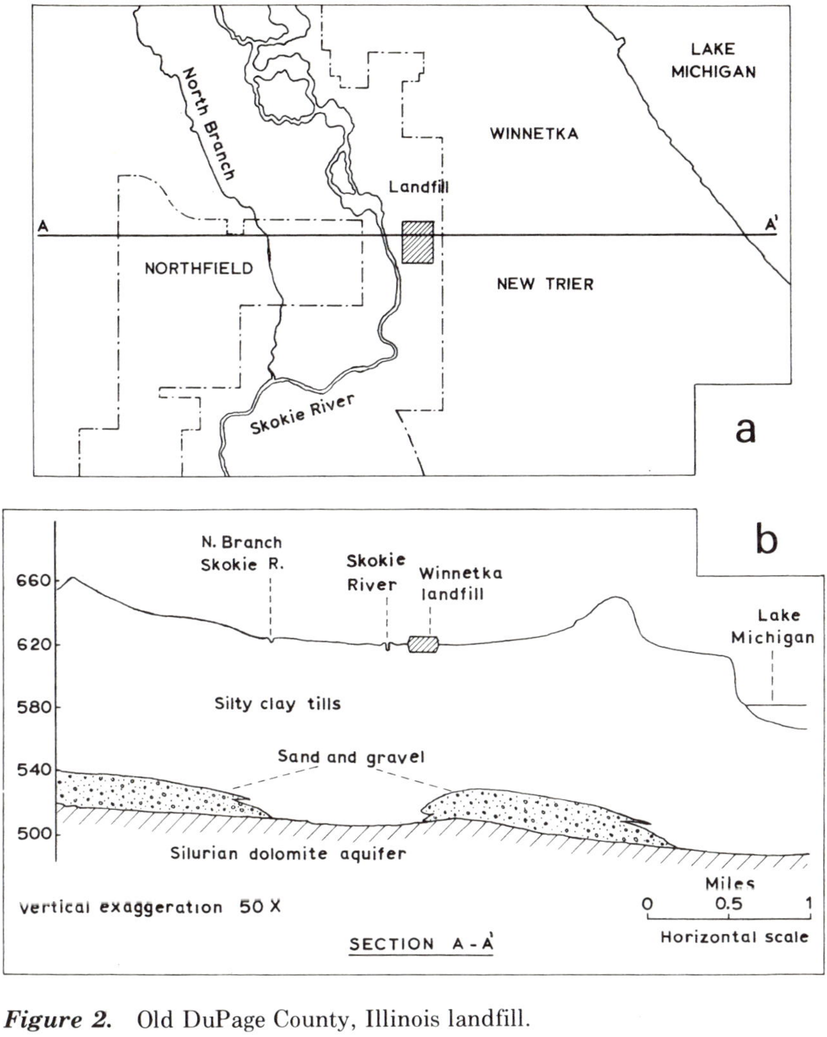

Figure 2. Old DuPage County, Illinois landfill.

and 3*b*). It is possible that contaminants from the landfill are moving downward through the surficial deposits to the aquifer and then down the groundwater gradient to the well. However, comparison of the rate of flow downward from the site with the rate of pumping of the well indicates a dilution factor so high that contamination would not be detected.

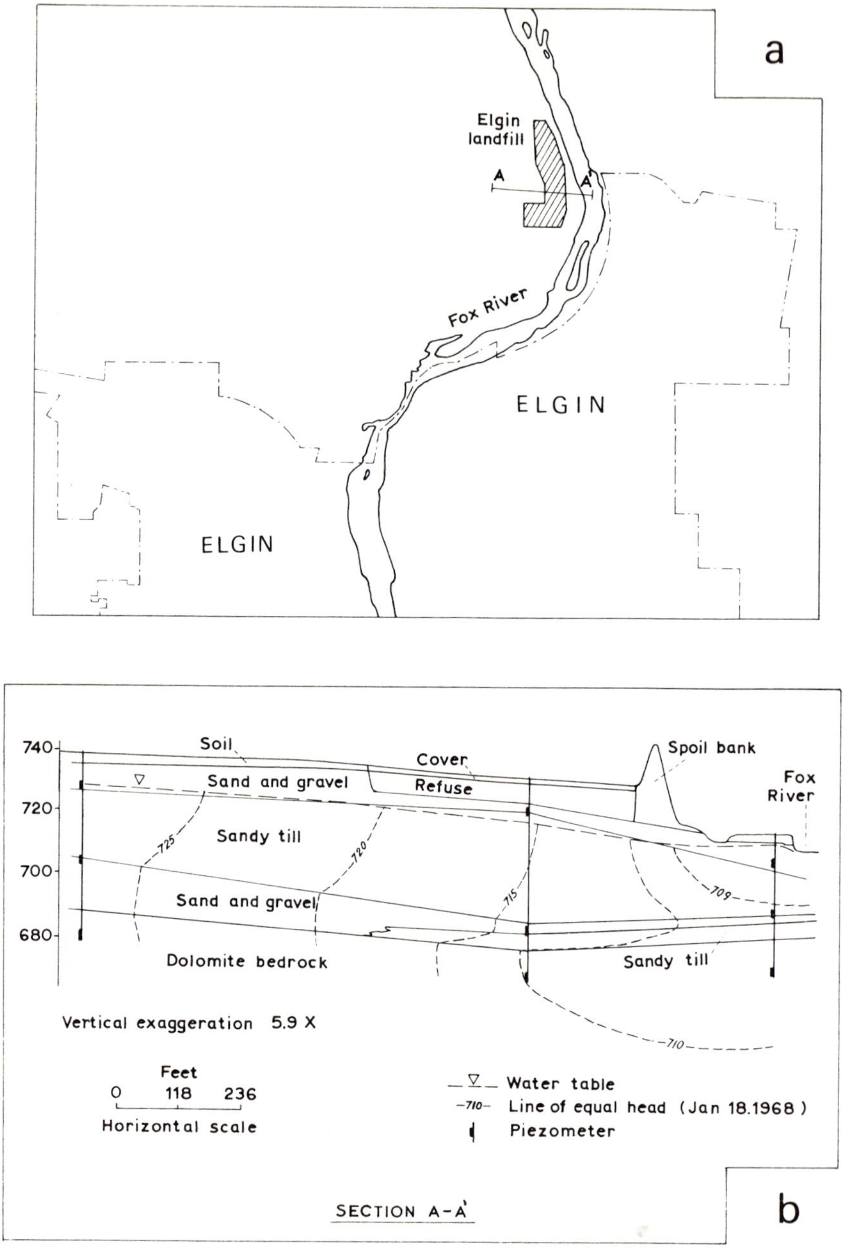

Figure 3. Elgin, Illinois landfill (after Hughes, Landon, and Farvolden). (*a*) Location, (*b*) generalized cross-section.

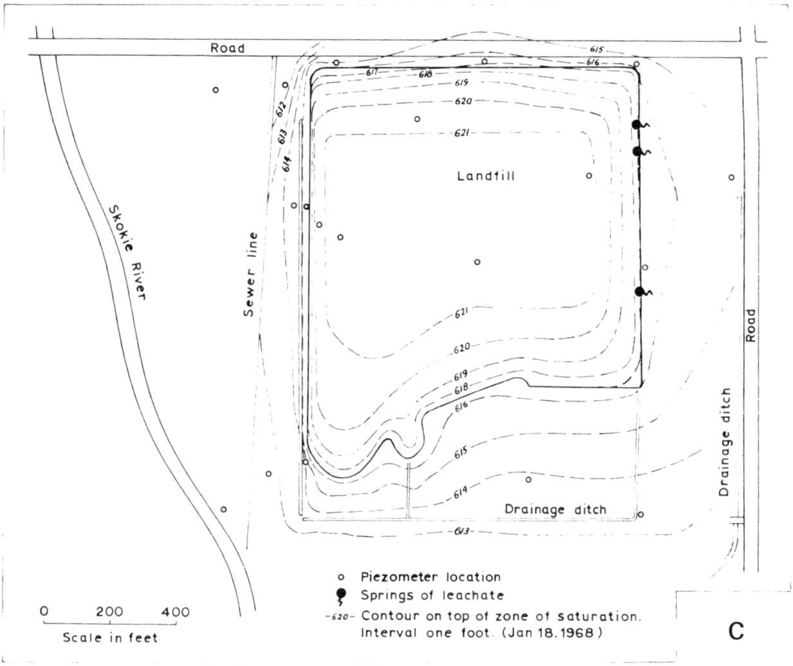

Figure 3. (*c*) Plan of site showing piezometer locations and ground water contours.

3. *Design to retain and collect the leachate within the fill.* It may be economic and convenient to use a site from which no migration of pollutants whatever is acceptable. In considering such a site, hydrogeologic factors are the most important controls on feasibility and cost.

Almost any plan that involves enclosing the waste within an impermeable liner or membrane is likely to be rejected on the basis of either cost or dependability. Even a trench in heavy clay is not proof against spread of leachate. Unless the cover on the fill is impermeable, water will infiltrate and eventually fill the trench and saturate the waste. As infiltration continues the leachate must overflow the trench, giving rise to springs of leachate and surface pollution rather than groundwater pollution.

A good example of this situation in the Chicago region is the city of Winnetka landfill (Figure 2*a*). The waste is buried in trenches about 8–10 ft deep and the fill is carried about 5–15 ft above original grade. No groundwater contamination was anticipated at this site and none was found from extensive system of sampling wells. However, a groundwater mound has formed under the fill (Figure 2*b*) and infiltration through the surface of the fill reappears as it seeps all along the east side where it forms offensive ponds on poorly drained ground. There are private residences nearby, so a situation that seemed acceptable in a rural area adjacent to the old DuPage County landfill is not acceptable here.

The solution to this problem is first to reduce the infiltration rate to a minimum, and then to prevent any of this reduced volume of fluid from leaving the site. The most dependable means of ensuring that all leachate is retained is to provide a hydraulic gradient toward the fill. This is best accomplished by pumping to create a cone of depression about the fill. Again, hydrogeology of the site locality is a critical factor in the feasibility of the operation. If the fill is in clay and the top cover is designed to reduce infiltration, then very little water must pumped to create the needed hydraulic gradient, because little water can migrate either from the surface or from the surrounding groundwater reservoir. However, if the fill is in permeable materials, a clay liner may be needed, just to reduce the volume of water to be handled. It is important to note that with this sort of hydraulic trap, failure of the liner will lead only to an increase in the pumping rate required to maintain the hydraulic gradient, and not to a release of pollutants from the site to the groundwater reservoir.

CONCLUSIONS

The most important physical requirements of any landfill site are the pertinent hydrogeologic parameters such as

1. properties and distribution of hydrogeologic units;
2. groundwater heads and water-table configuration for determining the natural flow systems of the locality;
3. Natural quality of the groundwater.

If this information is obtained by a thorough test drilling program, followed by sampling and monitoring, then a landfill that will meet an approved safety objective can be designed for almost any site. If the information is not obtained by a testing and monitoring program, the sanitary landfill cannot be considered safe.

REFERENCES

1. R. Farvolden, H. M. Hill, G. S. Fraquhar, and D. Weatherbe, "Sanitary Landfill Study—Phase One Report," University of Waterloo Industrial Research Institute Project, 8083, 1970, 68 pp.

2. R. N. Farvolden and G. M. Hughes, "Hydrogeologic Factors in Landfill Design," Proc. 16th Ann. Pollution Conf., Ontario Water Resources Commission, Niagara Falls, June 1969, pp. 43–50.

3. R. Allan Freeze and P. A. Witherspoon, "Theoretical Analysis of Regional Ground Water Flow: 1. Analytical and Numerical Solutions of the Mathematical Model," *Water Resources Res.*, **2,** No. 4, 641 (1966).

4. R. Allan Freeze and P. A. Witherspoon, "Theoretical Analysis of Regional Ground Water Flow: 2. Effect of Water-Table Configuration and Subsurface Permeability Variation," *Water Resources Res.*, **3,** 623–634, 1967.

5. R. A. Freeze and P. A. Witherspoon, "Theoretical Analysis of Regional Ground Water Flow: 3. Quantitative Interpretations," *Water Resources Res.*, **4,** 581–590 (1968).

6. G. M. Hughes, R. A. Landon, and R. N. Farvolden, "Hydrogeology of Solid Waste Disposal Sites in Northeastern Illinois—An Interim Report on a Solid Waste Demonstration," Grant No. D01-UI-00006), USSWM Cincinnati, 1969, 137 pp.

7. P. Meyboom, "Patterns of Ground Water Flow in the Prairie Profile," in *Proc. Hydrology Symp.*, No. 3, Dept. of Northern Affairs and National Resources, and the Geological Survey of Canada, 1963, pp. 5–33.

8. Irwin Remson, A. Alexander Fungaroli, and Alonzo W. Lawrence, "Water Movement in an Unsaturated Sanitary Landfill," *J. Sanitary Eng. Div., ASCE,* **94,** No. SA2, Proc. Paper 5904, 307–317, April 1968.

9. J. Toth, "A Theory of Ground Water Motion in Small Drainage Basins in Central Alberta, Canada," *J. Geophys. Res.*, **67,** 4375–4387 (1962).

10. J. Toth, "A Theoretical Analysis of Ground Water Flow in Small Drainage Basins," *J. Geophys. Res.*, **68,** No. 16, 4795–4812 (1963).

11. J. Toth, "Mapping and Interpretation of Field Phenomena for Ground Water Reconnaissance in a Prairie Environment, Alberta, Canada," Contribution 332 Research Council of Alberta, Edmonton, Alberta, Canada 1966.

CHAPTER II.6
aerobic landfill stabilization

RALPH STONE

BACKGROUND

Several potential problems are associated with completed sanitary landfills: combustible gas production, leachate groundwater pollution, and differential settlement. These problems, which may require extensive surveillance to reduce health hazards and various nuisances, are commonly associated with anaerobic decomposition of organic material in a landfill. Anaerobic decomposition of solid waste normally occurs slowly over a period of many years. In contrast, the maintenance of a suitable aerobic environment accelerates decomposition by promoting the rapid growth of cellulose-consuming microbes and by increasing the internal temperature, thereby increasing the rate of oxidation of the organic solid waste.

CHEMICAL REACTIONS OCCURRING IN SANITARY LANDFILLS

The aerobic chemical reactions that occur in an aerated landfill can be theoretically expressed as follows:

$$1.\quad \underset{\text{cellulose}}{(C_6H_{12}O_6)_n} \xrightarrow[\text{cellulase}]{\text{microbial}} \underset{\text{glucose}}{n(C_6H_{12}O_6)}$$

$$2.\quad \underset{\text{glucose}}{n(C_6H_{12}O_6)} + \underset{\text{oxygen}}{6n(O_2)} \xrightarrow{\text{microorganisms}} \underset{\text{carbon dioxide}}{6n(CO_2)} + \underset{\text{water}}{6n(H_2O)} + \underset{\text{heat energy}}{n(688{,}000 \text{ calories})}$$

$$\text{Sum:}\quad \underset{\text{cellulose}}{(C_6H_{12}O_6)_n} \quad \underset{\text{oxygen}}{6n(O_2)} \xrightarrow{\text{microorganisms}} \underset{\text{carbon dioxide}}{6n(CO_2)} + \underset{\text{water}}{6n(H_2O)} + \underset{\text{heat energy}}{n(688{,}000 \text{ calories})}$$

These rapid aerobic reactions can be contrasted to the relatively slow reactions in a normal anaerobic landfill, in which methane gas is a major by-product:

$$\text{1.}\quad \underset{\text{cellulose}}{(C_6H_{12}O_6)_n} \xrightarrow[\text{cellulase}]{\text{microbial}} \underset{\text{glucose}}{n(C_6H_{12}O_6)}$$

$$\text{2.}\quad n(C_6H_{12}O_6) \xrightarrow{\text{microorganisms}} \underset{\text{ethanol}}{2n(CH_3CH_2OH)} + \underset{\text{carbon dioxide}}{2n(CO_2)} + \underset{\text{heat energy}}{n(57{,}000\ \text{calories})}$$

$$\text{3.}\quad \underset{\text{ethanol}}{2n(CH_3CH_2OH)} + \underset{\text{carbon dioxide}}{n(CO_2)} \xrightarrow[\textit{omelianskii}]{\textit{Methanobacterium}} \underset{\text{acetic acid}}{2n(CH_3COOH)} + \underset{\text{methane}}{n(CH_4)}$$

$$\text{4.}\quad \underset{\text{acetic acid}}{2n(CH_3COOH)} \xrightarrow[\text{bacteria}]{\text{methane}} \underset{\text{methane}}{2n(CH_4)} + \underset{\text{carbon dioxide}}{2n(CO_2)}$$

$$\text{Sum:}\quad \underset{\text{cellulose}}{(C_6H_{12}O_6)_n} \xrightarrow{\text{microorganisms}} \underset{\text{carbon dioxide}}{3n(CO_2)} + \underset{\text{methane}}{3n(CH_4)} + \underset{\text{heat energy}}{n(57{,}000\ \text{calories})}$$

From the sums of the above theoretical reactions it is seen that cellulose can be completely oxidized to carbon dioxide and water under aerobic conditions with the liberation of 688,000 cal/mole of glucose, a building block of cellulose. Under anaerobic conditions, however, 50% of the carbon in the cellulose is oxidized to carbon dioxide and 50% is reduced to methane with the generation of only 57,000 cal/mole of glucose, assuming that all the biochemical reactions go to completion.

Most of the aerobic microorganisms are capable of oxidizing glucose completely, whereas under anaerobic conditions decomposition of cellulose to carbon dioxide and methane is brought about by groups of bacteria capable of utilizing only specific substrates. In the anaerobic process, conversion of ethanol to acetic acid and carbon dioxide, and conversion of acetic acid to carbon dioxide and methane are achieved largely by *Methanobacterium omelianskii* and methane bacteria, respectively. This results in the relatively slow decomposition of cellulose under anaerobic conditions. For example, in Southern California this writer has observed newspapers taken from anaerobic landfills, excavated after 40 years, which can still be read. The high concentrations of methane and the continuous differential settlement that occur in anaerobic landfills may persist for many years.

DESCRIPTION OF AEROBIC CELL

To illustrate the engineering aspects of aerobic landfill stabilization, a field demonstration of this process[1] will be described in this and the following sections.

Plan and section views of the aerobic cell are shown in Figure 1. The cell measured 50 by 20 ft at the bottom and 85 by 270 ft at the top. The earthen cell was excavated approximately 6 ft into natural ground. Earth berms, 11 ft high were constructed on the sides to increase the volume of the cell. Access ramps were provided at both ends.

The aerobic cell had an underdrain to distribute forced air throughout the cell. Valves provided control to regulate the flow of air. The underdrain consisted of 4-in.-diameter perforated vitrified clay pipes placed in parallel, with a distance of 6 ft between pipes (see Figure 1). The aeration piping was laid in trenches 12 in. wide by 12–15 in. deep and backfillled with select gravel. The electrical blower was rated to deliver 1400 ft^3/min at 12-in. static pressure and was capable of delivering 1200–1700 ft^3/min at 10–14-in. static pressure, respectively. The blower was controlled by a time clock which enabled variation of the operating cycle as required.

The cell had an access manway for physical observations of environmental conditions within the cell. It also permitted operation of the control valves for the aeration system.

A residue cell, into which the aerated solid waste was transferred for disposal, was constructed adjacent to the aeration cell. The residue cell was excavated approximately 5 ft into natural ground, with 7-ft berms constructed on both sides. The width of the cell as 60 ft, and the length was 300 ft.

An anaerobic cell was built as a control for the aerobic cell. The anaerobic cell was similar to the aerobic cell, except that it had no aeration equipment.

TEST PROCEDURES

The operating sequence of the aerobic cell consisted of the following steps:

1. Solid waste was deposited using normal sanitary landfill compaction procedures.

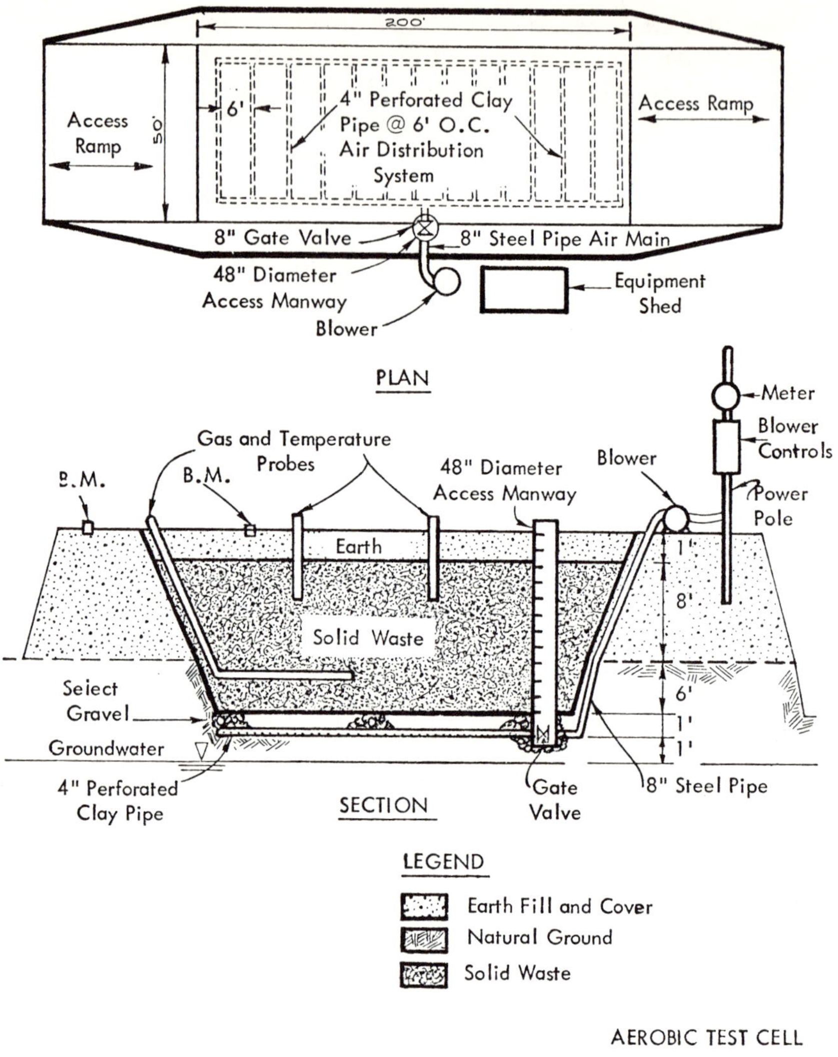

Figure 1. Aerobic test cell.

2. A 1-ft-thick soil cover was placed on the filled solid waste.
3. The solid waste was aerated for a period of 30–90 days.
4. The soil cover was removed and stockpiled.
5. The oxidized residue from the solid waste was excavated, transferred to a residue cell, and compacted.
6. A soil cover was placed on the oxidized residue.
7. The cell was refilled with solid waste and the sequence repeated.

Figure 2 illustrates the operating sequence of land reclamation by aerobic stabilization.

The aerobic and residue cells were designed so that about 20 cycles of the operating sequence could be carried out. The number of times that aerobic cells can be filled is limited by the life of the air distribution system, transfer costs, and by the disposal capacity of the residue cells.

Filling of the aerobic cell was monitored to determine the solid waste composition and density. All aerated solid waste was weighed. The volumes were measured by periodic field surveys. The apparent density of the solid waste in place varied from 935 to 1245 lb/yd^3 wet weight, and from 700 to 920 lb/yd^3 dry weight.

Aeration was started while the solid waste was being placed in the aerobic cell. This was done to stimulate the growth of aerobic organisms, to accelerate oxidation of the solid waste, and to prevent development of an anaerobic fill environment. The principal hazard in starting aeration before placement of the soil cover was the increased possibility of fire in the combustible waste.

The amount of mechanical aeration varied from 600 to 100 ft^3/min. The daily blower operating cycles ranged from 37 min on–43 min off to 98 min on–20 min off. On a volumetric basis, the daily rate of aeration ranged from 255 ft^3 air/yd^3 solid waste to 83 ft^3 air/yd^3 of solid waste. On a weight basis, the oxygen supplied ranged from 0.03 to 0.61 lb/lb of solid waste, dry weight.

Temperatures within the aerobic cell were measured by thermistors and thermometers embedded in the solid waste. Generally temperatures were measured 2–3 ft above the bottom of the aerobic cell and 2–3 ft below the surface of the solid waste.

Gas composition within the aerobic cell was determined by collecting gas samples from probes. While solid waste was being placed, probes were placed near the bottom of the fill, extended to the side walls of the aerobic cell, and then extended up the side walls of the cell to the surface. The probes were perforated and shielded. After the cell was filled with solid waste and covered with soil, portable surface probes were driven through the soil cover 2–3 ft into the solid waste. A small vacuum pump was used to extract the gas samples from the cell. Each probe was purged by operating the pump for approximately 5 min before collecting a gas sample. A water trap was installed between the air pump and the probe to reduce the quantity of condensate in the special gas sample container. Analyses in the field for CO_2 and O_2 were performed with an Orsat apparatus. Gas chromatograph analyses were carried out in the laboratory for CO_2, O_2, N_2, CH_4, CO, and H_2.

Gas samples and temperatures were obtained in the residue cell in a

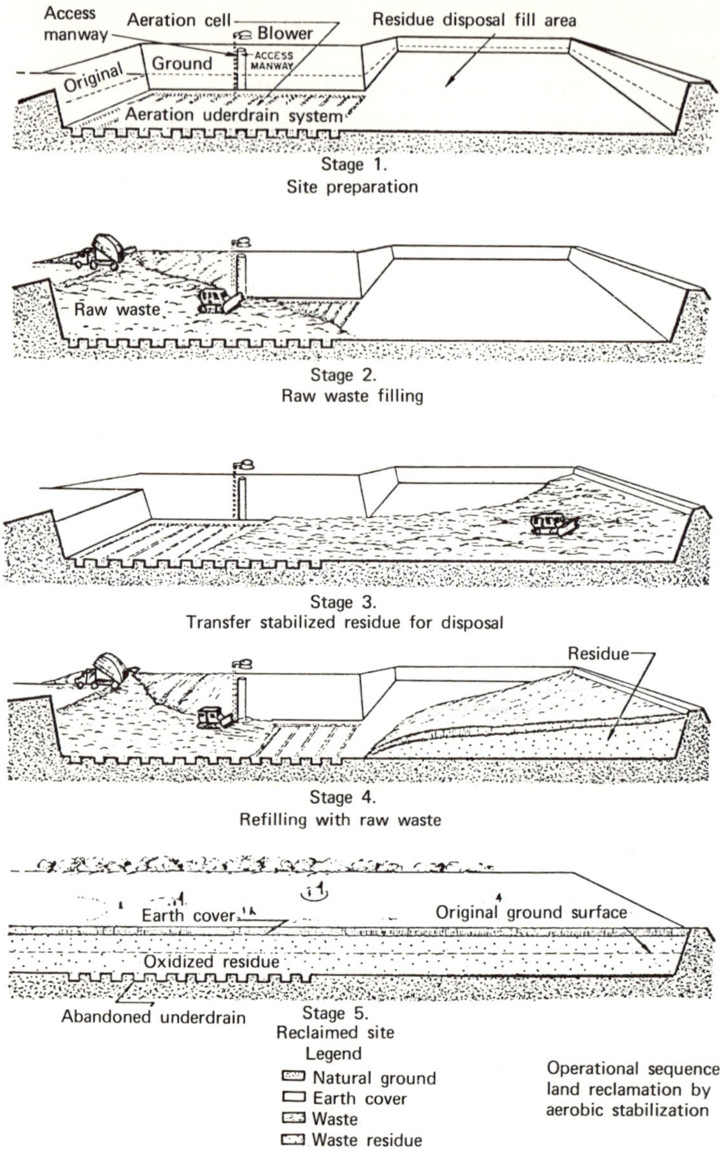

Figure 2. Operational sequence, land reclamation by aerobic stabilization.

manner similar to that of the aeration test cell. Probes were installed by driving ¾-in. diameter galvanized pipe well points through the soil cover approximately 2 ft into the solid waste residue. The pipes were perforated at the bottom. Thermistors were placed in the bottom of the probes.

The internal environment of the anaerobic cell was routinely sampled. Gas samples were analyzed by gas chromatograph, and temperatures were recorded by thermistors.

A limited field test was also conducted to determine the effects of aerobic and anaerobic landfills on the quality of groundwater. Three wells were dug for this purpose. One was located upstream from the landfill site, the second was located downstream from the aerobic test cell, and the third was located downstream from the anaerobic control cell. The wells penetrated the perched groundwater 1–4 ft. Four samples were taken from each well. The sampling procedure involved pumping each well for a short time to remove the standing water, letting the well recover, and then taking samples for analysis.

In situ tests were conducted to measure the concentrations of pollutants exhausted from the aerobic cell. Tests were conducted for sulfur dioxide, nitrogen oxide, and aldehydes using an inverted container method. This method consisted of embedding a sealed container 2 in. into the soil cover and drawing gases with a vacuum pump through the soil cover into the test apparatus. Air was exhausted and purged for approximately 10 min to evacuate static gases within the piping. The pumping was continued and the extracted gas samples were passed through moisture traps and reagents. Laboratory analyses were carried out following the procedures outlined in USPHS Publication Number 99-AP-11, except when modified to permit analysis of either excessive or low concentrations of pollutants.

The effect of the landfill environment on tensile strength of solid waste components was quantitatively measured in the laboratory. Three environments were simulated in glass jars: standard atmosphere, aerobic, and anaerobic. The standard atmosphere environment was maintained at 70°F, with 50% humidity. The aerobic and anaerobic atmospheres were maintained at 130°F and 90°F, respectively. Solid waste sample components were kept in their environments for 2½ months. The solid waste components were polyethylene, paper, and rubber.

AEROBIC FILL RESULTS

Values of Test Parameters

Measured data are presented below for the aerobic, residue, and anaerobic cells. Temperature data for the aerobic cell are presented for two filling cycles. Values of the parameters for the two filling cycles are presented in Table 1. After the first filling cycle, the aerobic oxidized residue was transferred to the residue cell, which was filled to a depth of 8 ft. The anaerobic control cell was filled with residential rubbish to a depth of 8 ft. A physical description of the types of solid waste used in the aerobic and anaerobic cells is presented in Table 2.

Temperature

Temperature data for the aerobic, residue, and anaerobic cells are presented in Figures 3–6. For the first filling cycle of the aerobic cell, maximum temperatures were reached 29 days after the beginning of aeration and then gradually decreased. The maximum temperature measured was 192°F, 2 ft below the surface of the solid waste. This temperature decreased to 172°F 82 days after the beginning of aeration. Temperatures measured 7 ft below the surface were generally lower than those measured 2 ft below the surface.

The variation of temperature with depth is seen in Figure 4, for the second filling cycle of the aerobic cell. During this cycle, probes were placed at depths of 3, 3½, 9, and 11 ft below the surface of the solid waste. The data in Figure 4 show that temperature decreased with depth and that there was a significant increase in temperature before the beginning of aeration. This indicated that the oxidation reactions began before mechanical aeration. At four of the five probe locations, temperatures decreased after the beginning of aeration. At one probe location, 3½ ft below the surface, the temperature increased to 172°F after 31 days had elapsed from the beginning of aeration. This temperature then decreased to 146°F after 80 days from the beginning of aeration.

The decrease in temperature with depth was due to two factors: (1) relatively cool air was supplied through the air distribution system at the bottom of the cell, and (2) this air displaced the warm gases which moved upward through the fill. As the cool air traveled upward, oxygen was consumed by the microorganisms; this resulted in an increased oxidation rate with its accompanying release of heat. The temperature variations between probes at the same depth may be attributed to the heterogeneity of the solid waste and to local air movements.

The temperature data of the oxidized residue from the first cycle of

Table 1 *Test Parameters for Aerobic Cell*

Cycle No.	Solid Waste Density (lb/yd^3 dry weight)	Period of Aeration (days)	Range of Aeration Cycles On/Off (min)	Average Rate of Aeration ($ft^3/min/yd^3$ of Solid Waste)	Air Added (ft^3/yd^3 of Solid Waste)	Oxygen Added (lb/lb of Solid Waste)	Depth of Solid Waste (ft)	Solid Waste Type
1	560	109	37/43 to 70/20	0.124	19,500	0.61	10	Residential rubbish
2	780	79	53/25 to 98/25	0.105	11,900	0.28	12	Residential rubbish, street sweepings, industrial and commercial solid waste

Table 2 *Solid Waste Description*

Cell	Type of Solid Waste	Properties	Solid Waste Components				Overall Values	
			Low Moisture Cellulose	High Moisture Cellulose	Inerts	Garbage (Foodstuffs)	Moisture Content (% dry weight)	Organic Matter (% dry weight)
Aerobic, first filling cycle Anaerobic	Residential rubbish	Percent by volume	53	42	2	3		
		Percent by wet weight	51	38	6	5		
		Percent by dry weight	67	20	9	4	46	73
		Moisture content, percent by dry weight	10	178	1	81		
Aerobic, second filling cycle	Residential rubbish, street sweepings, commercial and industrial solid waste	Percent by volume	72	25	3	1		
		Percent by wet weight	73	24	3	1		
		Percent by dry weight	85	11	4	1	29	71
		Moisture content, percent by dry weight	11	180	2	116		

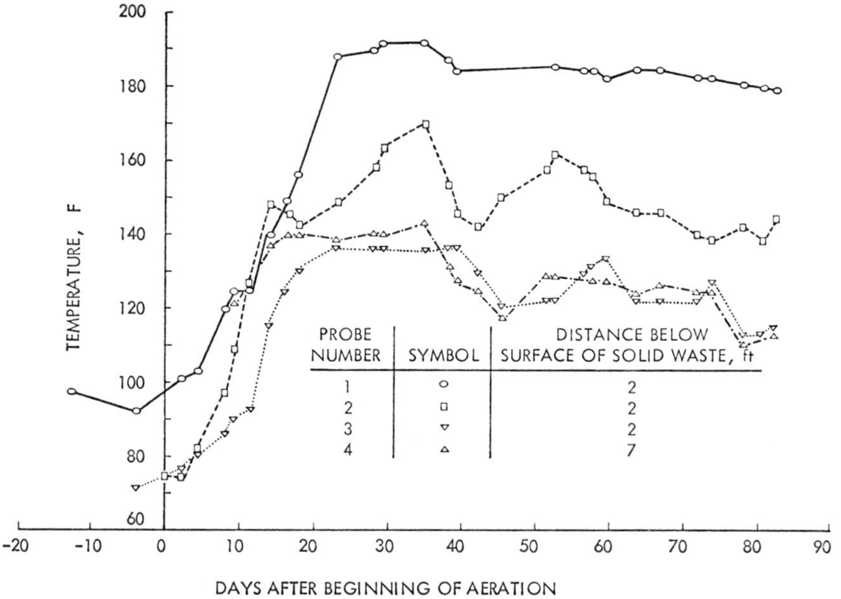

Figure 3. Temperature in aeration cell 2, first cycle.

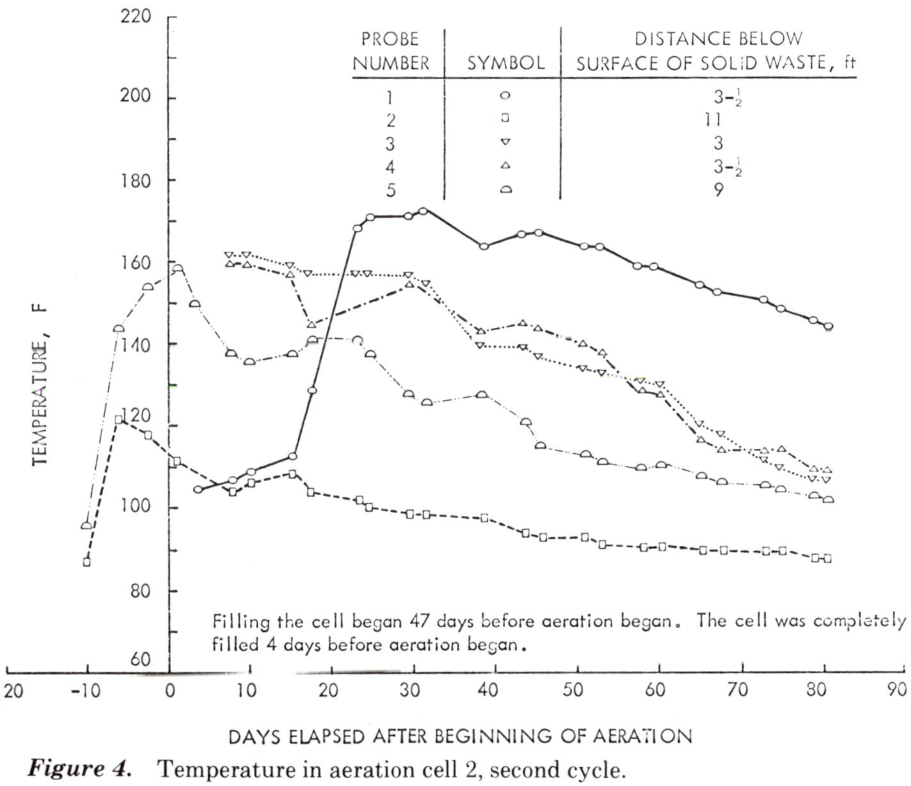

Figure 4. Temperature in aeration cell 2, second cycle.

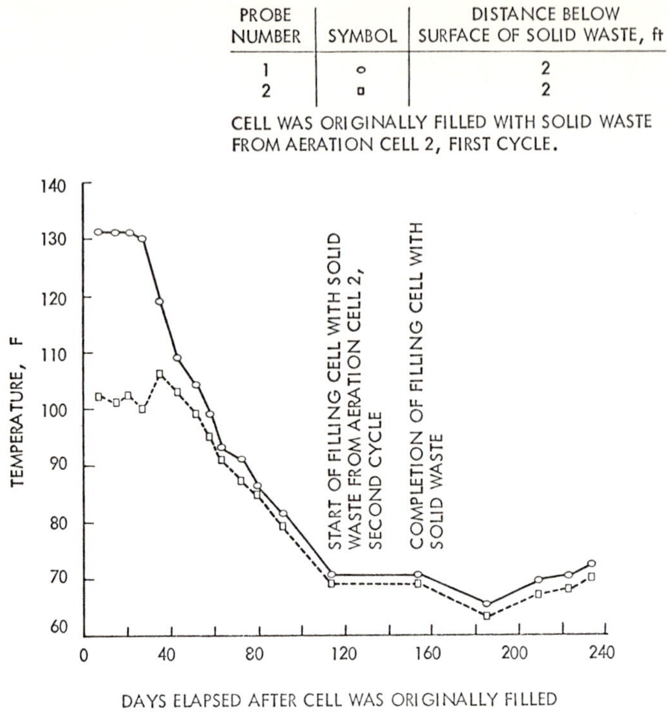

Figure 5. Temperature in residue cell 12.

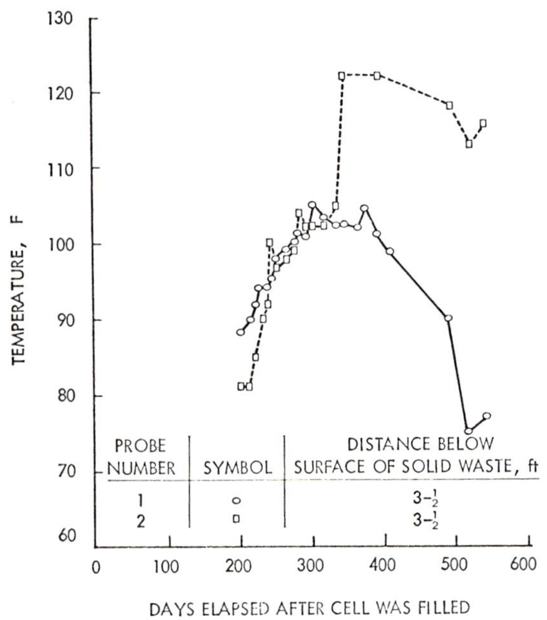

Figure 6. Temperature in anaerobic cell 3.

the aerobic cell are presented in Figure 5. As shown by the figure, temperatures decreased with time. The high initial temperatures, which lasted for 30–40 days, indicated that some anaerobic decomposition continued to take place.

The temperature data of the anaerobic control cell are presented in Figure 6, which shows there was a significant variation of temperature at the same depth. A maximum temperature of 122°F was attained 365 days after the cell was filled. Comparing this time with the time required (29–31 days) for the temperature to reach its maximum value in the aerobic cell and comparing the maximum temperature in the anaerobic cell (122°F) with the maximum temperatures in the aeration cell (171–192°F), it can be concluded that aeration causes the oxidation process to occur at a much higher rate and to go further to completion, thereby releasing more heat.

Gas Composition

The N_2, O_2, CO_2, and CH_4 concentration data for the aerobic cell are presented in Figures 7–10. As seen in Figures 7 and 8, the N_2 and O_2 concentrations were quite small before the beginning of aeration. After aeration began, the N_2 concentration rose to approximately 80% by volume, and the O_2 concentration rose to approximately 8% by volume.

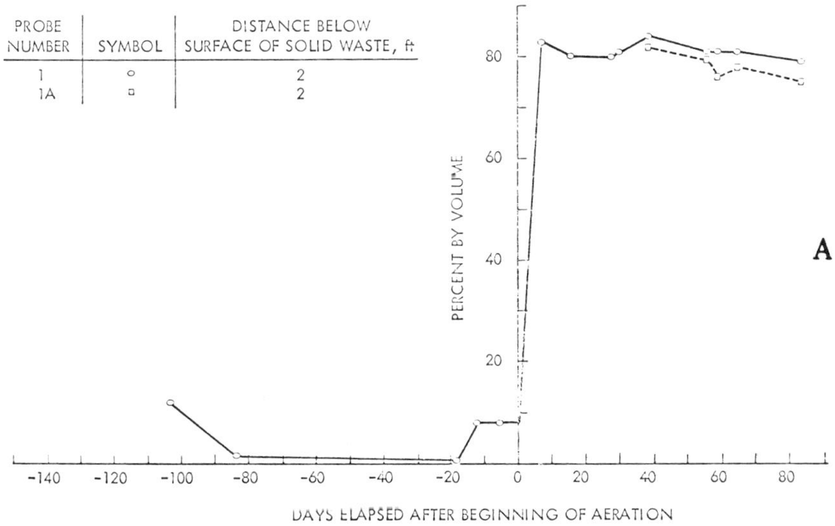

Figure 7. (A) N_2 concentration in aeration cell 2, first cycle. (B) N_2 concentration in aeration cell 2, first cycle. (C) N_2 concentration in aeration cell 2, first cycle.

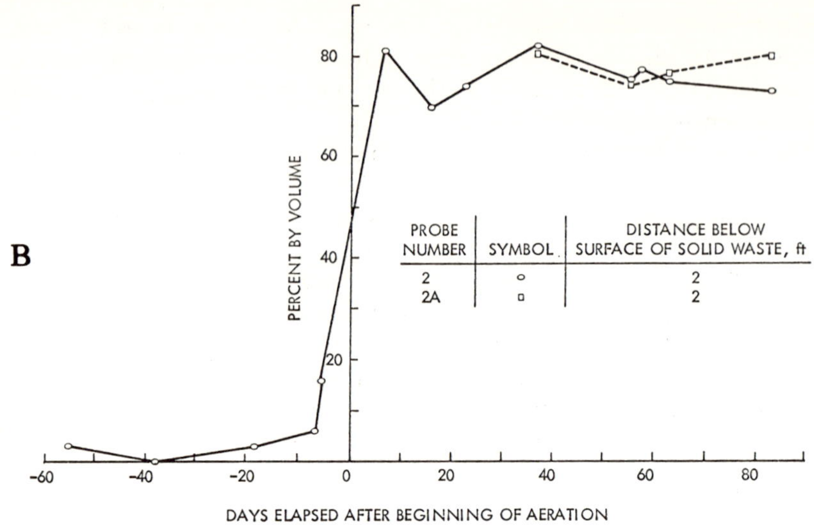

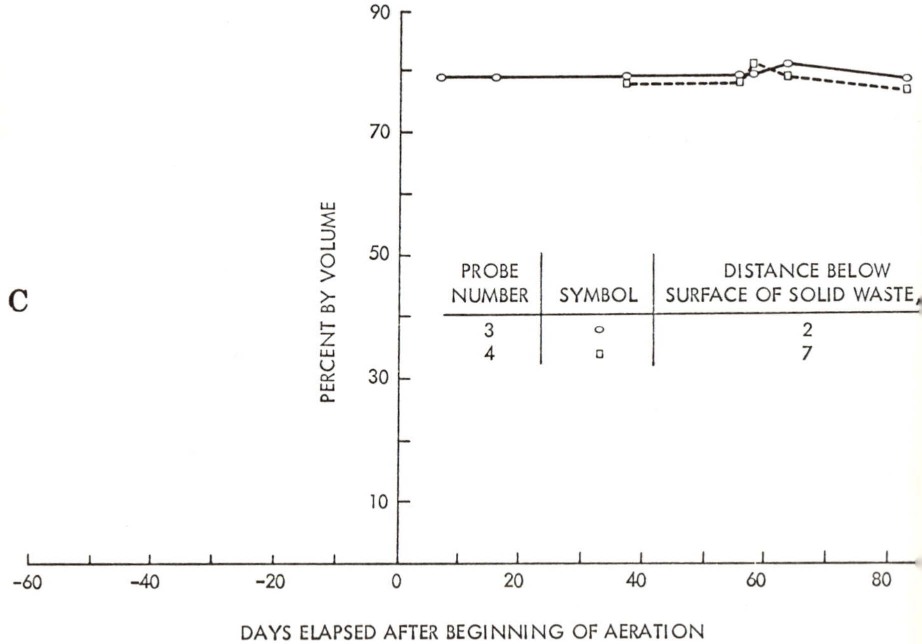

Figure 7. (Continued)

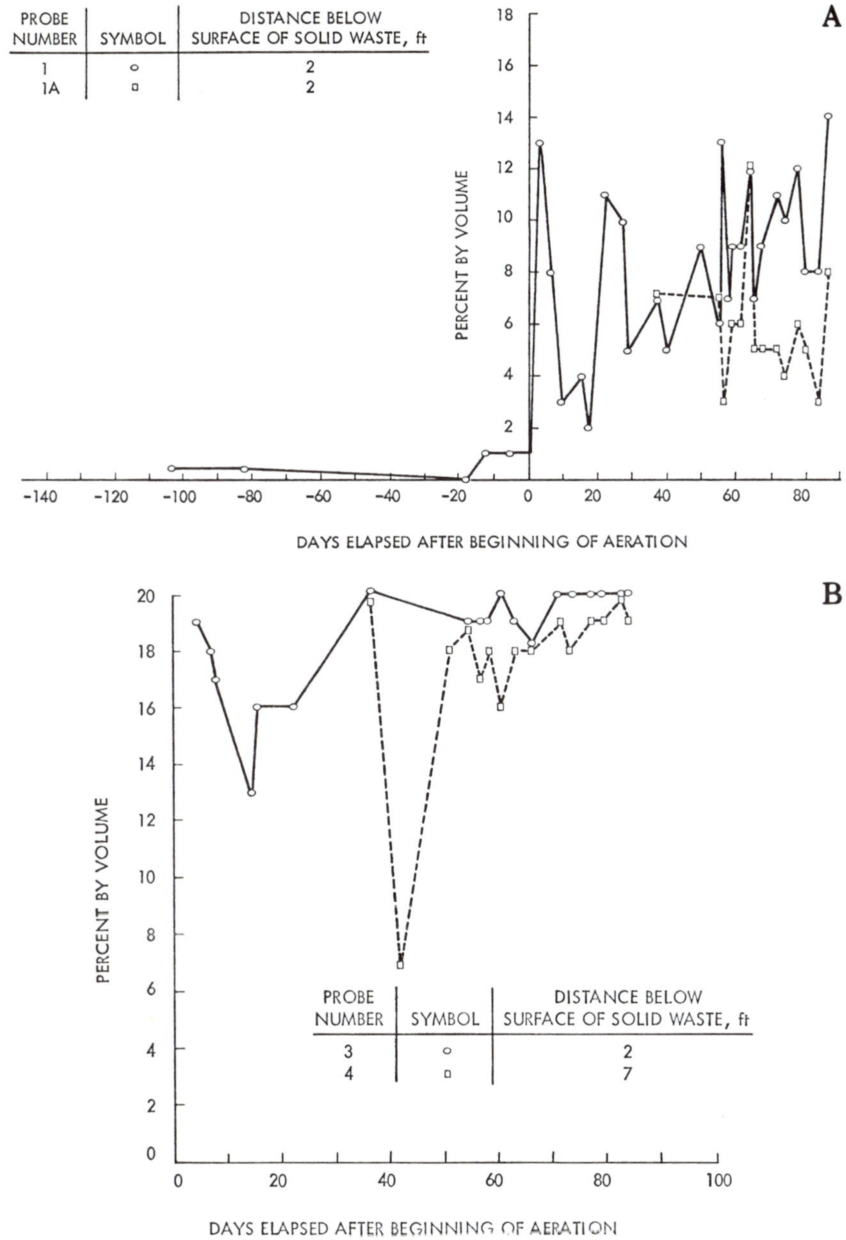

Figure 8. (A) O_2 concentration in aeration cell 2, first cycle. (B) O_2 concentration in aeration cell 2, first cycle. (C) O_2 concentration in aeration cell 2, first cycle.

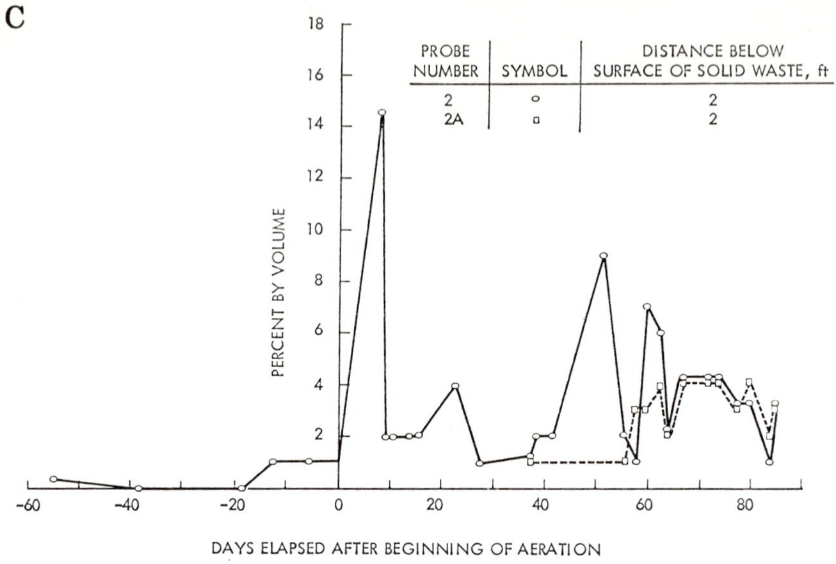

Figure 8. (Continued)

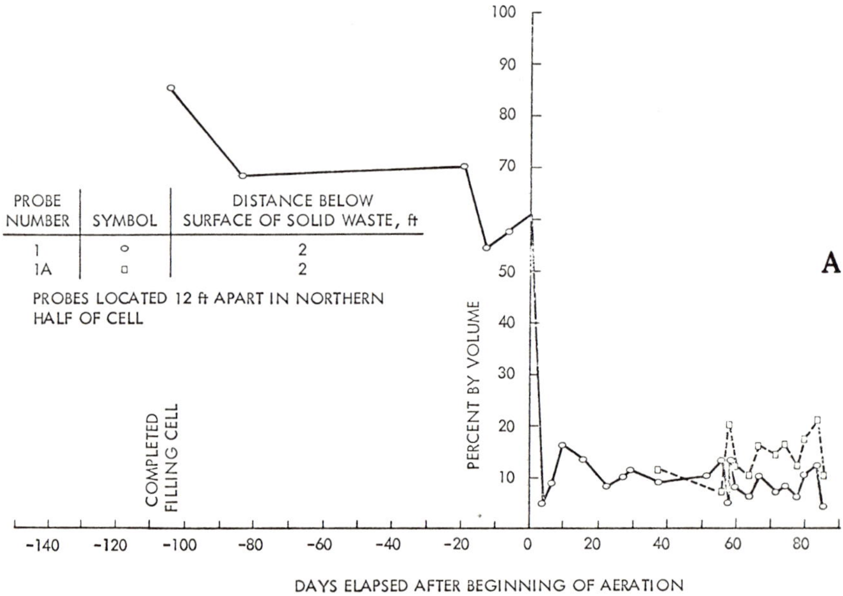

Figure 9. (A) CO_2 concentration in aeration cell 2, first cycle. (B) CO_2 concentration in aeration cell 2, first cycle. (C) CO_2 concentration in aeration cell 2, first cycle.

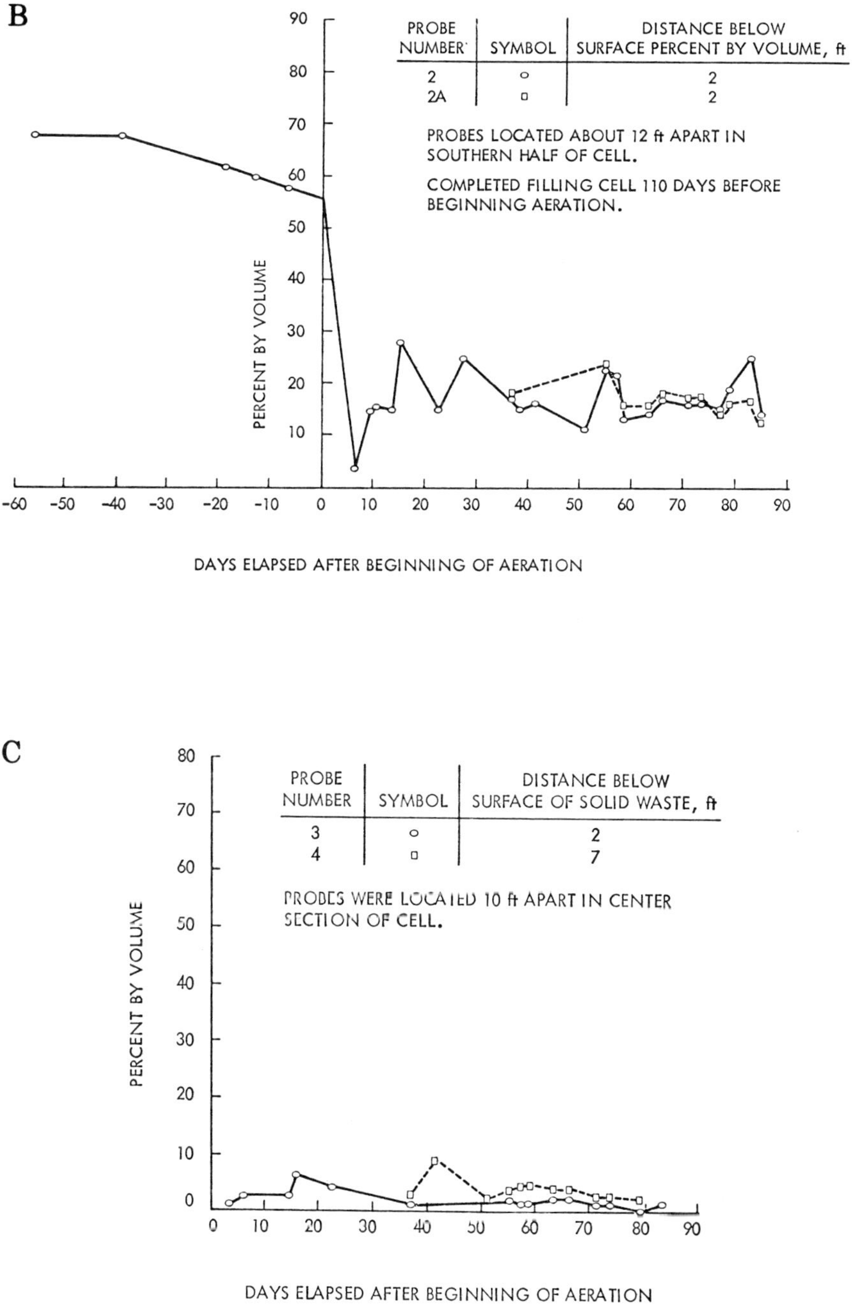

Figure 9. (Continued)

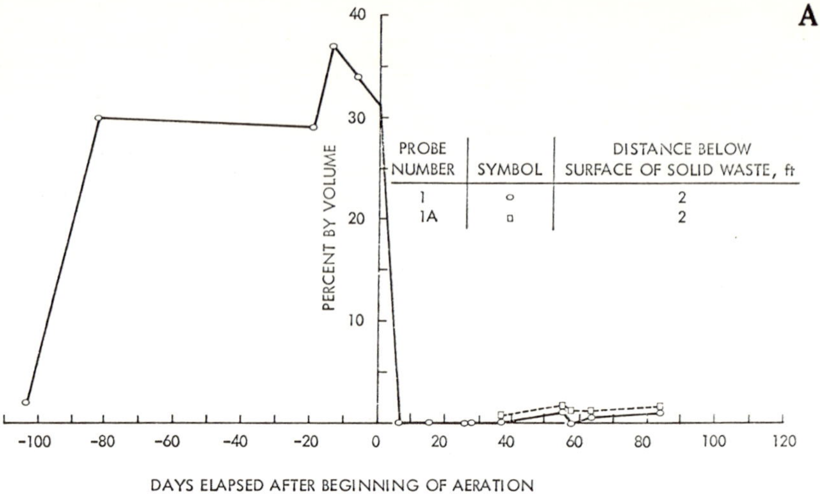

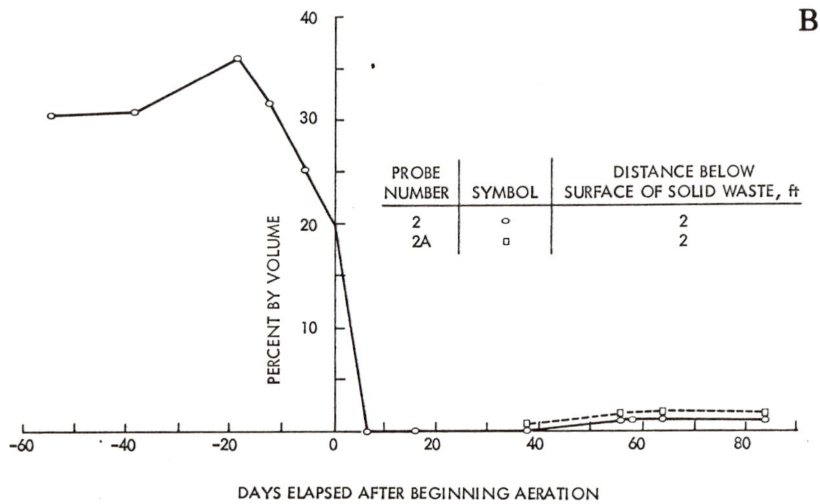

Figure 10. (A) CH_4 concentration in aeration cell 2, first cycle. (B) CH_4 concentration in aeration cell 2, first cycle. (C) CH_4 concentration in aeration cell 2, first cycle.

C

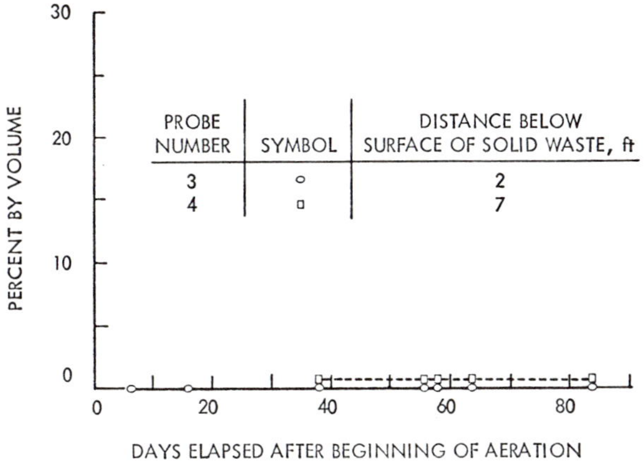

Figure 10. (Continued)

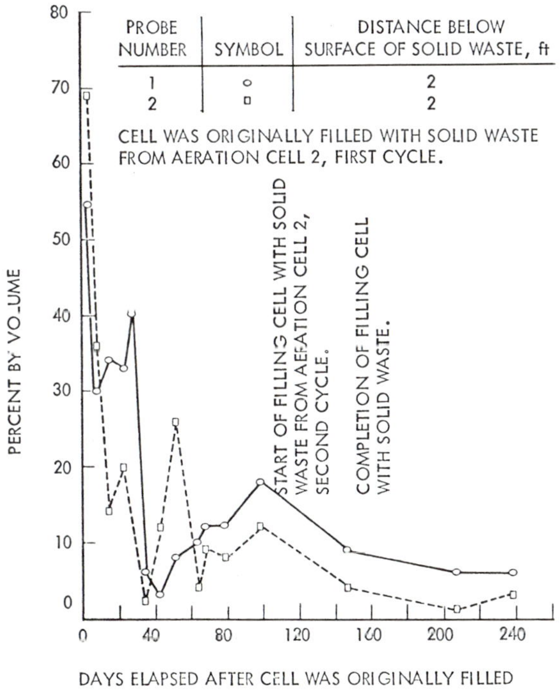

Figure 11. N_2 concentration in residue cell 12.

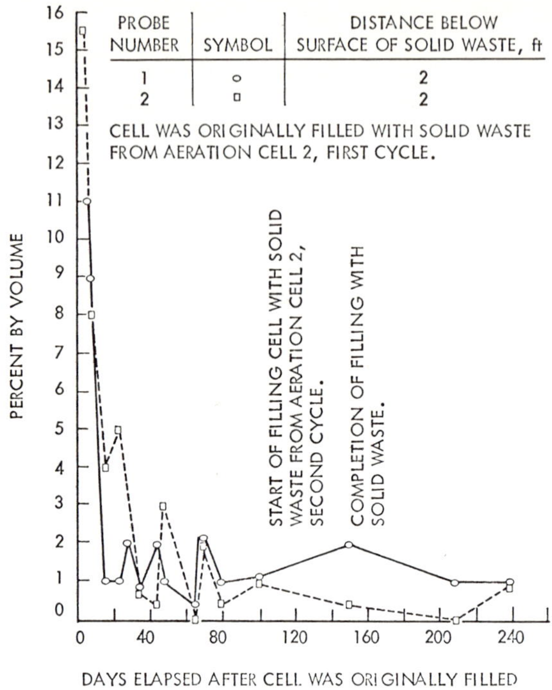

Figure 12. O_2 concentration in residue cell 12.

Before aeration, the CO_2 concentration was approximately 65% by volume, as shown in Figure 9. The introduction of air during the aeration process caused the CO_2 concentration to drop to approximately 10% by volume. Before aeration, the CH_4 concentration was approximately 30% by volume. The concentrations of CO_2 and CH_4 before aeration indicated that anaerobic decomposition was occurring. After aeration started, the CH_4 concentration dropped to nearly zero, as seen in Figure 10. The negligible concentration of CH_4 after aeration began was an indication that the anaerobic decomposition process had stopped and was replaced by the aerobic process.

The gas concentrations for the residue cell are presented in Figures 11–14. The N_2 and O_2 concentrations decreased from their values at the end of the aeration process to approximately 5% and 1% by volume, respectively. The concentrations of CO_2 and CH_4 increased from their values at the end of the aeration process to approximately 55% and 40% by volume. The concentration changes in the residue cell indicated that anaerobic decomposition took place in the residue cell. Ap-

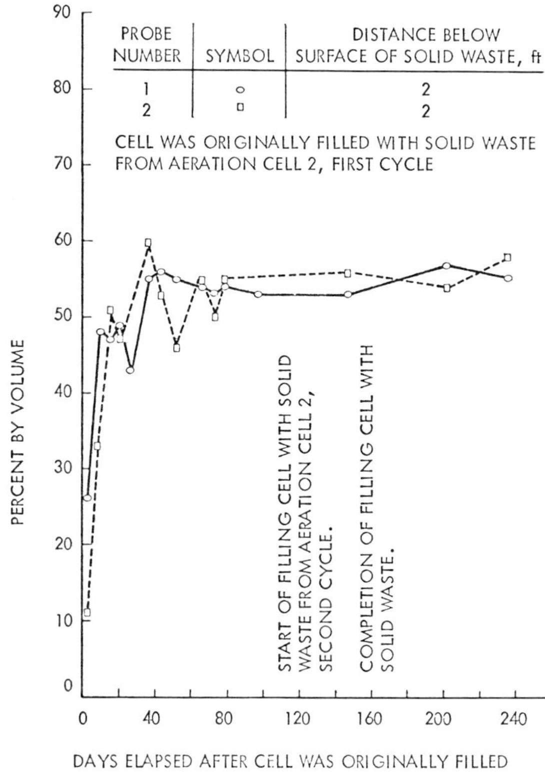

Figure 13. CO_2 concentration in residue cell 12.

proximately 40 days were required for the CO_2 and CH_4 concentrations to reach their peak values. These data are consistent with the temperature data in the residue cell (Figure 5), which also show that anaerobic decomposition occurred for approximately 40 days.

The gas concentrations in the anaerobic cell are presented in Figures 15–18. After 500 days the N_2 and O_2 concentrations fell to approximately 10 and 1% by volume, respectively. The CO_2 and CH_4 concentrations, also after 500 days, increased to approximately 55 and 25% by volume, respectively. The slight upward trend in the CH_4 concentration curve after 400 days indicates that anaerobic decomposition was still occurring.

Settlement

Settlement data are presented in Figures 19–21. For the aerobic cell, Figure 19, the average settlement rate prior to aeration was 0.00125

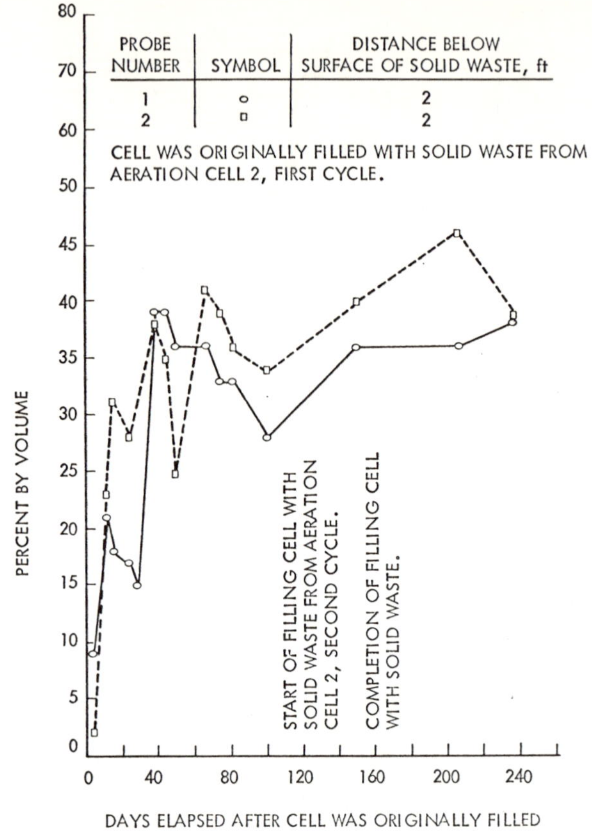

Figure 14. CH_4 concentration in residue cell 12.

ft/day. After aeration, the average settlement rate at the north benchmark was 0.00967 ft/day, nearly an eightfold increase. The total settlement varied from 0.60 to 0.79 ft 74 days after the beginning of aeration.

Settlement data for the residue cell are presented in Figure 20. A period of 7 days elapsed after the placing of the benchmarks before settlement was observed. For the next 139 days, the average settlement rate was 0.000856 ft/day. The total settlement, 146 days after the benchmarks were placed, varied from 0.12 to 0.13 ft. These data include the 0.06-ft settlement caused by the placement of a 1-ft soil cover.

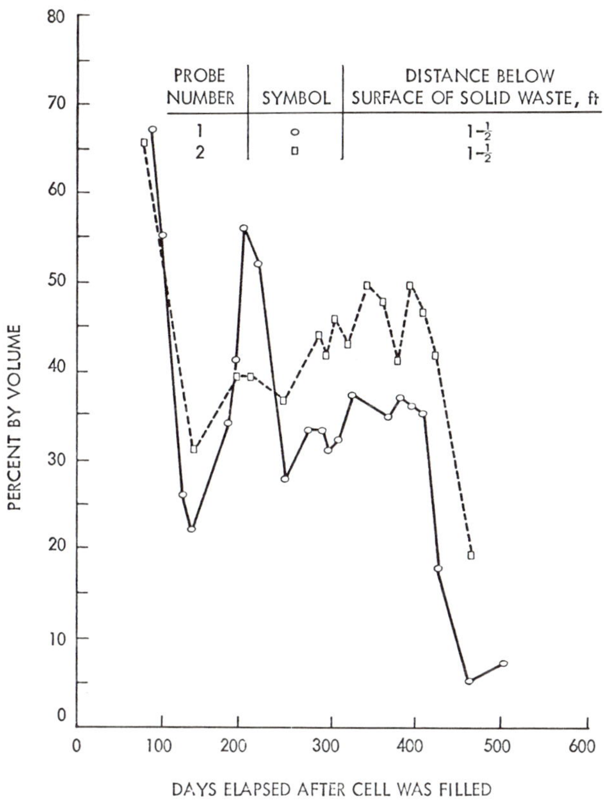

Figure 15. N_2 concentration in Anaerobic cell 3.

In the anaerobic control cell, Figure 21, settlement was not observed until 96 days after the placing of the benchmark. Thereafter the average settlement rate was 0.000537 ft/day. After 394 days from the placing of the benchmark, the settlement was 0.16 ft.

In the aerobic cell, the settlement observed produced a reduction in volume of approximately 7.0% 74 days after the beginning of aeration. In the residue cell, the reduction in volume was approximately 1.6% 170 days after the filling of the cell. In the anaerobic control cell, the reduction in volume observed was approximately 2.0% 394 days after the placing of the benchmark. These figures indicate that the surficial volume reduction in aerobic cells can be three to four times as great as that in anaerobic cells. In other tests, the final volume in the aerobic cell was 25% less than that in an anaerobic cell. The settlement and

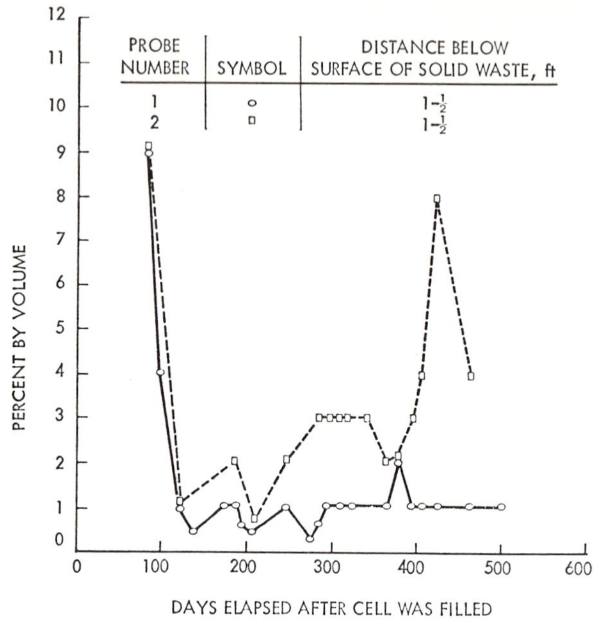

Figure 16. O_2 concentration in anaerobic cell 3.

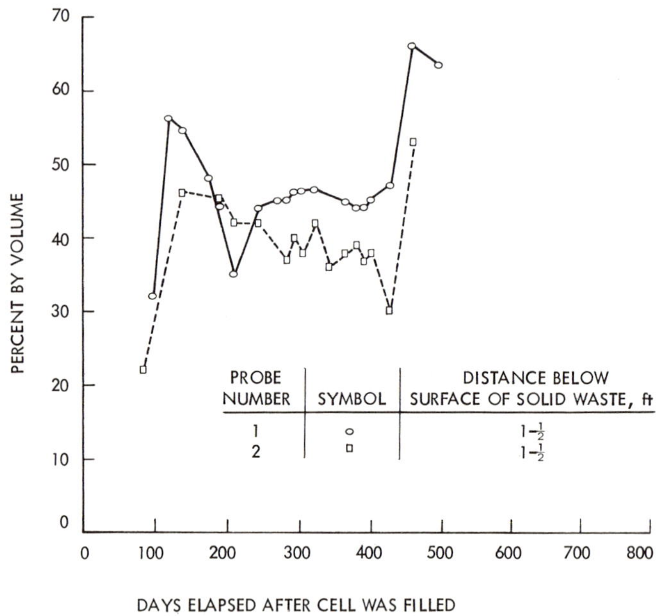

Figure 17. CO_2 concentration in anaerobic cell 3.

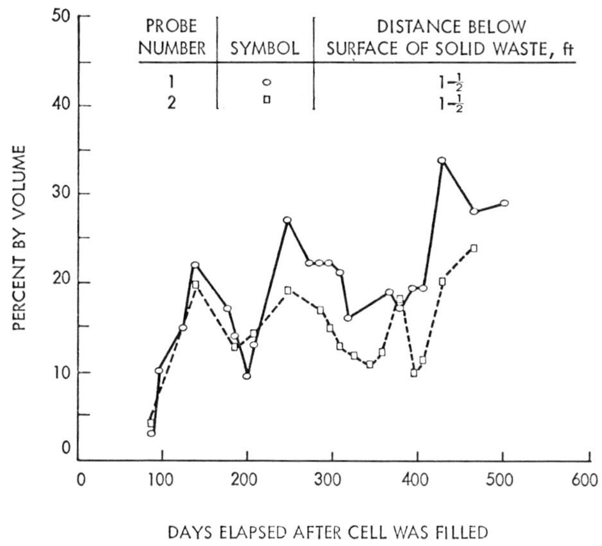

Figure 18. CH_4 concentration in anaerobic cell 3.

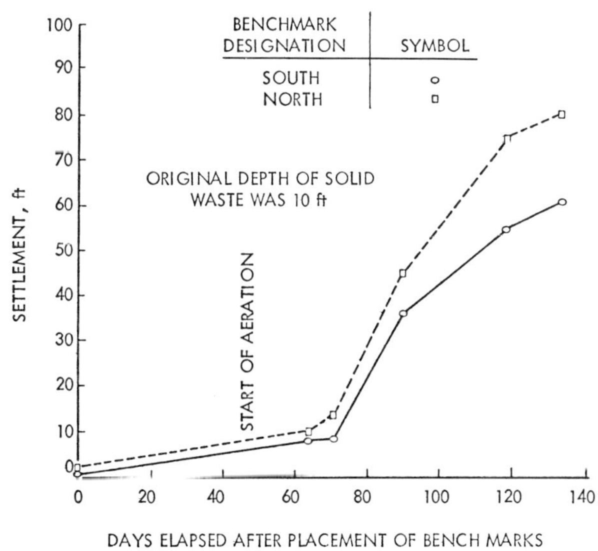

Figure 19. Precipitation in aeration cell 2, first cycle.

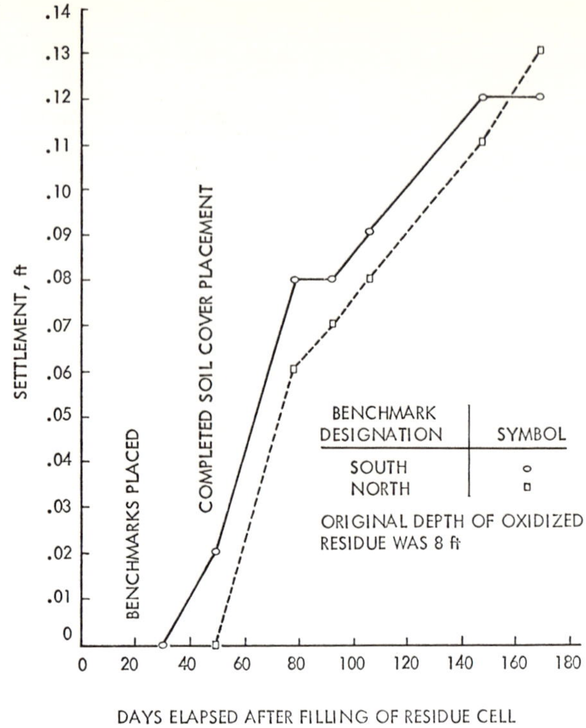

Figure 20. Precipitation in residue cell 7, first cycle.

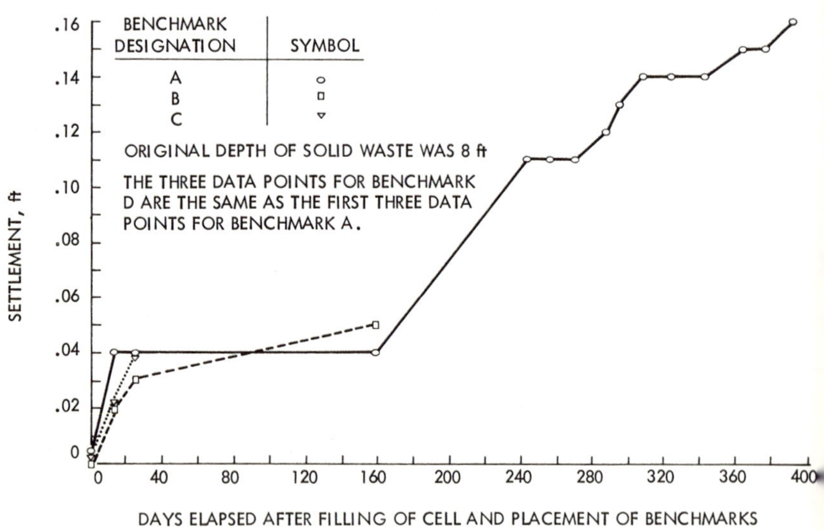

Figure 21. Precipitation in anaerobic cell 3.

volume reduction achieved depend on the percentage of inerts, moisture content, and degree of compaction. The aerobic residue compacts far better than raw solid waste, so that a volume reduction of over 25% may be achieved for the total decomposition, transfer, and disposal processes provided by the aerobic and residue cells.

Groundwater Pollution

The natural groundwater in the area was found to be of poor quality, to have a high mineral content, and to be nonpotable. It has been used only for limited agricultural irrigation.

The water from the well upstream from the landfill had the lowest mineralization, as measured by specific electrical conductance. The well downstream from the anaerobic control cell had the highest mineralization. The average values of the specific electrical conductance are given in Table 3.

Air Pollution

Pollutant concentrations in the gases exhausted from the aerobic cells are given in Table 4.

These results show that SO_2 and NO_x concentrations were within the upper threshold limiting values of 5.0 ppm.

Odors and Vermin

Septic odors were generally not encountered in the aerobic cell, covered or uncovered. A mild sweet tobacco type of odor was emitted from the open aerated solid waste and residue. Little or no odor was observed when the aerated solid waste was covered with 1 ft of soil. The sweet tobacco odor persisted after aeration was stopped and the cell had been allowed to become anaerobic, as well as during excavation and transfer of the solid waste residue to the residue cell. The significantly less noxious odor of aerated solid waste may be attributed to the

Table 3 *Electrical Resistance*

Location of Well	Average Specific Electrical Conductance (μmho/cm)
Upstream from landfill	3150
Downstream from aerobic cell	3800
Downstream from anaerobic cell	5500

Table 4 Concentration of Pollutants

Constituent	Range of Concentration (ppm)
SO_2	0.47–4.0
NO_x	0–0.11
Aldehydes	0.42–9.56

oxidation of putrescible wastes, and to the non-noxious odor generated by the extensive aerobic microbial populations.

Noxious septic odors were encountered in the anaerobic control cell when it was excavated. Similar odors have been observed at other covered sanitary landfills when there were surface cracks in the soil cover and at the sides of the landfill.

No vermin and very few flies were seen in the vicinity of the aerobic cell. This may be attributed to the relatively high temperatures of the aerated solid waste, which created an environment hostile to vermin, and to the rapid oxidation of the vector food supply.

Strength of Solid Waste Components

The laboratory tests of tensile strength of solid waste components indicated that both the aerobic and anaerobic environments significantly decreased the tensile strength of paper, but had little effect on polyethylene and rubber. The test results are presented in Table 5.

Table 5 Tensile Strengths of Solid Waste Components

Component	Tensile Strength (psi)		
	Standard Atmosphere	Anaerobic Atmosphere	Aerobic Atmosphere
Polyethylene	1586	1973	1316
Kraft paper	5000	2160	1960
Newspaper	2917	1083	900
Resilient rubber	42	38	—
Nonresilient rubber	28	28	28

ECONOMIC ANALYSIS

The costs of operating approximate 20-ft depth fills employing aerobic cells, anaerobic cells with good compaction, and anaerobic cells with normal compaction have been computed. The computations were carried out for the following conditions:

1. Net cost per acre of usable fill: $11,000.
2. Number of cycles for each aeration cell: 20.
3. Volume of residential solid waste: 6500 yd^3/cycle.
4. Weight of residential solid waste: 3250 tons/cycle.
5. Period of aeration: 90 days.

A cost summary of the additional expenses for the aerobic cell is presented in Table 6. Cost comparison data for the aerobic and anaerobic cells are presented in Table 7. The greater volume required for the anaerobic cell with normal compaction contributed to its high

Table 6 *Cost Summary for Aerobic Cell*

Item	Cost Per Cell ($)	Unit Costs	
		($/yd^3)	($/ton)
Field Construction			
1. Installation of air distribution piping	5,000		
Repairs	500		
2. Blower	500		
3. Irrigation system	200		
4. Electrical installation	500		
5. Power service connection	500		
6. Water service connection	300		
7. Earthwork	3,000		
Total field construction cost	10,500	0.10	0.16
Utilities			
1. Power consumed	2,800		
2. Water consumed	800		
Total utilities cost	3,600	0.03	0.05
Final disposal			
1. Residue transfer @ $0.24/yd^3	23,600		
2. Soil cover @ $0.24/yd^3	2,300		
Total disposal cost	25,900	0.24	0.40
Total cost	40,000	0.37	0.61

Table 7 *Cost Comparison*

Item	Unit Costs ($/ton)		
	Aerobic, Well Compacted	Anaerobic, Good Compaction	Anaerobic, Normal Compaction
1. Basic disposal cost	1.00	1.00	0.90
2. Additional cost for aeration and transfer	0.61	—	—
3. Soil cover cost @ $0.40/yd^3	0.09	0.11	0.14
4. Land cost @ $11,000/ acre, filled 14 ft deep	0.76	0.95	1.18
5. Cost for extended haul	—	0.18	0.33
6. Maintenance costs after redevelopment	—	—	0.15
Total costs	2.46	2.24	2.70

cost. Although the cost for the aerobic cell per ton of solid waste, $2.46, is slightly greater than the cost for the anaerobic cell with good compaction, $2.24, the aerobic cell has intangible advantages, the evaluation of which on a monetary basis would be a complex task. These advantages include the following:

1. Reduced long-term anaerobic decomposition, which allows aerobic stabilized landfills to be developed for parks or other purposes considerably sooner, since there is less continuous settlement and greater landfill stability than with anaerobic sites.
2. Reduced number of vermin, flies, and birds due to the elevated temperatures.
3. Reduced odor, and water and air pollutants.
4. The aerobic residue can be used for agricultural compost, metal, or glass reclamation.

Incineration costs, presented for comparison purposes, have a normal range of $7 to $10 per ton.

CONCLUSIONS

The aerobic landfill process is a simple, practical, and economic landfill operation that provides sanitary and environmental benefits.

Characteristics of the aerobic landfill process are accelerated landfill reuse, minimization of methane and other pollutants, fast rate of settlement and decomposition, and minimization of noxious odors, birds, and vermin.

REFERENCE

1. *Aerobic Landfill Stabilization;* Ralph Stone and Company, Inc., Los Angeles, California, 1970.

CHAPTER II.7

biologic oxidation

Biologic oxidation is essentially a unit operation, as that word is understood in chemical engineering. It is widely employed in solid waste disposal and is applicable to cannery wastes; animal manure and similar putrescible solids arising from poultry raising and slaughtering; packing and food processing; and the raising of swine and ducks. These subjects are discussed separately in this volume. Cattle feedlots have only experimented with this technique.

The organisms that consume both suspended and dissolved solids multiply and produce different solids and different problems. Primarily, biologic oxidation may be regarded as a treatment to produce potable water from waste water. Its application to all types of solid wastes requires the introduction of the subject here to acquaint the reader with the terminology.

INDUSTRIAL BIOLOGIC OXIDATION OF WASTES

Theories of biologic oxidation have been developed for the treatment of sewage; these theories have been adapted to waste treatment in general and have provided data for design. Treating wastes by aeration is not new; it has been tried extensively for pulp and paper waste streams and to some extent in the fruit and vegetable processing industries.

In 1960, the treatment of industrial waste in aerated lagoons[1] was being developed along with surface aerators, which consist of a gear motor driving an open impeller. These have been developed by several manufacturers. Although they are of different shapes, each manufacturer claims about the same efficiency of oxygen transfer. Aeration of open basins without fixed linings allowed the compliance to basin configuration that was necessary for diffused aeration. Surface aerators tended to give the basin's contents a more "completely mixed" nature when sufficient horsepower was available and allowed the application of "complete mix" theory for the biological treatment process. The entire contents of the aeration basin assimilated the waste flow as it entered the basin, giving an increased buffer capacity for high strength, variable strength, and high or low pH waste streams.

The development of these surface aerators by equipment manufacturers and their acceptance were not accompanied by development of theory or parameters for use in the treatment process.

The attainment of a completely mixed system by aeration allows steady-state removal of substrate (BOD and COD) by the biological mass measured as volatile suspended solids (VSS). The BOD and COD removal is equal to the maximum rate of removal of BOD or COD per unit mass of organism times the residual BOD or COD concentration in a solution, divided by the sum of this residual BOD or COD and the residual BOD or COD at one-half the maximum substrate removal rate:

$$R = \frac{Fs}{S + s} \tag{1}$$

where R is the BOD or COD removal rate, mg BOD, COD/mg MLVSS-day, F is the maximum removal rate, S is the BOD or COD concentration at $R = \frac{1}{2} F$, constant, and s is the BOD or COD concentration, substrate concentration.

When R is plotted against s, the concentration, the curve starts upward from zero and gradually decreases in slope until it is horizontal at the asymptote, which represents F. F is large as compared to a normally operating treatment system where R is much less than one. This equation simplifies to

$$R = fs$$

$$f = \text{BOD or COD removal coefficient} \times \frac{\text{mg BOD or COD removed}}{\text{mg MLVSS-day-mg/1 BOD or COD}} \tag{2}$$

This slope f, the BOD or COD removal coefficient, can be much more readily obtained from operating data of treatment systems than can the two constants F and S (1). The substrate BOD or COD concentration in a completely mixed system (s) is a constant since the entire contents are completely mixed and the effluent is also of this same concentration. Equations (1) and (2) become valid only in a completely mixed system.

There is a conversion rate of substrate to cell material that is a constant for a particular waste and waste system, and represents the milligrams of cellular material formed from the removal by organisms of a milligram of BOD or COD from the waste. This constant is always less than unity. The constant ratio of conversion apparently holds true regardless of removal or reaction rate and activity of the biological mass. There is a rate of destruction of cells, endogenous respiration, that occurs simultaneously with the cell growth; at is apparently not affected by the reaction rate of treatment system, but is primarily linked to the temperature of the system. The net sludge growth is

$$\text{Net sludge growth} = c \cdot \text{BOD or COD removed/day-}k \cdot \text{MLVSS} \tag{3}$$

where

$$c = \frac{\text{Sludge growth constant} - \text{mgVSS formed}}{\text{mg BOD or COD removed}}$$

$$k = \frac{\text{Endogenous respiration rate} - \text{mgVSS destroyed}}{\text{mgMLVSS-day}}$$

and MLVSS stands for mixed liquor volatile suspended solids.

The MLVSS is assumed proportional to the active biological mass in the system. Equation (3) can be divided by the MLVSS; (4) is the removal rate and (5) is the net sludge growth rate:

$$R = \frac{\text{BOC or COD removed/day}}{\text{MLVSS}} \tag{4}$$

$$G = cR - k \tag{5}$$

where G is the net sludge growth rate mg VSS/mgMLVSS-day.

MLVSS here refers to the total weight of volatile suspended solids in the treatment system and the net sludge growth rate (G) consists of cells or volatile suspended solids accumulated in the system, those which are wasted from the system as carryover in the effluent, and those intentionally removed by wastage of settled solids, divided by the total VSS in the system. In a steady-state system the effluent substrate would remain constant as would the MLVSS and, therefore, so would the net sludge growth. In a non-steady-state system, a portion of the net sludge growth could be accumulated in the system or excessive volatile suspended solids wastage could be carried on with the MLVSS increasing or decreasing, or the substrate could vary. The reciprocal of the net sludge growth rate is the "sludge age" of the biological solids in the system.

Oxygen is required for energy production for conversion of the substrate to cellular matter and for the maintenance of the biological mass or endogenous respiration. The oxygen required is given by

$$O_2 = \alpha \cdot \text{BOD or COD removed} + \beta \cdot \text{k} \cdot \text{MLVSS} \tag{6}$$

where O_2 is the oxygen required, α is a constant, mgO_2 required/mg

BOD or COD removed-day, and β is a constant, mgO_2 required/mg VSS-day.

By dividing (6) by the MLVSS we obtain Eq. (7), which evaluates the constants from operating data:

$$\frac{O_2}{\text{MLVSS}} = a \cdot R + b \cdot k \tag{7}$$

Most surface aerator manufacturers claim an oxygen transfer of 3.5–4 lb/hr per horsepower-hour expended, with clear, suspended solids and organic-free water, one atmosphere at 20°C and zero dissolved oxygen. Solids and organics in the water lower the ability of the water to take gaseous oxygen into solution as rapidly as clear water does, and these figures must therefore be reduced when being applied to a waste treatment system. In a treatment system a certain amount of dissolved oxygen must be maintained in order to sustain the biological growth and prevent oxygen deficient conditions in a portion of the system. This 1–2 mg/liter dissolved oxygen that must be maintained decreases the dissolved oxygen deficit, the driving force for solution of oxygen into the water, which also reduces the factor presented for their aerators. The corrections in Eq. (6), α and β, must be used for design. A capacity of 2 lb of oxygen per horsepower-hour is normally expected from surface aerators during actual operation in the field.

Activated sludge or contact stabilization treatment of fruit processing waste at low rates (below 0.4 mg COD removed per day per milligram of MLVSS) will provide greater than 90% removal of organic load and removal of solids from the effluent. Aerated lagoon treatment can provide greater than 70% BOD removal, but suspended solids remain in the effluent and are the principal source of effluent BOD and COD. "Completely mixed" aeration basins provide buffering to avoid pH fluctuation. Final clarification for activated sludge or contact stabilization is successful at low surface-loading rates (below 400 gal/ft^3/day), made necessary by the slow-settling sludge developed and with vacuum type sludge removal.

Nutrient addition is necessary to achieve successful treatment. Nutrient savings can be made by increasing the amount of biological sludge in the system, thus decreasing the waste loading rate on the sludge. This allows a greater destruction of sludge by endogenous respiration which returns nutrient to the system and cuts down the amount necessary to add. Activated sludge or contact stabilization treatment is necessary to accomplish this.

PVC sheeting is a suitable basin lining material provided that all joints are properly welded. The bottom should be covered with sufficient material to prevent it from floating even if gas should accumulate beneath it.

Surface aerators accomplished adequate mixing at 0.3 hp/1000 ft^3 provided that 2 lb of oxygen transfer per horsepower-hour under operating conditions was accomplished.

The biological sludge that developed seemed to be independent of the type of fruit processing waste and had the following characteristics:

(a) about 90% volatile,
(b) settled very slowly,
(c) settling improved at low COD and BOD removal rates,
(d) mg COD/mg VSS = 1.39,
(e) mg BOD/mg VSS = 0.3 + 0.28 × BOD removal rate,
(f) mg organic N:mg organic P:mg VSS = 0.087:0.016:1.

$$\frac{\text{Net sludge growth}}{\text{day}} = C \times \frac{\text{Substrate removed}}{\text{day} - k_t \times \text{MLVSS}}$$

where

$$c = 0.49\,\frac{\text{mg VSS formed}}{\text{mg COD removed}} \qquad \text{for pear waste}$$

$$c = 0.46\,\frac{\text{mg VSS formed}}{\text{mg COD removed}} \qquad \text{for peach waste}$$

$$c = 0.57\,\frac{\text{mg VSS formed}}{\text{mg COD removed}} \qquad \text{for apple waste}$$

II-7-217-50
II-7-217-F

$$k_t = 0.115 \times 1.14^{(T-20)}\,\frac{\text{mg VSS destroyed}}{\text{mg MLVSS-day}}$$

$$= \text{Endogenous respiration rate at } t^\circ\text{C}$$

$$\text{Substrate removal rate} = f_t \times \text{substrate concentration}$$

$$f_t = 0.017 \times 1.16^{(T-20)}\,\frac{\text{mg COD removed}}{\text{mg MLVSS-day-mg/1 COD}}$$

$$= 0.068 \times 1.16^{(T-20)}\,\frac{\text{mg BOD removed}}{\text{mg MLVSS-day-mg/1 BOD}}$$

$$= \text{Substrate removal rate coefficient}$$

$$\frac{\text{Oxygen required}}{\text{day}} - a \times \frac{\text{substrate removed}}{\text{day}} + b \times k_t \times \text{MLVSS}$$

$$a = \frac{0.34 \text{ mg } O_2 \text{ required}}{\text{mg COD removed}}$$

$$= \frac{0.46 \text{ mg } O_2 \text{ required}}{\text{mg BOD removed}}$$

$$b = \frac{1.2 \text{ mg } O_2 \text{ required}}{\text{mg VSS destroyed by endogenous respiration}}$$

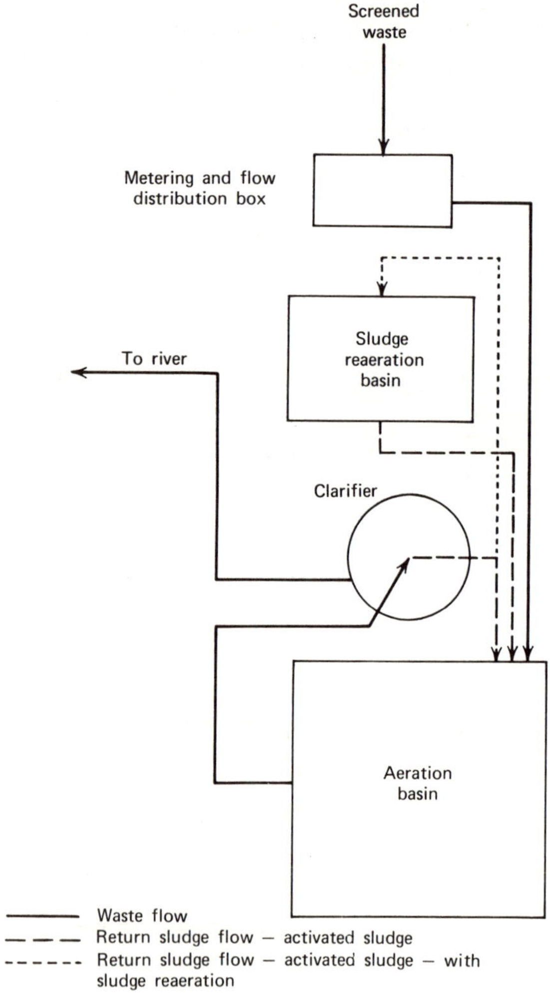

Figure 1. Typical system for cannery.

Cost of treatment is estimated to be

$0.041/lb BOD removed for aerated lagoon
$0.061/lb BOD removed for activated sludge
$0.67/lb BOD removed for contact stabilization

A typical system for a cannery is shown in Figure 1.

The waste load from a plant, screening solid wastes over 20 mesh vibrating screens, at 250 tons of raw fruit per two-shift day for pears, peaches, and plums or 100 tons of apples is given in Tables 1 and 2.

Thickened sludge is used for fertilizer or land disposal. The more concentrated the sludge can be made, the more economical will be the disposal costs since the fertilizer value lies in the solids only and not in the water that must be included.

Application of biologic oxidation through lagooning, air incorporation through impellers, while developed largely in water treatment, has been extended to cannery wastes,[1] solid wastes, and the like, by lagoons and oxidation ditches to cannery wastes,[4] citrus fruits,[2] poultry processing,[3] dairy plant wastes,[5] meat packing,[6] sugar beet wastes,[7, 8] hog production,[9] chicken raising,[10] and duck farming,[10] among others.

Table 1 *Wasteloads*

Product	Flow (gal/ton raw product)
Pears	5,800–6,600
Peaches	7,500–8,500
Purple plums	7,000–8,000
Tomato juice	3,200–4,000
Apples	8,200–10,000
Product	**COD (lb/ton raw product)**
Pears	105–120
Peaches	98–115
Purple plums	95–110
Tomato juice	Negligible compared to concurrent products run
Apples	71–90

Table 2 BOD per Ton

Product	BOD (lb/ton raw product)	Standard Deviation (lb/ton raw product)
Pears	78–90	10
Peaches	66–80	10
Purple plums	64–75	10
Tomato juice	Negligible compared to other products processed	
Apples	53–67	14

Product	BOD over COD	Standard Deviations
Pears	0.74–80	0.09
Peaches	0.67–77	0.10
Apples	0.75–72	0.07

Product	$\frac{N}{\text{BOD}}$	$\frac{N}{\text{COD}}$	$\frac{P}{\text{BOD}}$	$\frac{P}{\text{COD}}$
Pears	0.006	0.004	0.0012	0.0009
Peaches	0.011	0.007	0.0027	0.0016
Apples	0.0023	0.0017	0.0010	0.0008

REFERENCES

1. W. W. Rose, et al. "Production and Disposal Practices, Canning and and Freezing Fruits and Vegetables," Proc. 2nd Nat. Symp. on Food Processing Wastes, WPCRS, EPA, USGPO, Washington, D.C., Stock No. 5501-0167, pp. 109–128.
2. F. A. Eidsness, J. B. Goodson, and J. J. Smith, Jr., Ref. 1, pp. 271–886.
3. W. M. Crosswhite, R. E. Caraman, and John Amacon, Ref. 1, pp. 323–326.
4. C. D. Parker and G. P. Skerry, Ref. 1, pp. 25–270.
5. W. J. Harper and John L. Blaisdell, Ref. 1, pp. 509–546.
6. W. L. Paulson, D. R. Kueck, and W. E. Kranlich, Ref. 1, pp. 617–636.
7. W. J. Oswald, C. G. Golueke, R. C. Cooper, and R. A. Tsaugita Ref. 1, pp. 547–596.
8. Fisher, J. H. and E. H. Hungerford, Ref. 1, pp. 597–616.
9. This book, Chapter IV.6.
10. This book, Chapter IV.5.

CHAPTER II.8

the nitrogen cycle

New Jersey is an urban state with the greatest population density of any state in the union. Approximately 10% of its people have septic tanks for human waste disposal and in some instances for organic garbage disposal. These tanks are often on small lots with inadequate disposal fields, and they must be pumped out frequently. The septic tank contents have been dumped in ditches, storm sewers, sewage systems, on abandoned fields, or in landfill.

Both large and small municipalities are faced with the problem of disposing of sewage sludges from their municipal treatment plants.

Farmers throughout the nation have been litigated out of business because their animal wastes have created objectionable odors, polluted surface water, and have produced flies. There are tons of food processing and household wastes that are high in carbon content and an ideal energy source of biological systems. These biodegradable wastes could be incorporated in the soil media where they would be recycled and utilized by plants in natural biological systems.

Nitrogenous compounds from human and animal wastes are potential pollutants; therefore, one of the basic problems is protein conversion. The nitrogen cycle, Figure 1, indicates the complexity of the problem.

The two primary inputs are the wastes and the sites listed in the second block on Figure 2. In addition to agricultural and recreational land, there are thousands of acres of abandoned and unutilized sites now producing weeds and dust and contributing to flash flooding, soil erosion, and filling of lakes and rivers with sediment.

Consequently, farmers prefer to purchase and apply 100 lb of nitrogen from a 2- or 3-ft^3 clean package rather than to pay for the labor and equipment to transport and apply 100 lb of nitrogen in the form of a ton of an odoriferous, fly-breeding, slurried waste.

Odoriferous slurries are spread on land surfaces providing media for feeding and breeding flies and other pests. The wastes may be washed into streams and lakes before being tilled into the soil.

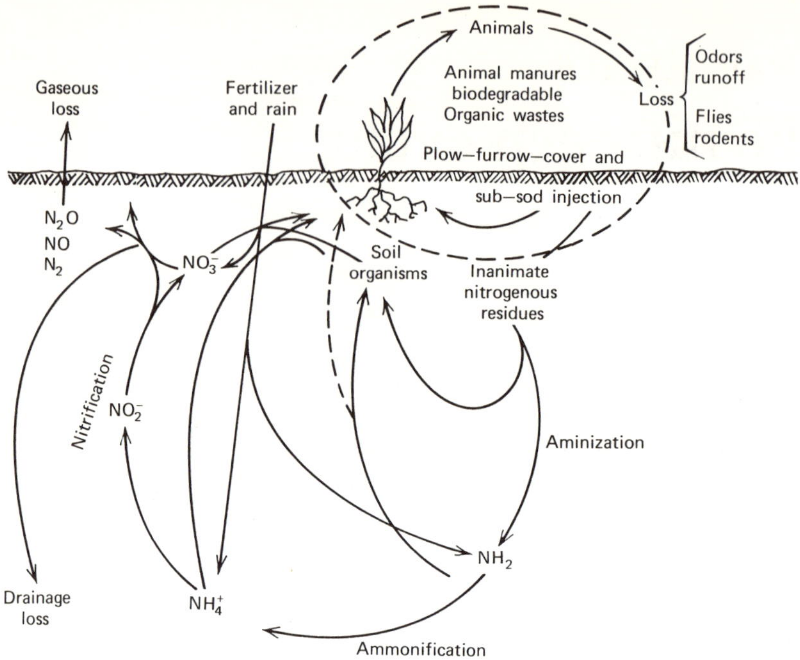

Figure 1. The nitrogen cycle. PFC and SSI organic wastes in slurry form in the upper 8 in. of soil—the aerobic, biologically active layer. Here soil microorganisms degrade the wastes to carbon dioxide, water, and nutrients.

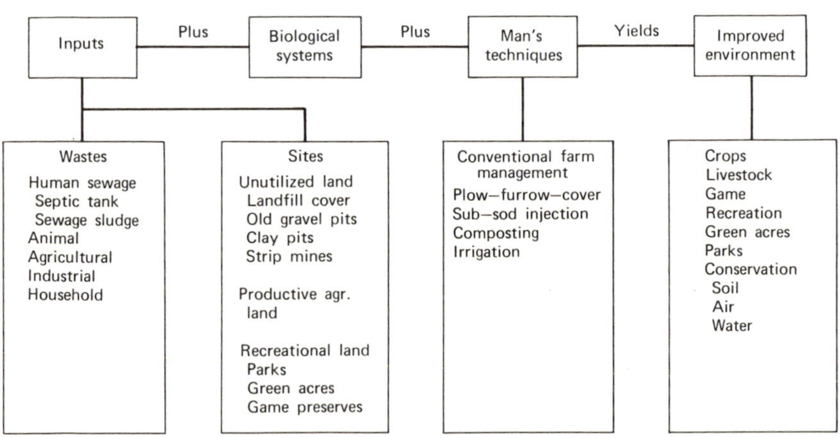

Figure 2. Recycling and utilization of biodegradable wastes in the soil.

The plow–furrow–cover (PFC) method of waste disposal consists of depositing a slurry or a semisolid waste in a 6- to 8-in.-deep plowed furrow. Immediately after deposition, in the same operation, a plow covers the waste and opens the next furrow. From 1½ to 2 in. of slurry can be completely covered, approximately 170–225 tons of slurry per acre. A furrow 16 in. wide, 7–8 in. deep, and 400 ft long, with 1½ in. of slurry, contains 500 gal or approximately 2 tons.

The subsod injector (SSI) will inject a band of slurry up to 1½ in. thick and 24 in. wide and 6–8 in. beneath sod or growing crop. Under normal growing conditions, turf recovers quickly and plant growth is stimulated.

The PFC and SSI techniques utilize the soil media in the aerobic region for degradation and green plants for the eventual utilization and recycling of organic wastes. With a closed system there are no odors and no opportunity for flies or other pests to breed or feed. Unless there is massive soil erosion there is no surface runoff of pollutants. With SSI there is no surface erosion or runoff of pollutants. The PFC and SSI equipment have been designed to operate at 3 mph and unload 200 and 400 gal/min, respectively.

Equipment for plow–furrow–cover and subsod injection techniques consists of the subsod injector and a tank mounted on a trailer with an adjustable gooseneck tongue. The tongue provides maneuverability of the equipment with either a 16-in. single-bottom moldboard plow, or the subsod injector mounted on a farm tractor with hydraulic controls and a three-point hitch. Instead of an adjustable tongue, the trailer axle frame could be adjustable for the required offset.

The vacuum pressure tank on the two-wheel trailer has been used for PFC and SSI of animal manure slurries. It has a capacity of 800 gal and will load, by vacuum, animal manure slurries with up to 20% solids and unload slurries with up to 30% solids (wet basis) by gravity.

A nonvacuum tank also has been designed to unload semisolid animal wastes with up to 30% solids, and sewage sludge with up to 15% solids. It is mounted on a four-wheel drive truck chassis with flotation tires and equipped with hydraulic controls and a three-point hitch. This assembly constitutes a self-propelled highway field transport for mixing and incorporating wastes in the soil by PFC and SSI.

The subsod injector with a 50-in.^{2} inlet, connected to a tank with a 6-in.-diameter hose, will inject up to 400 gal of slurry per minute with hen-size lumps. Except for the mushroom top and inlet, it has been assembled from available parts. It has a three-point hitch and a spring-trip release for passing over subsurface objects. It is heavy-duty piece of equipment comparable to a two-bottom plow in weight and durability.

CHAPTER II.9
fluid wastes and sludges

FLUID WASTE FOR ILLINOIS FARMS*

Processed sewage includes a great deal of water and other waste liquids from homes and industrial plants in addition to some solid materials such as hair and metal filings from the same sources. Processing at a municipal sewage disposal plant changes the physical form. The end products are effluent (the liquid material) and sludge (the solid material). These can be combined into "sluffluent" and spread on agricultural or "waste" land.

Several American municipalities, including Tucson, Arizona; San Antonio, Texas; and Springfield, Illinois, have been using land as a disposal site for either effluent or sludge. The former is most common, since it contains minimal solids and can be spread through conventional irrigation equipment. Most of the plant nutrients and other chemical elements are in the sludge.

So far, no harmful effects from heavy metals (e.g., mercury and lead), excessive micronutrients, or harmful bacteria have been documented. Data are not complete with respect to the mineral elements, because of the historically high costs of analysis.

Other cities (Milwaukee and Grand Rapids) sell "activated sludge," an oxidized dry material commonly containing 4–6% nitrogen.

The Soil Enrichment Materials Corporation (SEMCO) has successfully taken sluffluent to a site south of Arcola, a small town in east central Illinois, and "irrigated" over 150 acres with the material. A number of difficulties had to be overcome. The material in the lagoons was about 20% solids and had to be diluted in order to be flowable and pumpable. A well as dug and that water used to dilute the lagooned material.

After studies had shown that tank cars for transportation were feasible, new cars were designed and built by the North American

* Written by C. L. Mantell with the cooperation of Dr. S. M. King, Soil Enrichment Corp., Chicago, Illinois

Tank Car Co. and leased to SEMCO. A unit train, consisting of 40 of these 20,000-gal cars, began hauling in July 1970. A total of 3000 tons is hauled 150 miles each day. At the disposal site, the cars are emptied into a holding trench, from which the liquid sludge is pumped into a 35-acre storage lagoon. The material at this point is practically odorless and looks like crude oil. On the other side of the lagoon, another pump pushes the sluffluent into an irrigation system which spreads it with sprinklers on wheels.

In 1970, 140 acres of oats and alfalfa were irrigated with an average of 1½ in. of sludge (dry basis). After being plowed in the spring of 1971, the area was planted to corn. Soil and plant samples were taken for elemental analyses from areas set up as research plots. Water samples collected in the fall of 1970 showed no lead, mercury, nitrate, or other toxicities in wells or surface water.

In a 4-acre soybean demonstration south of Champaign, Illinois, sludge was applied in 1969 and plowed down. Soybean yields in 1970 were over 4 bushels per acre higher with sludge than with a conventional fertilizer treatment.

Acre plots were planted to corn near the lagoon site in the spring of 1970. Treatments were: no fertilizer, 140 lb/acre nitrogen as ammonium nitrate and about 600 lb/acre nitrogen as 2 in. (dry basis) of sludge. The fertilizer materials were surface-applied and lightly disked before corn planting. The soil bacteria were unable to convert a significant amount of the organic nitrogen in sludge to the available nitrate form. The corn plants were nitrogen-deficient in late July, but picked up more and more as the growing season ended.

The material transported to Arcola contains about 10% solids. With each acre-inch of water, about 10 tons/acre dry sludge are spread (see Table 1). An application of 20–25 dry tons per acre (2–2½ in. of water) per year is considered adequate. It is obvious that fluid waste, compared to a premium suspension formulation, is not a balanced fertilizer.

First, most of the nitrogen is in the organic form, and therefore only slowly available to crops. If sludge application is made in the crop year, additional nitrogen fertilizer is needed; the amount depends on the crop, the soil, and the type of tillage. Second, the sludge is low in potassium. Most crops and soils will require additional potash to balance the nitrogen and achieve high yields. Third, some micronutrients and nonfertilizer nutrients are present in relatively high amounts. It would be costly to remove them. They are, for the most part, tied up in organometallic form and therefore relatively unavailable to crops and relatively unleachable. On the other hand, sluffluent has physical as well as chemical advantages. The effluent

Table 1 Application Rates

Element	Application Rate per $2\frac{1}{2}$ in. Water or 25 Tons Dry Sludge (lb/acre)
Total nitrogen	900
Nitrate nitrogen	Trace
Ammonium nitrogen	Trace
Phosphorus (P)	1750
Potassium (K)	90
Calcium	2500
Magnesium	625
Sulfur	460
Boron	20
Copper	150
Iron	2000
Manganese	17
Molybdenum	Trace
Zinc	500
Arsenic	Trace
Cadmium	30
Chromium	250
Lead	115
Mercury	Trace

will add to the nutrient supply available and help ensure an adequate crop, even in dry weather.

Long-distance sluffluent hauling is new, so that no price range has yet been determined. Water content and distance always have to be taken into account. For the Arcola, Illinois, area, however, a farmer is charged $20 per acre for about 20 tons of sludge and 2 in. of water.

Fluid wastes will probably be a once-in-several-years application; they provide many nutrients in very slowly available form and must be supplemented with potash.

Current sludge production from the Metropolitan Sanitary District of Greater Chicago is sufficient to supply only 20,000 acres on a continuing basis. If fluid wastes from homes and industries throughout the whole country were applied to land, only 700,000 acres would be required. Much of this could be low-production areas that currently use no inorganic fertilizer. Sluffluent, then, will never be a threat to the chemical fertilizer business. Its application could, in fact, stimulate use of inorganic fertilizers in many areas.

SLUDGE

Sludge treatment and disposal present problems in the processing of sewage and waste, and frequently cost as much as all other treatment processes together. Concentration, stabilization, and disposal are difficult.

"Sludge" is the concentrated solid materials withdrawn from waste water. Untreated sewage sludge is a mixture of organic and inorganic solids, removed from the combination of domestic and industrial wastes, which becomes malodorous if left untreated. After treatment, the sludge loses its objectionable odor and assumes the properties of a plant fertilizer and soil conditioner.

Domestic sewage containing industrial wastes, including human wastes, garbage, sand, grease, and other particular matter has a suspended solids concentration of about 0.02–0.03%.

Sewage sludge may be disposed of by biologic oxidation, lagooning in ponds, wet combustion, multiple-hearth incineration, addition to composting procedures, and disposal on land as soil conditioner on farms and forest areas.

Incineration disposes of sludge from biological wastewater treatment. Sludge containing 60–75% moisture can be reduced to one-tenth its original weight, often without supplemental fuel. Typical combustion values are given in Table 2.

Table 2 *Typical Combustion Values of Waste Materials*

Material	Combustible (%)	Ash (%)	Btu/lb
Grease and scum	88.5	11.5	16,750
Fresh sewage solids	74.0	26.0	10,285
Fine screeings	86.4	13.6	8,990
Ground garbage	84.8	15.2	8,245
Rags	97.5	2.5	8,050
Digested sewage and garbage solids	49.6	50.4	8,020
Digested sludge	59.6	40.4	5,290
Grit	33.2	59.8	4,000

Note: Where organic polymers can be utilized to condition sludges rather than ferric chlorides and lime, the heat value of the sludge cake can be increased 1500–4000 Btu/lb of dry solids. The ash from the furnace will also be reduced by 5–20%.

The multiple-hearth furnace consists of a refractory-lined circular steel shell with refractory hearths located one above the other as represented in Figure 1. Sludge feeds through the furnace roof by a screw feeder or belt and flapgate. A rotating air-cooled central shaft, with rabble arms and teeth, plows the sludge across the top hearth to drop holes. It falls to the next hearth and then the next until ash discharges at the bottom.

The system has three operating zones: the top hearths, where feed is dried to about 48% moisture; the incineration–deodorization zone, which has a temperature of 1400–1800°F; and the cooling zone, where the hot ash gives up heat to incoming combustion air. Exhaust gases exist at 500–1100°F.

Sludge properties that have the most influence over incineration are moisture content, percent volatiles and inerts, and calorific value. Moisture is the principal property over which the plant operator has some control. Minimum moisture is important because of its thermal load on the incinerator. For example, with auxiliary fuel and an offgas temperature of 700°F, about 1800 Btu/lb are needed to evaporate water.

Wet oxidation and wet air oxidation are found in the Zimpro process, Figure 2. Sludge is oxidized in liquid phase, without mechanical dewa-

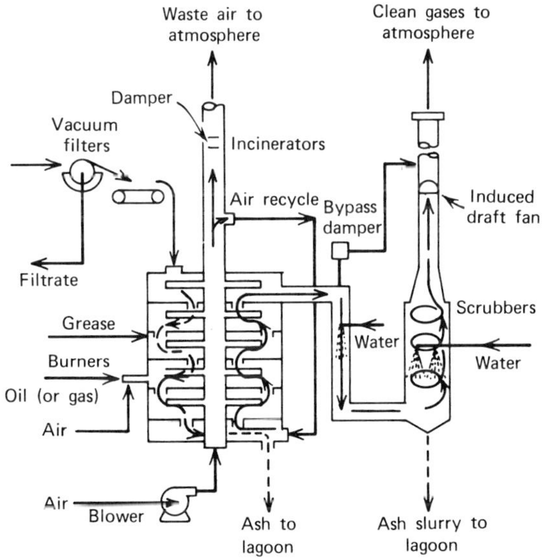

Figure 1. Multiple hearth system for sludge disposal.

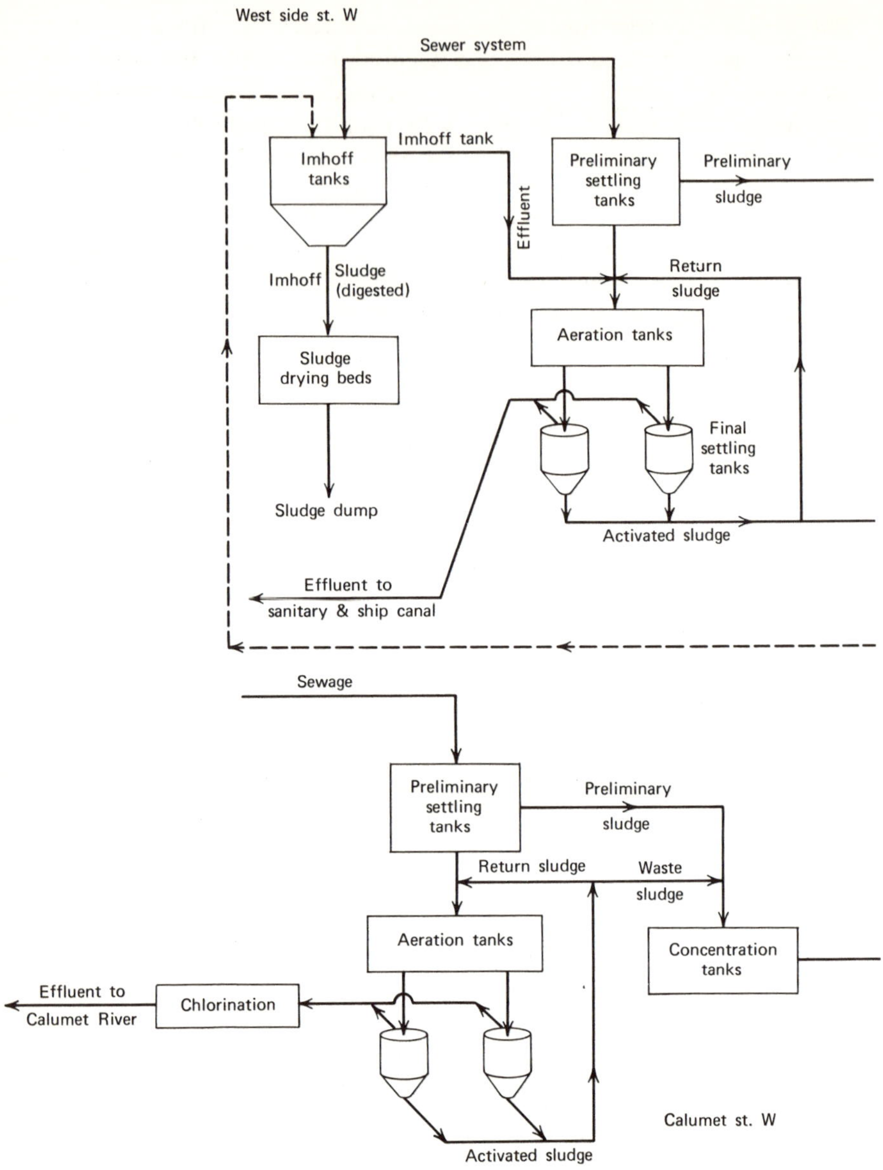

Figure 2. Zimpro wet oxidation process treats biological waste without mechanical dewatering.

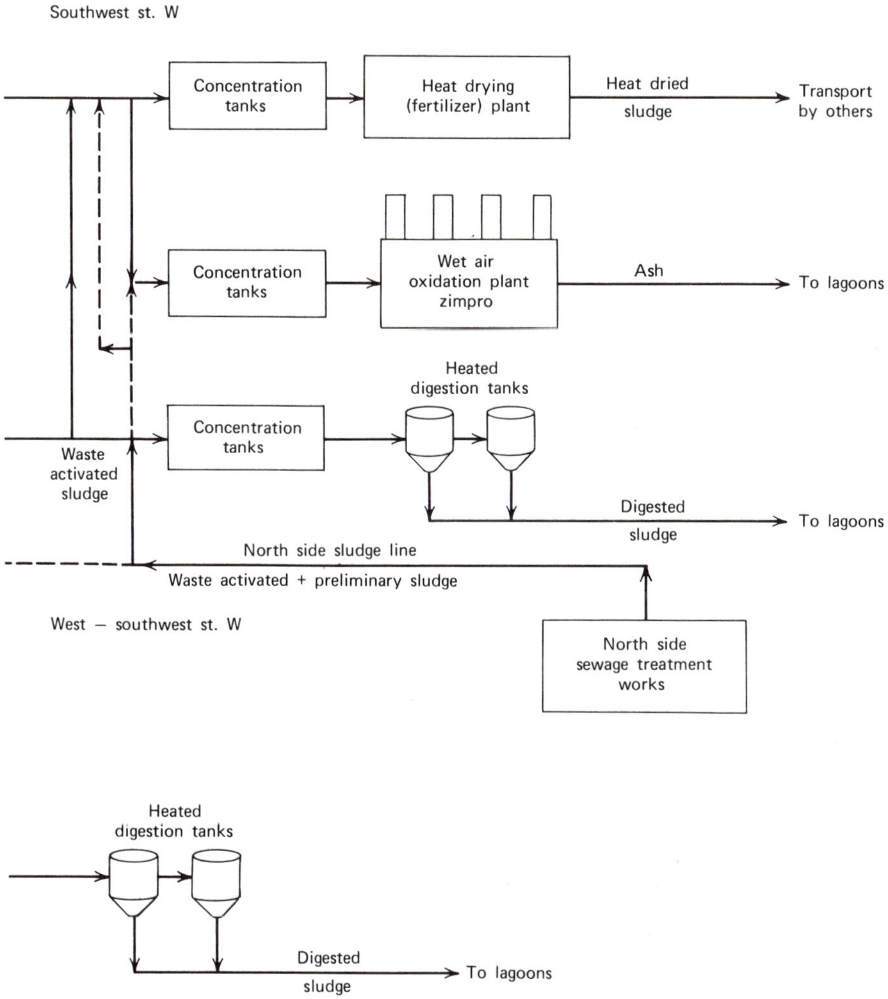

Southwest st. W
Concentration tanks
Heat drying (fertilizer) plant
Heat dried sludge
Transport by others
Concentration tanks
Wet air oxidation plant zimpro
Ash
To lagoons
Heated digestion tanks
Concentration tanks
Waste activated sludge
Digested sludge
To lagoons
North side sludge line
Waste activated + preliminary sludge
West – southwest st. W
North side sewage treatment works
Heated digestion tanks
Digested sludge
To lagoons

tering by heat and pressure, and in air to oxidize combustible matter. Degrees of organic destruction depend on temperature and pressure. The system includes a reactor, air compressor, heat exchanger, high-pressure sludge pumps, sludge grinders, power recovery equipment, and vacuum filters. The first installations operated at 1750 psi and about 525°F. The system is said to be thermally self-sustaining without power recovery at feed heat values of 1500 Btu/lb. Effluent liquid with about 5000 mg/liter BOD is returned to the activated sludge system for biological reduction. Solids discharged from the process are about 90% inorganic.

The Porteous steam-injection process is the oldest of the mechanical processes that change the physical structure of sludge. Sludge is pumped from a sedimentation basin or digestor through a heat exchanger to the reactor vessel. Here, steam enters through a steam-jet circulator, which mixes sludge and steam at 230–390°F and 180 to 230 psi. After 30–45 min, conditioned sludge passes back through the heat exchanger, giving up heat to incoming sludge and to the decanting vessel.

The decanter acts as a sedimentation tank in which solid material settles and the supernatant is drawn off. This reduces treated sludge volume by one-third. For dewatering, the effluent goes to vacuum filters, filter presses, or other mechanical equipment. The product may contain as little as 35% moisture and may be used for compost or landfill.

A flow diagram of the treatment processes at the Chicago Sanitary District Plants, 750 tons dry sludge equivalent per day, is shown in Figure 3. Raw sewage is 99.99% liquid and 0.01% solids; settled sludge is 97–95% liquid and 3–5% solids. The settled sludge is withdrawn from settling tanks with solids concentrations of 4–5%. Bacteria, protozoa, and viruses abound in this raw, settled sludge.

The sludge at the Sanitary District's activated sludge plants may be combined with incoming sewage and aerated, forming clumps of activated sludge floc. Protozoa and bacteria, present in these flocs, decompose organic matter and convert dissolved nutrients into cell matter. Activated sludge flocs are removed in the final settling tanks, and pumped into heated digesters.

The solids are classed in analyses as fixed (mineral) solids or volatile (organic) solids. Solids are also described as dissolved or suspended with the suspended solids further grouped into settleable and suspended classes. About 90% of the suspended solids will be removed from the sewage as untreated sludge.

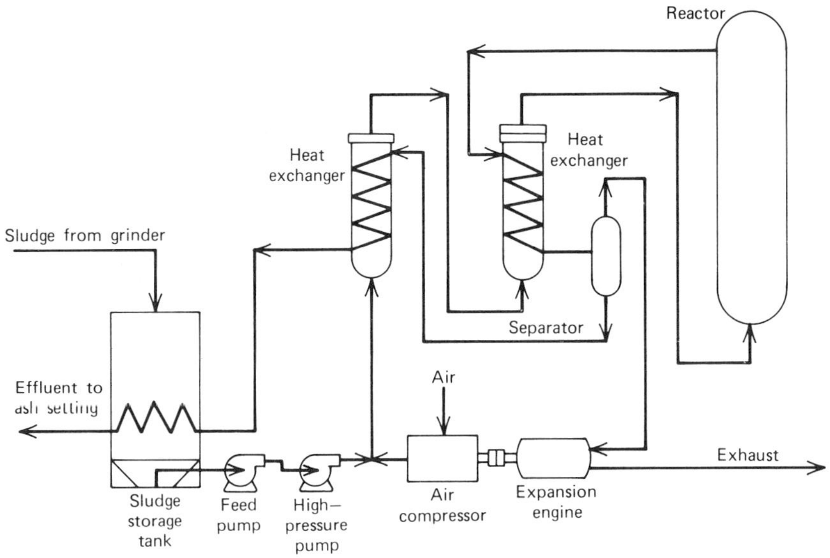

Figure 3. Flow diagrams of treatment processes, Chicago Sanitary District.

The weight and volume of untreated sludge are reduced by both heated and unheated digestion, the wet air oxidation process, or vacuum filters and heat drying.

Heated digesters (95°F) operate at the Calumet and West-Southwest plants. Bacteria act on the settled, untreated, activated sludge to convert about 30% of the organic matter to gas, and require about 2 wk to produce a stabilized sludge with a slight odor, similar to moist earth. The viruses and bacteria have been largely destroyed or combined in stable complexes. The digested activated sludge contains much water in the form of bacterial cells, difficult to dewater and concentrate.

At the West Side plant, untreated sludge is also collected in unheated Imhoff tanks. Organic matter is not converted into cell matter, as in the activated sludge process. Digested Imhoff sludge loses water faster, and contains less volatile matter, than digested activated sludge. Digestion is inefficient because the process depends on outside air temperatures and has no provisions for mixing. The time varies from 1 to 3 months. The digestion processes result in volatile solids reductions, which are summarized in Table 3.

Behn[11] measured viscous properties of digested sludge (see Table 4) for various concentrations. The shearing stress, S_s, is related to the

Table 3 Volatile Solids Reduction

Process	Reduction of Volatile Solids (%)	
Imhoff tank digestion	50	50
Heated digester digestion	30	30
Wet air oxidation	55	—
Composite	37	33

shearing rate, du/dy, by the equation

$$S_s = K\frac{(du)^n}{dy}$$

where

$$K = 0.0113\ C^{3.4}$$

and

$$n = 0.817 - 0.843 \log_{10} C$$

where C is the percent concentration of solids.

Table 4 Sludge Properties

Apparent density (lb/ft^3)			
3.5% solids		63	
10–20% solids		20–30	
Stabilization (%)		75	
(volatile solids destroyed)			
Average		50–60	
Chicago Sanitary District		33–35	
Odors		H_2S (hydrogen sulfide)	
Hydraulic properties			
Under 5% solids		As water	
Above 6% solids		Non-Newtonian	
Friction loss		Function of viscosity	
Chemical analysis	N	PO_4	K_2O
Commercial fertilizer	20	10	10
Wet sludge	6	3	0.6
Milorganite (dry)	6	4	0.5

For flow of non-Newtonian fluids in a pipeline the Darcy-Weisbach equation is applicable:

$$H_f = f\frac{L}{D}\frac{V^2}{2g}$$

The friction coefficient f is determined from the equation

$$f^{-1/2} = \frac{4.0}{0.75}\log_{10} N_R^{(1-0.5n)} - \frac{0.4}{n^{1.2}}$$

in which,

$$N_R = \frac{(\text{Density})\,(\text{diameter})^n(\text{velocity})}{K8^{(n-1)}}$$

For a pipeline velocity of 3.5 ft/sec and 1060 tons/day of dry solids, the effects of varying concentration on the Hazen-Williams coefficient (C_{HW}) and the power to overcome pipeline friction are shown in Table 5.

Since sludge is a concentrate of assorted domestic and industrial wastes is contains metals, salts, and organic matter in amounts greatly in excess of those present in drinking or irrigation waters. Several analyses of Sanitary District sewage sludge are included as Table 6.

Digested sewage sludge has been used for various agricultural and land reclamation purposes for many years and at many locations throughout the world; the literature on the subject is extensive.

Most installations apply liquid sludge by irrigation methods, although techniques range from fire nozzles to dump trucks. Sludge is frequently conveyed to the fields by tank truck or by pipelines, at rates of 0.5–500 tons of dry solids per acre per year (T/A/Y) in slurry concentrations of 1–10% dry solids. Application rates of 0.5–50 T/A/Y

Table 5 *Power Requirements; Variation in Power Requirements with Concentration*

C (%)	D (in.)	O (ft^3/sec)	C_{HW}	Hp/mile
3.5	24	10.5	160	11.5
5.0	20	7.3	160	9.6
6.0	18	6.1	160	8.0
8.3	16	4.4	110	16.0
10.5	14	3.5	100	17.0

Table 6 *Sanitary District Sludges*

	Units	Calumet Digester Sludge			W. Southwest Digester Sludge			S. W. Lagoon Digester Sludge
		7/66	7/66	6/67	7/66	7/66	6/67	6/67
Total solids	mg/liter	27,500	29,305	34,300	22,600	22,430	29,000	317,500
Mineral	mg/liter	13,200	13,200	17,800	10,200	10,200	14,000	201,500
Organic	mg/liter	14,300	16,105	16,500	12,400	12,230	15,000	116,000
Specific gravity, pH		—	—	1.011	—	—	1.002	1.126
		7.0	7.1	7.7	7.2	7.2	7.2	8.1
Maximum part size		—	—	10 mesh	—	—	20 mesh	—
Conductivity	μmho	3,800	3,800	6,000	3,600	3,600	5,500	2,000
Hexane soluble	mg/liter	1,727	—	4,200	1,774	—	2,700	17,600
Alkalinity as $CaCO_3$	mg/liter	2,745	2,750	2,460	2,615	2,600	2,130	—
Al	mg/liter	—	227	636	—	173	1,345	28,966
As	mg/liter	—	BDL′	—	—	—	BDL′	BDL′
B	mg/liter	0.9	1.9	5.1	1.2	1.1	0.9	14.0
Cd	mg/liter	1.9	1.0	2.4	10.6	9.3	8.4	120

Ca	mg/liter	1,180	1,240	—	811	845	—	
Cl	mg/liter	490	490	170	350	350	150	—
Cr	mg/liter	46	49	26.0	81	79	88	860
Co	mg/liter	—	—	BDL′	—	—	BDL′ [a]	BDL′
Cu	mg/liter	32	24	31.0	46	49	47.4	690
Fe	mg/liter	1,666	1,598	1,500	662	661	1,100	12,500
K	mg/liter	114	152	—	105	136	—	
Mg	mg/liter	332	291	446	228	249	194	240
Mn	mg/liter	143	14	—	45	5	—	—
Mo	mg/liter	—	0.06	BDL′	—	0.03	BDL′	BDL′
Total N	mg/liter	1,512	1,450	1,767	1,542	1,538	1,753	—
Ammonia N	mg/liter	528	603	790	608	657	736	1,338
Na	mg/liter	129	119	—	78	79	—	—
Ni	mg/liter	Trace	2	3	10	12	13	116
P	mg/liter	680	738	740	632	707	710	—
Pb	mg/liter	90	90	6	19	19	19	20
Si	mg/liter	—	9,800	2,773	—	5,200	3,129	43,138
S	mg/liter	45	67	288	52	82	306	1,660
Zn	mg/liter	92	91	90	147	148	191	1,670

[a] Prime denotes amount below detectable limits.

are reported where repeated annual loadings are used in conjunction with farm crops, which can utilize large amounts of nitrogen.

Application rates of 50–500 T/A/Y are reported where massive loadings are applied to sterile soil, such as dredged sand fill, for development of a top soil to support vegetation. Extremely heavy applications, over 200 T/A/Y, have been confined to dredged coastal areas where groundwater pollution is no problem.

No serious nuisance or health hazards have been reported as a result of using digested sludge. Slight or temporary insect or odor conditions were experienced occasionally. Toxic soil conditions occurred where loadings were excessive or other agricultural-management needs had been neglected.

Sludge has been used for agricultural or similar operations under many diverse climatic and cropping conditions. Walkenshaw[9] reports that several cities in the midwestern United States are operating farms for sludge utilization. Canton, Ohio, has the longest history of operation, with their farm having begun in 1900. Since that time Canton has been disposing of its sludge on 600 acres. The sludge is spread on approximately one-half this area, which is hilly and is used for pasture or is wooded. Sludge is not applied to the other, flatter land on which corn is planted. The land on which sludge is applied is from 10 to 20% slopes. The sludge has been distributed by pumping to ditches along the top of the slopes and releasing it down the slope. Some sludge ran off into natural drainage channels and off the land, but there were no complaints except after a heavy rain following the sludge distribution.

The sludge production figures indicate a loading of about 7 dry T/A/Y. The distribution was very uneven and small areas have probably received five to ten times this much. On a terraced area, 20 dry T/A/Y have been applied.

Troemper[10] reports that the Springfield, Illinois, Sanitary District disposes a portion of its secondary digested sludge on 22.5 acres of land. Liquid sludge is spread on 8.0 acres of terraced hillside; Bed dried sludge is spread on 14.5 acres. The soil is a dark colored loess. In 1965, 795 tons of bed dried sludge were spread on 14.5 acres. In 1966, 8.7 in. of liquid sludge were applied per acre, 4.9 in. during July and August and 3.8 in. during the winter. At a concentration of 6.8%, this was an application rate of 67 tons/acre. An application of about 107 tons/acre had been made in 1965, before the tract had been terraced.

Van Kleeck[1] reported that the principal value of digested sludges is the humus content resulting in increased moisture-holding capacity of the soil and a change in soil structure that results in a greater friability. Like all humus materials, sludge stimulates the growth and propagation of angle worms, which aid in soil aeration.

Experience using an elutriated digested sludge conditioned with ferric chloride is that liming is required about every 3 yr to maintain a soil pH between 6.5 and 7.0. Van Kleeck found that first crop yields are improved by low application rates of sludge to soils. He found a rate of 10 tons/acre was better than a rate of 20 tons/acre. Many smaller cities throughout the United States are able to sell or give away part, or all, of their sludge to farmers located near the treatment plant.

Merz[2] reported that primary sludge from the city of San Diego, digested at 85°C for 30 days, was used for barley grown on test sites. The studies indicated that increased fertility of soils was obtained at loadings from 10–100 tons of dry solids per acre; that a deleterious effect on the soil and interference with plant growth can be avoided by adequate leaching of the soil of intolerant salinity introduced with the sludge; that costs for sludge disposal were substantially decreased; and that composite soil samples showed an increase of saturation percentage from 27 to 50 for the 100-tons/acre loading. The sludge was applied to the land, allowed to dry for 30 days, then plowed under. No public health problem or nuisance developed. There was a slight musty odor present when the sludge was spread, but this disappeared after 2–3 wk. Coliform organism tests of water from nearby wells 20–40 ft deep gave negative results.

Experiments to determine if sprinkler irrigation of golf courses using sewage effluent created a public heatlh problem by releasing waterborne bacteria to the air were reported by Merz. Thirty-six air samples were collected of which 19 yielded positive results. Four of the 19 positive samples were collected within the limits of effective spray irrigation. The remaining 15 positive samples were all taken within the zone where mist from the sprinkler could be physically detected. No positive samples were obtained where mist was not detectable. The bacteria can be air-borne only in moisture droplets. The only possible health hazard is in the spray or mist zone.

Scanlon[3] reported that New York City disposed of digested sludge by disposal at sea, drying beds, spraying on sand covered landfill, and lagooning. Digested sludge was spread on a 180-acre area at Marine Park, Brooklyn, between 1936 and 1956, using the land as a drying bed. The dried sludge was removed in 1956 and plowed into sand-covered landfill to create topsoil in the order areas of the park. Topsoil was created by spraying sludge on sand-covered landfill, on 40 acres at a time at the rate of 1 gal/ft^2 and allowed to dry for 2 or 3 days before disking it into the upper 9 in. of sand. About 16 applications were required to form 4 in. of topsoil.

Sludge is also piped to lagoons and the dry sludge from the lagoon is spread on the sand. The estimated cost is \$3.50 per yd^3 to import topsoil, and only \$2.50 to form topsoil from sludge, including transport costs.

In Britain, liquid sludge is commonly used on farm land for its fertilizer value, particularly from the smaller treatment plants.

Coker[4] and Green and Cowling showed that the maximum amount of nitrogen that could be assimilated by grass was about 300 lb/acre/yr and this was usually applied in two or three applications of 100–150 lb nitrogen each. Since the nitrogen in liquid digested sludge is equal in its effect on increasing growth to that in fertilizers, this also defines the rate of application of liquid digested sludge for the most efficient production of grass. On the average the uptake of nitrogen from sludge was 86% of that of equivalent artificial fertilizers.

A study by the Natural Resources Committee, England,[5] showed that the dry solids recovered at sewage plants in England and Wales amount to 800,000 tons/yr, and 350,000 tons are applied to agricultural land. Approximately 400,000 tons more could be used, containing 10, 000 tons of nitrogen, 5000 tons of P_2O_5, and 1200 tons of K_2O. Other organic matter added to the soil in normal farming amounts to 17.5 million tons/yr.

Paris and Berlin both have irrigated farms with sewage. The farm in Paris includes about 10,000 acres and has been operated for 87 yr. Fouquet[6] reported that the application of sewage sludge for a soil supplement has been practiced in Paris since 1880. Until 1940, the sewage farms received all the sewage from Paris and with proper management of the irrigation, no ponding of the sewage took place and no inconvenience was experienced in the neighborhood. The principal crops grown are celery, beans, and leeks, but corn, rye, and oats are also grown as well as many other vegetables. The vegetables are noted for their size and color, but buyers are instructed to boil them before eating.

Musgrave,[7] Superintendent of Sewage-Treatment Plant, Fairfield, Illinois, reported that Fairfield has been applying sludge to the same farm for 10 yr. Two smaller farms have been added for standby, and other farmers are trying to get priorities for sludge.

The plant has a conventional settling and trickling filter, producing sludge by heated digestion without mixing. The output was 3000 and

1500 gal/day digested sludge. The digester operation is described as follows:

1. Try to hold in primary digesters for 45 days, then
2. pull off to secondary holding digesters, and
3. as it nears full, haul to farm.

Disposal is to a 200-acre general farm with 89–90 acres standby. Application was by tanker truck with deflector plate, before plowing and on plowed ground, on idle pasture land as well, at 0.46 T/A/T (dry solids).

San Diego hauled sludge by truck from the treatment plant to the land disposal area. Later, transportation costs were reduced to one-fourth of trucking costs by building at 7.5-mile pipeline to the disposal site. On the San Diego site, furrows are plowed and filled about 18 in. deep with liquid sludge. The process is repeated after drying until the sludge, up to about 410 tons/acre, has been applied.

New York City uses barges and pipelines to transport large quantities of liquid sludge to several hundred acres of sandy soil. The sludge is applied to produce a depth of 18 in. and allowed to dry.

Sludge is conveyed to fields by pipelines and applied by surface spreading at Springfield, Illinois, and Canton, Ohio. Sprinkler irrigation equipment is connected by light-weight pipe to pumps mounted on the tanker trucks at some English operations. Sprinkler equipment also has been used for application of animal-manure slurries. Myers[15] states that sprinklers have been used at Pennsylvania State University for several years, applying treated effluent under all extremes of weather.

Klute[12] reports on long-term tests conducted at Riverhead, Long Island, on level plots of Sassafras Silt Loam. Manure loadings were applied at rates up to 40 tons dry solids annually for periods up to 40 yr. Soil organic levels were increased to 2.5–5.0% and moisture holding capacity was increased by about 25%. However, available moisture was increased by less than 20%, since the wilting point had been increased also. The stability of 2–5 mm aggregates was improved significantly.

Leaver,[13] referring to sludge disposal practices at several northwestern communities, stated that liquid sludge is disposed for fertilizer at four communities. At Olympia, a nurseryman pays $2.00 per 1000-gal load ($4.80 per dry ton) and hauls it away in his own truck a distance of 2½ miles. Chehalis received $400 per year for 142, 000 gal of sludge in 1954 and at 7% solids the return was about $10 per dry ton. Issaquah has approximately 15,000 gal of thin digested sludge hauled each 6 wk by a farmer who uses it on hay fields. At Shelton, private haulers have taken most of the available liquid digested sludge since 1951. A reference shows the constituents of various fertilizers. In Seattle, manure costs about $16.60 per ton dry. Milorganite costs $114 per ton. Portland, Oregon, and Tacoma, Washington, lagoon liquid digested sludge. Some sludge was removed when the cities advertised its availability. The Washington State Health Department requirements state that raw or improperly digested sludge cannot be used nor can digested sludge be used on root crops or on low growing vegetables, fruits, or berries. When used for hay fields, cattle must be excluded for 45 days. Air-dried sludge must be dried for 30 days or until it can be handled with a pitchfork. Sludge may be used on root crops and vegetables if it is plowed into the field 6 months before planting.

Pennypacker[14] reported on studies made at Pennsylvania State University. The effluent from the University Sewage Plant was applied by a sprinkler system to university agricultural land and forests. The BOD of the effluent was 17–18 mg/liter. Total suspended solids of the effluent was 15–32 mg/liter. The sprinkler system applied 0.25 in./hr; the application rates were 1 and 2 in./wk. The agricultural production revealed a very good increase in yields from effluent plus irrigation: hay showed a yield increase of 139%; corn silage, of 39%; corn grain, of 78%; and out grain, of 70%.

The rates of application commonly used depend on whether the liquid sludge is being applied for (a) nitrogen-phosphate fertilizer properties, (b) soil conditioner value, (c) disposal as the primary objective. The literature does not indicate any consistent approach to determine loading rates.

Several controlled tests have shown that application rates of 2–25 tons/acre are satisfactory in supplying the nitrate and phosphate need of a subsequent crop. Lunt[16] recommends a rate of about 10 tons/acre, for a sludge containing about 2.5% total nitrogen. Trials were carried out on both sandy and medium textured soils.

Liquid sludge is reportedly used for the reclamation of dredged-sand fill in the cities of New York and San Diego. Applications are made at rates of 300–1000 tons/acre. In New York, sludge is dried before plowing, but in San Diego, wet sludge is plowed into the previously sterile sand to develop a "topsoil" capable of supporting vegetation. Topsoil had been trucked into these areas previously at greater cost. Sludge is applied to dredged-shell fill at the Virginia Key Plant in Miami. A relatively heavy load is applied initially, and repeated annual loadings of about 14 tons/acre are maintained after grass cover is established.

Few operating problems have been reported for the utilization of sludge. Small plants report inconvenience owing to holding sludge when fields or roads are impassable. Modification of the New York City transportation barges for sea disposal was needed so that unloading could be accomplished at the reclamation sites when the barges were diverted from their normal sea disposal circuit.

Rohde[17] attributed decreased yields on sewage-irrigated farms at Berlin and Paris to the accumulation of high concentrations of copper and zinc in the soil.

After many years of sewage disposal by irrigation on cropland near Berlin and Paris, the lands began showing signs of exhaustion and decreased crop yields following earlier high yields. Applications at Paris have been made for 87 yr. In 1957 soil samples were collected from land on which healthy plants were growing and also from lands showing exhaustion at both Berlin and Paris. Soils at the Berlin sewage farm are sandy and generally acid (pH near 6.7), while those at Paris were near neutral (pH generally near 7.3) and rich in lime. Soils on land showing signs of exhaustion contained more organic matter than soil from land on which healthy plants were growing, suggesting that the amount of sewage sludge in the soil was an important factor. Some soil organic matter contents reported are 2.2% (in soils with healthy plants) versus 7.1% (adjacent soils with unhealthy plants), 3.7% (healthy) versus 8.9% (unhealthy), and 1.8% (healthy) versus 3.45% (unhealthy).

Rohde[17] used weight loss on ignition to estimate organic matter content. This method causes loss of water from hydrated clay minerals; thus values for organic matter are somewhat high. He also[17] investigated the status of several "trace elements," but he concluded that zinc and copper were the chief depressants to plant growth on the sewage farms. Table 7 shows zinc and copper concentrations.

The normal total zinc content in soils ranges from 10 to 300 ppm; a typical value is 100 ppm. Zinc is usually more concentrated in surface than in subsurface soil horizons. Available zinc normally is in the range of about 0.5–5.0 ppm. Water-soluble zinc is in the parts-per-billion range. Zinc in soils is largely in organic complexes; some is held as exchangeable zinc on cation exchange sites of clay minerals. The latter is probably the form most available to plants. Deficiencies occur when available zinc falls below 0.5 ppm.

Zinc may be fixed in small amounts by silicate clays, and it may precipitate as the phosphate, carbonate, or hydroxide at slightly acid to alkaline pH. Hence, liming is often used to relieve zinc toxicity to plants. Liming can, however, create other problems for plant growth,

Table 7 Zinc and Copper Concentrations in Soils at the Berlin and Paris Sewage Farms

	Extracted with 10% HCl		Extracted with Boiling H_2O[a]	
Soil Sample	Copper (ppm) Berlin	Zinc (ppm) Berlin	Copper (ppm) Paris	Zinc (ppm) Paris
Average				
Healthy	118	197	8.71	6.48
Unhealthy	508	609	20.06	19.12
Average				
Healthy	132	386	8.5	3.43
Unhealthy	291	1032	11.2	10.48

[a] Water extracted copper and zinc are readily available to plants.

since the availability of other necessary plant nutrients is curtailed at alkaline pH values. If the plants are to be used for forage, excessive molybdenum uptake may result from alkalinity of the soil.

Plants vary widely in their normal content of zinc, from less than 20 to as much as 10,000 ppm.[18] Those that can assimilate large amounts of zinc, e.g., ragweed, can be used as indicators of the zinc content of the soil. Plants showing symptoms of deficiency will contain less than 20–25 ppm zinc, dry basis; satisfactory levels are in the range 25–150 ppm. Little is known about excess zinc, but indications are that amounts greater than 400 ppm may be harmful. In an area of Illinois near zinc mines, corn showed stunting and other symptoms of toxic levels; analysis showed zinc levels of 200–300 ppm in the tissue of this plant.

A summary of the effects of sludge application in soil is given in Table 8.

PETROLEUM SLUDGES

The American Oil Refinery at Mandan, North Dakota, found the problem of disposing of oily sludge and spent caustic most perplexing. Oily sludge is characterized by the presence of oil, solids, and water. Its consistency varies from that of a fluid slurry to a greasy semisolid. It is encountered when cleaning oil storage vessels, removed from the waste water oil–water separator, and it is collected as a residue from oil–water emulsion-breaking operations.

The disposition had been to dump the sludge into open earthen pits. To conserve land, the volume of sludge was reduced by burning the hydrocarbon that migrated to the surface, with a billowing black cloud of smoke. This method was unsatisfactory.

Spent caustic results from treatment of petroleum products for the removal of naphthenic acid, cresols, mercaptans, and hydrogen sulfide by caustic soda. These compounds emit intense disagreeable odors. Disposal of spent caustic by neutralization and by trickling into the waste water treating system released odors during neutralization and the waste overtaxed the bio-oxidation pond.

When the spent caustic was dumped into an earthen pit, the caustic leached through the earth with contamination of the natural ground water. When the pit was lined with plastic sheets, it rapidly became full. The volume of spent caustic was increased by rainfall. The only reduction of the waste occurred from evaporation which produced an undesirable odor.

Various incineration systems were considered. The oxidation of spent caustic alone would require a prohibitive quantity of supplemental fuel. The incineration of oily sludge appeared more feasible. When oily sludge is subjected to high temperatures, the oil evaporates. Should the resulting hydrocarbon vapors fail to ignite, they could create hazardous explosive mixtures with combustion air.

Conditions offered by the fluidized bed incinerator were high heat-transfer efficiency, mixing, and stable combustion conditions. A bed of inert fluidized solids acting as a heat sink reduced the hazard of unburned hydrocarbon vapor and air.

The fluid bed incineration process[19] has been demonstrated to be practical and effective for the disposal of petroleum refinery generated spent caustic and oily sludge. The process creates no atmospheric pollution problems, emitting only carbon dioxide, nitrogen, and water vapors, the odor of the off gas being slight to nonexistent. The ash produced contains sodium carbonate, sodium sulfate, other soluble inorganic salts and inert material such as sand, clay, and rust, and is readily buried. Operations are frequently terminated prematurely by the loss of bed fluidity caused by excessive bed particle size. The particle size growth rate can be controlled by including proper design features and operating techniques.

Table 8 *Effects of Sludge Application on Soil*

Property	Sand Soils	Silty Clay Loam Soils
	Physical Effects	
(1) Soil Structure		
Degree and type of aggregation	Slight improvement owing to some aggregation of particles from cementation by organic compounds.	Moderate improvement due to aggregation of clay particles makes soil more friable, less cloddy. Stability of aggregates will be greater.
(2) Infiltration Rate		
Penetration of water into surface soil	Decreased rate owing to addition of fine particles and grease from sludge.	No change or increase owing to improved aggregation.
(3) Permeability		
Movement of water within soil	Decreases permeability if sludge mixed into soil; if placed as a layer may cause temporary perching of water.	Increases permeability if sludge mixed into soil.
(4) Maximum Retentive		
Capacity: Percent H_2O at saturation	Increases substantially, e.g., 30–50%	No change or may decrease slightly.
(5) Available Moisture		
Capacity: Moisture available to plants	Increases somewhat; field capacity ($\frac{1}{3}$ atm) is increased substantially but wilting coeff. (15 atm) increases proportionately, so net gain in available moisture is not great.	No significant change, or slight increase.
(6) Free Water		
Superfluous to plant needs.	No effects if underdrainage is provided.	May cause brief periods of water logging even though underdrainage is provided. Dependent on loading rate of the wet sludge and tim ng of application.
(7) Trafficability		
Vehicular traffic on field surface.	No change or delay due to sludge application.	May delay traffic over field for 2–3 days depending on rate of sludge application and natural rainfall.

(8) Susceptibility to Wind and Water Erosion Assuming no cover crop.	No appreciable change or may be somewhat less susceptible to wind erosion, particularly if sludge applications are frequent enough to keep surface moist.	May lessen water erosion by increasing infiltration rate.
(9) Soil Texture	It is assumed that a substantial part of the insoluble ash in sludge is of silt and clay size (<50 microns). Therefore, it is expected that an admixture of 10 tons sludge (or about 2–3 tons ash) per acre per year will gradually produce a finer textured soil. For example, a loamy sand may be changed to a sandy loam.	The effect of sludge on silty clay loam soils would tend to coarsen the texture somewhat.
	Chemical Effects	
(1) Cation Exchange Capacity (CEC) (Also buffering capacity or resistance to pH change)	Should increase substantially, sandy soils have CEC of about 5 meq/100 g, with sludge it might be increased to 10–15 assuming organics in sludge are comparable to soil humus which has a CEC of about 150.	May increase somewhat: clay component in these soils and native organic matter have CECs of 20–40 meq/100 g
(2) Exchangeable Cations	The normal order of abundance of exchangeable cations held on exchange sites of Illinois soils is Ca > Mg > K > Na. This is determined by factors other than soil texture. The order of abundance in the sludge is the same although the proportions may be quite different than in soils. Assuming that these elements are in ionic form (readily dissociated salts), it does not appear that sludge applications to soils will upset the balance of exchangeable cations except that occasionally Na may surpass K in abundance. This probably will not have any serious side effects, i.e., Na should not become so abundant on the exchange sites of clay particles as to cause dispersion ("puddling"); also, K can be readily applied as a commercial fertilizer should it become a limiting plant nutrient.	
(3) pH	Owing to the high pH of sludge (7.0–7.5) and the low buffering capacity of sands the pH should be increased to near neutrality in a short period of time.	Sludge will increase pH but the change will be more gradual than in sands; also greater amounts of sludge would be required to cause changes in pH comparable to those in sands.

Table 8 *Effects of Sludge Application on Soil—Continued*

Property	Sand Soils	Silty Clay Loam Soils
(4) Ability to Adsorb Heavy Metals (Cu, Zn, Ni, Pb, Cr, Mo.	Sands in original condition have virtually no capacity to adsorb heavy metals. Any absorption capacity for heavy metals would be associated with the sludge itself. It is not known at this time to what extent this capacity is saturated with respect to the heavy metals already present in the sludge. Research is needed to determine this.	Soils with fair amounts of silicate clays (montmorillonite, vermiculite, and illite) have a fairly high capacity to adsorb heavy metals. However, little is known about the long-term effects, e.g.: is the adsorption capacity saturated with respect to Cu, Zn, etc., in a few years, in a hundred years of continued sludge application?
(5) Ability to Adsorb Ammonium Ion	Sands have no ability to adsorb ammonium ion. The addition of sludge will increase the adsorption capability of the soil–sludge mixture.	It is reported that as much as 50% of the total N present in certain illite-containing soils has been found to be in the "fixed" ammonia form*. In this fixed form, the nitrogen is not subject to rapid oxidation, although in time it may become available.
(6) Ability to Adsorb Phosphates	Sands have little or no ability to adsorb phosphates. Adsorption of phosphates will be related to organic matter content. From the standpoint of phosphate-fertility the addition of sludge should therefore be quite beneficial on sandy soils, particularly since Ca is also added and the pH of the sludge is near neutral to slightly alkaline.	As in the case of sandy soils much of the soil phosphorus in silty clay loam soils is associated with the organic matter. Although the anion adsorption properties of the chief clay minerals present in Illinois soils are low some adsorption of this type will undoubtedly take place. Also, if soil pH is acid, fixation of phosphorus by Fe and Al compounds will take place (most of these compounds occur in the clay fraction); at alkaline pH values phosphates will be precipitated as Ca-phosphate.

	Bacteriological Effects	
(1) Bacteria Population	Owing to the droughty nature of sand soils organic residues are quickly oxidized. Hence, existing populations of microorganisms are generally quite low. The addition of the organics in the sludge to sand soils may increase existing populations as well as introduce new colonies and species. California data[a] suggest that a 10–50-fold increase in bacterial populations in sandy soils from sludge application may be expected. A pH increase to about neutral due to sludge application may further stimulate bacteria growth.	Good agricultural soils with moderate moisture holding capacities and fairly high organic matter contents usually carry a great number of microorganisms. Populations introduced with the sludge and the addition of organic materials in the sludge may or may not increase bacteria populations. The possibility exists that the competition between various kinds of microorganisms may depress total populations; also certain inorganic substances, e.g., copper salts may act as bactericides. It is likely that sludge applied to the silty clay loam prairie soils of Illinois will not be beneficial to soil bacteria either in kinds or numbers.

[a] Report on Continued Study of Waste Water Reclamation and Utilization. State Water Pollution Control Board. Sacramento, California, Publication No. 15, 1956.

REFERENCES

1. LeRoy W. Van Kleek, "Utilization of Sewage Sludge," *Water and Sewage Works,* Ref. Edition, 1953.
2. Robert C. Merz (Ed.), *Third Report on Study of Waste Water Reclamation and Utilization,* California Water Pollution Control Board, Publication No. 18, 1957.
3. Arthur Scanlon, "Utilization of Sewage Sludge for the Production of Topsoil," *Sewage and Industrial Wastes,* August 1957.
4. E. G. Coker, Science Department, Hertfordshire Institute of Agriculture, "Experiments in East Hertfordshire on the Use of Liquid Digested Sludge as a Manure for Certain Farm Crops," *J. Inst. Sewage Purification, 1965.*
5. Anon., "Use of Towns' Refuse in Agriculture." A study by the Natural Resources (Technical) Committee; *Her Majesty's Stationery Office,* London, England, 1954; *Sewage and Ind. Wastes,* 28 March 1956.
6. P. Fouquet, Chief Engineer, Paris Sewage Farm, personal communication, October, 1967.
7. Irs Musgrave, Superintendent of Sewage-Treatment Plant, Fairfield, Illinois, personal communication, May 6, 1967.
9. G. B. Walkenshaw, Superintendent, Sewage Treatment Plant, Canton, Ohio, personal communication, May 23, 1967.
10. A. Paul Troemper and Jerry, Peters, Springfield, Illinois, Sanitary District, personal communication, 1967.
11. V. C. Behn, *J. Sanitary Engineering Division, ASCE,* **88,** May, 1962.
12. A Klute and W. C. Jacobs, "Physical Properties of Sassafras Silt Loam as Affected by Long-Time Organic Matter Additions," *Proc. Soil Sci. Soc. Am.,* **14,** 1950.
13. Robert E. Leaver, "Sludge Disposal Practices in the Pacific Northwest," *Sewage and Industrial Wastes,* **23,** 323, March 1956.
14. Stanley P. Pennypacker, William E. Sopper, and Louis T. Kardos, "Renovation of Wastewater Effluent by Irrigation of Forest Lands." *J. Water Pollution Control Fed.,* February 1967.
15. E. A. Myers, "Engineering Problems in Year-Round Distribution of Waste Water," *Proc. Natl. Symp. Animal Waste Management,* ASAE, May 1966.
16. Herbert A. Lunt, "Digested-Sewage Sludge for Soil Improvement," Connecticut Agricultural Experiment Station, Bulletin 622, April 1959.
17. G. Rohde, "The Effects of Trace Elements on the Exhaustion of Sewage-Irrigated Land," *J. Inst. Sewage Purification,* **6,** 581–585 1962.
18. H. D. Chapman, "Diagnostic Criteria for Plants and Soils," University of California, Division of Agricultural Sciences, 1966.
19. "Fluid Bed Incineration of Petroleum Wastes," EPA WPC Research, *R*12050 EKT 03/71 USGPO 5501–5502.

CHAPTER II.10
composting

Composting is a method of converting organic material into a drier, nonodoriferous form through bacterial action, primarily to supply humus to soil. It is unlikely that this treatment method will become common in the United States in the near future.

About three decades ago Europeans, led by the Dutch, began development of composting plants for the disposal and utilization of urban solid wastes. At first the objective was to produce compost for a soil builder or conditioner; later, the objective was primarily to provide a sanitary method of waste disposal and secondarily to utilize the resulting compost. Although there were many composting processes, two types of plants predominated: the rasping system-windrow composting plants developed by the Dutch N.V. Vuilafvoer Maatschappij (VAM) and the Dano biostabilizer plants developed by the Dano Works in Denmark. These European methods are briefly reviewed below.

The German farmer has recognized the need to maintain the organic content of his oil by operating an agricultural system that resulted in materials moving from the soil to crops to animals and back to the soil. The efficiency was such that there was a need to supplement the organic content by a source outside the system. Compost could produce results that were comparable to those obtained by commercial fertilizers applied to mineral-deficient soils. The results created confusion as to the role of compost but studies have shown that compost is beneficial as a source of humus and not of minerals.

Compost has short- and long-term advantages when added to the soil for crop production. The exact mechanisms are unknown. Compost increases the biological activity in the soil; compost-produced crops experience less plant disease. Compost is not a fertilizer, yet when applied to the soil it produces a crop response similar to that with commercial fertilizer. One study related to compost and soil effects involved earthworms. Two windrows of ground refuse were placed side-by-side, one as control, the other inoculated with earthworms. After 2 months the control pile was anaerobic at the center and decomposing slowly while the inoculated windrow had decomposed rapidly into compost.

The high population density of Germany requires maximum production from the agricultural land available, and sustained high-level productivity over a long period of time, so that soil losses must be minimized. One crop with erosion potential is grapes grown on hillside vineyards. The grapevine growth is such that there is no ground cover, leaving the soil with exposure to the erosion. The hillsides are terraced and gradually the terraces have been replaced by the open mountainside. Compost has been found to be an effective material for erosion control, and the wine growers are probably the largest group using compost in Germany.

In the United States the soils are young and fertile, although often suffering from poor management. A decreased land area per capita may require the preservation of lands for intensive agricultural production by transforming wastes into compost or humus for the soil, but this transformation is, generally, not yet necessary.

Composting plants receive only the domestic and some part of the commercial refuse from a community and have a small intake, ranging from 40 to 200 tons per day. The storage bunkers (pits) receive and hold four collection vehicle loads of refuse. When a breakdown occurs, an alternative disposal system must be available. Most of the bunkers are of a trapezoidal design with the two side walls sloping to a conveyor belt along the bunker bottom. The sloping walls help move the refuse onto the conveyor belt but allow bridging over it. The solution has been to construct a vertical false wall along one side, producing one straight and one sloped side. On new installations, a wider conveyor bottom would be recommended (conveyors 6 ft wide or more).

Most of the installations on the continent once had dust control equipment for the bunker area, but it was either not used, broken, or considered ineffective in operation. Where the storage bunker was too small, refuse would be discharged on the ramp and pushed in by a tractor as the bunker was emptied. A steel conveyor belt at the bottom of the bunker fed a rubber conveyor belt that moved the refuse to the next stage.

REFUSE PREPARATION

As the refuse moved, some salvage and sorting operations were performed. Most of the plants were eliminating picking or sorting because of increasing labor costs and the small income from salvage.

Pickers removed rocks, tires, large pieces of iron, bottles, large sheets of plastic, and other materials undesirable for composting. It was considered important to remove as much plastic as possible; glass was of little concern, and no attempt was made to salvage paper.

An over-bank magnet, a drum-head magnet, or both, were standard equipment. The drum-head magnet is an electromagnetic roller at the end of the conveyor belt. The magnetic field places the ferrous material in a trajectory different from that of the other materials leaving the belt. The over-band magnet is an electromagnet directly over the conveyor belt. An endless belt positioned between the refuse and the magnet sweeps the material out of the magnetic field into a storage container. The iron removed was baled for storage and was sold for whatever price it would bring, or else given away for the removal.

Following intake, the refuse could be processed by grinding, rasping, or by the Dano drum. Commercial grinders are supplied by the Buhler Company (double- or single-rotor grinders), the Hazamag Company (double-rotor grinder), or the Gondard Company. The ballistic separation action of the Gondard grinder provides a mechanical sorting or removal of much of the noncompostable material. The Hazamag grinder has a high-speed double rotor with swinging arms of flat steel. The Buhler single-rotor grinder uses flat steel arms with pivoting sledge-hammer-type heads at the end. In either, the grinding surface can be reversed for additional use. The double-rotor Hazamag and the single-rotor Buhler mill are primary grinders; the double-rotor Buhler is a secondary unit.

As is standard, both grinders were considerably overpowered. The excess power is required for starting and for overcoming slugs of material or materials difficult to grind. Hazamag seemed to be relying on the inertia and grinding action of the double rotor to overcome the hard spots. The Hazamag mill was more heavily constructed than the Buhler. In the double-rotor Buhler mill the rotors turn at different speeds, and the hammers of the faster rotor have about one-third the life of the slower rotor hammers.

Developed by the Vuilafvoer Maatschappij (VAM) firm, the rasp is marketed by the Dorr-Oliver Company. It is a vertical cylinder with a diameter of 15 ft and a total height of 7 ft, containing both a grinding and a discharge floor. The grinding floor is formed from wedge-shaped sections with alternate sections being constructed with round holes (about 0.5–1.5-in. diameters) or with 0.5-in. steel pins projecting from the section. The pins also extend part way up the wall all around the inside. There are eight heavy hinged arms that can rotate at various speeds up to 25 rpm. The arms move the material over the floor that serves to alternately screen and rasp the material. The arms can lift up over ungrindable material until a trap door can be opened and the rejects swept out the door by the arms. The material passing through the grinding floor falls on the discharge floor and is moved out a discharge opening by a rotating arm attached to the center spindle. The rasp is physically larger than a grinder, is quieter in operation, requires less maintenance, but has a higher initial cost. The grinding and rasp systems of composting appears to be competitive.

The Dano drum, usually 9–12 ft in diameter and 60 ft or more in length, is rotated at a speed of about 1 rpm during the working day and at about ¼ to ½ rpm at night. The raw refuse is usually magnetically separated before being discharged into the drum, where a slight inclination requires about 2–5 days for the material to pass through, detention time being a function of the speed of rotation. The drum is operated half full, and the slow rotation provides aeration and the self-grinding action that is effective in the reduction of the raw refuse. When the material reaches the end of the drum, a short section of 1.5-in.-diameter screens removes large pieces. The rejects are periodically discharged from the drum via a trap door in the screen. A characteristic of the drum compost is the presence of small balls of paper.

COMPOST CURING

Compost curing is by windrow. Windrow turning is by an end-loader and elevating conveyor that piles in the triangle-shaped windrow. No test has been developed for the frequency of windrow turning. Temperature is measured by a thermometer and moisture is evaluated by the compost *meister's* nose, sense of touch, and experience. This reliance on the compost meister has developed a fraternity of these men, and they circulate from plant to plant.

When the compost is judged to be finished, it will be screened once after final grinding to provide an evenly blended and uniform texture. Glass is the most objectionable material in compost. In general, the machines are of three types: (1) Ballistic separation is used for glass and other hard particle removal. Compost falling into one end of the machine strikes a high-speed horizontal rotor. The rotor paddles are often short railroad tracks. The paddles or impellers strike the compost and propel it in a horizontal or slightly upward trajectory. Hard, dense, and resilient particles will obtain a high speed and long trajectory, while the soft, light, and inelastic particles travel only a short distance. The trajectory paths are confined in a horizontal enclosure, the length and partitioning of which can be adjusted. (2) The secator is cheap and mechanically reliable. Compost falling from the end of a conveyor strikes a shield or bounce plate that directs the material vertically downward onto a revolving drum. The strike plate may direct the material onto the top center of the drum or slightly off center in the direction opposite that of the drum rotation. The heavy and resilient particles will bounce off the drum into a collection bin, while the light and inelastic materials are carried on the drum under the strike plate and into storage. (3) With the inclined plate conveyor, compost falls onto a steel plate conveyor that moves at an angle upward against the falling compost. Heavy and resilient particles bounce down and off the conveyor while the lighter material is carried upward.

For improved quality the compost should be exposed to two or more of the systems, or to the same system twice. The problem is that greater glass removal results in loss of compost, and an operational balance between a maximum amount of glass removal and a minimum amount of compost rejection must be attempted.

Screening or sieving of compost is universal at composting plants to provide a gradation of the final product. The screening of the compost and its cost are related. Most of the screens are the vibrating type, although a few plants use small drum-type screens. An experimental sieve for glass removal is an adaptation of a coal sieve that operates on the counterflow principle. The compost enters the top and falls onto a fine-mesh, inclined, vibrating screen. Air, blown at high velocities upward through the screen, suspends the compost a short distance above the screen while at the same time moving the material up the slope of the screen and eventually out of the machine. The heavier particles remain and are moved downward and off. A cross-flow air current for large particle removal is directed across the compost falling from a conveyor. The installation is effective in removing wine corks, bottle caps, some plastic, and bits and pieces of other material.

Most of the compost plants were located adjacent to the sewage treatment plants. The sewage sludge was recognized as useful in the composting process, and composting was considered a good method for sludge disposal. The cost accounting system usually included the expense of the equipment needed to transport and incorporate the sludge with the refuse but seldom recognized, by means of charges to the sewage treatment plant or something similar, the benefits of using composting for sludge disposal.

In most of the operations the "expenses" exceeded the "income," but the necessary subsidy was justified as a reasonable expense. When questioned why composting was the system used, some operators sooner or later admitted that a high-powered salesman had influenced the judgment of the local political body. Even though the system did not prove profitable, as long as the plant subsidy was kept low, the operation was considered acceptable. In most cases composting was justified on the basis that incineration capital costs were too great, that there was a lack of room for tipping, or that the watershed laws forbade the deposition of any type of raw refuse on the ground. Some questions might be raised about a law that forbids depositing "raw" refuse but fails to establish standards or quality control regulations established for compost produced unless the compost is marketed as fertilizer. There are definite quality control standards for fertilizer production.

MARKETS

There is no problem with producing compost; the problem is what to do with it. One organization for the utilization of compost is the previously mentioned Vuilafvoer Mattschappij (VAM) of Amsterdam, Holland. One of the largest stockholders in the company is the Dutch government. The company has three major sections: technique, agriculture research, and sales. VAM's first plant was built at Wijster in 1932 when there was a large demand for coarse, low-quality compost in land reclamation. At that time about 80% of production went to the land reclamation companies. By 1940, 90% went to farms and only 10% to the reclamation companies. During World War II commercial fertilizers were almost impossible to obtain, and it was necessary to utilize compost to maintain crop production. By 1950 compost had found several new uses. Cattle preferred grazing on pasture that had been covered with compost rather than stable manure; mixing compost with farm manure increased the organic content of soil; compost could be used as bedding material for chickens and pigs, and was effective in curing anemia in pigs. These uses increased the demand. In the 20 years between the middle of the 1930s to the 1950s only one composting plant was in operation, but since then a total of 15 additional plants produce compost.

During the 1950s vegetable growing, flower cultivation under glass, and flower bulb cultivation became large industries in Holland. A finer quality product was required and the processing of the compost became more complex. A market opened for "refuse heating manure," freshly ground refuse devoid of any sharp particles, in demand by greenhouses. The attractive feature is the large amount of heat produced during the decomposition of the organic matter.

Flower bulb cultivation along the sandy North Sea coast created a large market when hand-placed reeds and straw were replaced by mechanized application of compost. Thus, while most compost is sold in bulk, there is an increasing demand for small quantities in attractively packaged bags. Although the compost *meister* curses plastic as one of his biggest problems, the sales people find it the most attractive way to market the product.

The VAM was established to develop and maintain a market for compost. The government is a strong partner because the sale of compost provides a solution to refuse and sewage sludge. The VAM company sets forth the following points necessary for compost sales: (1) a good outlet in the neighborhood of compost producing plants; (2) absence of competitive products such as farm manure, heating manure, and the like, unless they have long supply distance; (3) cooperation by a large and well-stocked government information center; (4) the production of quality products that can be guaranteed; (5) favorable trading results with the buyers and no government restrictions; (6) very dry years when crops have had bad yields or died early because of lack of organic matter; (7) marketing with a trademark.

COMPETITIVE PROCESSES

VAM states that compost is not profitable. Several other countries are faced with problems similar to those in Holland, and in some the federal government is involved; yet none has a sales group comparable to VAM and none enjoys the success in compost utilization that Holland does.

A composting system can be considered to have two parts: one for the preparation of the material and the other for biological environment. Composting systems are of four categories: (1) fermentation cells with prior crushing of the refuse and windrow curing; (2) fermentation cells and windrow curing; (3) windrows without prior preparation and product preparation; (4) windrows with prior treatment of refuse.

The Van Maanen System, established in 1932, is the *open-dump composter.* Most of the refuse is railhauled from Den Haag, Holland, about 175 km from the plant at Wijster. The rail cars discharge the refuse onto the ground from rail spurs built on viaducts. Water sprinkling systems moisten the refuse. Schedules provide up to 7 months' storage in the windrow. After curing, the transfer is via gantry cranes and mining cars to processing and screening by successively coarser screens. Fines first removed and treated by magnetic separation are nearly free of glass and make the best quality. The coarse-screen product is treated by magnetic separation, then either by a roller crusher or a hammer mill. This lower grade can then be refined by ballistic separation and stockpiled.

The operation is isolated from nearby farms. There are vector and odor problems and a potential groundwater pollution problem. Efforts to use chemical masking agents have not been successful in solving the odor problem. An underdrain system designed to collect the leachate has been destroyed by corrosion, and some work is now being done to develop a new collection system and provide some measure of treatment for the water collected. A barge canal transported some of the compost to market, but trucks have now replaced the barges.

The Baden-Baden system is the *modified tip composter.* Similar to the Van Maanen system, the operation is on a much smaller scale and provides better environmental control. Refuse is discharged from the collection vehicles into a storage bunker, then lifted by crane to a receiving hopper, feeding a rotating drum. The refuse passes from the conveyor into a second drum and is mixed with sewage sludge (from sludge drying beds) until a 50–60% moisture content is developed, and then discharged into a truck or wagon for transport to windrows.

The windrows are constructed on two lines of large-diameter, perforated concrete pipe. Formerly air was blown into the pipes and out through the windrow, but odors became a very serious problem, which was corrected by drawing air into the pipes through the windrow; the air drawn in is used as the air supply in the bulky waste burner on the site. Water is applied to the windrows and drainage is discharged into storm drains or to the surrounding marshy area. After curing, the compost is ground and sieved for sale. The purpose of the plant is reported to be the disposal of sewage sludge and not the production of compost. The compost not sold raises the level of an adjacent marsh.

The Caspari-Meyer Brikollare system, the *cookie-cutter composter,* replaces windrow curing with a "block" or "controlled" environment curing. Refuse is processed by magnetic separation and some hand sorting before discharge into a Dorr-Oliver rasp. After further magnetic and ballistic separation, it is transferred via conveyor to the "brick" machine. Raw sludge, dewatered by a vacuum filter, is conveyed to a pug mill blending tank where three parts refuse and one part sludge is blended to 50%–60% moisture and conveyed to a holding tank, placed in a mold, vibrated for compaction, extruded from the mold, and stacked on a wooden pallet. The brick, measuring about 15 × 15 × 20 cm, has the appearance of dark, raw papier-mâche. A half-moon slot parallel to the long axis runs along the middle of the bottom of the brick. While the bricks are stored for about 2 wk in the shed, temperatures of about 60°C are maintained and large quantities of mold form.

The group of stacked bricks provides a body of heat that maintains the temperature near the optimum for the desired biological activity. The exterior surface of the brick is open to the atmosphere and decomposition is aerobic, while inside the brick it is anaerobic. As the products of the anaerobic activity are transferred outward, they encounter the anaerobic–aerobic interface, where they are transformed into unobjectionable products. The brick moisture moves outward and air moves inward and the microorganisms maintain the balance. When the bricks are to be sold, they are passed through a hammermill and the compost is then screened.

The Humboldt system processes by magnetic separation material being conveyed from the storage bunker to a pug mill for shredding and mixing with sewage sludge. The pug mill includes shearing knives spaced at intervals along the twin rotors. The homogenized and shredded material discharges into a tall conical drum. The material moves downward by gravity and is discharged at the bottom by a scraper. At the center of the drum are pipes for air and moisture. Nylon stockings present a problem in the pug mill because they do not shear and then become wrapped about the rotor and entangle pieces of refuse. The result is a huge ball of material that jams the machine.

SLUDGE DISPOSAL

One function of compost is to dispose of sewage sludge. In an experimental installation at St. Gallen, Switzerland, sludge was processed from the Imhoff tanks of the sewage treatment plant. The sludge was pumped from a storage pit to a metering control into an auger conveyor, which provided some mixing of the sewage sludge with sawdust. The mixture (approximately 70% sawdust) passed from the auger through a mixing or agitating storage hopper and up an elevating conveyor into a rotating drum. Heat in the rotating drum was supplied by a gas burner placed at the drum inlet. The drum was designed with internal baffles arranged to pass the mixture the length of the drum three times before reaching the exit where the material was transferred by a blower into the top of a cyclone separator. The hot gases exited at the top of the cyclone while gravity moved the heavier particulate matter to the bottom where a short conveyor moved the material onto a vibrating sieve. The material that passed through the sieve was sacked for sale to a local fertilizer company while the sieve rejects, mostly sawdust, entered the auger at a point upstream from the sludge entry.

The sawdust served as a blotter and absorbed most of the moisture of the sludge. When the mixture was passed into the drum, the heat dried the sawdust and burned the sludge. In the cyclone, the fine particulates went out with the exit gases, but the inert particles of the sludge and a large proportion of the sawdust were recovered by the cyclone. If too much material passed through the sieve, the sawdust would be lost, and if too many rejects from the sieve were recycled, heat would be wasted.

A different sludge processing system at Dübendorf, Switzerland, centers around a multiple-hearth Nichols roaster. Refuse in the storage pit was moved by a crane and conveyor belt into a primary grinder. An auger conveyor mixed the refuse with raw sludge, and the mixture was discharged into the top of the Nichols roaster. As the mixture was transferred from one level of the tower to the next by slowly rotating arms, hot gases circulating through the tower first dried and then burned the material. The product was a fine, brown powder. Temperature control could be provided by waste-oil-burning nozzles or jets at the fifth or sixth level of the tower and exhaust gas vents at the top of the tower.

A Nichols system was used for sludge disposal at Stuttgart, but no refuse was included. The Stuttgart system was sensitive to variations in the moisture of the sludge.

Lausanne, Switzerland, used the fluidized bed technique. Thickened sewage sludge was taken from the storage tanks and dewatered in a vacuum filter. The cake was transported into the burning tower with auxiliary fuel. Air was blown upward through a sand bed on the tower bottom. The sludge ignited and burned as it moved downward through the hot gases and flames. The system was equipped with an electrofilter and a waste-heat boiler to recover the heat content of the sludge in the form of steam or hot water.

COMPOSTING IN THE UNITED STATES

In the United States, Waksman and his associates conducted research from 1926 to 1941 on aerobic decomposition of plant residues and manures. The first basic studies on aerobic composting of community solid wastes were conducted by the University of California, Sanitary Engineering Research Project during 1950–1952. One conclusion of this work was that composting should be considered as a means of disposal and reclamation for municipal refuse. Gotaas later prepared a comprehensive text on this subject for the World Health Organization but copies are not available.

From 1953 to 1962 Wiley and others of the Public Health Service conducted studies at Savannah, Georgia, and at Chandler, Arizona, and indicated that composting of municipal refuse, with or without sewage sludge, is a feasible and sanitary method of treatment.

Combined treatment of refuse and raw sewage sludge by composting will improve the economy of the process, enhance the decomposition, and augment both structure and nutrient content of the compost. There are several examples of refuse-sludge composting in Europe, and three plants have used refuse and sludge in this country (Williamston, Michigan; Largo, Florida; and experimentally at Altoona, Pennsylvania). The added expense of incorporating sludge with refuse is small compared to the cost of sludge treatment in a conventional wastewater treatment plant. Since moisture must usually be added to mixed refuse for optimum composting, sludge can be added to provide moisture along with other benefits. However, combining all of the sludge with the refuse from the same contributory population will introduce excessive moisture unless sludge thickening is practiced.

Most plants have relied on the sale of compost and salvaged materials for their primary income, coupled with a nominal payment by the municipality for disposal. Inability to dispose of large quantities of compost at a favorable price greatly affects the economy of the process and is probably a major factor in the closing of six U.S. composting plants during from 1962 to 1964. Although compost contains some plant nutrients and trace elements, its principal value is in its use as humus for soil building or conditioning. It can be fortified with additional chemical nutrients to form an organic-based fertilizer. There is no widespread demand for compost in commercial agriculture now and no increase in demand is expected.

A second problem in composting relates to the hygienic aspects in plant operation and in compost utilization. Cleanliness and prompt handling of the wastes are essential to avoid the production of odors and propagation of flies and rats. In normal composting, the aerobic thermophilic decomposition produces temperatures of 140°F or higher for several hours or even days. Such a time–temperature exposure is adequate to destroy most, if not all, pathogens.

For a number of years the Altoona FAM, Inc. plant composted garbage with combustible rubbish by the windrow method. In December 1963 the Fairfield Engineering Company began operation of a Fairfield-Hardy composter at the Altoona plant. This was the second installation of a Hardy digester.

The refuse is ground in a wet pulper followed by two dewatering presses. The ground material is composted for about 5 days with intermittent stirring and aeration in the digester. Stirring is provided by augers suspended from a rotating bridge in the circular tank. Air is provided by a blower and air pipes in the floor of the tank. Operating temperatures are 140–160°F. The plant had a capacity of about 45 tons/day but operates at about 18–23 tons/day. Sewage sludge was added to the ground refuse. The compost is reported to sell as a soil conditioner or as an organic base for fertilizer. However, the company buying compost from the Altoona plant ships most of it in bulk a distance of more than 500 miles.

The original Largo, Florida, plant was built by Peninsular Organics, Inc.; it operated for a few years. In 1963, National Composters Company refinanced Peninsular Organics Company and rebuilt the Largo plant, which had a capacity of 50 tons/day. Operating 5 days a week it treated 40–50 tons/day of mixed refuse to which was added 1000–1500 gal/day of raw sewage sludge.

The steps involved sorting and magnetic separation, double hammermilling, mixing of ground refuse and sludge, composting for 6 days in a long narrow open tank with turning and aeration, regrinding, screening, and bagging. The digester was 12 ft wide by 6 ft deep by 280 ft long having a metal false bottom with air holds. The turner was a rail-mounted bucket elevator moving in either direction the length of the digester. Material was turned twice during composting. The final compost had a market in bulk or when fortified to a 3:2:2 analysis and bagged.

The Norman, Oklahoma, plant was built by the Naturizer Company of Norman, Oklahoma. This plant and the later one at San Fernando, California, were taken over by SACS, Inc. (Salvage and Conversion Systems), which entered into a marketing agreement with Westinghouse Electric Corporation. In February 1965, these companies, together with Allstate Insurance Company, formed a new company, International Disposal Corporation, which incorporated the assets of SACS. The plant treated the 35 tons/day of mixed refuse from the city of Norman. Operations were discontinued in 1964.

The digester system provided 6-day retention in two 3-story buildings with a movable floor in each story. Compost was sold either in bags or bulk. Table 9 gives an incomplete listing of composting plants in the United States.

The Phoenix composting plant was the second Dano-type plant in the United States and was built by Arizona Biochemical Company to process 300 tons/day of refuse. Originally two Dano biostabilizers were provided, but two additional units were added in March 1963. By the European practice of 3-day digestion in the biostabilizers, the capacity of these rotating drum units is 175 tons/day. The plant provides sorting, magnetic separation, salvage, hammermill grinding, composting in the biostabilizers for 24–30 hr, screening, regrinding and deglassing, and further composting in windrows. When an adequate

Table 9 *Municipal Solid Waste Composting Plants in the United States*

Location	Company	Process	Capacity (tons/day)	Began Operation	1973 Status
Altoona, Pa.	Altoona FAM, Inc., Fairfield Engg. Co.	Fairfield-Hardy	45	1951 1963	Operating
Largo, Fla.	Peninsular Organics, Inc., National Composters Co.	Peninsular Organics	50	1963[a]	Closed
Norman, Okla.	International Disposal Corp.	Naturizer	35	1959	Closed 1964
Phoenix, Ariz.	Arizona Biochemical Co.	Dano	300[b]	1963	Closed 1963[c]
Sacramento Co., Calif.	Dano of America, Inc.	Dano	40	1956	Closed 1963
San Fernando, Calif.	International Disposal Corp.	Naturizer	70	1963[d]	Closed 1964
Springfield, Mass.	Springfield Organic Fertilizer Co.	Frazer-Eweson	20	1954[e]	Closed 1962
Williamston, Mich.	City of Williamston	Riker	4	1955	Closed 1962
Wilmington, Ohio	Good Riddance, Inc.	Windrow	20	1963	Operating
Brooklyn, N.Y.	Ecology, Inc.	Windrow	20	1972	Closed 1973

[a] Date of initial operation of the original Hardy plant is unknown; plant rebuilt in 1963.
[b] According to European Dano practice, the plant capacity is about 175 tons/day.
[c] Reported to be back in limited operation, March 1965.
[d] Date of initial operation is unknown; continuous operation began in July 1963.
[e] Partially burned in 1958; rebuilt in 1961.

market for compost had not been developed by late 1963, the plant was shut down.

The plant in Sacramento County, California, built by Dano of America, Inc. in 1957, was the first Dano-type plant in the United States. It had a capacity of; about 40 tons/day of refuse and, except that only one biostabilizer unit was employed, operated in much the same way as the Phoenix plant. The plant was closed in 1963.

The San Fernando, California, compost plant was built by Lockheed Aircraft Corporation and operated by General Conversions System Corporation. Later, the plant was taken over by SACS, Inc. and finally by International Disposal Corporation, Shawnee, Oklahoma. The plant had a capacity of 70 tons/day of refuse but operated at only about 40 tons/day. It was a version of the process at the Norman, Oklahoma, plant. Appreciable salvage operations were performed. It was necessary to provide an afterburner on top of the composting unit to comply with Los Angeles County Air Pollution Control District requirements to ensure that no odors would be discharged.

The plant was shut down in October 1964. Compost was sold in bags and in bulk. "Sta-Soil" was a water suspension of compost, chemical fertilizer, grass seed, and shrub seed. It was sprayed over denuded slopes, or cuts and fills, to provide a blanket against soil erosion and a "soil" in which grass and shrubs would quickly take root.

The Springfield, Massachusetts, plant was built in 1954, partially burned down in 1958, and was rebuilt in 1961. It was owned by the Springfield Organic Fertilizer Company. The plant had a capacity of 20 tons/day of garbage from the city of Springfield. The garbage, being too wet to compost aerobically, was held for about 2 days to dewater. This frequently created fly, rat, and odor problems. The city failed to renew the contract in 1962 and the plant was closed.

The Williamston, Michigan, plant was built in 1955 and operation was taken over by the city in 1956. The 4-ton/day plant treated garbage, vacuum-filtered raw sewage sludge, and corncobs. Ground garbage and corncobs, mixed with the sludge cake, were composted for 21 days in two 4-compartment vertical composters. The plant was closed in 1962 primarily because a nursery in Detroit stopped purchasing the compost.

The Wilmington, Ohio, 20-ton/day refuse composting plant was built by Good Riddance, Inc. A minimum of sorting was provided and two hammermill grinders were used in series. The ground refuse was composted in windrows where it was turned weekly by a front-end loader. Difficulties were experienced in selling the compost.

There is another windrow composting plant at Oahu Prison,

Honolulu, which handles all of the solid wastes of the prison in addition to some food processing wastes. There was also a small plant near Reading, Pennsylvania, built by Keystone Organic Corporation, which was shut down in October, 1961.

At Houston, Texas, three companies (National Organic Corporation, Biochemical Sales, Inc., and Metropolitan Waste Conversion Corporation) submitted contracts; for plant construction at their own expense with each company to receive \$2.75 to \$3.50 per ton from the city for treating the refuse. One plant was reported to be of 350-ton/day capacity, and the others are expected to be in the 300-ton/day range.

The production of Hou-Actinite (heat-dried activated sludge sold by the city) was greater than the market. The amount of Hou-Actinite sold during the period 1937–1944 varied from 846 to 2136 tons/yr, whereas the compost on a dry weight basis would be expected to be 40–100 times this amount if the raw refuse amounts to 950 tons/day.

A compost plant to treat refuse and sewage sludge (about 300 tons/day) was reported under construction at Mobile, Alabama. All of the refuse of Mobile (225 tons/day) will be processed but only a portion of the sludge (digested sludge from a plant serving about half of the city). In this case the city is to build and operate the plant, selling all of the compost to a private concern at a guaranteed price plus a share of the profits.

A contract has been signed by the city of St. Petersburg, Florida, for a composting plant to treat 105 tons/day of refuse. The Naturizer system plant was to be privately built and operated with payment by the city amounting to \$3.24 per ton.

NORCO (National Organic Corporation, Atlanta) reports that four U.S. plants were planned, in Elmira and Irondequoit, New York, Houston, Texas, and New Orleans, Louisiana. The system is the same as that used at McKeesport, Pennsylvania, and Kingston, Jamaica. It includes double grinding and windrow composting, much the same as at Wilmington, Ohio. The amount of refuse from Kingston and St. Andrew, Jamaica, has been reduced from about one-half to one-tenth owing to lack of compost demand.

A survey revealed that Johnson City, Tennessee, had a well-operated refuse collection system, a good accounting system, and a site of adequate size suitably located adjacent to a primary sewage treatment plant. In addition, Johnson City is located in that area of the Tennessee Valley where there already existed the highest ratio of organic to chemical materials used as soil amendments.

A flow diagram of the plant arrangement is given in Figure 1. Based

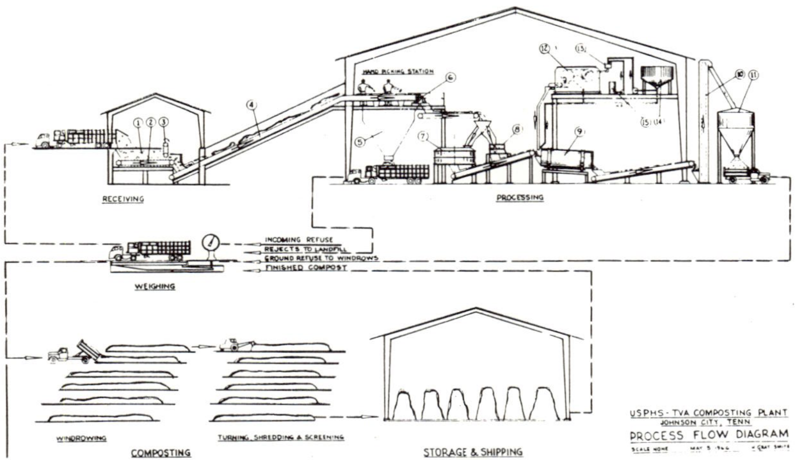

Figure 1. Composting plant flow diagram.

PROCESS FLOW DIAGRAM

① RECEIVING HOPPER
② RECEIVING HOPPER CONVEYOR
③ LEVELING & METERING GATE
④ ELEVATING BELT CONVEYOR
⑤ REJECTS HOPPER
⑥ MAGNETIC SEPARATOR
⑦ RASPER
⑧ GRINDER
⑨ MIXER
⑩ BUCKET ELEVATOR
⑪ GROUND REFUSE STORAGE BIN
⑫ SLUDGE THICKENER
⑬ SLUDGE COAGULATING TANK
⑭ SLUDGE HOLDING TANK
⑮ CHEMICALS MIXING TANK

on records of refuse collection quantities from this city of 33,000 population, the plant was designed to process 10 tons/hr with an allowance of 1–2 hr for cleanup at the end of the shift.

Although some items of painting, landscaping, and so on, and the final grinding and screening equipment still were incomplete, the principal components of the plant were sufficiently completed to permit operation in 1967.

The operating crew consists of ten men including the foreman, all on annual salaries. The operating crew is divided into the following classifications: Foreman, Operator A (acts for Foreman in his absence), Operator B (operates Payloader, compost turner, etc.), one building maintenance mechanic, two truck drivers, and four laborers. Most of these men are dually classified to permit them to perform duties with higher rates of pay as needed to replace absent workers. Initially, working hours were set at 10.00 a.m. to 6.30 p.m. with a half-hour lunch period. After a few months, the city was able to revise its delivery schedules so that working hours were changed to 8.30 a.m. to 5.00 p.m. Refuse is normally delivered by the city in two compaction trailers with capacities of 53 and 65 yd^3.

The initial plan was to deliver all refuse collected in the city to the composting plant. Experience soon demonstrated that the commercial collection during the night was 95% paper, mostly large cardboard boxes. Since refuse with paper content this high must be rejected, it was apparent that the commercial collection should be diverted directly to the landfill. Another problem was experienced with the wastes from certain industries such as a textile mill, a wire mill, and a hospital. Cooperation by the city was helpful in diverting much of this directly' to the landfill.

Initially, refuse materials were dumped directly into the receiving hopper. Placing 50–75% of each load on the ramp in front of the receiving hopper allowed sorting the noncompostable materials. A 6-ft-wide plate conveyor located in the bottom of the hopper moves refuse toward the discharge end at a rate of 2–10 ft/min. Refuse discharged from the receiving hopper travels to the sorting station in the processing building on a 3-ft rubber belt moving at the rate of 60 ft/min. Extension of the skirts for the entire 150-ft length of the belt substantially reduced cleanup time.

At the sorting station two or three laborers remove glass bottles, nonferrous metals, carpeting, rags, and some plastics. The ferrous metals are removed by a magnetic pulley. Rejects, including ferrous metals, amount to approximately 25% of the total refuse received. Rejects from the picking station and the magnetic pulley drop into a 20-yd^3 bin located below the picking station. A 12-yd^3 open-bed truck hauls rejects to the landfill.

From the end of the feed conveyor, refuse can be shuttled by a short cross conveyor into the rasper or the hammermill. The rasper (Dorr-Oliver Company in Holland) consists of a circular bin 18 ft in diameter and 4½ ft high in which materials are moved by radial sweeps, revolving about a vertical axis at 8.3 rpm, and are rasped against protruding cast steel pins on the walls and bottom of the chamber. The rasper is driven by two 40-hp motors connected through a hydraulic coupling. The floor of the bin has plates perforated with 1¼-in. holes. As material becomes fine enough from the rasping action it drops through these holes into a lower chamber. It takes about 30 min for the charge to build up in the rasper so that output equals input.

The 32-in. hammermill (Williams Patent Pulverizer and Crusher Company) is driven at 1800 rpm by a 125-hp motor. Four sets of three 50-lb hammers, mounted to swing freely on horizontal rods, grind the material against a breaker plate and cage bars.

Sludge is pumped from the digestion tanks at the sewage treatment plant some 1300 ft away to a 2600-gal storage tank located beside the Permutit DCG (Dual Cell Gravity) Solids Concentrator on the second floor of the processing building. Partially digested sludge was used initially, but only raw sludge is now used. Moisture content of the sludge as received from the treatment plant varies from 91 to 95%. With the aid of polymers it is further thickened in the DCG machine to a moisture content of 85–88%. Filtrate is returned to the sewage treatment plant.

Sludge is mixed with the ground refuse in a rotary drum mixer 10 ft long and 48 in. in diameter. Ground refuse is hauled in 6-yd^3 dump trucks from the silo and placed in windrows in a 5-acre composting yard. This area was graded to slope in two directions, then surfaced with crushed stone and compacted. Each windrow should be 5–5½ ft. high and 7–8 ft wide at the base.

Windrows are turned by a revolving drum mounted on a power drive chassis which straddles the windrow. The drum is 10 ft long and 30 in. in diameter, and it revolves at 450 rpm. Angular-shaped teeth, 3 in. long, are mounted on the drum and faced toward the center so compost is thrown toward the center to reshape the windrow.

Following the period in windrows the compost is moved to an open, 60 × 200-ft curing shed, the floor of which is surfaced with bituminous concrete. Here the moisture content was expected to drop to approximately 25% in a 2-wk period. The largest amount processed in a single day was over 58 tons.

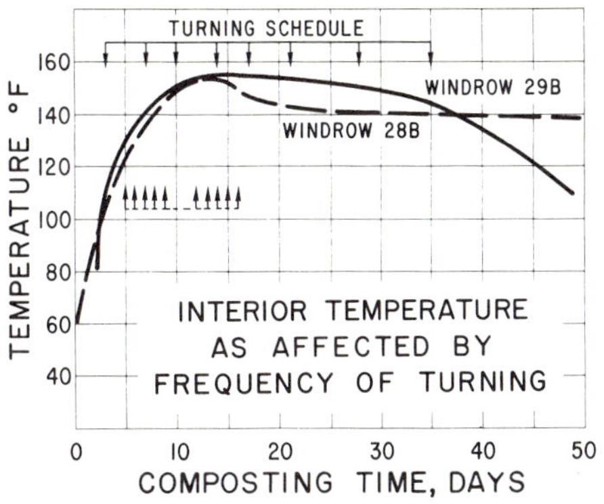

Figure 2. Interior temperature as affected by frequency of turning.

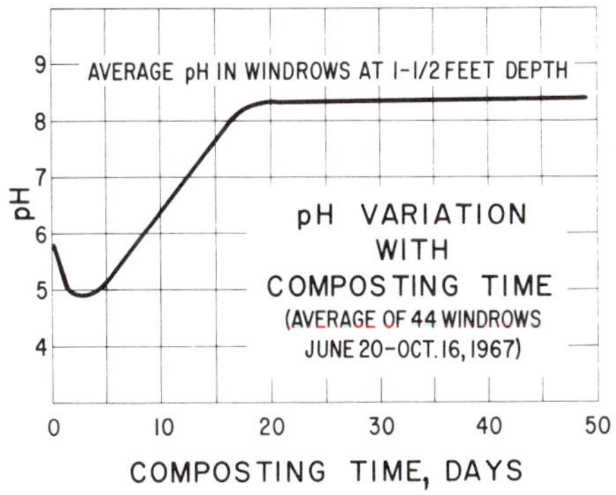

Figure 3. pH variation with composting time.

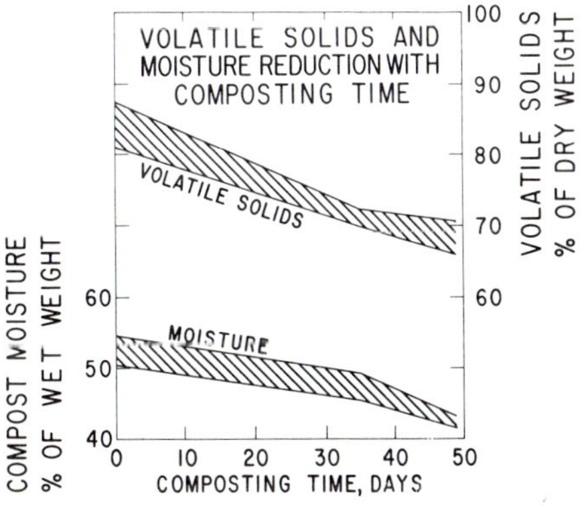

Figure 4. Volatile solids and moisture reduction with composting time.

There are no local markets for salvaged materials such as paper, glass, and metal. Consequently, private collectors have not found this operation attractive and the municipal refuse in Johnson City contains a much higher proportion of paper than might be expected.

The variation of temperature with frequency of turning is given in Figure 2; pH variation with composting time is given in Figure 3, and the variation of volatile solids and time is given in Figure 4.

REFERENCES

1. Samuel A. Hart, "Solid Wastes Management in Germany," Report of the U.S. Solid Wastes Study Team Visit, June–July 1967, PHS #1812, USHEW, PHS, NCUTH, SWP, 1968, 18 pp.
2. Eliassen, "Solid Waste Management," prepared by Ad Hoc Group for Office of Science and Technology, Executive Office of the President, May 1969, USGPO, Washington, D.C.
3. John S. Wiley and O. W. Kochtitzky, "Composting Developments in the U.S.," Region IV Vector Control Conference, Biloxi, Miss., June 3, 1965, 5 pp.
4. John S. Wiley, F. E. Gartrell, and H. Gray Smith, "Concept and Design of the Joint USPHS TVA Composting Project," Johnson City, Tenn., 5th Annual Sanitary and Water Resources Engineering Conference, Nashville, Tenn., June 3, 1966; *Compost Science,* Autumn 1966, 14 pp.
5. "Experimental Composting Research and Development," PHS and TVA Composting Project, Johnson City, Tenn., undated, 2 pp.
6. Samuel A. Hart, "Solid Waste Management, Composting—European Activity and American Potential," PHS 1826, USHEW, PHS, CPEHS, ECA, SWP, 1968, 40 pp.
7. John S. Wiley, "Some Specialized Equipment in European Composting Systems," *Compost Science,* Spring 1963, 4 pp.
8. Maurice Franz, "Municipal Garbage into Mushroom Soil," *Organic Gardening,* Oct. 1972, Emmaus, Pa., pp. 60–63.
9. Richard P. Lonergan and Eugene M. Herson, "Solid Wastes—A Natural Resource," Western Resources Conference Book, 1967: *Man and the Quality of His Environment,* Boulder, Colo., 1968, 15 pp.
10. "Tezuka Refuse Compression System," APWA Research Foundation, Project Do1-U1-00073, USHEW, PHS, CPEHS, SWM, 1969, 41 pp.
11. M. E. Jansen, "Observations of Continental European Solid Waste Management Practices," PHS 1880, USGPO, Washington, D.C., 1969, 46 pp.

12. O. W. Kochtitzky W. K. Seaman, and J. S. Wiley, "Municipal Composting Research at Johnson City, Tenn.," *Compost Science,* **9** (4) pp. 5–16, Winter, 1969.

13. Gainesville Compost Plant, Gainesville Municipal Waste Conversion Authority, Cincinnati, 1969, USHEW Grant Do1-U1-00030, 345 pp.

CHAPTER II.11

optimum conditions for composting

JOHN S. JERIS and RAYMOND REGAN

The results reported here are based on four years of laboratory scale composting studies using municipal refuse, paper, and a synthetic refuse as the solid waste. The main difference between this laboratory study and other small-scale studies was the degree of laboratory control of many environmental factors affecting the composting rate.

EXPERIMENTAL

Three major experimental systems were used: (1) three-bench scale composters, (2) a shake flask apparatus, and (3) a Warburg respirometer. The composters were fed continuously while batch studies were performed with the other equipment. All equipment operation allowed for close control of desired environmental parameters and was sufficiently flexible to operate over wide ranges of conditions.

A schematic drawing of the experimental bench scale composter is shown in Figure 1. Each of the three composters used was a horizontal stainless steel cylinder, 12 in. in diameter, 2 ft long with a total volume of 1.5 ft^3

Each composter was loaded with a total of 1000–3000 g, which filled about one-half the cylinder volume. Continuous heat was supplied by an electric heating rod suspended in the air space directly over the composting mass. The temperature was maintained constant by controlling the power to the heating rod with a rheostat. In this manner, the control of the temperature was not dependent on microbial activity. The temperature could be adjusted to any desired level and used as a controlled variable for studying its effect on composting efficiency. Mixing and aeration of the material was provided by intermittent rotation of the cylinder.

Air was supplied to each unit through a rotameter at a rate sufficient

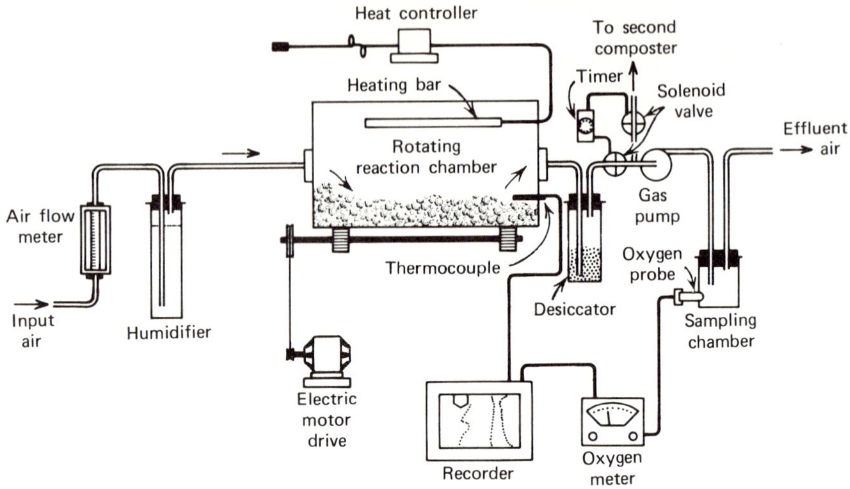

Figure 1. Laboratory composting system.

to maintain an oxygen residual of at least 10% in the effluent gas. The residual oxygen of each unit was measured automatically every 4 hr by an oxygen probe placed in the sampling chamber and monitored by a multipoint strip chart recorder. Daily samples of gas were also taken and analyzed by gas chromatography to determine the carbon dioxide content of the effluent.

The shake flask apparatus was composed of four incubators each containing eight shake flasks, which were mounted on the arms of two Burrell wrist-action shakers. Each incubator was 25 × 20 × 16 in. constructed of masonite with 2 in. of rock wool insulation on five sides. The sixth side was an observation door. A heating rod mounted in the incubator provided heat, which was regulated by a rheostat.

The flasks used were 500-ml Erlenmeyers, sealed with a rubber stopper containing a glass tube which served as a gas sampling port. Approximately every 3–4 hr during a run, a gas sample for carbon dioxide analysis was taken from each flask and the flask was completely aerated. The Burrell shaker provided a tumbling action for a 2-min period every 2 hr. A maximum of 32 flasks, at four different temperatures, could be set up at one time, ensuring coverage of a wide range of conditions.

While using paper as a substrate, the shake flask apparatus provided satisfactory results of the carbon dioxide production; however, when the municipal refuse study was initiated, the high rates of gas production and oxygen depletion necessitated too frequent handling of the shake flasks, and the Warburg Respirometer was used. The Warburg apparatus in effect was a small-scale shake flask apparatus provided with eighteen 125-ml flasks but having manometers for superior gaseous measurement and for simpler operation. Potassium hydroxide was placed in the center well of the Warburg flasks for carbon dioxide absorption; thus the manometer readings indicated the oxygen consumption.

The three composters provided an environment for continuous composting under a given set of conditions. This composting mass was used as seed for further study in the shake flask or Warburg apparatus. As these apparatuses had many flasks, a wide range of environmental conditions could be scanned quickly and effectively to locate optimum conditions. The results were used to adjust the operating conditions of the lab scale composters.

Solid Waste Substrate

The soild waste feed included newsprint and typical municipal and synthetically prepared mixed refuse containing 60–70% paper. The cellulose content was significant in each of these substrates. In Table 1 the typical characteristics of the solid wastes used are given.

Table 1 *Solid Waste Characteristics*

	Newsprint	Municipal Refuse	Synthetic Municipal Refuse
Lipids (%)	0.5	3.0–9.0	9.7
Crude fiber (%)	72.0	41–61	58.0
Ash (%)	0.5	4–20	11.0
Volatile matter (%)	99.5	96–80	89.0
Crude fiber/ash	144	2–15	8.1
Bulkweight (lb/ft^3 @ 65% moisture)	5	10–15	9
Moisture (%)	5.0	30–60	47
Average total weight in composters (g wet)	1000–1400	3000	2000–3000
pH	4.2	5–8	7.0
Ammonia (mg/g) N as NH_3	0	0.5–1.4	0.7
Ammonia (mg/g) added	2.0	2.0	2.0
Organic nitrogen (mg/g)	—	4.8–14	24.3
Total nitrogen (mg/g)	—	7–17	27.0
Protein (0.625 × org. N) (%)	—	3.1–9.3	16.5
Phosphorus added (mgP/g)	1.3	1.3	1.3
Potassium added (mgK/g)	1.3	1.3	1.3

Two methods were employed for obtaining mixed refuse. Initially, batches of urban refuse were collected bi-weekly and stored in metal containers until needed. This method was inadequate because the chemical characteristics of the refuse thus obtained varied from batch to batch. Furthermore, the refuse deteriorated during storage, making comparison of the experimental results difficult. To improve the degree of control, a synthetic refuse, with characteristics similar to municipal solid waste was prepared in the laboratory. It consisted of wastes from the college cafeteria, supermarket wastes, and newsprint. Typically meat and fish scraps, lettuce and other vegetable trimmings were used.

The solid waste was prepared by cutting and grinding the waste with a garden shredder, to a particle size no greater than ¼. Since the waste materials contained insufficient nutrients, nitrogen in the form of ammonium chloride was added. The wastes were also supplemented with phosphorus. The moisture content was adjusted to levels suitable for composting with tap water. It was assumed that the tap water supplied necessary trace elements.

Results

The objective of the investigation was to obtain optimum stabilization in the laboratory using a high degree of experimental control. The environmental conditions recommended for a maximum composting rate of solid wastes are summarized in Table 2.

Table 2 *Environmental Conditions for Optimum Composting*

Substrate	Parameter Controlled	Optimum Value
Newspring	Temperature	48°C
Mixed refuse	Temperature	59 ± 5°C
	Moisture	67 ± 3%
	Free air space	32 ± 3%
	Seed recycle	50–75%
	Mass turnover time	4 to 5 days
	pH	8.0 ± 0.5
	Nitrogen added (as NH_3–N)	2.0 mg/g dm[a]
	Phosphorus added (as P)	1.34 mg/g dm
	Air requirements	16.7–20.3 ft_2/day-lb VM[b]
	Air residual	10%

[a] dm, dry mass.
[b] VM, volatile matter.

The average reductions of the components in the refuse after composting are summarized in Table 3. The results represent 24 runs, the data of which were obtained after 10–30 days of continuous operation under the desired conditions. The mass turnover time was normally 4–5 days.

Temperature

The laboratory experiments indicated that temperature has a significant effect on the composting rate of solid wastes containing large amounts of paper. For freshly collected municipal refuse, containing 60–70% paper, an optimum composting rate of 3.7–4.5 (mM) CO_2 and O_2 per day-gram volatile matter was measured at 59° ± 5°C. The composting rate of paper was less than 10% of the rate obtained for mixed refuse.

In Figure 2 a plot of temperature versus the rate of carbon dioxide production is given for mixed refuse (municipal and synthetic), newsprint, and composted refuse; results obtained by Schulze are included for comparison.[1] The results of newsprint and composted refuse obtained from the Altoona, Pennsylvania, composting plant are approximately the same, indicating high paper content and the slow biodegradability of paper. Schulze's results were obtained under nonequilibrium conditions using a waste high in readily degradable solids with less than 10% paper.

Moisture Content and Free Air Space

The optimum moisture content and free air space for composting typical mixed refuse were 67 and 30%, respectively. Calculations

Table 3 *Average Reduction of Mixed Refuse by Composting*

Refuse Component	Reduction (%)
Total matter	40
Volatile matter	45
"Other" volatile matter (carbohydrates)	65
Lipids	86
Crude fiber	30
Protein	22
Bulk volume[a]	42–50

[a] Mass adjusted to 65% moisture.

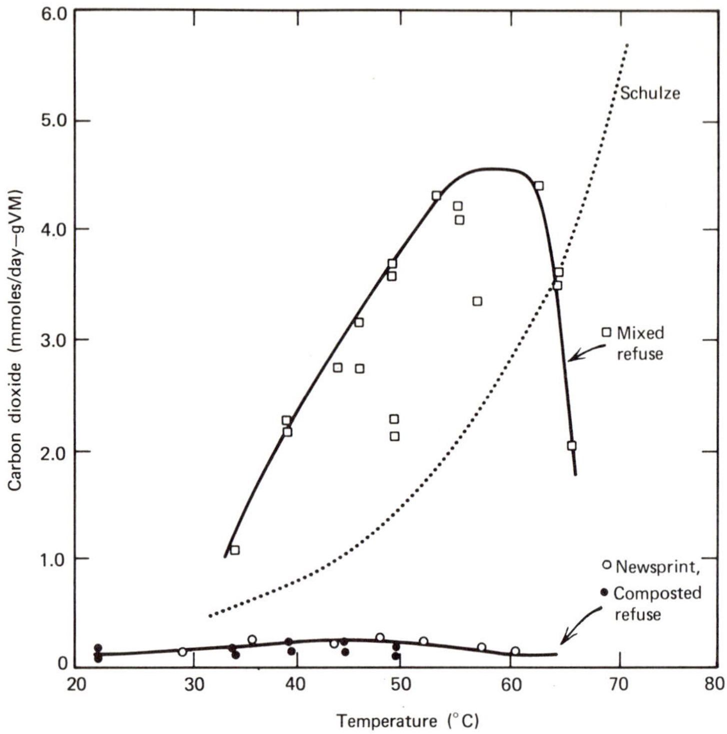

Figure 2. Carbon dioxide production rate versus temperature for cellulose derivatives.

performed using information in the literature indicate optimum free air space at 29–35% for many other organic wastes having significantly different optimum moisture requirements. The free air space concept suggests a more fundamental and universally applicable parameter for composting not only municipal refuse, but many other types of organic solid wastes. The "free air space" concept has been adapted from soil science and relates the specific gravity (sp gr) bulk weight (wB) and dry mass (dm = 1-moisture fraction) of a waste to the air-filled voids within the material. The working equation to determine the percent free air space for any given waste is

$$FAS = 100\,(1\text{-}[wB \times dm/sp\ gr]) \times dm \qquad (2)$$

This equation shows that as paper content increases, the most suitable moisture content for composting an organic waste increases. For example, the moisture requirement for composting increased from 53% for beef feedlot wastes to 65% for wastes containing significant quantities of paper. The corresponding free air space remained constant at 32–36%.

Figure 3 presents the data of three runs using a mixed refuse, they indicate an optimum of about 30% free air space. Runs 1, 2, and 3 represent data collected using a fresh refuse, 10-day-old refuse, and partially stabilized refuse, respectively.

Seed Recycle

The need or value of recycling seed in composting operations has long been open to debate.[3] Most composting studies have indicated that any type of "starter" material is unnecessary.[4]

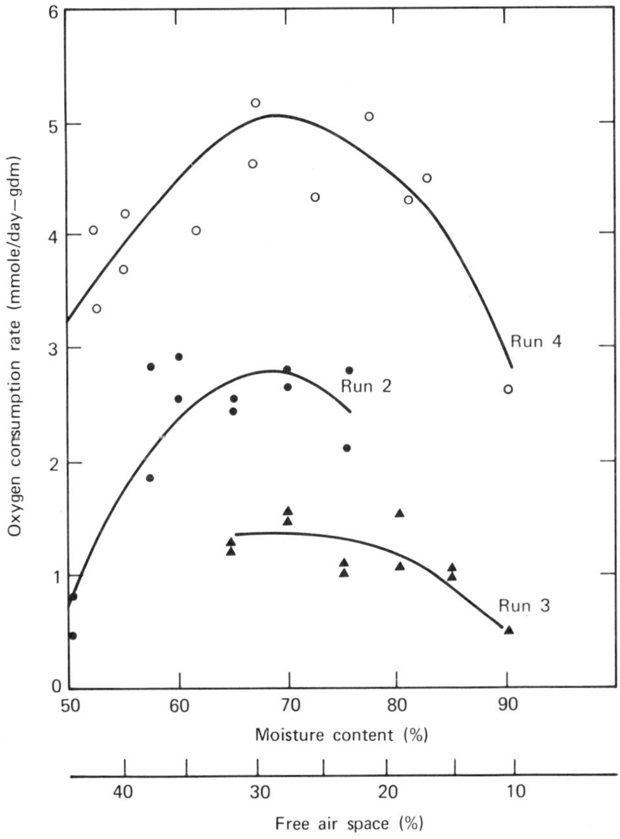

Figure 3. Oxygen consumption rate versus moisture and free-air space.

The amount of seed recycle required for optimum composting in the closely controlled laboratory experiments was found to range between 50 and 70% and was affected by the following practical considerations.

Temperature

Microorganisms indigenous to refuse are stimulated by warmer temperatures. During the summer months, seed recycle may be decreased because the incoming refuse is more biologically active. During the winter, seed recycle may be increased to compensate for the lower biological activity of the incoming refuse.

Storage Time

The development of microbial growth in refuse begins soon after the waste is placed in the sanitary container. Microbial activity may be significant when the refuse is brought to the composting plant. Depending on the sanitation pickup schedule, refuse may be stored in a trash can for 2, 3, or more days, thus affecting the biological activity.

Shredding

When the refuse is shredded, the composting rate is increased as the surface area for microbial growth is greater, provided that aeration is not restricted. Since the seed material is more biologically active, less recycle would be required if the incoming refuse were well shredded. Shredding also provides mixing and better contact and distribution of the food and microorganisms.

Recycling under thermophilic conditions with accompanying decreased treatment times may result in reducing the total cost of composting municipal refuse. However, information is required on a pilot plant scale to ascertain the benefits of recycling.

pH

Most species of microorganisms, including bacteria and fungi, have specific pH preferences, although less desirable conditions may be tolerated. The controlled bench scale composter studies indicated maximum thermophilic composting rates were obtained at pH 7.5–8.5. Stabilization efficiency was significantly decreased at lower or higher pH levels. Generally the pH of the refuse was below 7, causing a decrease in pH of the composting mass after mixing with recycle seed. However, the pH returned to 7.8–8.2 within 4 hr. The highest composting rate, as measured by oxygen consumption rate, occurred 6–10 hr later at alkaline pH levels. In Figure 4 the effect of pH on composting rates is given based on data obtained from the bench scale composters.

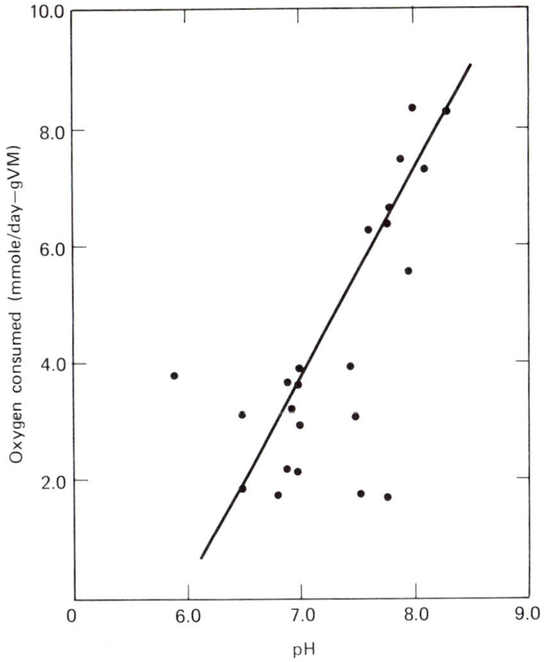

Figure 4. Effect of pH on the oxygen consumption rate bench scale composter.

Paper Content

The composting rate of newsprint as measured by oxygen consumption was 0.2–0.3 mmole/day-gVM, while for municipal refuse containing 60–70% paper the rate was 4 to 5, or an order of magnitude greater. As typical municipal refuse of American cities contains large amounts of paper and other cellulistic materials, its affect on composting must be considered. The laboratory results indicated a significant decrease in composting rate when the amount of paper was greater than 75–80%, as shown in Figure 5.

It was also quite evident from these experiments that the major component remaining in the final composted mass was paper. The readily degradable fraction of the solid refuse was consumed leaving the paper, and it does not appear possible to degrade paper significantly in 4 days or less.[5] Since the composting in effect concentrates the paper fraction, the composted mass could serve as a prime material for reclaiming paper, fertilizer filler, clean landfill, and building materials as well as other uses.

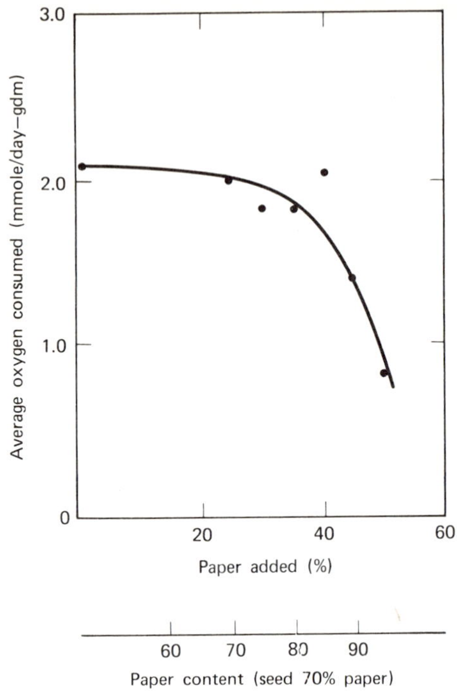

Figure 5. Effect of paper content on oxygen consumption rate.

REFERENCES

1. K. L. Schulze, "Aerobic Decomposition of Organic Waste Materials," Final Report Project RG-4180 (C5R1) National Institute of Health, 1961.
2. J. S. Wiley, "High Rate Composting," *Trans. ASCE,* No. 2895 (1957).
3. C. Goloeke, O. Card, and P. McGauhey, "A Critical Evaluation of Inoculums in Composting," *Appl. Microbiol.,* **2,** (1954).
4. H. Gotaas, "Composting Sanitary Disposal and Reclamation of Organic Wastes," WHO, Geneva, 1956.
5. J. S. Jeris et al., Final Report, "Cellulose Degradation in Composting," USPHS, SWP, Research Grant EC00161, Manhattan College Bronx, New York, January 1971.

CHAPTER II.12
pyrolysis

From the engineering viewpoint, pyrolysis has been widely studied in the form of low- and high-temperature coal carbonization, preparation of adsorbent carbon on a tonnage scale from lignite, rice hulls, nut shells, and related organic matter; removal of organic insulation from wire in secondary metal plants; and partial distillation of rubber from waste tires. In these cases the equipment engineering has been developed to a high order, tonnage plants created and operated, and products of the operation disposed of at a profit.

Trash and waste are seldom disposed of at a profitably, and therefore pyrolysis for this purpose commands few financial resources, although much publicity.

Pyrolysis is destructive distillation, often with carbonization, in the absence of air by the application of indirect heat; often, the term thermal decomposition is applied. Pyrolysis can convert trash, cellulose, wastes after drying, and contaminated paper into usable fuels, or it may be employed as modified incineration to produce ash, slag, and ferrous metals from solid or, in chemical plants, moist and liquid residues. Gases generated may be useful as fuel or they may be consumed, with a gas holder added to the system, as process energy.

The Bureau of Mines' Pittsburgh Energy Research Center has been studying pyrolysis in one form or another for years. Monsanto's Enviro-Chem Systems, Inc. has developed and field tested a 35-ton/day pilot pyrolysis plant and guarantees that systems capable of pyrolyzing up to 1500 tons/day of municipal refuse will meet all existing pollution control standards at the time the contract is signed.

Work on pyrolysis at BuMines started nearly half a century ago with basic studies on the carbonization of coal. BuMines' pyrolysis unit, built in 1929, was designed to be a research tool, and features a recovery train for trapping pyrolysis products for analysis. With minor alterations, the unit remains in operation today.

Among the first waste material to be studied at BuMines was rubber, in the form of shredded tires. The products recovered were gas with a high Btu content, oil, and a soild residue with a high heating value. Each of the products could be burned cleanly. From a ton of tires with smoky heating value of 32.5 million Btu, about 95% of the heat could be recovered as products.

The BuMines team explored a variety of waste products, including plastic battery cases and shoe soles. Municipal refuse seemed a likely next choice.

For pilot studies, BuMines collected samples of municipal trash from Altoona, Pennslyvania. Altoona was chosen because glass and cans must be separated from garbage by its citizens in compliance with a local ordinance. Furthermore, the refuse is taken to a local mulch plant which shreds the trash, so it is easy to obtain reasonably representative samples.

Pyrolysis of the refuse yielded a solid residue and gas and oil fractions, each with significant thermal value. Water in the feedstock and any water produced in pyrolysis were recovered as an aqueous fraction containing pyroligneous products—aldehydes, alcohols, acids, ketones, and so on. Product distribution depended on water content of the refuse and pyrolysis temperature.

The heating value of garbage is less than that of tires, about 10 million Btu/ton. Since it takes 1–2 million Btu in electrical energy to pyrolyze a ton of refuse, the net energy yield will fall between 6 and 8 million Btu/ton of municipal refuse or some 9–13 billion Btu/yr nationwide.

The BuMines unit consists of a furnace heated with nickel-chromium resistors and a recovery train to trap products. Shredded refuse is sealed inside a retort to charge the furnace and heated to the desired

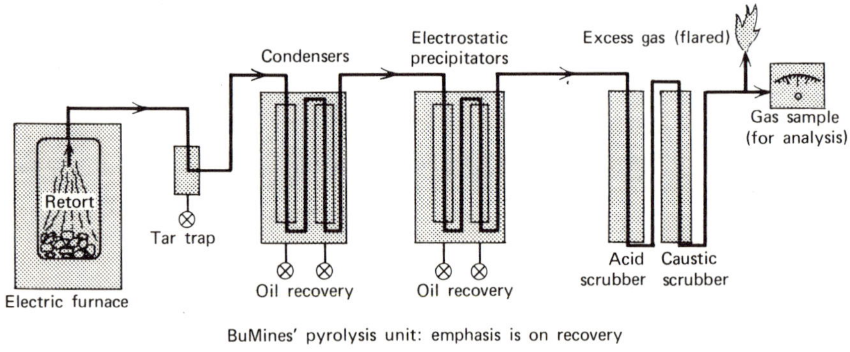

Figure 1. BuMines pyrolosis installation.

Table 1 *Yields from 1 Ton of Municipal Waste*

Char residue	150–230 lb
Tar and pitch	0.5–5 gal
Light oil	1.2–2 gal
Gas	$11\text{–}17 \times 10^3$ ft^3
Aqueous liquor	80–133 gal
Ammonium sulfate	18–25 lb
Thermal values	
Char	8–13,000 Btu/lb
Oil	15,000 Btu/gal
Gas	500 Btu/ft^3

temperature. Gases given off first pass through an air-cooled trap (Figure 1) where tars and heavy oils condense. Uncondensed vapors pass through a series of water-cooled columns where additional oils and aqueous liquors are removed. The gas stream passes through electrostatic precipitators to remove oil mist and from there to acid and caustic scrubbers. The scrubbed gases are passed through meters which direct 99% of the gas to flaring and 1% to analytical instruments. Table 1 shows yields.

For 500, 1500, and 2500-ton/day plants, disposal costs are estimated to be \$6.00, \$3.00, and \$2.00 per ton, respectively.

The Monsanto Landgard system (Figure 2) is designed to be a totally

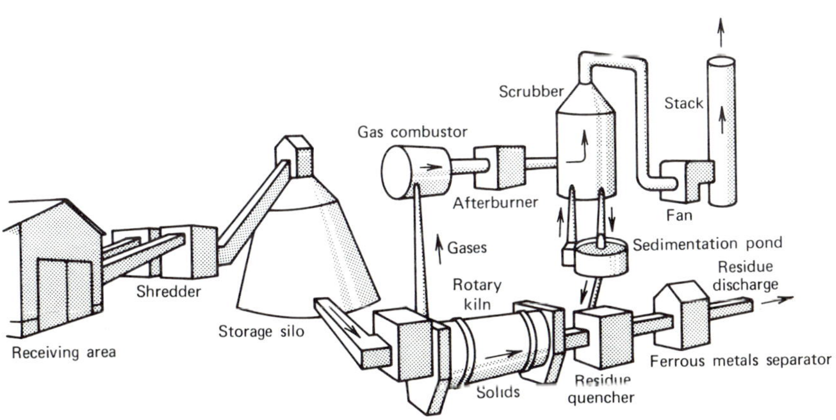

Figure 2. Enviro-Chems Landgard System.

Table 2 *Pyrolysis*

	H_2	Gas
Municipal waste	1.87 scf	9.3 scf
Oil shale	1.54 scf	6.0 scf
Lignite	1.54 scf	9.9 scf
Transport cost	2¢/1000 ft³/100 miles	
Per capita daily waste	5.72 lb; 1.5 lb C	
Cost of plant	$100 × 10^6/200 × 10^6 scfd/ of gas	
Disposal cost	25¢/ton waste, gas at 40¢/1000 ft³	

enclosed, self-contained operation. Refuse is trucked to an enclosed receiving area (kept under slight negative pressure to control dust and odors) where it is dumped into a pit. Enclosed conveyors carry refuse to shredders where rubbish is reduced to a mixture with particle size about 3–4 in. in diameter. A second conveyor system removes refuse from the shredder and dumps it into a storage silo.

From the silo, refuse is conveyed to a rotary kiln. The kiln is lined with refractory materials and there are no grates or moving parts. The kiln is inclined from exit to entrance so charred material falls by gravity into a water quenching unit. The residue is water-cooled before passing to a magnetic separator that removes ferrous metals. The wet residue, reduced in volume by 94%, is then loaded into trucks and taken to landfill. Cooling water is diverted to sedimentation basins for recirculation.

Gases driven off from the kiln are passed through a combustion chamber where hydrocarbons are oxidized. Gases are passed through an adiabatic spray scrubber and then released. Steam plumes are eliminated by heating the stack gases. Total operating costs are estimated to be $7.50 to $8.50 per ton.

There is a cyclical pattern by which contamination builds up as material is recycled and recycled again. Recycling uses must be carefully selected to ensure necessary quality and to fit the economic picture.

Pyrolysis for trash treatment has been proposed along with hydrogenation to produce pipline gas.

Municipal waste theoretically could yield 9.3 standard cubic feet (scf) of methane compared to 6 scf for oil shale (+1.54 scf H_2) (+1.54 scf H_2) (Table 2).

CHAPTER II.13

incineration

LEO J. COHAN and J. H. FERNANDES

MUNICIPAL WASTE GENERATION

To obtain data regarding solid waste generated in municipalities in the United States, a survey of six cities was made for the U.S. Government by Combustion Engineering, Inc.[24] Cities were classified by population into two groups, (1) 25,000–50,000 persons and (2) 50, 000 persons or more.

The survey developed data on (1) past and present per capita refuse production, (2) quantity and characteristics of municipal and industrial waste handled by the municipality, (3) trends in vacant land, and (4) amount of waste disposed in sanitary landfill, open dump, or other methods. In urban areas there are four basic sources of solid waste: residential, commercial, institutional, and industrial.

Although there is a small amount of self-disposal of residential waste, it is not considered significant since it is well within the norm for standard deviations based on geographical and socioeconomic conditions. A statistical analysis of the waste generated in urban areas appears in Table 1.

To arrive at a figure that would indicate the total solid waste generated for the United States, the amount of commercial waste, 1.4 lb/cap/day, would be reduced because certain nonurban areas do not generate much commercial waste. It is suggested that this can be weighted by multiplying the commercial waste rate by the percentage of urban population of 70%. The weighted national figure would then be revised as shown in Table 2.

Figure 1 indicates the distribution of the disposal facilities by municipalities over 25,000 population.

INDUSTRIAL WASTE GENERATION

Very few data have been published on the quantities of industrial waste streams, particularly with regard to liquid and semisolid wastes

Table 1 *Solid Waste Generated in Urban Areas*[24]

	lb/cap/day
Residential	2.4
Commercial	1.4
Bulky waste	0.3
Subtotal	4.1
Industrial	3.2
Total	7.3

(sludges and colloids). Industrial waste generation and disposal is frequently masked in municipal inventories and incineration and landfill practices. In investigating industrial waste inventories by means of surveys and interviews with personnel in 320 plants in 24 selected industries, attempts were made to establish the amount of waste generated for disposal by each of the industries, its method of disposal, and an estimate of the quantities of waste in 1975. Table 3 indicates some of the data collected during the 320 interviews. The "Waste for Disposal" column refers to a waste-to-employee ratio expressed in thousands of pounds per year per employee. By means of change ratios (the ratio of physical production for 1965), the 1975 wastes for ultimate disposal, adjusted where possible by recognizable trends expected in this decade, are predicted.

The waste generated in the 24 basic industrial groups is 350,500 million lb/yr, of which sawmills account for approximately 55%. Because of the high utilization, only 181,500 million lb/yr are disposed of,

Table 2 *Solid Waste Generated in United States*[24]

	lb/cap/day
Residential	2.4
Commercial	1.0
Bulky waste	0.3
Subtotal	3.7
Industrial	3.2
Total	6.9

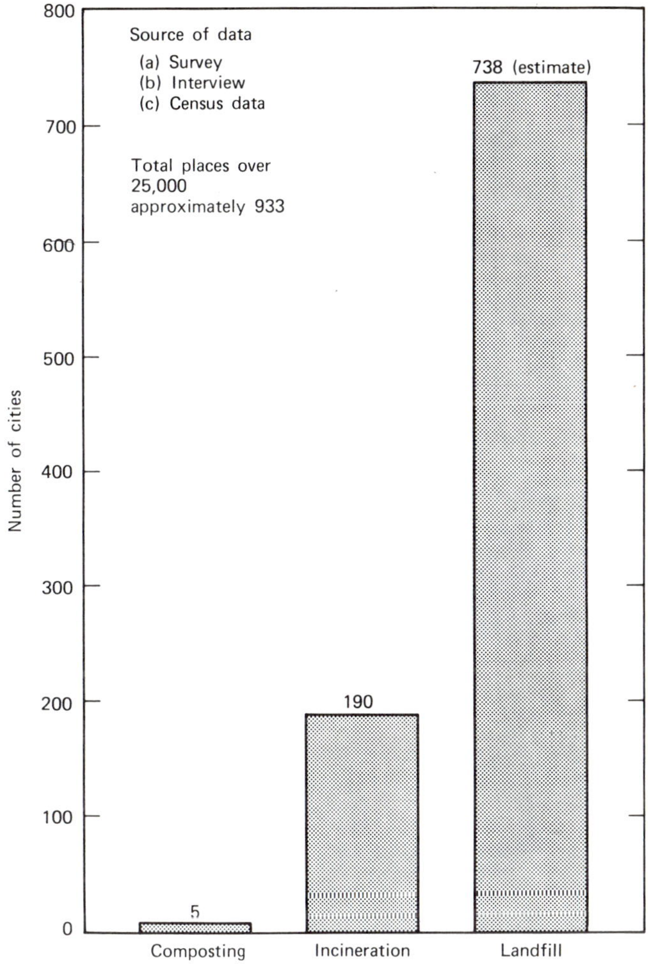

Figure 1. Types of solid waste disposal used by municipalities with over 25, 000 population.

but nearly 65% of this is accounted for by industry exclusive of sawmills. This is based on an employee population of 10,146,000. The nonprojected codes within these industries represent an employee population of 5,900,000, and waste generated by this segment of the industrial population was estimated and added to the previous figure giving a total industrial waste for the 24 codes of 229,229 million lb/yr.

The dispersion (disposal results) is known with somewhat less ac-

Table 3 *Summary of Waste Ratios, Projections, and Predictions (Adapted from Ref. 24)*

Industry	Number of Interviews	Known Scrap and Waste 1965 (million lb/yr)	Quantity, Waste For Disposal (million lb/yr) 1965 Estimated	1975 Projected	Total Employees Projected on	Waste For Disposal (klb/yr/employee)
Manufacturing						
Food	14	11,500	10,584	14,076	982,220	10.8
Meats	19	2,963	1,650	2,400	225,000	7.3
Textile mill products	10	2,629	1,706	2,132	677,600	2.52
Apparel and related products	9	981	696	1,037	1,238,354	0.562
Wooden containers	15	4,996	2,470	2,190	29,737	85
Sawmills	30	192,600	65,600	23,000[a]	236,910	277
Mill work	9	1,072	570	890	65,919	8.65
Wooden furniture	16	4,197	3,090	5,170	268,735	11.5
Paper	29	11,100	9,950	14,700	570,000	17.5
Printing and publishing	10	7,243	2,318	3,222	140,900	16.5
Chemicals	13	2,625	2,625	4,900	439,000	13.07
Paints	8	324	324	390	62,000	5.3
Asphalt roofing	8	1,180	1,148	1,538	14,300	80.5
Rubber	13	2,900	2,900	3,900	247,000	11.9
Tanning	7	608	598	670	30,800	19.4
Glass	7	3,200	2,680	3,936	118,000	98.49
Metal cans	3	973	183	258	53,745	3.40
Fabricated metal products and machine except electrical	18	12,337	6,020	8,758	2,056,286	2.93
Electrical machinery	20	3,651	2,760	4,968	1,327,581	2.08
Auto and aircraft	9	5,786	2,910	3,660	1,361,144	2.14

Total 20 manufacturing code groups	267	272,865	120,782	101,795	10,146,132	11.9
Total 19 manufacturing code groups (except sawmills)	237	80,265	55,182	78,795	9,909,222	5.56
Nonmanufacturing						
Cotton ginning	13	3,836	1,572[b]	1,665	>25,000	<63
Demolition	16	38,100	38,100	44,300	8,449	4510
Stockyards (including auction)	11	1,919	779	779[c]	>3,000	<260
Super markets	13	33,800	20,310	26,400	568,000	35.7
Total nonmanufacturing	53	77,655	60,761	73,144		16.9
Total 24 codes manufacturing and nonmanufacturing		350,520	181,543	174,939	10,750,581	11.0
Total 23 codes manufacturing and nonmanufacturing (except sawmills)		157,920	115,943	151,939	10,513,671	

[a] Assuming all sawdust and shavings sold in 1975.
[b] 1962 to 1963.
[c] Assumed same as 1965.

curacy than the waste quantity figures, but within the 24-code study 48% of the waste was utilized in some manner. Excluding sawmills, only 27% is utilized in the remaining industry.

Fifty percent of the industrial waste surveyed was disposed of in private facilities, with 36% going to municipal facilities. Excluding sawmills, 22% goes to private and 56% to municipal facilities. The disposal method distribution for the total surveyed is 40% in incinerator for open burning and 40% to dumps. Excluding sawmills, 57% goes to dumps.

This industrial inventory has shown that a large percentage of industrial waste is of uniform character (exclusive of liquid and semisolids) consisting of shipping wastes, plant trash, and office waste. Two industries, food and chemical, are characterized by process wastes peculiar to the process involved. Many industries are presently disposing of quantities of waste to streams and sewers. This waste will have to be collected, separated, and treated for disposal by more acceptable means in the future.

WASTE CHARACTERISTICS

A systematic classification for types of refuse has been developed by The Incinerator Institute of America (IIA). This has received a large measure of acceptance.

MUNICIPAL WASTES

Municipal wastes have general applicability but are not specific for direct combustion calculations. The combustion of refuse follows the same physical laws as do other fossil fuels and therefore the same qualitative characteristics must be known. Trends in food processing and packing have affected the moisture characteristics of refuse. Garbage—the vegetable and animal food wastes from the preparation, cooking and serving of food, and market wastes—has been greatly reduced. At one time it accounted for 65% or more of the volume of municipal refuse with an overall moisture content of 50% This figure in certain areas is seasonal because of a small introduction of coal ash during the winter months. Today, garbage in municipal waste may average 10% or less, with a moisture content of 15–25% (see Table 4).

Physically, mixed municipal refuse contains large quantities of paper bags, crates, and similar dry combustible materials. Densities vary from 200 to 500 lb/yd^3. Wide variations result from the heterogeneous characteristics, geographical locations, and methods of

Table 4 Refuse Heading Value

	Moisture	BTU as fired
Domestic Refuse		
Paper, cardboard, cartons, bags	3	7,660
Wood crates, boxes, scrap	7	7,825
Brush, branches	17	7,140
Leaves	30	4,900
Grass	50	3,820
Garbage	75	1,820
Greenstuff	50	3,470
Greens	50	4,070
Rags, cotton, linen	10	6,440
Industrial Scrap Refuse		
Boot, shoe trim and scrap		8,500
Sponge waffle and scrap		8,500
Butyl soles scrap		11,500
Cement wet scrap		11,500
Rubber		12,420
Tire cord scrap		12,400
Rubber scorched scrap		19,700
Tires, bus and auto		18,000
Gum scrap		19,700
Latex coagulum		19,700
Latex waste, coagulum waste		12,000
Leather scrap		10,000
Waxed paper		12,000
Cork scrap		12,400
Paraffin		16,803
Oil waste, fuel oil residue		18,000
Plastic and Synthetic Refuse		
Cellophane plastic		12,000
Polyethylene		19,840
Polyvinyl chloride		17,500
Vinyl scrap		17,500
Aldehyde sludge		18,150
Solvent naptha		18,500
Carbon disulfite		8,000
Benzine		10,000
Miscellaneous		
Carbon to CO_2		14,093
Carbon to CO		4,347
Sulfur		3,983
Methane		23,879

collection (compaction). While highly variable in appearance and density, the chemical analysis is quite uniform, consistent with cellulose by-products. Major variables are moisture, ash, and other solid inerts (see Table 3).

AVERAGE REFUSE COMPOSITION

Based on the works of Kaiser,[23] and EPA contracts,[25] sufficiency analysis for combustible calculations are provided in Tables 5–9. Average 1968 refuse as computed from the above by Arthur D. Little may be summarized in Table 10.

INDUSTRIAL WASTE

Industry, because of profit motivation, is constantly seeking ways to recycle or reprocess industrial waste streams. Nevertheless, there is 1 constant stream of waste that must be disposed. While most industrial waste, on a statistical basis, is cellulosic material, there are many waste streams unique to specific processes or products. The proprietary nature of many processes has hampered the collection of specific data relative to the physical and chemical properties. Tables 11 and 12 compiled by Heschles give some insight into many of these materials. [Note: With one exception (Table 8), all is solid waste.]

Liquid industrial wastes vary considerably in viscosity, moisture, heat content, and flash and fire points. Some of those frequently incinerated and/or used as fuel include solvents (alcohols, ketones, light aromatics, etc.) waste oils (mineral and cutting), oil sludges, oil–water emulsions, monomers, polymers, resins, chlorinated hydrocarbons, phenols, creosols, tars, combustible chemicals, greases, and fats. Many of these are high in moisture and will not support combustion without blending or support fuel. Many of these liquid wastes can be neutralized and precipitated to burnable sludges for disposal.

COMBUSTION CALCULATIONS[6]

Combustion of refuse involves the usual parameters but in a more complex manner because of the heterogeneous nature of the fuel. The most difficult factors to be accommodated are the amounts of moisture and noncombustibles in the refuse. The average of extended measurements is quite consistent, but from instant to instant, the heat content in the refuse can vary substantially. Therefore, a certain amount of premixing is required to ensure that a minimum of moisture and

Table 5 *Proximate Analyses in Percent by Weight, as Discarded (A.D.) by Householders (Adapted from Ref. 24)*

Refuse Component	Moisture	Volatile Matter	Fixed Carbon	Ash	Btu/lb A.D.	Btu/lb Dry Basis
1. Newspaper	5.97	81.12	11.48	1.43	7974	8480
2. Brown paper	5.83	83.92	9.24	1.01	7256	7706
3. Trade magazine	4.11	66.39	7.03	22.47	5254	5480
4. Corrug. paper boxes	5.20	77.47	12.27	5.06	7043	7429
5. Plastic coated paper	4.71	84.20	8.45	2.64	7341	7703
6. Waxed milk cartons	3.45	90.92	4.46	1.17	11327	11732
7. Paper food cartons	6.11	75.59	11.80	6.50	7258	7730
8. Junk mail	4.56	73.32	9.03	13.09	6088	6378
9. Veget. food wastes	78.29	17.10	3.55	1.06	1795	8270
10. Citrus rinds and seeds	78.70	16.55	4.01	0.74	1707	8015
11. Meat scraps, cooked	38.74	56.34	1.81	3.11	7623	12443
12. Fried fats	0.00	97.64	2.36	0.00	16466	16466
13. Leather shoe	7.46	57.12	14.26	21.16	7243	7826
14. Heel and sole composition	1.15	67.03	2.08	29.74	10899	11026
15. Vacuum cleaner catch	5.47	55.68	8.51	30.34	6386	6756
16. Evergreen shrub cuttings	69.00	25.18	5.01	0.81	2708	8735
17. Balsam spruce	74.35	20.70	4.13	0.82	2447	9541
18. Flower garden plants	53.94	35.64	8.08	2.34	3697	8027
19. Lawn grass	75.24	18.64	4.50	1.62	2058	8312
20. Ripe tree leaves	9.97	66.92	19.29	3.82	7984	8869

Table 6 *Ultimate Analysis, Dry Basis in Percent by Weight (Adapted from Ref. 24)*

Refuse Component	Carbon	Hydrogen	Oxygen	Nitrogen	Sulfur	Ash
1. Newspaper	49.14	6.10	43.03	0.05	0.16	1.52
2. Brown paper	44.90	6.08	47.34	0.00	0.11	1.07
3. Trade magazine	32.91	4.95	38.55	0.07	0.09	23.43
4. Corrug. paper boxes	43.73	5.70	44.93	0.09	0.21	5.34
5. Plastic coated paper	45.30	6.17	45.50	0.18	0.08	2.77
6. Waxed milk cartons	59.18	9.25	30.13	0.12	0.10	1.22
7. Paper food cartons	44.74	6.10	41.92	0.15	0.16	6.93
8. Junk mail	37.87	5.41	42.74	0.17	0.09	13.72
9. Veget. food wastes	49.06	6.62	37.55	1.68	0.20	4.89
10. Citrus rinds and seeds	47.96	5.68	41.67	1.11	0.12	3.46
11. Meat scraps, cooked	59.59	9.47	24.65	1.02	0.19	5.08
12. Fried fats	73.14	11.54	14.82	0.43	0.07	0.00
13. Leather shoe	42.01	5.32	22.83	5.98	1.00	22.86
14. Rubber composition heel and sole	53.22	7.09	7.76	0.50	1.34	30.09
15. Vacuum cleaner catch	35.69	4.73	20.08	6.26	1.15	32.09
16. Evergreen trimmings	48.51	6.54	40.44	1.71	0.19	2.61
17. Balsam spruce	53.30	6.66	35.17	1.49	0.20	3.18
18. Flower garden plants	46.65	6.61	40.18	1.21	0.26	5.09
19. Lawn grass, green	46.18	5.96	36.43	4.46	0.42	6.55
20. Ripe tree leaves	52.15	6.11	30.34	6.99	0.16	4.25

Table 7 *Average Refuse Composition, As-Discarded Basis*

Category	wt %	Description
Glass	8.3	Bottles (primarily)
Metal	8.2	Cans, wire, foil
Paper	35.6	Various types, some with fillers
Plastics	1.1	Polyvinyl chloride, polyethylene, styrene, etc. as found in packaging, housewares, furniture, toys and nonwoven synthetics
Leather and rubber	1.5	Shoes, tires, toys, etc.
Textiles	1.9	Cellulosic, protein, and woven synthetics
Wood	2.5	Wooden packaging, furniture, logs, twigs
Food wastes	23.7	Garbage
Miscellaneous	1.7	Inorganic ash, stones, dust
Yard wastes	15.5	Grass, brush, shrub trimmings
Total	100.0	

maximum of combustibles are available to the process and that the resulting mixture is capable of supporting combustion.

Once mixing has been accomplished, the material must be dried and raised to the ignition temperature. At this point in combustion, air must be available in sufficient quantity to support combustion. The air and refuse must be mixed in a turbulent manner, allowing for initial

Table 8 *Percent Moisture in Refuse Categories on "As-Discarded" and "As-fired" Bases*

Refuse Category	Percent Moisture	
	"As-Fired"	"As-Discarded"
Glass	3.0	2.0
Metal	6.6	2.0
Paper	24.3	8.0
Plastics	13.8	2.0
Leather and rubber	13.8	2.0
Textiles	23.8	10.0
Wood	15.4	15.0
Food wastes	63.6	70.0
Yard wastes	37.9	50.0
Miscellaneous	3.0	2.0

Table 9 *Estimated Annual Average of Municipal Refuse Composition (1968) ("As-Discarded" Basis)*

Category	Unseasonal State (e.g., Florida)	Semiseasonal State (e.g., Alabama)	Seasonal State (e.g., Massachusetts)
Glass	7.6	8.1	8.8
Metal	7.5	8.1	8.7
Paper	32.6	35.1	38.2
Plastics	1.0	1.1	1.1
Leather and rubber	1.3	1.4	1.5
Textiles	1.8	1.9	2.0
Wood	2.3	2.4	2.7
Food wastes	18.2	19.5	21.1
Yard wastes	26.1	20.7	14.1
Miscellaneous	1.6	1.7	1.8

pyrolysis and finally complete combustion. These two operations take time, which means that certain furnace velocities and volume are required to properly complete the combustion and release the heat available in the refuse.

Mixing, and the need for it, has been discussed but drying must also be accomplished. Drying intimately affects furnace temperature and proper burnout. Therefore, it must be given attention in any incinerator combustion calculations.

Drying is accomplished by convective and radiant heat transfer, convective transfer from air or gas flows and radiant transfer from the flame or furnace surfaces. The material can be either predried external to the furnace or dried within the furnace itself as a result of the burning systems design.

Predrying ahead of the furnace may be accomplished in many ways, including flash drying or drying in rotary, tray, chute, or fluid bed dryers. All of these bring air or gas flow in contact with the material to

Table 10 *Heating Values and Moisture*

Heating value	4450 Btu/lb as fired (HHV)
% Moisture	28.6
% Ash	20.8

Table 11 Waste Fuel Analysis

Type of Waste	Heating Value (Btu/lb)	Volatiles (%)	Moisture (%)	Ash (%)	Flash Point (COC°F)	Fire Point (COC°F)	Sulfur (%)	Dry Combustible (%)	Density (lb/ft³)
Coated fabric—rubber	10,996	81,20	1.04	21.20	265	270	0.79	78.80	23.9
Royalite	20,299	81.90	0.37	9.62	270	280	0.04	90.38	24.3
Coated felt—vinyl	11,054	80,87	1,59	11.39	165	170	0,80	88.61	10.7
Coated fabric—vinyl	8,899	81.06	1.48	6.33	155	175	0.02	93.67	10.1
Ensolite—expanded	10,216	88.90	0.35	9.99	265	300	0.32	90.01	5.7
Missile—rubber scrap	12,238	71.36	1.69	24.94	340	360	1.17	75.06	28.9
Fuel-cell spray booth	12,325	79.10	1.74	20.67	125	130	0.25	79.33	9.5
Missile—rubber dust	9.761	62.36	0.87	36.42	250	260	1.06	63.58	9.9
Banbury—rubber scrap	13,242	60.51	1.74	4.18	145	180	0.53	95.82	34.9
Polyethylene film	19,161	99.02	0.15	1.49	180	200	0	98.51	5.7
Foam—cloth backed	10,185	55.06	0.37	31.87	215	235	1.44	68.13	11.3
Uppers—cloth	7,301	70.73	1.25	24.08	295	300	0.40	75.92	13.5
Foam—scrap	12,283	75.73	9.72	25.30	185	240	1.41	74.70	9.1
Tape—resin-covered glass	7,907	15.08	0.51	56.73	300	330	0.02	43.27	9.5
Fabric—nylon	13,202	100.00	1.72	0.13	625	640	0	99.87	6.4
Fuel-cell bladder and tire cord	15,227	87.35	1.24	3.57	270	290	0.55	96.43	19.5
Vinyl scrap	11,428	75.06	0.56	4.56	155	165	0.02	95.44	23.4
Liquid waste	13,140	100.00	3.2	1.04	68	68	0.07	95.76	53.0

Table 12 *Waste Fuel Analysis*

Type of Waste	Heating Value (Btu/lb)	Volatiles (%)	Moisture (%)	Ash (%)	Sulfur (%)	Ignition Temperature (°F)	Ash-Fusion Temperature (°F)	Density (sp gr)
Batch stock (raw)	14,177	60.29	0.46	13.36	1.07	220	2200	1,112
Uncured frictioned duck	9,343	76.44	2.57	9.16	0.52	200	2580	0.628
Cured duck	16,454	69.94	1.47	3.76	1.28	280	2220	0.957
	11,306	78.08	1.47	11.95	0.86	260	2270	1.004
Wire-braid hose	12,846	52.13	0.40	11.67	0.87	280	2480	2.109
	7,820	68.83	2.18	15.30	0.29	220	2500	2.047
Uncured C. I.	9,842	55.76	1.73	24.25	1.62	300	2360	1.415
Cured C. I.	10,150	49.89	1.66	20.31	0.60	320	2240	1.478
Cured matting trim	7,228	51.13	0.99	41.47	0.84	300	2240	1.593
Packing trim and C. I.	10,413	53.77	0.99	27.79	0.92	300	2260	1.430
Cured flash and molded goods	15,442	62.38	0.91	9.13	0.97	320	2800	1.232
Miscellaneous yarn (rayon and cotton)	7,759	90.28	6.59	0.61	0.15	220		0.504
Semicured tubes	15,392	49.70	1.24	2.03	0.70	320	2720	1.279
Miscellaneous tubing and die strip	11,352	64.15	0.49	30.54	2.23	300	2120	1.209
Rubber tape	9,629	54.51	0.54	31.63	0.56	280	2460	1.698
Ensolex trim	9,139	65.22	0.40	22.99	0.72	260	2800	0.170
Miscellaneous rubber	11,335	59.42	0.37	23.37	1.50	260	2240	1.252

be dried. The term drying has no accepted limitations and will be used here to the exclusion of other terms for moisture removal. External drying can be accomplished with gases drawn from the furnace exit, but they must be returned and reheated for deodorizing. Drying can be done with preheated air from a gas to air heat exchanger in the incinerator passage, a fired air heater, or in a steam coil air heater. If drying is accomplished with air, it must enter into the combustion process or the odors produced will have to be taken care of separately. Without mixing, the load on the drying system would be erratic and make for an impossible control system.

The furnace can be designed so that drying is accomplished within the enclosure itself; most modern incinerators incorporate furnace drying rather than predrying ahead of the furnace. An exception is where one is attempting to burn material such as sewage or industrial sludges. These materials have a moisture content in excess of 50%. This is, however, not the case with normal U.S. municipal and industrial refuse. The moisture content is approximately 20%, but it may range over 30% at certain times of the year and under adverse weather conditions. Once again, with mixing, the refuse can be quite satisfactorily burned with drying accomplished in the furnace.

If the drying is external to the furnace, the refuse is brought to about 220°F and the rest of the sensible heat addition required to bring the refuse to ignition temperature is added in the furnace. This would require that 1200 Btu/lb of moisture evaporated be added to the refuse in the external predrying system.

This 1200 Btu was determined by comparing with the requirements for evaporating a pound of water and some of the factors that affect drying. It requires 1122.3 Btu to evaporate 1 lb of water from a mass of refuse at an ambient temperature of 60°F. Since drying equipment is not 100% thermally effective, the heat input to accomplish the required evaporation must be greater. The effectiveness of the thermal drying process is governed by four fundamental factors:

A. Moisture dispersion within the refuse mass, allowing maximum exposure of moisture surface to the heated gas.

B. Large temperature differentials—high gas temperatures are required for rapid drying (high rate of heat transfer).

C. Maximum refuse agitation to increase the rate of heat transfer.

D. Particle size must be minimum to increase the surface from which moisture must be evaporated and allow the moisture to reach the surface for drying by the normal capillary action of the material.

This last requirement, coupled with the fact that oxidation as well as evaporation take place on the surface, indicates that a decrease in particle size of refuse will improve not only its drying but its combustion, once in the furnace. This argues for some form of preparation if maximum efficiency is to be obtained in both the drying and burning operations.

Internal drying, if the furnace is of the mass burning type, is usually accomplished on a stoker, the first grate system of which must dry and must ignite the refuse. This is accomplished by designing the furnace so that the drying stoker is gas swept and exposed to the extremely high-temperature furnace surfaces so that convective and radiant drying will be accomplished rapidly. If, on the other hand, a form of suspension firing can be incorporated, it will allow further drying and burning before settling on the grate for final burning and ash disposal. Heated combustion air will also assist drying. Air heating is limited in its application where pure grate burning is used, since stokers usually require that the grate be cooled by the combustion air. If excessive air preheating is incorporated, grate maintenance will be prohibitive. In either case, preheating recovers waste heat while promoting more rapid and complete combustion.

Once the mixed refuse has been dried, sensible heat from the furnace flame and enclosure must be added to bring the material to its ignition temperature and begin the rapid oxidation known as combustion, releasing the heat contained in the refuse fuel.

SUPPLYING THE BURNING PROCESS WITH SUFFICIENT AIR[7]

This topic will be considered from the standpoint of a continuous-process type of incinerator rather than the batch type. The continuously fed incinerator lends itself to accurate control and efficient burning. The larger incinerator sizes being considered in this section are almost exclusively of the continuous-feed type and will be discussed in detail. The requirements for the batch-type incinerator are similar, but special emphasis must be placed on the start-up and burn-down conditions, which are a major factor in the batch incinerator, and specific requirements are very difficult to establish.

While refuse appears highly variable, the typical refuse is quite uniform in chemical analysis and fits into the "family tree" of the fossil fuels. Refuse, both industrial and municipal, bears close relation to wood, bark, and bagasse, all of which have been burned sucessfully for many years. Considering that paper, cartons and crates are wood products, this chemical relationship is easily understood and it can be used to our advantage in computing air requirements. In addition to the predominantly cellulosic material, an increasing amount of plastic containers and wrappers are found in refuse. This is particularly true of industrial refuse. This results in an increased heating value, but not enough at present to prevent the application of calculations for basic cellulosic material. One can therefore apply the latest combustion design and practices that have made modern industrial furnaces so much more efficient than their predecessors.

One might suggest that although the material is by nature cellulosic, variables in the refuse could upset the computation. Nevertheless, experience has shown that the variables in refuse are mostly the noncombustible residues and moisture. Chemical composition of the combustibles is quite uniform. Design is therefore only slightly more difficult than for other fuels such as wood and bagasse, about which a great deal is known. Tests have shown that the heat of combustion on a moisture ash-free basis is very close to that of cellulose, 8000 Btu/lb, for normal municipal and most industrial refuse fuel. This fact simplifies the calculations in the heat process and makes it possible to predict the behavior of the burning material. The most important variable is the moisture in the refuse, which is considered one of the main energy-consuming loads on the incinerator.

The heating values of the main constituents of municipal and industrial refuse on a moisture and ash-feed basis are wood, 8420; brush, 8600; paper, 7900; garbage, 7280; and plastics, about 18,000.

Although plastics could upset the situation as they grow in importance, the "family tree" of combustible materials in Figure 2 is further proof of refuse's cellulosic nature. This chart can also act as a check on the more detailed calculations that will be presented later. The computations will take plastics into account as well as the cellulosic components and allow the proper design of an incinerater. It should be noted that the "family tree" (Figure 2) is based on a moisture and ash-free percentage of the material. To assist in appreciating the heating value and offer an opportunity for estimating the heating value of refuse, Table 13 is offered. This table illustrates the quantities of air required and heat released from the various constituents in the combustion of refuse. With the assistance of this table, one can arrive at an approximate heating value for refuse. In order to accurately determine the necessary design parameters, the air and gas weight must be determined from an ultimate analysis. The ultimate analysis used in these sample calculations will be as follows

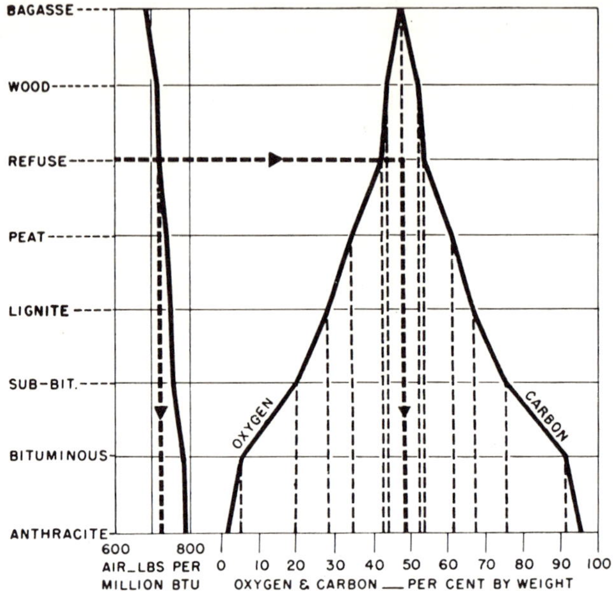

Figure 2. Refuse fits into a family tree of combustible material in relationship to wood and peat.

Table 13 *Combustion Equations (Adapted from Ref. 7)*

Combustible	Molecular Weight	Equation		Heat Release (Btu/lb) (High)	Theoretical Air (lb/lb of Fuel Element)
Carbon	12	$C + O_2$	$\rightarrow CO_2$	14,100	11.52
Hydrogen	2	$H_2 + 0.5\ O_2$	$\rightarrow H_2O$	61,100	34.50
Sulfur	32	$S + O_2$	$\rightarrow SO_2$	4,000	4.32
Hydrogen sulfide	34	$H_2S + 1.5\ O_2$	$\rightarrow SO_2 + H_2O$	7,100	3.10
Methane	16	$CH_4 + 2\ O_2$	$\rightarrow CO_2 + 2\ H_2O$	23,880	17.28
Ethane	30	$C_2H_{61} + 3.5\ O_2$	$\rightarrow 2\ CO_2 + 3\ H_2O$	23,320	16.12
Propane	44	$C_3H_{18} + 5\ O_2$	$\rightarrow 3\ CO_2 + 4\ H_2O$	21,660	15.68
Butane	58	$C_4H_{10} + 6.5\ O_2$	$\rightarrow 4\ CO_2 + 5\ H_2O$	21,300	15.48
Pentane	72	$C_5H_{12} + 8\ O_2$	$\rightarrow 5\ CO_2 + 6\ H_2O$	21,090	15.38

(this table represents an approximation for an average municipal refuse): carbon, 30%; hydrogen, 4%; oxygen, 22%; moisture, 24%; and ash, metals, and glass, (noncombustibles), 20%.

It may be of interest to compare this selected composition with the ranges found by various investigators: carbon, 25–35%; hydrogen, 3.5–4.5%; oxygen, 15–30%; moisture, 15–30%; and ash, metals, and glass 15–25%.

These averages represent some mixing. Otherwise, there would be excursions outside the range of these limits. However, one can readily see that present U.S. municipal refuse very closely approximates a cellulose fuel, and a heating value of 5000–6000 Btu as received is a logical expectation. Design should be based on the highest average heat value that may be expected. Otherwise, the designer could wind up with an undersized furnace volume or, if a mass burning system, insufficient grate area to adequately burn the refuse.

The heat unit generally used is the million Btu/hr, which is the product of the fuel burned, in pounds per hour, times the heating value in Btu per pound, as fired. In the following example the heating value is found to be approximately 5000 Btu/lb as received. On the moisture and ash-free basis, it comes out to be 8940 Btu/lb. If the ultimate analysis is adjusted to the moisture and ash-free basis, the following results are obtained: carbon, 53.5%; hydrogen, 7.2%; and oxygen, 39.3%. With this information, enter the "family tree" Figure 2 of combustible material and find that the 53.5% carbon and 39.3% oxygen come out exactly on the refuse line for a combustible material, requiring 720 lb of air per million Btu for proper combustion. This will now be checked more accurately below.

AIR AND GAS WEIGHT DETERMINATION

The refuse ultimate analysis chosen for these sample calculations requires the following theoretical air (from Table 13):

$$\begin{aligned} C \rightarrow 0.30 \times 11.54 &= 3.46 \text{ lb air} \\ (\text{H-0/8}) \rightarrow (4 - \tfrac{22}{8}) \times 34.56 &= 0.43 \text{ lb air} \\ \text{Total} &= 3.89 \text{ lb dry air} \end{aligned}$$

Therefore, the amount of theoretical atmospheric air to completely react with 1 lb of this refuse is 3.89 lb. To ensure complete combustion, 50% excess air is assumed. This can vary with the type of firing and whether the furnace is water cooled or not. In suspension firing, less excess air may be used, but in mass burning at least this amount is normally used. Greater amounts may be used to dilute the products of combustion and control furnace temperatures, but this air is not truly part of the combustion process and is not included here. The total combustion air to be used with 50% excess air, regardless of how it is admitted to the burning zone (under- or overfired air), is computed as $1.5 \times 3.89 = 5.84$ lb dry air per pound of refuse. The computation can be checked against Figure 3 if the 5000 Btu/lb. refuse is converted to millions Btu by dividing 5000 into 1,000,000. It is then determined that there are 200 lb of refuse per million Btu. Introducing this into the above calculations (5.84×200), 1168 lb of air per million Btu are to be used with 50% excess air. This compares with 1070 lb air per 10^6

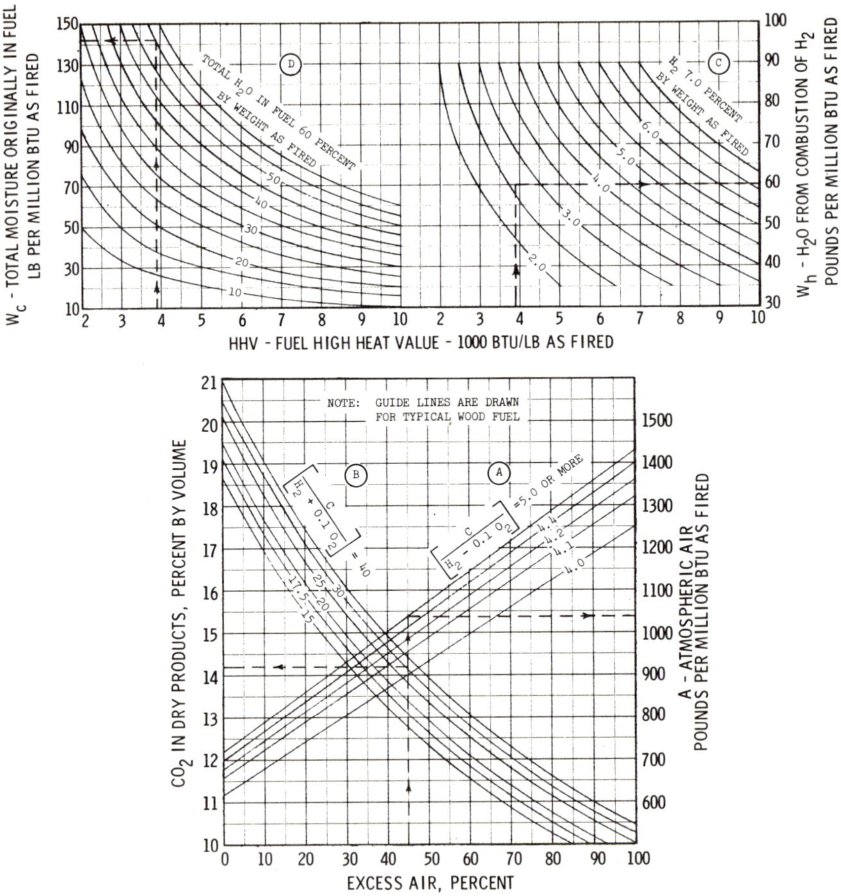

Figure 3. C, H_2, and O_2 on a combustible basis; chart for wood and bagasse.

Btu from Table 13, indicating a required air quantity 8.4% below the computed estimate. Another quick method of estimating the theoretical air for a refuse fuel is to determine the Higher Theoretical AIP (T.A.) Heating Value (HHV) and substitute it into the formula $-\text{T.A.} = 7.7 \times 10^{-4} \times \text{HHV}$ = pounds of air per pounds of refuse. The value obtained, 3.85 lb/lb, compares well with the computed value above. The constant in this equation (7.7×10^{-4}) varies somewhat with variations in refuse composition. The results, as with both the other methods given above, are good for rough design estimates. The air requirements remain at best a design estimate because of the peculiar hydrogen bonding of the complex cellulose material and its effect on both the air requirements and heating value. To add to the uncertainties, some metal oxidation takes place, increasing both the heating value and the air requirements. This completes the estimate of air required for fan selection and air ducting design.

Returning to the combustion calculations, the next step is to compute the flue gas volume. The quantity of flue gas (in pounds per pound of refuse) produced from this example is determined from the foregoing results:

Carbon burned*	0.300
Hydrogen	0.040
Oxygen	0.220
Moisture	0.240
Dry combustion air (50% excess)	5.840
Weight of flue gas	6.640 lb gas/lb of refuse (plus air moisture)

The quantity of flue gas is frequently expressed as pounds per million Btu; for this example the value would be 1328 lb per 10^6 Btu.

PRODUCTS OF COMBUSTION[7]

The 6.64 lb of flue gas per pound of refuse indicated is the total produced minus air moisture. When this is added the total wet products of combustion are obtained. The quantity of moisture in the

* If combustion is incomplete, assume that the combustion exhibits a hydrogen preferential and the percent unburned combustible is entirely attributable to the carbon in the refuse. This is usually the case. Once the unburned carbon is computed, correct the carbon burned to reflect this fact.

flue gas can be computed as follows: Assume* 0.020 lb moisture per pound of dry air used, $0.020 \times 5.84 = 0.117$ lb/lb refuse. Therefore total products of combustion for this example is $6.64 + 0.12 = 6.76$ lb per pound of refuse.

Moisture, 0.24	= 0.24
Hydrogen, 0.04×9	= 0.36
Air moisture from above	= 0.12

Total moisture in flue gas 0.72 lb/lb of refuse

The dry products of combustion can now be computed:

$$6.76 - 0.72 = 6.04 \text{ lb/lb of refuse fired}$$

The composition of the gaseous products of the refuse can be computed as follows from the assumed ultimate analysis:

$$CO_2 = 0.30 \times 3.67 = 1.10 \text{ lb/lb of refuse fired}$$
$$H_2 \text{ (as computed above)} = 0.72 \text{ lb/lb of refuse fired}$$

$$O_2 \left(\frac{3.89}{2} \times 0.232\right) = 0.45 \text{ lb/lb of refuse fired}$$

$$N_2\ (3.89 \times 0.769 \times 1.5) = 4.49 \text{ lb/lb of refuse fired}$$

Total wet products of combustion of 1 lb of the assumed refuse equals 6.76 lb/lb of refuse.

Therefore

$$CO_2 = \frac{1.10}{6.76} = 0.163 \text{ lb/lb of flue gas}$$

$$H_2O = \frac{0.72}{6.76} = 0.106 \text{ lb/lb of flue gas}$$

$$O_2 = \frac{0.45}{6.76} = 0.067 \text{ lb/lb of flue gas}$$

* This assumption depends on the inlet air conditions. They were assumed in this case to be 80°F and 90% relative humidity. The moisture value is then read from a psychrometric chart.

$$N_2 = \frac{4.49}{6.76} = 0.664 \text{ lb/lb of flue gas}$$

$$\text{Total} \quad \overline{1.000} \text{ lb/lb of flue gas}$$

This can be expressed on a dry basis as follows:

$$1.000 = 0.106 = 0.894 \text{ lb of dry gas per pound of refuse}$$

$$CO_2 = \frac{0.163}{0.894} = 0.182 \text{ lb/lb of dry flue gas}$$

$$O_2 = \frac{0.067}{0.894} = 0.075 \text{ lb/lb of dry flue gas}$$

$$N_2 = \frac{0.664}{0.894} = 0.743 \text{ lb/lb of dry flue gas}$$

$$\text{Check total} \quad \overline{1.000} \text{ lb/lb of dry flue gas}$$

Because of the frequent use of the Orsat apparatus to analyze the products of combustion, it is usual to compute the volumetric composition of the dry products of combustion:

$$CO_2 = \frac{0.182}{44} = 0.00413 \qquad \frac{4.13}{32.87} = 0.126 = 12.6\%$$

$$O_2 = \frac{0.075}{32} = 0.00234 \qquad \frac{2.34}{32.87} = 0.071 = 7.1\%$$

$$N_2 + \frac{0.743}{28} = 0.02640 \qquad \frac{26.4}{32.87} = 0.803 = 80.3\%$$

$$\text{Moles of dry gas} \quad \overline{0.03287} \qquad \text{Check} \quad \overline{100.0\%}$$

The value of an Orsat analysis on incinerator flue gas becomes quite evident. If the carbon dioxide is less than indicated above, the oxygen will be greater, indicating greater than 50% excess air being used in the system. Conversely, a greater volume of carbon dioxide and less oxygen means that less than 50% excess air is being used.

FURNACE TEMPERATURE

An important parameter in the design of incinerators is the temperature of the flue gases leaving the furnace. This can be approximated by determining an average specific heat and using it to compute the temperature.

The furnace exit temperature can be estimated from Figure 4. Using the 24% moisture and 50% excess air used in the illustration above, the temperature is found to be approximately 2500°F.

If the designer now refers to the gas tables, he finds the specific heat at constant pressure of carbon dioxide varies from 8.89 to 14.16 Btu/lb-mole-°F between 80°F and 2500°F for an average of 11.53 Btu/lb-mole °F or 0.262 Btu/lb-°F.

Similarly for

Moisture	0.545 Btu/lb °F
Oxygen	0.247
Nitrogen	0.274

These are admittedly only estimates of the specific heat of these materials over the applicable range. They could have been arrived at more accurately and from many other sources. When all the influences affecting the furnace exit temperature are considered, however, this degree of accuracy appears sufficient.

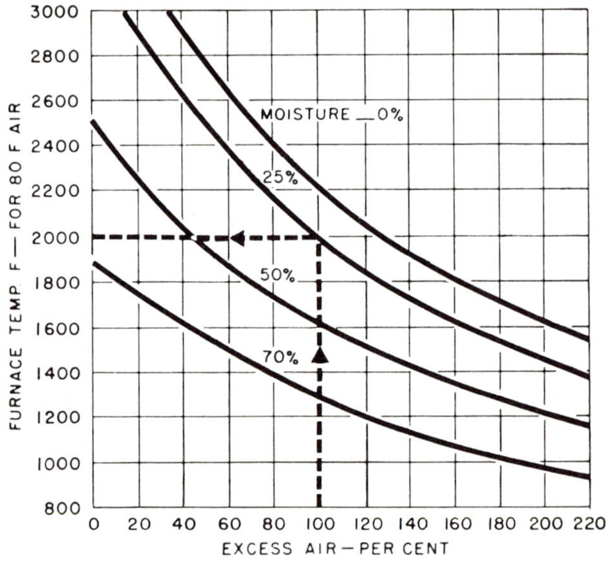

Figure 4. Furnace gas temperature in a refractory furnace as a function of excess air and moisture control.

The temperature is now computed as follows:

$$\underset{CO_2}{[(wCp)} + \underset{H_2O}{(wCp)} + \underset{O_2}{(wCp)} + \underset{N_2}{(wCp)]} \times [T_2 - T.]$$

$$\begin{aligned} CO_2 &= 0.163 \times 0.262 = 0.0427 \\ H_2O &= 0.106 \times 0.545 = 0.0577 \\ O_2 &= 0.067 \times 0.247 = 0.0166 \\ N_2 &= 0.664 \times 0.274 = \underline{0.1820} \\ \textstyle\sum wCp &= 0.2990 \end{aligned}$$

The furnace inlet temperature has been assumed as 80°F which will be used here.

The heat released and made available to the furnace per pound of refuse will be assumed as 4,750 Btu/lb. This assumes a 5% heat loss in incomplete combustion and through the furnace walls. Divided by the 6.76 lb of flue gas per pound or refuse, this is 703 Btu/lb of flue gas.

Estimated furnace exit temperature is therefore

$$\underset{\text{exit}}{[T - 540]} = \frac{703}{0.299} = 2350$$

$$T \text{ exit} = 2350 + 540 = 2890R = 2430°F$$

This value is now available to the remaining furnace calculation.

FURNACE CALCULATION

The following example shows how these values are used in incinerator design.

Assumed rated capacity in tons per day (this is a convenient multiple for larger units)	100 tons/day
Pounds per hour for 24-hr operation	8340
Millions Btu/hr – 8340 × 5000	41.7×10^6 Btu/hr

If the design is to incorporate mass burning on a grate, the grate area is determined by dividing the total Btu input by the allowable grate heat release expressed in Btu/ft³/hr. This is an empirical value that takes into account cost as well as performance. For typical U.S. municipal refuse, 300,000 Btu/ft²/hr is considered satisfactory. This is equivalent to 60 lbs/ft²/hr of 5000 Btu/lb refuse.

$$\text{Grate area} \quad \frac{(41.7 \times 10^6)}{300,000} \quad 139 \text{ ft}^2$$

In a similar manner, the furnace heat release rate for grate firing of refuse has been found to fall between 15,000 and 20,000 Btu/ft^3 for refractory lined furnaces and can be as much as double this value if water-cooled walls are used. For this example 20,000 Btu/ft^3 will be used.

$$\text{Furnace volume} \quad \frac{(41.7 \times 10^6)}{20{,}000} \quad 2085 \text{ ft}^3$$

$$\text{Furnace height required} \quad \frac{2085}{139} \quad 15 \text{ ft}$$

(Therefore 15 ft clearance above grate is needed to complete combustion and allow introduction of overfired air.)

Total air weight at 50% excess air:

8340 × 5.84	48,700 lb/hr
48,700 ÷ 60	812 lb/min
Air volume to be supplied at 80°F	11,050 ft^3
Furnace exit quantity	
6.76 × 8340	56,400 lb/hr
56,400 ÷ 60	940 lb/min

Gas volume at furnace exit is computed using the 2430°F furnace exit temperature computed above. Then calculate the furnace exit gas flow as 68,500 ft^3/min.

The furnace exit opening can be computed from this volume using a normal furnace exit velocity of 20 ft/sec.

The exit area is therefore

$$\frac{68{,}500}{60 \times 20} = 57 \text{ ft}^2$$

From the above data the grate can be selected, the fans can be specified, the various furnace gas velocities can be computed, and other necessary design data obtained.

STORAGE OF REFUSE

The logistics of refuse collection are independent of the incineration of the material. Refuse collection is usually structured to accommodate the public. It is usually on a periodic basis and arrival at the in-

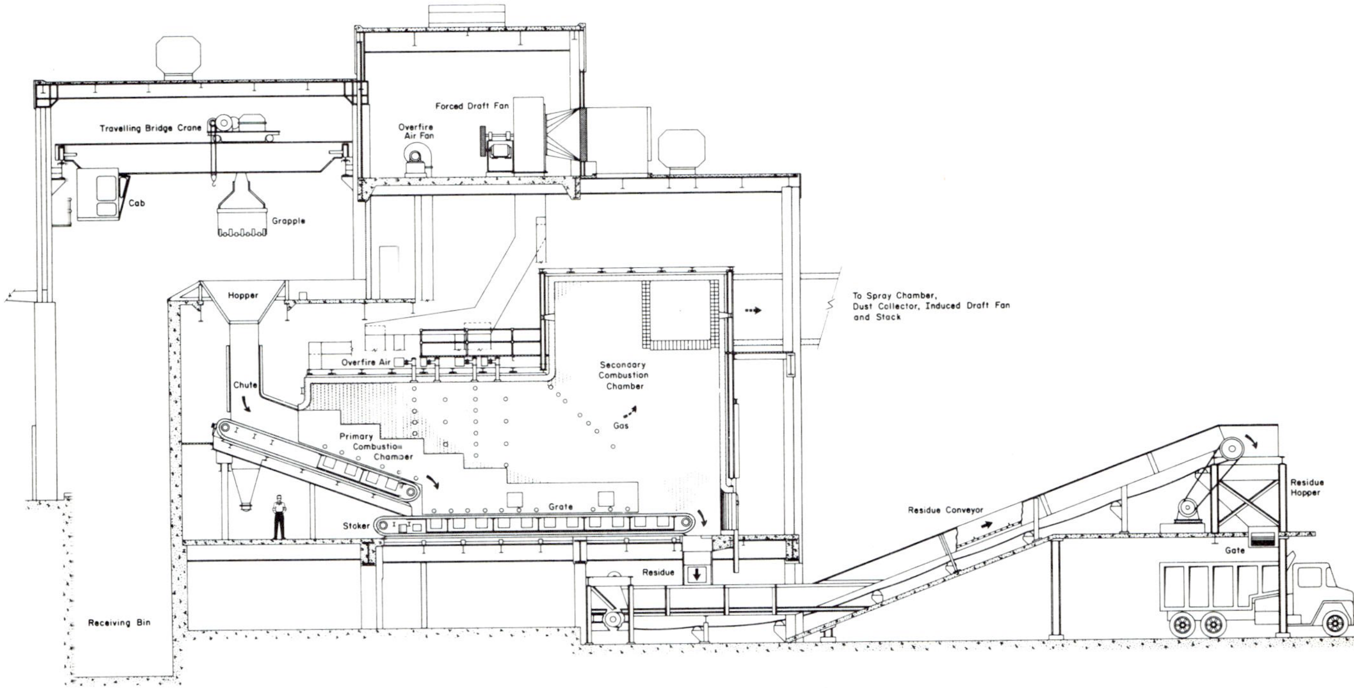

Figure 5. Elements of a municipal refuse incineration system.

cinerator facility is cyclic with peak periods. Where the incinerator operates more than one shift, collected refuse must be stored for burning over longer periods. The same is true during holidays, weekends, or periods of planned or unplanned maintenance.

Storage is divided into two general classifications: (1) bin and silo storage and (2) floor storage.

Bins are usually large, concrete rectangular enclosures set below ground. Frequently, incinerators are constructed on marginal land leading to expensive, complicated construction. The storage capacity may be set by state or local codes. A 1½-day storage capacity is not uncommon. This can have a major influence on plant design and is frequently difficult to obtain in larger plants (1000 ton/day and up). Storage bins of this type are shaped to permit full access by overhead cranes (see Figure 5, elements of a municipal refuse incinerator system).

Silo storage is a method that is currently being explored. It does require use of preconditioned refuse (sized).

There are many types of live bottom bins currently being used for storage in other industrial processes. It is contemplated that refuse can also be stored in similar bins and work has been done to establish this. These are usually vertical tanks or structures; the bottoms are usually divergent and open. Reclaim devices such as drags or screws are used to continuously undercut the refuse. In principle this method conforms to the accepted material handling concept "first-in-first-out." This is not true with bin storage, which utilizes the crane to remove material.

FLOOR STORAGE

Floor storage, as the name implies, consists of a surface sheltered or enclosed floor. Refuse is usually handled on the tipping floor by tractors with bulldozing blades and lifting buckets. This method is set satisfactory for smaller incinerators. Since the below-grade bin is currently the most widely accepted method of storing refuse, the overhead crane with a grab bucket or grapple is used to remove refuse from the pit and feed it to the furnaces. Monorail cranes have also been used in smaller plants. This is a single fixed overhead rail supporting a trolley so arranged as to discharge into the feed openings of the furnaces.

Large incinerators have traveling bridge cranes. The operators cab may move with the crane. This type is common in the United States, whereas in Europe, the cab is really a small control room located in the side of the bin at a point that provides observation of the bin for the operator. The operator does not move with the traveling bridge but remains stationary.

The crane, besides removing material from storage, provides a function unique to the incineration process. Cranes provide a means of sorting and mixing refuse to condition the refuse feed for burning. This is a cyclic action of acceleration and deceleration and casting. Cranes are subject to severe duty in an incinerator plant because of this process and must be highly reliable, since a crane stoppage can but down an entire plant. They also require special provisions in the building which houses them, such as strong mounting, head room, and well-protected electrical source.

PRE-PREPARATION

Suspension fire requires pre-preparation or size reduction of refuse. This method of firing offers many potential advantages, and while it has not been applied on a wide scale to incinerators, all of the components and techniques have been extensively used in the firing of other cellulosic and fossil fuels. Several installations are under construction and will be operating shortly. This method opens the door to feed methods in storage not currently being used in incinerators. The role of the crane is primarily to remove it from a small receiving area, the large below-grade storage bin being replaced by a storage silo and reclaim system. The crane is used to feed a shredder. Refuse is shredded and conveyed by either high-pressure pneumatic systems or belts to a storage silo.

STORAGE AND RECLAIMING OF SHREDDED SOLID WASTE

Controlled combustion of shredded solid wastes (SSW), as with other fuels, requires a fuel feed rate that ensures positive response to the demands of the combustion control system However, meeting this requirement with SSW is considerably more difficult than with other fuels, but much easier than with unprepared refuse.

Even after being shredded, SSW may contain fibrous and stringy materials in sufficient quantity to cause serious storage and feeding problems. In storage, SSW tends to compact and mat under its own weight, creating an interlocking mass that is very difficult to reclaim from storage and feed reliably at a smooth and controllable rate. Attempts to reclaim from storage and feed this material with conventional augers, drag chains, vibrators, and the like, therefore, have not always been successful.

One concept of reclaiming and feeding that has proven effictive with SSW is shown in Figure 6. Reclaiming of SSW from storage is accom-

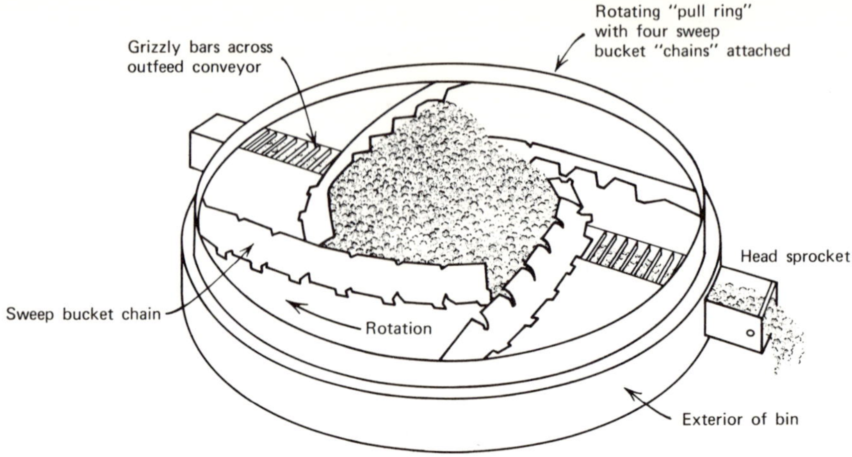

Figure 6. The Atlas storage and reclaim system.

plished by "chains" of "sweep buckets," each chain having one end pinned to a powered rotating pull ring that rotates around the periphery of the bin; the sweep chains automatically trail toward the center. The sweep buckets contact the stored SSW at the outside of the pile and, as the pull ring continues to rotate, the buckets fill and the SSW is swept through the floor openings onto a drag-chain type of conveyor recessed in the floor of the bin. This outfeed conveyor then delivers the SSW at a controlled rate to the plant pneumatic or other type of conveying system for feeding to the combustion unit.

METHODS OF PREPARATION

In principle, materials are sized by pressure, impact, shearing, tearing, and friction. Since municipal refuse is a man-made composite material, every item will respond to some principle of size reduction. Unfortunately, a compromise must be sought to select the equipment that will best handle the mixture.

Tables 14–16 have been prepared to show the fitness of disintegration for waste processing, the raw material properties, and their suitability for various processes and some of the technical processes and solutions. No machine for size reduction can satisfy all of the needs, but basically a compromise is based on ability to reduce, first cost, maintenance and operating costs, noise level, and space. There are few published data of comparative shredding methods.

Table 14 *Raw Material Properties: Typical Representives in Municipal Waste and Adequate Disintegration*

	1 Raw Material Property and Wear and Tear Characteristics	2 Typical Representative in Waste	3 Pressure	4 Friction	5 Hammering	6 Shearing	7 Tearing	8 Impact
A	Hard, much wear and tear	Natural stones, baked stones (tiles)	□	▽	○	□	▽	○
B	Hard, wear and tear, if outsize troublesome	Metals: tins, plate, bits of cast iron	▽	▽	▽	○ ▲	▽	○
C	Wear and tear brittle	Ceramics, hard-baked clinker	○	▽	○	▽	▽	○
D	Soft and fibrous; hardly any wear and tear	Textiles, leather, paper	▽	▽	▽	○	○	□
E	Elastic, hardly any wear and tear for metals	Metal springs, rubber, synthetic materials with softeners foil	▽	▽	▽	○	○	▽
F	Sludgy, much wear and tear	Sludge	▽	▽	▽	▽	○ ●	▽

▽, in appropriate; ○, appropriate; □, conditionally appropriate; ▲, provided adequate shear strength is available; ●, processed with equal amounts of garbage.

Table 15 *Methodology: Operation and Fitness of the Disintegrators for Waste Processing*

Type of Disintegration	Disintegrators	Fitness for Disintegration
Pressure	Jaw crusher, conical breaker, roller crushers	No
	Edge runners	No
Pressure and friction	Ring mills or pendulum mill (slow rotating)	No
	Roller breakers, roller mills	Conditioned
Hammering	Roller breakers, roller mills (fast rotating)	Conditioned
Hammering and friction	Ball drop hammer device, Symons breaker	No
	Hammer crusher, ball drop hammermill	No
	Ball drop swinging mill, stamping mill	No
Hammering, shearing and tearing	Hammermills	Yes
Shearing	Dust mill with shear blades	No
	Waste shears, press shears	Yes
Friction and shearing	Grinding, slow rotating	No
	Pebble mill, ball crusher	No
	Rasp	Yes
Impact and tearing	Impact hammer breaker, impact bar crusher	Yes
	Impact plate breaker, disk attrition mill	No

A summary of an analysis by Battelle Memorial Institute was prepared by A. D. Little. This summary recognized that few definitive data existed on refuse size reduction and suggested that "caution should be exercised in the use of the results for definitive planning or budgeting."

There is no reason to doubt that the background in material preparation and size reduction, which is more extensive than the "state-of-the-art" of incineration, can successfully be applied to effect cost reductions. The basis limitation on the capacity of mass burning equipment in itself is sufficient motivation. Figure 7 is typical of an accepted shredding device for municipal refuse.

WASTE REDUCTION

The prime function of any waste reduction facility is to achieve a minimum residual volume and sterile end products. The by-products such as gas and water must be discharged to the atmosphere within acceptable pollution limits. Two methods of thermal reduction which provide a sterile ash or a useable by-product are incineration and

Table 16 *Disintegration and Screening: Technical Processing Problems and Their Solutions*

Process Number	Steps in Processing					Process Target
	A		B		C	
1			Disintegration			Incineration
2	Prescreening		Disintegration			Incineration volume reduction
3	Prescreening		Disintegration		Postscreening	Composting
4	Prescreening	Predisintegration	Intermediary screening	Postdisintegration		Composting
5	Prescreening	Predisintegration	Intermediary screening	Postdisintegration	Postscreening	Composting rapid composting
Aim of Process	Elimination of bulky waste and finest particles	Surface enlargement	Facilitate post-disintegration	Production of end-size	Elimination of outsize and foreign substances	

Figure 7. Municipal waste shredder.

pyrolysis. Incineration satisfies more of the basic needs for bulk reduction than other methods developed to date, reducing the bulk by oxidation to a residue usually 20–25% by weight and 10–12% by volume. With normal municipal incineration, only one collection is necessary; this includes most noncombustibles as well as combustibles.

For a successful incineration system, whether large or small, municipal or industrial, it is necessary that the following provisions be included for carrying out certain essential functions:

1. Receiving loads of mixed refuse at varying rates of supply.
2. Measuring the quantity of refuse received.

3. Storing in a sanitary, accessible, nuisance-free manner.
4. Sorting, mixing or otherwise preparing refuse for incineration.
5. Feeding the refuse to the system at a controlled rate.
6. Drying the refuse sufficiently to permit ignition of combustibles.
7. Burning the refuse to produce essentially inert solid residue and clean gases.
8. Dissipating the heat of combustion.
9. Cooling, collecting, and removing noncombustibles.
10. Cleaning and discharging effluent gases and liquids in an acceptable form.
11. Controlling the process for safety, efficiency, economy, and community or plant acceptance.
12. Protecting personnel and equipment from the elements, dangerous refuse, and careless operation.

Refuse is handled in a batch-feed or a continuous-feed system. The heart of the reduction system is the burning of the material, which is accomplished by mass burning or suspension burning. Mass burning takes unprepared refuse and burns it with a minimum of mixing on a fuel-supporting and air-admitting surface called a stoker. Stoker firing is mechanical firing that provides continuous or intermittent feed, drying, ignition, and burning of combustibles, and discharge of noncombustibles. All stokers are classified as underfeed, crossfeed, or overfeed. Stokers for refuse burning may incorporate more than one of these basic classifications. The classification of stokers is based on the method by which fuel and air are introduced and their relative direction.

In the *underfeed* principle of firing, the fuel and air have the same relative direction. Both enter the active burning zone from beneath the fuel bed. In the *crossfeed* principle of firing, the fuel enters the furnace at right angles to the main (primary) air flow. In the *overfeed* method, the fuel enters the active combustion zone from above and opposite to the direction of primary air flow. A further breakdown of the specific types of mechanical stoking devices used for mass burning in incinerators is as follows.

Traveling-Grate (crossfeed) stokers consist of a grate surface comprising a series of cast iron sections, or keys, cast on carrier bars which in turn are fastened to two or more drive chains to form an endless conveyor, the upper surface of which supports the fuel. These are usually arranged in multiple levels in the furnace enclosure. Refuse enters the feed end, is burned and conveyed to the discharge end where the ash is continuously removed. See Figure 8.

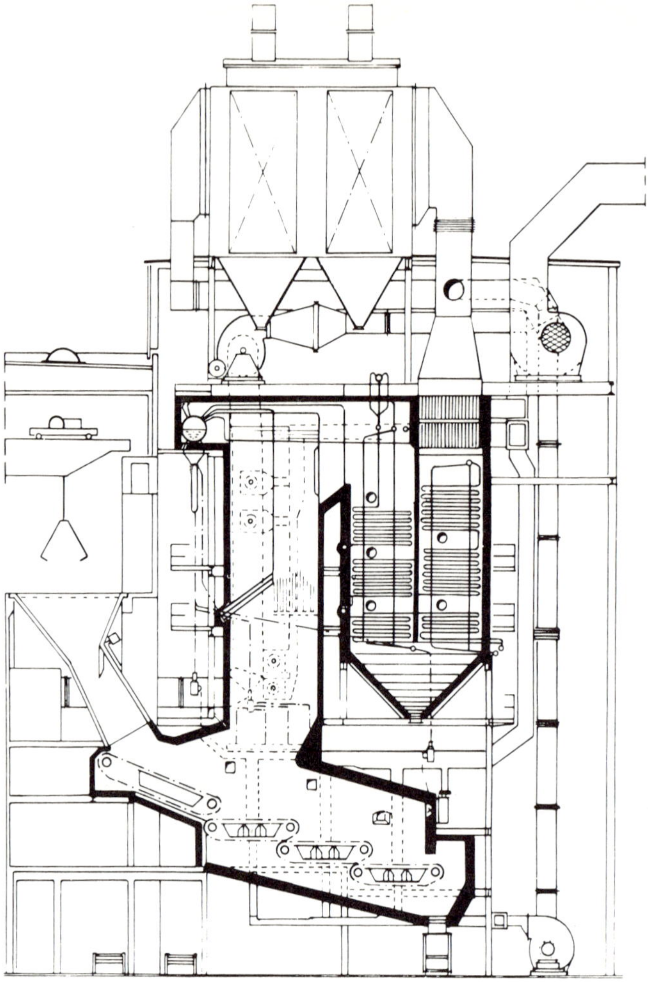

Figure 8. Boiler with multiple grade stoker.

Rotating-drum (crossfeed) stokers consist of an inclined row of rotating individually driven cylinders. The drum surface is made up of segmental keys. In principle, this method is an adaptation of the traveling grate in that the traveling grate is in reality a stretched out circle. See Figure 9.

Reciprocating (crossfeed–underfeed) stokers have a reciprocating grate that is either direct acting or reverse acting. In a direct-acting grate, the surfaces reciprocate in alternate rows towards the discharge end. In the reverse-acting grate, the grates reciprocate towards the entrance of the furnace. Instead of horizontal movement, the alternate surfaces are arranged at an angle of approximately 35°. While predominantly embodying crossfeed principles because the fuel enters the furnace at right angles to the air flow, the reciprocating action, moves the fuel at the bottom to the top in the same relative direction of air

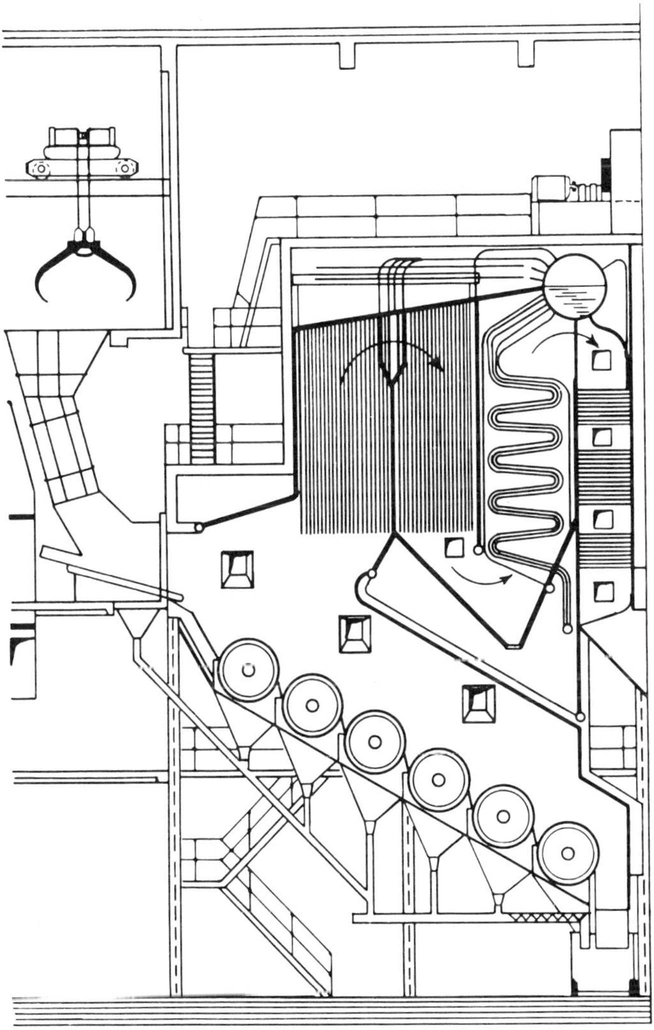

Figure 9. Boiler with Düsseldorf or barrel type stoker.

flow; hence it combines some of the characteristics of the underfeed principle. See Figure 10.

Rocking grate (crossfeed–underfeed) stokers consist of a series of movable steps (grates) each of which is rocked in a coordinated manner, lifting the refuse and advancing it to the discharge end. This embodies principles of both crossfeed and underfeed firing. The underfeed principle is more dominant than in the reciprocating stoker because of the positive fuel bin turnover. See Figure 11.

The inertial grate (crossfeed) stoker is a fixed bed either in a single plane or in steps; it is carried on rollers or flexures and activated in a manner so as to draw the entire bed back and then release it so that the bed moves forward until abruptly stopped by a checking device. The

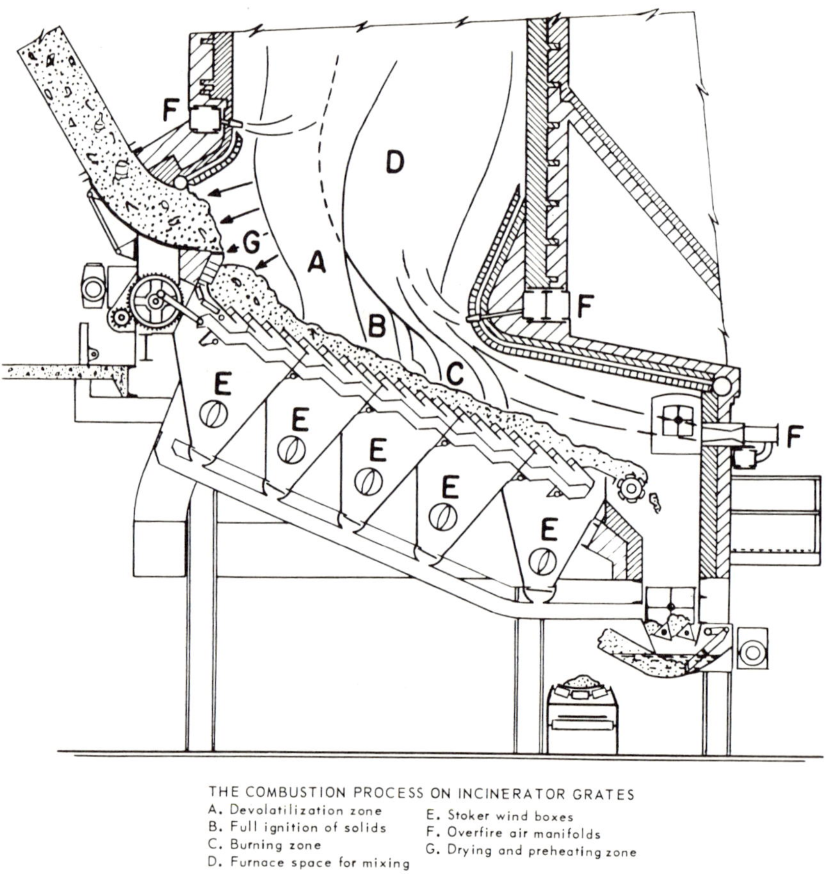

Figure 10. The combustion process on incinerator grates.

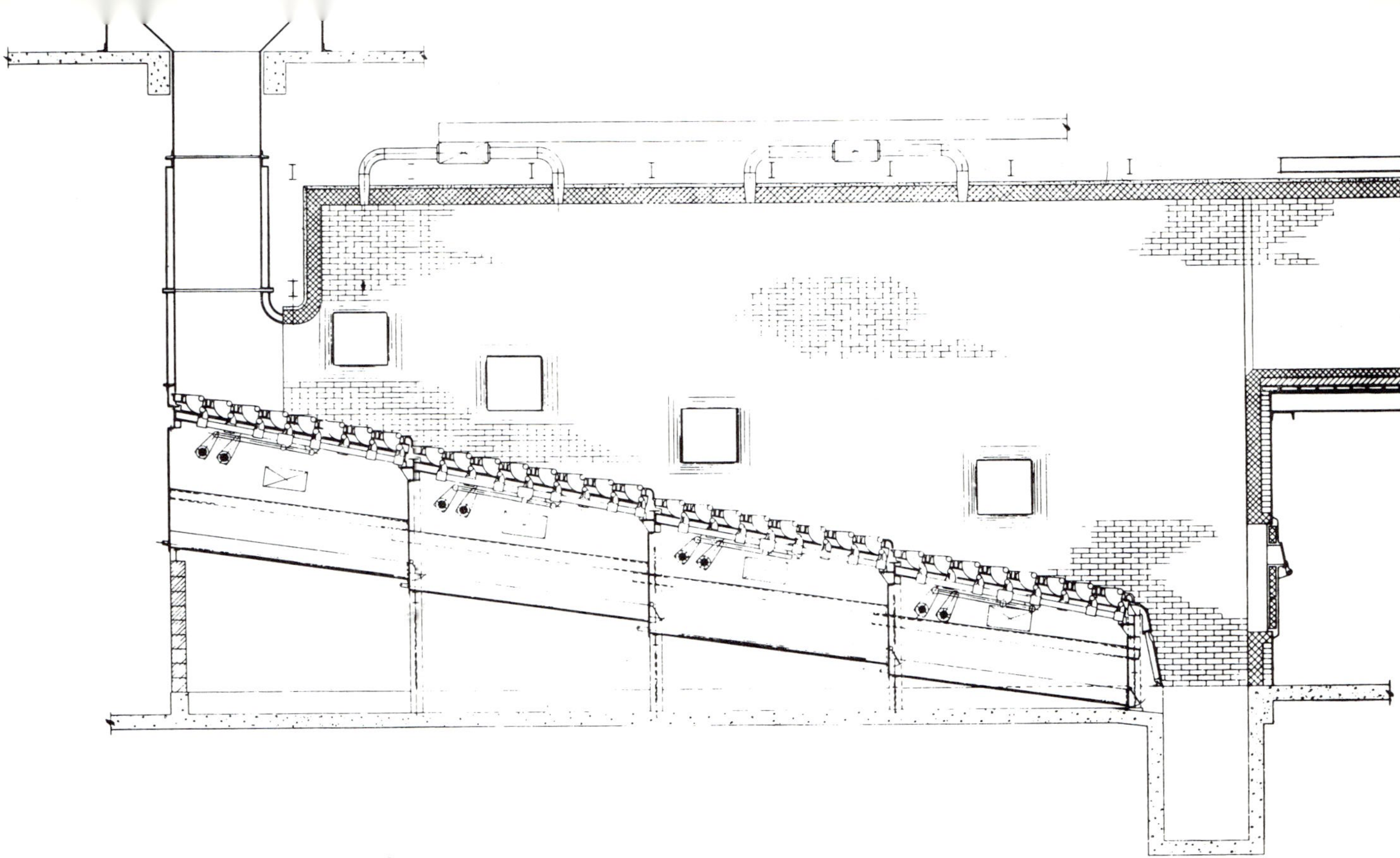

Figure 11. The rocking grate stoker.

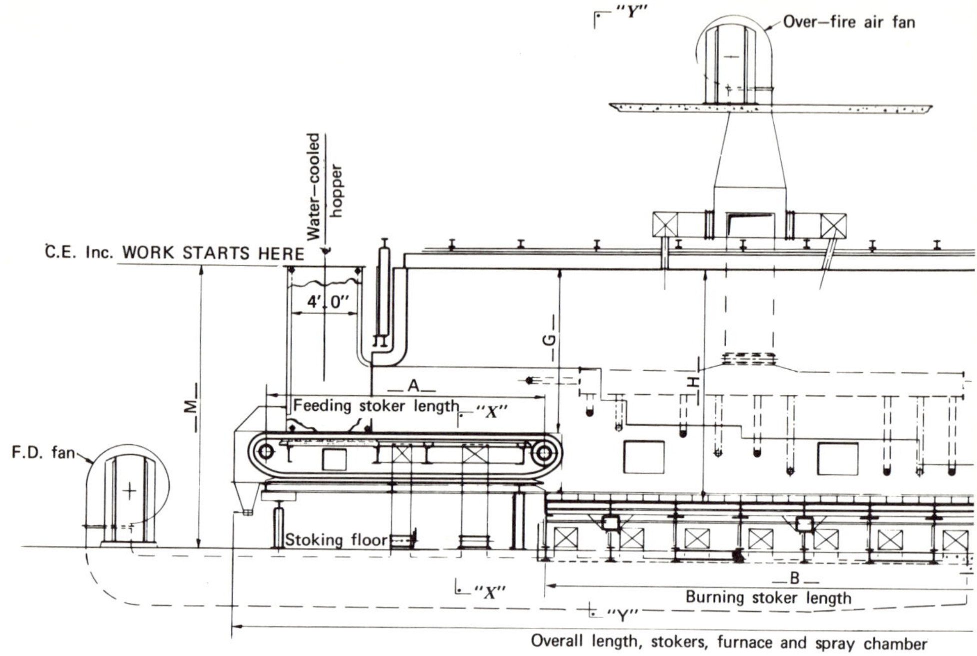

Figure 12. Standard arrangement of travelling grate and Combustikator incinerator stokers.

inertia of the refuse carries it a small distance along the surface of the stoker, repeating the cycle and discharging the burned-out residue. See Figure 12.

The rotary kiln is really not a true stoker but consists of a combination of stokers usually reciprocating (crossfeed–underfeed) to feed the refuse into a rotating refractory-lined cylinder. The cylinder is horizontal or slightly inclined. Most of the drying and devolatilization takes place on the stoker with burnout occurring in the kiln.

The fluid bed, while not a true stoker either, does use the system of crossfeed. This technology has the advantage of high heat transfer and reaction rates and has been primarily used for sludges and semisolid wastes. Investigators are currently exploring the use of this principle for solid waste reduction. Intimate contact of solid particles and gas promotes good mixing and burnout. Lack of uniform size and density distribution presents problems, and closely controlled pre-preparation to produce a uniform feed mixture will be required and is currently being investigated. The fluid bed system operates most efficiently

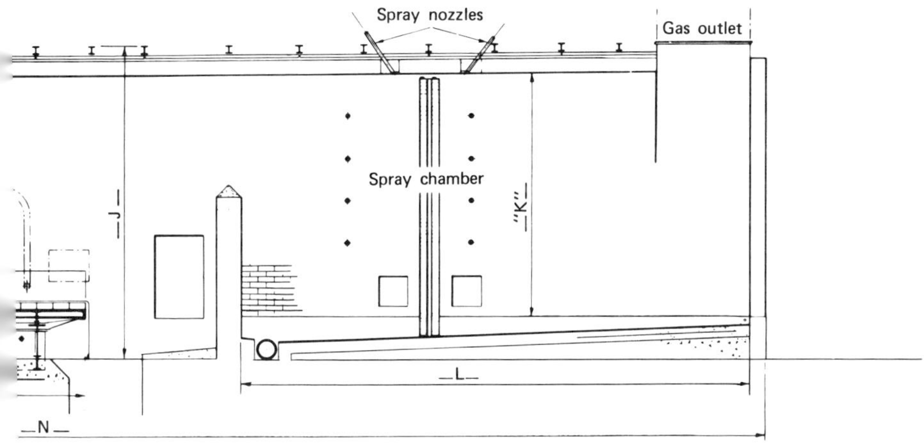

within narrow limits of combustible contents and requires close control of the process. It is anticipated that the limitations and throughput (capacity) will minimize acceptance within the current known state of the art.

Table 17 has been prepared indicating advantages and disadvantages of these various methods.

In the case of all mass burning methods, the furnace configuration is dictated by the firing method.

Suspension burning of the refuse is a new development and is classified as the overfeed principle (fuel introduced above and in the direction opposite to primary air flow). It differs from mass burning in that it requires preconditioning where sizing of the refuse is required. It combines the principles of pulverized fuel and stoker firing. Fine, dry, volatile material burns in suspension, and larger particles burn on a grate. Prepared refuse is introduced through the front or side walls by mechanical or pneumatic distributors or introduced pneumatically by a high-pressure conveying system and fired tangentially.

In the tangential system, the fuel is introduced into the furnace high enough to allow drying of the refuse before it reaches the grate. Turbu-

Table 17 *Firing Classification*

Type	Classification		Siftings	Maintenance	Fly Ash Carryover	Residue Quality	First Cost	Power
	Crossfeed	Underfeed						
Traveling grate	×		Low	Low	Low	Good	Medium	Low
Rotating drum	×		Low	Low	Low	Good	High	Low
Inertial	×		Low	Medium	Medium	Excellent	Low	Low
Reciprocating direct acting	×	×	High	High	Medium	Excellent	Medium	Low
Reverse acting	×	×	High	Medium	Medium	Excellent	High	High
Rocking	×	×	High	High	High	Excellent	Medium	Low
Rotary kiln	N.A.[a]	N.A.	N.A.	High	High	Excellent	High	Low
Fluid bed	×	×	N.A.	High	High	Excellent	High	High

[a] N.A., not applicable.

lence is provided by blowing tangentially directed streams of preheated air at high velocity through nozzles at various furnace elevations. Overfire and underfire air are proportioned in accordance with the volatile content of the refuse. All the refuse passes the rows of nozzles at various elevations. When refuse enters the highly turbulent high-temperature zone, a portion will burn rapidly in suspension, and the larger particles will be dry for complete reduction on a grate. Tangential firing provides longer residence time in the hottest furnace zone. Unlike other stokers, the grate area is independent of the fuel feed rate.

Suspension burning provides more effective use of furnace volume, and unit capacity no longer need be restricted to the physical dictates of the stoker. Similarly, air temperatures are not limited by mechanical and material considerations of the stoker thus improving the combustion process. Close control of fuel–air ratios is possible. The economics of preparations is favorable when the entire system is considered.

Because of the advantages of low excess air operation and lack of financial constraints (stoker area) this method should be considered for use in large steam generation. See Figure 13.

Incinerator furnaces are divided into refractory furnaces and water-cooled furnaces. Refractory furnaces are described in Refs. 26 and 27.

Water-Cooled Furnaces

The application of incinerator equipment to steam generators has been brought about by the potential of economic trade-offs by steam generation. Attempts to apply convection boiler sections to refractory furnaces have met with limited success. The disadvantages of this arrangement are primarily limited to the fact that the refractory furnace sets the design, i.e., furnace temperature is controlled by excess air. All too frequently, the convection section was a standard selection for fossil fuels and did not take into consideration the slagging characteristics of refuse. Tube spacing was inadequate, velocities were high and caused erosion.

In the 1920s, the first completely water-cooled furnace was installed. When bare-tube water cooling was first proposed for furnace linings, many engineers felt that ignition would be retarded to a point where conplete combustion would be impossible. They failed to realize that, once ignition temperature is reached, the completeness of combustion then depends on the rapidity with which it is possible to supply the demand of the burning fuel for oxygen. The furnace temperature, above that required for ignition, is of minor importance to the combustion process, nor does it greatly affect the mechanical mixing of the fuel and

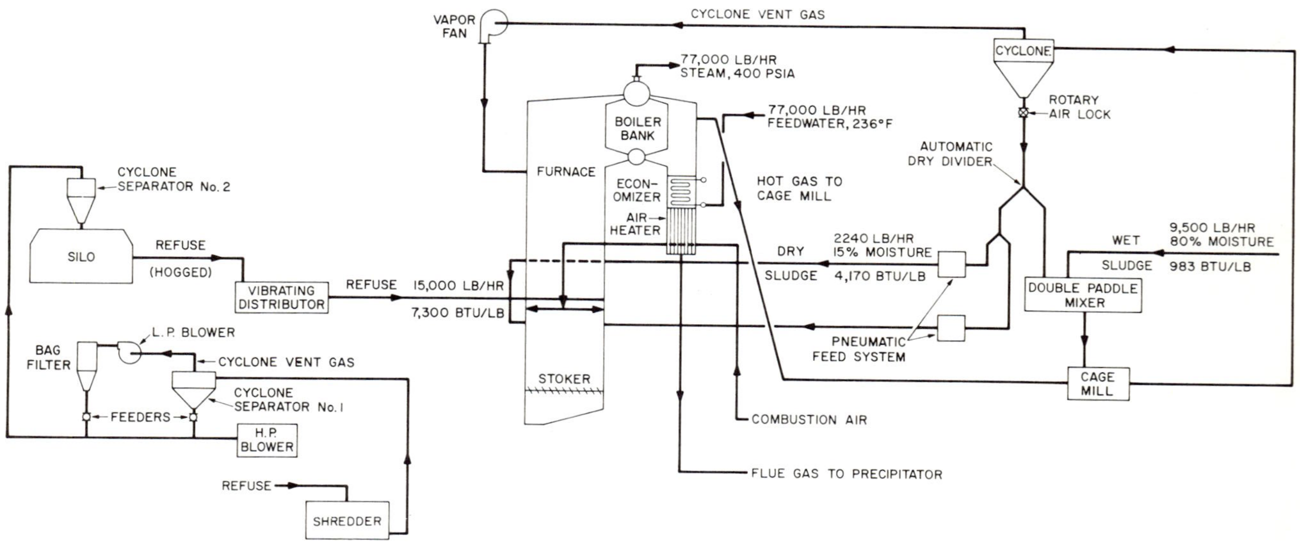

Figure 13. System flow diagram.

its air aupply. Rapid combustion is usually accompanied by high temperature. Because of this, the conclusion is frequently drawn that high temperature is the cause of this rapid burning. Actually, intense and thorough mixing causes rapid combustion and results in high temperature. Therefore, in furnaces having a combination of refractory and water-cooled walls, or with complete water cooling, there is no danger of incomplete combustion because of excessive cooling. The temperature is always high enough so that combustion is maintained, so long as there is a correct supply of air mixed with the fuel.

Practically all modern power boilers are equipped with waterwalls. In large central station boilers, waterwalls completely cover the interior surfaces of the furnace, thus providing practically complete elimination of exposed refractory surface. In addition, the waterwalls serve as the sole means of heating and evaporating the feedwater supplied to the boiler from the economizer.

Waterwalls usually consist of substantially vertical tubes arranged tangent or approximately so and connected at top and bottom to headers. These tubes receive their water supply from the boiler drum by means of downcomer tubes connected between bottom of drum and lower headers. In a typical central station boiler, roughly 50% of the heat released by the combustion of fuel in the furnace is absorbed by the waterwalls, as shown in Figure 14. Heat so absorbed is utilized in evaporation of all or a large percentage of the water supplied to the boiler. The steam generated and a substantial quantity of accompanying water are discharged from the top of the waterwall tubes into the upper waterwall headers and thence passed through riser tubes to the boiler drum. Here the steam is separated and the accompanying water together with the incoming feedwater is returned to the waterwalls through the downcomers.

Waterwalls serve as a durable furnace lining. By their use, a continuous metallic surface is provided that can absorb and transfer a large quantity of heat without deterioration, thus minimizing furnace maintenance and appreciably increasing availability of the boiler.

Furnace design must take into consideration water heating and steam generation in the water tubes as well as the process of combustion. All large modern boilers are designed with walls made of water-cooled tubes to form complete metal coverage. Present practice is to use tube arrangements and configurations that permit practically complete elimination of refractory in all areas that are exposed to high-temperature gases.

Tube diameter and thickness are of concern from the standpoints of

Boiler service	Fuel	Capacity (lb/hr)	Pressure	Steam temperature (°F)	Reheat temperature (°F)	Feedwater temperature (°F)	Boiler efficiency
Heating	Oil	60,000	125	Sat	–	212	80.20
Small industrial	Oil	100,000	650	750	–	300	85.60
Medium industrial	Coal	220,000	650	750	–	265	87.25
Large industrial	Coal	600,000	1500	880	–	275	88.00
Small utility	Oil	1,580,000	1890	1005	1005	472	85.20
Large utility	Coal	3,785,000	2820	1000	1000	480.5	90.04

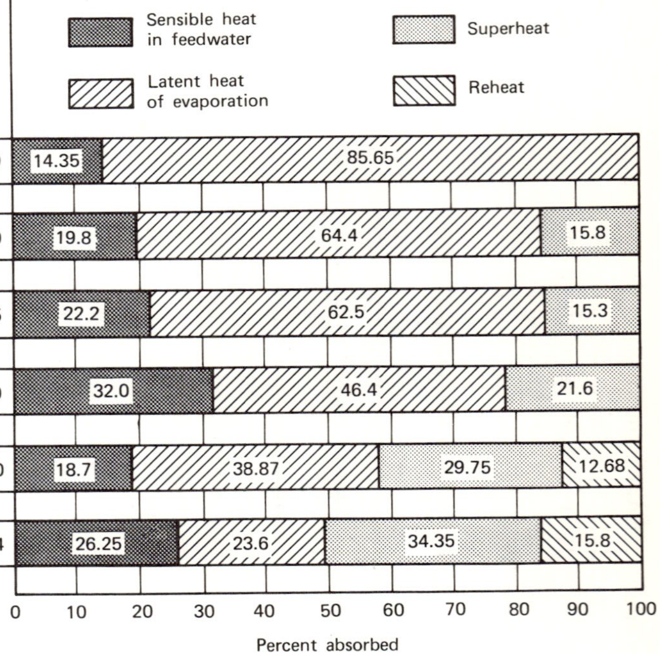

Figure 14. Percentage of heat absorbed by components.

circulation and metal temperatures. Natural circulation boilers generally use larger-diameter tubes than controlled circulation or once-through boilers. This practice is dictated largely by the need for more liberal flow area to provide the lower velocities necessary with the limited head available. The use of small-diameter tubes becomes advantageous in high-pressure boilers since the smaller tube thickness required results in lower outside tube metal temperatures.

Factors such as velocity of water and water–steam mixtures and quality of water and steam mixtures are of particular concern in high heat absorption areas of high-pressure boilers. Although it is not possible for the designer to provide complete freedom from corrosion and plugging, there is much that can be accomplished to minimize these objectionable features. In the furnace, factors such as size, shape, amount of water cooling, and elimination of all refractories are given due consideration with respect to fuel analysis. Provisions for continuous cleaning of surfaces and areas by the installation of proper types, numbers, and location of soot blowers, and the installation of ash hoppers in proper locations are means of maintaining continuously clean surfaces. In connection with the superheaters and reheaters, particular importance is placed on the arrangement and spacing of tubes and selection of materials for tubes and selection of materials for tubes and supports.

Waterwalls arranged with tangent tubes have generally been provided with skin casings consisting of thin steel sheets welded together in the form of an envelope placed behind and in contact with the waterwall tubes. It is this welded enclosure that provides the seal against furnace leakage. However, this type of construction requires a considerable amount of field fabrication and welding. In order to lessen the amount of this work and decrease the time required for boiler erection, an improved design referred to as panel wall construction has been developed.

Panels consist of a number of tubes joined together by a process of fusion welding and are made in widths of approximately 5 ft. The tubes are spaced about $\frac{3}{8}$ to $\frac{1}{2}$ in. apart by bars that are then fused together with the tubes to form a continuous metal furnace lining.

The Incinerator Boiler

When there is an in-plant demand or a nearby market for steam or electricity it is usually a simple matter to justify the cost of installing a boiler system to reclaim the heat generated from refuse incineration. Even when there is no immediate use for the steam other than for the boiler auxiliaries, it may be possible to justify a boiler when one considers the saving in the cost of flue gas cleaning that results from lower excess air in the furnace and elimination of spray-water cooling of the flue gas. The heat in the unused steam would be dissipated in air- or water-cooled condensers. Figure 15 illustrates the potential for steam generation from refuse and the effect a boiler has on the volume of flue gas to be cleaned.

Two boiler design problems must be given special consideration when firing refuse, namely, fouling of heating surfaces and potential corrosion. It is imperative that the boiler designer treat these problems correctly to ensure a high availability factor for the refuse disposal unit. In a refuse-fired boiler, all sections of heating surface (waterwall, superheater, generating bank, economizer, and air heater tubes) are subject to fouling from slag and flyash deposition. Proper furnace sizing, arrangement of heating surfaces, and correct use of soot blowers can reduce fouling to an acceptable level.

Proper furnace sizing means providing adequate volume and residence time to ensure complete burnout. The required volume and

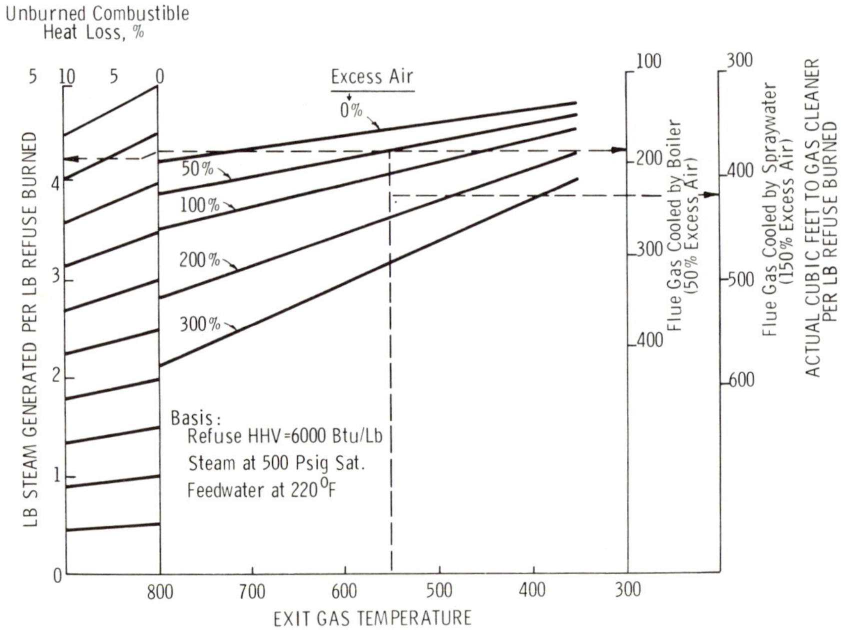

Figure 15. Relationship among excess air, stack temperature, steam capacity, and flue gas.

dimensions for flame travel will vary with the method of fuel preparation, burning mechanism (i.e., fixed-grate, reciprocating-grate, traveling-grate, suspension-firing, etc.), overfire air and underfire air system, and the amount of water cooling in the furnace. Proper furnace sizing also means providing sufficient water cooling to reduce the temperature of the products of combustion to the point where the ash is not fluid and thus will not "freeze" on boiler tubes. This type of deposit is difficult to remove with soot blowers. Also, the furnace design must provide for a temperature in the burning zone that is high enough to destroy bacteria and ensure proper burning.

The next step in preventing serious slag and fly ash deposition is proper arrangement of the heating surface. This begins with adequate furnace cooling of the flue gas followed by using wide convection tube spacing, espcially in the high-gas-temperature zones, and in-line tube rows. Staggered tube rows should not be used in a refuse-fired unit. Also, the gas velocity should be low and retractable soot blowers should be used to clean convection heating surface.

The second problem mentioned above is that of potential corrosion. When a high percentage of the boiler heat input is from refuse, the problem can be serious and inattention to it or overoptimism during the design stages can result in extremely high maintenance costs and greatly reduced incinerator availability. There are at least four recognized types of corrosion which must be considered by the designer: (1) high-temperature, liquid-phase corrosion, (2) corrosion due to nonuniform furnace atmosphere, (3) corrosion by hydrochloric acid, and (4) low-temperature of dew-point corrosion. A brief discussion of each follows.

High-temperature, liquid-phase corrosion is probably caused by molten alkali-metal sulfates. Much has been written about this type of corrosion and there is disagreement among investigators on the exact mechanism and the temperature range in which the corrosion occurs. However, it is generally agreed that this type of corrosion is likely to occur at metal temperatures above 900°F. Recent data indicate that high gas temperatures aggravate the problem. One obvious way to avoid high-temperature corrosion is to eliminate high metal temperatures that occur in superheaters of units where refuse contributes a high percentage of the heat input.

The second type of corrosion is caused by the products of partial combustion in a reducing atmosphere. A reducing atmosphere can be present locally in a unit supplied with 200% or more excess air as a result of stratification or improper air or fuel distribution. Thus, carbon monoxide or hydrogen sulfide can be produced. It is thought that these compounds attack and cause failure of waterwall tubes by reducing the iron oxide on the surface of the tubes. It is important, therefore, that the burning system provide not only the correct fuel–air ratio, but also proper distribution of air to the fuel and sufficient turbulence to prevent stratification.

The third type of corrosion, resulting from hydrochloric acid and chlorine, has been recognized for years. Chlorine is introduced in the form of polyvinyl chlorides in plastics. There is much disagreement on the mechanism and temperature range of this type of corrosion. It is the author's opinion that metals operating at temperatures above the dew point and below 550°F should not suffer severely from hydrochloric acid and chlorine attack.

Finally, low-temperature or dew-point corrosion occurs when the flue gas contacts surfaces which are at temperatures below the dew point of corrosive constituents in the gas. Temperatures low enough to cause acid condensation may be found at the water inlet end of an economizer if the feedwater temperature is too low, in the cold end of an air heater, and on boiler outer casings. For this reason, welded-wall construction (previously referred to) as shown in Figure 16 is considered desirable. The tubes and fins are welded to form a self-cased, pressure-tight envelope to prevent flue gas from contacting a cold casing. Low-temperature corrosion could also be a problem during unit outages. Some deposits are corrosive, and where they are hygroscopic, the problem becomes more severe as the length of the outage increases. If a lengthy shutdown is contemplated, the fireside of the unit should be water washed or, alternatively, the unit could be kept hot by using an external source of heat.

Furnace Performance

Furnace exit gas temperature must be established and is determined by empirical curves relating furnace temperature to net furnace heat release (total heat input less moisture loss). Furnace net heat release is obtained by dividing net heat release by effective projected radiant surface (EPRS) of the furnace, which is the total envelope in square feet of the furnace multiplied by appropriate effectiveness factors for the various furnace walls, i.e., tangent tube construction, fin tube construction, and welded-wall construction.

ENVIRONMENTAL CONTROL

Every operation must be sensitive to environmental requirements if

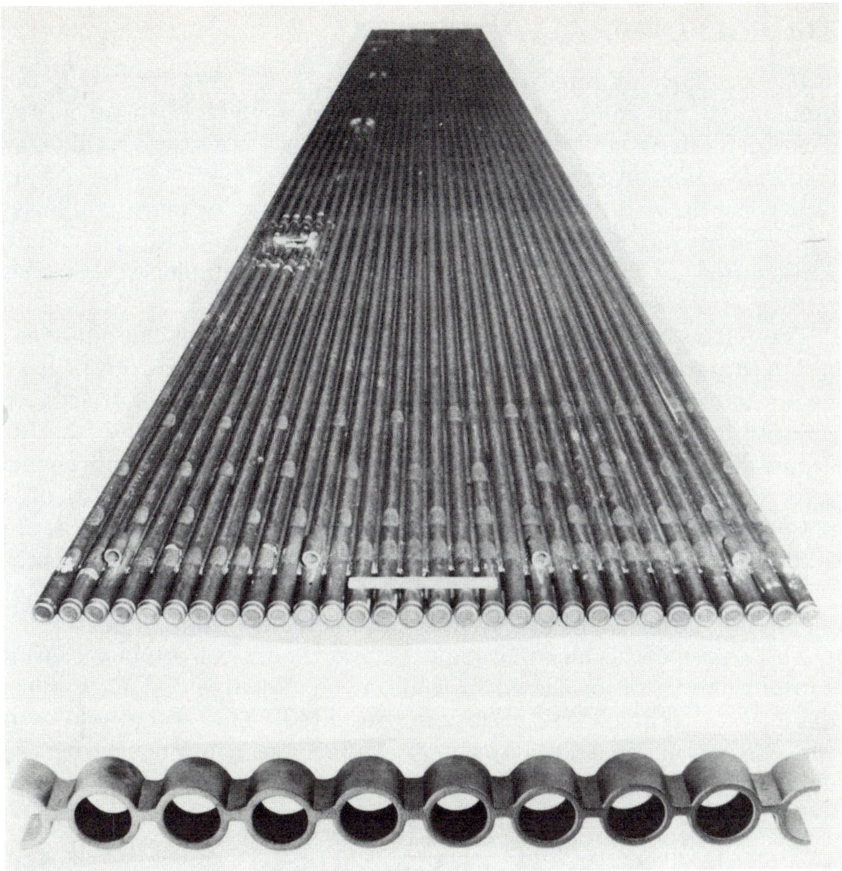

Figure 16. Welded wall construction.

it is to be a successful venture. Incineration is particularly susceptible to these requirements because in the past open dumps and inefficient incinerators were quite common. Because of this, some will immediately react negatively to the proposal of incineration, claiming that it is not a true solution but merely a transfer of the problem to another medium, such as water or air. This is a legitimate reaction. Open burning dumps and inefficient incinerators have built a legacy that is difficult to live down. Equipment is available, however, to incinerate solid waste efficiently and cleanly with only a sterile ash remaining and without polluting either the air or the water; these topics will now be covered.

AIR POLLUTION

The problem of refuse disposal is inextricably involved with the problem of air pollution, regardless of the method of disposal. Incineration offers the opportunity to reduce refuse to a sterile residue for disposal to landfill and to remove offensive odors, but it can be significant contributor to the air pollution problem in an urban community. The primary air pollution concern in incineration is with particulate emission rather than gases or odors; therefore, the emphasis here is on particulate emission.

There have been comments that a properly operated incinerator does not need particulate collection equipment. Many systems with little or no air pollution control equipment have been represented as effectively meeting dust emission requirements when they actually do not. This occurs because the excess air used for combustion and cooling is so great (200–500%) that it dilutes the effluent to the extent that it does not appear objectionable, although excessive quantities of dust are actually emitted. With the trend toward larger, more efficient incinerators located close to the population centers served, effective control of incinerator atmospheric pollution is extremely important.

The degree of gas cleaning required and the cost of primary control equipment will be discussed in this section. Relative cost for various degrees of control will also be presented. The discussion will concentrate on air pollution control of the large, continuously fed incinerator now common in modern municipal practice.

INCINERATOR PARTICULATE EMISSION

The quantity and size of particulate emission leaving the furnace of an incinerator varies widely, depending on such factors as the refuse being fired, method of feeding, operating procedures, and completeness of combustion. The information presented here includes published data from a number of sources (see References).

The rate of furnace dust emission has been found to vary from less than 10 to as much as 60 lb dust per ton of refuse burned. High-performance, compact, turbulent incinerators operate close to the lower end of this range, or about 20 lb/ton. In practice, dust loadings are reported in various ways. This 20 lb/ton may be reported as

2.0 lb/million Btu (assuming 5000 Btu/lb of refuse).

1.7 lb of dust per 1000 lb of flue gas adjusted to 50% excess air (corrected to a flue gas condition equivalent to burning with 1.5 times the theoretical air required).

0.90 grains/standard ft^3 adjusted to 50% excess air.

These dust loadings refer to conditions "leaving the furnace" (leaving the combustion zone, including any afterburner or secondary furnace, but before the quench chamber).

Dust sizing, like dust loading, varies widely. Most factors that affect dust loading also affect dust sizing. Improved incinerator performance, which reduces dust quantities, also decreases the size of the individual particles. The dust is always quite heterogeneous, consisting of typical incinerator fly ash combined with large, low-density flakes. Dust density varies from an average of slightly over 2 g/cm^3 (125 lb/ft^3) to as high as 3 g/cm^3 (187 lb/ft^3). The dust size as determined in the BAHCO centrifugal classifier, using the methods and procedures of ASME Performance Test Code No. 28, indicate a size-range distribution as presented in Figure 17. It is evident that about 35% of the average dust leaving the furnace is below two microns (1 micron = 3.94×10^{-5} in.) and is difficult to collect. Simple settling chambers and spray chambers do not remove sufficient quantities of this dust to

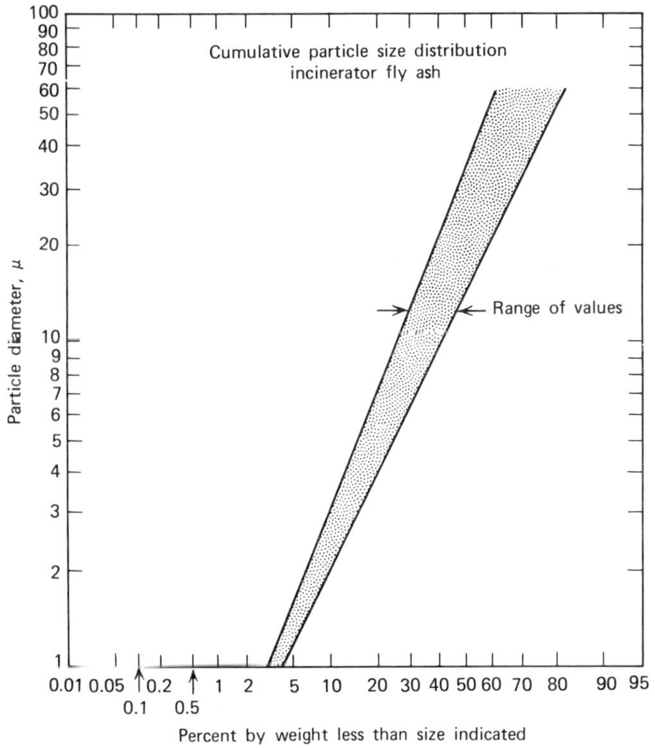

Figure 17. Cumulative particle size distribution—incinerator fly ash.

meet even lenient air pollution regulations, and therefore more sophisticated equipment must be used.

Fly ash from incinerators is a specific case of the general problem of particulates, discussed in Part 5 of the present volume.

Air pollution control equipment performance is significantly affected by the completeness of combustion, operating procedures, and the adequacy of maintenance. To ensure proper collection performance with any of the systems discussed, frequent and thorough inspection and maintenance of air pollution control equipment are required. Without proper maintenance, design performance as an operation criterion is meaningless. As a further check on proper performance, stack emission checks should be performed occasionally.

Incinerator Air Pollution Control Equipment Performance

Particulate collection equipment performance may be classified in a number of ways, but the most widely accepted criterion is the weight efficiency. The weight efficiency relates the quantity of the dust collected to the dust that enters the collector with the gas. This number may be meaningfully applied only under conditions similar to those entering the collector during the test, including the given dust density and size distribution, the entering gas dust loading, the collector energy level, and the inlet gas temperature. The results can sometimes be related to other applications if the dust density, size distribution, dust resistivity (if a precipitator), collector energy level, and gas condition are known.

Another important collector performance criterion is the fractional efficiency curve (Figure 18), sometimes called the size or grade efficiency curve. It represents the performance of the particular collector on each size of dust particle of a given density for a given collector energy level, gas temperature, and dust resistivity (if a precipitator).

The two efficiencies are related and one can be computed from the other if the dust size distribution is known. This is very important, since most air pollution control equipment manufacturers would prefer to guarantee the known fractional efficiency performance for their equipment and allow each purchaser to compute the efficiency for his particular dust.

The size and composition of incinerator fly ash and the extremely large quantities of air used in incineration mask the real pollution potential. As a result, stack observations are no measure of an incinerator's pollution control. An accurate determination of stack emissions can be obtained only by actual tests based on samples taken in the dust leaving the air pollution control equipment. It is suggested

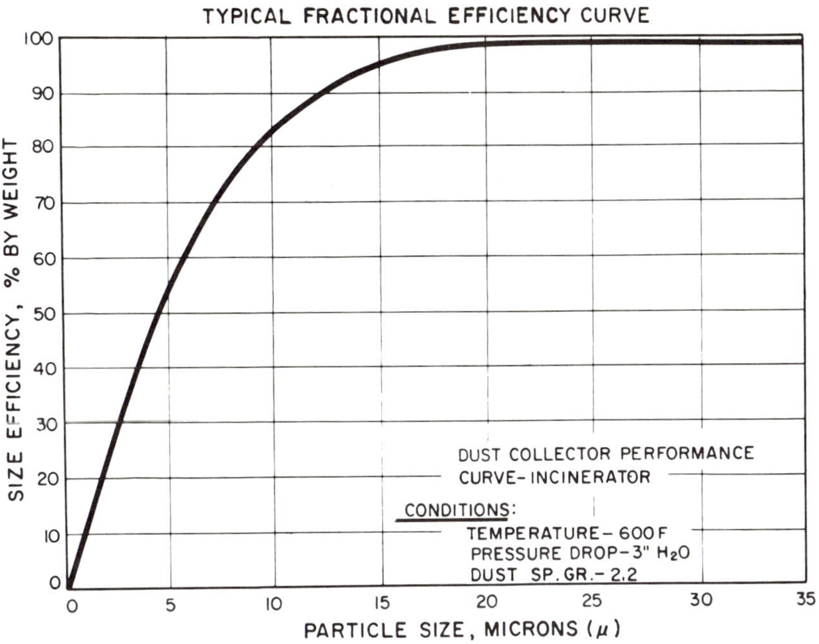

Figure 18. Dust collector performance curve incinerator.

that test connections be designed into the ducting before and after the primary dust collection equipment. A make-shift arrangement to accommodate sampling at a later date is, at best, a compromise. Preengineered sampling points will permit accurate measurement of particulate emission from the stack and the testing of the primary air pollution control equipment to determine if it is functioning properly.

Collector performance and the resulting emission to the atmosphere are summarized in Figure 19. This can be compared to the local emission standards, which may be used as an entry to the graph. The efficiency required is read on the left ordinate while the right ordinate presents the class of air pollution control equipment that could be designed to meet this requirement. As an example, if the ASME 1966 maximum emission lever were used, one can enter Figure 19 with the 0.8 lb of dust per million Btu and read 77% efficiency on the left ordinate and on the right ordinate note that a mechanical collector could be designed for this service. Once again, the reader is cautioned that these data assume a properly designed and maintained collector and an incinerator with good combustion conditions. The ranges of perfor-

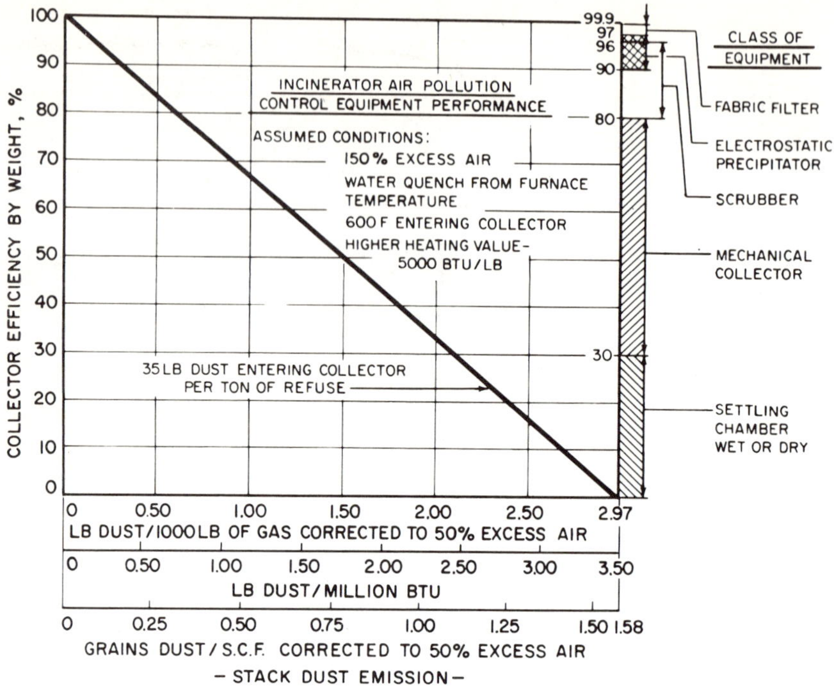

Figure 19. Collector performance and emissions.

mance presented on the right ordinate of this figure indicate areas in which it is reasonable to expect each class of equipment to perform. In most cases, the 35 lb of dust per ton of refuse leaving the furnace, assumed as a basis for this graph, is a satisfactory starting point. If, for a certain type of incinerator, the designer knows the furnace emission is greater or less than the assumed 35 lb/ton, a second line can be drawn (from the 100% efficiency and zero-emission point to the expected furnace emission on the zero-efficiency line) and the graph used as before for these new conditions. See the dashed 20 lb/ton emission line).

This graph illustrates that today's technology can provide good pollution control on modern incinerators. As mentioned earlier, the highest efficiency collectors may require additional development to achieve their full potential, but they are available to the industry today.

Cost of Air Pollution Control Equipment for Incinerators

It is extremely difficult to precisely pinpoint the price of a particular class of air pollution control equipment. Prices vary substantially from one vendor to another, and with market conditions. In addition to these factors, certain improvements in performance and reliability cost more than less sophisticated designs of the same class. Therefore, all values presented in this section must be considered estimates representative of a range of possible values. The local architect-engineer involved in a particular design is best equipped to estimate the cost of the air pollution control equipment for a given plant.

The approximate cost of air pollution control equipment per ton per day of incinerator capacity can be developed on the basis of the following assumptions: (1) 600°F inlet gas temperature, (2) 150% excess air, and (3) 5000 Btu/lb refuse.

The volume of gas handled is approximately 520 ft^3/min/ton/day of capacity if the gas cooling is accomplished with water quenching to 600°F. If the cooling is performed indirectly by water heating, steam generating surfaces, or an air heating device, the quantity of gas to be handled is substantially reduced. (See Figure 20.) In this case, the volume of gas to be handled is approximately 365 ft^3/min/ton/day of capacity. If combustion is completed with 50% excess air and a steam generating surface is used to control furnace temperatures and cool the gas to 600°F, the volume of gas is reduced to approximately 220 ft^3/min/ton/day of capacity (Figure 20).

Many sources can be quoted for the cost of air pollution control equipment. The data given in Table 18 were obtained from statements given before the U.S. Senate Committee on Public Works, Subcommittee on Air and Water Pollution, May 18, 1967 by Earl L. Wilson, president of the Industrial Gas Cleaning Institute, Inc. These data are included here since all other reliable information falls within these broad ranges.

Table 18 *Cost of Air Pollution Removal Equipment*

Type of Collector	$/$ft^3$/min[a] Equipment Cost	$/$ft^3$/min Erection Cost	$/$ft^3$/min Yearly Maintenance and Repair Cost
Mechanical collector	0.07–0.25	0.03–0.12	0.005–0.02
Electrostatic precipitator	0.25–1.00	0.12–0.50	0.01–0.025
Fabric filter	0.35–1.25	0.25–0.50	0.02–0.08
Wet scrubber	0.10–0.40	0.04–0.16	0.02–0.05

[a] Dollars per actual cubic foot of gas handled.

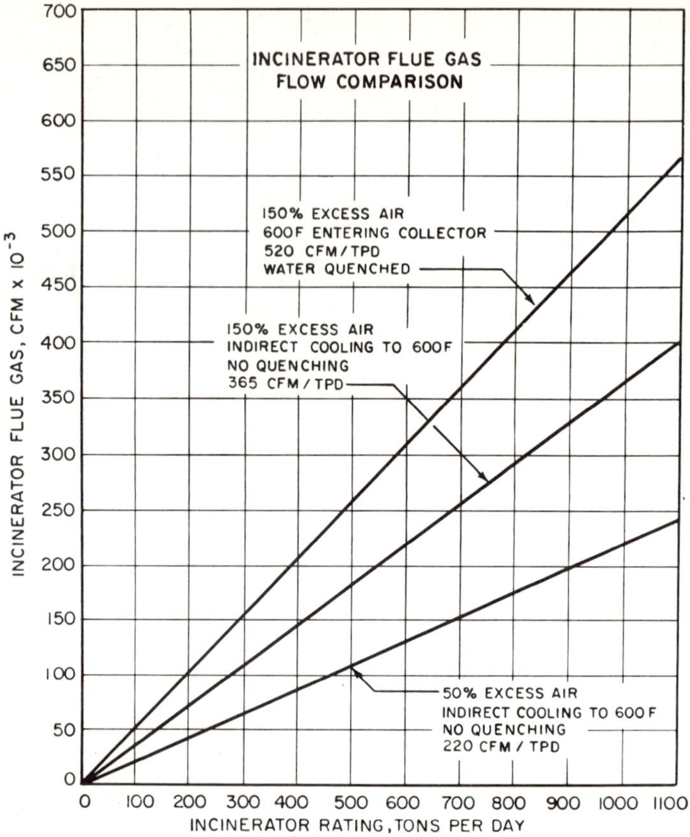

Figure 20. Incinerator flue gas flow comparison.

Because of the difficulties involved with the collection of incinerator fly ash, the actual cost of incinerator air pollution control equipment would probably be found at the higher ends of the ranges given. These cost ranges also account for the decrease in price with increase in unit size.

By way of illustration, a 95% efficient electrostatic precipitator on a 400-ton/day incinerator operated in accordance with the foregoing assumptions (520 ft^3/min/ton/day) would have the following cost in 1967:

$$520 \times 400 = 208{,}000 \text{ ft}^3/\text{min}$$
$$208{,}000 \times \$.80 = \$166{,}400 \text{ fob manufacturer}$$
$$208{,}000 \times \$.40 = 83{,}200 \text{ delivery and erection cost}$$
$$\text{Total delivered and erection cost} = \$249{,}600$$

The capital cost is reduced for air pollution control equipment, ID fan, ducting, and stack when the flue gas is cooled indirectly (see Figure 2). This difference justifies consideration of the utilization of the heat generated in an incinerator. If the energy released was used to supply the power to run the incinerator plant or to heat the plant and surrounding buildings, the saving in capital cost due to the reduced size of the air pollution control and gas handling equipment plus the energy saving could offset the additional cost of the heat conversion equipment. The operating savings could possibly reduce the cost of incinerating the refuse. There are excellent opportunities in the field of incinerator heat utilization, but additional research is needed if it is develop to its full potential.

A reduction in flue gas temperature to 600°F was assumed for the examples presented here, but an additional reduction in size and cost of gas handling equipment is possible if the flue gas is indirectly cooled even further. It may be possible to reduce the incinerator flue gas temperature to 350°F or less before release to the atmosphere. This would decrease the gas volume to be handled by nearly 15%. The additional heat recovery could make energy utilization even more attractive.

Table 19 summarizes the interrelationship and comparison of the various air pollution control equipment systems with respect to performance, size, and cost. The second column introduces a new and important parameter: the space required by each class of system. Column six presents a very important factor that is frequently overlooked: a comparison of the relative operating cost between the various systems. Many organizations buy their units on a lowest capital cost basis without regard for the continuing operating expense. On a capital cost basis, it would be difficult to justify a unit with improvements such as indirect heat exchange. Units of this type, when energy credits are not included, are at best only on a cost par with the simpler water quench systems.

Summary

Although an incinerator stack may appear reasonably clean, the fly ash in the gas may be excessive. Large quantities of air used in incineration often mask the real pollution potential. As a result, stack observations are no measure of emission from an incinerator. An accurate determination of stack emissions can be obtained only by actual tests based on samples taken in the duct leaving the air pollution control equipment.

This section has presented a survey of the performance capability of air pollution control equipment. More knowledge of incinerator

Table 19 *Comparative Air Pollution Control Data for Municipal Incinerator*

Collector	1 Relative Capital Cost Factor (F.O.B.)	2 Relative Space (%)	3 Collection Efficiency (%)	4 Water to Collector (Per 1000 ft^3/min)	5 Pressure Drop (in. Water Column)	6 Relative Operating Cost Factor
Chamber	Not applicable	60	0–30	2–3 g/min	0.5–1	0.25
Multicyclone	1	20	30–80	None	3–4	1.0
Cyclones to 60 in. dia. tangential inlet	1.5	30	30–70	None	1–2	0.5
Scrubber[a]	3	30	80–96	4–8 g/min	6–8	2.5
Electrostatic precipitator	6	100	90–97	None	0.5–1	0.75
Fabric filter	6	100	97–99.9	None	5–7	2.5

[a] Includes necessary water treatment equipment.

pollutants is required, and some research must be performed if all classes of high performance air pollution cotrol equipment are to be applied to incinerators with optimum results. Regardless of whether lax or stringent air pollution regulations are in effect, collectors of good design, properly installed, operated, and maintained can be selected and the relative cost estimated.

Water Pollution

Nearly all incinerators require water. The residue is quenched to facilitate handling except for small units and some batch type units where the refuse is allowed to burn down and the residue is removed after it has cooled. The gas leaving the incinerator furnace is water quenched except where steam generation or other heat exchange is employed. Therefore, extensive use of water is a normal feature of the present incinerator and is potentially a water pollution problem. Since water is required, many incinerator systems call for a scrubber as the method of gaseous effluent cleanup. It is a natural extension of the already extensive water requirements indicated above. When water is used in this manner, a small percentage must be discarded to keep the quantity of fines from increasing and being reintrained when the water is recirculated. However, this can normally be taken care of by connecting the scrubber sump with the residue quench tank, so that in removing the residue, some of the water is removed. Because a wet sludge is taken from the residue tank, this should remove a sufficient quantity of the collected fines so that it will not be reintroduced in the quenching chamber or scrubber and be reintrained in the gas flow, adding to the burden on the air pollution control equipment.

The sludge collects in an open container awaiting removal and the basic character of the ash itself neutralizes the material. The water involved will take on a neutral pH after only a short time in contact with the residue. If the residue for some reason was unable to neutralize the sludge, chemicals would have to be added. Therefore, the incinerator residue, after a short period of holding in the final residue receptacle, is a sterile inert material, capable of being used in construction or land fill without offending the environment. If a scrubber is used and the water is recycled back to the quencher-scrubber area, the water may have to be treated to ensure a neutral uncorrosive pH value that would not be detrimental to these units. The scrubber-and-quench chamber should be lined to protect it from exposure to condensing gases, which in themselves can quite readily attack the metal before they are sufficiently diluted by the water to reduce their activity.

Another aspect of the water, when it is being reused and recycled within the system, is that it becomes heated and can generate considerable steam around the unit. It may be advisable to use a small cooling tower to cool the water before reintroducing it into a scrubbing system. In the quenching area, the amount of makeup is sufficient so that no cooling is necessary. The steaming problem occurs only in cold weather when the high moisture of the gases creates a considerable steam plume. This is not considered an environmental hazard unless the incinerator is in a location where under certain meterological conditions the plume may be brought down to fumigate a nearby hillside. When the air temperature is low, this can cause icing problems and is a potential danger to nearby activities. This, however, can be taken into consideration when designing the plant. Such things as the exit gas velocity and height of the stack can be designed to avoid the problem.

The foregoing assumes water recycling. If fresh water is plentiful, the decision may be made not to recycle and thus reduce the potential for corrosion. In these cases, water treatment must be employed before the water is returned to the water course. The water treatment would include chemical neutralizing and clarifying. Filtering and aeration might also be recommended. The services of a water consultant would be advisable when designing this subsystem. Operation must be as simple and automatic as practicable to ensure that acceptable water is introduced to the receiving water course.

Residue

An incinerator's primary objective is to decrease the volume of the refuse supplied and to produce a residue that is inoffensive to the environment. It may have as a secondary objective the generation of steam or in some other manner make use of the heat released, but the primary consideration is the reduction in volume to a sterile residue. The potential offense against the environment is measured in two ways. One is the putrescible content of the residue and the other is the combustible content. Tests for these elements are mentioned elsewhere and unfortunately are not as well standardized as might be hoped. The putrescible is the organic material remaining in the residue after it has passed through the incinerator. An accepted standard for putrescible at the present time is an amount less than 0.5% of the residue produced in a properly operating incinerator. Tests for putrescible involve chemical analysis or fermentation studies. Probably the more severe requirement is the meaningful sampling of the residue to ensure representative results. These representative samples must also be tested for their carbon or combustible content. Certainly, the putrescibles are burnable and will constitute part of the combustibles, but the principal constituent of the combustibles in the residue is carbon and many specifications state the combustibles as the percentage of carbon in the residue. A reasonable percentage for a properly operating incinerator is a combustible value for the residue of less than 5%.

If the residue meets the foregoing requirements, it is essentially sterile and can be disposed of in many ways, such as a valuable fill in construction. The metallics in the residue can be separated out for salvage, making the residual ash even more valuable as a fill or in place of sand and aggregate in construction and road work. If these profitable opportunities are not available to the incinerator operator, he can bury the residue ash without contributing to an environmental offense. Unlike normal refuse, the sterile residue compacts well and allows for the immediate reclamation of the land used as the filling site.

All incinerators should be designed to meet the above-mentioned putrescible and combustible residue content. Not only must they be designed to meet them, but they must be properly operated or they will not meet them. The best designed incinerator, poorly operated, can result in unsatisfactory residue and problems. Treatment of the residue must be done in the incinerator, not attempted after it has left the residue quench chamber.

MUNICIPAL PRACTICES

The nature of municipal purchasing requires an understanding of many of the current practices. Since the bulk of the waste reduction facilities will involve municipal activities, an understanding of the many problems is required. Public works projects cannot generally be purchased without competitive bidding. To establish the quality of the equipment required and its relation to the rest of the system is a difficult matter, rendered no less so by the complexities introduced by the various governmental bodies concerned.

Figure 21 to illustrates the various amounts of information required. Basically, municipal incinerators are purchased either by performance standards or construction specifications.

In the case of performance standards, minimum requirements are prescribed and the approach to the process utilizes the system concept. The need for adequate test procedures has made it difficult to achieve sound purchasing and assurance that the performance requirements

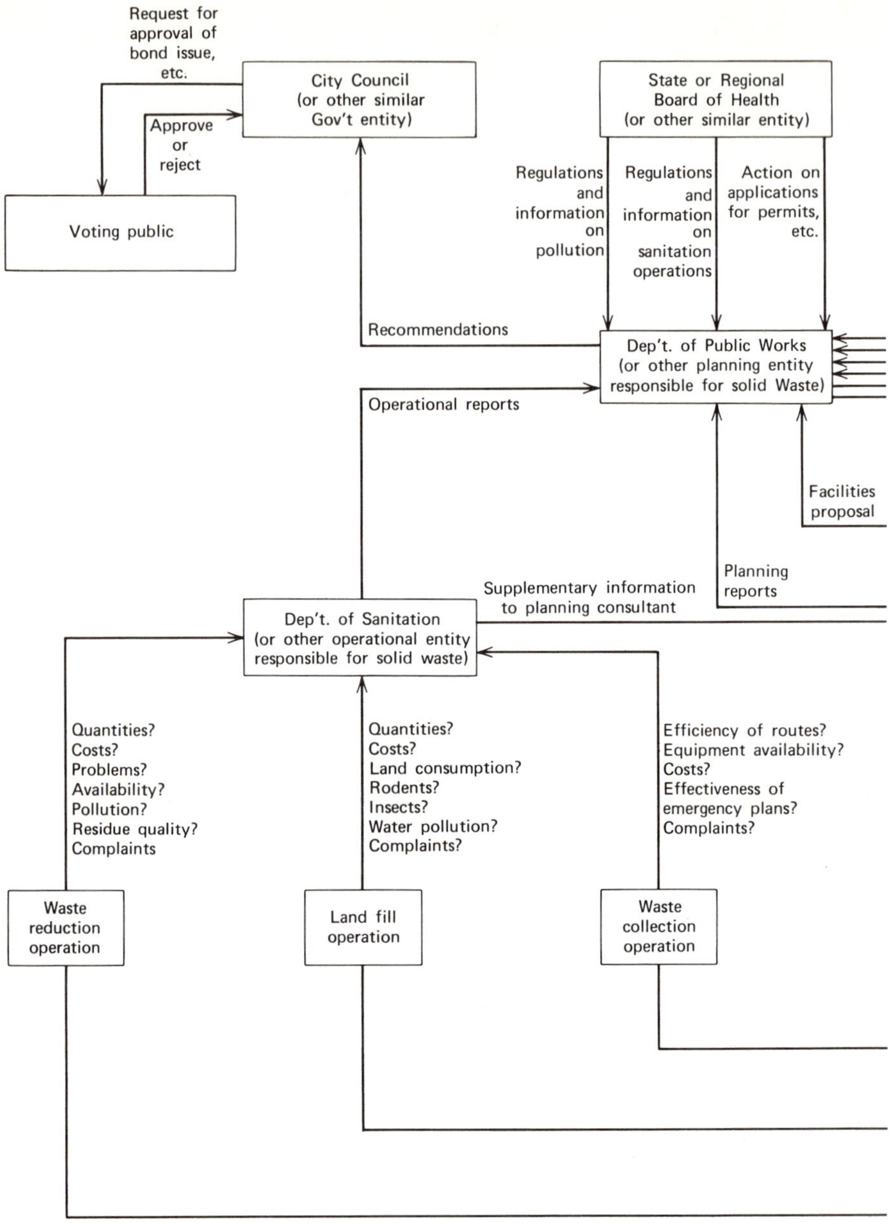

Figure 21. The flow of information in municipal refuse activities. Each box denotes an entity, as labeled; the arrows indicate the flow of data and information.

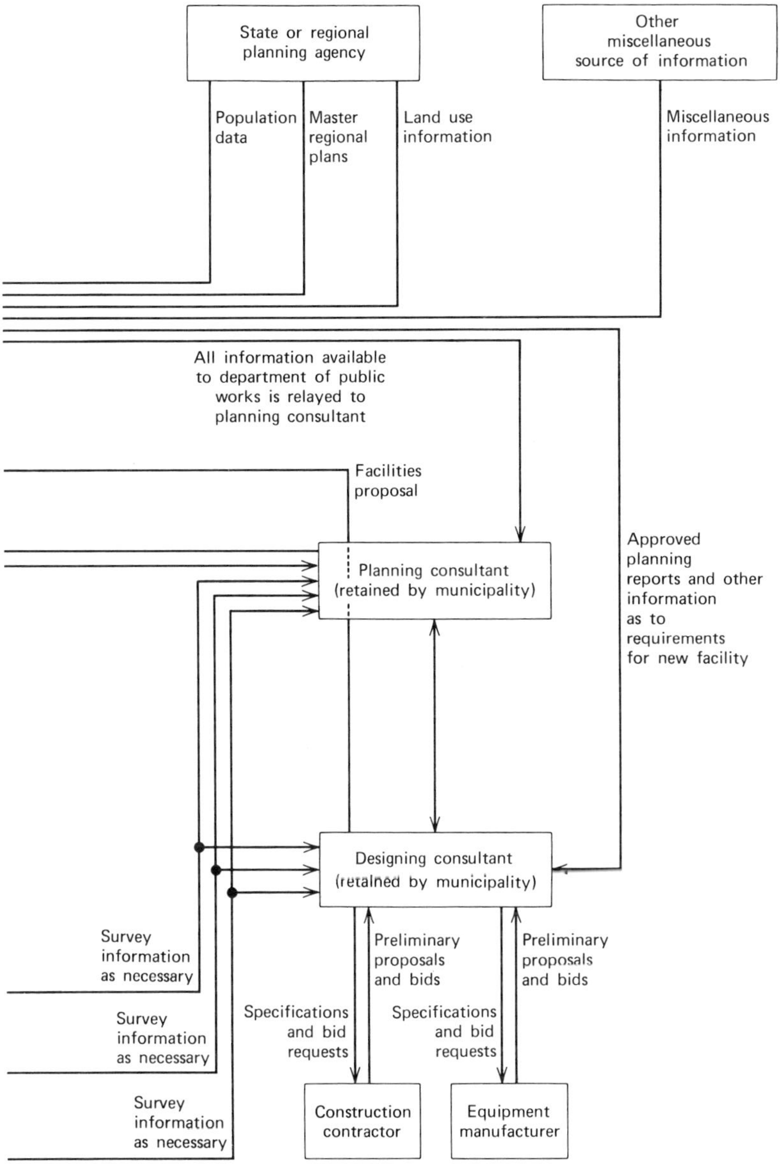
State or regional planning agency
Other miscellaneous source of information
Population data
Master regional plans
Land use information
Miscellaneous information
All information available to department of public works is relayed to planning consultant
Facilities proposal
Planning consultant (retained by municipality)
Approved planning reports and other information as to requirements for new facility
Designing consultant (retained by municipality)
Survey information as necessary
Survey information as necessary
Survey information as necessary
Preliminary proposals and bids
Preliminary proposals and bids
Specifications and bid requests
Specifications and bid requests
Construction contractor
Equipment manufacturer

are being met. Many agencies are currently directing their activities towards the development of measurement standards that will enable sound evaluation of equipment performance.

A construction specification is more commonly used. It defines in detail a minimum set of standards and requirements. It tends to be a "hardware" type of specification, and frequently does not take into consideration the coordination of the system. Major components purchased on this basis are frequently judged on price alone, without regard for their performance and relationship with other components.

In addition to the various methods by which specifications are written, the type of approach is often dictated to by the opinion of the various buying influences. Municipalities are necessarily structures in the same manner, and the buying influences and the control exercised by these influences frequently differs. The role of all parties must be completely understood if improved performance and progress are to result.

Incinerator Testing

Incinerator performance is difficult to measure and as a result data obtained are often controversial and frequently inaccurate and misleading. Recognizing this, the American Society of Mechanical Engineers has established a committee to develop a performance test code for incinerators (PTC 33). When this is available, it should be broadly adopted; specifically, the procedures will be adaptations of standard acceptable test practices modified to suit refuse characteristics.

Certainly, the heterogeneous nature of refuse complicates the problem, and comparisons between installations can probably best be made on some statistical basis, but data used for comparisons must at least be obtained under similar conditions and tested under identical test procedures. Valid performance criteria are essential to the understanding and ultimate resolution of the problems confronting us. To this end, a procedure should be directed to test methods that will enable us to establish an acceptable level of burnout. Any specification and test method must meet certain basic requirements. These are

1. Clearly stated and understandable test objectives.
2. A base operating condition and a refuse standard.
3. A method to contract fuel including refuse.

4. Procedures fo sampling input material.
5. Procedures for sampling output material.
6. Procedures for determining qualitatively the input.
7. Procedures for determining qualitatively the output.
8. Method for evaluating results including necessary calculation and report.
9. Test duration.

Before specifying a required value for burnout, the engineer must determine what the test objectives are:

1. Bulk reduction only.
2. Acceptable bulk reduction and some specified minimum acceptable value of organic in residue.
3. Minimum organics in residue and some special emphasis on throughput.
4. Other pertinent data.

Tests normally conducted on an incinerator plant are to determine

1. Throughput (capacity).
2. Percentage of reduction.
3. Organic in residue.
4. Efficiency or performance of steam generation if included.
5. Other purposes of tests, allied with or separate from the primary objectives, include air pollution as revealed by dust loading and others, that is, SO_X, NO_X etc. efficiency of dust collection, data on water consumption, quantity and composition of waste water, efficiency of waste water treatment equipment, and density of residue.

The most frequently used method of determining overall performance for incineration is "burn-out."

Specification

The specification must clearly state what is required, that is, it should state that the residue leaving the unit will not contain more than the agreed percentage of organics by weight of the residue, organic putresibles being defined as the sum of the protein, carbohydrates, and lipids. Limit of other combustibles should be specified.

The lipid, carbohydrate, and protein content will be analyzed in accordance with the accepted test procedures and subject to the errors outlined in the evaluation of results.

Corrections to Design Conditions

There must be some pre-agreed-upon basis for correcting results to a "standard refuse" condition.

Procedure for Sampling Refuse

It is imperative that every sample be collected and prepared carefully and conscientiously and in strict accordance with procedures prescribed; for if the sampling is done improperly the sample will be in error, and it may be impossible or impractical to take another sample. However, if the analysis is in errror, another analysis can easily be made of the original sample.

Because of the wide variations in conditions under which the material must be sampled and the nature of the material to be sampled, it is important that the samples be collected by a trained and experienced sampler.

A sampling method:

1. Establish a length of sampling time (usually the test duration).
2. Obtain a 50–75-lb representative sample every 10 min until approximately 2200 lb (about 130 ft^3) is accumulated.
3. Remove large, *clean* bulky objects such as bicycles, auto wheels, and stoves, until a net sample of 2000 lb is obtained.
4. Pass all of sample through size reduction machine and reduce to ½ to ¾-in. size.
5. Reduce size further until it all passes through a 10-mesh screen.
6. Mix total sample thoroughly, divide in successive steps until four 500-lb samples are obtained and seal two of the samples for arbitration.
7. Further divide one of the 500-lb samples until five 100-lb samples are obtained. Seal each of the 100-lb samples and send to the laboratory.
8. Send the other 500-lb sample to the laboratory for preliminary tests for moisture, ash, and heating value.
9. Each of the five 100-lb samples shall be further divided in the laboratory to obtain five 20-lb samples. We now have twenty-five 20-lb samples.
10. Each of the 20-lb samples shall be analyzed for the following constituents and be reported as a percentage of the dry sample. Twenty-five analysis for each constituent:

    Water soluble fraction
    Organic soluble fraction

Volatile fraction
Chemical water
Silica
Iron
Free carbon
Ash
Moisture
Heating value

11. The results of each constituent shall be averaged to determine final average of each constituent.

Figure 22 illustrates this procedure.

Procedure for Sampling Residue

As much care and effort should go into the sampling of the residue. This residue sampling procedure is indicated as follows:

1. Sampling of residue shall begin when sufficient time has elapsed to allow sampled refuse to appear in residue.
2. Residue sampling time shall be the same as the refuse sampling time.
3. Obtain a 25-lb sample every 10 min; 5 lb from each equally divided section across the width of the discharge grate until 400 lb have been accumulated.
4. Pass all material through a machine to reduce the size to ½–¾ in.
5. Further reduce size until all residue passes through a 10-mesh screen.
6. Mix final sample and divide until four 100-lb samples have been obtained.
7. Seal two of the 100-lb samples for arbitration purposes.
8. Send two 100-lb samples to the laboratory, one sample for preliminary analysis of moisture, ash, and heat value.
9. The remaining 100-lb sample shall be further mixed and divided in the laboratory to obtain four 25-lb samples.
10. Each of the 25-lb samples shall be analyzed for the following constituents and be reported as a percentage of the dry sample. Five analyses from each sample for each constituent. Twenty analyses for each of the following constituents:

Water soluble fraction
Organic soluble fraction
Volatile fraction

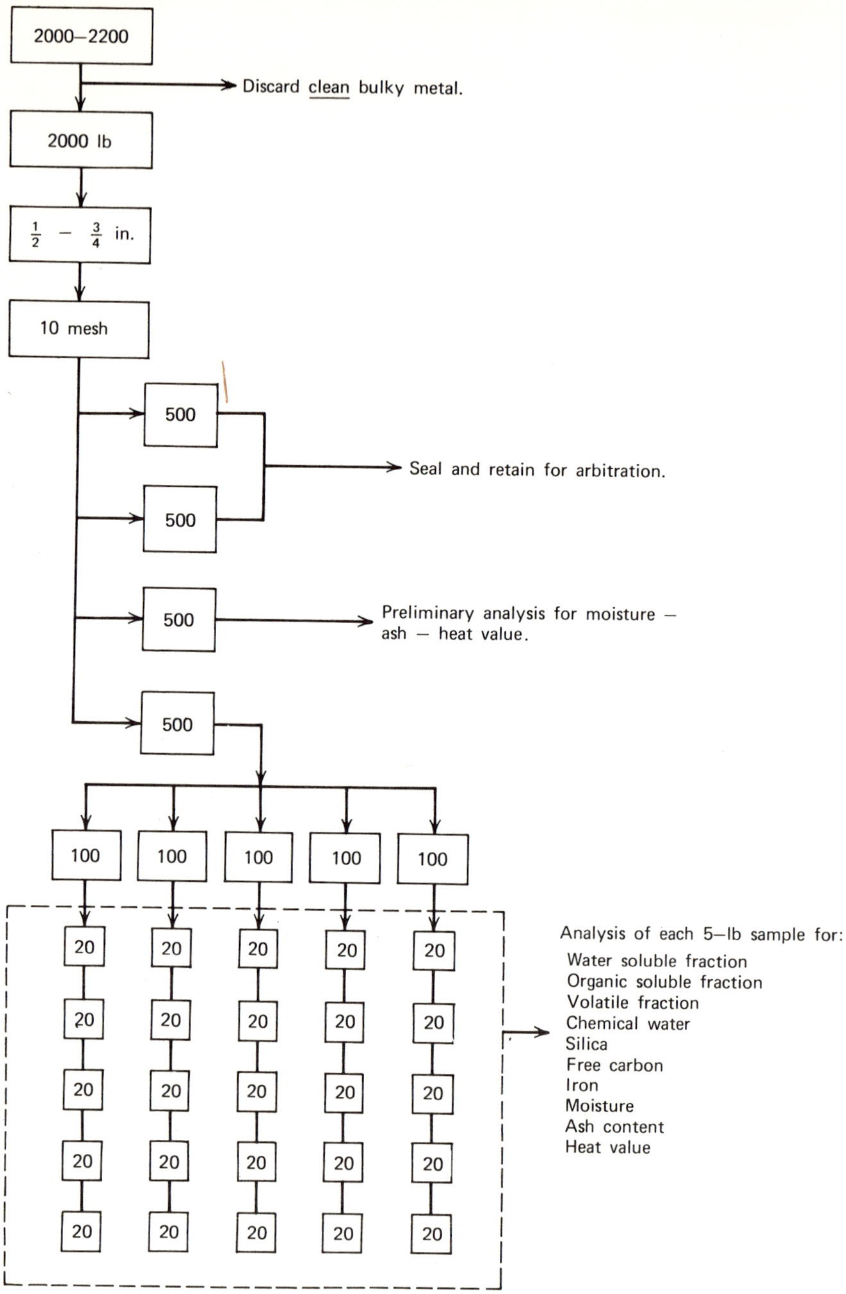

Figure 22. Sampling procedure for refuse.

Chemical water
Silica
Iron
Free carbon
Ash
Moisture
Heating value

11. The results of each constituent shall be averaged to determine final average of each constituent.

Figure 23 illustrates this procedure.

While this may appear unduly complicated, it is necessary if accurate and meaningful results are to be obtained.

The sampling recommendations of the ASTM Standards on Coal and Coke may appear tedious and costly. Coal is a relatively homogeneous material and its properties are well known; the sampling of refuse and residues should accordingly be subject to at least as stringent requirements. Variations in minimum weight increments and minimum weight of gross, the sample and number of sampling increments are specified for eight size classifications ranging from ⅝ in. and under, to over 6 in., and for four ash classifications from 8% ash and under to 15% ash and over. Weight increments vary from a minimum of 2 lb to 30 lb. Minimum weight of gross sample varies from a minimum of 30 lb to a maximum of 1050 lb. Minimum number of increments vary from 15 to 35. All of these vary in accordance with the size and ash content.

ANALYTICAL PROCEDURES

Standard pre-agreed-upon chemical analytical procedures should be applied for specific tests.

Evaluation of Results

Since methods of evaluating or comparing results are important technically as well as legally, acceptable limits or variations in results should be defined in advance. Points to be considered should be

1. Variation in results if two different laboratories are engaged to make determinations on the same sample.
2. Variation if different technicians are employed in the same laboratory.
3. Variation if two different laboratories analyze two different samples of the same test.

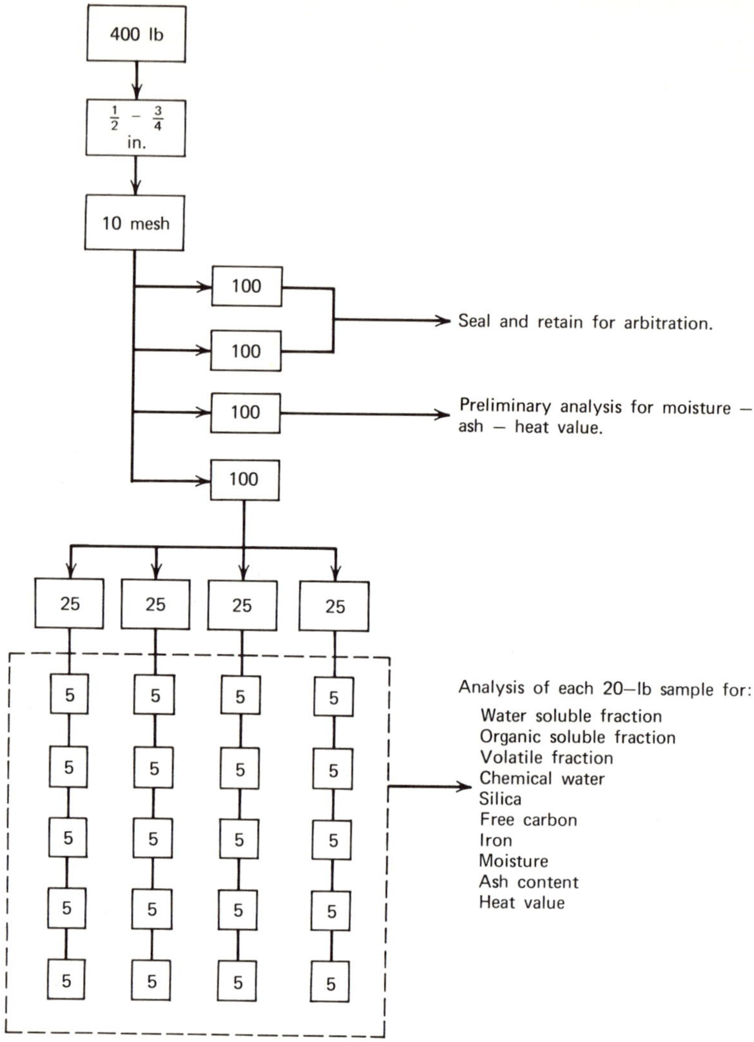

Figure 23. Sampling procedure for residue.

Steam Generation

Where steam generation is combined with incineration the steam generator should be tested in accordance with accepted ASME test procedures (Performance and Test Code PTC 4.1).

REFERENCES

1. J. M. Bell, "Development of a Method for Sampling and Analyzing Refuse," Purdue University, 1963.
2. J. M. Bell, "The Physical and Chemical Composition of Municipal Refuse." (Ref. 1)
3. L. J. Cohan and A. P. Selker, "The Design and Development of a Compact Incinerator," presented at 62nd Annual Meeting of APCA.
4. L. J. Cohan and J. H. Fernandes, "Industrial Waste Disposal," Presented at New England Plant Engineering and Maintenance Conference, Newton, Massachusetts, 1970.
5. L. J. Cohan and J. H. Fernandes, "Potential Energy Conversion Aspects of Refuse," presented at Winter Annual Meeting ASME, Pittsburgh, Pa., 1967.
6. Otto DeLorenzi, *Combustion Engineering,* Combustion Engineering Inc., 1957, Windsor, Conn.
7. J. H. Fernandes, "Incinerator Air Pollution Control," presented at National Incinerator Conference, ASME, New York, 1968.
8. J. A. Fife, "Control of Air Pollution from Municipal Incinerators," presented at National Conference on Air Pollution, Washington, D.C., December 1966.
9. J. V. Fitzpatrick, "Solid Refuse Disposal Practices as Related to Air Pollution Problems," presented at National Conference on Air Pollution, Washington, D.C., December 1966.
10. G. R. Fryling, *Combustion Engineering,* Rev. ed., Combustion Engineering, Inc., 19th ed.
11. P. B. Hawk, and O. Bergeim, *Practical Physiological Chemistry,* Blakeston Company, Philadelphia, Pa.
12. W. B. Hickman, unpublished correspondence.
13. E. R. Kaiser, "Combustion and Heat Calculations for Incinerators, Refuse Composition and Flue Gas analysis from Municipal Incinerators." Proceedings of the National Incinerator Conference, ASME, New York, 1964.
14. J. F. Mullen, unpublished correspondence.
15. H. Munde, unpublished correspondence.

16. W. R. Niessen, "Systems Study of Air Pollution from Municipal Incineration," Report to National Air Pollution Control Administration, USHEW Contract CPA-22-69-23.

17. W. R. Niessen and A. F. Sarofin, "The Emission and Control of Air Pollutants from Incineration of Municipal Solid Waste," presented at 2nd International Clean Air Congress, Washington, D.C., December 1970.

18. G. V. Prible, "Refractory Characteristics and Trends in Incinerator Furnaces," presented at ASME, Incinerator Division, New York, November 1968.

19. P. W. Purdom, Director, Environmental Engineering and Science, Drexel Institute, Philadelphia, personal communication.

20. J. W. Regan, J. F. Mullen, and R. D. Nickerson, "Suspension Firing of Solid Waste Fuels," presented at American Power Conference, Chicago, Ill., 1969.

21. J. W. Regan, "Generating Steam from prepared Refuse," presented at National Incinerator Conference, ASME, Cincinnati, Ohio, May 1970.

22. A. B. Walker, and F. W. Schmitz, "Characteristics of Furnace Emissions from Large, Mechanically Stoked Municipal Incinerators," Proceedings of the National Incinerator Conference, ASME, New York, 1966.

23. G. Stabenow, "European Practices in Refuse Burning," National Incinerator Conference, ASME, New York, May 1964.

24. Technical-Economic Study of Solid Waste Disposal Needs and Practices," Combustion Engineering Inc., for U.S. PHS PH-86-66-163.

25. Quad City Solid Wastes Project," USHEW, PHS, NCUIH, Cincinnati, Ohio, 1968, Grant Do1-U1-00026.

CHAPTER II.14
rail transportation

Hauling solid waste by rail would potentially make many landfill sites available. Solid waste rail haul landfilling differs from existing methods primarily in scale of operations. Present methods of sanitary landfilling could be used with minor modifications. Owing to the kind of equipment needed, a trench, for example, might be 100–200 ft wide and 50–100 ft deep.

Pits and quarries are available almost everywhere but are not appropriate to rail hauling of solid wastes because they usually lack the required capacity and sufficient inexpensive cover material.

Open pit mines are large but few; for example, there are iron ore mines in Minnesota and the copper mines in Utah and other states. These mines would probably be adequate for the disposal of solid wastes in their areas, but they are not ordinarily available. Technological advances extend the life of a mine by making it profitable to exploit lower-grade ores; hence the owners would be reluctant to forego this opportunity by having the pits filled with wastes.

Solid waste disposal in abandoned strip mines might be considered as a special case of ordinary sanitary landfilling. Abandoned strip mines differ from other sites primarily in topography. One disposal opportunity is offered by the abandoned last trench, another in the gaps between the spoil banks. The latter fits well with the reclamation requirements of many states, and would help the owners to defray their reclamation costs. It might also be possible to work abandoned strip mines backwards in order to use the total area better.

Solid waste disposal could also be combined with the normal activities of an operating strip mine. This procedure would add little or no cost to the mine operations and would eliminate the costs of providing disposal space and of covering material for the wastes.

No existing landfill is as large as one that would be appropriate to a solid waste rail haul system. The largest present system disposes of 8000 tons/day of refuse at Fresh Kills, Staten Island, N.Y. On the other hand, a rail haul system would be capable of delivering tonnages that require a single site capable of handling 10,000–30,000 tons/day, or even more.

As in many other fields, costs for sanitary landfills follow the economics of scale: The larger the amount of wastes to be disposed of, the lower the unit cost, as shown in Figure 1. The costs shown exclude the cost of land and site development, but as a rule these amount to less than $0.05 per ton.

Rail haul could provide access to active strip mines, thus opening heretofore unused disposal opportunities, and at the same time providing positive benefits to the community through a land reclamation program. Such a program would require integration of strip mine operations with disposal activities in order to realize the economies.

Strip mining equipment is adaptable to large-scale disposal operations; some equipment, for example, might be used in reworking abandoned strip mines, thus enabling solid waste rail haul to aid in the reclamation of waste lands.

The USDI estimates that about 3.2 million acres or 5000 miles^2 of land had been disturbed by surface mining up to January 1, 1965. Only

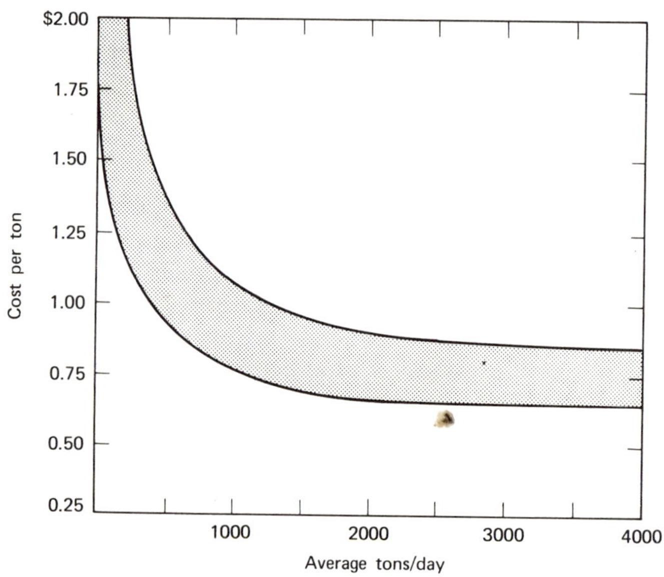

Figure 1. Range of sanitary landfill operating cost. (Source: American Public Works Association Research Foundation Project 67–1.)

about one-third of this acreage has been reclaimed, leaving about 2 million acres requiring reclamation. Although it is difficult to estimate the annual increase in the area disturbed by surface mining, the 1964 figure is estimated at 150,000 acres.

Figure 2 suggests that about 41% of the land disturbance is caused by the surface mining of coal; the second largest factor is sand and gravel, 26%. The mining of seven commodities accounts for 95% of all the land disturbed by surface mining.

The amount of coal mined by stripping increased by about 9% between 1965 and 1966, from about 165 million to about 180 million tons. *Coal Age* forecasts that by 1985 the production of bituminous coal by strip mining will be at least 380 million tons and could be as high as 520 million, or about two to three times the present level. Active coal strip mines thus appear to offer ample potential sites for the disposal of solid wastes for many years to come.

Geographical location of coal strip mines is favorable for consideration as disposal sites. States east of the Mississippi accounted

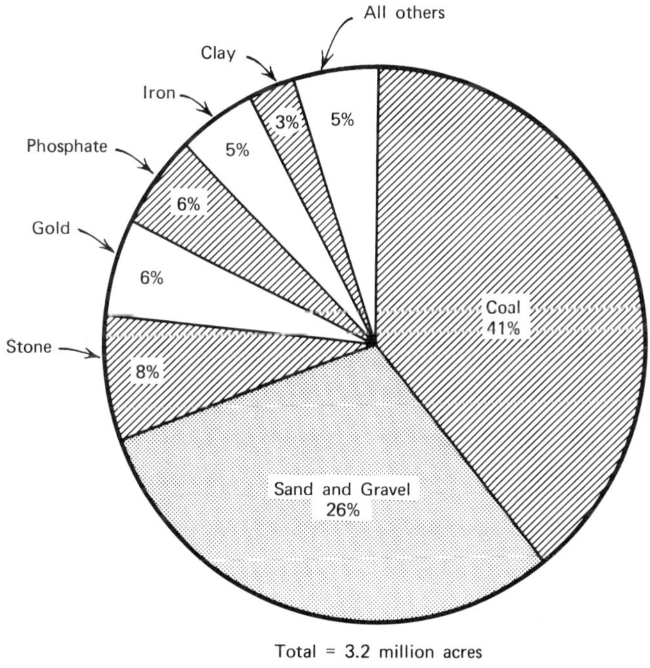

Figure 2. Percentage of land disturbed by surface mining. (Source: U.S. Department of the Interior.)

for 95% of the total coal produced in 1967, or about 549 million tons. Kentucky, West Virginia, Virginia, and Pennsylvania rank highest in number of operating coal mines; see Table 1. Thus mining operations are found either in or close to most industrial states.

Three methods are used in surface mining of coal.

Area stripping refers to digging a series of parallel trenches in relatively flat or rolling terrain. The spoil material is placed in a previously made cut, and the mine then resembles the ridges of a washboard with an open trench where the last cut was made.

The second major method is *contour stripping,* which refers to digging around a hillside in steep or mountainous country. It creates a shelf bordered on the inside by an edge that forms a wall that may be as high as 100 ft and on the outside by an edge that forms a rim with a steep outslope covered by loose spoil material.

The third method is *auger mining,* which refers to boring horizontally into a seam to get more coal after the stripping is finished. A cutting head is drilled into the seam and coal is scraped out along the auger threads.

Although there are fewer strip mines than underground mines, their productivity, in average output per man per day, is twice as great. There has been significant growth in coal mining by the strip method. Owing to the rising costs of mining operations, those methods having high productivity will also have a continued growth.

The investigation of the feasibility of solid waste disposal in active strip mines concentrates on area strip rather than contour mine operations, because of the multiple moving and handling costs involved in contour mining. It is also felt by mining companies that the relatively high operating costs of contour mines might make them unlikely sites for the disposal of solid wastes within the context of ongoing operations. An example of area strip mine operations is illustrated in Figure 3.

In considering the cost structure of area strip mines, it must be emphasized that each mine is operated somewhat on an ad hoc basis. There are many local variables which might affect cost data rather substantially. These variables include the amount of stripping that needs to be done, the overburden soil conditions, the length of haul, the demands on the roads that have to be built, coal yields, and the number and types of trucks that can be used.

To identify the unit operations which might be affected by solid waste disposal, it is necessary to gauge the general cost structure of area coal strip mines. The cost range of ten major cost centers can be identified (Table 2).

Five to eight of the ten cost centers might be affected in varying

Table 1 *Major Coal Production Activities by State, USBM*

States East of Mississippi River	Total Coal Production (Deep, Auger, and Strip) 1967 tons (mm)	Number of Bituminous Coal and Liguite Mines, 1965	States West of Mississippi River	Total Coal Production (Deep, Auger, and Strip) 1967 tons (mm)	Number of Bituminous Coal and Liguite Mines, 1965
Alabama	15.2	206	Arkansas	0.3	8
Illinois	65.7	90	Colorado	5.4	79
Indiana	18.0	61	Iowa	1.0	28
Kentucky	96.3	1,827	Kansas	1.2	6
Maryland	1.4	69	Missouri	3.8	16
Ohio	45.7	417	Montana	0.4	13
Pennsylvania (anthracite)	11.6	N.A.	New Mexico	3.6	8
Pennsylvania	79.0	1,140	North Dakota (Lignite)	4.0	29
Tennessee	7.3	230	Oklahoma	0.8	15
Virginia	37.9	1,271	Utah	4.5	31
West Virginia	152.2	1,660	Washington	0.06	5
			Wyoming	3.7	14
Total:	519	6,971		30	252

Source: Keystone Coal Buyers Manual—1968, Minerals Yearbook, 1966, U.S. Bureau of Mines.

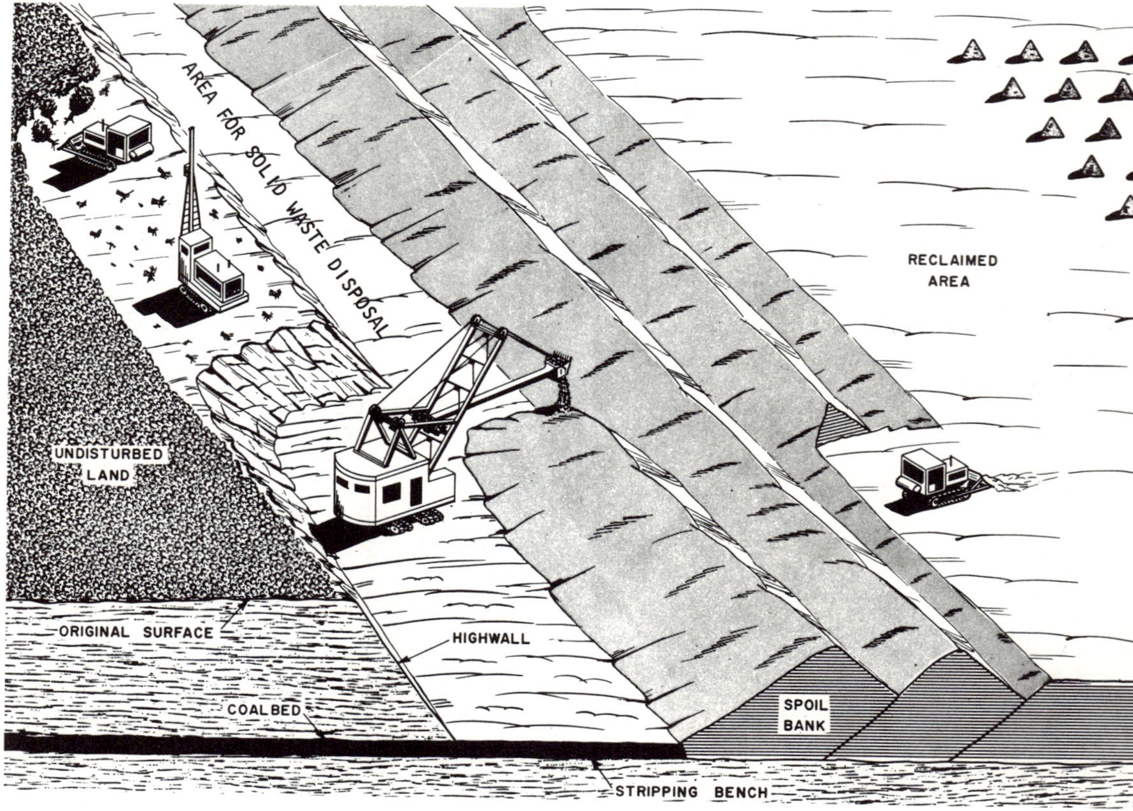

Figure 2 Overview of active strip mining operations

Table 2 *General Distribution of Revenue in Area Coal Strip Mines by Major Cost Center*

Major Cost Centers	Percent of Total Revenue[a]
1. Stripping of the overburden	35–45
2. Coal loading, preparation, prospect drilling	5–10
3. Haulage by truck	3–19
4. Roads	0.5
5. Supervision on site	1–2
6. Land reclamation	0.5
7. Noncontrollable expenses, e.g., UMWA welfare, taxes, insurance	10–15
8. Depreciation	15–20
9. Royalties, general overhead	5–10
10. Profit before taxes	8–10

[a] Total revenue, or mine realization, runs currently at about $4.50 per ton.

degrees by solid waste disposal operations. Excluded are the stripping of the overburden and the operations dealing exclusively with the mining of coal. Based on a marginal cost analysis, the depreciation, the noncontrollable expenses, and the royalty, sales, and general overhead expenditures appear to be only slightly influenced by the solid waste disposal operations.

REFERENCES

1. Rail Transportation of Solid Wastes," APWA Research Foundation under Grant Do 1-U1-00073, USHEW, PHS, CPEHS, ECA, SWM, Cincinnati, Ohio 1969.

CHAPTER II.15

disposal at sea

Sea disposal of U.S. radioactive wastes from nuclear reactors and medical research activities was discontinued by the Atomic Energy Commission in 1967. However, some European nations still dump their radioactive material into the ocean off their coasts.

After every war, all over the world, every nation with outmoded arms and waste material has loaded these on ships, taken them out to sea, and sunk the whole mass. In late 1970 considerable attention was directed in the press to the case of the S.S. LeBaron Russell Briggs, an outmoded freighter, which the Army loaded with tons of nerve gas and sank in a deep trench off the southern coast of the United States. The furor was so great that the military services assured the world it would never again dump chemical warfare agents into the sea.

Smith and Brown,[2] of the Dillingham Environmental Company, La Jolla, California, conducted a study for the Bureau of Solid Waste Management in reference to ocean dumping. They studied 20 U.S. cities: Seattle, Washington; Portland, Oregon; San Francisco, Los Angeles, and San Diego California; Galveston, Texas City, Houston, Port Arthur, and Beaumont, Texas; New Orleans, Louisiana; Pascagoula, Mississippi; Mobile, Alabama; St. Petersburg, Florida; Charleston, South Carolina; Norfolk, Virginia; Baltimore, Maryland; Philadelphia, Pennsylvania; New York, and Boston. The quantitative aspects are given in Table 1.

By the end of the 1960s, it is estimated that 50,000,000 tons of waste were annually dumped into the sea with disposal costs of approximately $30,000,000. About 60% of the tonnage was dredging spoils from harbor deepening operations, carried out mainly by the U.S. Army Corps of Engineers. Some of the spoils include amounts of residues from municipal and industrial discharges.

Industrial wastes are about 5,000,000 tons discharged at a cost of about $8,000,000.

Slurries of solid material remaining after municipal waste water treatment, residual human wastes, untreated sewage, and other organic

Table 1 *Dredging Spoils Constitute Bulk of Wastes Dumped at Sea*

Type of Waste	Pacific Coast[a] Annual Tonnage	Atlantic Coast[a] Annual Tonnage	Gulf Coast[a] Annual Tonnage	Total Tonnage	Estimated Cost ($)
Dredging spoils	7,320,000	15,808,000	15,300,000	38,428,000 (80%)	15,583,000
Industrial wastes[b]	981,300	3,013,200	696,000	4,690,500 (10%)	8,193,000
Sewage sludge[c]	0	4,477,000	0	4,477,000 (9%)	4,433,000
Construction and demolition debris	0	574,000	0	574,000 (<1%)	430,000
Refuse and garbage	26,000	0	0	26,000 (<1%)	392,000
Explosives	0	15,200	0	15,200 (<1%)	235,000
Total	8,327,300	23,887,400	15,996,000	48,210,700	29,266,000

[a] Figures for 1968 for 20 coastal cities.
[b] Includes chemicals, acids, caustics, cleaners, sludges, waste liquors, and oily wastes.
[c] Tonnage on wet basis.

Source: Dillingham Corp.

and inorganic wastes are greater than 4.5 million tons at a disposal cost of about $4.5 million, or roughly a dollar per ton. Construction and demolition debris and refuse was less than a million tons and was disposed of at less than a dollar a ton. Other wastes are debris, refuse, and garbage, chiefly paper, food wastes, garden wastes, steel, and glass containers, and outdated military explosives and chemical warfare agents.

Along the Atlantic seacoast many of our summer resort areas in the north and the winter resort areas in the south find it most convenient and least costly to pump either their untreated or treated sewage wastes a mile or so out into the ocean; this is stated to be the situation even at Palm Beach, Florida, where the wealth of the inhabitants is legendary. There have been many proposals for pumping the waste into the ocean 10–15 miles off the coast on the assumption that the enormous volume and buffering capacity of the ocean would tolerate a large amount of residues. This has now been judged objectionable by the courts.

In the eastern part of the United States industrial waste dumping more than doubled between 1959 and 1968. The total tonnage of industrial waste, sewage sludge, construction demolition debris, refuse, and garbage dumped into the ocean increased more than fourfold in the 20 years between 1949 and 1968, with the expectation that it will increase considerably in the next decade.

Increasingly water and air pollution laws, loss of land areas now used for sanitary landfill and ponding of industrial waste, and increase in population and industry have already caused cities like San Francisco, New York, and Philadelphia to investigate sea disposal of municipal waste and garbage.

Industrial wastes include those from petroleum refining, steel and paper production, pigment processing, insecticide, herbicide, fungicide production, chemical manufacture, oil well drilling, and metal finishing, cleaning, and plating. Annually, barges dumped 2.7 to 3 million tons of waste acid, more than half a million tons of refinery wastes, a third of a million tons of pesticides waste, and 140,000 tons of paper mill wastes into the sea. There are apparently about 126 present disposal areas, 42 in the Pacific, 51 in the Atlantic, and 33 in the Gulf of Mexico.

Off-shore fishermen are quite familiar with the acid waters off the New York Bight in the vicinity of Sandy Hook, New Jersey. It is estimated that this area has received 11 million yd^3 of sewage sludge and dredge spoil in the years 1966–1970. The fishermen call this the "acid waters." and go to the areas because the catch is large on the periphery of this disposal area.

The dumping place of New York City's sewage sludge (Figure 1) and other waste from the metropolitan area was studied at close range by scientists in a minisubmarine. The sewage sludge has remained at its original resting place, a little more than 8½ miles due east of Sandy Hook. Samples of the bottom, water current velocities, make videotapes, and photographs were taken.

A second phase of the project was to see whether waste disposal in the underwater valley tended to disperse rapidly away from the continent. The Hudson Shelf Valley is a natural conduit for sediment transport, and it is possible that wastes dumped in the area would be rapidly carried away.

The area comprises three major dumping sites, one of sewage sludge, one for material collected in dredging operations, and one for debris called "cellar dirt"—broken concrete, brick, dirt, and other debris from building excavations.

The largest volume dumped was the mud from dredging. This came to about 6.65 million yd^3 in the year ending June 30, 1970. In that same year, sewage sludge dumping totaled 3.88 million yd^3, while the "cellar dirt" figure was 486,000 yd^3

The waste material comes from New York City, Westchester and Nassau Counties, and parts of New Jersey. The city's contribution to

Dumping area

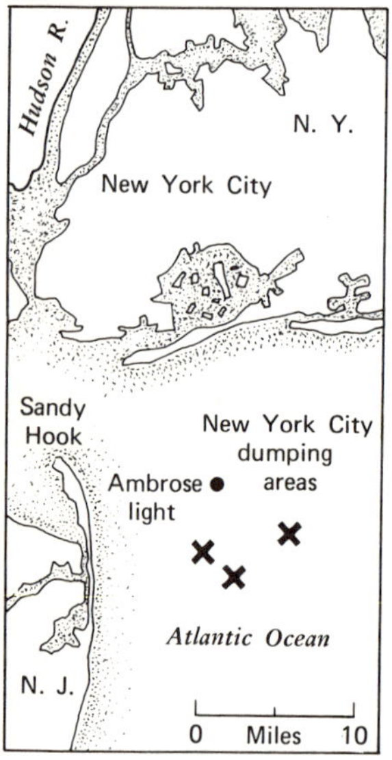

Figure 1. New York's ocean dumping area. (Source: *The New York Times,* Sept. 30, 1971.)

the total was about 25 barge loads of sewer sludge totaling about 1.5 million ft^3/wk.

A third phase of the diving project involved descents along the New Jersey coast to look at offshore underwater sand ridges to determine whether they are shaped primarily by existing and recent seafloor currents or are the remains of ancient beaches that have been overrun by the ocean.

Another project was to learn about the movements of sand and other materials in the New York Bight—the area of the continental shelf from Nantucket, Massachusetts, to Cape May, New Jersey.

The testing Philadelphia carried out in its $100,000 evaluation of the Atlantic Ocean area, where municipal sewage sludge is being disposed of, may be a forerunner of the ecological testing chemical plants will have to perform if they expect to use similar waste disposal methods.

At a session on waste disposal in the ocean, Samuel Baxter, Philadelphia's Water Commissioner, defended his city's disposal of digested sludge into the Atlantic. He strongly refuted critics who would ban all ocean dumping.

Philadelphia has been transporting anaerobically digested sludge to sea since 1961, disposing of it in an area about 12 nautical miles off Cape May. Disposal volume is now about 100 million gal/yr of a sludge that consists of about 90% water, 5% organics, and 5% inert materials. After the Food and Drug Administration banned shellfishing in a 100-mile area that included Philadelphia's sludge disposal site, the city began its study in cooperation with Franklin Institute.

There is no sludge layer in the dumping area. Rather, the bottom material consists of clean sand and gravel. Both vetebrates and invertebrates abound at the disposal site, and they are in good health. Dissolved oxygen, at all water depths, shows no signs of depletion. Coliform measurements are zero at all water depths. No significant heavy-metal concentrations could be measured in macrofauna taken from the disposal area. "It appears that no harm has been done to the ocean after 10 years of sludge disposal by Philadelphia," Baxter says. "Despite this, there are strong demands that we phase out this practice and replace it with an untried alternate, which has the potential of increased sludge-disposal costs," referring to incineration.

Much of the criticism of Philadelphia's disposal program was engendered by findings in 1969 on sludge dumped by New York City into the New York Bight. The New York sludge was largely undigested and besides, the New York Bight is not as well flushed as Philadelphia's disposal area. Each ocean disposal project, Baxter asserted, has to be evaluated separately, taking into account the dumping location, material to be handled, and actual impact on the marine environment.

For example, the Atlantic Ocean site into which NL Industries has been disposing of its titanium pigments wastes has become a fishing spot. The present program seems fully compatible with the ocean environment, and barging beyond the continental shelf would put the sludge beyond the reach of effective biotic recycling.

Bulk industrial wastes are normally discharged from tank barges in motion propelled by seagoing tugs, but containerized disposal methods are increasing in application. Dredged spoils may cost as much as \$0.40 per ton for disposal to as high as \$24.00 per ton for containerized industrial wastes, though these in bulk shipment average about \$1.70 per ton for disposal.

At present little is known about the long-term environmental effects of dumped wastes on the marine environment and on marine organisms, except the testimony of fishermen.

Certain inert waste materials, however, may be discarded at sea to advantage. Worn out automobiles, tires, and other stable solid wastes have been used successfully to make more than 100 experimental artificial reefs for fish sheltering, feeding, and spawning along the U.S. coasts. Discarded automobiles and tires have also been used in fighting the depredation of polyp and coral-eating starfish and related animals that are destroying the Barrier Reef in Australia. There solid residues serve as living areas for the destructive starfish and provide food for game fish, which have increased in numbers and reduced the ravages of the starfish. Structural uses of solid wastes have been proposed for artificial islands, breakwaters, off shore bars, jetties, and the like.

It is expected, particularly in the shore cities, that there will be growing pressures to dispose of municipal and industrial wastes in the sea, since with decreasing availability of waste disposal sites on land, the economics of sea disposal become increasingly attractive.

The Bechtel Corp.[1] in a report on the disposal of sludge, recommended a land disposal system broadly, but for coast cities the recommended ocean disposal; see Table 2.

A regional waste management system, involving ocean disposal for the Baltimore–Washington region, is presnted in Figure 2. The digested sludge system shown, capable of serving the needs of this region through the year 2000, is based upon pipeline collection and transportation with an 80-mile ocean outfall utilized for disposal. Although significant economic gain would result from the use of a shorter outfall to the continental shelf waters, substantial questions remain concerning the long-term effects on the marine environment. A major re-

Table 2 *Projected Quantities for the Atlantic Coast Region (in Tons)*

	1968	1975	2000
Digested sludge	2,190	2,740	6,120[a]
Maintenance dredgings	51,000	54,800	67,500
Power plant fly ash	7,320	8,520	3,040
Filter plant residue	173	197	323

[a] Includes waste chemical sludge from tertiary treatment.

search and development program will be necessary before the potential savings can be realized.

The estimated initial capital investment for the system is $53.5 million and future additions will total $8.1 million. The unit disposal cost for the period 1975–2000 will range from $55.00 per ton for the initial year, down to $23.10 for the final year, with an average cost of $27.80 per ton of dry solids.

These costs compare very favorably with costs projected for other disposal methods. For example, sludge incineration costs recently projected for Washington, D.C., are in the range of $52 to $58 per ton of dry solids. In addition, the pipeline-ocean disposal system will completely eliminate pollution of local waterways, due to loss of digester solids and nutrient discharge.

The suggestion has been made in light of the theory of continental drift and plate tectonics that garbage be buried in selected areas of the sea to find its way to deep trenches.

Although it was once thought that the ocean floor has remained stationary and relatively unchanged for billions of years, scientists are now certain that it is continually spreading and being renewed with fresh material welling up from a 47,000-mile-long chain of volcanically active undersea ridges. As this new rock from deep within the earth's mantle moves slowly away from the mid-ocean ridges, it carries the continents along, thereby providing the mechanism for continental drift. Millions of years after leaving the ridges, the material reaches the extremely deep trenches that lie just off several of the continental coasts. There it plunges back into the earth.

The rate of movement into the trenches is no more than an inch or so a year. Some of the more conveniently located trenches could be used as geophysical garbage dumps. The trick would be to dump packaged

LOCATION MAP – YEAR 2000

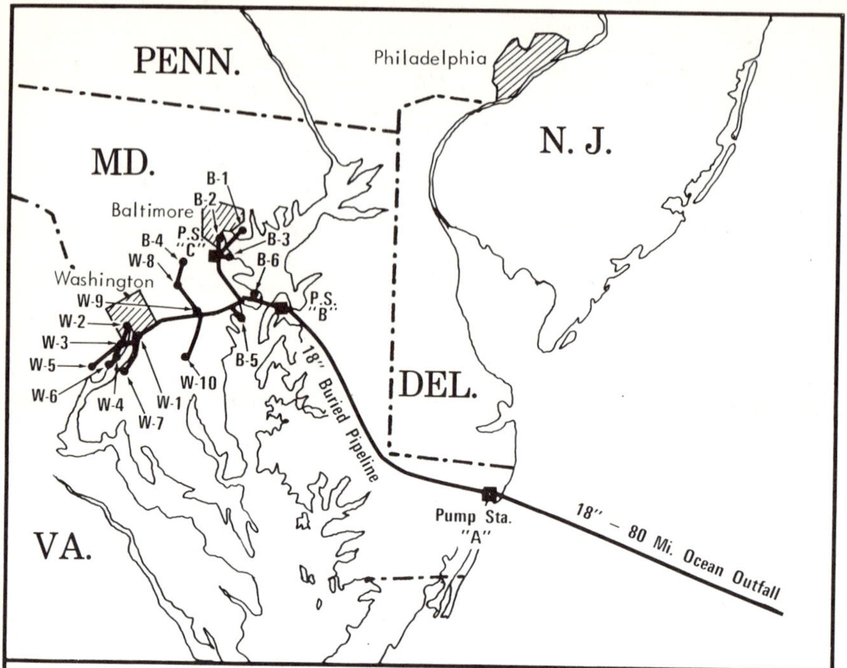

SEWAGE TREATMENT PLANTS

W - 1 DIST. OF COLUMBIA	B - 1 BACK RIVER
W - 2 ARLINGTON	B - 2 PATAPSCO
W - 3 ALEXANDRIA	B - 3 COX CREEK
W - 4 WESTGATE	B - 4 SAVAGE
W - 5 POHICK	B - 5 ANNAPOLIS
W - 6 LITTLE HUNT. CREEK	B - 6 BROADNECK
W - 7 PISCATAWAY	
W - 8 PARKWAY	
W - 9 BELAIR–BOWIE	
W - 10 WESTERN BRANCH	

Figure 2. FWPCA Waste Management Study. Ocean disposal system summary. Digested sludge—80-mile outfall case.

FACILITIES DIAGRAM

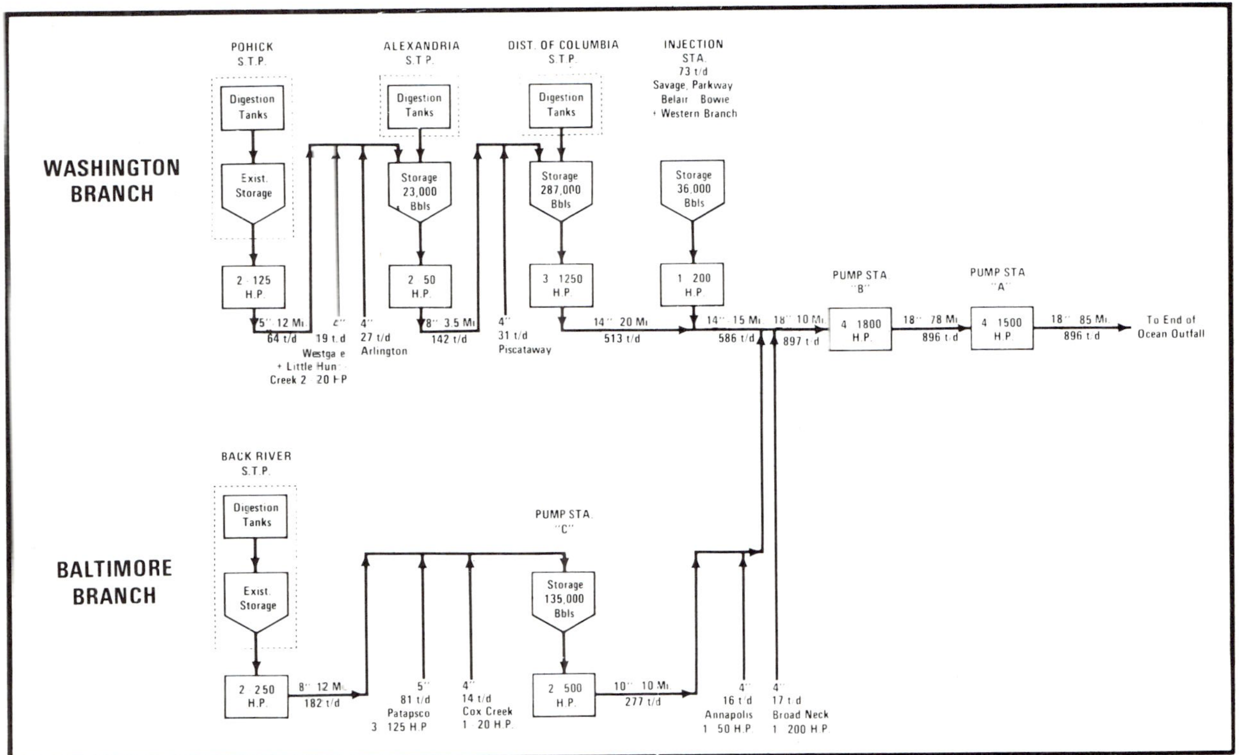

Figure 2. (Continued)

COST BENEFITS

ECONOMIC BENEFIT

	Present Worth of Capital & Operating Costs @ 6% Years 1975 to 2000 * (Millions of Dollars)
Present Methods:	$103.0
• Assumes Average Cost of $40 /ton	
• Assumes Capture of all Solids from Digester	
Proposed Regional Ocean Disposal System:	81.5
Economic Benefit (difference)	$21.5

Environmental Benefits :

- Regional Solution to Sewage Sludge Problem
- Reduction in Region's Air Pollution
- Complete Solids Capture from Digesters
- Complete Nutrient Removal from Digesters

* 1971 Dollars

COSTS

CAPITAL COSTS

	Costs (Millions of Dollars)*
Original Installation: 1974	$53.5
Additions in Later Years	8.1
Total Capital Costs	$61.6

ANNUAL COSTS	AVERAGE	1975	2000
Direct Operating Costs	$2.02	$1.39	$2.72
Capital Charges @7.82%	4.32	4.19	4.83
Total Annual Costs	$6.34	$5.58	$7.55
Tons Transported /Year	228,000	101,500	327,000
Unit Costs in $/ton	$27.80	$55.00	$23.10

* 1971 Dollars

Figure 2. (Continued)

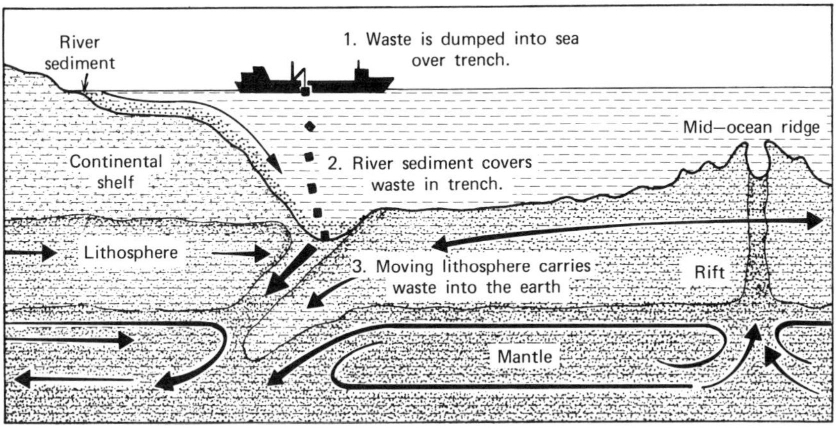

Figure 3. Disposal of solid waste in ocean trench.

waste into the sea off the mouths of fast-flowing rivers, which annually wash vast amounts of mud into continental trench areas. Though the garbage would not be drawn far into the earth for many years, it would soon be buried so deep in mud that there should be little danger of pollution during the interim.

It would have to be determined, for example, whether a river deposits mud quickly enough to accommodate the projected garbage load. The garbage would have to be compressed into dense, sinkable packages, and transported by barge to the disposal site. Figure 3 indicates the concept proposed.

REFERENCES

1. "Bulk Transportation of Waste Slurries to Inland and Ocean Disposal Sites," Bechtel Corp. Contract 14-12-156, FWPCA, USDI, December 1969.
2. D. D. Smith and R. D. Brown, Dillingham Corp., LaJolla, Calif., for SWP, EPA, 1970.

PART III agricultural and food processing waste

Chapter III.1 deals with world production and waste from fresh, canned, and frozen fruit. The United States contributes half of the world total. Chapter III.2 is short, on burning for insect control, now mostly prohibited by legislation. Apples, discussed in Chapter III.3, are only a mild seasonal problem; citrus, also discussed in Chapter III.3, has received much attention. Coffee processing, discussed by Pfluger of General Foods Corp., the largest coffee producer, in Chapter III.4, is an engineering success. Corn and corncobs (Chapters III.5 and II.6), have been exhaustively studied and are discussed by Arnold, who was one of the workers in the field; they are related as an economic failure and are still a problem. Cotton is treated in Chapter III.7, from the field through ginning to baled fiber, and from seeds from the gin to cottonseed oil. Chapter III.8, by Holmes and Potter, discusses tung oil, whose wastes have toxic problems. Nut shells and fruit pits, treated in Chapter III.9, are very large in tonnage and generally become boiler fuel. Peanuts, discussed in Chapter III.10, by Penny, have an aflotoxin problem, and waste is generally used as boiler fuel.

Potatoes, discussed in Chapter III.12, are processed, and large amounts of waste result, but these wastes become animal feed and are often integrated with feeding stations. Rice is discussed in Chapter III.12; it has a large volume, high ash stubble, and a hull problem, still solvable, in the main, only with economic distress. Sugar beets are treated by Pugsley in Chapter III.3. After sugar extraction, the solids become animal fodder. Sugar cane, discussed in Chapter III.4, has large amounts of pressed cane stalks or bagasse, which are made into board, sometimes paper, but often burned as fuel. Fish and shellfish wastes are treated in Chapter III.5 by Soderquist; wastes are huge and seeking application. Some, after drying, become poultry feed. Reconstituted tobacco, treated by Mellen in Chapter III.6, is an outstanding engineering success. Tobacco is too costly to waste, so there is no waste. Wheat, treated by Miller in Chapter III.17, once had its straw and stalk waste converted into strawboard for packaging, but much cheaper corrugated paper has driven strawboard off the market. Chapter III.18 is a brief summary of food processing wastes.

Since discussion in this part is detailed rather than general, as in Part I, references are at the end of each chapter.

CHAPTER III.1

world production and waste from fresh, canned, and frozen fruit

The Commonwealth Secretariat[1] estimates that the world production of fresh, canned, and frozen fruit is approaching 4.5 million tons (4.423 million long tons in 1968) as compared to 3 million tons in the late 1950s. The United States produces half the world's canned fruit. Table 1 gives the statistics.

The resultant wastes at the canneries—pits, kernels, leaves, and cores—are seasonal and present a waste disposal problem not solvable by lagooning and biologic oxidation, or by washing and drying for disposal as cattle fodder; but after drying these wastes are useful for in-

Table 1 *Annual World Production*

U.S. Production (long tons)	
Peaches	1,000,000
Pineapples	650,000
Pears	291,000
Apricots	14,000
Cherries, mangoes, berries, etc.	440,000
Total	2.32×10^6
Exporting Countries (long tons)	
S. Africa	206,000
U.S.A.	192,900
Australia	114,700
Importing Countries (long tons)	
United Kingdom	408,000
Federal Germany	262,300
U.S.A.	441,500

Table 2 *Solid Wastes—Canned and Frozen Fruits and Vegetables*

Number of plants	1838
Workers, thousands	167
Dollar value added to raw crop, billions	2.2
Water useage, billion gallons	100
Raw product, million tons	26.4
Solid waste, million tons	8.9
Percentage solid wastes	33.4

cineration or boiler fuel. Conversion to chars, activated carbon, or adsorptive materials consumes only small amounts of the residues.

Waste from vegetable canning, or—leaves, trash, culls, and washings—are processed by lagooning and biologic oxidation; waste from root crops such as potatoes and sugar beets are processed to animal feed for which there are large markets.

In the individual discussions in this section, row crops and truck farm produce are omitted in that either the waste is left in the fields (asparagus, broccoli, cabbage, celery, tomatoes, and the like) and in later cultivation is plowed under, or returned to the fields, after separation (peas) or with wash water to the growing areas. They present no waste problems to the farmer.

Table 3 *Solid Wastes—Fruit According to Crop*

Fruit	Thousands of Tons of Raw Fruit	Suspended Solids (lb/long ton)	Solid Residuals (lb/long ton)	Solid Residuals Thousands of Tons Waste	% Solid Wastes
Apple	1,000	5	600	320	32
Apricot	120	12	360	21	17.5
Cherry	190	6	300	27	14.2
Citrus	7,800	7	880	3390	43.4
Peach	1,110	10	500	270	24.5
Pears	400	20	660	120	30.0
Pineapple	1,000	8	900	450	45.0
Other	400	10	440	80	20.0
Total	12,250			4680	38.2

Table 4 *Solid Wastes—Canned and Frozen Vegetables*

Vegetable	Thousands of Tons of Raw Vegetables	Suspended Solids (lb/ long ton)	Solid Residuals		
			lb/long ton	Thousands of Tons Waste	% Solid Wastes
Asparagus	120	7	720	45	37.5
Beans, lima	120	80	280	16	13.3
Beans, snap	630	4	420	130	20.6
Beets	270	50	820	110	40.7
Carrots	280	40	960	150	53.0
Corn	2,500	10	1340	1660	66.4
Peas	580	10	260	78	13.0
Pumpkin and squash	220	15	1240	150	68.0
Sauerkraut	230	3	640	75	32.6
Spinach, green	240	10	320	37	15.0
Potato, sweet	150	80	88	60	40.0
Potato, white	2,400	50	760	910	37.9
Tomato	5,000	4	200	400	8.0
Others	1,300	30	770	500	38.5
Total	14,040			4321	30.8

Despite years of research, other wastes present massive problems of disposal (cornstalks, corncobs, cereal grains stalks, straw, plant residues, soya beans, wheat, barley, and sorghum); and the most important of these problems are discussed. Others, such as peanuts, with their aflotoxin problem, and tung nuts cannot be employed for animal foods, and their residues have to be disposed of by whatever method is available.

Rose et al.[2] gave most of their attention to liquid wastes but coordinated data from the USDA Reporting Service, the Census of Manufacturers, USDC, and the National Canners Association; these data in its solid wastes aspects are given in Tables 2–4.

The solid wastes from fruit and vegetable processing have relatively little economic value and are not marketable.[3] Disposal often consists of distribution to farmers who use them as feed for cattle, swine, or poultry, or spread there on the land. However, from citrus, corn, sugar beets, and potatoes, the tonnages of solid wastes are so large they are employed for cattle and animal feeding in confinement and become useful. Thus about three-quarters of all the residuals from fruits and vegetables are actually by-products.[7]

The USDI[4, 8] estimates the waste loads (in millions of pounds) by the biologic methods employed in treating them and calculated by various testing methods:

Method	1963	1968	1972	1977
BOD	660	785	845	905
SS	750	890	960	1,035
TDS	710	845	910	980

In this section major crops are individually discussed from the solid waste viewpoint. Many crops are seasonal, but the disposal and by-product recovery from citrus, coffee, cotton, potatoes, sugar beets, sugar cane, fish and shellfish, and tobacco are successful and economic. Corn and corncobs, tung oil, nut shells and fruit pits, peanuts, rice, and wheat have problems, limited markets, or many studies but uneconomic application; there are disposal problems in areas where burning is limited or prohibited, despite vast sums spent in research and development, particularly by agricultural development laboratories.

Eidsness, Gordon, and Smith[5] have reported on treatment of citrus food processing wastes at large plants in Florida.

REFERENCES

1. The Commonwealth Secretariat, Marlborough House, Pall Mall, London, S. W. 1, England, 1972, 358 pp.
2. W. W. Rose, W. A. Mercer, A. Katsuyama, R. S. Sternberg, G. W. Brauer, G. W. Brauer, N. A. Olsen, and K. G. Weckel, "Production and Disposal Practices for Liquid Wastes from Cannery and Freezing Fruits and Vegetables," Proceedings of the 2nd National Symposium on Food Processing Wastes, March 1971, Denver, Colo., EPA Research Series 12060, 03/71, pp. 109–127.
3. H. Hudson, "Solid Waste Production in the Food Processing Industry," Ref. 2.
4. "Industrial Waste Profiles," FWPCA No. 6, USDI, 1967, Cincinnati, Ohio.
5. F. A. Eidsness, J. B. Gordon, and I. J. Smith, Jr., "Biological Treatment of Citrus Processing Wastes," Proceedings of the National Symposium on Food Processing Wastes, FWPCA, EPA, USGPO stock 5501-0167, pp. 271–287.
6. W. E. Boyle and L. B. Polkonski, "Observations of Aerated Lagoons for

Treatment of Vegetable Cannery Wastes," Wisconsin Canners and Freezers Association, 1968.

7. National Canners Association, "Preliminary Summaries of a Survey of Solid Wastes from Canning and Freezing," SWM, USHEW, 1970.

8. Secretary of Agriculture and Director, Office of Science and Technology, Agricultural Related Pollution, Washington, D.C., 1969, personal communication.

9. "Cost of Clean Water Industrial Waste Profiles," No. 6, "Canned and Frozen Fruits," 1967, Cincinnati, Ohio, USDI, FWPCA.

CHAPTER III.2

burning for insect control

Vast quantities of sugar cane trash, cereal straw, stubble, cotton stalks and trash, cornstalks, soya bean wastes, and the like are burned or allowed to decay in the field. Apple, apricot, cherry, peach, and pear tree prunings are burned in the winter, while grapevines and prune and walnut tree trimmings are burned in the late winter and spring months. Burning is carried out as a means of disease and insect control, since it appears to be the only safe way for eliminating plant pathogens in the wastes. As a result of legislation, burning has been largely displaced by disking as a disposal procedure for grapevine prunings. Shredding, followed by the incorporation of the wastes into the soil, has been tried for orchard wastes, with the advantages of the avoidance of air pollution from burning as well as returning organic matter to the soil.

Disinfection by chemicals is one way of retarding the spread of fungus diseases. Greenhouse wastes are hauled to landfills for disposal or incinerated on the site. Pesticide containers accumulate in agricultural areas and they constitute a hazard to humans and live stock, but many are attractive for reuse for other purposes. Residues when burned are a source of smoke; when not burned they are reservoirs of plant diseases and pests, so that most agricultural waste, when permitted by law, is burned as a practical procedure. In individual counties in California, 40,000 50,000 tons of orchard wastes were burned, as was the stubble of approximately 140,000 acres of rice.

There are about 2 tons of residue per acre from the approximately 1 million acres of grass grown for lawn and pasture seed each year, and about one-third of this residue is burned as a sanitation measure. Prescribed burning in forests each year produces about 32 cubic miles of smoke, while generating 6½ tons of particulate matter and 68,000 tons of hydrocarbons. This burning again is performed for the sake of insect control.

Contributions of burning agricultural waste to photochemical air pollution have been evaluated. The burning of fruit tree prunings, rice straw, barley straw, and dry native range brush produce 14, 9, 1, 18, and 7 lb, respectively, of hydrocarbons per ton, compared with 130 lb of the same hydrocarbons in auto exhaust per ton of fuel consumed. Except for its aesthetic factors, burning of agricultural waste does not appear to add significantly to the burden of pollutants in community air, yet there is increasing legislation and local resistance to such burning. In the western United States, large quantities of agricultural waste from orchards, grain fields, and range lands are burned as the most practical means of ridding the land of wastes. The potential of agricultural wastes to affect health is a basis of concern, but any demonstrated effect is a matter of judgment, rather than fact. Suburban areas collect prunings, grass clippings, brush, and the like, mixing them with their trash and disposing of them at local incinerators.

CHAPTER III.3

apples and citrus fruits

APPLES

Apples not marketable as commercial fruit are pressed for apple juice, which in turn might be processed or fermented to cider and in turn to vinegar. The solid residues or pomace, to the extent required, are processed for the pectin content after which the residues go for cattle feed, incorporated with other animal feeds. The markets are seasonal, being important only at harvest time or the storage period between crops.

Cuttings, prunings, dead trees, fallen fruit, and the like are cut up, shredded, and incorporated in the orchard soil whereas formerly they were burnt in the fall season to restrict the growth of pests.

CITRUS FRUITS

Citrus production, processing, and marketing comprises a $2 billion industry in Florida alone. More than 800,000 acres of the peninsula, extending some 250 miles from north to south, are planted in citrus trees. This area yields about 75% of the oranges and 80% of the grapefruit grown in the United States. The total Florida citrus crop for the 1967–1968 season amounted to about 9 million tons. Since about one-third of the orange acreage had not attained bearing age by the 1966–1967 season, and since citrus acreage is constantly expanding, a steady increase in production is anticipated for years to come.

Most Florida citrus is processed within the state to produce single-strength or concentrated juice for marketing in containers. Surveys showed more than 52 processing plants in the state, which processed about 82% of the crop. The industry has made substantial progress in disposing of wastes through recovery of valuable by-products. Florida plants produced more than 350,000 tons of cattle feed and 48,000 tons of citrus molasses during the season. Cattle feed production amounted to about 573,000 tons annually. Some 130 million gallons/day of wastewater, with a 5-day biochemical oxygen demand (BOD) loading of about 319,000 lb/day, were discharged from Florida plants.

The industry continues to make progress in solving the solid waste problem through recovery of by-products such as peel, pulp, citrus molasses, employment of fermentation or biologic oxidation, and recovery of essential oils by steam and distillation.

Most of the problems of citrus plant processing are those of processing water treatment.

CHAPTER III.4

soluble coffee processing

RICHARD A. PFLUGER

Total green coffee consumption in the United States amounts to about 2.8 billion lb/yr. Approximately 500 million lb of this total is used by the U.S. coffee processing industry for the manufacture of soluble coffee. More than 90% of the soluble processing is done in less than two dozen instant coffee plants located in various parts of the country. Since only about one-third of the green coffee bean ends up as actual soluble powder, one question of major importance to the coffee processor lies in the handling of the remainder, which consists primarily of solid wastes. Airborne wastes from the roasting and drying operations, although a relatively small percentage by weight of the total, also present a significant challenge to the manufacturer in that pollution levels must be controlled in keeping with established standards.

THE SOLUBLE COFFEE PROCESS

A schematic flow sheet for the soluble coffee process is shown in Figure 1. The desired blend of green coffee is first roasted to a predetermined color in large commercial roasters under conditions controlled for proper flavor development. After cooling and tempering of the roasted beans to equalize moisture, the coffee is ground to a relatively coarse particle size for use in the commercial percolators. For large-scale percolation of coffee, it is common practice to use a battery of six to eight vertical columns operated in series countercurrently under 100–200 psig pressure. Capacity of such units ranges from several hundred to 2000 lb/hr of soluble solids. Percolation temperatures vary from about 350°F in the spent stage to 200°F at the fresh coffee end of the extraction train. At a concentration of about 25% solids, the coffee liquor is drawn off from the extractors in batches and immediately cooled to prevent flavor deterioration. After clarification in a

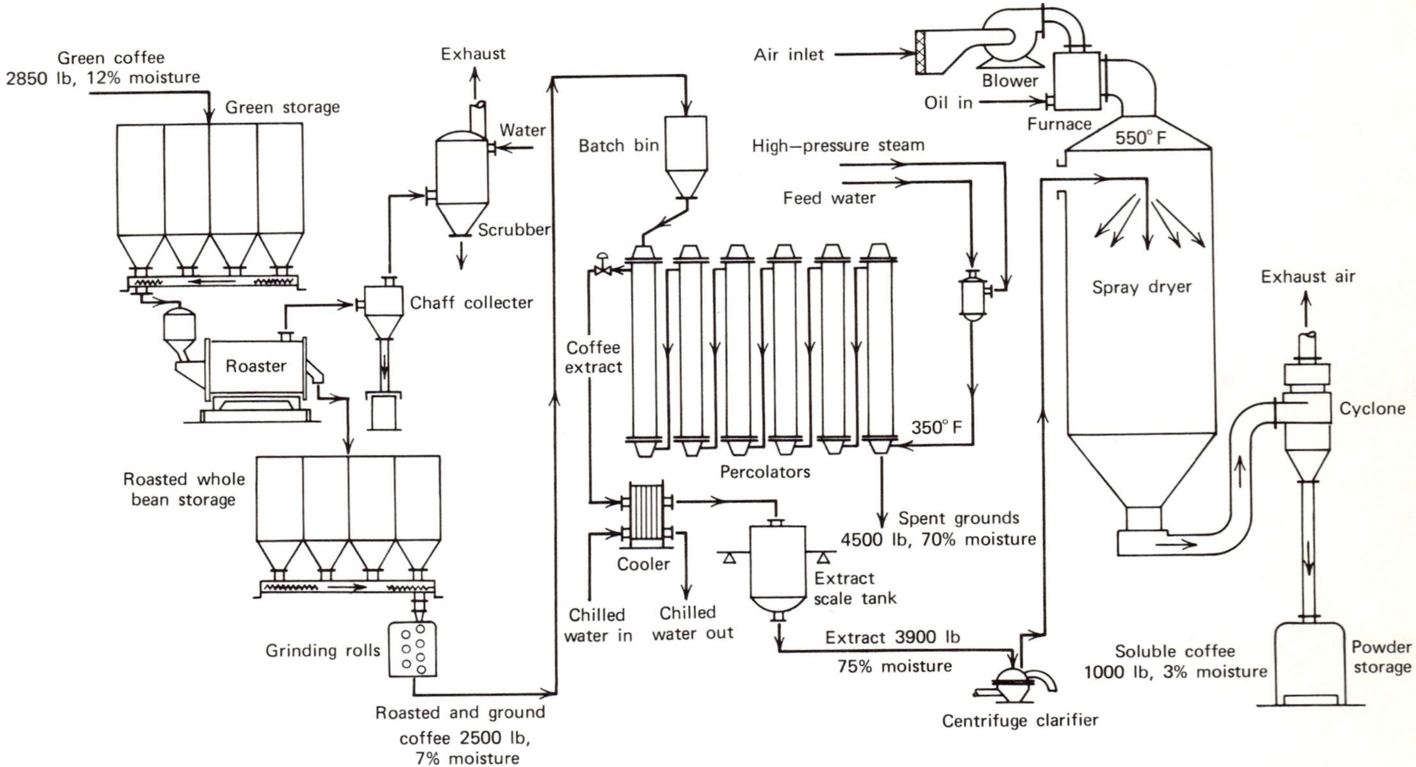

Figure 1. Soluble coffee process flowsheet.

Table 1 Composition of Spent Grounds

Gross Analysis	
Ash	0.5%
Oils and fats (ether extraction)	23.0%
Protein (6.25 N)	14.0%
Carbohydrates	62.5%
Total	100.0%
Individual Components	
Nitrogen	2.2%
Reducing sugars after acid hydrolysis	17.5%
Crude fiber	37%
Alpha-cellulose	37%
Furfural	1.21%

continuous solid-bowl centrifuge, the extract is spray dried to produce the basic soluble coffee powder of commerce.

It is these basic steps in soluble coffee manufacture which are of most significance when considering the generation and handling of industrial wastes. Other finishing steps such as plating of aromas on the finished powders, or agglomeration of the powders to produce certain appearance changes or physical attributes do not, as a rule, contribute to waste generation.

WASTE GROUNDS

Much thought and effort, and many dollars, have been spent on the problem of the utilization of waste coffee grounds. For every saleable pound of soluble powder produced, more than 2 lb of wet grounds remain for disposal. Possible uses must take account of the chemical analysis, Table 1, and the physical form of these grounds. Suggested uses have included mulch and soil conditioners; material for fuel cells; ingredient of pipe tobacco; source of oils; animal feed supplement; source of fatty acids; carrier for insecticides and herbicides; filler for paper; ingredient of polishes, paints, and resins; specialty lubricant; antibiotic growth medium, and many others.

Without exception these possibilities have proved to be uneconomic, or to require long and expensive development programs.

By far the most satisfactory solution known for disposing of waste grounds, after more than 25 years of searching, is to burn them as fuel. A schematic flowsheet of this seemingly inelegant, but highly ac-

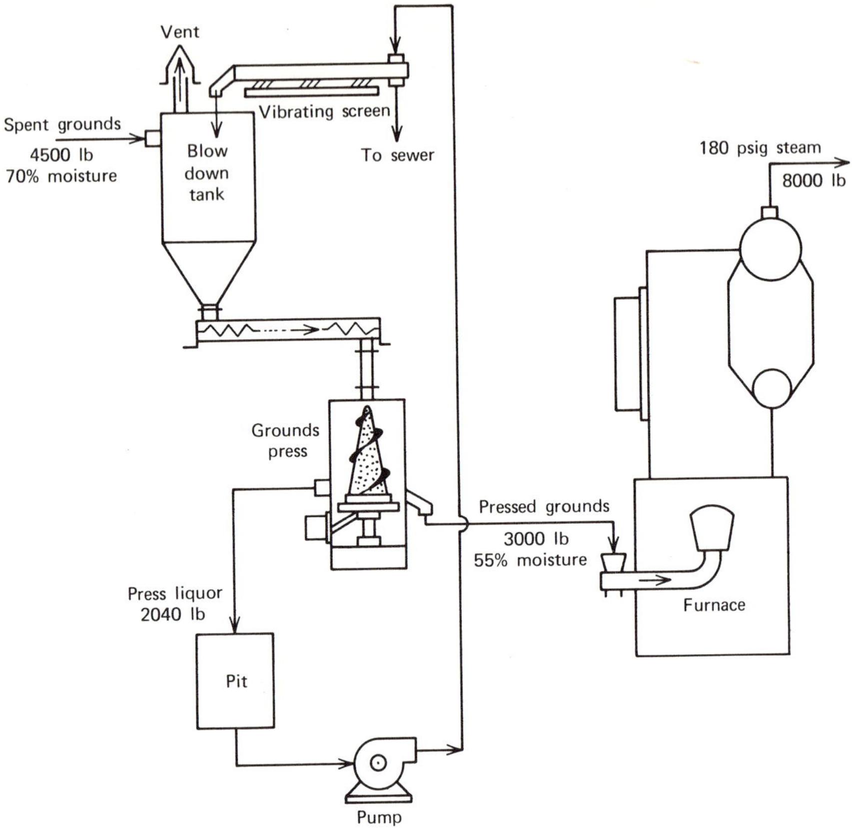

Figure 2. Waste grounds incineration flowsheet.

ceptable approach, is shown in Figure 2. With the advent of strict pollution controls, which have in recent years been heavily focused on power plant emissions, the use of spent grounds for steam generation has proved highly desirable because of the very low sulfur content. Waste grounds handling and incineration equipment have been developed to the point where grounds can be burned very cleanly with essentially no objectionable poolution.

Waste grounds as discharged from the percolation step contain 65–75% moisture. In order to burn the spent grounds adequately, they are first dewatered by pressing to about 50–55% moisture. Figure 3 shows typical equipment used for this purpose. At the reduced moisture content, it is usually necessary to burn a small amount of

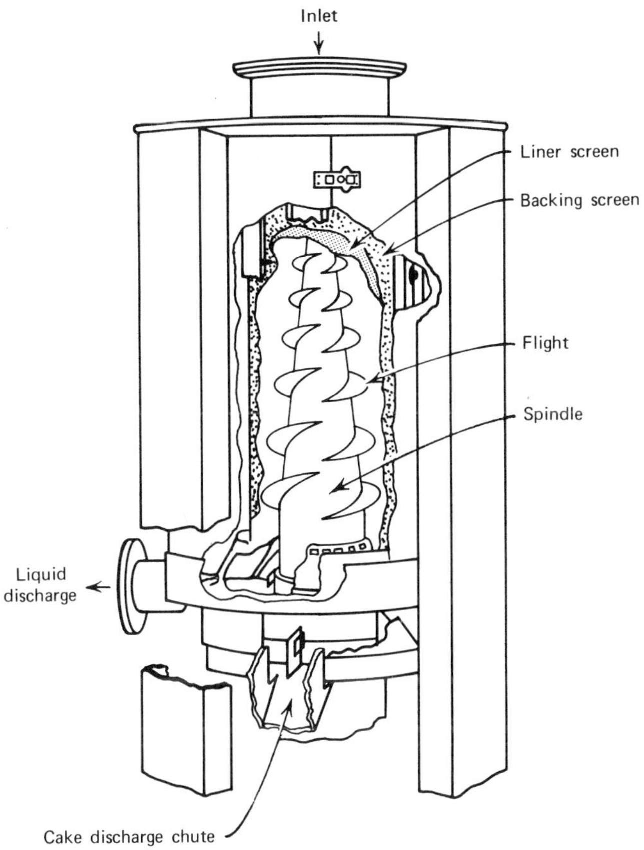

Figure 3. Waste grounds press—Beloit Corp.

auxiliary fuel, approximately 5% of the total heat demand, in order to maintain furnace temperatures at desired levels (1900–2200°F). Because of the burning characteristics of spent grounds, they are usually underfed to a Dutch oven type of stoker with pressurized overfire air. A typical furnace arrangement of recent design is shown in Figure 4. Boiler efficiencies when operating on spent grounds are generally 10–15% lower than when operating on oil or gas (80–85%).

ECONOMICS OF WASTE GROUNDS INCINERATION

In general, modern soluble coffee plants are designed to include waste grounds incineration equipment where highest efficiency and

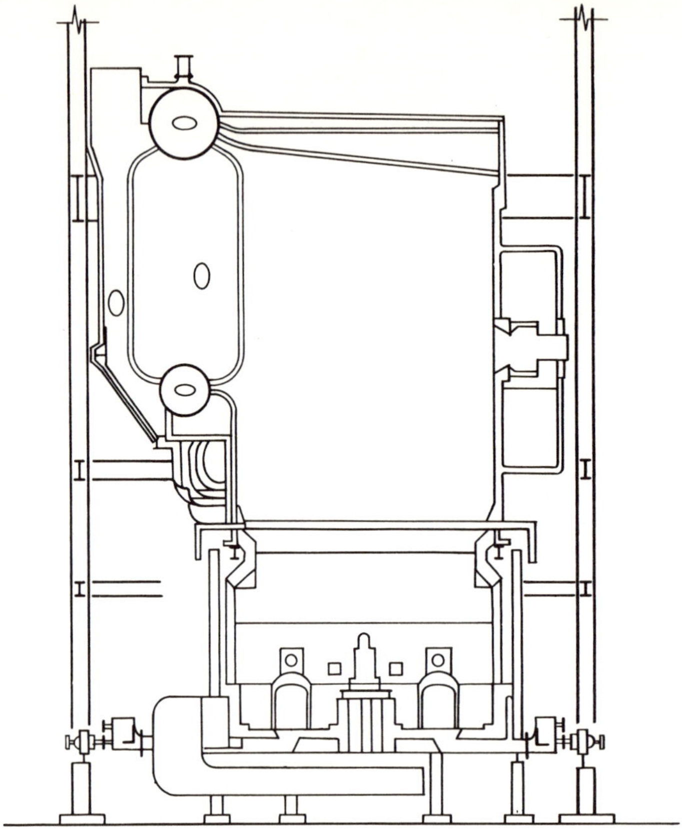

Figure 4. Waste grounds steam boiler.

Table 2 *Heating Values of Spent Grounds*

% Moisture, Wet Basis	Heating Value of Grounds, Btu/lb	
	Dry Basis	As Is
0	10,000	10,000
25	9,600	7,200
50	8,900	4,450
75	6,700	1,680

lowest operating costs are desired. The Btu value of spent grounds varies with moisture content with approximate values given in Table 2.

About 40 lb of spent grounds at 50% moisture produces the same amount of steam as 1 gal. of #6 fuel oil. For the typical soluble process, spent grounds can supply more than 75% of the total heat requirements of the plant.

The most frequently used alternative to incineration of waste grounds is to truck them to some suitable dumping site where they are mixed with other wastes and used as land fill. This alternative is not only costly ($3–6 per ton of wet grounds; trucking costs in metropolitan New Jersey), but is becoming increasingly difficult in most locations as adequate dumping areas become scarce.

Aside from savings in trucking costs, other items of importance when considering incineration are fuel prices, costs of additional capital installations for grounds incineration, and additional operating and maintenance requirements. A cost breakdown for a large incineration installation (40,000–50,000 tons of grounds per year at 55% moisture) might typically be as given in Table 3.

The value of the spent grounds as fuel based on oil costs of 7–10¢ per gallon is $4–5 per ton. Consequently, the payback on a grounds incineration system can be quite attractive. Where the only alternative to incineration is to truck the grounds away to a dumping site, the savings in trucking charges, as previously noted, can also be substantial. Needless to say, the economics of waste grounds incineration, although usually quite favorable, can vary widely by plants and should be evaluated for each specific installation.

PREDRYING OF SPENT GROUNDS

In some cases, it has proved economical to predry the spent grounds prior to incineration. This not only results in a higher heating value of

Table 3 *Incineration Costs*

Investment and Operating Costs	$/ton at 55% Moisture
Depreciation on investment for additional combustion equipment (10-yr straight line)	1.50–2.50
Labor and maintenance (one operator per shift)	1.50
Miscellaneous power, supplies, etc.	0.10
Total	3.10–4.10

the grounds as fed to the furnace, but enables somewhat smoother combustion. The trade-off that must be made is the cost of external water removal versus the cost of removal and handling of the excess water within the furnace. Commercial dryers operate in a cost range of 0.25–0.50 cents per pound of water removed and their use for spent grounds predrying normally would not be economically attractive unless local conditions exist that act to reduce this cost. Availability of spare drying equipment for savings in manpower scheduling and utilization are factors of this nature.

COMPOSTING OF WASTE GROUNDS

Many people have investigated ideas relating to the use of waste coffee grounds as mulch and soil amendment. Indeed it is not at all rare to find a neighbor who has, at one time or another, sprinkled spent grounds from a home percolator onto the soil surrounding a favorite shrub or plant. Controlled experiments have in fact shown spent grounds to be beneficial for certain types of shrubs such as azaleas and rhododendrons. As a rule, it is necessary to compost the grounds, preferably with other materials such as soil, manure, or garbage, to achieve desired results. When handled in this fashion, the grounds produce an excellent humus having a small amount of nitrogen (less than 1%), and an acid character beneficial to certain types of plants.

In view of the fuel value, however, and the additional cost of trucking, handling, and processing of the grounds, the economics of composting as a way of handling spent coffee grounds usually proves unattractive.

CONTROL OF ATMOSPHERIC POLLUTION

The primary sources of atmospheric pollution in soluble coffee processing are the roasting and spray drying operations. Chaff, or the outer covering of the green bean, flakes off during the roasting operation and is carried into the exhaust stream from the roasters. In addition, aromas produced during the roasting operation, although quite pleasant under certain circumstances, are considered by the conscientious coffee processor as a pollutant. In the spray drying operation, fine particles produced by the atomizers can be carried out of the dryer exhaust stack to an objectionable degree unless proper control measures are taken.

The amount of chaff produced during roasting is about 0.5–1.0% by weight of green coffee. Most of the chaff produced is separated from the roaster exhaust stream by means of a cyclone separator. Where a waste grounds incineration system of the type described is being used, the chaff can easily be disposed of by burning with the grounds. In direct fired oil or gas furnaces, this is usually not possible because the chaff particles are carried out of the combustion zone before they can be completely burned. As a result, it is usually best to truck the chaff to a dump when a grounds incinerator is not available.

Several means are available for handling and deodorizing the gases from the chaff separator. Direct oxidation has been used extensively by many processors for smoke and odor control. This is usually accomplished by means of an afterburner in the roaster exhaust stack which heats the gases to the temperature required for combustion (1200–1400°F). This method of control has the disadvantage of increasing fuel requirements for roasting by 50–150%.

An alternative method that requires less fuel is catalytic oxidation which makes use of a catalyst such as alumina-platinum alloy (Houdry Oxycat) plated onto a ceramic support which is adapted to the roaster exhaust stack. Preheating of the exhaust gases to 600–800°F is required just prior to contact with the catalyst. It has been claimed that fuel penalties can be controlled to about one-half the requirement for an afterburner system. Widespread commercial application of this technique has not evolved because of the high replacement costs of the catalyst elements.

A third method of controlling roaster emissions is to scrub the exhaust gases with water. A fairly effective system of recent design is shown schematically in Figure 5. This equipment is of relatively simple construction and has low operating costs. The gases pass through two stacks of open-pore polyurethane foam filters (Scott Paper Co.) that have a high surface-to-volume ratio for collecting small particles. The filter stacks have a graded porosity, with the coarsest filter being on the inlet side of the primary stack. The filters are continuously sprayed with water to scrub out objectionable odors and smoke. A unit of the type shown, having a cross-sectional area approximately 4 × 4 ft uses 40 psig water at a rate of 8 gal/min and can handle the exhaust from one batch roaster with an output of 2000–3000 lb/hr of roasted coffee. FOB cost of scrubber in stainless is $2500 (1973). A fan is usually used on the exhaust side of the scrubber (3000–4000 ft^3/min at 3 in. W.G. pressure drop). Such scrubbers may not be directly applicable to all types of roasters, especially those having relatively high stack discharge

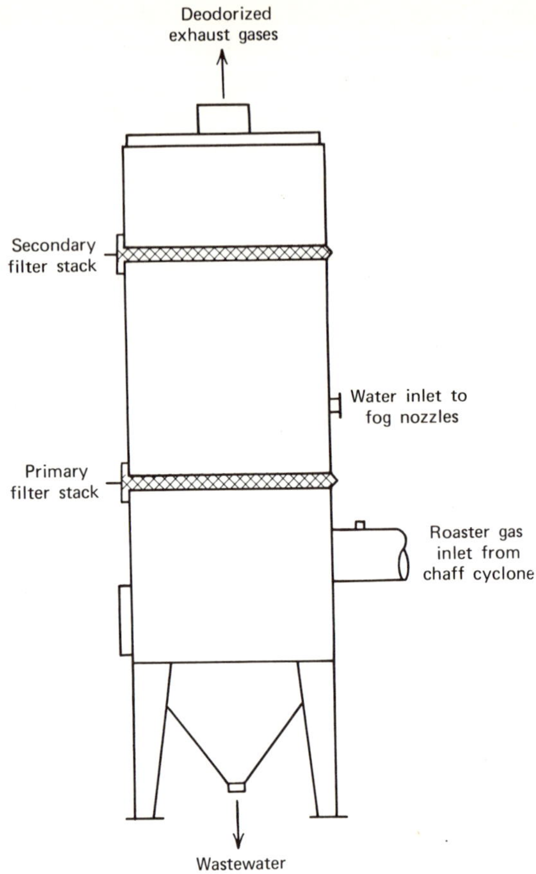

Figure 5. Roaster exhaust gas scrubber.

temperatures. Each installation should be carefully evaluated on the basis of actual field tests.

A soluble coffee spray dryer of 1000–2000 lb/hr capacity can produce an exhaust stream containing up to 75 lb/hr of fine coffee dust. At this level, the environment in the vicinity of the plant becomes noticeably affected and neighborhood complaints rise to a high level. Dust levels, however, can be controlled to less than 15 lb/hr by use of well-designed cyclone separators that are properly maintained. The value of the powder recovered (approximately $1–2 per lb) is sufficient to justify the best dust collection equipment available. Because of its simplicity and trouble-free operation, the cyclone separator is heavily favored as the primary device for separation of coffee particles from spray dryer exhausts. As the diameter of a cylone is reduced, its separation efficiency increases. When handling large volumes of exhaust gas in small-diameter cyclones, it is necessary to use the cyclones in multiples and to arrange them in such a way so as to not interfere with basic simplicity and ease of operation. A unit of this type is illustrated in Figure 6. The installed cost of such units for large spray dryers of 1000–2000 lb/hr capacity is on the order of $50,000–100,000.

Where the plant is close to residential areas, it is sometimes necessary to further treat spray dryer exhausts such that essentially all of the particulate matter is removed. This usually requires use of a

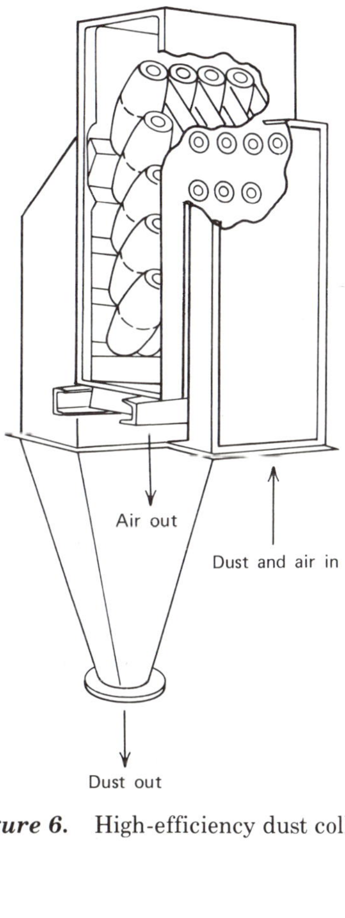

Figure 6. High-efficiency dust collector.

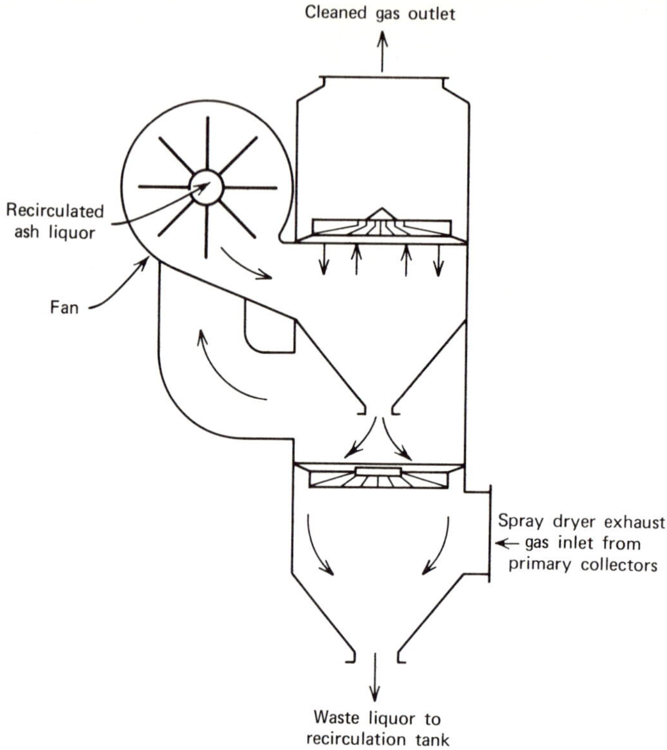

Figure 7. Spray drier exhaust gas scrubber.

scrubber capable of handling large volumes of air (20,000–30,000 ft^3/min). One such device is shown in Figure 7. Attempts at use of electrostatic precipitators, steam scrubbers, and incineration, have been largely unsuccessful. The device has dimensions of 9 ft D × 27 H ft when design for 30,000 ft^3/min of exhaust air. Liquid rates are 90–100 gal/min with the liquid being recycled via pump from a recirculation tank mounted under the waste liquor discharge. The tank contains an overflow to the sewer and a fresh water makeup line. The system is operated to control concentration at desired levels by regulating the makeup water rate to the tank. Installed cost of the above unit is 75,000–100,000.

CHAPTER III.5

the commercial utilization of cornstalks

LIONEL K. ARNOLD

CORNSTALKS

Indian corn, *Zea mays,* is an annual plant raised for its grain in the temperate and subtropical zones. Of the seven varieties of the species, only one dent variety is raised extensively in the United States for grain. The stem ranges in height commonly from 5 to 8 ft in the Corn Belt; it is made up of an outer wall or cortex of a smooth glossy epidermis enclosing a center of pith. The cortex contains a layer of closely united fibrovascular bundles, with other similar bundles running vertically through the pith. At the point where the long blade like leaves are attached to the stem, the fibrovascular bundles unite to form hard nodes that constitute about 26% by weight of the stem proper. The outer woody portion of the stalk makes up about 54% of the stem. The pith, although much greater in volumn than the cortex, constitutes only about 20% of the stem proper. Leaves, tassel, and the husks that cover the grain are ordinarily left on the stem when the cobs and grain are removed.

The corn grain grows in concentric rows around a cylindrical cob about 6–8 in. long and 1 in. in diameter forming the "ear," which is covered by several leaflike husks which are heavier and tougher than the leaves.

Although it is possible to grow corn in every state, except possibly Alaska, most of it is grown in the states listed in Table 1. It is estimated[1] that with 100 bushels of corn per acre there will be produced 2.6 tons of dry stalks, leaves, and cobs per acre. Assuming 14 lb of cobs per bushel this leaves 1.9 tons of stalks and leaves. Because of variations in yield, it is probably more realistic to use a lower figure such as 1.5 tons/acre.[2]

Corn was originally harvested by cutting off the stalks near the ground and tying them in bundles; these bundles were then *stacked* in

***Table 1** Corn Production in Selected States*[a]

State	Acreage in 1000s	Grain in 1000s bushels	Stalks in 1000s tons	Cobs in 1000s tons
Illinois	10,130	901,570	15,195	6,308
Iowa	9,696	901,728	14,544	6,312
Indiana	4,898	416,330	7,347	2,914
Minnesota	4,449	360,369	6,673	2,525
Nebraska	4,149	302,877	6,223	2,120
Missouri	2,958	245,514	4,437	1,719
Ohio	2,884	242,256	4,326	1,696
South Dakota	2,399	110,354	3,598	772
Wisconsin	1,754	163,122	2,631	1,142
Georgia	1,455	58,200	2,182	407
Total (U.S.)	55,707	4,374,840	83,560	30,624

[a] Acreage and corn yield data from USDA, Agricultural Statistics, USGPO. Stalks calculated as 1.5 tons/acre; cobs calculated as 14 lb/bushel of grain.

shocks. The ears were then broken off or snapped from the stalks. Later the ears were simply snapped from the standing stalks and thrown into a wagon that was driven through the fields. If stalks were to be harvested, they were then cut by hand or by a binder; otherwise they were simply plowed under in preparation for the next crop. Mechanical corn pickers were later developed; these were drawn

***Table 2** Chemical Composition of Cornstalks and Several Woods*[2]

Constituent (%)	Cornstalks	Pine: West. Yellow	Pine: Long Leaf	Douglas Fir	White Spruce
Moisture	9.3	6.4	—	—	—
Ash	3.2	0.5	0.4	0.4	0.3
Pentosan	28.0	7.4	7.5		10.4
Methyl pentosan	0.1	1.6	3.6	6.0	3.6
Lignin	31.3	26.7	—	4.4	—
Cellulose	45.0	57.4	58.5	61.5	61.9
Alpha cellulose	24.0	35.6	—	—	—

through the field by a tractor and they snapped off the ears depositing them in a following wagon. Most of the stalks remained standing although many were broken off by the picker.

Modern harvesting employes a *combine,* which cuts off the stalks, removes the grain from the cobs, and deposits the grain in a wagon. The stalks and cobs are dropped onto the ground at the rear of the machine. Equipment for shredding the stalks before dropping onto the ground is also available.

Cornstalks need to be baled for transporting and storing. Probably the most practical way to harvest the stalks would be with a machine that would bale them as it harvests the grain. Whole stalks store well if properly baled and piled unless the moisture content is too high.

PAPER FROM CORNSTALKS

The principal chemical constituents of cornstalks are cellulose, pentosans, and lignin as shown in Table 2 in comparison with several woods; the compositions of the parts of the stalk are shown in Table 3. Because of their cellulose content, cornstalks were early considered as a possible raw material for paper.

One of the earliest patents on the manufacture of paper from cornstalks was granted to Allison and Hawkins in 1802.[3] Activity in this area has been continuous ever since: in 1828 paper was made from corn stalks and cornhusks in England,[4] Cooper was granted a patent in 1829,[5] and Harcourt in 1838.[6]

Other attempts were recorded by Diamont[7] in 1856 and Harshberger[8] and Dodge[9] in 1893. Cornhusks were advertised as a source of paper in 1865.[10] Welsbach[11] patented a caustic soda process in 1863, and Roth[12] in 1864. Apparatus was patented by Dixon.[13] Other early attempts are recorded in Refs. 14–24.

Table 3 *Compositions of Parts of Cornstalks*[2]

Constituent	Outer Shell	Vascular Bundles	Pith	Total Stalk
Pentosan	25.9	26.4	27.7	27.6
Lignin	33.5	35.2	32.0	34.3
Cellulose pulp	55.9	50.2	50.1	52.6
Pentosans in pulp	16.7	13.1	12.2	14.2
Cellulose (by diff.)	39.3	37.1	37.9	38.4

In 1908–1909 the Bureau of Plant Industry of the USDA investigated cornstalks as a papermaking material.[25] Researchers digested the cornstalks in water and then in caustic soda solution. The best results were secured using 10–15% caustic soda and cooking 2 hr at 110 lb pressure.[26] The pulp was screened to separate the long fibers from the pith. They secured 12–18% long fiber suitable for book, writing, and other papers. The pith yield was 13–30% of pulp suitable for pie plates, grease proof wrappers, fiber boxes, and insulating material. Semicommercial runs were made in mills of the S. D. Warren Company[27-30] producing enough paper from cornstalks and other waste materials to print a bulletin[30] describing the work. The water extract was evaporated and fed to dairy cattle[31] and hogs[32] with promising results. In a report to the House of Representatives on the research[33] it was stated that better paper can be made from cornstalks than from wood and at a lower cost. The yield was given as 1000 lb of pulp from 1 ton of cornstalks.

In 1910 a patent[34] was issued to Deiss on a process in which the stalks were treated with an acid followed by neutralization with an alkali and fermentation with a yeast. The pulp was then digested with caustic soda. In 1911 Heckel[35] in France digested cornstalks 6 hr at 59 lb pressure with 2.4% caustic soda using 200 gal to each 2200 lb of stalks. The yield after bleaching was 51%.

In 1914 Bache-Wiig patented[36] a process for making a substitute for ground wood pulp for newsprint. The stalks were digested under pressure for 2–3 hr in sodium chloride and sulfur dioxide. Burton received two patents[37, 38] in 1914 and 1916 covering a digester and methods of digested with caustic soda. In 1917 Drewsen[39] patented a process for digesting stalks 4–8 hr at 40–120 lb steam pressure with 25% sodium monosulfite. Marsden[40] in 1924 softened cornstalks by steaming and then fermenting with yeast. The remaining fibers were digested in alkali or acid to produce a paper pulp. In a process patented by Markoczi and Adams[41] anaerobically fermented cornstalks were digested in water followed by digestion in a low concentration (0.1–0.7%) of caustic soda.

In Dorner's[42] process the cornstalks were first treated with dilute acetic acid under pressure, followed after washing by digestion with dilute caustic soda under pressure. Experimental work was done under Jackson[43] by the Euromerican Cellulose Products Corporation. The Cornstalk Products Company[44] was organized to put Jackson's process (referred to as the Dorner Process) into operation at Danville, Illinois, to manufacture paper pulp and alpha cellulose in 1928. Their first commercial paper, a blend of wood sulfite and cornstalk pulp, was too hard and nonabsorbent for newsprint or for magazine work, although it was said to be a good bond paper. Rotogravure paper was made of 50% cornstalk pulp and 50% ground wood pulp. Cornstalk pulp was also used in tissue, bond, and book paper. The exact process was not disclosed but it was understood to have been a modified soda process. The plant produced 50 tons of pulp a day but failed to make a profit and quit operation in 1930.

Considerable research by the chemical engineering staff at Iowa State University was done on the production of paper from cornstalks.[45] Equipment consisted of several small digesters, with capacities of approximately 3, 5, and 7 ft^3, a small Hollander paper beater, a Jordan refiner, a paper screen, a hand sheet machine, and a sheet dryer. Much of the work used sodium hydroxide as the digesting chemical. Some of the results are shown in Table 4. When only the outer fiber was used the results in Table 5 were secured; results of the work using sulfate digestions are shown in Table 6. A limited number of sulfite digestions were run. Results are shown in Table 7.

Table 4 *Paper from Shredded Cornstalks Digested in a Caustic Soda Solution*

No.	NaOH (%)[a]	Yield (%)	Bursting Strength Ratio	Breaking Length (yd)	Folding Endurance Ratio
1	7.5	58.5	31.5	3950	1
2	12.5	47.4	43.9	479	41
3	15.0	45.5	45.6	5060	51
4	20.0	45.2	51.3	5200	195
5	25.0	48.6	50.0	4730	264
6	7.5	34.5	51.8	—	—
7	10.0	30.9	54.6	—	—
8	15.0	39.0	48.0	4645	37.8
9	20.0	53.5	42.1	4880	67.6
10	25.0	43.9	38.2	4883	28.1
11	5.0	47.1	33.4	4225	14.1
12	10.0	36.1	41.2	3990	22.8
13	15.0	42.4	34.9	4290	21.9

[a] Based on weight of stalks.
Note: All digestions at 280°F. No. 1–5, digested 1 hr in caustic; No. 6–10, digested in water before caustic, with one caustic cook; No. 11–13, same as 6–10 except digested 4 hr in caustic.

Table 5 *Paper from Cornstalk Outer Fiber Digested in Caustic Soda Solutions*

No.	NaOH (%)[a]	Yield (%)	Bursting Strength Ratio	Breaking Length (yd)	Folding Endurance Ratio
1	5	55.6	32.3	4650	2
2	10	52.5	57.9	4480	60
3	15	55.0	47.1	4220	96
4	20	56.2	43.9	4580	38
5	25	48.1	42.6	4900	89
6	5	59.4	28.2	3580	4
7	10	50.1	46.3	5270	52
8	15	—	42.5	5102	59
9	20	54.0	34.8	3340	40
10	5	50.5	39.4	5180	10
11	7.5	62.6	43.4	4850	24
12	10	57.4	47.4	4800	60
13	13.7	55.5	37.3	4555	29
14	15	50.5	41.2	4940	52
15	20	57.6	39.3	4250	31
16	25	51.4	43.7	4645	105

[a] Based on fiber weight.

Note: No. 1–5 digested 1 hr at 262°F; No. 6–9 same as 1–5 except digested in water first; No. 10–16 digested only in caustic solution for 4 hr.

Table 6 *Cornstalk Paper by the Sulfate Process on Outer Fiber*[45]

NaOH (%)	Na_2S (%)	Temp. (°F)	Time (hr)	Yield (%)	Bursting Strength Ratio	Breaking Length (yd)	Folding Endurance Ratio
13.7	7	240	4	55.0	29.5	2860	17.2
13.7	7	280	4	54.5	42.0	4350	60.0
13.7	7	285	4	47.1	23.8	1550	10.6
0	7	262	4	60.0	26.4	3930	3.0
4.2	7	262	4	60.0	29.0	2550	25.7
8.3	7	262	4	58.0	29.6	2850	25.7
15.0	7	262	4	54.7	48.4	5250	79.8
13.7	0	262	4	55.5	33.2	4220	20.8
13.7	2	262	4	61.0	38.1	4015	85.1
13.7	4	262	4	61.3	39.8	4350	37.5
13.7	10	262	4	52.6	31.9	3800	10.5

Table 7 Paper from Shredded Entire Stalks by the Sulfite Process

Time (hr)[a]	Free SO_2 (%)[b]	Ream Wt. (lb)[c]	Bursting Strength Ratio	Breaking Length (yd)	Folding Endurance Ratio	Yield (%)
6	2.5	53	22.5	1520	2.5	52.6
12	2.5	59	20.5	1170	3.3	49.0
18	2.5	51	18.2	1160	4.0	44.3
6	3.3	67	22.0	1800	2.8	52.5
4	5.0	62	23.2	1200	3.2	48.2

[a] Temperature 280°F.
[b] Combined SO_2 1.1%.
[c] Basis of 500 sheets 25 × 40 in.

A process for producing a pulp by fermentation following a chemical cook was patented by Sweeney.[46]

Although good paper pulp has been produced from cornstalks by many research workers, none is being produced commercially. Several objections to cornstalks as a paper making material have been stated by various critics. The paper is said to be dirty, indicating the necessity of keeping the stalks clean when harvested. Paper from the entire stalk is said to be objectionable because of "rattle." Perhaps the answer is to blend the pulp with wood pulp to overcome this. Removing the pith is expensive unless a use for the pith can be developed. The cost of cornstalks is high compared to wood.

BUILDING BOARD FROM CORNSTALKS

Insulating Board

Although a great deal of experimental work has been carried out on a variety of materials for building or rigid insulating boards, the use of cornstalks has been studied mainly at the Engineering Experiment Station laboratory at Iowa State College*[47-52] the most significant work was in a pilot plant of digesters, refiners, forming machines, and a dryer. The original forming machine was a modified continuous cylindrical filter. In each of the two later machines the mat of board was formed on a wire screen or belt as it emerged almost horizontally from a pool of pulp. Refiners included a swing hammermill, a rodmill, two

* Now Engineering Research Institute, Iowa State University of Science and Technology.

Table 8 *Typical Cornstalk Board Characteristics*

Type of Pulp	Weight (lb per 1000 ft², ½ in. thick)	Modulus of Rupture lb/in.²
Mechanical	539	217
Mech. plus 10% paper	600	487
Steamed, 18% paper[a]	805	618
Cooked[b]	662	471

[a] Digested at atmospheric pressure in water for 3 hr.
[b] Digested in water at 45 psi for 3 hr.

paper beaters, a Jordan refiner, a Claflin refiner, and an attrition mill.

Soft but rigid insulating board was made as follows. For "mechanical" pulp the shredded stalks passed in a water suspension through two refiners, such as the rod mill followed by the Claflin refiner, to storage tanks. Three percent rosin paper size was added between refiners. In some cases 10–20% repulped old newspapers were also added to increase the strength of the board. Instead of the newspaper pulp, a cornstalk paper pulp made by a caustic cook could be added. Before forming the refined pulp had added to it enough alum to bring the pH to 4.5. Sulfuric acid was substitued for part of alum in some batches. The sized pulp was next pumped to the forming machine. Here it was continuously formed into a mat which would trim to 2 ft in width. As it emerged from the forming machine, the mat passed between three pairs of smooth rolls to remove part of the water. It then passed into a roller type of tunnel dryer. The length of the dryer limited the length of board made to 40 ft.

If a "steamed" pulp was desired the shredded stalks were boiled in water at atmospheric pressure for about 3 hr before refining. "Cooked" pulp was made by digesting the stalks in water at 50 psi steam pressure for about 3 hr.

The board was light in weight and rather springy although rigid. It was formed from a heterogeneous mixture of fiber bundles up to about 1 in. long together with the pith particles and the added paper pulp. The pith and the hydrated paper pulp cemented the fiber bundles together at their points of contact while still leaving an abundance of air films, which give the board its thermal insulation properties. Some typical weights and strength values are given in Table 8. Thermal conductivities range from 0.34 to 0.55 Btu/hr°F for a square foot 1 in. in thick. The board was suitable for sheathing in buildings, as a plaster base, and for other purposes where both structural strength and heat insulation are desired.

A factory for the production of insulating board from cornstalks was built in Dubuque, Iowa, about 1928. The product was marketed under the trade name of "Maizewood" and competed well with similar boards from other raw materials. After several years the use of cornstalks as a raw material was discontinued and a switch made to another raw material said to be cheaper. Part of the higher cost was that commercial equipment suitable for harvesting and baling the stalks was not available, and part was the cost of rathr long-distance hauling.

Acoustical Board

The acoustical properties of the cornstalk insulating board have been studied.[53] The board was made slightly oversize and then sanded to ½-in. thickness. It was cut into "tile" 1 ft^2 in size and then two pieces were glued together to give one 1-in.-thick tile. Tests of the coefficient of absorption at various frequencies were made. The effect of the type of board on absorption is shown in Table 9. Bevelling the edges and cutting slots into the surface increased the absorption to as much as 60%. Similar results were secured by drilling holes into the board as is done in commercial acoustical tile. One-inch-thick tiles surfaced with

Table 9 *Effect of Pulp Composition on the Coefficient of Absorption of Acoustical Board*[53]

		Coefficients of Absorption at Frequencies (in cps) (%)				
Kind of pulp	Percent	256	512	1,024	2048	NRC[a]
Mechanical	0	37.5	54.9	56.8	51.5	50.3
Mechanical	9	40.7	48.4	44.4	45.1	44.7
Mechanical	18	39.3	44.6	41.1	43.5	42.2
Steamed	0	38.1	49.6	48.1	48.5	46.6
	9	34.8	43.8	42.0	45.1	41.5
	18	—	34.9	32.1	41.8	—
Cooked	0	48.3	59.0	49.9	49.3	54.1

[a] Noise reduction coefficient—that is, the average of data given.

⅛-in. hard board with holes drilled through it ¾-in. depth gave good absorption.

Hard Board

"Pressed" board or "hard" board was produced experimentally at Iowa State University[54-56] from both the "steamed" and "cooked" pulps used in insulating boards. The mat of board was formed on the same machine used for insulating board. After passing through the press rolls the mat cut into lengths to go into a hydraulic press with stream-heated platens. One of the presses was a split platen press. The two plates of each platen were separated about ¼ in. to allow a continuous spiral weave screen belt to pass between them. The mat of board was loaded on top of the upper plate of one platen. The lower plate of the platen above had attached to its lower surface a plate of smooth stainless steel. Pressures up to 500 psi were applied to the mats while they dried in the press. The top surface of the board in contact with the stainless steel plate had a very smooth surface. The bottom surface showed the screen marks. The screen on the bottom allowed the moisture to escape from the board. Boards were made from cooked pulp with densities as high as 65 lb/ft^3 and having a modulus of rupture of 7500 psi. The boards were similar to products of the market made from wood waste; there were no commercial developments.

Molded Fiber Products

Molded fiber products can be made from cornstalks pulp. For example, ordinary two-plane basement window sashes were molded at Iowa State University.[57] Cooked pulp of the type used for insulating and hard board was made up in water to 3% consistency. Flaked paraffin wax was added to the slushed pulp to act as a waterproofing size. The slurry was poured into a steel mold having an exaggerated vertical dimension and equipped with drainage holes and a screen. Pressure was applied to the cold mold by a plunger. After removing as much water as would press out, the mold was placed in a steam-heated press and the molded piece dried under pressure of about 65 psi and a temperature of about 300°F. At this temperature the wax melted and was distributed over the fibers. After drying the sashes were conditioned in a hot moist chamber to prevent warping. Strength tests indicated adequate strength for normal use. The sashes were varnished and painted satisfactorily. The experimental results indicate that other and larger units can be made. No commerical developments are known.

MISCELLANEOUS PRODUCTS

Xylose

When a pentosan such as xylan is heated with a dilute mineral acid it hydrolyzes to form the corresponding five carbon sugar, in this case xylose. Cornstalks contain about 20% xylan. A crude xylose syrup can be made by digesting cornstalks or corncobs with dilute acid and evaporating to the desired consistency. Firstenberger[58] found it desirable to use a preliminary digestion in water to remove nonsugar water solubles first. He found the best conditions for hydrolysis were 1 hr at 40 psi in a 0.2 *N* sulfuric acid solution. Successive extractions with fresh acid solution were not as practical as three stages of countercurrent hydrolysis. The cook liquor was neutralized with lime. Partial purification of syrup was obtained by treatment with phosphoric acid and activated carbon. Firstenberger secured 18.4 lb of xylose in a syrup from 100 lb of stalks. He did not develop a satisfactory method of crystallizing the xylose. Crystalline xylose has been made from cottonseed hull bran.[59]

Maizolith

If a pulp like that used for paper making is refined continuously in a closed circuit of a Hollander beater and Jordan refiner, a jelly-like pulp can be produced. This dries to a hard, dense material similar to vulcanized fiber and may be machined to a desired shape. Such a material was made at Iowa State University by Hartford[60] and named Maizolith.

The material had a modulus of rupture of 20,000 psi, a tensile strength of 7,000 psi, and a dielectric constant of 7. The material is difficult to dry and probably will never compete with certain plastic products.

Fuel Gas

Work at the University of Illinois[61, 62] has shown that a mixture of cornstalks and sewage may be fermented to produce a mixture of 64% methane, 28% carbon dioxide, 3.4% hydrogen, and 4.3% nitrogen. This gas has a calorific value of 640 Btu/ft^3. The sewage furnishes the nitrogeneous material needed. The residue consisted of the harder, long fibrous portions of the cornstalks that are not readily attacked by the bacteria.

Usually the husks are harvested with the stalks and considered a part of them. One of the most extensive uses of cornhusks in the past was in the manufacture of buffing wheels for polishing hard rubber articles. The buffing wheels were made up of a series of layers of cornhusks clamped between circles of canvas and held between steel disks. The husks were arranged with the butts outward.

The production of cornstalks in the United States is in excess of 83 million tons annually. There is practically no utilization of cornstalks. As with the corncobs, modern corn harvesting methods would need some modification to ensure clean stalks at low cost. The stalks would need to be dried before baling to prevent deterioration during storage.

Cornstalks have very little value as fertilizer but on some soils are useful as a source of organic material to be plowed under. Many cornstalks are raked into piles and burned.

REFERENCES

1. E. E. Ferlemann, "Total Corn Harvest; A Combine Attachment for Shelled Corn and Silage," M. S. thesis, Iowa State Univ., 1966.
2. O. R. Sweeney and L. K. Arnold, "Cornstalks as an Industrial Raw Material," Bulletin 98, Engineering Experiment Station, Iowa State Univ., 1930.
3. Mentioned in Ref. 4. Not available at Patent Office.
4. Anon, "Cobbett's Corn Paper," *World's Paper Trade Rev.,* **66,** No. 20, 38–9, 1916.
5. L. H. Weeks, *A History of Paper Manufacture in the United States,* Lockwood Trade J. Co., New York, 1916.
6. Homer Holland, "Process for Preparing the Lignia of Fibrous Portion of Cornhusks, etc.," USP 878, 1838.
7. Moritz Diamont, "Die Verwethung der Maisplanze als Spinn-und Webestoff sowie zur Papierzeugnung" ("The Utilization of the Corn Plant as Spinning and Wearing Material, As Well as for Paper Production"), *Dinglers Polytech.,* **116,** 413–415, 1862.
8. J. W. Harsberger, "Maize; A Botanical and Economic Study." Univ Pennsylviania, Contribution from Botanical Laboratory, **1,** No. 2, 1893.
9. J. R. Dodge, "Maize-paper and Maize-cloth." Report of U.S. Commission of Agriculture, 1863, pp. 436–438.
10. Anon, "Cornhusks Wanted," *Am. Agriculturist,* **24,** 143, 1865.
11. A. C. Welsbach, "Improvements in the Manufacture of Paper, etc. from the Husks of Indian Corn," USP 38,220, 1863.
12. J. A. Roth, "Improved Fibrous Material from Cornstalks," USP 41,642, 1864.

13. J. W. Dixson, "Process for Making Paper pulp from Cornstalks," USP 51,432, 1865.

14. J. R. Barber, "Made Paper from Cornstalks Years Ago," *Paper Trade J.*, **68,** No. 25, 28, 1909.

15. Anon, "Paper Fiber from Wood and Plants," *Sci. Amer.*, **39,** 212, 1878.

16. G. R. Sherwood, "Machine for Pithing Stalks," USP 627,882, 1899.

17. M. W. Marsden, "Leak Protector for Vessels," USP 534,611, 1895.

18. G. R. Sherwood, "Process of Manufacturing Products from Cornstalks or Analogous Plants," USP 720,850, 1903.

19. G. R. Sherwood, "Process of Manufacturing Products from Cornstalks or Analogous Plants," USP 720,851, 1903.

20. G. R. Sherwood, "Process of Preparing Fiber Stock from Cornstalks and Analogous Plants," USP 905,374, 1908.

21. Viggo Drewsen, "Process of Manufacturing Products from Cornstalks, Sugar Cane, Sorghum or Analogous Pithy Stalks and Paper Produced Thereby," USP 789,417, 1905.

22. Viggo Drewsen, Ref. 21, pp. 418.

23. Viggo Drewsen, "Process of Manufacturing Soft Absorbent Pulp from Sugar Cane, Cornstalks, Sorghum and Analogous Pithy Stalks and Products of Such Processes," USP 789,887, 1905.

24. Viggo Drewsen, "Process of Manufacturing Products from Cornstalks, Sugar Cane, Sorghum or Analogous Pithy Stalks and Paper Produced Thereby," 853,943, 1907.

25. U. S. Bureau of Plant Industry, Report of Acting Chief, 1910. Annual Report, USDA 1910, pp. 301–302.

26. Anon, "The Process for Cornstalks," *Paper Making,* **28,** No. 5, 1914.

27. U. S. Bureau of Plant Industry, USDA, "Method of Taking Food Extract from Cornstalks and Reducing the Residue to Fiber for Paper Making," 1927.

28. Anon, "Manufacturing Paper from Cornstalks," *Sci. Amer.*, **99,** 274, October 24, 1908.

29. Anon, "Investigating Substitutes for Wood," *Paper,* 10, No. 3, 19, 1913.

30. C. J. Brand, "Crop Plants for Paper Making." USDA Bureau of Plant Industry, Circular 82, 1911.

31. Bureau of Plant Industry, Report of the Chief, USDA 1911, pp. 283–284.

32. Bureau Plant Industry, Report of the Chief, USDA 1912, p. 417. Also see Report of Bureau of Animal Industries, 1912, p. 341.

33. A. O. Stanley, "Substitute for Wood in the Manufacture of Paper Pulp." A Report from Committee on Agriculture to House of Representatives, 60th Congress, 2nd Session Report 2142 (Ser. No. 5-384).

34. Auguste Deiss, "Process of Manufacturing Cellulose," USP 967,001, 1910.

35. L. Vidal, in "Moniteur ete la Papeterie Francaise," April 15, 1911, abstract from M. Vidal.

36. Carl Bache-Wiig, "Method of Preparing Pulp," USP 1,084,244, 1914.

37. G. D. Burton, "Method of Treating Cornstalks and Similar Substances," USP 1,123,166, 1914.

38. G. D. Burton, "Process of Treating Cornstalks and Similar Substances," USP 1,199,241, 1916.

39. Viggo Drewsen, "Process of Making Fibers for Paper, USP 1, 229,422, 1917.

40. M. W. Marsden, "Method of Making Fibrous Pulp from Low Cost Vegetable Material," USP 1,510,855, 1924.

41. E. Markoczi and S. J. Adam, "Industrial Utilization of Strawlike Material," USP 1,534,236, 1925.

42. Bela Dorner, "Treating Maize Straw and the Leaves," USP 1,038,730, 1912.

43. J. E. Jackson, private communication, 1927.

44. E. F. Hulbert, "Paper from Cornstalks," *Paper Trade J.*, **87,** No. 13, 51–58, 1928.

45. O. R Sweeney and L. K. Arnold, "The Production of Paper from Cornstalks," Iowa Engineering Experiment Station Bulletin 100, Iowa State Univ., 1930.

46. O. R. Sweeney, "Process for Elaborating Cornstalks into Pulp for Use in Paper Making and the Like," USP 1,639,152, 1927.

47. O. R. Sweeney, C. E. Hartford, R. W. Richardson, and E. R. Whittemore, "Experimental Studies on the Production of Insulating Board from Cornstalks," *Iowa Eng. Exp. Sta. Bull.* 102, Iowa State Univ., 1931.

48. O. R. Sweeney and L. K. Arnold, "Studies on the Manufacture of Insulating Board," *Iowa Eng. Exp. Sta. Bull.* 136, Iowa State Univ., 1937.

49. O. R. Sweeney and L. K. Arnold, "Moisture Relations in the Manufacture and Use of Cornstalk Insulating Board," *Iowa Eng. Exp. Sta. Bull.* 163, Iowa State Univ., 1948.

50. O. R. Sweeney, "Process Relation to the Production of Synthetic Lumber from Cornstalks," 1,803,737, 1931.

51. O. R. Sweeney, "Machine for Producing Insulating Lumber," USP 1,814, 268, 1931.

52. O. R. Sweeney, "Process and Apparatus for the Manufacture of Synthetic Lumber," USP 1,846,406, 1932.

53. L. K. Arnold, H. J. Plagge, and D. E. Anderson, "Cornstalks Acoustical Board," *Iowa Eng. Exp. Sta. Bull.* 137, Iowa State Univ., 1937.

54. Baker Wingfield, T. R. Naggziger, E. R. Whittemore, C. B. Overman, O. R. Sweeney, and S. F. Acree, "Production of Pressboard from Cornstalks," NBS Misc. Pub. 123, USGPO, 1936.

55. O. R. Sweeney, "Method of Making Hard Lumber Substitute," USP 1, 772,502, 1930.

56. O. R. Sweeney, "Matboard Handling Machine," USP 2,084,980, 1937.

57. L. K. Arnold, "Pressure-Molded Fiber Products from Cellulosic Material," *Paper Ind.*, **21,** 760–764, 1939.

58. B. G. Firstenberger, "The Data, Design and Specifications for a Plant to Produce Xylose from Cornstalks," Ph.D. Thesis, Iowa State Univ., 1942.

59. W. E. Emley, "Xylose," *J. Chem. Ed.*, **7,** 2403–4, October 1930.

60. C. E. Hartford, "Manufacture and Properties of a Cellulose Product (Maizolith) from Cornstalks and Corncobs," NBS Miscellaneous Bulletin 108, 1930.

61. C. S. Boroff and A. M. Buswell, "Fermentation Products of Cellulose," *Ind. Eng. Chem.*, **21,** 1181–2, December 1929.

62. A. M. Buswell, "Production of Fuel Gas by Anaerobic Fermentation," *Ind. Eng. Chem.*, **22,** 1168–72, December 1930.

CHAPTER III.6

the commercial utilization of corncobs

LIONEL K. ARNOLD

THE CORNCOBS

The corn grain is arranged around the outside of the cob. The cobs of the common dent varieties of corn generally are 7–10 inches long[1] and 1 to 1⅜ in. in diameter at the butt. The cob tapers to about ¾ in. near the opposite end. The outer layers of the cob consist of fine and coarse chaff. Under these is a hard tough woody ring surrounding a center of pith. The physical composition, by weight, is given by Sweeney[2] in Table 1. The chemical composition of corncobs is similar to that of the cornstalks.

When corn was picked by hand or mechanical pickers, piles of corncobs accumulated near grain elevators or on the farms, depending on where the corn was shelled. When corn is harvested by combine the cobs, together with the stalks, drop to the ground behind the machine. Unless the ground is dry or frozen it is difficult to secure clean stalks and cobs. Some cobs are segregated in piles at hybrid seed corn processing plants.

FURFURAL

When a material, such as corncobs, containing pentosans is heated with a dilute mineral acid the corresponding pentoses are formed. Further heating results in the elimination of three molecules of water and the production of 2-furaldehyde or furfural. Furfural was discovered by Dobereiner[2] in 1830. Stenhouse in 1840 prepared it in laboratory quantities by treating oat meal and sawdust with sulfuric acid. He gave it the formula of $C_5H_4O_2$, which was later shown to be correct. LaForge, USDA,[3] in 1920 listed furfural as one of the products he obtained from corncobs. The first commercial production of furfural

Table 1 *Physical Composition of the Corncob*

	Average %
Chaff	24.97
Woody substance	73.36
Pith	1.67
Total	100.00

using oat hulls was by the Quaker Oats Company in 1922. While theoretically furfural can be made from any plant material containing pentosans, most of it is produced from corncobs, oat hulls, cottenseed hulls, and sugar cane bagasse. A furfural plant at Omaha, Nebraska uses corncobs and another at Memphis, Tennessee uses some corncobs.

For the manufacture of furfural[4] from corncobs, coarsely ground cobs are put into large rotary digesters lined with carbon bricks. Dilute sulfuric acid is added and the cobs digested under steam pressure. The furfural is removed by steam distillation. The vapors from the digesters containing largely furfural and water are condensed. A small amount of low boiling heads are removed in a methanol column. The water layer goes to a stripping column for recovery of the furfural. The furfural layer, containing about 6% water, goes to a dehydrating column where the water comes off the top and the furfural off the bottom. The product is marketed in 55-gallon drums and tank cars. The commercial product has a purity of 99.5%.

Furfural is a selective solvent used to dissolve gums and other undesirable constituents from lubricating oil. It is used to dissolve the unsaturated portions of vegetable oils, such as soybean oil. The unsaturated fraction is superior to the original oil as a drying oil while the saturated fraction can be used as a food oil. Furfural dissolves out color bodies in rosin and is used in the purification of tall oil.

When furfural replaces formaldehyde in phenol-formaldehyde resins the resin has a longer flow time before setting up thermally. Furfural also reacts with urea, lignin, casein, sulfamides, and amines to form resinous materials. Furfural is a starting point for the manufacture of hexamethylene diamine and adipic acid, which are intermediates for nylon production.

Smaller amounts of furfural are used as fungicides, disinfectants, and herbicides. Among its derivatives are furfuryl alcohol, tetrahydrofurfural, furan, tetrahydrofuran, hydrofuramide, and furoic acid.

SUGARS AND SOLVENTS FROM CORNCOBS

Work at the Northern (Peoria) Laboratory of the USDA showed that it is possible to separate 85% of the pentosans in the corncobs as solution of the pentoses in dilute sulfuric acid, to neutralize the acid in the solution and crystallize out the pentoses, mainly xylose; the latter could be converted to furfural. The pentoses are fermented to butanol, acetone, and alcohol. The cellulose from the residue may be hydrolyzed to dextrose. The dextrose may be recovered as the sugar or the solution fermented to ethyl alcohol.

To make these utilizations of the cobs practical, it is necessary to find uses for the lignin that is left behind. Lignin has a variety of uses in such items as plastics, rubber, soil stabilizers, leather manufacture, adhesives, ceramic plasticizers, cement dispersants, and core binders. Large amounts of cheap impure lignin are recovered from paper mill wastes, making an oversupply available.

PLASTICS

Plastic products were made experimentally at Iowa State University.[2, 5] It was necessary to add hexamethylene tetramine to corncobs to produce a thermosetting product. The best samples compared favorably with phenol-furfural products on the market, but the process was never commercialized. Corncob flour was used as a filler in phenolic plastics but produced a product with higher moisture absorption than that made using wood flour.

BOARDS FROM CORNCOBS

The production of insulating and hard boards similar to those made from cornstalks has been found to be impractical because of the short fiber length in the cobs. When ground cobs were moistened with concentrated sodium hydroxide solution and ground in a ball mill for several days, a plastic gummy mass was produced. After the caustic was washed out, the material was made into a mat and dried with a hot roll. The product showed promising properties but was not developed further.[2]

It is also possible to produce a good board by mixing ground corncobs with sufficient resin, such as phenol formaldehyde, to bond the cob material together. However, the cost of the resin makes this board too expensive for most uses.

A hard fire-resistant board has been named "Corncrete" and

***Table 2** Composition of Corncob Gas*[2]

Constituent	Percent by Volume
Carbon dioxide	53.3
Oxygen	0.4
Illuminants	3.0
Carbon monoxide	22.1
Methane	13.1
Hydrogen	5.3
Nitrogen	2.8
Total	100.0

patented.[6] Corncobs are digested in 5–30% phosphoric acid and dried. The corncobs are crushed to give a mixture of approximately 40% retained on a ⅜-in. scrren, 10% on a Number-10 screen, and 40% on a Number-40 screen with the balance passing through a Number-40 screen. Furfuryl alcohol and spent sulfite paper liquor are added to the corncobs. The mixture is molded into panels and dried in a heated press at from 80–200°C. Good results were secured with 65% cobs, 18.5% ammonium lignosulfonate (sulfite liquor), and 16.4% furfuryl alcohol, mixed together and then formed into panels and dried at 100°C and 15 psi for 10 min. These panels had a flexural strength of 1500 psi.

FUEL GAS FROM CORNCOBS

The destructive distillation of corncobs is similar to that of wood. The corncobs are heated out of contact with air to about 700°C and the gas, methanol, acetone, acetic acid, and tar are collected as the heating continues. Charcoal is left behind in the retort. The gas varies somewhat between runs and during individual runs. Table 2 shows the composition of the gas obtained on early experimental runs in the Chemical Engineering Laboratory at Iowa State University[2] using a batch operation. The yields on the various products are shown in Table 3. Later the USDA at Ames, Iowa used a continuous still of pilot plant size.[7] The results were promising but never commercialized.

The production of acetic acid, aceton, and methanol by synthetic methods together with the substitution of coke for charcoal in iron smelting have made the destructive distillation of wood impractical. It

Table 3 Destructive Distillation Products[2]

Products	Percent of Air Dried Cobs[a]
Charcoal	38.30
Crude acid	37.75
Tar	4.10
Gas (diff.)	19.85
Methanol	0.62
Acetone	0.37
Acetic acid	3.46

[a] Total time of heating, 7.5 hr; final temperature 360°C.

does not appear that substitution of corncobs for wood would be economically practical.

When mixed with household sewage to supply the necessary protein, corncobs and cornstalks can be fermented to produce a mixture of gases composed of approximately 50% carbon dioxide, 25% methane, and 25% hydrogen.[8] Thus 50% of the gas is composed of flammable gases; 1 ton of corncobs will yield 10,000 ft^3 of gas having a heating value of 500 Btu/ft^3. It was considered too complicated a process for home use and was not developed beyond the experimental stage.

Producer gas from wood and charcoal has been used in many countries as a fuel. Work at the Northern Regional Research Laboratory of the USDA has shown that the quality of the gas produced from corncobs equals that from wood.[9] Some study has been made using the Fishcer-Tropsch process but apparently nothing practical has been developed.

CLEANING METAL PARTS

A mixture of ground corncobs and rice hulls has been used for cleaning carbon, oil, and combustion products from such airplane engine parts as cylinders and pistons by airblasting. The process was developed during World War II by the Northern Regional Laboratory of the USDA.[10] The cobs were ground minus 12 and plus No. 35 screen. The use of cobs eliminated the possible damage when sand was used.

Corncob grit blasting may be used to clean large electric motors and generators rapidly and economically. Corncob grits under an air pressure of 20 to 30 psi remove oil, grease, and surface dirt without damaging insulation or coil coverings. Cleaning with corncobs eliminates the hazards of solvent cleaning. Ground corncobs have also been used in the barrel method of deburring, burnishing, and polishing such products as metal stampings, molded plastics, and metal die castings. The products are tumbled with the ground cobs in a barrel. Cob grits are also used for absorbing oils, greases, and dirt from metal stampings and heat-treated metal pieces.

OXALIC ACID FROM CORNCOBS

Strong nitric acid oxidizes sugars and cellulose to oxalic acid. Since corncobs are composed mainly of pentosans, lignin, and cellulose, it should be possible to oxidize them to oxalic acid by nitric acid.[2] Early work at Iowa State University[2] showed this could be done, and later Webber[11] recommended the following process.

> Add slowly with constant stirring one part by weight of ground cobs (¼ to ½ inch size) to three parts by weight of cold 95 to 100 per cent nitric acid. When the cobs have become well wetted by the acid and the initial reaction has moderated, heat the mass until all of the cobs dissolve. Cool the mixture and add 0.1 per cent of freshly prepared vanadium pentoxide (V_2O_5) and three parts (based on the original weight of cobs) of 50 to 55 per cent nitric acid. Allow the mixture to stand at room temperature 2 to 3 days to complete the oxalic acid formation and crystallization.

The corncobs would be converted to oxalic acid in acid-resistant vessels. The mother liquor from the oxalic acid crystals would flow down an absorption tower and the nitrogen oxides formed in the process would go to the tower and react with the mother liquor. The nitrous oxide in the mother liquor would be oxidized by air to nitric acid. Nitric acid would be recovered from the tower for reuse.

Experimental work was carried out at Iowa State University[2, 4] to produce oxalic and acetic acids by fusing corncobs with sodium hydroxide; the best results were obtained with 3:1 ratio of alkali to cobs. Cornstalks and corncobs give identical yields but the stalks are bulkier to handle. The optimum conditions were heating 1–20 hr at 220–250°C. The yield of acetic acid was about 20% and oxalic acid 35% of the weight of the air-dried corncobs. Since the product was a mixture of the oxalates it was necessary to add sulfuric acid to convert it to a mixture of acids. Less drastic conditions are necessary when using corncobs than when using soft wood.

MISCELLANEOUS USES

Corncob Concrete

Corncobs as an aggregate for low-density concrete of higher than ordinary thermal insulating value have been studied.[1, 12] The cobs were ground in a hammer mill to sizes of ⅛–1 in. and the loose pitch and chaff blown out. The cob particles were soaked in water before adding to the concrete mixture. Compressive strengths of 500–600 psi were obtained with ⅛–¼ in. cob particles. With ½-1-in. pieces the strength rose to over 1300 psi. The concrete took nails and screws well.

Adhesive Extenders

To reduce the cost of synthetic resins used as adhesives in the production of plywood and furniture, extenders such as flours from corncobs, nut shells, and fruit pit shells are added.[1, 13] The extenders improve both the spreading and adhesive properties of the resins, but corncob flour is less resistant to moisture than the nut and fruit pith flours. Freight costs to the northwest plywood plants are also higher for the corncob flour than for the other flours.

Abrasives for Soap

Corncob meals in screen sizes from No. 40 to No. 100 are used as a mild abrasive in powdered hand soaps.[1] Only the hard part of the cob is desirable for this purpose, the chaff and pith being removed. As much as 60% corncob meal is used in some soaps.

Dynamite

Corncob grits have been used in dynamite in place of wood flour.[1] Size requirements vary but screen ranges No. 10/16 and No. 40/60 are common.

Light-Weight Bricks and Ceramics

Ground corncobs are mixed with the clay. When products are fired the cob particles burn leaving a porous product.

Floor Sweeping Compounds

A common sweeping compound contains 35% fine sand, 40% fine sawdust, 10% paraffin oil, 10% water. All parts are by weight. Corncob particles of suitable size may be substituted for the sawdust.

Rug Cleaners

Rug cleaners commonly consist of about two-thirds by weight of corncob meal and one-third of a nonflammable petroleum fraction. Sometimes trichloroethyene is added as a solvent for the petroleum fraction. This solvent evaporates and leaves the dirt and oil on the cob fraction.

Based on the yields of corn produced in the United States in 1968, it is calculated that in excess of 30 million tons of corncobs were produced (see Table 1 in Chapter III.5). It is difficult to secure accurate data on the utilization of corncobs, but it is estimated at less than 1 million tons/yr or about 3% of the production.

Apparently there are plenty of corncobs available; however, modern corn harvesting methods do not provide for harvesting the cobs efficiently. Increase in utilization will depend largely on the development of equipment or attachments to current equipment to separate the corncobs economically at the corn harvesting stage. The cobs, like the grain, will need to be dried before storing in large quantities.

Limited quantities of corncobs are added to animal feeds purely for their roughage value. When plowed under on the fields they decay very slowly are of very little if any value.

REFERENCES

1. T. F. Clark and E. C. Lathrop, "Corncobs—Their Composition, Availability, Agricultural and Industrial Uses," Mimeo. Circular AIC-177 (Revised), Northern Regional Research Laboratory, USDA, 1953.
2. O. R. Sweeney, "The Commercial Utilization of Corncobs," *Iowa Eng. Exp. Sta. Bull.* 73, Iowa State Univ., 1924.
3. F. B. LaForge, "Practical Study of Corncob Utilization," *Chemical Age,* **28,** 332, 1920.
4. A. R. Dunlop, "Furfural and Other Furan Compounds," *Kirk-Othmer Encyclopedia of Chemical Technology,* 2nd ed., Vol. 10, pp. 237–251, Wiley, New York, 1966.
5. O. R. Sweeney and L. K. Arnold, "Plastics from Agricultural Materials," *Iowa Eng. Exp. Sta. Bull.* 154, Iowa State Univ., 1942.
6. Turgut Demirel, "Acid-Treated Corncob Construction Material and Method of Producing the Same," USP 3,301,800, 1967.
7. P. B. Jacobs, "Destructive Distillation of Agricultural Wastes," *Ind. Eng. Chem.,* **32,** 214–226, 1940.
8. P. B. Jacobs, "Heat and Light on the Farm," *Architectural Rec.,* **75**(4), 358–368, 1934.

9. J. C. Porter and R. Wiebe, "Gasification of Agricultural Residues," Circular AIC 174, Northern Regional Research Laboratory, USDA, 1948.

10. E. C. Lathrop and S. I. Avonovsky, "Soft Grit Blasting of Metals," *Compressed Air Mag.* **50,** 268–272, 1945.

11. H. A. Webber, "The Production of Oxalic Acid from Cellulosic Agricultural Materials," *Iowa Eng. Exp. Sta. Bull.* 118, Iowa State Univ., 1934.

12. J. S. Boyd and T. J. Breuik, "Use of Corncobs as a Filler for Concrete Used in Farm Construction," *Michigan State College Agr. Exp. Sta. Quart. Bull.,* **35** (1): 87–94, 1952.

13. R. V. Williamson and E. C. Lathrop, "Agricultural Residue Flours as Extenders in Phenolic Glues for Plywoods," *Mod. Plastics,* **27** (2): 111–112, 169–170, 172, 174, 1949.

CHAPTER III.7

cotton processing, by-products, and wastes

GLENN H. GLOVER

Cotton, one of the important crops produced by American agriculture, has contributed to the well-being and comfort of people throughout the world for centuries. It was produced and woven into cloth in the Indus River Valley of Pakistan and the Nile River Valley of Egypt 3000 years before the birth of Christ. Trading of cotton took place between Arab merchants and Europeans about 800 A.D. The fiber's importance in world trade continued to increase and was known throughout the world by 1500.

Production of cotton on a planned basis in the United States was practiced by settlers at St. Augustine, Florida, in 1556 and in Virginia in 1607. It had become an important crop for the colonists along the James River in Virginia by 1616.

Hand processing of cotton into cloth continued until 1730, when mechanization was introduced in England. Invention of the cotton gin by Eli Whitney and the Industrial Revolution in England marked the starting point for the position that cotton now holds. Within 10 years after Whitney developed the cotton gin, the annual cotton crop in the United States increased in value from $150,000 to more than $8,000, 000.

Cotton grows in most warm climates. Major producers are the United States, the Soviet Union, The People's Republic of China, and India. Four other countries—Brazil, Egypt, Pakistan, and Turkey—produce at least 1 million bales each year.

It is produced in 19 states in the United States (Figure 1). Texas produces the largest quantity each year, with Mississippi, California, Arkansas, Alabama, Arizona, Tennessee, Georgia, Louisiana, South Carolina, Missouri, North Carolina, Oklahoma, New Mexico, Florida, Virginia, Kentucky, Nevada, and Illinois following in descending order. These states now produce an average yield or more than 1 bale per

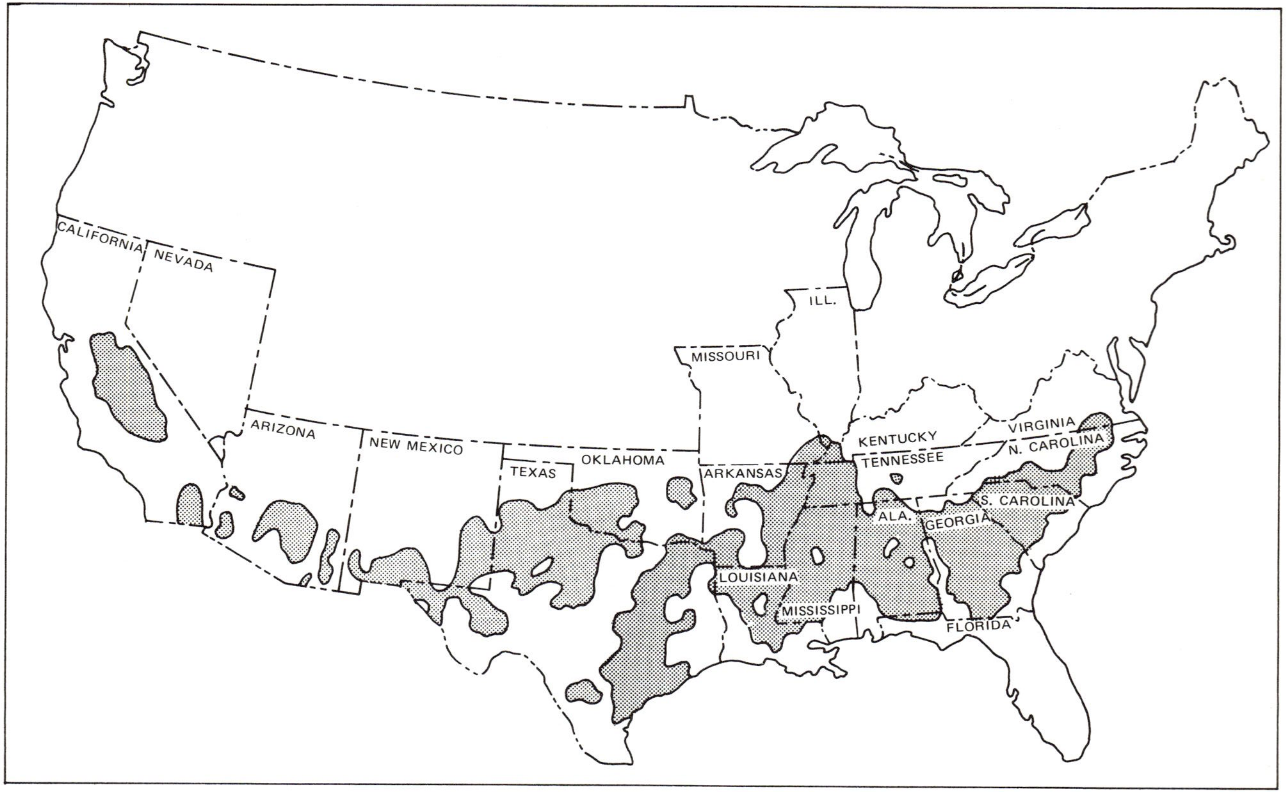

Figure 1. Areas in the United States where cotton is produced.

acre, compared with average production of about ½ bale per acre in 1945.

Waste and by-products are generated from the production and processing of cotton. The increased quantity now grown makes the waste and by-products more significant than they were during earlier periods.

In this chapter, we first define the terminology associated with this cotton and describe methods of its production and processing. Second, the quantity of by-products and wastes that result from the production and processing of cotton is estimated. How these by-products and wastes are utilized and the significance of the wastes as pollutants are reviewed.

TERMINOLOGY

Specific terms applied to by-products and waste particles removed from cotton lint as it is ginned and processed into cloth are as follows.

BURS
: Burs are the remaining parts of mature, open cotton bolls that hold the lint and seed on the cotton stalk. While harvesting, many burs are pulled from the plant mixed with the cotton lint and seed. They are a waste product and sometimes used as a mulch by nurseries.

GRASS
: Cotton harvested from fields where noxious grasses were not adequately controlled usually contain grass as a foreign material. Grasses are usually more difficult to remove from cotton lint than other types of foreign matter.

LEAVES
: Leaves from cotton plants are usually mixed with the fibers during harvest. They are found in larger quantities in lint picked from fields where a defoliant was not properly used.

STEMS
: Short stems from cotton plants are sometimes mixed with the lint during harvest. This occurs more often with mechanical picking than when the lint is harvested by hand.

PEPPER TRASH
: Small leaf particles which are very difficult to remove from lint cotton. Pepper trash is usually made up of ground, dry leaf particles.

GREEN BOLLS

When cotton is harvested with a mechanical striper, green bolls are usually found mixed with cotton lint and seed. These are removed from the lint and seed before ginning.

SHALE

The lining of open cotton burs, which is harvested with cotton lint and seed from the field..

MOTES

Immature, undeveloped seeds found in cotton lint are referred to as motes.

By-products and wastes from the processing of cotton fibers into fabrics and seeds into meal and oil are

PICKER WASTE

Undesirable particles found in cotton lint after the bales have been opened and blended for processing. Picker wastes are usually not foreign matter, but undesirable cotton fibers such as neps and naps. Neps are small, tangled knots of fibers visible as dots or specks when a thin web of fibers is held to the light or against a dark background. Neps in lint are undesirable because, if not removed, they will appear as defects in yarn or fabrics. The removal of neps from lint is difficult, costly, and frequently impossible. Naps are large clumps, or matted masses of fibers that contribute to the rough appearance of ginned cotton. Their formation is influenced primarily by the condition of the seed cotton at the time of ginning. Cotton that is ginned green or wet is usually "nappy." Naps do not lower the grade of cotton as much as neps.

CARDING WASTES

Carding wastes are also items that are not foreign to the cottom fiber, primarily neps and naps, which are further removed in the carding process. Carding is normally performed immediately after the picking process, completing the cleaning action, removing short fibers, the first step toward bringing the fibers into a parallel order, and producing a thin, filmy sheet of web that is condensed into a ropelike, untwisted strand roughly 1 in. in diameter.

COTTON LINTERS

Linters are the downlike fibers that adhere to the cotton seed through the ginning process. Linters must be removed before the seed is processed into cotton seed meal and cotton seed oil.

When cotton is harvested from the field, it is difficult and uneconomical to remove only the cotton lint and associated seed from the mature cotton plant. Dry leaves, stems, burs, and other foreign materials from the plant usually become mixed in with fiber and seed when mechanical harvesters are used. Ginning removes these and seed from the lint cotton.

After ginning, the seed, lint, and foreign materials are held separately. The combination of burs, leaves, grasses, stems, and green bolls is waste and disposed of; some are used productively as mulch by nurseries. The seeds are transferred to plants where they are processed into meal and oil after the linters are removed.

COTTON PRODUCTION AND PROCESSING

Most cotton produced in the United States today is harvested with mechanical equipment—the mechanical picker and the mechanical stripper. The picker removes the cotton fibers from the open bolls and the stripper removes the fibers and bolls from the stalks. The trash, or foreign material content, in mechanically picked cotton is usually much less than in mechanically stripped cotton.

Approximately 1450 lb of mechanically picked cotton is required to obtain 480 lb of ginned lint, compared with a yield of about 2000 lb of mechanically stripped cotton. The picked cotton is ready for ginning, seed and foreign materials are separated from the lint.

Ginning

Cotton is transported from the farm to the gin in large open farm trucks or trailers. When cotton arrives at the gin, it is moved into the ginning equipment with air suction equipment. In most gins, cotton first passes through dryers to facilitate processing, then into equipment that removes burs, sticks, dirt, leaf trash, and other foreign matter from the lint and seed. It then passes into the gin stand, which is the heart of the ginning process. In the gin stand, lint is separated from seed. This is accomplished by circular saws mounted very close together on a revolving shaft. These saws project through narrow slits in a metal screen and pull the lint cotton through the slits away from the seed. The screen slits are designed so that seeds cannot pass through them with the lint. The lint is then removed from the gin saws by brushes or by air blasts. With existing technology in the gin, 100% of the lint is not removed from the seed in the process. Fine fibers, called linters, adhere to the cotton seed and are passed on through the gin with the seed. After ginning, cotton seed and lint are handled separately.

Cotton harvested with a mechanical stripper normally has larger quantities of foreign material than that harvested with a mechanical picker or by hand. Equipment adjustments in a gin set-up for machined-picked cotton are usually required to process machine-stripped cotton. The machine-stripped cotton is cleaned before it is ginned to remove burs, bolls, and other excess foreign material. Some machine-picked cotton requires cleaning before ginning; however, in most cases, it can be ginned without substantial prior cleaning. After the seed and lint are separated in the ginning process, the lint is usually cleaned and then pressed into bales. Running bales are often made weighing roughly 500 lb. The normal 500-lb bale contains 480 lb of lint cotton and 20 lb of baling materials. Gin capacities range from six to 24 bales per hour.

After ginning, seeds go to oil mills, where they are processed into oil, meal, and by-products; The major by-products include linters and hulls. Linters are removed through a process similar to ginning and, in some cases, with acid. Oil is then extracted from the seeds by pressing or through a chemical solvent extraction process. The oil is used primarily for human consumption after refining, and the meal is usually used for animal feeds. Linters are used by various industries in the manufacture of synthetic fibers, films, plastics, paints, gun powder, paper, and other items.

The bales of ginned cotton are transported from gins to textile mills. When received at textile mills, the bales are opened and cotton from several is blended together and cleaned to obtain the desirable average staple lengths and other quality characteristics. The cleaning and separation of individual fibers takes place in picking and carding machines. When cotton fibers emerge from the carding machine, they are formed into untwisted, ropelike strands, normally referred to as "slivers," which are then sometimes placed through combs, which comb out short fibers. Output from the carding and combing processes is then drawn to blend fibers and arrange them in parallel order.

BY-PRODUCT AND WASTE YIELDS FROM COTTON

The quantity of by-products and wastes removed from cotton in ginning and processing depends, to a great extent, on the condition of the cotton plant when harvested from the field and the harvesting techniques used. Approximately 1450–1500 lb of cotton harvested give a 480-lb batch of cotton. For each 480-lb net weight bale of cotton lint produced from machine-picked cotton, there are usually about 79 lb of field trash, which consists of burs, grass, leaves, stems, and green bolls removed in the ginning process. In addition an average of 827 lb of seed and 64 lb of linters are removed. Thus, approximately 891 lb of by-

Table 1 *Average Nonlint Content and Processing Picker and Card Waste for American Upland Samples of Cotton From Selected Gin Points, United States*

Crop Year	Total Nonlint Content (%)	Picker and Card Waste (%)
1957	2.6	8.0
1958	2.6	7.7
1959	2.8	8.5
1960	2.6	8.2
1961	2.7	7.3
1962	2.7	7.9
1963	2.4	6.9
1964	2.6	5.0
1965	2.8	5.2
1966	3.2	5.8
1967	3.3	6.0
1968	3.1	6.2
1969	3.2	6.8
1970	3.2	6.7
1971	3.2	6.8
1972	3.3	6.8

products and seed and 70 lb of wastes are removed in ginning a 480-lb bale of machine-picked cotton. The amount of waste is usually somewhat higher in machine-stripped cotton.

From 1957 through 1969, the total nonlint content in bales of ginned cotton ranged from 2.4% to 3.3% (Table 1). This percentage has increased over the period owing to changes in harvesting practices and ginning techniques. The nonlint content consists primarily of pepper trash, shale, and other fine nonlint particles that were not removed in ginning. The percentage of picker and card wastes, which is part of the lint content, has ranged from 8.5% to 5.0% during the same period (Table 1). This reflects an improvement in the quality of cotton harvested from the field for ginning. Picker and card wastes consist primarily of neps and naps.

The average yield of by-products and waste from each ton of cotton

Table 2 *Average Yield (lb) of Cottonseed Products per Ton of Cottonseed Crushed in the United States*

Year Ending July 31	Crude Oil	Cake and Meal	Hulls	Linters	Loss or Waste	Total
1941	324	888	504	171	113	2000
1942	312	874	495	186	133	2000
1943	311	887	482	190	130	2000
1944	313	928	469	185	105	2000
1945	311	919	463	182	125	2000
1946	312	879	480	182	147	2000
1947	315	882	471	191	141	2000
1948	312	930	452	186	120	2000
1949	320	897	463	183	137	2000
1950	323	895	469	176	137	2000
1951	321	896	461	185	137	2000
1952	320	930	451	185	114	2000
1953	328	961	431	184	96	2000
1954	332	946	444	184	94	2000
1955	331	976	434	187	72	2000
1956	339	942	447	180	92	2000
1957	340	964	433	181	82	2000
1958	339	922	455	176	108	2000
1959	342	928	463	181	86	2000
1960	339	928	458	181	94	2000
1961	338	936	455	177	94	2000
1962	337	905	478	176	104	2000
1963	333	937	464	171	95	2000
1964	336	948	455	171	90	2000
1965	338	936	458	176	92	2000
1965–1970	338	934	460	176	92	2000

seed crushed in the United States from 1941–65 is presented in Table 2. These figures reveal tha approximately 173 lb of linters and 93 lb of wastes are generated from each ton of cotton seed processed.

The cotton linters are of necessity removed from the cotton seed before the seed is processed into meal and oil. Most mills run the seed through delinting machinery twice and produce first-cut and second-cut linters. First-cut linters, which are the longer, more resilient fibers removed in the delinting process, are usually used for felting, while the second-cut linters are used principally for chemical production. First-cut linters are employed in the manufacture of absorbent cotton, wicks, twine, gauze, carpet yarns, and batting in mattresses, other bedding

Table 3 *United States Production of Cotton Lint, Seed, and Linters and Estimated Wastes*

Year	Cotton Lint (lb)	Cotton Seed (tons)	Cotton Linters (lb)	Estimated Wastes (tons)
1959	6,987,633	5991	999,088	575,024
1960	6,850,579	5886	957,308	563,746
1961	6,872,618	5978	983,367	565,559
1962	7,135,948	6096	994,067	587,229
1963	7,360,187	6192	964,395	605,682
1964	7,287,146	6227	996,841	599,671
1965	7,186,835	6087	1,001,650	591,417
1966	4,595,983	3960	677,400	378,211
1967	3,578,199	3135	538,800	294,456
1968	5,255,040	4640	784,200	432,446
1969	4,807,200	4186	705,600	395,593

products, and in upholstery for furniture and automobiles; an increasing volume of first-cut linters is being used in the manufacture of fine paper.

Second-cut linters as raw material for the chemical industry must first be purified. The resulting linters pulp is practically pure cellulose, suitable for explosives, photographic and X-ray films, plastics for automobile instrument panels and steering wheels, lighting fixtures, radio cases, flexible pipe, and many other molded products. Another market is in viscose fibers for tire cord and acetate fiber for clothing, draperies, and other household articles.

SUMMARY

A substantial amount of by-products and wastes are generated in the production and manufacture of cotton into finished fabrics; however, most of these wastes and by-products are used for the manufacture of other products (Table 3). Few of them could be justifiably classified in the category of pollutants, since those that are not useful in the production of other products are, in many cases, returned to the soil and serve a useful purpose in improving its fertility.

CHAPTER III.8

by-products and wastes from the tung oil industry

RAIFORD L. HOLMES and GEORGE F. POTTER

Tung oil is one of the best of the vegetable drying oils. It is produced from the seeds (nuts) of the tung oil tree (*Aleurites fordii*), which was introduced from China to the United States in 1905. The fruit is about the size of a black walnut but contains four or five seeds. The oil is contained in the kernels of the seeds. The kernels are surrounded by thin hard shells, which in turn are surrounded by thin papery inner hulls; the whole is surrounded by a fibrous outer husk about ¼ in. thick.

Commercial plantings of tung trees were first made in the United States in 1924, and the first tank car of oil was shipped from Gainesville, Florida in 1932. The industry expanded through Florida, the southern parts of Georgia, Alabama and Mississippi, and through southeastern Louisiana until by the 1950s 200,000 acres of orchards had been planted. In 1964 the production was 123,000 tons of fruit yielding about 32,000,000 lb of oil valued at $7,600,000. The production for 1964 was good, but the annual production is erratic because of damage from late spring frosts.

Tung oil has met with increasing competition from synthetic materials used in the coating industry and from cheap oil imported from South America causing the price of tung oil to drop until it was necessary for the USDA to support the price of tung oil to make its continued production feasible. Also, Hurricane Camille in August 1969 did terrific damage to the industry, destroying 50,000 acres of orchards mostly in southern Mississippi. It remains to be seen how well the industry will recover from this disaster considering the low price at which tung oil has been selling.

The tung fruits mature and drop in the fall. Harvesting starts in November. The whole fruits are picked up after partially drying and carried to disk hullers where the inner and outer hulls and about half the shell are removed from the kernels. If the partially hulled kernels

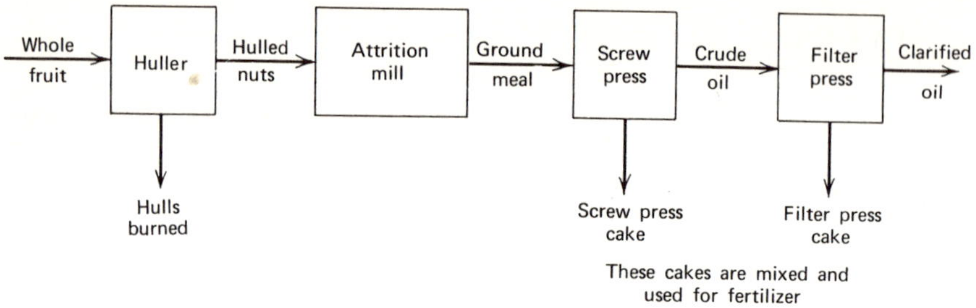

Figure 1. Tung oil processing.

are not already dry enough (9–10 moisture), they are passed through driers, ground to a coarse meal in attrition mills, and the meal pressed in steam-jacketed screw presses. The hot oil is filtered through plate and frame filter presses and is ready for sale to the protective coatings industry. The cake is discharged at the end of the screw press (Figure 1).

The by-product of the tung industry is the mixture of the screw-press and filter-press cakes. This mixture contains about 6% tung oil, 25% protein (4% nitrogen), 1% phosphoric acid, 3% potash, and about 54% crude fiber plus pentosans.[1] Because of its oil content, care has to be taken in storing tung cake to avoid spontaneous combustion. Tung cake cannot be used for stock feed as are most other oil seed cakes because it is toxic to animals, including man. Considerable experimental work has been done trying to characterize the toxic substance and develop a method for detoxifying the meal,[2, 3] but no commercially feasible method has been developed. It is used only as an organic nitrogen fertilizer, most of its being scattered over the tung orchards.

The tung hulls constitute a waste product. In the early years of the industry some ground hulls were used as a fertilizer conditioner, as a mulch for plants, and as a fuel under the boilers. At present all hulls are burned in the open. The hulls contain about 3% potash, 0.3% phosphoric acid, and 0.6% nitrogen. The remainder consists of cellulose, lignin, and the like.[1] When the hulls are burned the potash is concentrated in the ash, which is used as a fertilizer.

REFERENCES

1. Raiford L. Holmes and R. S. McKinney, "Tung Hulls and Press Cake," USDA AIC-357, June 1953, 14 pp.

2. Raiford L. Holmes, and E. T. Rayner, "Isolation of Two Nitrogen-free Toxins from Tung Kernels," *J. Am. Oil Chem. Soc.,* **35,** 586–589, November 1958.

3. Raiford L. Holmes and R. T. O'Connor, "Chemical and Physical Characteristics and Possible Configuration of Toxins from Tung Kernels," *J. Am. Oil Chem. Soc.,* **38,** October 1961.

CHAPTER III.9
nut shells and fruit pits

Nut shells and fruit pits accumulate in quantity about cracking plants, canneries, and food-processing plants. Under favorable conditions and with proper handling, these residues can be converted into profit. Economically, the residues include only those materials that accumulate in quantities large enough to makes their processing and use worth while—the pits of apricots, cherries, peaches, and dates, and the shells of almonds, coconuts, filberts, peanuts, pecans, and both English and black walnuts. More than 1 billion lb are available annually within the United States.

Attempts to use agricultural residues in commercial operations are not new. Many scientific papers, patents, and proposed methods for utilization of agricultural residues are found in the literature. Failure of many of the past developments and methods has resulted from the false assumption that these agricultural materials were available at little or no cost and that low-cost substitute products might be made from them.

Collection or purchase and the preservation of the nut shells and fruit pits must be economic, with sound merchandising and business management. Uses must be based on superior or unusual chemical and physical characteristics. In the early 1930s a plant for grinding walnut shells was established on the Pacific Coast, with a line of products including flours for plastics, adhesives, and insecticides, and grits for blast-cleaning, burnishing, and polishing, for nonskid paints and dynamite manufacture. Uses for processed materials from nut shells and fruit pits have expanded.

Structurally and physically the shells (endocarp) of nuts and fruit pits have characteristics in common. During growth of nuts and fruits, the pits or shells have an outer covering or husk (mesocarp) that often covers the entire shell. In fruits this husk is the portion for which the plant is cultivated. Although this fleshy material is removed at can-

Table 1 *Proximate Analyses of Shells of Nuts and Fruit Pits, Dry Basis*

		Solubility in					Constituents							
Type of Shell	Crop Year	1% Sodium Hy-droxide (%)	Alcohol-Benzene (%)	Acetone (%)	Hot Water (%)	pH of Hot Water Extract	Ash (%)	Silica in Ash (%)	Nitrogen (%)	Pentosans (%)	Cutin (%)	Lignin (%)	Lignin, Corrected for Cutin (%)	Alpha Cellulose[a] (%)
Almond, hard	1946	17.8	4.2	2.4	4.4	5.2	1.0	5.8	0.19	34.2	0.4	27.5	27.1	23.6
Almond, soft	1946	27.9	2.6	3.9	11.9	5.0	3.1	3.2	0.32	31.0	0.0	24.9	24.9	22.9
Apricot pit	1946	14.1	1.1	1.3	2.1	5.0	0.7	14.9	0.12	32.8	0.9	31.0	30.1	27.2
Black walnut	1945	13.7	1.9	1.5	2.5	5.0	0.3	51.7	0.20	28.3	12.7	43.5	30.8	28.1
Cherry pit	1948	19.9	4.4	3.2	6.5	4.5	0.8	—	—	27.9	—	34.8	—	22.8
Coconut	1946	22.0	3.0	3.7	3.3	5.4	0.7	2.9	0.12	31.6	0.3	34.2	33.9	24.3
Date pit[b]	1946	37.4	11.1	11.3	12.0	5.8	1.1	6.5	0.96	2.9	1.5	10.2	9.1	39.6
English (Persian) walnut	1946	18.5	1.1	3.2	3.5	4.1	0.8	12.0	0.17	29.5	0.7	38.7	38.0	25.2
Filbert	1944	13.3	1.8	1.8	2.2	4.8	0.7	31.7	0.16	29.4	10.0	44.5	34.5	25.8
Peach pit[5]	1946	14.8	1.5	1.7	2.3	4.6	0.4	21.9	0.15	31.1	1.7	38.9	37.2	22.6
Peanut	1943	44.0	6.4	6.4	11.2	6.6	8.8	59.8	1.49	16.8	3.7	28.7	25.0	25.1
Pecan	1946	25.5	1.1	1.9	7.8	5.0	1.8	27.1	0.24	26.0	0.5	46.2	45.7	19.7
Pecatraannin extracted	1946	24.0	2.9	2.6	1.4	4.9	1.5	2.3	0.26	25.2	0.3	43.6	43.3	22.7

[a] Monoethanolamine method, ash-free and corrected for pentosans.
[b] Averages for pits obtained from California and Washington.

neries or prior to delivery to shell-processing plants, some of this material, generally fibrous, may still be attached to the shell's outer surfaces.

One difference between nut shells (or outer covering) and fruit pits (from inside the fruit's flesh) is the kernel inside the fruit pit. Nuts are cracked for removal of the meats or kernels only; broken fragments of the meats are not purposely left in the shells. With fruit pits, however, the kernel remains within the shell unless, in processing at a cannery, the shell is split and the kernel released. Apricot pits in California are cracked to recover the kernels, which are marketable.

The outer surfaces of shells vary from very rough to smooth, soft to hard, and fibrous to essentially nonfibrous. The physical structure of the shells varies considerably from their outer to inner surfaces. The shell of the English walnut has three layers. The outer or hardest is composed of densely packed cells, essentially isodiametric and colorless. The cell walls are highly lignified. The middle layer is composed of irregular-shaped cells, larger, more spongy and with thinner walls. Cells of the inner layer are thin-walled, loosely packed, and serve as a lining. Shells of pecans and filberts also contain an appreciable quality of spongy lining material.

Many nuts also contain parchmentlike or woody tissues that act as partitions or septa between the cotyledons or segments of the kernels. These materials remain with the shells after cracking and shelling.

The shape of the original nut shell is not significant. However, fruit pits frequently accumulate without rupture of their shells and thus still contain kernels. Shape and size of the pits influence the manner in which pit shells may be cracked and the kernels removed. Peach pits of a flattened ovoid shape may be as large as 1½ in. long, 1 in. wide along the suture, and ¾ in. high across the suture. Cherry pits are approximately ⅜ in. long, 5/16 in. wide, and ¼ in. across the suture. Broken sections of coconut shells may be as large as 3–4 in. in one dimension.

The shells of nuts and fruit pits are quite similar in composition. Proximate analyses are presented in Table 1.

These analyses do not include tannins, which in the outer shells ranged from 0.4% to 1.1%, while that of the lining material ranged from 25.6% to 47.9%, depending upon variety and the locality.

The USDA estimates that 1,000,000 tons annually of nut shells and fruit pits are available; the major ones are in Table 2.

Each year a considerably smaller amount actually becomes available for commercial processing. Materials are used where grown, by canning, or as seed, or sold fresh without processing, or the location is not conducive for shipping to processing plants, or volume is too small or there are culls or unharvested crop.

Table 2 *Tonnage and Colorific Value of Nut Shells and Fruit Pits*

	Tonnage	Percent Shell	Tonnage Shell	Percent Pits	Tonnage Pits	Percent Fresh	Percent Canned	Percent Dried	Percent Frozen	Percent Brined	Btu/lb (Dry) Heating Value of shells or pits
Almonds	45,000	50–60	25,000								8360
Black Walnuts		88									8000
Coconuts		10–15									6000
English Walnuts		58									8000
Filberts		60									8300
Peanuts	1,000,000	34	340,000								8800
Pecans	80,000	60	48,000								8950
Apricots	200,000			4–8	16,000	15	50	33	2		8520
Cherries	240,000			4–8	19,200	24	42		22	13	7800
Peaches	2,000,000			4–8	166,000	60	29	8	2		8200
Chestnuts											7900

Canneries, cracking, drying, freezing, and other packing plants become the sources of these residues. Continuous year-round operation (one shift) of shell-grinding plants requires 5000 tons of shells. Frequently, the shells are burned as fuel in the plant boilers. This practice reduces the expense of power. Shells are often discarded on a dump or used for ground fill.

Processing of shell wastes affords a profitable means for disposing. Where processing is to follow cracking, the size of processing operations can be geared to handle the major portion of the waste material; thus, facilities for storing raw material can be kept at a minimum. However, when processing demands are greater than the supply of shells from a plant, the local shells must be augmented by similar material obtained from other cracking plants. Facilities must be provided for unloading, short-term storage, and cleaning prior to processing.

The problem of disposal of fruit pits at canneries and fruit-packing plants differs to some extent from that for nut shells. The pits have residual fruit flesh or pulp on their surfaces. Often these pits are ground in a hammermill and flushed into the sewer. In canning and packing the fruit is often halved by sawing. This operation releases the kernels and also mashes then so that separation and recovery of shells or pits is difficult. The shells are converted to charcoal or ground for industrial purposes. In many cases no separation of kernels is made prior to disposal. Custom grinders often do pit cleaning at canneries. Transportation of residual fruit pulp is avoided. No pits accumulate on the cannery site.

The cost of unground shell and fruit pits depends upon the source, value as fuel, moisture and impurities, transportation, storage facilities, and the price the consumer will pay for finished products. The cost to grinders is their value as fuel, being approximately one-half that of coal. Processors of apricot pits can afford to procure these materials from areas within a 40-mile radius. Peanut shells may require baling in order to reduce their bulk for shipping.

Almond and peanut shells are used for poultry litter; their absorbent character keeps poultry house floors dry. Baled peanut shells are shipped more than 500 miles in competition with corncobs, bagasse, vermiculite, and other materials. Poultry litter can be used for months by periodic additions of fresh material to the 4–5 in. initially spread. The shell material should be larger than ¼ in. and dust-free.

Charcoal prepared by destructive distillation of nut shells and fruit pits is used in poultry and animal feeds.

Shells from peanuts, almonds, pecans, and filberts are used as mulch. The fines (less than ¼ in.) and light fractions are removed to prevent packing and to avoid lighter materials being blown around. These mulching agents conserve moisture for the plants and help check weed growth. Competition from sawdust and wood chips, corncobs, bagasse, vermiculite, and compost is intense. Shell material from almonds and peanuts is ground for mixing with commercial fertilizer to provide bulk, retard caking, and improve free-flowing characteristics in such products. Flour from walnut shells is used in dusting materials.

For years corn meal was used for cleaning furs, and was replaced by hard maple sawdust and the woody portion of corncobs, ground and classified to No. 12/20, the beeswing and dust, and chaff being removed on ground nut shells and fruit pits, with size classification of pit materials similar to the cob products.

During World War II, the military forces developed processes for removing carbon, oil, and products of corrosion from cylinder heads, pistons, valves, and other parts of aircraft engines, by air-blasting with cob grits or mixtures of cob grits and rice hulls. Shells of almonds, apricot pits, cherry pits, coconuts, filberts, peach pits, pecans, black walnuts, and English walnuts are competitive.

Parts to be blasted may be degreased first if oil is excessive. Induction or pressure-tank-type blast units may be used, depending upon the size and nature of the pieces to be cleaned. The grits are recovered continuously and recirculated until worn out or contaminated excessively with oils and greases. Grits are contained in a hopper beneath the booth and fall by gravity to an induction chamber from which they move by suction to the blast nozzle. Spent grits drop through a grating or screen within the booth to the storage hopper, to be disposed of as fuel under boilers. Air pressure of 80–90 psi is required.

The method has been extended to the finishing of molded plastics and hard rubber parts. Flash and fins along parting lines are removed from the molded pieces without harm to the molded surfaces if air-blasting is done with soft grits classifying as small as No. 50. Such cleaning can be done in a continuous operation.

Tumbling finishing includes the removal of burrs from the edges of metal stampings, flash and fins from molded plastics, hard rubber, and metal die castings, and the burnishing and polishing of these. For years hardwood sawdust and wood chips, pegs, and balls, as well as a variety of metal shapes or jacks, were used. Ground corncob, shells, and fruit pits are well adapted to metal finishing.

Finely ground shells of most nuts and fruit pits are excellent fillers in phenolic molding compounds. Flours for molding should be finer than No. 100 Tyler. Some compounders of plastics prefer flours finer than No. 200, and even as fine as No. 325. These finer grades of flour filler impart superior surface characteristics. The lighter-colored flours are preferred because of the economy and ease with which dyes and pigments can be incorporated. Only the inner woody portion of most fruit pits and nut shells should be used. The dark outer surfaces, together with lining material and septa, should be removed in grinding and classifying operations. In some cases, as with pecan shells, tannins that are concentrated in the inner linings and septa are interfering substances; kernel residues and nut meat fragments likewise interfere. Some processors of shells and pits expose the incoming stock to weathering during which worms and other small organisms consume the kernel and meat residues. These are removed by screening and aspiration prior to grinding operations.

Charcoal, gases, methyl alcohol, phenolic compounds, acetic acid, tar acids, other chemicals, and pitch, may be produced by the destructive distillation of nut shells, fruit pits, other agricultural residues and wood. Manufacture of chemicals from natural gases, coal, air, limestone, and water has been found more economical than the destructive distillation processes. High-quality charcoal is obtained by the destructive distillation of dense shells from black walnuts, coconuts, English walnuts, pecans, and pits of apricots and peaches. Activated carbon prepared from black walnut shells is of excellent quality and effective in hypersorption units purifying the constituents in natural gas, as an absorptive agent in gas masks and industrial respirators.

In general, nut shells and fruit pits, being hard, friable materials, undergo size reduction by impact and attrition without much difficulty. The few shell materials that have fibrous characteristics may require cutting action to reduce their bulk during the initial stages of processing. Thorough consideration is given to these factors in a bulletin, "Dry Grinding of Agricultural Residues, A New Industrial Enterprise," AIC 336, published in May 1952 by the Northern Regional Research Laboratory, Peoria, Illinois. In addition to giving data on actual grinding and on commercial as well as pilot-plant equipment, many difficulties and hazards of a grinding business are discussed. Dust explosions and fires in grinding plants are also given detailed consideration. Studies on the grinding of black walnut shells to fine flours for the Tennessee Valley Authority are reported.

Preliminary reduction of nut shells and fruit pits is made with a high speed hammermill or with an attrition-type of disk mill. The plate pattern selected for the preliminary reduction depends on shell characteristics and the work to be accomplished at this stage. The reduction to fine particle size and flour takes place in attrition mills. Although mills of this type run hot in dry grinding, the possibility of the fine products catching fire is avoided by the use of water, steam, air, and inert gases in maintaining nonhazardous conditions in the grinding system. One method of using steam in a dry-grinding system is claimed in a patent granted to Ayers. At the Northern Regional Laboratory steam is introduced into the pneumatic collecting system at the discharge of the attrition mill. This has been very effective in preventing explosions and fires.

A perennial suggestion with nut residues and pits is to convert them to "activated" carbon by destructive distillation. Destructive distillation can be traced back to antiquity. In Egypt distillation of wood was practiced for fluid wood tar and pyroligneous acid for embalming, for char alone or, when resinous coniferous wood was used, for tar and pine oil.

The demands of the dye intermediate and synthetic organic industries for wood alcohol became so great that wood was distilled for this solvent. In 1870, the manufacture of acetic acid and acetone via calcium acetate from wood was developed.

For many years wood distillation was the only source of acetic acid, methanol, and acetone, and furnished tar, pitch, methyl, acetone, and wood oil. During World War I, however, a process for acetone by fermentation was worked out competitively, displacing acetone from wood distillation. In 1928 acetic acid from calcium carbide and acetone from propylene further decreased the wood distiller's market. Synthetic methanol displaced wood alcohol. In 1953–1954 by-product acetone from the cumene phenol process completed the demise of the wood chemicals industry.

From 100 hardwood-distillation plants, the number in operation dropped to 46 in 1932. In 1947 all but 15 of those had been shut down permanently, and in 1954 this number was reduced. There is no synthetic counterpart for charcoal for barbecues or fuel.

Wood distillation is practiced principally in Pennsylvania, Michigan, New York, Arkansas, and Tennessee, where the wood is grown as a crop and harvested for timber, the scrap going into retorts.

Almost any vegetable substance can be destructively distilled to produce a char. The patent literature mentions pine, birch, logwood, quebracho, hemlock, cedar, and other woods, particularly in waste forms such as small pieces, shavings, and sawdust; cornhusks, corncobs, cornstalks, cane trash, bagasse, peat, coals of various kinds, lignite, flue ashes, rice hulls, coffee, cocoa, mate, molasses, alcohol slop, waste liquors and extracts from paper manufacture, ivory-nut, shavings, coconut hulls, peach-nut shells and pits, spent olive pulp, surgical cotton, flax waste, moss, hydrogenated oil residue, rubber waste, extracted cotton hulls, and so on.

Mantell[1] has described commercial manufacture of carbon adsorbents in detail, but most of those from agricultural wastes have been commercial failures, such as Carborox from rice hulls, Konigsfeld from spent coffee, Actibon from sawdust, Carbex from cellulose, Eponit from wood wastes, Kelpchar from seaweed, Magnechar from oak, and Kuri from chestnuts, all of which are out of production, but there are successful operations based on pine wood, paper mill waste liquor, lignite, and bituminous coal.

With the increasing use of tertiary water treatment employing activated carbon, perhaps large markets (Mantell,[1] p. 205 ff) will develop which will be based on peanut shells, corncobs, nutshells and fruit pits. Competitive marketing will be most important for commercial success.

REFERENCES

1. C. L. Mantell, *Carbon and Graphite Handbook,* Wiley, New York, 1968, Chapters 9 and 13.

CHAPTER III.10

peanuts

NEWTON W. PENNEY

Annual average peanut production in the United States amounted to 1,233,398 tons during the 5-yr period 1965–1969. This tonnage was produced in three major producing areas Figure 1. The Virginia–Carolina area produced an annual average of 306,500 tons, the southeastern area produced 629,898 tons, and the southwestern 297,000 tons.

Three types of peanuts account for nearly all production. The Virginia–Carolina area produces most of the Virginia type, commonly known as "ballpark peanuts," roasted in the hull. However, Runners and some of the Spanish type are also grown in the area. In the southeastern area both the small Spanish type and the Runner type are produced. These are utilized for peanut butter manufacturing, roasted peanuts, and in confections such as candy bars. The roasted nuts are usually shelled and dry or oil roasted and sold in packets, as cocktail peanuts, mixed nut packages, and other ways. In the southwestern area the Spanish type is produced almost exclusively. The Spanish type is used for every purpose except roasting in the hull.

Most of the peanut crop is sold as farmer's stock, a trade term for uncleaned peanuts as they come from the farm in bulk truckloads. Peanuts are harvested by plowing up and are left in the field to partially dry. Harvesting is continued with a pickup combine that removes the nuts from the vine and deposits the vines on the land. Unless the field drying reduces the moisture content to 8% the peanuts are transported to a drying plant for lowering the moisture to at least the 8% level. Obviously, peanuts so harvested will have some loose dirt, stems, sticks, pebbles and trash remaining commingled with the peanuts. In some cases the farm truckloads of peanuts are precleaned before selling and storing, especially in cases where foreign material content is high.

If peanuts are sold and unloaded at shelling plants or independent warehouses, they are sampled and graded for foreign material content,

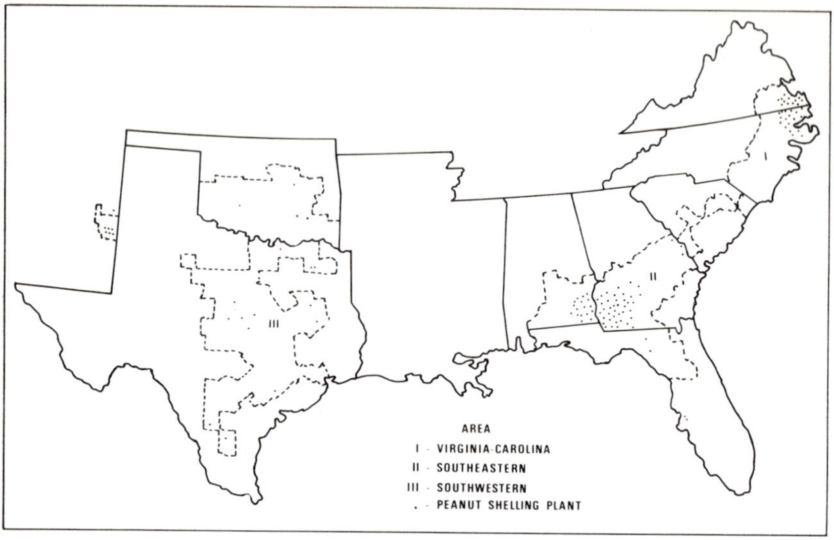

Figure 1. Peanut producing areas of the United States.

moisture, sound mature kernels, splits, and damaged kernels. They are segregated and stored according to grade in common storage in-shell. Except those loads of peanuts that are unmerchantable because of foreign material, the nuts are stored with the foreign material commingled. In the latter case, the foreign material is removed in the shelling process. Those loads of peanuts that are unmerchantable have to be precleaned before selling or storing.

During the 5-yr period 1965–1969 the peanut supply going into commercial channels was utilized in the following foods: candy making, 21%; shelled roasted peanuts, 23%; peanut butter, 53%; and all other uses 2%.

In recent years a rather high percentage of the entire production of peanuts has been utilized for crushing. Aside from 3–5% of the shelled edibles removed as undesirable for the food products mentioned above, some loads of farmer's stock peanuts have sufficiently high damage that they are declared unmerchantable for shelled edible uses and have a value only for oil stock. Additionally, near the beginning of the new crop harvest each year the Commodity Credit Corporation disposes of stocks held on loan by competitive bids for oil uses. The amount of peanuts utilized for crushing would be about 5–10% with no surplus disposal program by the federal government.

Assuming that the hull content of farmer's stock peanuts averages

26.15%, the 5-yr annual average volume of hulls (including the waste and foreign material content) amounts to 318,000 tons. Hulls constitute a serious waste problem for shelling plants. Their bulk and light weight coupled with low value cause the hulls either to have no economic value or perhaps a negative economic value. Most shelling plants are powered by electricity, but originally many of them were powered by steam; these steam-powered utilized some or all the hulls. However, hulls were never a good fuel because of their peculiar burning and heating characteristics. Moreover, burning the hulls as a means of disposal or as fuel is undesirable from the environmental point of view: the ash fallout pollutes the air and utilizes oxygen from the air.

Commercial uses of hulls include litter for poultry houses; bedding for cattle, horses, sheep and other animals; mulch for shrubbery; soil conditioner; roughage for feeding ruminant animals; as a carrier for pesticides and insecticides; and as a conditioner or as a filler for fertilizers. These applications are sporadic and the supply is highly seasonal. Demand depends largely on weather conditions. Unless ground or compressed, transportation costs are prohibitive, and cost of long-distance hauling prohibits sales in distant markets due to low unit value.

Considerable research work has been done on hull utilization. Some possibilities seem to be as follows: char made from hulls reportedly has variable heating characteristics that may make the char useful for industrial purposes in addition to uses for cooking; activated charcoal for decolorizing and deodorizing; abrasives, sweeping compounds; and for making various kinds of building materials such as particle board. However, the latter has not yet been proven successful. Another application is as a component of presto logs for fire place heat. One other possibility is the use of hulls as a raw source of cellulose to serve as energy for conversion into edible protein substance. The process involves subjecting the cellulose material to a microorganism discovered by Han and Srinivasan.[1] If this possible use should prove economically and commercially feasible it would provide an outlet for the entire peanut hull production of the United States.

THE AFLATOXIN PROBLEM

In 1960, more than 100,000 young turkeys died within a few months in England from an apparently new disease. Almost simultaneously, an outbreak of trout hepatoma was discovered in the United States. Although there was no known relationship between the poultry and trout diseases at the time of the outbreaks, it was later shown that a common causative agent was indeed involved. The oilseed meals used in the poultry and trout diets were found to be contaminated with mycotoxins (toxic metabolites produced by fungi).

The mycotoxins were later isolated from toxic peanut meal and identified as aflatoxins, a group of closely related difuranocoumarin compounds produced primarily by two common saprophytic molds, *Aspergillus flavus* and *A. parasiticus*. Agricultural products, particularly oilseeds, may become contaminated by these metabolites during harvesting, storage, or transportation if the environmental conditions are favorable for the growth of molds. A profusion of spores is produced which are widely disseminated over the environment; the most common reservoir of *A. flavus* is the soil. The fungus requires for growth a certain minimum concentration of organic compounds, a temperature of 20–30°C for near-optimum growth rate, and adequate moisture. When a spore germinates in an optimum environment, a profusion of new spores may be produced within 3 days. In growing, the fungus extends itself by means of microscopic filaments that penetrate and permeate the substratum, for example, a corn grain, peanut kernel, or superficial coconut tissue. All structures of the fungus are moderately resistant to environmental hazards so that if moisture, for example, becomes too low, the fungus will cease growing actively. It will remain dormant for relatively long periods until conditions for growth again become favorable or until it is killed by heat, chemicals, the activity of other microorganisms, etc. For this reason, a physical procedure like dry milling of corn does not kill the fungus but rather fragments the substratum and leaves the fungus alive in finer particles of corn.

As a natural consequence of growth, especially of rapid growth in a favorable medium, metabolic by-products are produced. Four such products, in the case of *A. flavus,* are aflatoxin B_1, B_2, G_1, and G_2. Unfortunately, these products are toxic, even in low concentrations, to many animals, unlike other metabolic fungal products that are less toxic or even beneficial, such as antibiotics. As the fungus grows, aflatoxin is deposited in the tissue of the supporting medium. Physical, chemical, or other agents such as heat or organic solvent may kill the fungus but leave the aflatoxin unchanged.

Seeds of some crops, for example, sorghum, soybeans, and rice, are apparently resistant to the rapid growth of the fungus and attendant aflatoxin formation. Distribution of aflatoxin is such that a few seeds tend to have a very high concentration while most seeds contain little or no toxin. This is clearly the situation in peanuts. This fact makes feasible the reduction of toxin in finished peanut products by physical or electric sortings.

In general, *A. flavus* and many deteriorating microorganisms do not gain a foothold in actively growing plant tissue. However, when senescence and maturity occur, these fungi may rapidly degrade the crop. In present studies, many other fungi have been detected in the seed assayed. The total activity of all fungi in grains, peanuts, and coconut is to reduce the quality that was possible at full maturity. Control of deterioration fungi can be accomplished with fungicides, but this is often hazardous because the result may be a pesticide residue problem rather than an aflatoxin problem. The most economical and effective method of control is rapid moisture reduction to the point at which *A. flavus* and other fungi, although not killed, will not grow actively. This state of control depends, moreover, on the maintenance of this low degree of moisture. In the case of peanuts, 7–9% is satisfactory. With corn, 12–13% would be a relatively safe moisture, and copra is apparently safe at a moisture of about 6%. Since the aflatoxins are acutely toxic and chronically carcinogenic to many animal species, they represent potential public health hazards. Where they are not controlled the oilseed production cannot be used, but must be disposed of as waste.

Although the aflatoxin problem first came to public attention in an oilseed, namely the peanut, the problem in not unique to oilseeds. It is, however, of particular importance in this group of crops since they are excellent sources of high-quality protein. The frequent occurrence of aflatoxins in peanuts and other oilseeds and their corresponding products thus presents serious problems in utilizing these sources to supplement low-protein diets throughout the world.

The concurrent outbreaks of disease in poultry and fish in different parts of the world in 1960, then, focused worldwide attention upon diet contamination by mycotoxins. Paradoxically, because of the aflatoxin problem, the public is getting much higher quality oilseed products than ever before.

PEANUT SHELLING

The Shelling Process

The shelling plant process includes cleaning, shelling, separating, sacking, and weighing. The shelling plant is usually operated over the entire year, but they are operated nearer full capacity during fall months when peanuts are being received and stored. The marketing period for farmers begins in late August and usually ends in about 2 or 3 mo. Much of the shelling plant machinery now in use was designed originally for other uses and has been adapted to peanut cleaning, shelling, and separating. Likewise many of the buildings were originally constructed for other purposes and have been adapted to the present use. The shelling plants are located in the producing areas.

Cleaners and Cleaning

The cleaning job involves removal of foreign materials such as pebbles, stems and sticks, dirt, and miscellaneous matter. Peanuts have varying amounts of foreign material, whose nature varies with areas of production, harvesting practices, weather during harvesting, and so on.

Peanuts are usually moved to the cleaning machine from the unloading point or storage facility by bucket elevators and conveyor belts. Most cleaners have flat, slightly inclined screens that are vibrated so that the stems and sticks tail over at the lower end while the peanuts pass through perforations and are moved into conveyors. This cleaning removes the largest sticks, stems, and pebbles, but some remain in the unshelled peanuts. In some cases the peanuts are double-screened; that is, the peanuts ride a second screen and are aspirated at the lower end while the rocks fall off because of higher specific gravity. Peanuts are deposited in a bin or chute or on a conveyor and pass on to the shellers. The cleaners do not remove all of the foreign material; this creates problems in the succeeding steps throughout the process. Pebbles and stems are the most difficult to remove. Consequently, several kinds of stoners are used, some before shelling and some after shelling. Some foreign material remains in the unshelled peanuts as they pass to the shelling machines. The speed of cleaning limits the shelling capacity of the shelling plant as a whole.

Stoners and Stoning

Stoning of peanuts means the removal of rocks, pebbles, or stones. This cleaning job is done at several places in the shelling process. Many pebbles are removed by the first cleaning in most plants. However, many of the small pebbles move on to the shelling units and through the separating machinery. Some pebbles are removed in separating the shelled and unshelled peanuts. A few small ones are removed at the picking tables, which afford the last opportunity for their removal. A few pebbles get by and remain in the shelled goods shipped to processors.

Shellers and Shelling

Peanuts are shelled in machines consisting of a straight-bladed rotor and a stationary grid or screen. The screen holds the unshelled nuts allowing the shelled kernels, unshelled peanuts, and hulls to pass through after the rotor has cracked the hulls.

Peanuts that pass through the principal shelling machines without being shelled are collected in the separating machines and returned to the "nubbin" and "return" shellers. The shelled peanuts removed in the separation pass through another separating machine for removal of the unshelled "nubbins." The "nubbins" are small round unshelled peanuts with only one kernel. They are produced when only one kernel develops in a pod or when a two-kernel pod is broken into two parts.

Peanuts are not sized for shelling, and the machines do not do a complete job of shelling. Consequently, unshelled peanuts must be routed to the "return" shellers. The grates of the shellers wear because of friction and the presence of sand, gravel, etc. Breakage of grates because of rocks and metal is a considerable expense, and the adjustment of the grates and their degree of wear affects the efficiency of shelling. A small pod will go through unshelled and an oversize kernel will be broken. Sizing the nut before shelling would lower the breakage.

Breakage decreases as the moisture content increases. The ideal condition would be to have clean, sized peanuts, with the kernels at about 7% moisture and the shells quite dry and brittle. Under storage conditions, the shell usually contains a higher percentage of moisture than the kernel.

Separators and Separating

The mixture of shelled peanuts, hulls, and unshelled peanuts passes from the shellers to separators. The hulls are removed by aspiration. The shelled and unshelled peanuts are separated by screening. The unshelled peanuts are routed to one or more of the shellers, usually referred to as the return sheller. Unshelled peanuts are recirculated in this manner until shelled. The shelled peanuts are screened a second time for removal of small shriveled kernels, which are routed to bags or bins holding oil stock. The larger kernels, those riding the screens, are routed to a specific-gravity separator.

When the shelled peanuts reach the gravity separator some unshelled peanuts, including a high proportion of nubbins, small rocks, and some other foreign material remain in them. The gravity separator separates shelled peanuts from unshelled peanuts and rocks. Often as the rocks pass from this machine some shelled peanuts also are removed.

The nubbins are returned to one of the shelling machines, the nubbin sheller. The shelled peanuts leaving the gravity separator are routed to a holding bin before going to the picking tables.

Grading Shelled Peanuts

After removal of the hulls and foreign material from the shelled peanuts, it is necessary to remove the split kernels (Number 2's) and the shriveled kernels. Several machines are on the market for separating the splits from the whole kernels. The shriveled kernels are usually removed by moving the peanuts over a slotted or round hole shaker screen commonly known as a scalping shoe. In some instances the remaining sound mature kernels are separated by size after the damaged kernels have been removed by handpicking or electric eye sorting machines. Grading by size is usually done only to fill an order specifying size.

Sacking, Weighing, and Loading

From the storage bins the peanuts pass through a chute with a manually operated shut-off and pour into a sack. When they are full, the sacks are weighed on portable platform scales and are sewed by hand or machine. From this point the bagged peanuts are loaded for shipment.

BLANCHING

Peanuts are blanched, which removes their red skins, by two methods: the most common is dry blanching; the other is water blanching. Only about 1% of the blanched peanuts are processed by the water method. Almost all end-use products are processed from blanched peanuts; at least 95% of the edible peanuts are blanched before final processing.

The dry skins are bagged and sold for use in feed mixtures and seldom constitute a problem of disposal. However, the skins that come from the water-blanched process do constitute a problem of disposal; they are now disposed through sewage systems and in city dumps.

The removal of skins from peanut kernels causes a loss in weight of about 60 lb from a ton of shelled peanuts. This results in about 22,000 tons/yr of skins to dispose of by the blanchers or end product manufacturers.

SUMMARY

The 1965–1969 annual average peanut production, farmer's stock basis, in the United States was approximately 1¼ million tons. Of this quantity 98.7% (1,200,000 tons) was sold by producers and moved into commercial channels. About 71% of the peanuts marketed (745,000 tons) was utilized as food products—peanut butter, confections, and roasted peanuts. About 29% (298,000 tons) was crushed for oil. All the farmer's stock peanuts that arrive in the local markets in-shell are dried to 8% moisture if necessary and some foreign material is removed by precleaning if above moisture tolerance. The peanuts are stored in common storage in-shell prior to shelling and processing.

The hulls and kernels of the peanuts during the same 5-yr period averaged 26% and 74% respectively. During the processing stages hulls and skins are separated from the kernels. The annual average yield of hulls was approximately 320,000 tons. These hulls constitute a problem in solid waste disposal in the local markets in the producing areas where all the shelling plants are located.

The yield of shelled edibles amounted to an annual average of 745, 000 tons; they were processed into end-use products mostly outside the producing areas. The skins were removed from most of these peanuts. Roughly 60 lb of skins per ton were removed in the processing stages. Skins accounted for about 22,000 tons out of the shelled edible volume. When skins were removed by the water-blanch process they constituted a solid waste problem at the local plant, but if dry blanched then the skins were sold for utilization in feeds.

The average annual volume of peanuts crushed for oil amounted to 298,000 tons farmers stock basis. Assuming the average shellout as 74% kernels, 220,000 tons of kernels were crushed for oil and cake. Peanuts crushed yielded 46%, crude oil or 101,000 tons and 54% cake or 119, 000 tons. The cake was used for feeds and the crude oil was refined. The yield of refined oil was 95% of 96,000 tons and was utilized in food products. The remaining 5% of the original crude oil consisted of about 2.5% "foots" and was utilized in feeds (2,500 tons) and 2.5% acid oils, or free fatty acids, in the same amounts as "foots." This acid oil was utilized in soaps and the like.

REFERENCES

1. Y. W. Han and V. R. Srinivasan, "Purification and Characterization of B-Glucosidase of Alcaligenes Faecalis," *J. Bacteriology,* **100,** No. 3, December 1960, pp. 1355–1363.

2. The author is indebted to the following persons and firms for their generous help in providing materials, data, and advice in the preparation of this manuscript: Mr. John Penner, Goldkist Div., Cotton Producers Association, Atlanta, Ga.; Mr. Marvin Shaw, Seabrook Blanching Corporation, Edenton, N.C.; Mr. W. M. Birdsong, Jr., Birdsong Storage Co., Suffolk, Va.; Mr. A. D. Bubp, The Bower Bros. Co., Springfield, Ohio; Mr. H. R. Strop, The V. D. Anderson Co., Cleveland, Ohio; Dr. Ben

Doupnik, Coastal Plains Experiment Station, Tifton, Ga.; Dr. Curtis R. Jackson, Georgia Experiment Station, Experiment, Georgia; Dr. J. G. Woodroof, Retired, Georgia Experiment Station, Experiment, Georgia.

3. B. Doupnik, "Aflatoxins produced on peanut varieties previously reported to inhibit production." *Phytopathology,* **50,** 1554, 1969.

4. L. Goldblatt *Aflatoxin,* Academic Press, New York, 1969, 72 pp.

5. Curtis R. Jackson, "Report to AID on Aflatoxins in the Philippines," 1970.

6. "Fats and Oils Situation," USDA, FOS-244, September 1968.

7. Census of the United States, Bureau of the Census, USDC.

8. N. M. Penny, "Physical Aspects of Peanut Marketing," Univ. of Ga. Exp. Stations, Georgia Station, Experiment, Ga., Bulletin 284, 1953.

CHAPTER III.11
potatoes

The annual potato production in the United States is more than 15 million tons; a per capita consumption of 110 lb/yr, more than half of which is processed as flakes, chips, flour, canned and peeled, and starch. Expected per capita consumption in 1975–1980 is 18–20 lb/yr as frozen potato products.

A significant fraction of the potato is rejected as waste during processing. The amount wasted depends on the quality of the potato as well as the particular type of process, and many vary from as low as 20% to more than 50%. The waste streams from various types of processing plants were determined as population equivalents based on biochemical oxygen demand (BOD) as shown in Table 1.

The average population equivalent of an untreated waste stream is 315 per ton of raw potatoes processed; an untreated waste load is equivalent to a population of about 5.5 million people. Projections for frozen and dehydrated potato products indicate 7.4 million tons of raw potatoes will be used in these two processes alone in 1975–1980. An untreated waste load is equivalent to a population of almost 8 million people. The industry is located primarily in Idaho, Maine, and the Red River Valley of the North, concentrating the problem in three small areas.

Today, the potato is grown in almost every important agricultural country in the world. In the United States potatoes comprise 7% of the 1488 lb of food consumed per capita annually. They are available for specified purposes and environments. Varieties with characteristics such as resistance to disease, high solids content, low reducing sugar content, and thick or thin skin have been developed. New varieties are derived mainly from cross-pollination. A potato tuber is a modified stem, with its axis greatly shortened and its lateral members weakly developed, the latter froming the "potato eyes." The outer layer, the skin, consists of a corky periderm. When the potato is cut or wounded,

Table 1 *Population Equivalents per Ton of Raw Potatoes Processed*[a]

Flakes (lye peeling)	420
Chips (no treatment)	171
Chips (with clarifier)	84
Flour (no treatment)	318
Flour (with clarifier)	110
Starch	353

[a] Based on 0.17 lb of 5-day 20°C biochemical oxygen demand (BOD_5) per population equivalent.

a wound periderm is formed which apparently is more effective in protecting the tissue than the original layer. Underlying the periderm is the cortex, a narrow layer of parenchyma tissue. Between the cortex and the pith, which forms the central core of the tuber, is the vascular area. This is part of the tissue and is high in starch content. The vascular storage parenchyma tissue is divided in two unequal parts by a discontinuous vascular ring, consisting of large polyhedral parenchyma cells in which small islets of phloem are embedded. The pith or "water core," connected to each of the eyes by lateral branches, consists primarily of large cells, containing less starch than the cells in the vascular area and the inner part of the cortex.

The eyes show the relation of the tuber to the stem. Each eye is a leaf scar, arranged in a spiral around the tuber, with 13 eyes to each five turns of the helix.

The composition of the potato is given in Table 2 and mineral content in Table 3.

The quality and yield of potatoes is determined by the environmental conditions during growth, including date of planting, soil type, soil reaction, soil moisture, season, location, mineral nutrition of the plant, cultivation and weed control, spray program for control of insects and diseases, temperature during the growing season, time and method of vine killing, and time of harvest. Since different varieties of potato are affected to different degrees by these factors, the choice of variety is of course very important.

High-quality raw potatoes are important to the processor. Potato quality affects not only the final product, but also the amount of waste produced. Generally, potatoes with high solids content, low reducing

Table 2 *Proximate Analysis of White Potatoes*

	Average (%)	Range (%)
Water	77.5	63.2–86.9
Total solids	22.5	13.1–36.8
Protein	2.0	0.7–4.6
Fat	0.1	0.02–0.96
Carbohydrate		
Total	19.4	13.3–30.53
Crude fiber	0.6	0.17–3.48
Ash	1.0	0.44–1.9

sugar content, thin peel, and uniform shape and size are desirable for processing. Cull potatoes sometimes are sorted for starch and flour production, but more often low-quality potatoes must be wasted and thereby represent a loss to the processor.

Raw potatoes must be washed thoroughly to remove sand and dirt prior to processing. Sand and dirt carried over into the peeling operation can damage the peeling equipment. Often the incoming potatoes are washed in the water used for transport from the storage or receiving area into the plant. Fluming is an economical and effective method causing a minimum of damage. Sand and gravel are removed in stone traps in which these heavier solids settle out while the potatoes pass on. Tubers transported into the plant by other means or washed inadequately in the flume are passed through washers, of which there

Table 3 *Mineral Content of Potato Ash*

	Average (%)	Range (%)
K_2O	56	43.95–73.61
P_2O_5	15	6.83–27.14
SO_3	6	0.44–10.69
MgO	4	1.32–13.58
Na_2O	3	0.07–16.93
CaO	1.5	0.42–8.19
SiO_2	1	0.16–8.11

are a number of types. Barrel-type washers are quite common. Potatoes are tumbled and rubbed against each other and against the sides of the barrel while immersed or sprayed with water. Other types use rubber brushes, rolls, or paddles to loosen the adhering dirt. Following washing, the potatoes drain while being carried to the next step of inspection where trash and rotten potatoes are removed.

Water consumption for fluming and washing varies from 600 to 2700 gal/ton. Of this volume, two-thirds is used for fluming and one-third washing. The waste water contains 100–140 lb of solids per ton of potatoes. Organic degradable substances are predominantly in dissolved or finely dispersed form, and amount to 2–6 lb of BOD_5 per ton of potatoes.

Peeling requirements vary from process to process and from plant to plant. Potato chips, for example, require less complete peel removal than French fries. Peeling methods are abrasion, steam heating, and lye processing, either as a batch or as a continuous process.

Abrasive peelers have abrasive disks or rolls that remove the peel by uniform contact with the surface of the potatoes. Peelings and potato tissue are removed from the abrasive surfaces by water sprays, which also reduce the tendency of potatoes to darken through enzymatic action. The potatoes must be spun and rotated so that all surfaces are equally exposed to the abrasive surfaces. Abrasive peeling is used especially in potato chip plants where complete removal of the skin is not essential, with losses of 1–8%. Abrasion peeling has been used in the production of pre-peeled potatoes, since the heat ring, which results from high-temperature exposure, does not occur with this method. High peeling losses, possibly as high as 25–30%, may be necessary to produce a satisfactory product, however. Up to 90% of the settleable solids appear to consist of coarse potato matter, the rest being fine starch particles.

Steam peeling yields thoroughly clean potatoes. The entire surface of the tuber is treated, and size and shape are not important factors as in abrasion peeling. The potatoes are subjected to pressures of 3–8 atm, with an exposure time of 30–90 sec in a pressure vessel. The surface tissue is hydrated and cooked so that the peel is softened and loosened. After discharge from the pressure vessel, the softened tissue is removed by brushes and water sprays similar to the ones used for prewashing.

Because the gelation point of potato starch is surpassed in steam peeling, there will always be a heat ring. This is not always objectionable, as, for example, in the production of French fries, where it is not noticeable in the finished product.Steam peeling of pre-peeled potatoes may result in enzymatic darkening under the heat ring. The ring may harden during storage, seriously affecting the quality of the product.

The peelings and solids usually are removed by screens before the waste water is treated. These screenings are simply cooked potato tissue and can be used for cattle feed without additional treatment except for possibly some drying or mixing with other feed materials. Because of the relatively high temperature of the waste waster, a considerable amount of starch dissolves and cannot be removed by plain sedimentation.

Lye peeling is a popular method in the United States. The combined effect of chemical attack and thermal shock softens and loosens the skin, blemishes, and eyes so that they can be removed by brushes and water sprays. With lye temperatures of 190–220°F above the gelation point, immersion for 2–6 min in lye concentrations of 15–25% is required. Below 140°F immersion times are obtained at lye concentrations of 15–20%.

The peel, after having been softened and loosened in the lye bath, is removed by brushes and water sprays as in steam peeling. Sometimes a short period of time is allowed for the lye to work on the potato tissue before removing the peel. The tubers then must be thoroughly washed to remove any lye and lye-affected surface that will harden and cause a yellowing of the surface.

Lye-peeling waste water is the most troublesome potato waste. Because of the lye, the waste water pH is very high, between 11 and 12. The solids are colloidal, and the organic content is higher than for the other methods. The temperature, usually 50–55°C, gives a high dissolved starch content. The waste water has a tendency to foam.

Because this peel is extremely caustic, the pH is lowered through microbiological action before the solids can be used for cattle feed. For example, the pH of the settled solids can be lowered by controlling solids detention time in the tank so that the primary sludge can be fed directly to cattle after dewatering.

Abrasion-peeling wastes do not have the high content of dissolved starch and fine and colloidal solids associated with wastes from lye- and steam-peeling installations, and thus can be treated to a higher degree by sedimentation. The amount of solids that must be removed from abrasion peeling waste is large, however, since the peeling losses generally are high.

In comparing lye- and steam-peeling wastes, the greatest difference is the pH of the two wastes. While steam-peeling wastes are usually near neutral, lye peeling wastes have pH values from 11 to 12. This high pH may cause problems in receiving waters as well as in biological

Table 4 Waste Streams from Peeling

	Abrasion	Steam	(ppm)	Lye	(ppm)
Flow, gal/ton	600	625	10000	715	
COD, lb/ton	—	52.5	10000	65.7	11000
BOD, lb/ton	20	32.6	6750	40.0	6730
TS, lb/ton	—	53.2	10200	118.7	20000
VS, lb/ton	—	46.8	9000	56.4	9500
SS, lb/ton	—	26.8	5150	49.7	8350
SS, ft^3/ton	9015	—	—	—	—
pH	5–6	3		12.6	

treatment. The organic content of lye-peeling waste is generally higher than that of steam-peeling waste.

The characteristics of the total waste water for abrasion, steam, and lye peeling are given in Table 4.

The processing to potato chips involves slicing, washing the slices in cool water, rinsing, partially drying, and frying in fat or oil. White skinned potatoes with high specific gravity and low reducing sugar content are desirable for high-quality chips.

After conventional prewashing, the potatoes are peeled, usually by the abrasive method, washed, inspected, and trimmed prior to slicing. Losses vary over a wide range. A survey of more than 50 plants in the United States showed that peeling and trimming losses for old potatoes varied from 2% to 20% and for new potatoes from 0.3% to 13%. Another survey showed the losses ranging from 0.5% to 25% and from 0% to 8% for old and new potatoes, respectively. Trimming losses alone may vary from negligible amounts of 10%. Trimming contributes primarily to the amount of solid waste.

Slicing yields 15–20 slices per inch. Considerable amounts of starch are released in the operation and must be washed off to avoid matting and sticking of the chips during frying. Most of the starch and slivers are removed in a washer with cool, fresh water.

Losses during slicing and washing amount to 0.05–1.0% of the raw potatoes. After washing, the chips are rinsed in one or two water baths to further remove starch. The waste from these steps is negligible in both volume and organic content. Drying of the chips prior to frying reduces the frying time and thus increases the capacity of the kettle. The chips also may be treated by hot water or chemicals prior to frying to prevent darkening. Chemicals used include sodium bisulfite, hydrochloric acid, phosphoric acid, and citric acid.

After frying, the chips are passed over a shaker screen to remove free oil, then salted or flavored, and finally cooled and inspected before packaging. A survey by Porges and Towne[2] on four chip plants showed an average of 50 lb of BOD_5 and 66 lb of suspended solids discharged per ton of potatoes processed. The average waste flow was 3980 gal/ton.

Annually 32% of the potatoes processed are used for frozen French fries. Including other frozen potato products, the figure amounts to 37%. The grade and quality of potatoes processed into frozen products vary over a wide range. For French fry production large potatoes of high specific gravity and low reducing sugar content are most desirable. Large potatoes result in lower peeling losses and larger yield of long French fry cuts. High specific gravity ensures higher yield per ton of potatoes processed and low reducing sugar content improves the color of the finished product.

After prewashing, the potatoes are peeled by the steam or lye method. The peeling requirements on potatoes for French fry processing are high, and the abrasion peeling would result in excessive losses. Peeling and trimming losses vary with potato quality and range from 15% to 40%. On the trimming belt low-quality potatoes are diverted from the process line and small tubers may be removed for processing into coproducts like potato patties, hash brown, or mashed potatoes.

The strip cutter produces, besides French fry cuts, slices and nubbins that must be removed. This is usually accomplished by shaker screens or rotating reels having slots through which the smaller pieces pass. These pieces may then be diverted to the coproduct processing lines. The removal of slivers and nubbins may amount to an additional loss of 10%, reducing the overall yield of raw French fry cuts to within the range of 50–75% of the tonnage of potatoes processed.

The by-products contribute a minor part of the waste. This of course depends on the number and kinds of by-products and upon the fraction of the pototoes diverted to the by-product line.

After conventional prewashing the potatoes are peeled. Lye, steam, and abrasion peeling as well as combinations of these are used in the canning process. Abrasion peelers sometimes are used rather than lye or steam peelers to give the potatoes a smoother surface. Because the heat ring may present a problem, low- rather than high-temperature lye peeling appears to be more suitable for canning.

Since canned potatoes are a low-cost product, only a minimum of trimming is allowed. Some processors prefer to sort out the potatoes that need trimming and send them through the peeler a second time.

If the potatoes have not been size graded previously, this is done following the trim table inspection. Large size tubers are usually not

Table 5 *Operating Values*

	Chips	French Fries	Dehydrated Diced	Mashed	Flakes	Starch	Flour	Canned Potatoes	Pre-peeled
Flow, gal/ton raw potatoes	1140	2310	70–2000	70–2000	1540	1740		700–1000	800–1200
COD, lb/ton	75.5	10	50–70	50–70	56.2	50–90	75	50–70	50–70
BOD, lb/ton	21.9	22–30	30–40	30–40	38.2	30–50	45	30–40	30–40
TS, lb/ton	118.5	50–60	70–100	50–90	125	35–50	70	60–100	60–100
VS, lb/ton	63.6	78–80	70–80	65–75	54		70	50–70	50–70
SS, lb/ton	53.8	25–30	30	32–40	16.5		50	30–50	30–50
pH	7.4	11.1	7–10	6–10	5.2	4–6	4.2	3–11	4–12
Waste, solid, %	15–28	20–40	25–26	25	25–30	3–10	5–10	15–27	15–27

canned whole but cut to make diced, sliced, shoe string, or julienne potatoes.

As a result of the demand for convenience foods and the high cost of hand labor, restaurant operators generally find it more economical to purchase pre-peeled potatoes than to invest in labor for hand peeling or in peeling machinery. Pre-peeled potatoes also have gained popularity in the household.

Potatoes are pre-peeled for sale as whole potatoes as well as for French fry cuts, chips, hash browns, and others. The selection of raw material depends to a great extent on the final product.

The pre-peeling processes do not differ significantly and the waste streams are similar. Abrasion and low-temperature lye peeling are used.

Discoloration of pre-peeled potatoes is the enzymatic formation of melanin, a dark pigment, or the nonenzymatic "after-cooking darkening." Black-spot or black-heart may also occur. Microbial spoilage furthermore limits the shelf life of pre-peeled potatoes.

To prevent discoloration, the common additive is suflur dioxide, 50 ppm of which delays browning of peeled potatoes up to 2 days at 20°C and up to 14 days at 4°C. Microbial deterioration may be controlled for 10–14 days by keeping the peeled potatoes under refrigeration.

Alcohol production from potatoes in the United States is small compared with the volume produced from grain. In Europe potato alcohol production is important.

In fermentation plants, the potatoes are washed, steamed at about 135–140°C and cooled to 62–65°C. The starch is hydrolyzed by barley malt or by a mold amylolyticenzyme. The mash, of 17–20% dry matter, is subjected to alcohol fermentation for 2–3 days. The residue after distillation (120–140 kg/100 kg potatoes, of 4.5% dry matter) is a valuable feed for cattle. Waste water from distilleries has been reported to have a BOD_5 of about 530 mg/liter. Joint production of starch and alcohol is carried out in Poland, Russia, and Czechoslovakia. The pulp and protein water from the starch production are used for alcohol fermentation, and the distillery residue is fed directly to livestock. This allows for disposal of all wastes (see Table 4).

The flow sheets for peeled and canned potatoes, flour, starch, chips, French fried, flake and granules are given in Figure 1. Operating values for processed potatoes are given in Table 5.

A treatment facility at the R. T. French Company, Shelly, Idaho, demonstrated the feasibility of a complete mix activated sludge system for secondary treatment of potato processing wastes. The facility was designed for a daily flow of 1.25 million gal/day and a BOD loading of

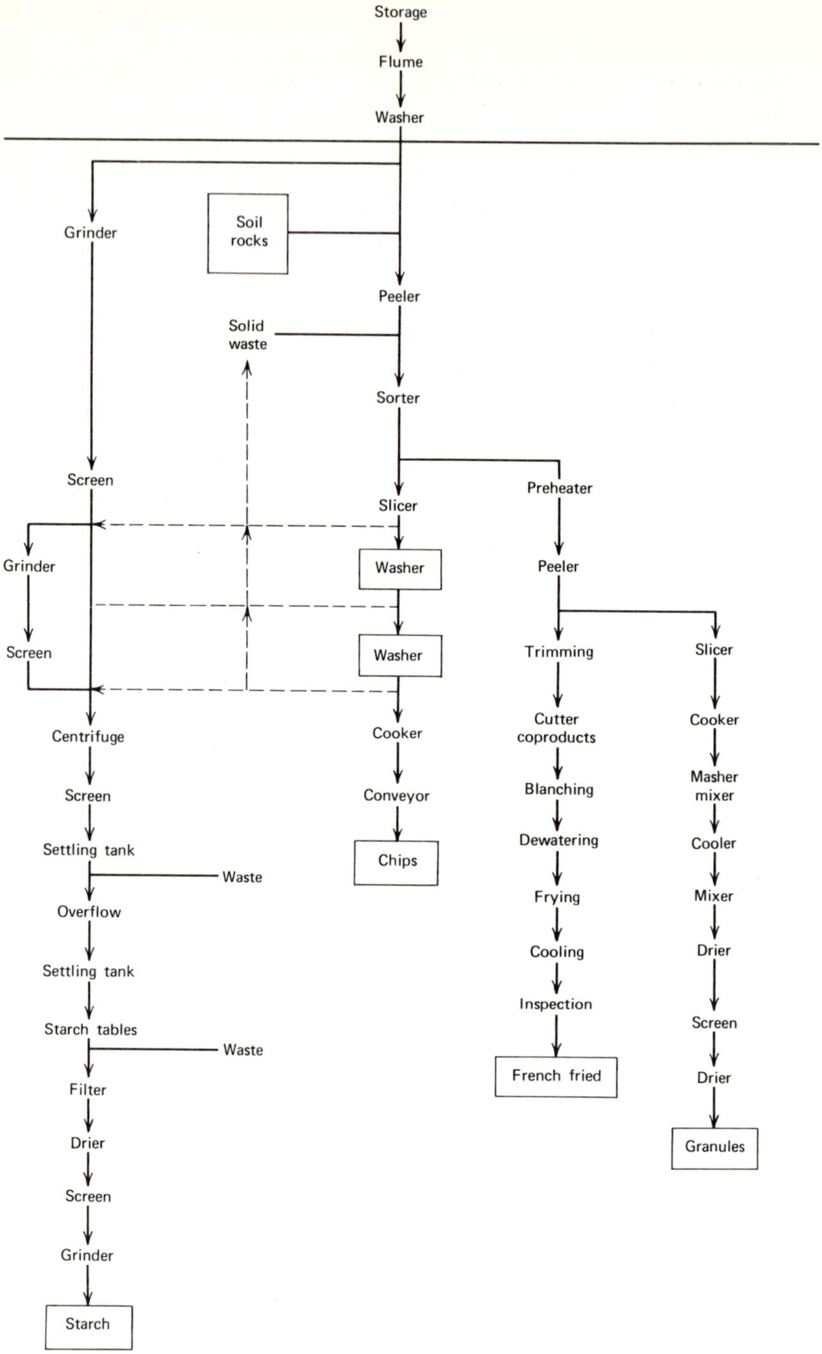

Figure 1. Flowsheets of potato processing.

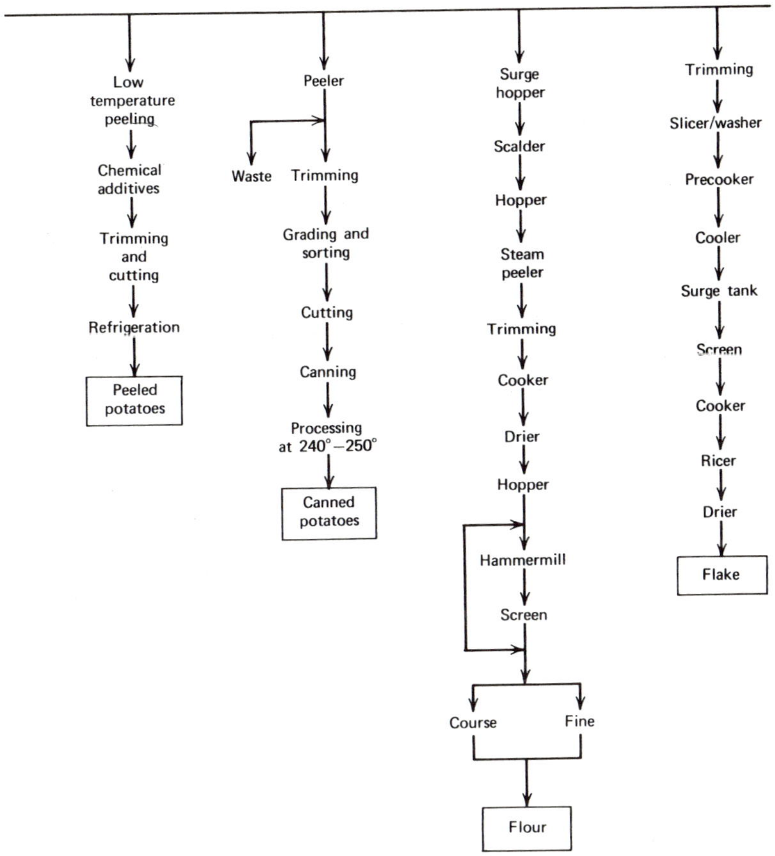

Low temperature peeling
Chemical additives
Trimming and cutting
Refrigeration
Peeled potatoes
Peeler
Waste
Trimming
Grading and sorting
Cutting
Canning
Processing at 240°–250°
Canned potatoes
Surge hopper
Scalder
Hopper
Steam peeler
Trimming
Cooker
Drier
Hopper
Hammermill
Screen
Course
Fine
Flour
Trimming
Slicer/washer
Precooker
Cooler
Surge tank
Screen
Cooker
Ricer
Drier
Flake

14,100 lb/day. BOD removals of over 90% were obtained for periods of time. These removals have been obtained with (1) MLSS concentrations between 2000 and 8,000 mg/liter, (2) aeration basin DO concentrations between 0.3 and 5.2 mg/liter, (3) aeration basin temperatures of 45–67°F, (4) aeration basin pH between 7.1 and 8.4, (5) organic loadings between 10 and 120 lb BOD/1000 ft^3/day, (6) hydraulic detention times of 0.9–8.7 days, and (7) BOD/MLVSS ratios of 0.15–0.47.

During the 1969–1970 potato processing season, aerobic secondary treatment of potato processing wastewater at R. T. French Company[1] was studied. Coliform removals in excess of 96% were obtained; high pH values of influent process wastewater were buffered in the aeration basins and were not detrimental to treatment efficiencies; low air temperatures have not caused failure; the quantity of excess biological sludge, X_v, was a function of the BOD removal rate, S_r, and the average MLVSS (mixed liquor volatile suspended solids), X_d; the substrate removal rate coefficient, k, was a function of temperature, $k = 0.016 \times 1.17^{(T-20)}$; system oxygen requirements are a function of the BOD removal rate, S_r, and the average MLVSS, X_d, with the equation $O_2 = 0.48\ S_r + 0.03\ X_d$; total annual costs are estimated to be $0.038 per pound of BOD applied and $0.021 per pound of COD applied (these costs do not include waste activated sludge disposal); the clarifier-thickener approached 100% suspended solids removal when operated as a straight silt removal system; operation of the clarifier-thickener for dewatering secondary sludge in combination with silt was not successful.

About 300,000 tons of potatoes per year are processed by three J. R. Simplot Company processing plants near Burley, Idaho.[2]

The Burley Processing Company processes about 28% of the national total. It is a highly automated plant and produces instant mashed potatoes and related specialties. The potatoes are washed, lye peeled, cut, automatically sorted for flakes or granules, dehydrates, and packed for shipment.

The Heyburn plant, one of the world's largest in this field, processes about 59% of the national total, or 180,000 tons of potatoes per year. Products include French fries and potato specialties. Here, too, the potatoes are washed and lye peeled; then trimmed, blanched, processed, and packed. The remaining 13% of the Simplot Company's total potato tonnage at Burley is processed in a starch plant.

Waste streams from the three processing plants are piped to a primary waste treatment plant. They first enter a receiving tank and then are passed through two 5-ft-diameter, 10-ft-long rotary drum screens. All solids retained on the +20 mesh screens are removed and stored in bins.

The screened waste water enters a 100-ft-diameter clarifier with an average overflow rate of about 800 gal/day/ft^2. Most of the settleable solids are removed.

Solids collected in the clarifier are pumped through a centrifuge for thickening. The thickened sludge is also stored in bins. Both the screenings and the sludge from the centrifuge are trucked to a cattle feedlot for use as part of the animals' diet.

REFERENCES

1. 12060 EHV 12/70, EPA, "Aerobic Secondary Treatment of Potato Processing Wastes," R. T. French Co., Shelby, Idaho.
2. EPA 07/69, "Secondary Treatment of Potato Processing Wastes."

CHAPTER III.12

rice

In 1968, Louisiana growers reaped over 2.67 billion lb of rice from their more than 669,000 acres. Adding to this Arkansas's production of 2.54 billion lb and Mississippi's 0.27 billion gives more than half the amount produced by all the 50 states.

More than a dozen varieties of short, medium, and long grain rice flourish in the middle South states of Arkansas, Louisiana, and Mississippi, with production methods that undoubtedly bear little resemblance to those incorporated in coaxing the initial Louisiana crop to maturity. Airplanes are employed first, to plant many fields, and later to distribute insecticides. Machines cultivate fields before planting, and others harvest the ripened grain. The result: middle South farmers spend less than 2 man-days per acre from field preparation to final harvesting, while production methods outside the United States may require as many as 400.

Furthermore, the spectrum of products reaped from the slender stalk or rice today has been vastly altered since the 1700s. Most familiar still is the milled white rice, but supplementary varieties have combined to increase Americans' consumption of the staple grain, including parboiled rice, subjected to steam under pressure to increase vitamin and mineral retention; precooked rice, when is actually cooked then dehydrated to minimize preparation time; brown rice, produced simply by not removing the grain's outer layers of bran. The list goes on—canned rice, baby foods, breakfast cereals, and snacks.

Hulls are incorporated in the production of hand soap, fertilizers, and furfural. The bran layer provides an ingredient for livestock feed; similarly, the straw, after milling, often becomes roughage or bedding for farm animals. Small kernel bits broken during milling become "brewers rice."

In Fall 1967, climatic conditions dovetailed well into the needs of California rice growers. Dry, warm weather made it possible to harvest in record time 360,000 acres of late planted rice, which could have been severely damaged if the rains had been even in moderate amounts in late October.

A dry autumn made it possible to promptly burn rice straw and stubble after harvest, concentrating heavy palls of smoke from Red Bluff southward. Warm weather and calm were conditions teamed up to result in an air inversion condition over the entire Sacramento Valley that confined the smoke to within 2500 ft of ground level. Even though rice stubble smoke contains none of the irritants common to auto-exhaust-induced "smog," the public throughout the entire area became very much aware of the visibility problem during the burning period. The Sacramento Metropolitan jet airport was closed for periods by the smoke problem. This in turn helped to underline the fact that new California legislation makes provision for statewide air quality control. Rice farmers may find that fire as a tool for disposal of about 200,000 tons of hulls and 1,250,000 tons of straw will be greatly restricted.

No way has been found to economically utilize rice straw for industrial or livestock feed purposes. Thus rice farmers long have found stubble burning a convenient tool. In the past available farming equipment could not cope with the incorporation of so much partially dry material into the dense mat of roots in the surface 12 in. of riceland soils. By partially reducing the straw mass by burning, the available comparatively light and low-powered equipment could be used. But heavier disks and plows and tractors with greater horesepower are now available. Harvesters equipped with straw spreaders or shredders now distribute the straw more evenly over harvested fields, and it is mechanically possible now to incorporate residues from a rice field. What will be the long-term effects of such regular annual incorporation of organic materials?[2] Will there be a build-up of insect and disease pests currently not a problem? Soil structure may be so modified that the soils will no longer be best suited for rice production. Many answers are needed before burning of rice straw is no longer necessary.

Department of Agronomy workers during 1914–1949 showed that a program involving plow-down of 3000 lb of straw per acre plus some nitrogen fertilizer increased rice yields about 800 lb/acre, whereas buring the straw cut yield over the period about 650 lb/acre if no compensating nitrogen fertilizer was applied.

If the straw contains 0.7% nitrogen or more, on an average the straw plow-down plots produced 800 lb/acre more rice than in those where the straw was burned. Rice straw must contain at least 0.55% nitrogen not to adversely affect rice yield.

Incompletely incorporated organic matter can cause stand establishment failures. From experience, growers know that trashy fields are especially subject to seedling disease problems. Scum management is also more of a problem in trashy fields.

Mosquito build-up is much more severe in flooded, trashy rice paddies than in rice fields where residues have been completely covered by disking and deep plowing. Breeding mosquitoes prefer the microclimate and environment of flooded, decomposing organic matter.

Rice hull utilization or disposal is a particularly acute problem for the California rice industry. On milling, for each 100 lb of paddy rice produced, 20 lb of hulls accumulate on an average. Thus with an annual 20-million-bag (cwt) crop, the industry must somehow dispose of 200,000 tons of hulls. At 8 lb/ft^3 of hulls, this amounts to 50 million ft^3 of material. Even after burning, the hull ash amounts to 5 million ft^3. Although several California firms have developed a limited market for hulls and hull products, about 75% of the rice hulls are being burned near Sacramento just to get rid of them. The remaining 25% are used either as (1) ground livestock feed, (2) poultry and livestock litter, (3) boiler fuel, (4) diluent for other products, (5) manufacture of plywood glue, (6) insulating material, (7) abrasive material for household cleaners and other special cleaning compounds, and (8) soil conditioner.

Rice hulls can be utilized for soil conditioning and reclamation. Rates of up to 100 tons of hulls per acre have applied to entire fields. Special heavy duty disk plows and "slip" plows have been used to mix the material to a depth of 3–5 ft.

Rice and corn produce tremendous weights and volumes of waste. Paper can be made from the rice straw as it was with corn straw, but the project never met commercial success. Flax straw is made into cigarette and special papers.

CHAPTER III.13
beet sugar

E. B. PUGSLEY

In 1969, 1.5 million acres of sugar beets were harvested in the United States, producing 28.5 million tons of beets. These beets were processed to produce approximately 3.2 million tons of sugar. The yield per acre of beets varied from a low of 14.4 tons to a high of over 25 tons. (These are state averages.) The amount of sugar produced per ton of beets varies from region to region and on whether Steffens or non-Steffens processes are used. In the former case, yields of over 300 lb of sugar per ton of beets are common; in the latter, the yield is about 50 lb less of sugar per ton of beets.

As an average, approximately 5% dry weight of beets processed is plain pulp equivalent. This material is used as livestock feed. When sold as dried pulp (either directly or pelletized), the feed is used by cattle, sheep, pigs, chickens, etc. Sometimes, molasses is added to the plain pulp, for both nutritional and economic reasons. At least one sugar company adds the end liquor from its monosodium glutamate plant to the pulp, resulting in an appreciably enhanced feed—but only for ruminant animals. Only these animals, principally cattle and sheep, can fully utilize the food value of this additive, so this type of pellet is used only by cattle and sheep. In any case, most of the pellets are used as cattle feed, if for no other reason than that cattle constitute the bulk of livestock on feed in the United States. On a dry-weight basis, approximately 1.4 million tons of dried plain pulp were thus utilized annually as livestock feed in the United States in the late 1960s.

BEET SUGAR FACTORIES

The extraction of sugar from beets and the packaging of the sugar for market involve the unwanted production of waste products, as with most industries. The management of some of these wastes has been solved to a high degree of both technical and economical success, while that of others is much less successful. The wastes include air pollutants, water pollutants, and solid wastes.

It must be kept in mind that practices necessarily vary from factories in warmer areas, such as California, to those in the colder climates, such as the Northern Plains and the Rocky Mountain states.

Factory Processes

Although production of sugar from cane has been practiced since antiquity, production from sugar beets is relatively recent. The first factories of Europe date from soon after 1800 and the first successful factory in the United States was built in 1870.

Sugars beets are generally dug in the fall, though some southern grown beets are overwintered in the ground and dug in the spring and summer. The beets are topped in the field and delivered to the factory area. In most areas, the beets are stored in open piles, 15–25 ft high, about 120 ft wide, and as long as space and convenience permit. As factory production demands, the beets are taken from the piles, by various means, to hoppers. From the hoppers, they are transported to the factory proper by water flume.

Fluming is especially effective for moving the beets. They have almost the same density as water and are easily moved with little bruising. The flume water also does a considerable amount of cleaning of the beets as they are carried along. At the factory, the beets are thoroughly washed and trash (mostly weeds and tops) is separated.

The manufacturing process consists, in outline, of four steps, with most factories adding from one to three additional processes. The four common to all are diffusion, juice purification, evaporation, and crystallization. The additional three processes are pulp drying, recovery of sugar from molasses (Steffens process), and the concentration of the Steffens filtrate. All of these seven steps are described below, as well as being illustrated schematically in the Figures 1, 2, and 3. It should be remembered that the figures are intended only to show the general order of the beet sugar factory, with emphasis on typical flows for factories practicing basically no water recirculation, factories practicing some water recirculation, and factories practicing total water recirculation. Probably, no individual factory has a flow exactly like the figures, but most are similar to one of them.

If a factory is located where water is scarce or where thermal pollution is closely watched, the condenser water is frequently recirculated, as shown, within its own circuit, with a spray pond or water cooling tower added. The relative degree of water recirculation throughout the factory depends on many factors, among which the more important are availability of and cost of water, the size of a possible receiving water body, and the strictness of the applicable water pollution control authority.

Diffusion

The beets are sliced into thin strips, called cosettes. The cosettes are put into one of two types of diffuser, a continuous diffuser (more modern) or a battery of cells. (It is believed that no cell battery diffusers remain in operation in the United States, though some may still be used abroad.) In one type of continuous diffuser, the cosettes are propelled up a sloping trough by a rotating screw, mechanism, against the downward flow of extracting hot water, which removes the sugar by diffusion. In the battery type, cells are filled with cosettes and the extracting water is fed first into the oldest cell, then successively through the next oldest to the newest cell. In rotation, the exhausted cell is dumped and filled with fresh cosettes and it then becomes the last cell in the process.

In either method, the exhausted cosettes, now pulp, go to a screen, for preliminary draining, then to a press for extraction of as much of the liquid as practicable. From the press, the pulp goes to a dryer for eventual sale as a dry animal feed. In the United States, almost all factories have at least some drying facilities, but where no drying is done, the pulp normally is taken from the screen and sold as wet pulp. Pending sale, it is stored in open silos; the drainage from this pulp constitutes a major source of water pollution.

Juice Purification

The juice from the diffuser contains sugar and nonsugars. The sugar is in solution, but the nonsugars are both dissolved and in colloidal form. In order to purify the juice, the customary process is one called carbonation. This involves adding lime and carbon dioxide to the juice, generally simultaneously. The lime precipitates as calcium carbonate. The colloidal material is coagulated in the process, and some of the colloids adhere to the calcium carbonate floc.

The lime is added either as an aqueous solution and suspension, known as milk of lime, or as a slurry of calcium saccharate and sweetwater, called saccharate milk, which is received from the Steffens process of removing additional sugar from molasses. The pH of the juice leaving this step is about 11.2. The carbonated juice is settled or filtered before proceeding further.

After this carbonation and filtration, the juice is carbonated a second

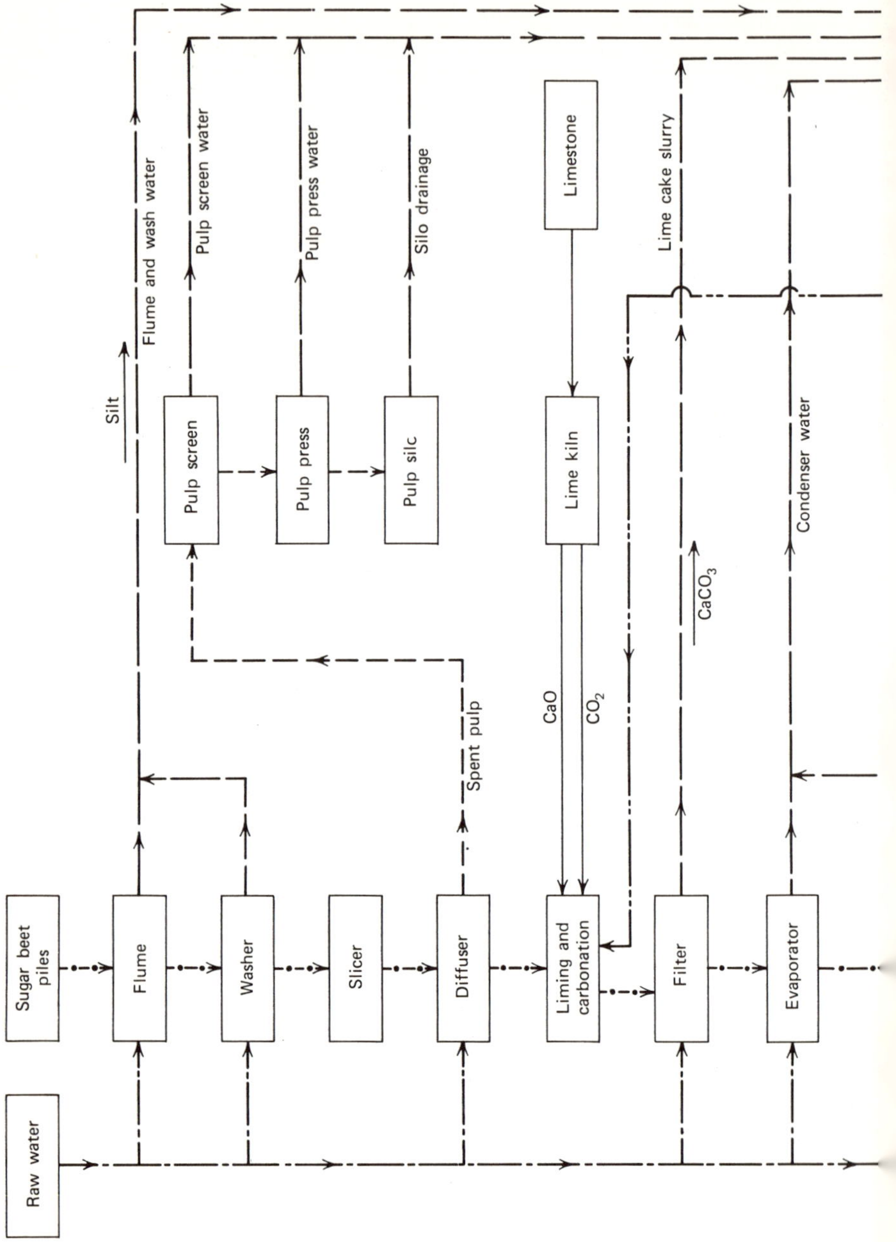
Raw water
Sugar beet piles
Flume
Washer
Slicer
Diffuser
Liming and carbonation
Filter
Evaporator
Silt
Flume and wash water
Pulp screen
Pulp press
Pulp silc
Pulp screen water
Pulp press water
Silo drainage
Spent pulp
Limestone
Lime kiln
CaO
CO_2
Lime cake slurry
$CaCO_3$
Condenser water

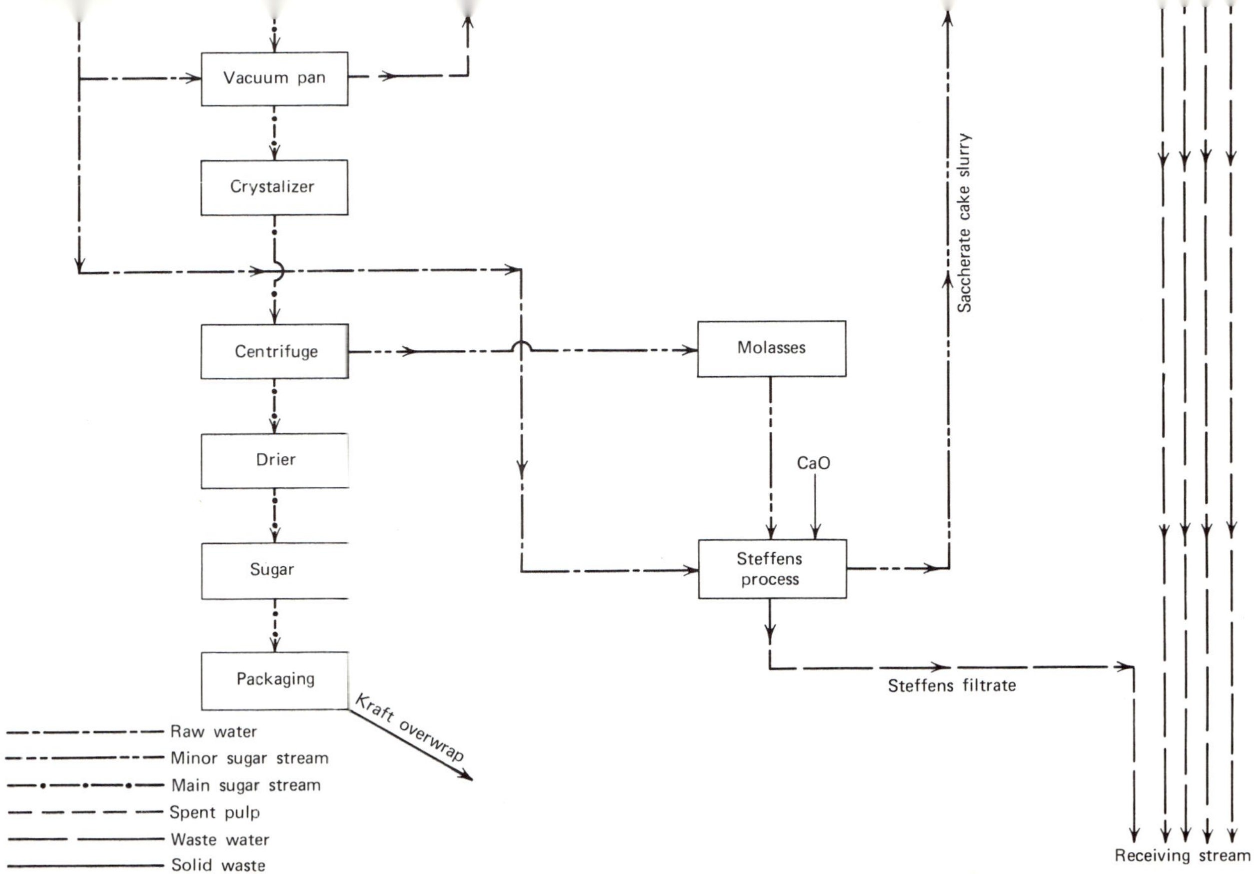

Figure 1. Sugar beet processing with no water circulation.

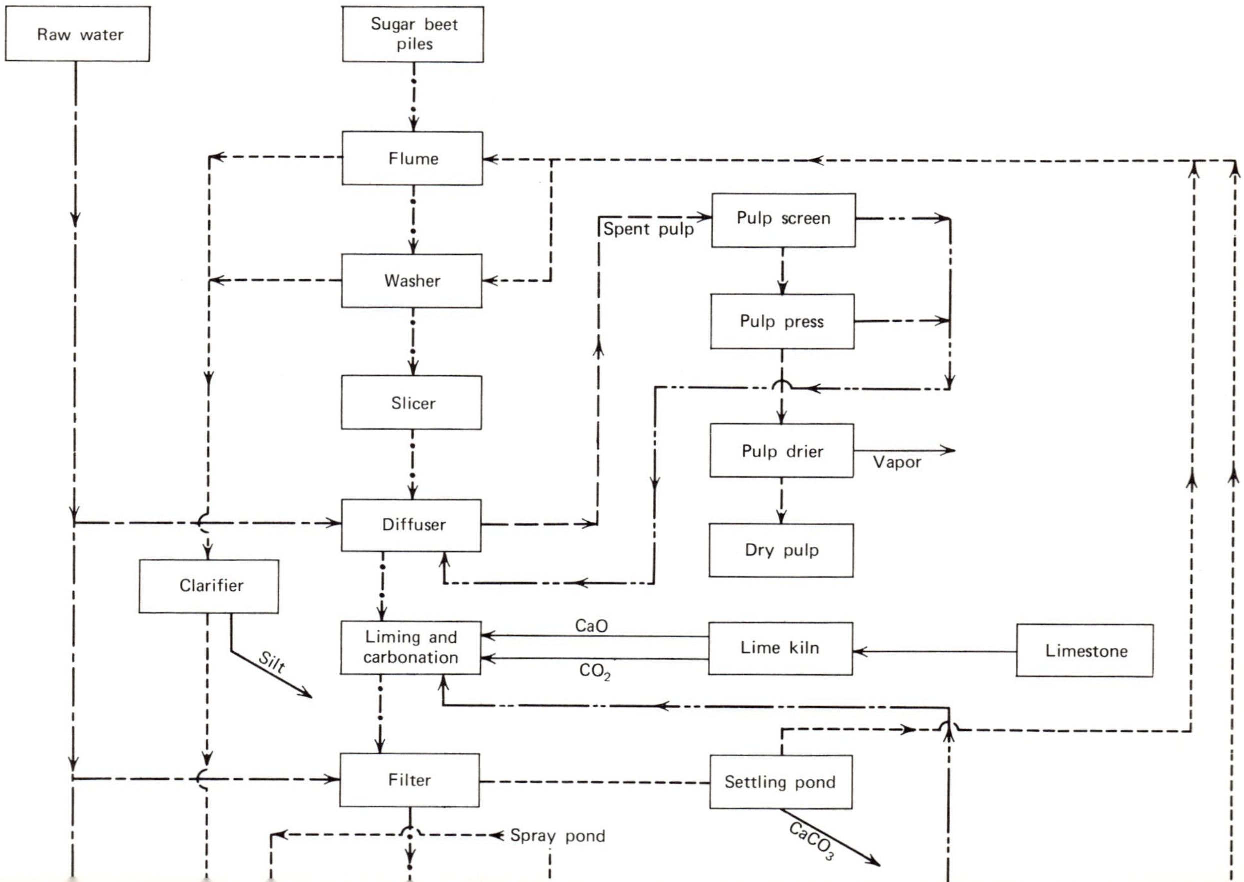
Raw water
Sugar beet piles
Flume
Washer
Slicer
Diffuser
Clarifier
Silt
Liming and carbonation
Filter
Spray pond
Spent pulp
Pulp screen
Pulp press
Pulp drier
Vapor
Dry pulp
CaO
CO_2
Lime kiln
Limestone
Settling pond
$CaCO_3$

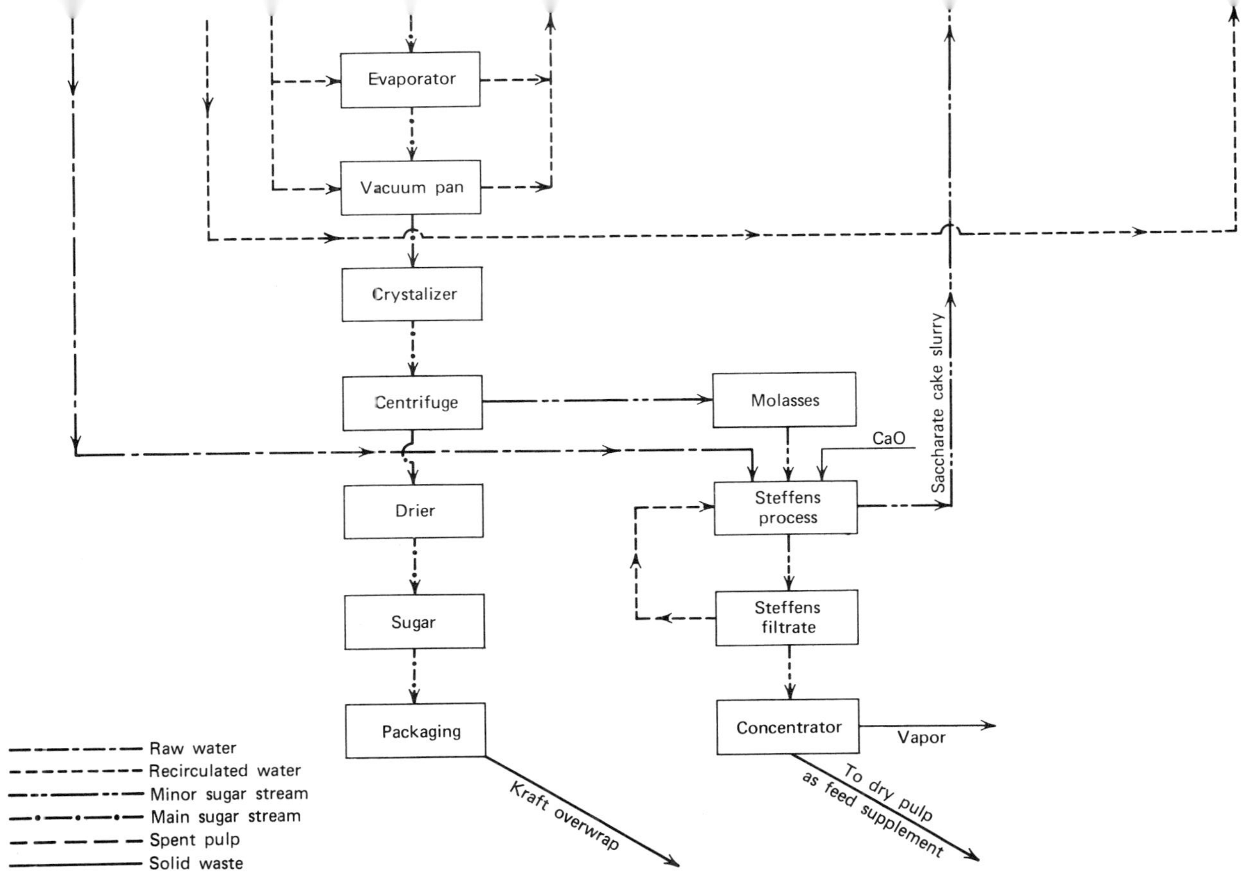

Figure 2. Sugar beet processing with complete water circulation.

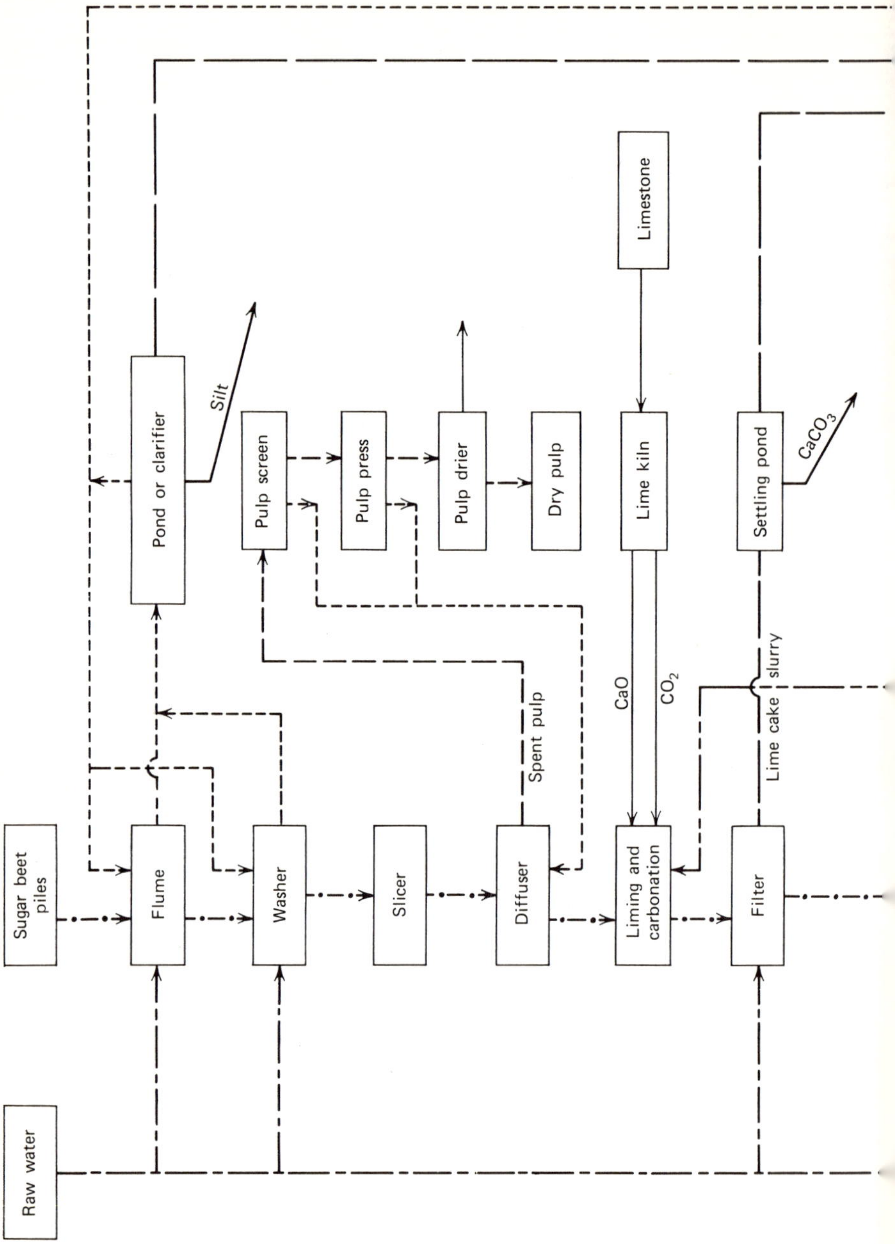

Sugar beet piles
Raw water
Flume
Washer
Slicer
Diffuser
Liming and carbonation
Filter
Pond or clarifier
Silt
Pulp screen
Pulp press
Pulp drier
Dry pulp
Limestone
Lime kiln
CaO
CO_2
Settling pond
$CaCO_3$
Spent pulp
Lime cake slurry

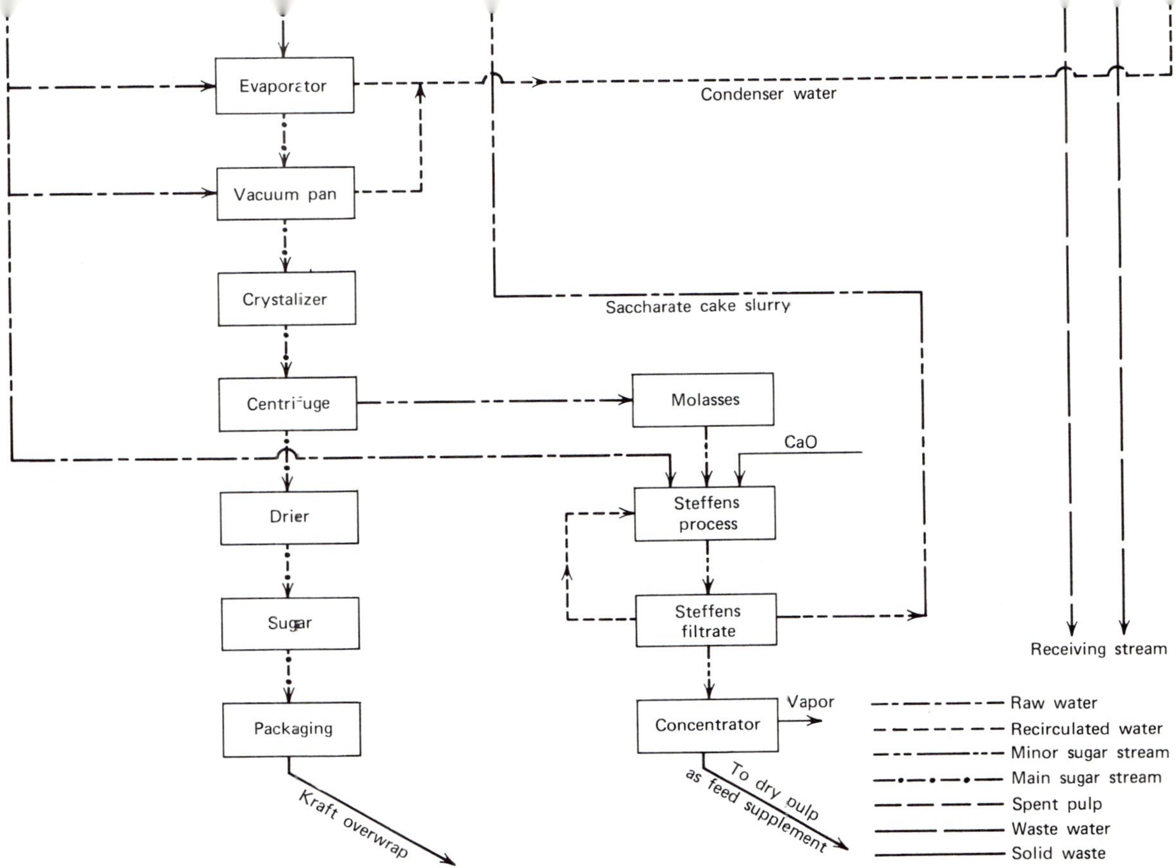

Figure 3. Sugar beet processing with partial water recirculation.

time to remove as much of the remaining lime as practicable, at a temperature near the boiling point. Additional calcium carbonate is formed and the juice filtered again. The pH of this juice is about 9.2. The principal reason for the two carbonations at different pH values is that some of the nonsugars that are soluble at pH of 9.2 precipitate and can be removed at pH of 11.2.

The juice is treated with a small amount of sulfur dioxide and sometimes boiled for a few minutes to remove any bicarbonates that may have formed and to partially remove any ammonia gas that may have formed from the decomposition of amines. The liquid product is now called thin juice.

Settled or filtered solids are removed and washed free of sugar on vacuum filters. The wash is called sweetwater. It is used either to slake lime in a "straight house" (a factory that does not use the Steffens process to remove sugar from molasses) or to make saccharate milk in a "Steffens house." Any surplus sweetwater is normally returned to the first carbonation station.

Evaporation

The thin juice is pumped through heaters and sent through a multiple effect evaporator, where the sugar content is raised from 10–15% to 50–65% percent. This product liquid is called thick juice.

Crystallization

Sugar from the second and sometimes the third boilings (see next paragraph) is mixed with the thick juice, filtered, and fed into vacuum pans. It is from this liquid, called standard liquor, that white sugar is crystallized. This sugar is formed by pan boiling (first or white boiling) at low temperature and pressure, to avoid burning and caramelization. The boiling continues to supersaturation, when the liquid is seeded by addition of powdered sugar or fondant. Crystal nuclei form and, by proper control of pressure, temperature, and supersaturation, crystals are grown to proper size.

The mixture of liquid and crystals, massequite, is fed to a centrifuge, where the liquid is forced out through a perforated screen, leaving the crystals behind. Hot wash water is sprayed onto the crystals to remove the remaining liquid. If the product is to be sold as white sugar, the crystals are dried and then packaged in a desired size or sold as bulk. Other products from this step are simple sugar syrup (white sugar dissolved in water) and invert syrups (a mixture of equal amounts of glucose and fructose, plus varying amounts of sucrose or white sugar.)

Sometimes, the syrup is completely inverted, with no white sugar. Inverted syrups are normally formed from simple syrup.

The liquid passing through the centrifuge is called syrup. The syrup serves as feed liquor for the second or intermediate boiling (sometimes called High Raw), carried out in the same manner as the first boiling. The sugar from this step is called intermediate or High Raw Sugar. The intermediate syrup serves as feed material for the third or Low Raw boiling. Sugar from this third boiling is known as Low Raw or raw sugar. Syrup spun from this step is molasses.

Molasses is an end product, and no more sugar can be recovered from it by direct crystallization. It is either sold as a cattle food, yeast food, used in various bacteriological processes, or treated by the Steffens process for further sugar recovery. With the production of molasses, the straight house stops processing. The Steffens house continues.

Recovery of Sugar from Molasses (Steffens Process)

Molasses is diluted, cooled to a low temperature, and finely powdered quick lime is added. The result is an insoluble precipitate of sugar and lime, called saccharate, which is filtered out. The filter cake is called cold saccharate cake. When the filtrate is heated, a second precipitate is formed, hot saccharate or hot cake. This, too, is filtered out. The mixture of the hot and cold saccharates is slurried in sweetwater and known as saccharate milk. It is used as a liming agent in the first carbonation. The filtrate is very high in BOD.

Pulp Drying

In order to avoid the difficulties of handling and storing wet pulp, pulp drying facilities are installed. Wet pulp loses food value in storage; drainage from the pulp constitutes a major water pollution problem; odors from the stored wet pulp frequently stored in open silos, are rather offensive; and considerable physical space is required. In the drying process, much of the water is removed by presses, with the remainder removed by heat. If molasses is added to pulp during the evaporation process, molasses-dried pulp is produced; if not, plain-dried pulp is the product. In some factories, the pulp is further processed into pellets.

Concentration of the Steffens Filtrate

The filtrate from the Steffens process is generally concentrated by evaporation, yielding concentrated Steffens filtrate (CSF). This is added to the wet beet pulp, prior to drying. In at least one company, the CSF is received from several factories and processed into monosodium glutamate, potassium sulfate, and a residual concentrated solution for animal feed mixture.

Waste Origin

Water

The origin of the various polluted waters involved in beet sugar factories includes flume and wash waters and process waters. The process waters include pulp screening filtrate, pulp press water, pulp silo drainage, lime cake slurry, Steffens filtrate, and condenser water from evaporators and vacuum pans. (An additional polluted water which might be listed is power plant blowdown, but, since this is similar to that of any power plant, it will not be discussed here.)

The pollution in the flume and wash waters consists of suspended and dissolved organic and inorganic material. The suspended material is largely dirt and trash and is discussed in the following section. The dissolved material is largely sugar and other organic material from the beets. The concentrations vary, depending on the degree of bruising and cutting of the beets as they are handled from the field through the washing process. Normally, the amount of sugar extracted from the beets in these processes is not great and the flume water tends to build up in dissolved sugar concentration to a certain point and not increase further. Hence, the amount of sugar lost in fluming and washing is not a major factor in the economics of the sugar factory.

Sugar is lost (and thus becomes a water pollutant) in the pulp screening filtrate, in the pulp press water, in pulp silo drainage, in the lime cake slurry, and in the Steffens filtrate. Typically, a straight house will recover about 81% of the sugar in the beets and a Steffens house will recover about 91% of the sugar. The proportion of the nonrecovered sugar in the water effluents of the factory varies considerably from factory to factory, depending principally on the amount of recirculation the system contains. The percent of sugar recovered has dropped somewhat in recent years, principally due to farm practices. However, the amount of sugar produced per acre has increased, principally due to these same farm practices. The main factor has been increased fertilization, which has introduced additional soluble organics into the diffused juice and prevents crystallization of sugar to the degree formerly possible.

Dissolved organic and inorganic materials other than sugar are, of course, present in the waters. However, as this work is not concerned primarily with the management of water pollution, more detailed treatment of quantities and concentrations of water pollutants is left to other sources of information.

Solid Wastes

The solid wastes involved in a beet sugar factory include beet pulp (even though it is completely and profitably utilized), dirt (from washing action in the flume and washing steps), trash (tops, weeds, bits of beets), lime cake slurry, and kraft paper overwrap of sugar packaging materials.

Beet pulp is listed as a solid waste only because it is a by-product and not the principal product. It typically amounts to 0.03–0.05 tons (dry) per ton of beets processed.

The amount of dirt removed from the beets in the factory process varies, depending on the weather before to and during harvesting, on the manner of harvesting, and on the soil in which the beets are grown, to mention some of the affecting conditions. Heavy rains and the resulting mud generally cause larger amounts of dirt to cling to the beets; different mechanical methods of digging the beets will likewise cause varying amounts of dirt to remain on them. One factory with a season's production of 200,000 tons of sugar from approximately 1,400, 000 tons of beets, estimated that it removed approximately 8000 tons of dirt. One source has stated that in some areas of Europe, the amount of dirt has reached 25% of the net beet tonnage.

Trash, consisting of tops, weeds, and broken bits of beets, is separated from the beets both in the flume water and in the washing step. The principal factors governing the amount of trash are the cultivating of the beets (to lessen the amount of weeds in the fields and, thus, in the delivered beets), the care with which the beets are topped in the fields, and the care with which the beets are harvested and handled en route to the factory. The weather can play an important part in this sector, also. For instance, heavy freezes just before the beets are harvested have resulted in subsequent rotting. This, in turn, has caused the beets to be easily broken up, with increased solid waste in the flume and washing steps. Freezing and thawing also cause a diminution of the sugar content of the beets—sometimes to the extent that it does not pay to harvest them. Also, frozen tops wilt and result in poorer elimination of the tops during harvest.

Lime cake slurry is the largest single solid waste to be found in the sugar beet factory. Limestone solids normally amount to about 3% of the tonnage of beets processed in a straight house. The figure rises to about 5% in a Steffens house. Considering a Steffens factory with a handling of approximately 1,400,000 tons of beets, the lime cake slurry discarded amounts to about 70,000 tons/yr (dry material).

Kraft paper overwrap can easily constitute a major item in solid waste management. A factory of the above size may use 1,000,000 100 lb paper sacks, in packaging its refined sugar. An even greater proportionate overwrap occurs in connection with small sugar boxes—1–5 lb. No estimates have been found concerning actual amounts of wrapping paper, but industry sources have stated that it amounts to tons per season per factory.

Solid Waste Collection

Beet pulp is collected at the pulp screen, from where it is partially dewatered in the press (unless the pulp is to be sold in the wet form, in which case the pulp is normally not pressed, but stored wet). Dirt is collected from the flume and wash water at the clarifier or settling pond. Trash is collected at the trash screen or at a similar location. Lime cake is collected at the filters that follow the first and second carbonation steps. Kraft paper overwrap is collected at the packaging stations.

Utilization of Solid Wastes

Beet pulp, as been pointed out, is utilized profitably in the beet factory. Although modern methods have improved its desirability, through drying and pelletizing, apparently the excellent feed qualities of the material have been known for as long as the sugar beet has been used as a source of sugar. Increased profitability of the product—for it really is product and not a waste—appears to hinge on two factors:

1. Reduction of the cost of extraction of water, possibly through improved presses.
2. Reduction of the necessity of removing so much water, while still retaining the desirability of the material as a feed.

The dirt, removed during fluming and washing, is not utilized, except to build up dykes on the retaining ponds of the factory installation.

The trash—beet tops, weeds, and broken bits of beets—collected on screens is generally utilized as animal feed. In some locations this material is sold to livestock producers as feed, while in other areas the factory has to pay to have it hauled away. As an aside, the tops cut off in the field are also used as a feed, and a highly desirable one. For comparison, the tonnage of tops, per acre, is approximately the same as the tonnage of alfalfa that could be grown on the same land. This, of course, is in addition to the production of the beets themselves. Fifteen tons of green wet tops, per acre, is a not unusual harvest. Tops, cured on the ground, are the nutrient equal of alfalfa. Tops, dried and made into silage, are the equal, pound for pound, of corn silage.

Lime cake slurry is not normally utilized, but is put into retention ponds which, over the years, build up to substantial heights, with dykes built ever higher. However, in some areas, where the soil is acid, the lime slurry is spread on the ground as a counteractant. Much more important are the attempts being made to reuse the lime.

Some factories reburn the lime cake, but dust problems are considerable, as most of the kilns used are similar to cement kilns. Where this is done, about 20% of the lime is still wasted, as impurities build up in the cake. The dust problem is the one situation in which air pollution (outside of odors and the power plant) might be a factor in a beet factory.

At present, interested groups in the industry are experimenting with briquetting the cake. Should this become practical, a vertical shaft type of limestone kiln might be used and the air pollution problem virtually eliminated. At this writing, briquetting is still in the laboratory. For the future, both economic and political aspects will have to be considered in deciding whether a factory should continue to build up its dykes or recover the lime.

Kraft paper overwrap is not generally utilized, though some factories have arrangements with waste paper firms to take the material. As with the trash, some factories sell the material and some have to pay to have it hauled away. Probably the main reason is that most factories are at a considerable distance from metropolitan centers, where waste paper reclaiming firms are located, and transportation costs are too great for profitable sale of the paper.

Disposal

Water

The suspended solids in flume and wash waters are disposed of by eventual settling in ponds, though in some installations, the solids are more or less concentrated in a clarifier first.

The sludge that is separated in the clarifying or settling action is transported to ponds or drying beds. Depending on the net evaporation rate of the area and the size of the ponds, there may be an overflow to lagoons, streams, or other water bodies. Likewise, depending on the permeability of the bottom of the pond or sludge drying bed, there may be considerable percolation of the water in the sludge to the ground. Even the term "net evaporation rate" must be viewed with caution, as the major evaporation normally takes place in the summer, while sugar beet factories normally operate from early October through January—the time of minimum evaporation rate. Hence, unless the ponds or drying beds are properly sized, overflow to water bodies will occur and the dissolved solids in the flume and wash water will at least partially be disposed of in this manner. If the ponds are of sufficient size, though, the dissolved solids will eventually precipitate out to the bottom of the pond. Of course, in the warmer areas where the factories operate into the warmer season, evaporation will be a more positive factor that it is in the Northern Plains.

Silo drainage, where it still occurs, is extremely high in pollutants, particularly in BOD. Where such drainage is discharged to a public water body, the practice has a short life expectancy, as such contamination is being prohibited in most jursidictions, with varying lengths of time allowed before actual prohibition.

Organic matter dissolved in lime cake slurry is carried to the lime retention ponds. This matter is acted on by bacteria—one hopes aerobically, but not infrequently anaerobically, with its attendant odors. As with the ponds that receive the sludge from the flume water clarifier or pond, there may be an overflow from these lime ponds to some body of water. As the ponds are filled during the cold weather, bacteriological action is at a relatively low level until spring.

As the season ends, the discharges to the ponds stop. In the colder areas, ice forms on the ponds and the ponds become anaerobic for lack of dissolved oxygen. In the spring, on melting of the ice, the ponds turn over and a period of a few days to a few weeks of disagreeable odor will occur.

The filtrate from the Steffens process is of low volume but very high BOD concentration; it must be treated in some manner. In some factories, typical mechanical sewage treatment processes are used, in others, oxidation ponds. It is not known whether anaerobic processes, such as mixed anaerobic reactors or submerged filters, are used in any factory for the breaking down of the organic matter in this filtrate, but it appears that this process might be an excellent one for the purpose. Low ambient temperatures, however, would make the process slow, unless economical means of keeping the anaerobic reactor warm (preferably, well above 70°F) were available.

Solid Wastes

No disposal problem exists with beet pulp, as it is completely utilized as animal feed. In most sugar beet areas, little problem exists with trash (as used in the present context), since it is normally utilized as an animal feed, also. Where local conditions are such that the trash is not utilized as a feed, it must be handled as an unwanted solid waste and is normally landfilled. It is probable, though no specific instances have been found where it is a common practice, that some factories incinerate this material.

Normally, the term solid waste signifies such material as garbage, waste paper, tires, and packing. In this context, the major solid waste material of a beet sugar factory is the Kraft paper overwrap. As mentioned above, this paper is generally not sold, but is considered a completely waste product. In some factories, it is incinerated and in others it is landfilled. A few factories are able to sell the overwrap and thus recover some of the cost.

The two solid waste problems of major consideration in a sugar beet factory are the dirt from the flume and wash waters and the lime cake slurry. The control of these two waste products is generally considered to be a function of water pollution control. However, since modern practices have shown that there need be no water discharge from a beet factory, the problem resolves itself into what is to be done with the dirt and the lime cake slurry.

The sludge (dirt) that settles out from the flume and wash waters contains more than 95% water—water which contains dissolved organic matter, with resultant BOD. Because of this, the sludge cannot simply be spread on the ground. Odors would be pronounced and the water would drain off, uncontrolled. Hence, the sludge is taken to ponds, which preferably are sealed and nondischarging. Each season, the silt builds up in the ponds. When the silt has lessened the available depth to the point where the capacity is less than that desired, it is excavated and put on the dykes that surround the ponds. While water is in the ponds, they act as oxidation ponds for the bacteriological breakdown of the organic matter. If sufficient oxygen is not available for aerobic action, anaerobic bacteria become dominant and disagreeable odors result.

One problem that exists in some of these ponds is the variation in the water; depth. Generally, depth of between 3 and 5 ft is optimal or proper functioning of an oxidation pond. However, typical evaporation ponds at a sugar factory have depths varying from a few inches to 8 or 10 ft. The shallower depth is not sufficient for proper action, while the deeper depth is too great for aerobic bacteria and anaerobic action takes place. Good management dictates that the ponds be excavated and maintained at a level bottom and that they be segmented so that each part could be filled to at least 3 ft rather rapidly and to no more than 5 or 6 ft. Successive discharges throughout the season could be routed to separate segments to keep all parts of the system at optimal depths.

A program for proper operation of these ponds is easy to establish in theory but somewhat difficult to carry out. The principal difficulty is that the liquid entering the ponds is heavy with silt, which tends to settle out near the entry point, creating shallow areas. Conditions at each installation dictate how to accomplish the desired configuration of the ponds, in size, shape, and depth, plus the schedule for routing the liquid to the various segments.

The lime cake slurry presents a similar situation, with a large amount of suspended inorganic matter, plus a substantial concentration of BOD. The water pollution control aspects were discussed previously, and the solid precipitates are handled in a manner similar to that of the silt in the flume and wash water circuits. In some areas, it is found that the excavated lime sludge is not good dyke material, so it is left and the dykes heightened with other material.

Summary

Air Pollution

Except for the odors frequently present around some of the lagoons, settling ponds, evaporative ponds, and so on, of a sugar beet factory, most such factories are concerned with air pollution only to the extent that any factory having a power plant is concerned. It is possible that some factories will have a dust problem, should they decide to recalcine the lime cake. Briquetting of the cake appears to offer a means of avoiding this dust, provided research furnishes a substance useful in the regular limestone kilns.

Water Pollution

Older factories discharged practically all of their streams directly to some water body. As recovery methods became economical and as regulatory bodies became stricter, these effluent streams have been successively more completely treated or recirculated. Some factories now have no discharge to the environment of water streams, except for evaporation and percolation. Other factories approach this situation. There appears to be no technical reason why all sugar factories cannot practice complete water recirculation and produce no water discharge to public water bodies.

Solid Wastes

Most sugar beet factories today have three principal solid waste products. Silt from the flume and washing processes appears to offer no hope of treatment and no means of disposal other than discharge to evaporation ponds and eventual use as dyke buildup or similar use. Lime cake, carried to disposal lagoons as a slurry, is now the largest single item in solid waste management of these factories. It appears only a matter of time before methods are developed whereby the lime is recalcined, with about 20% wasted, due to increased impurity concentrations. This will eliminate a major part of the solid waste products of the factories. Such reuse would be most desirable both for conservation of natural resources and for eliminating a waste disposal problem. In some areas the lime cake is spread on agricultural lands for better soil management. The third solid waste product of the factories is the Kraft paper overwrap. Whether this is sold as waste paper, incinerated, or landfilled is apparently a matter of the economics of the individual factory location.

CHAPTER III.14

sugar cane

The processing of sugar cane starts with the harvesting of the crops, either by hand or mechanically. In mechanical harvesting it is standard practice to burn the cane before harvesting to remove the greater part of the foliage. There are approximately 4 tons/acre of unwanted foliage on the cane and 11 tons/acre of grass and weed, all of which is combustible.

Following harvesting the cane is washed to remove larger amounts of dirt and trash, crushed and shredded to reduce the size of the stalks, and then the juice is extracted by milling or diffusion. In milling the cane is pressed between heavy rollers to press out the juice and in diffusion the sugar is leached out by water and thin juices. The raw sugar then goes through clarification, evaporation, and crystallization to produce the final fine sugar.

The largest sources of emissions from sugar cane processing are the open field burning in the harvesting of the crop and the burning of bagasse as fuel. In crushing, evaporation, and crystallization some particles are emitted but in relatively small quantities.

The fibrous residue consisting of 30% pith, 10% water solubles, and 60% fiber (1.5–1.7 mm) is bagasse. Treatment through magnetic separator eliminates tramp iron, through a depithing mill gives pith for fuel, pulp for paper and wall board (Celotex) after separation of dirt, fine fibers, weeds, and nonfiber. There are more than 30 bagasse pulp mills throughout the world. Since the crop is seasonal, most of the bagasse is burned on low-density grates to produce steam for sugar mill process and for power. The paper products are insulation board, bleached printing paper, and unbleached kraft. Residues from combustion in boilers are employed as fertilizer in the cane fields.

Competition for Celotex-type material is strong, as indicated by paper board, particle board, secondary boxboard, bark board, and interior grades of plywood.

CHAPTER III.15

fish and shellfish wastes

M. R. SODERQUIST and K. J. WILLIAMSON

The present world harvest of fish and shellfish is approximately 60 million tons/yr[1] and has been increasing at an annual rate of 5%. However, the annual catch in the United States has been declining; since 1962 U.S. fish harvests have decreased by 20%. The reasons given for this decrease include low harvesting efficiencies, inadequate and expensive labor supplies, and governmental restrictions.[2]

The annual U.S. catches (cleaned) average approximately 4 billion lb.[3] These fish are utilized as follows: 35% are rendered, 30% are marketed fresh, 20% are canned, 10% are frozen, 1% are cured, and the remaining 4% are handled by miscellaneous means.[4, 71]

Significant portions of the fishes and shellfishes processed are wasted. These percentages of wastage range from zero for whole-rendered fish such as menhaden to 85% for some crabs. The average wastage for all fish and shellfish is about 30%. In addition to the large volumes of solid wastes, large wastewater flows result from the butchering, washing, and processing of the product. The volumes of solids and wastewater vary widely with species and processing method.

The Washington State Water Pollution Control Commission[5] characterized (in general terms) fish processing wastewaters are shown on Table 1. The listed biochemical oxygen demands (BOD_5) and solids concentrations are very high compared to domestic sewage. These values can be considered only crude estimates at best, since neither the products, processes, nor plant sizes are listed.

THE INDUSTRY BY SPECIES

Bottom Fish

The most important bottom fish species were listed by Slavin and Peters[6] as haddock, cod, ocean perch, whiting (silver hake), flounder,

***Table 1** Fish Processing Wastewater Characteristics*[5]

Parameter	Unit	Value
Volume	gal/ton fish	465–9100
BOD_5	mg/liter	2700–3440
BOD_5	lbs/1000 gal effluent	2.6–29
BOD_5	lbs/ton fish	8–120
BOD_5	lbs/ton product	21–24
Suspended solids (S.S.)	mg/liter	2200–3020
Total solids (T.S.)	mg/liter	4198–21820
Population equivalent (P.E.)[a]	BOD/based ton fish	47–706

[a] Assuming 1 P.E. = 0.17 lb BOD_5 per capita per day.

hake, and pollock. Halibut are regarded technically as bottom fish, but will be considered separately. Approximately 80% of the industry is located in the North Atlantic region. The 1968 catch exceeded 465 million lb and had a value of $46 million.[3]

Processing

The fish are usually caught in otter trawls. In a typical operation the fish are spread upon the trawler deck, sorted, and iced. Perch, flounder, and whiting are stored whole, whereas cod, haddock, and pollock are sometimes eviscerated on deck. The viscera and blood are washed overboard.

At the wharf, unloading is usually accomplished manually or with conveying systems, the method depending on production volume. The fish are then weighed, washed, and iced in tote boxes.

Most plants processing fillets use mechanized equipment. First, the fish are washed by water sprays in large rotating tumblers. Next the fish pass to filleting machines or hand-filleting tables. Filleting machines only operate on certain fish sizes and shapes, but considerably reduce labor costs and increase yields over hand filleting. The skin is removed from the fillet by hand or machine. The solid wastes from filleting and skinning operations are usually rendered for pet food or animal meal.

Figure 1 outlines a typical bottom fish filleting operation. On this figure (and subsequent flow sheets) the product is depicted as flowing through the unit operations from the top of the page to the bottom. The water, wastewater, and solid waste flows are depicted on the diagram as flowing from left to right. Where one flow is indicated as dividing and moving in two different directions, as are the cases with liquid wastes and with some solid wastes, this is meant to illustrate that either route (or, in some cases, both) may be followed. Wastewaters from a bottom fish filleting process, for example, may be treated or may pass directly to the receiving waters.

The skinned fillets are transported by conveyor belt through a washing tank and, in some cases, a brining tank. After inspection the fillets are packed into containers by hand or frozen and then packed.

Waste Quantities

In most filleting operations, the fish are not eviscerated. The unfilleted portions are discarded or recovered for by-products. This fraction of the fish has been estimated to range from 35 to 40% by weight.[7] Water is used continuously in the spray washers and during filleting and skinning for bacteriological control. Blood and small pieces of fish flesh are entrapped in this flow. Other waste flows include the packing ice and the cooling water (see Figure 1).

Claggett and Wong[8] listed the wastewater flow from a bottom fish plant as 0.65 mg/day with 750 mg/liter total solids. A report by the firm of Stevens, Thompson, Runyan and Ries, Inc.[9] listed wastewater

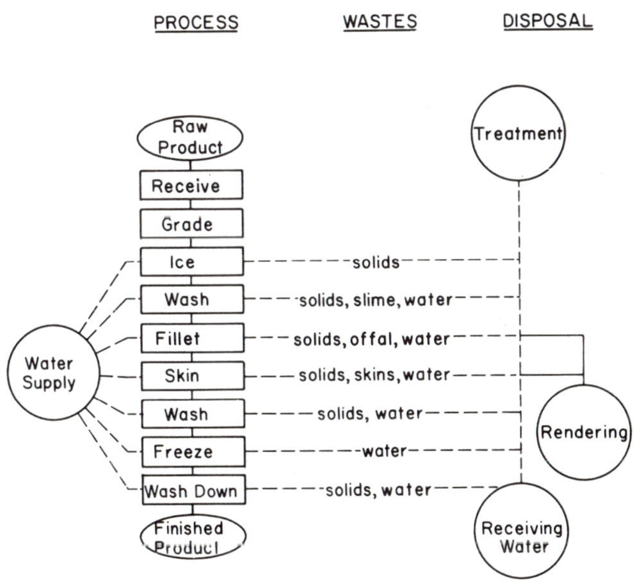

Figure 1. The bottom fish filleting process.

Table 2 Bottom Fish Processing Wastewater Characteristics

Parameter	Unit	Value	Ref.
Flow	gal/min	450	11
	gal/min	320–410	9
	gal/min	105	10
	gal/min	132	11
BOD	mg/liter	192–640	9
	mg/liter	640	10
	mg/liter	1726	11
	lb/ton of product	74	9
Suspended solids	mg/liter	300	10

flows from a bottom fish processing plant as 0.46–0.59 mg/day. This process included water transport of the fish to the filleting tables. The BOD_5 concentration varied from 192 to 640 mg/liter and the BOD_5 per hundred pounds of product averaged 3.7 lb. The organic loading ranged from 298 to 1100 lb/day of BOD_5.

In 1969 the firm of Cornell, Howland, Hayes and Merryfield[10] measured the waste loadings from this same fish processor after the fish flume had been replaced by a conveyor belt. The flow had decreased to 0.15 mg/day; the average BOD_5 concentration was 640 mg/liter, and the suspended solids, 300 mg/liter (see Table 2). The average waste loadings were 800 lb/day BOD_5 and 375 lbs/day suspended solids. The fish scaling operation produced significant quantities of wastewater. The wastewater flow from the scaler was 0.23 mg/day with average concentrations of 400 mg/liter BOD_5 and 290 mg/liter suspended solids[10]; however, the scaler was only operated for approximately 2 hr per average processing day.

Limprich[11] reported on the waste flows from a bottom fish processing plant at Cuxhaven, Germany. The average discharge was 132 gal/min with a BOD_5 concentration of 1726 mg/liter. This waste was clarified and the sludges were centrifuged and then processed for fish meal.

Catfish

Since 1965 the production of farm catfish has increased steadily. Four species (channel catfish, blue catfish, white catfish, and brown bullhead catfish) have been grown and managed successfully in ponds. Catfish are considered a delicacy in the southern and south-central states and markets have continued to expand. Sixty-four million pounds of catfish were harvested in a recent year.[12]

Processing

Several authors[13-16] have described in detail the raising of catfish in ponds. The process involves planting 6-in. fingerlings that are fed a commercial feed ration until maturity. The fish are harvested by draining the ponds and are shipped alive in tank trucks to processing plants. Live hauling eliminates the need for meat preservation before processing, but generates the problem of disposal of the feces-contaminated holding water.

Figure 2 depicts the processing method and the wastes resulting. The fish are held in live tanks until processing, which results in more feces-contaminated water.

The fish are first stunned, commonly with electric shock, and then butchered. The butchering process, which includes skinning, beheading, and eviscerating, can be either manual or mechanical. Only the outer layer of pigmented skin is removed for esthetic reasons. This process results in solid wastes containing skins, heads, viscera and wastewaters containing blood, slime, and flesh.

Waste Quantities

Jones[12] estimated 45% of the whole catfish to be waste and the Bureau of Commercial Fisheries,[17] 40%. Several methods have been suggested for catfish offal disposal,[18] and each should be considered on its

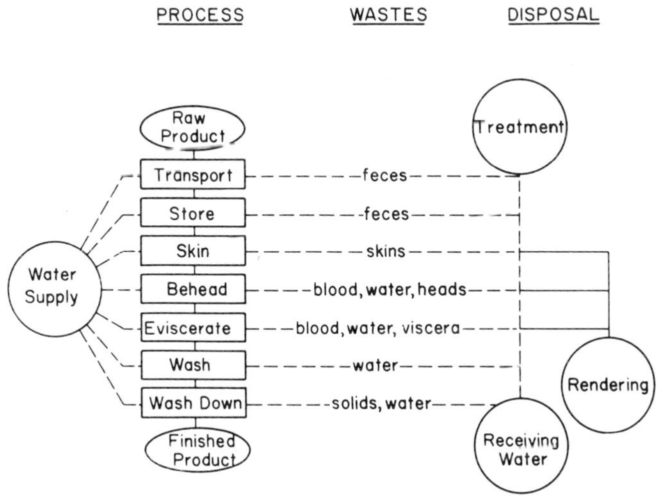

Figure 2. Catfish processing.

economic merit. These methods include rendering for pet food, catfish feed, fish meal, and burial. Catfish offal has been rendered to a meal containing over 45% protein.[19]

Crabs and Lobsters

The blue crab, which comprises 70% of the U.S. crab production, is harvested on the Atlantic Coast, principally in the Chesapeake Bay area.[20] The remaining crab harvest takes place mainly on the Pacific Coast, where Dungeness crab is the leading species, followed by Alaskan king crab. Present catches exceed 240 million lb annually.[3]

The lobster fisheries include the catch of the northern lobster of the North Atlantic region and the spring or rock lobster of the South Atlantic and Gulf states. The 1968 harvest was listed as 45 million lb.[3]

Processing

Crabs are harvested from shallow water in baited traps. Rapid and careful handling is necessary to keep the crabs alive; dead crabs must be discarded because of rapid decomposition.

At most plants, the whole crabs are steam cooked in retorts for 20–30 min.[20] Pacific Coast Dungeness crab operations first butcher the crabs (remove the backs), and then cook them for 12 min or less. Cooked crabs are marketed in the shell, butchered or whole, or the meats, picked from the shell, are marketed fresh, frozen, or canned. Most of the Atlantic blue crab meat is marketed fresh or frozen, but most Pacific Coast crab meat is canned.[21] A large quantity of Dungeness crab is sold in the shell and large quantities of king crab are butchered at sea[22]; both practices minimize the volumes of butchering wastes be handled at the processing plant.

The crabs are water cooled after cooking to facilitate handling. The backs are removed if the crabs were not butchered before cooking, and the remaining viscera are washed free. The cooking, cooling, and washing waters contain considerable solids and organic pollutants (see Figures 3 and 4). The meat is picked from the shells by hand with a small knife. Mechanical methods have only recently been developed to extract the meat from the shells.[23]

Crab meat quickly degrades in quality and must be chilled, frozen, or canned. Chilled meats can be stored for only a few days; even frozen meats lose texture and flavor qualities rapidly. Canning of crab meat results in additional wastewater flows: retort and can cooling waters.

Lobsters are caught in large traps and must be kept alive until processed; Many lobsters are marketed live, and some are shipped

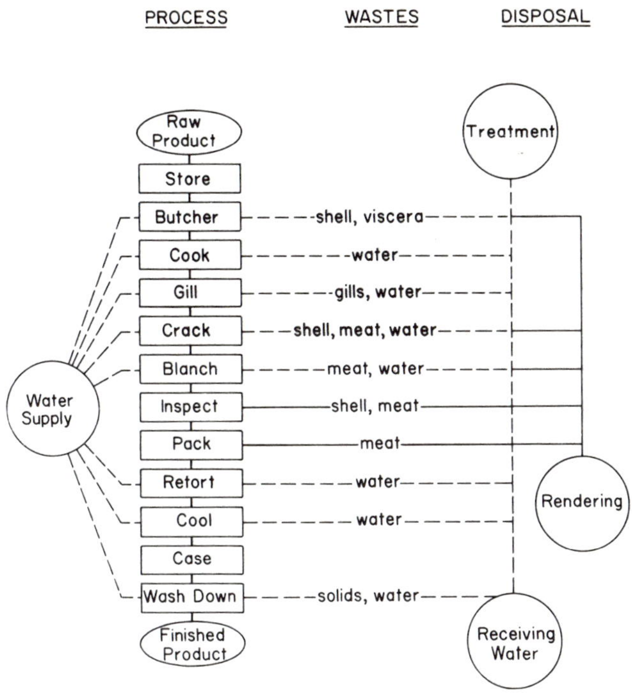

Figure 3. Crab canning process.

alive thousands of miles, carefully packed in moist seaweed and sawdust.

Lobsters are cooked, cooled, and butchered in a manner similar to crabs. The cooking, cooling and washing waters are normally highly polluted (see Figure 5). A small number of cooked lobsters and meats is frozen for later marketing. Low storage temperatures and quick turnovers are necessary for maintenance of high quality. Little lobster meat is canned because of the rapid degradation of texture and flavor quality of the canned product.

Waste Quantities

The major portion of the crab is not edible, and, as a result, is wasted in processing. This waste consists of the shell and entrails, amounting to approximately 80% of the crab by weight. Large quantities of water are necesssary for cooking, cooling, and washing of the entrails from the body. The wastage of the total crab has been listed for blue crab[20] as 86%; king crab,[24] 80%; and Dungeness crab,[25] 73%.

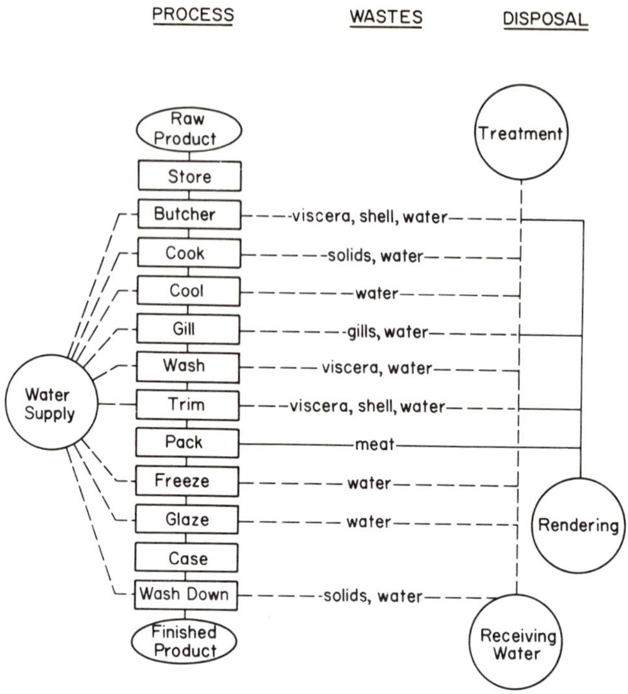

Figure 4. Crab freezing process.

The composition of shellfish waste is largely determined by the exoskeleton. The exoskeleton is composed primarily of chitin (a polysaccharide structural material), protein bound to the chitin, and calcium carbonate. While the major portion of the waste generally consists of exoskeletal materials, varying significant amounts of attached or unrecovered flesh and visceral materials are included. Crab wastes are typically 20% protein, 30% chitin, and 50% calcium carbonate.[26] Hoalihan[27] reported the cleaning loss of lobsters to be 80%. However, only a small percentage of the lobsters are cleaned before marketing; most are sold alive or cooked in the shell.

Halibut

The halibut is a large fish; the commercially-landed sizes vary from 20 to 50 lb. They are caught near the sea bottom using baited longlines. The major halibut fishery is centered in the Pacific Northwest with the commercial season extending from April or May through October. The halibut catch in 1968 was approximately 26 million lb.[3]

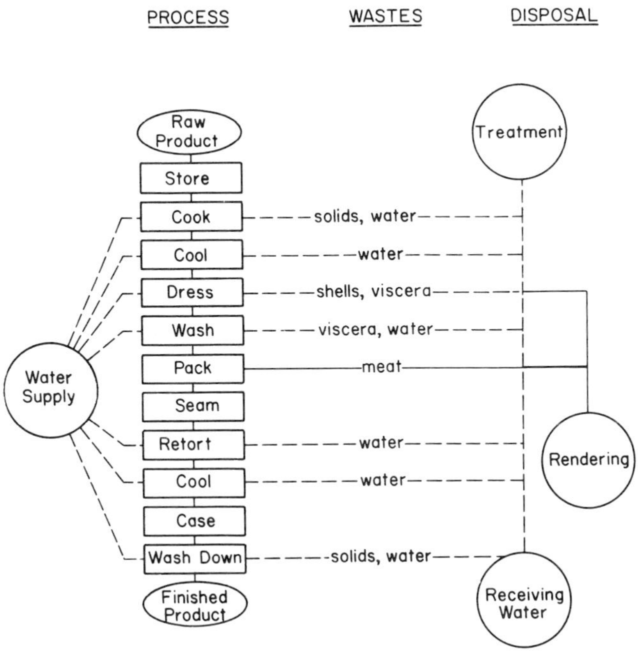

Figure 5. The lobster canning process.

Processing

After being landed on the vessel, the halibut are dressed by removing the viscera and cutting away the gills. They are then packed in ice in the hold. Halibut are ordinarily processed in relatively small plants. The fisherman usually unload and behead the fish before sale to the processors.

If the fish are not to be processed immediately, they are re-iced in the fish house (some are sold fresh, but most are marketed frozen). A continuous-belt washer sprays the fish before freezing. The fish are frozen with a glaze protection at approximately $-20°F$. The filleting and freezing operations are diagrammed on Figures 6 and 7.

Waste Quantities

Jensen[24] estimated that 35–40% of the halibut is wasted. The viscera and gills are usually disposed of at sea; Dassow[28] estimated the remaining waste to be approximately 8% of the total weight. This included heads, skins, and fins.

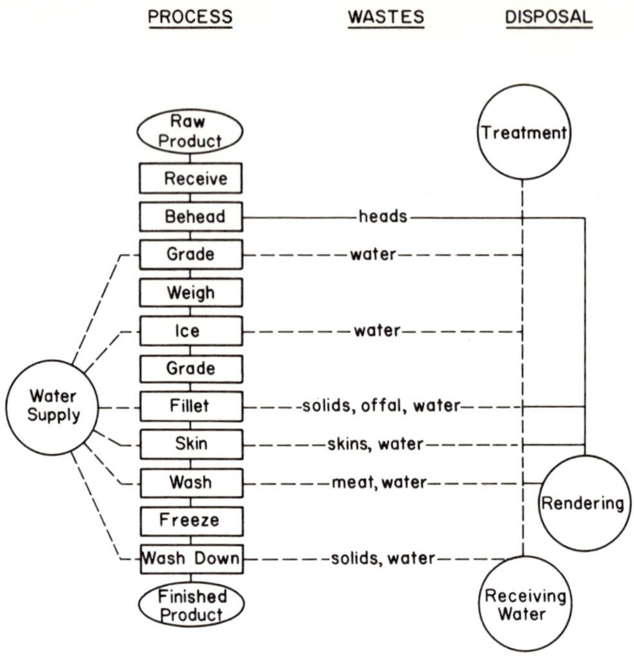

Figure 6. The halibut fillet freezing process.

Menhaden, Sardines, Mackerel, Anchovies, Herring, and Alewives

Menhaden, sardines, mackerel, anchovies, herring, and alewives are all classified as small, oily fishes. Sardines, anchovies, and mackerel are used mainly for human consumption, whereas menhaden, herring, and alewives are most frequently used in oil and fish meal rendering and for bait. Herring are sometimes canned in the North Atlantic region as "sardines."

Of these six species, menhaden are harvested in greatest quantity; the yield was 1.1 billion lb in 1968.[3] Dried scrap and meal are the most highly valued products from menhaden, although oil production was the initial reason for processing. Most of the scrap and meal is used as an animal feed supplement.

Processing

Sardines, anchovies, and mackerel are stored in water in the hold of the fishing vessel and are unloaded by pumping. The catch is then weighed and transferred to dockside holding tanks.

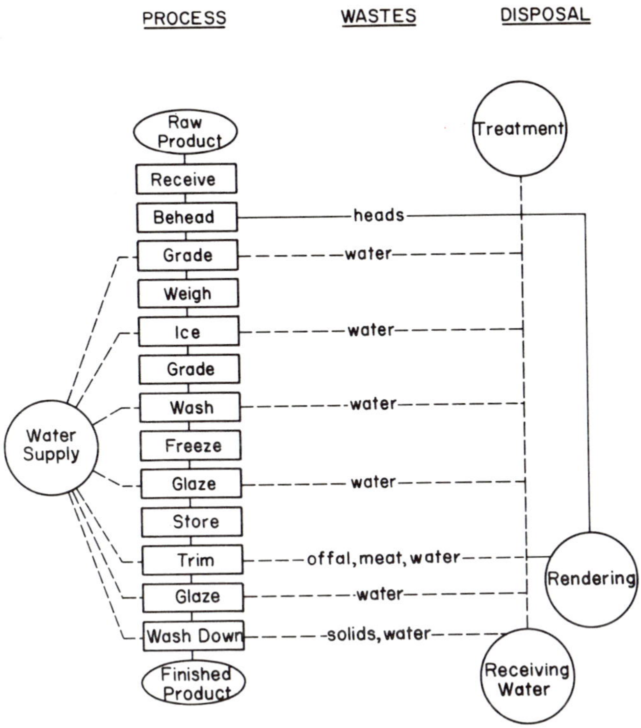

Figure 7. The whole halibut freezing process.

From the receiving tanks the fish are pumped onto the cutting tables, where the workers insert them into the slots of a conveyor belt. Revolving knives trim and slit the fish, and the viscera are removed by a suction process. The cleaned fish are then washed and canned (see Figure 8).

Menhaden are caught in purse seines and loaded into the holds. Ice or refrigeration is used to preserve the fish if the length of trip exceeds 1 day.

The fish are pumped from the holds, washed, automatically weighed, and conveyed into the plant; they are cooked by a continuous steam method. The fish are then pressed to remove the oil and most of the water. This press water is screened to remove solids and centrifuged to separate the oil. The remaining water, called stickwater, is discharged or evaporated to produce condensed fish solubles. The solid residual from which the water and oil have been pressed is known as pressed

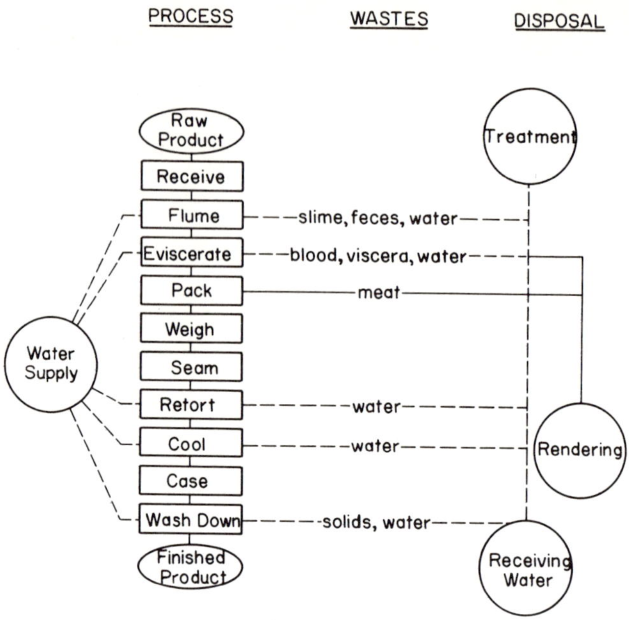

Figure 8. The canning process for sardines, anchovies, and mackerel.

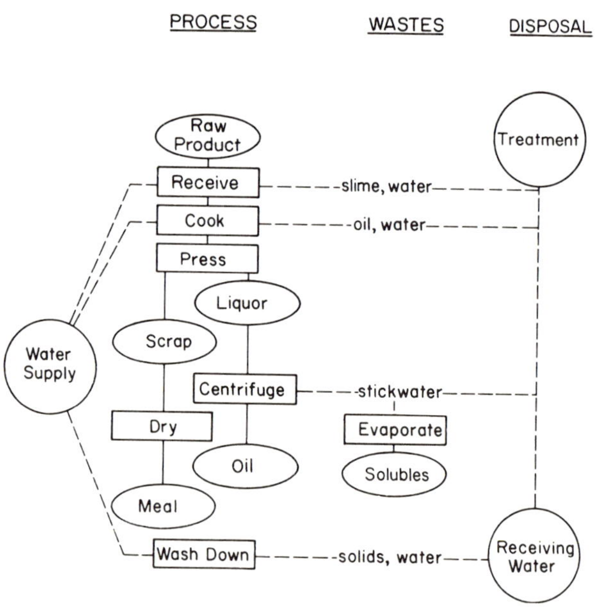

Figure 9. The menhaden rendering process.

cake. The pressed cake is dried to about 10% moisture and then ground for fish meal.[29] Figure 9 shows the process as described.

The majority of the herring and alewives are rendered in a manner similar to the menhaden process.

Waste Quantities

The Bureau of Commercial Fisheries estimated sardines and anchovies to be 15% waste; and mackerel, alewives, and herring, 30%.[25]

Limprich described in detail the wastes from a German fish meal plant using approximately 450 tons/day of raw material. The resultant flows and strengths are summarized on Table 3.

The wastewaters from the production of fish meal, solubles, and oil (diagrammed in Figure 9) from herring, menhaden, and anchovies can be divided into two categories: high-volume low-strength wastes and low-volume high-strength wastes.

The high-volume low-strength wastes consist of the water used for unloading, fluming, transporting, and handling the fish plus the washdown water. Davis[30] estimated the fluming flow to be 2000 gal/ton of fish with a suspended solids content of 5000 mg/liter. The solids consist of blood, flesh, oil, and fats. Claggett and Wong[8] listed a herring pump water flow of 250 gal/min with a total solids concentration of 30, 000 mg/liter and oil concentration of 4500 mg/liter. Pump water is used to transport the fish from the holds of the boats to the processing plant. The bilge water in the boats was estimated at 400 gal/ton of fish with a suspended solids content of 10,000 mg/liter.[30] Other wastes come from leakage from holding tanks, wash-down, evaporators, and the drier air scrubbers. Paessler and Davis[32] described in detail these wastes and process modifications that can be used to reduce the waste loads.

The National Canners Association conducted a 1-week study of the wastes from four Maine sardine packers. These wastes were divided

Table 3 *Fish Meal Processing Wasetwater Characteristics*[11]

Parameter	Cooling Waters	Other Wastewaters
Discharge	1440 gal/min	680 gal/min
	240–480 gal/ton fish	510–1020 gal/ton fish
BOD_5	621 mg/liter	1005 mg/liter

Table 4 *Sardine Packing Wastewater Characteristics*[38]

Source	COD (mg/liter)	BOD_5 (mg/liter)	Suspended Solids (mg/liter)	Oil and Grease (mg/liter)	Flow (gal/min)
Flume water	500–1400	200–1150	400	300–360	130–300
Hold water	800	370	—	—	—
Pump water	170–340	10–45	—	—	800–1000
Waste flume water	240–1700	100–2200	100–2100	60–1340	40–180

into four categories: pump water, flume water, hold water, and processing wastes flume water. Ranges are listed on Table 4.

The pump waters transferred the fish from the shipboard holds to a screen separator in the plant. These wastes were lightly polluted, as shown by a BOD_5 of under 50 mg/liter, but comprised the largest flow. The flume water conveyed the fish through the plant and became heavily polluted. The hold water resulted from the storage of fish in the boats. The small volumes made this wastewater of lesser consequence. The processing wastewaters were used to transfer the solid wastes to a truck for disposal. This waste flow could easily be eliminated through the use of dry capture techniques. All the liquid wastes were discharged untreated.

Other high-volume wastewaters include the condensate from the cooking operation and the cooling water from the condensers. The volumes are large, but the organic loads are small. These wastes contain small amounts of soluble and insoluble solids, including fats. The chemical oxygen demand is usually about 300 mg/liter and the wastes are not readily putrefiable.[33] Tanzler[34] recommended that this wastewater be first passed through a separator before discharge and the water draining from the fish be passed through a vacuum thickener. Condensate from the thickener would still contain suspended matter, which should be removed before discharge to a stream. The final wastewaters generated in the process come from the drying of the fish meal. These wastes are similar to the condensate wastes.

The drainage and rinsing waters from the storage bunkers are usually thick, slimy, highly colored, and strongly malodorous.[33] These wastes have large amounts of insoluble and soluble solids with high nitrogen and phosphate levels. Active decomposition begins rapidly, resulting in the production of hydrogen sulfide. Limprich[11] listed this waste volume as 36–48 gal/ton of fish.

The strongest segment of the fish meal wastes in the stickwater. In most instances stickwater is now evaporated to produce condensed fish solubles, but in previous years it was discharged untreated. The volume was estimated by Kempe et al.[35] and Davis[30] to be about 120 gal/ton of fish processed. Paessler and Davis[32] reported stickwater strengths of 56,000–112,000 mg BOD_5/liter and grease concentrations of 4200–24, 400 mg/liter for menhaden. California rendering plants using sardine scrap produced an average BOD_5 of 42,000 mg/liter. Jordan[33] stated that stickwater decomposes rapidly, evolving hydrogen sulfide, and leads to nuisance conditions if discharged to waterways or sewage treatment plants. Davis[30] estimated stickwater to be 6% solids, consisting almost entirely of protein, with very little oil. Knowlton[36] and Tetsch[37] agreed that separation of fats, greases, oils, and protein emulsions should take place at the fish processing plant before discharge to municipal sewers.

Oysters, Clams, and Scallops

Oysters, clams, and scallops are all bivalve mollusks. Harvesting results in large quantities of wastes and small quantities of highly valued meat. Excluding shell weights, present harvests approximate 140 million lb annually.[3]

Processing

Oysters are marketed shucked or unopened.[39] If marketed unopened, only washing, packing, and chilling of the shellfish are required. Prior

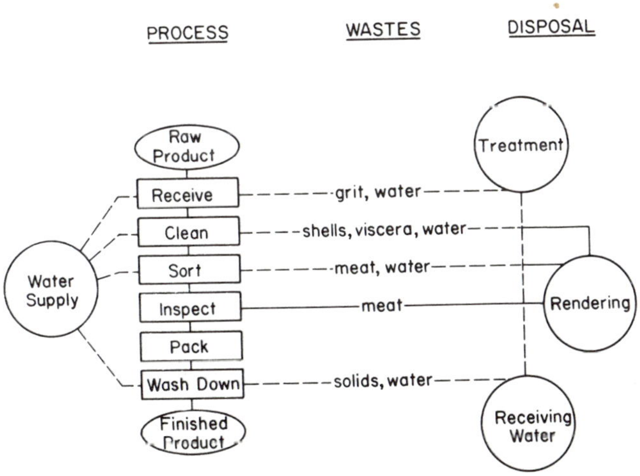

Figure 10. Oyster processing.

to shipment, the oysters may be stored in chlorinated water to minimize bacterial growth.

Most oysters are sold as shucked meats. The meats, when removed from the shells, are aerated in water to remove the sand and silt. After washing, the meats are graded and then packed into tins or glass containers (see Figure 10).

Before cleaning, clams must be washed free of sand and silt. Cleaning involves removing the shell and the visceral portion and trimming dark portions from the siphon tips. Clam meats are marketed canned as whole meats or minced clams or fresh as whole meats.

Scallops are shucked on board the fishing vessel and the large adductor muscles are removed. The adductor muscle is the only portion marketed; the remaining portions are discarded at sea. Scallops are marketed frozen, chilled, or precooked.

Waste Quantities

The Oregon State Department of Environmental Quality[7] estimated oysters to be 75% waste by weight. However, this waste consists mainly of shells and can be used for several by-products. In most cases, the oyster is not eviscerated; thus the organic portion of the waste is small.

The Bureau of Commercial Fisheries[25] estimated clams to be 65% waste. This waste also includes the shells, but contains a much higher organic content than oysters because clams are totally eviscerated in processing.

Scallop wastes present no terrestial disposal problems since they are discharged at sea. Utilization as fish food or crab bait could be considered if the solids were to be brought to the cannery.

Salmon

The five main species of salmon found in this country are chinook (king), sockeye (red), silver (coho), pink, and chum. The major portion of the catch is canned. The 1968 catch exceeded 300 million lb.[3]

Processing

The fish are caught fairly close to the canneries and are often stored in the boats without refrigeration. Canning operations are conducted, for the most part, employing standard cannery equipment in a conventional manner. The principal exception is the use of the "iron chink." The iron chink performs several functions in one operation by mechanically removing heads, fins, and viscera. During all phases a strong stream of water continuously washes the blood away.

The remaining operations are somewhat standard, as shown on

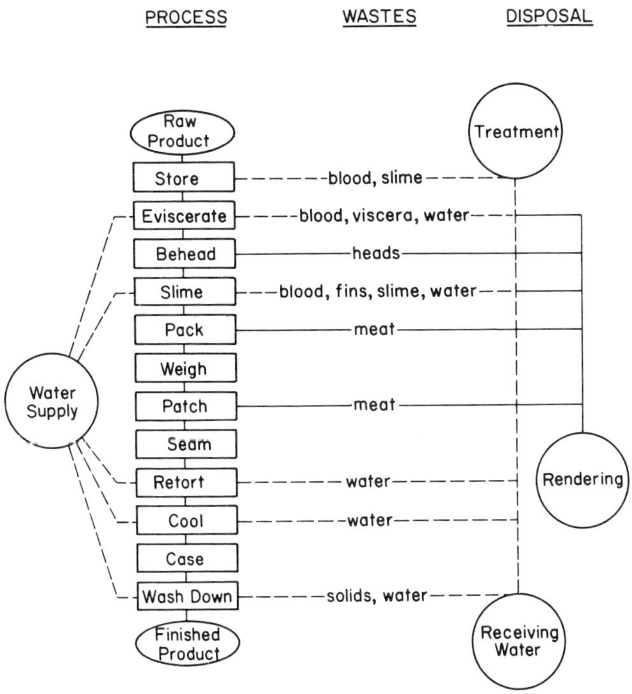

Figure 11. The salmon canning process.

Figure 11. The fish are washed, inspected and cut into controlled-length pieces. These pieces are cut into can-size portions and the cans are filled mechanically. Finally, the cans are automatically sealed and retorted.

Waste Quantities

The quantities and possible uses of salmon wastes have been rather thoroughly researched. Magnusson and Hagevig[41] reported salmon processing waste to be 34%. Other estimates include Brody's,[42] 33%; Jensen's,[24] 37%; and the Oregon State Department of Environmental Quality's,[7] 30–35%. The Bureau of Commercial Fisheries[25] listed waste fractions by species as follows: chinook, 30%; sockeye, 33%; silver, 33%; pink, 35%; and chum, 33%. Salmon waste is composed of the various body portions excluding the flesh. The relative amounts of each portion for the five species are shown on Table 5. Each portion has distinct by-product possibilities and recovery value. The quantities of milt and roe vary with the time of year.[41]

Table 5 Composition of Salmon Waste[43]

Portion	Percent of Total Salmon Cannery Waste by Species				
	Pink	Red	Chum	King	Coho
Head and collar	57	61	54	50	60
Tail and fins	16	14	11	11	11
Liver	5	5	5	3	4
Roe	8	9	16	15	8
Milt	5	5	6	4	6
Digestive tract	9	6	8	18	11
Heart	0.8	0.8	0.7	0.7	0.7

The wastes from the salmon canning process include butchering water, viscera, wash water, retort water, and cooling water. Also included is cooking water when oil recovery from the heads is practiced.

Claggett and Wong[8] listed the flow from a salmon canning line as 300 gal/min with a total solids concentration of 5000 mg/liter and oil concentration of 250 mg/liter. The firm of Stevens, Thompson, Runyan and Ries, Inc. reported on the effluents from several salmon processors.[9] Later studies of the same firms were reported by Foess.[10] The wastewater characteristics are listed on Table 6. The values for all parameters are quite variable; the strengths depend on the efficiency of solids removal. The BOD_5 concentrations range from 200 to 4000 mg/liter; suspended solids, 40–5000 mg/liter; total solids, 90–7000 mg/liter; and volatile solids, 70–7000 mg/liter. Mild curing was reported to produce considerably weaker wastes than canning.

Caviar production results in extremely strong wastes, but waste volumes are small. These wastes should be recovered and not discharged to the waterways or sewage treatment plants.

Shrimp

The shrimp industry is the most important seafoods industry of the Gulf of Mexico and South Atlantic areas. Shrimp are also found off the Pacific Coast in significant numbers. In 1968 the catch exceeded 290 million lb with a value of more than $110 million.[3]

Processing

Shrimp are caught commercially in otter trawls to a distance of approximately 50 miles offshore. The shrimp are separated from the trash fish and stored by various methods. When short storage times

Table 6 *Salmon Processing Wastewater Characteristics*

Process	Flow (mg/day)	COD (mg/liter)	BOD_5 (mg/liter)	BOD_5/ raw product (lb/ton)	Suspended Solids (mg/liter)	Total Solids (mg/liter)	Volatile Solids (mg/liter)	Ref.
Canning	0.043–0.046	5920	3660–3900	6.5–178	508–4780	1188–7444	1048–7278	219
Canning	0.33	—	3860	—	2470	—	—	220
Caviar	—	—	270,000	—	92,600	386,000	292,000	219
Mild curing	0.018–0.066	—	173–1320	10–80	44–456	258–2712	98–2508	219
Mild curing and fresh	0.011–0.036	—	206–2218	3.2–36	112–820	484–2940	184–1756	219
Mild curing or freezing	0.014–0.046	—	397–3082	3.8–19	40–1824	88–3422	67–2866	219

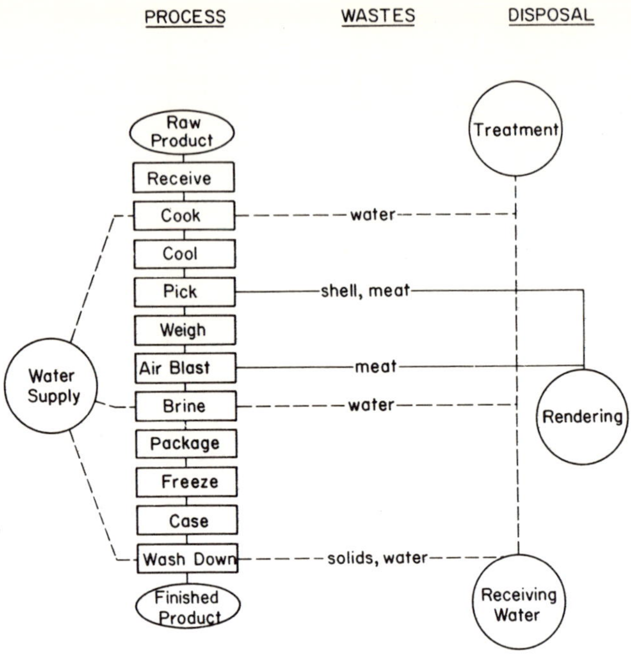

Figure 12. Freezing of manually picked shrimp.

will suffice, no preservation methods are used; the shrimp are taken directly to a processing plant or to a wholesale marketing vessel. When longer storage times are necessary, the shrimp are iced in the holds and re-iced every 12 hr. In some cases, notably the Gulf states, the shrimp are beheaded at sea and the heads discarded. Since the heads contain most of the active degradative enzymes, this practice retards spoilage. If the shrimp are beheaded within 30 min after being caught, the intestinal vein is readily removed with the head. This increases the value of the product.

The shrimp are unloaded from the vessel into a flotation tank to remove the packing ice, conveyed to a rotary drum to remove surplus water and bits of debris, and then weighed. In some areas (Texas and the South Atlantic states), the shrimp are iced after the initial preparation to optimize peeling conditions.

Next the shrimp are peeled and picked, if the head is still attached, manually or by machine. Machine-peeled shrimp are used mostly for canning.[24] The machine-peeled shrimp are paler in color, and have poorer flavor and texture than the hand-picked product. By hand, a picker can peel 100–400 lb/day of shrimp, as compared to a machine's capacity of 4000–12,000 lb/day.[44]

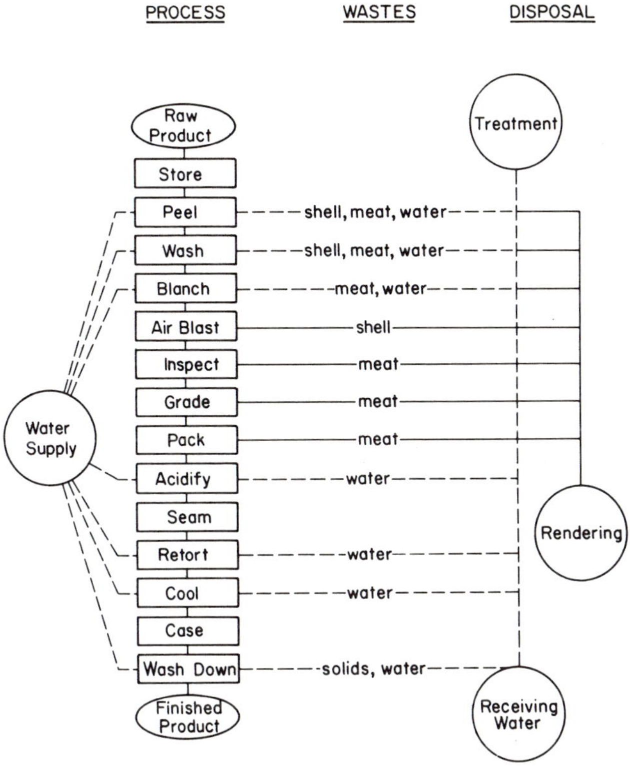

Figure 13. Canning of mechanically picked shrimp.

After peeling, the meats are inspected and washed. They are then blanched in a salt solution for about 10 min and dried by various methods to remove surface water. Again the shrimp are inspected and then canned. The process is outlined on Figures 12 and 13.

Waste Quantities

Jensen[24] estimated that 78–85% of the shrimp is wasted in mechanical peeling and 77–84% in hand picking. The Oregon State Department of Environmental Quality[7] estimated 75% wastage for hand picking and the Bureau of Commercial Fisheries listed a cleaning loss of 55%.[25] The low value of the Bureau of Commercial Fisheries was apparently due to ignoring the blanching loss, which ranges from 30% to 35% of the picked weight.[44] Vilbrandt and Abernethy[45] mentioned that the shrimp heads comprise 43–45% of the whole raw shrimp. A typical shrimp waste consists of 25% protein, 50% chitin, and 25% calcium carbonate.[26] Crawford[22] reported the effluent from mechanical shrimp peelers to average 29,000 mg/liter total solids and 185 mg/liter total nitrogen.

Tuna

Tuna ranks as the primary seafood in the United States; Americans consume over one billion cans of tuna per year.[46]

Tuna are large, migratory fish. They feed on whatever small sea animals are most abundant and easiest to catch. Their distribution in the oceans is still nearly completely unknown, although research in this field is under way. The major runs are found off the Pacific Coast. The annual tuna catch averages approximately 300 million lb, of which the majority is canned.

Processing

Most tuna which are canned in the United States are caught in distant waters. A modern tuna vessel can hold 150–300 tons of fish and has a range of 1000 miles.[47] Because of the long transport times, the fish are normally frozen aboard the fishing vessels.

The fish usually are unloaded (while frozen) by mechanical hoists and conveyed to the weighing station. After weighing, the fish are inspected and thawed.

Tuna are eviscerated by hand in several steps. The body cavities are flushed with fresh water and all adhering viscera carefully removed. The viscera are used for fish meal or pet food and the livers are sometimes recovered for oil and vitamins.

After butchering, the fish are precooked in large, open chambers. The time of cooking varies with the body size, but is usually about 3 hr. Weight loss during cooking (attributable to oil and moisture loss) averages 22–26%.[47]

The cooked fish are cooled for approximately 12 hr to firm the flesh. The meat is separated by hand from the head, bones, fins, and skin. All dark meat is removed and usually recovered for pet food. The meat to be canned is placed on a conveyor belt and transferred to the "Pak-Shaper" machine.

The tuna slices are arranged lengthwise in the Pak-Shaper. This device molds the loins into a cylinder, fills the cans and trims the meat after filling. The machine can fill 125–150 cans per minute.[47]

Salt and vegetable oils are next added to the cans and they are vacuum sealed and retorted by standard procedures. The entire process is diagrammed on Figure 14.

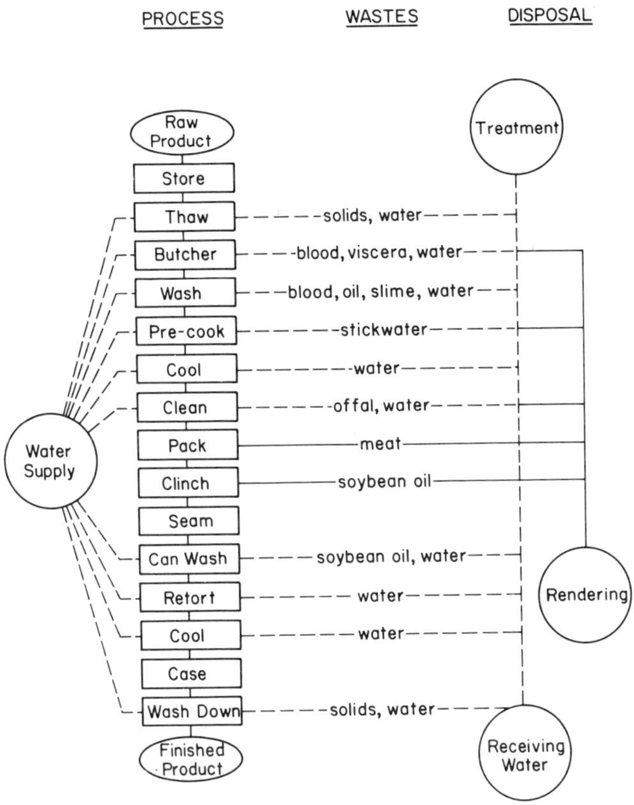

Figure 14. The tuna canning process.

Waste Quantities

The Oregon State Department of Environmental Quality[7] estimated that 65% of the tuna is wasted in the canning process. Tuna processing wastes (listed on Figure 14) include water from butchering, cooking, canning, retorting, and cooling and the eviscerated solid wastes. The Kennedy Engineers[48] reported on tuna cannery wastes in American Samoa. The waste concentrations averaged 5100 mg/liter BOD_5, 5890 mg/liter grease, and 1730 mg/liter suspended solids, of which 85% was volatile.

Chun et al.[49] studied in detail the wastes from a tuna canning and rendering plant in Hawaii. Their results are listed on Table 7. The average waste flow was 6800 gal/ton of fish. Phosphorus and nitrogen levels in excess of those needed for biological waste treatment were present in the wastes.

Table 7 *Tuna Wastewater Characteristics*[49]

Parameter	Concentration	
	mg/liter	lb/ton of fish
COD	2,273	129
BOD_5	895	48
Total solids	17,900	950
Suspended solids	1,091	58
Grease	287	15

The 5-day BOD was only approximately 40% of the COD value. Due to the high nitrogen levels and high proportion of particulate matter in the waste, a considerable BOD would be expected to be exerted after 5 days. In this case at 22 days the BOD exerted was 3525 mg/liter and was still increasing. It is important to realize that the waste will exert a considerable nitrogenous BOD in excess of the 5-day value.

BY-PRODUCT UTILIZATION

Fish Meal

Fish meal is one of the products of fish rendering; the others are fish oils and solubles. In 1968, 127 plants produced over 469 million lb of fish meal and scraps with a value of $30 million.[50]

Fish meal was used primarily as a fertilizer prior to the 1930s, when research showed its possible value in feed rations.[51] Fish meal has been shown to be a source of concentrated protein with essential and nonessential amino acids, B vitamins, an unidentified growth factor, and trace elements, including phosphorus.

The most common species used for fish meals include menhaden, alewives, anchovies, Pacific and Jack mackerel, herring, and wastes from groundfish, crabs, and shrimp.

Methods of Manufacture

Kempe et al.[35] listed four classes of currently used rendering processes: wet, dry, solvent extraction, and digestion. The wet process is over 100 years old and is still the most prominent. The dry process is used only in small operations. The solvent and digestion processes are not used extensively.

In the wet rendering process, shown on Figure 9, the fish or fish waste is removed from the boat, weighed, and cooked by steam. This material is then pressed to give a solid press cake and liquid press liquor.

The press cake is dehydrated in one of several different types of dryers to a meal containing about 8% moisture.[35] The dried meal is then sold directly or further ground and powdered prior to marketing.

In the dry rendering process, the fish are first dried in large steam-heated dryers. The dried product is then pressed to release the oil and the press cake sold as a fish meal.

The solvent and digestion processes are used to obtain a fish protein concentrate with a protein level exceeding that of the normal fish meal (i.e., 40–50%).

Fish Oils

Fish oils are the most valuable of the fish rendering products. Fish oils in food products are currently prohibited by the Food, Drug and Cosmetics Act of 1938.[52] They are considered nonedible because the raw material from which the oil is extracted is not completely edible. Fineberg and Johanson[52] estimated that 75% of the fish oils produced outside the United States are used for human consumption.

Methods of Manufacture

Fish oils are obtained from the press water in the wet reduction process. The press water is screened to remove any solids and then passed through a gravity separator or a centrifuge. The sludge from this process is returned to the press cake and the clarified press liquor is passed to an oil centrifuge. The oil centrifuge yields two products; the fish oil and the stickwater. The concentrated oil is marketable.

In the dry rendering process, the liquid emerges from the presses as fish oil; no recovery steps are necessary.

The press liquor yield and composition vary; an average yield is 5–10 lb of oil per 100 lb of press liquor.[6] Presently, oil is manufactured from ground heads and in some cases, ground offal in some Alaskan salmon canneries.[53] The recovered oil is added back to the low-oil-content sockeye salmon pack when necessary. Only fresh heads and wastes are used and the quality of the oil is carefully monitored.

Condensed Fish Solubles

Condensed fish solubles are the by-products produced from the stickwater generated in fish meal and oil plants. In 1968, 32 plants yielded over 143 million lb of solubles with a value of approximately $4 million.[50]

Methods of Manufacture

Fish solubles contain mainly water-soluble substances from stickwater. The stickwater is transferred from the holding tanks to an evaporator, which reduces the water content from 95% to 50%.[54] Solubles are produced in the form of a brown, somewhat viscous liquid with a mild, fishy odor. They may be sold in the feed trade, but normally are dried as press cake for meal.

Animal Feed

The use of whole fish and fish scraps for animal feed has been studied thoroughly. The fresh wastes or whole fish are usually processed into fish meal, fish oil, or cooked and canned pet food. In addition to meal and oil, the Bureau of Commercial Fisheries[50] listed, for the United States in 1968, ten plants producing crushed shells for poultry feed, four plants producing miscellaneous animal feeds, and two plants producing pelletized fish hatchery feed; all using fish wastes and whole fish.

STANDARD WASTE TREATMENT METHODS

The liquid wastes resulting from fish processing are most commonly discharged to adjoining waters. This practice has been restricted in many areas in recent years as plants have consolidated and enforcement of water pollution control regulations has intensified. The resultant action in many cases has been to discharge the wastes to the municipal sewers.

The specific difficulties encountered in the treatment of fish processing wastes are attributable in large part to the characteristics of the wastes. These usually include high flows, medium to high BOD_5 and suspended soilds, and high grease and protein levels (when compared to domestic sewage). The high grease and protein levels probably produce the most serious treatment problems due to difficulty of removal. The frequently short processing seasons, high peak loadings, and rapid biodegradability of the wastes are also important considerations.

Screens

Claggett and Wong[55] studied the effectiveness of screening salmon canning wastewaters. The two specific types tested were rotary and tangential screens. A 34-mesh rotary screen, rated at 100 gal/min, was investigated. Solids were removed with a screw conveyor and blinding was prevented through the use of high-pressure nozzles.

The tangential screen employed two screening surfaces, each 1 ft^2 in area, sized at 20 and 40 mesh. The resulting operating capacities were 35 and 20 gal/min, respectively.

Both screen types were judged to be successful on salmon canning wastes.[55] The results, shown on Table 8, indicate that with the low capital and operating costs associated with screening, a processor could expect removal of about one-half of the total solids in his waste stream.

Centrifuges

Jaegers and Haschke[56] stated that centrifuges can be effectively used to remove fish pulp from waste streams. Fats and proteins can also be recovered by this method.[57] However, centrifuging entire waste streams is very expensive when compared to other methods, owing mainly to high capital costs.

Gravity Clarifiers

Large quantities of fats and greases are present in the wastewaters from the processing of oily fishes (sardines, herring, etc.). Knowlton[36] reporte the fats and grease content of sardines canning wastewaters to be 1000–30,000 mg/liter compared to 50–200 mg/liter in domestic sewage. These organics are present as flotables or as emulsions. When the untreated wastes are discharged, serious problems can result if the emulsified grease coalesces and rises to the surface of the receiving water.[58]

The greases can be removed by two methods in clarifiers: flotation and sedimentation. Flotation will be described in the following section. Sedimentation of the fats and greases is enhanced by various coagulants. Lumprich[11] reported that the application of 2.5 g clay plus 2.5 g lime and 100 mg of ferric chloride per liter gave an optimal precipitation, with a BOD_5 decrease of 75%. A similar procedure was

Table 8 *Solids Removal (g/liter) from Salmon Wastewater by Screening*[55]

Screen	Mesh Size	Raw Waste	Underflow	Overflow
Rotary	34	4.2	2.4	105.1
Tangential	40	4.5	2.5	164

described by Schulz[59] using 2.5 g/liter of Al_2O_3, 2.5 g/liter of lime, and 100 mg/liter of $FeCl_3$. Griffen[60] also mentioned that the high fat and protein wastes can be treated with lime. Chlorination before sedimentation is recommended to prevent serious odor problems from rapid degradation.[37, 59, 60] It should be noted that the coagulant dosages recommended above would lead to sludge volumes of at least an order of magnitude greater than those normally encountered in wastewater treatment practices. Thus, in many cases, these dosages would prove impractical.

Buczowska and Dabaska[62] listed sedimentation results for fish processing wastewaters. In 2 hr of quiescent settling, 32% of the suspended solids were removed with 25% of BOD_5. About 58% of the organic matter in the wastewaters was in solution or colloidal suspension. Limprich[11] stated that 58% of the suspended matter settled out in 2 hr for "fish wastes." The resulting sludge was described as being "very voluminous." Corresponding values for sewage after 2 hr of settling would be approximately 70% suspended solids and 40% BOD_5 removals.[61]

A partially successful gravity clarification system was developed using large quantities of a commercial coagulant called F-FLOK.[55] F-FLOK is marketed by the Georgia Pacific Corporation and is derived from lignosulfonic acid. The floc formed slowly, but after formation, sedimentation rates of 4 ft/hr could be achieved. The summary for a large scale test on salmon wastewaters showed a maximum solids removal of about 70%. The underflow was judged to be quite dilute.

Claggett[63] mentioned that normal detention times in gravity clarifiers may lead to strong odors due to rapid microbial action.

Flotation Clarifiers

The flotation technique relies on the entrainment of minute air bubbles which float particles to the water surface. The resulting sludge blanket is continuously skimmed from the surface. Two methods are used to entrain the air bubbles in the flow, each method having definite advantages over the other.

The first method uses mechanical aerators to "whip" the air bubbles into solution. Dreosti[64] reported that good laboratory results were obtained using fish wastes with suspended solids levels of up to 8000 mg/liter. Higher suspended solids concentrations led to sludges that did not consolidate well on the surface.

For optimum results, Dreosti[64] recommended a small quantity of air for flotation and agitation times of only 1 or 2 sec. Centrifugal pumps could be used if air were bled into the pump chambers. Coagulants improved the removal efficiency; however, no mention was made of types or quantities used. The minimum detention time was estimated to be 5 min.

Hopkins and Einarsson[65] reported on a fish waste treatment installation using a "whip-type" air flotation unit. A flow of 0.065 mg/day was passed through tanks with an "Air-O-Mix" aeration unit. The resulting sludge, including 15.5 lb/day of grease and 35 lb/day of fish solids, was flash incinerated.

The second method involves flow pressurization. The total influent flow or a part of the flow is pressurized and then passed into the flotation unit, which is at ambient pressure. The now-supersaturated solution begins to release air, forming many tiny bubbles. These bubbles then float the suspended solids to the surface. This method requires pressure pumps and containers. However, greater efficiency is usually obtained than with the "whipping" method.

Claggett[63] and Claggett and Wong[8, 55] described in detail pilot scale tests of the flow pressurization method of solids removal from fish processing wastes. Water was pressurized by a centrifugal pump to about 40 psig. Air was added at the rate of about 2% by volume. The pressurization tank had a 1-min detention time. The recovery sludge was heated and the protein and oil fractions removed by centrifuges.

All tests were conducted using a coagulant aid. These aids act by breaking the oil–water emulsions, coagulating small particles, and reducing the solubility of protein fractions. The specific aids tested were alum, ferric chloride, F-FLOK, aluminum hydroxide, Zetol A (trade name for an animal glue), and lime.

In the first tests on salmon processing wastewaters, alum, ferric chloride, and F-FLOK were compared. The partial results in Table 9 show that alum and ferric chloride performed well as coagulants, but large dosages of F-FLOK were necessary to achieve comparable results. Ferric chloride-treated and recovered solids showed signs of extreme oxidation in the oils. In all cases a significant carry-over of the floc was noticed in the effluent.

In the second test on salmon processing wastewaters, precipitated aluminum hydroxide, with and without Zetol A and F-FLOK, were compared. The results in Table 10 show that precipitated aluminum hydroxide was only partially effective. The F-FLOK gave similar results, but dosages over 2000 mg/liter were used, leading to large sludge volumes.

Claggett and Wong[8] concluded that flotation cells could be used effectively on fish processing wastes. Alum treatment was judged the most promising of the methods used. Feeding tests showed that alum

Table 9 *Effects of Flotation with Coagulant Aids on Salmon Processing Wastewaters*[8]

Coagulant Aid	Dosage (mg/liter)	pH	Influent Total Solids (mg/liter)	Effluent Total Solids (mg/liter)	Solids Removal Efficiency (%)	Dry Solids Recovered at 50 gal/min (lb/hr)
Alum	47	6.1	5400	1560	71	11.5
Alum	47	6.0	2290	1200	48	8.5
Ferric chloride	60	5.5	5580	2400	57	11.0
Ferric chloride	60	5.5	2860	950	67	15.0
Ferric chloride	133	4.1	1800	1180	34	10.5
F-FLOK	1000	4.0	5900	1200	63	—

Table 10 *Effects of Flotation with Coagulent Aids on Salmon Processing Wastewaters*[55]

Coagulant Aids	Influent			Effluent			Removal Efficiencies		
	T.S. (mg/liter)	S.S. (mg/liter)	BOD_5 (mg/liter)	T.S. (mg/liter)	S.S. (mg/liter)	BOD_5 (mg/liter)	T.S. (%)	S.S. (%)	BOD_5 (%)
1. 375 mg/liter alum + 75 mg/liter NaOH	2685	640	1775	1505	180	475	44	72	73
2. (1) + 1 mg/liter Zetol A	2441	697	1275	1625	200	331	33	71	74
3. 375 mg/liter alum + 75 mg/liter lime	4268	1993	2833	2162	397	633	49	80	78

Table 11 *Wastewater Characteristics of a Japanese Fish Sausage Plant*[69]

Parameter	Units	Value
pH		6.9–7.1
Total solids	mg/liter	1560–2450
Volatile solids	mg/liter	1120–1900
Suspended solids	mg/liter	320–695
BOD_5	mg/liter	1000–1900
Total nitrogen	mg/liter	70–311

could be included in the recovered solids up to the 1% level without altering the growth rate of chickens.

A method has been developed to remove fish oils down to the 0.008% level (80 mg/liter) by acidification of the waste stream followed by flotation.[66] This method would require neutralization after treatment. Specially coated treatment equipment would be needed to avoid corrosion.

Aerobic Biological Treatment

Buczowska and Dabaska[62] concluded that the carbon-to-nitrogen ratio of fish processing wastewaters indicates that biological treatment should be successful. The biochemcial oxidation rate was said to be similar to sewage, but nitrification begins sooner and is more significant. Assuming primary stage removal of reasonable levels of solids, grease and oils, no special problems should be encountered, the authors said. Without this pretreatment several problems can develop. Matusky et al.[67] mentioned that oil and grease can interfere with oxygen transfer in an activated sludge system. Czapik[68] reported on a trickling filter that clogged because of high solids and oil levels in the wastewaters from a fish processing plant.

A Japanese activated sludge plant has been especially designed for fish wastes.[69] The wastewater flow is approximately 0.41 ft^3/sec (0.27 mg/day) and the BOD_5 concentration ranges of 1000–1900 mg/liter (see Tables 11 and 12). Pilot plant studies were conducted using a 10-hr separation time and the organic and hydraulic loadings listed on Table 12. The results showed adequate removals using conventional biological waste treatment. Bulking occurred when the organic loading rate exceeded 0.31 lb/ft^3/day.

Table 12 Activated Sludge Pilot Plant Results[69]

Parameter	Raw Waste	Effluent Characteristics; BOD loading (lb/ft^3/day) 0.075	0.14	0.21	0.26
pH	6.9	5.5	6.0	6.0	6.2
S.S. (mg/liter)	320	—	—	12	70
BOD_5 (mg/liter)	1000	5	10	13	27
(% removal)		99.5	99.0	98.7	97.3
Total N (mg/liter)	70	—	—	35	51

Anaerobic Biological Treatment

Fish wastes were judged to present no unusual problems in digester operations, assuming that large waste solids are first removed at the processing plant.[5] Matusky et al.[67] stated that fish solids and oil digested readily and the resultant sludge dewatered easily. The digester loading rates varied from 0.1 to 0.36 lb/ft^3/day of volatile solids (VS).

In the system described by Hopkins and Einarsson,[65] the clarified wastewater was effectively treated in a series of septic tanks.

Borchardt and Pohland[70] investigated the feasibility of anaerobic treatment of the stickwater from the wet rendering of alewives. Their laboratory studies led to the conclusion that the wastes could be treated by anaerobic digestion in a municipal system if (1) the supernatant were recycled through the plant; (2) the loading rate didn't exceed 0.05 lb VS/ft^3/day; (3) sufficient acclimation time were allowed; and (4) the ratio of stickwater to domestic sludge were no greater than 1:1.

REFERENCES

1. S. J. Holt, "The Food Resources of the Ocean," *Sci. Amer.*, **221,** 3, 178–194, 1969.
2. H. L. Wilcke, "Potential of Animal, Fish, and Certain Plant Protein Sources," *J. Dairy Sci.*, **52,** 409–416, 1969.
3. C. H. Lyles, "Fisheries of the United States . . . 1968," C.F.S. No. 5000. BCF FWS USDI, 1969, 83 pp.
4. G. Borgstrom and C. D. Paris, "The Regional Development of Fisheries and Fish Processing," in G. Borgstrom, Ed., *Fish as Food,* Vol. III Academic Press, New York, 1965, pp. 301–397.

5. D. Nunnallee and B. Mar, "A Quantitative Compilation of Industrial and Commercial Wastes in the State of Washington." 1967–69 Research Project Study 3F, Washington State Water Pollution Control Commission, Olympia, Washington, 1969, 126 pp.

6. J. W. Slavin and J. A. Peters, "Fish and Fish Products," in C. F. Gurnham, Ed., *Industrial Wastewater Control,* Academic Press, New York, 1965, pp. 55–67.

7. J. W. Slavin and J. A. Peters, "A Survey of Waste Disposal Methods for the Fish Processing Plants on the Oregon Coast," Oregon State Sanitary Authority, Portland, Oregon, 1969, 77 pp.

8. F. G. Claggett and J. Wong, "Salmon Canning Wastewater Clarification, Part I, Flotation by Total Flow Pressurization," Circular No. 38, Fisheries Research Board of Canada, Vancouver, B. C., 1968, 9 pp.

9. F. G. Claggett and J. Wong, "Industrial Waste Survey for Port of Bellingham" (unpublished), Stevens, Thompson, Runyan and Ries, Inc., 1966, 38 pp.

10. J. Foess, "Industrial and Domestic Waste Testing Program for the City of Bellingham" (unpublished), Cornell, Howland, Hayes and Merryfield, Inc., Seattle, Washington, 1969, 17 pp.

11. H. Limprich, "Asbwasserprobleme der Fischindustrie" [Problems Arising from Waste Waters from the Fish Industry], IWL-Forum 66/IV, 1966, 36 pp.

12. W. G. Jones, "Market Prospects for Farm Catfish Production," presented at the Commercial Fish Farming Conference, Univ. of Georgia, Athens, Georgia, 1969, 24 pp.

13. R. A. Grizzell, Jr., E. G. Sullivan, and O. W. Dillon, Jr., "Catfish Farming: An Agricultural Enterprise" (unpublished), SCS, USDA, 1968, 16 pp.

14. M. F. Boussu, "Harvesting Pond-Raised Catfish," presented at the Commercial Fish Farming Conference, Univ. of Georgia, Athens, Georgia, 1969, 11 pp.

15. J. E. Greenfield, "Some Economic Characteristics of Pond-Raised Catfish Enterprises," Working Paper No. 23, Division of Economic Research, BCF, USDI, Ann Arbor, Michigan, 1969, 19 pp.

16. B. Barksdale, "Catfish Farming," *The Farm Quarterly,* **23,** 4, 62–67, 1968.

17. B. Barksdale, "Master Plan for Fisheries: Cultured Catfish Fishery," Region 4 Catfish Committee, BCF, FWS USDI, Ann Arbor, Michigan, 1968, 35 pp.

18. T. J. Billy, "Processing Pond-Raised Catfish," Ms. AA-61 Final Draft, Technological Laboratory, BCF, FWS, USDI, Ann Arbor, Michigan, 1969, 13 pp.

19. T. J. Billy, "Station Experiments with Catfish By-Products," *Feedstuffs,* **41,** 19, 38, 1969.

20. M. E. Stansby, "Processing of Seafoods," in M. A. Joslyn and J. L. Heid, Eds., *Food Processing Operations,* AVI Publishing Co., Westport., Conn., 1963, pp. 569–599.

21. J. A. Dassow, "The Crab and Lobster Fisheries," in M. E. Stansby, Ed.,) *Industrial Fishery Technology,* Reinhold, New York, 1963, pp. 193–208.

22. D. L. Crawford, 1968, personal communication.

23. D. L. Crawford, "Extracts Crabmeat Automatically," *Food Eng.,* **42,** 2, 36, 1970.

24. C. L. Jensen, "Industrial Wastes from Seafood Plants in the State of Alaska," Proceedings of the 20th Industrial Waste Conference, Purdue Univ., Eng. Ext. Series No. 118, 1965, pp. 329–350.

25. C. L. Jensen, "Operations Involved in Canning," Fisheries Leaflet No. 79, in *Research Report No. 7, Commmercial Canning of Fisheries Products.* BCF, FWS, USDI, 1944, pp. 93–111.

26. C. L. Jensen, "Preservation and Processing of Fish and Shellfish—Quarterly Progress Report." Technological Laboratory, BCF, FWS, USDI, Ketchikan, Alaska, 1968, 2 pp.

27. J. Hoalihan, 1969, personal communication.

28. J. A. Dassow, "The Halibut Fisheries," in M. E. Stansby, Ed., *Industrial Fishery Technology,* Reinhold, New York, 1963, pp. 120–130.

29. F. C. June, "The Menhaden Fishery," in M. E. Stansby, Ed., Reinhold, New York, 1963, pp. 146–159.

30. H. C. Davis, "Disposal of Liquid Waste from Fish Canneries, from the Viewpoint of the Fish Canning Industry," *Sewage Works J.,* **16,** 947–948, 1944.

31. S. Lassen, "The Sardine, Mackerel and Herring Fisheries," in M. E. Stansby, Ed.,. *Industrial Fishery Technology,* Reinhold, New York, 1963, pp. 131–145.

32. A. H. Paessler and R. V. Davis, "Waste Waters from Menhaden Fish Oil and Meal Processing Plants," Proceedings of the 11th Industrial Waste Conference, Purdue Univ., Eng. Ext. Ser. No. 91, 1956, 371–388.

33. G. Jordon ["Fish Meal Factories and Their Waste Waters"], *Kleine Mitt. Ver. Wasser, Boden-u. Lufthyg.,* **13,** 308, 1937.

34. K. H. Tanzler ["Waste Waters from Fish Metal Factories"], *Gesundheitsing,* **65,** 157, 1942.

35. L. L. Kempe, N. E. Lake, and R. C. Scherr, "Disposal of Fish Processing Wastes," Office of Research Administration, Univ. of Michigan, Ann Arbor, 1968, 46 pp.

36. W. T. Knowlton, "Effects of Industrial Wastes from Fish Canneries on Sewage Treatment Plants," *Sewage Works J.,* **17,** 514–515, 1945.

37. B. Tetsch ["Separators for Light Material in Waste Waters, Technique With Special Reference to Waste Waters from the Fish Industry"], *Ber. Abwass. Techn. Verein.*, **9,** 278–286, 1954.

38. B. Tetsch, "Maine Sardine Waste Survey," Research Report 2-69, Washington Research Laboratory, National Canners Ass., Washington, D.C., 1969, 6 pp.

39. L. G. McKee, "Planting and Marketing Oysters in the Pacific Northwest." Fishery Leaflet 52. FWS, USDI, 1948, 6 pp.

40. F. Wheaton, "Engineering Approach to Oyster Processing," presented at the Annual Meeting of the ASAE, Purdue Univ., West Lafayette, Indiana, 1969, 15 pp.

41. H. W. Magnusson and W. H. Hagevig, "Salmon Cannery Trimmings, Part I—Relative Amounts of Separated Parts," *Commercial Fisheries Rev.,* **12,** 9, 9–12, 1950.

42. J. Brody, *Fishery By-Products Technology,* AVI Publishing Co., Westport, Conn., 1965, 232 pp.

43. G. I. Jones, and E. J. Carrigan, "Possibility of Development of New Products from Salmon Cannery Waste: Literature Survey," in M. E. Stansby et al., Eds., *Utilization of Alaskan Salmon Cannery Waste,* Special Scientific Report: Fisheries No. 109, FWS, USDI, 1953, pp. 7–35.

44. E. B. Dewberry, "How Shrimps are Canned at a New Orleans Factory," *Food Manufacture,* **39,** 7, 35–39, 1964.

45. F. C. Vilbrandt and R. F. Abernethy, "Utilization of Shrimp Waste," in *Report of Commissioner of Fisheries,* BF Document 1078, 1930, pp. 101–122.

46. F. C. Vilbrandt and R. F. Abernethy, "The U.S. Tuna Industry," *Good Packaging,* **24,** 7, 56–76, 1963.

47. E. B. Dewberry, "Tuna Canning in the United States," *Food Trade Rev.,* **39,** 37–42, 1969.

48. E. B. Dewberry, "Report on Waste Disposal in Pago Pago Harbor, Tutuila, American Samoa," Kennedy Engineers, San Francisco, California, 1964.

49. M. J. Chun, R. H. F. Young, and N. C. Burbank, Jr., "A Characterization of Tuna Packing Waste," Proceedings of the 23rd Industrial Waste Conference, Purdue Univ. Eng. Ext. Ser. No. 132, 786–805, 1968.

50. M. J. Chun, R. H. F. Young, and N. C. Burbank, Jr., "Industrial Fishery Products, 1968 Annual Summary," C.F.S. No. 4950, BCF, FWS, USDI, 1969, 9 pp.

51. F. B. Sanford and C. F. Lee, "U.S. Fish-Reduction Industry," Commercial Fisheries TL 14. BCF, FWS, USDI, 1960, 77 pp.

52. H. Fineberg, and A. G. Johanson, "Industrial Use of Fish Oils," Circular 278, BCF, FWS, USDI, 1967, 16 pp.

53. R. M. R. Smith, 1970, personal communication.

54. S. Lassen, "Fish Solubles," in G. Borgstrom, Ed., *Fish as Food,* Vol. 3, Academic Press, New York, 1965, pp. 281–299.

55. F. G. Claggett and J. Wong, "Salmon Canning Wastewater Clarification. Part II. A Comparison of Various Arrangements for Flotation and Some Observations Concerning Sedimentation and Herring Pump Water Clarification," Circular No. 42, Fisheries Research Board of Canada, Vancouver, B.C., 1969, 25 pp.

56. K. Jaegers and J. Haschke ["Waste Waters from the Fish Processing Industry"] *Wasserw.—Wass. Techn.,* **6,** 311, 1956.

57. K. Jaegers and J. Haschke, "Protein Recovery from Fish Filleting Waste Waters," *Effluent Water Treatment J.,* **9,** 46–47, 1969.

58. R. W. Stenzel, "Treating Oily Waste Water Such as Those from Fish-Canning or Vegetable-Oil Plants," *Chem. Abs.,* **37,** 2500, 1943.

59. G. Schulz ["Purification of Waste Waters from Fish Ponds"], *Wasserwirtsh. Wassertech.,* **6,** 314–316, 1956.

60. A. E. Griffen, "Treatment with Chlorine of Industrial Wastes," *Engineering Contract Record,* **63,** 74–80, 1950.

61. G. M. Fair and J. Geyer, *Elements of Water Supply and Wastewater Disposal,* Wiley, New York, 1958, 597 pp.

62. Z. Buczowska and J. Dabaska ["Characteristics of the Wastes of the Fish Industry"], *Byul. Inst. Med. Morsk. Gdansk.,* **7,** 204–212, 1956.

63. F. G. Claggett, "Clarification of Wastewater Other Than Stickwater from British Columbia Fishing Plants," Technical Report 14, Fisheries Research Board of Canada, Vancouver, B.C., 1967, 8 pp.

64. G. M. Dreosti, "Fish Solids from Factory Effluents," *Fishing News International,* **6,** 11, 53–54, 1967.

65. E. S. Hopkins and J. Einarsson, "Water Supply and Waste Disposal at a Food Processing Plant," *Indus. Water and Wastes,* **6,** 152–154, 1961.

66. A. Aminodan, "Recovery of Oil and Protein from Wastewater," Br. Pat. 1,098,716, 1968, 4 pp.

67. F. E. Matusky, J. P. Lawler, T. P. Quirk, and E. J. Genetelli, "Preliminary Process Design and Treatability Studies of Fish Processing Wastes," Proceedings of the 20th Industrial Waste Conference, Purdue Univ. Eng. Ext. Ser. No. 118, 60–74, 1965.

68. A. Czapik, "Fauna of the Experimental Sewage Works in Krakow," *Acta. Hydrobiol.,* **3,** 63–67, 1961.

69. Magasawn ["The Achievement of Wastewater Treatment in Public Nuisances Prevention Corporation"], *Water and Waste,* **10,** 66, 1968.

70. J. A. Borchardt and F. G. Pohland, "Anaerobic Treatment of Alewife-Processing Wastes," *J. Water Pollution Control Fed.*, **42,** 2060–2068, 1970.

71. M. R. Soderquist, K. J. Williamson, G. I. Blanton, Jr., D. C. Phillips, D. K. Law, and D. K. Crawford, "Current Practice in Seafoods Processing Waste Treatment," Final Report, Project 12060ECF, FWQA, USDI, 1970, 119 pp.

CHAPTER III.16

tobacco recovery by reconstituted sheet processing

WILLIAM J. NELLEN

Although the literature and patents concerned with reconstituted tobacco sheet extend back to the nineteenth century, the large-scale commercial use of tobacco sheet for cigarettes and cigars dates from the early 1950s.

At that time, manufacturers were forced by rising costs to look for economies in the use of raw materials. With the price of tobacco rising sharply, they sought methods of utilizing the factory fines, leaf scrap, stems, and other leaf fragments that could not be used without further treatment and were discarded by dumping or burning at their expense. One procedure was to cut stems, roll them to paper thinness, then shred and incorporate them in blends at relatively low percentages. Smoking taste preferences limited their use to a small percentage of the blend. Reconstitution processes offered the possibility of producing a more acceptable product that could be incorporated in the blend at higher levels.

By 1960, all of the major American cigarette manufacturers were producing or purchasing filler sheet. In 1961 an estimated 100 million lb of filler sheet was used in U.S. cigarettes. During the 1960s, cigarette filler sheet came into commercial use in Europe, particularly in France and Italy. The widespread use of filler sheet in the United Kingdom, Germany, Canada, and other major cigarette producing countries is expected to occur in the next 5–10 years. Current annual production of cigarette filler sheet is about 200,000,000 lb in the United States and 40,000,000 lb in Europe. Tobacco sources are shown in Figure 1.

TOBACCO SOURCES

After picking, tobacco leaf is subjected to numerous processing operations including curing, threshing to separate the stem from the leaf, moisture-conditioning, cutting, blending, and handling between

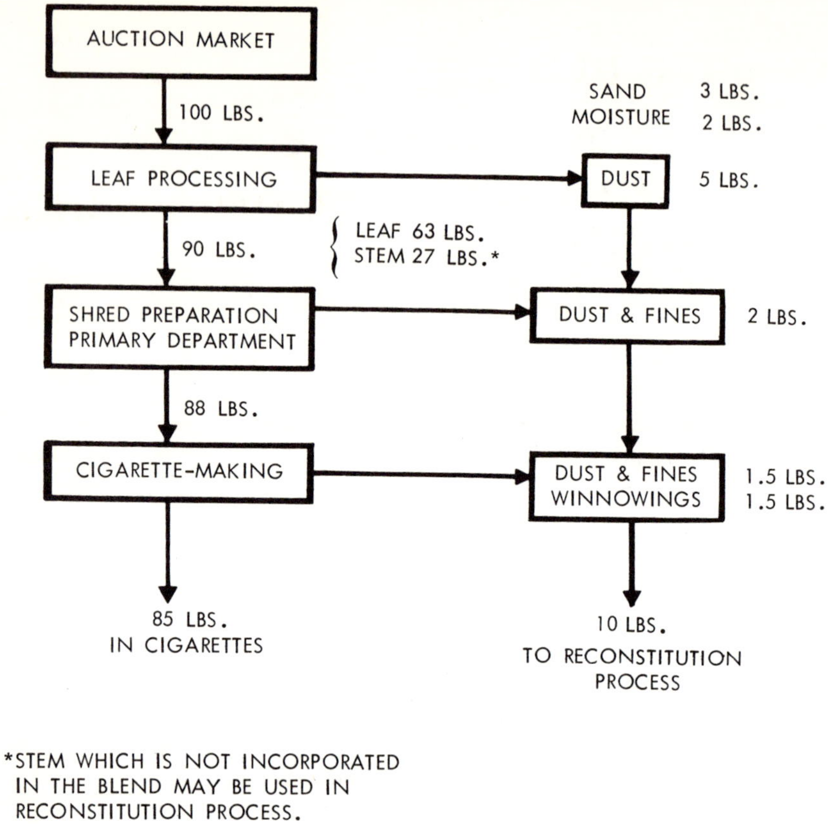

Figure 1. Tobacco sources.

operations. In these operations, tobacco fragments, including dust, fine particles, and short lengths, are formed. These tobacco fragments are of full quality in all respects except for their size; they are not retained in cigars and cigarettes. They can often be used directly in the production of other tobacco products such as snuff and chewing and smoking tobacco, but this provides only a small market for the large volume of tobacco particles generated in the making of cigarettes and cigars. Thus there is the need for reconstitution processes to upgrade the size of the fragments to the point where they are fully usable in the production of cigarettes and cigars.

The overwhelming proportion of tobacco goes into cigarettes. For example, in the United States in 1968, 580 billion cigarettes were produced, compared to only 8 billion cigars. The tobacco required to

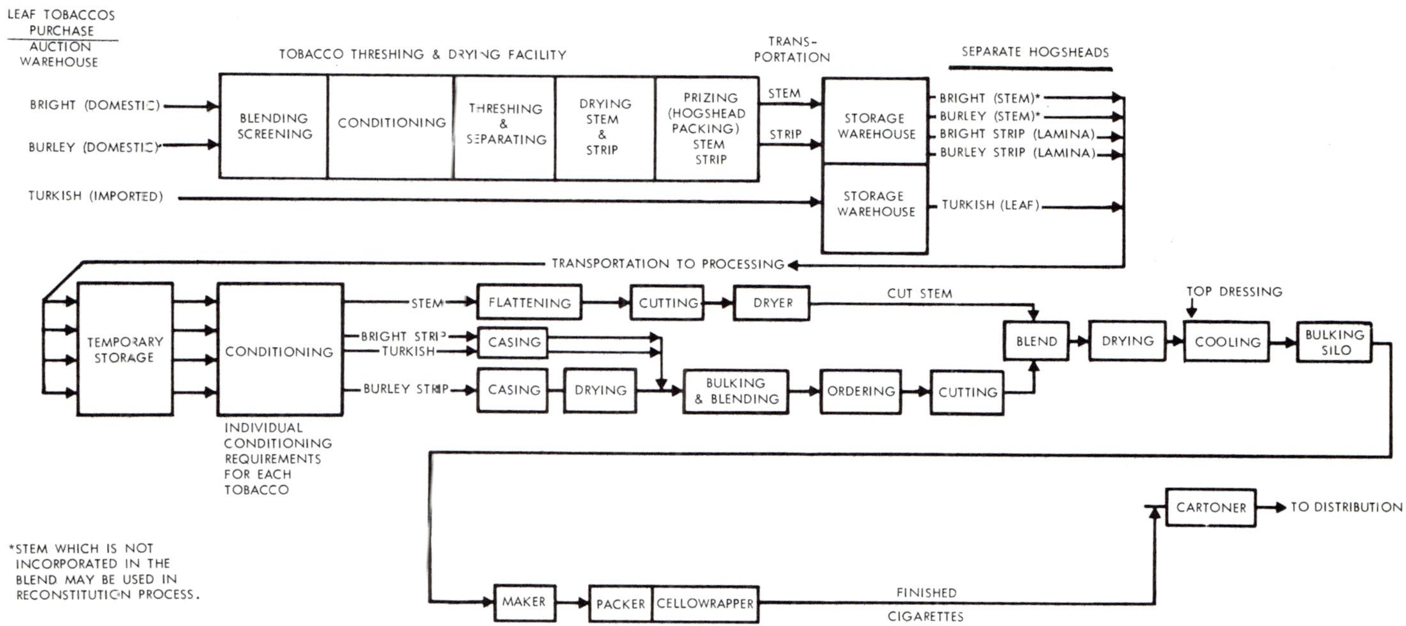

Figure 2. Tabocco processing—cigarette manufacturing.

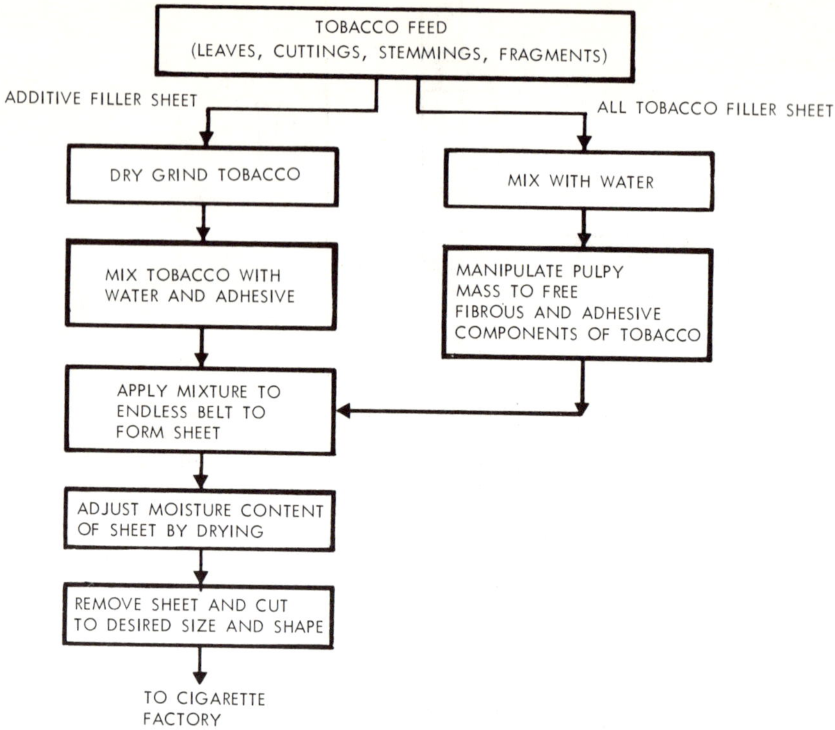

Figure 3. Slurry process.

fill this quantity of cigarettes is about 1.3 billion lb. Thus a discussion of the sources of tobacco raw materials for sheet making in the cigarette industry will be representative of the handling of the bulk of tobacco wastes. A flow diagram of a representative tobacco processing and cigarette manufacturing operation is shown in Figure 2.

Tobacco leaf cured at the farm is brought to market and auctioned in lots of about 200 lb. From the auction the tobacco is transferred into hogsheads, wooden cylinders holding about 800 lb of leaf. The hogsheads are taken to a threshing plant where the stem is separated from the leaf. The leaf and stems are separately packed in hogsheads for aging in storage warehouses until ready for use. Small particles of leaf and stem (dust and fines) are produced in the threshing operation. Sand is also separated from the leaf. The final yield in a representative operation is approximately leaf strip, 63%; stem, 27%; dust and fines, 5%; sand, 3%; and moisture loss, 2% (total: 100%).

After aging, the various types of tobaccos are blended, conditioned and cut into shreds. An additional 2% of dust and fines are obtained from these operations. The shreds are transported to cigarette making machines that produce a continuous cigarette rod that is cut to length. Filters, if used, are attached and the cigarettes are packaged for shipment. In the cigarette making operation, an additional 3% of tobacco

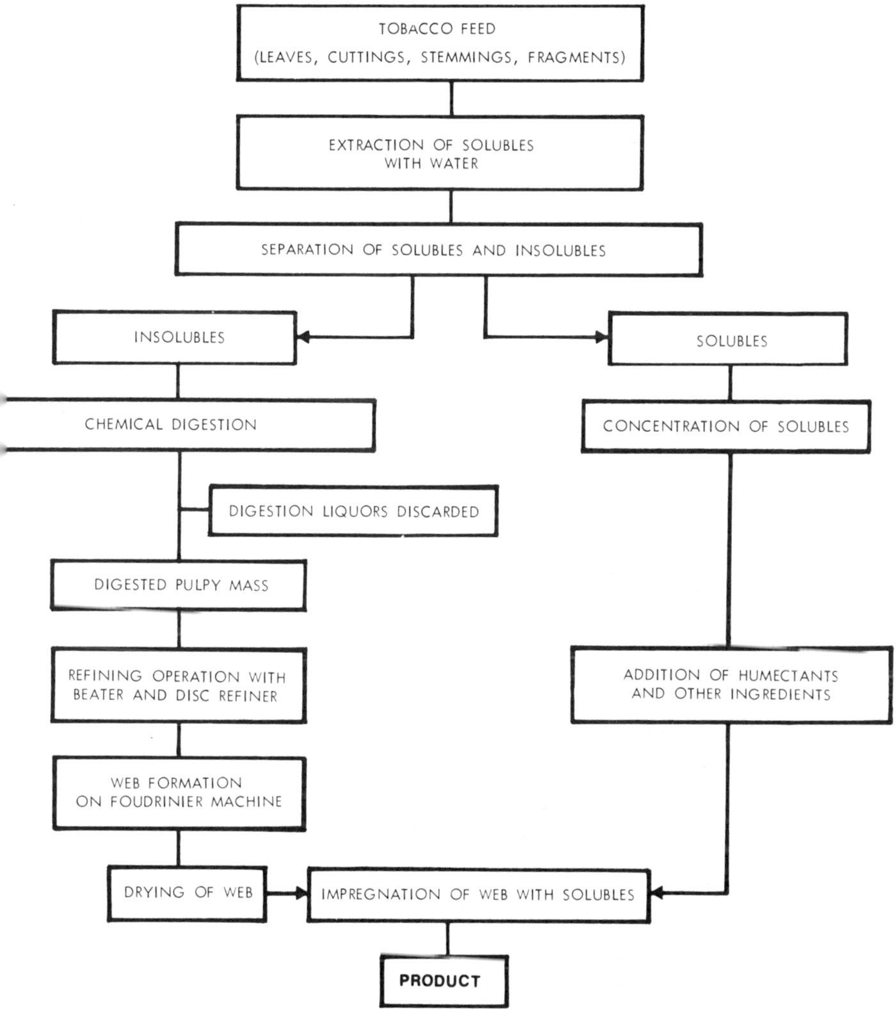

Figure 4. Paper process.

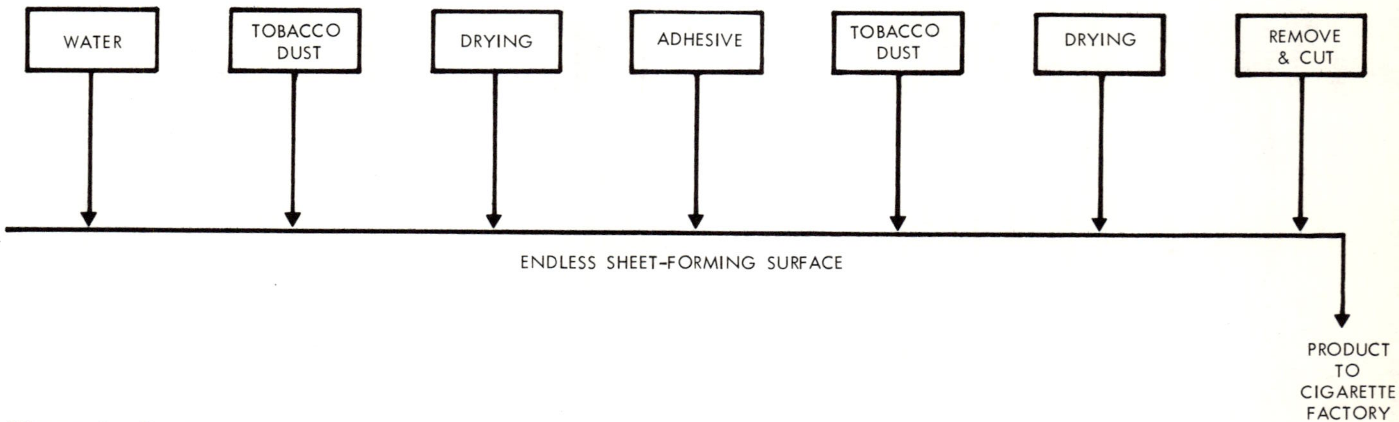

Figure 5. Dust process.

dust, fines, short shreds, and leaf vein pieces known as winnowings are produced.

Figure 1 summarizes the sources of raw materials for reconstituted tobacco sheet.

RECONSTITUTION PROCESSES

A number of processes have been developed to produce tobacco sheet from tobacco fines, leaf scrap, and stems. The major types are slurry process, Figure 3; paper process, Figure 4; dust process, Figure 5; and extrusion or calendaring process. Within each category there are numerous possible variations in process steps, ingredients, and equipment. The brief descriptions that follow are intended only to be indicative of the basic concept of each.

Slurry Process

The slurry process is a major process for the manufacture of reconstituted tobacco sheet. It involves the preparation of a viscous dispersion, or slurry, of tobacco dust, fibers, humectants, and other formulation additives, as necessary. This tobacco slurry is shaped into sheet form, generally by metering it onto an endless stainless steel belt. It is then dried to form a tobacco sheet.

An all-tobacco sheet can also be made by the slurry process. The sheet structure is obtained by manipulating a tobacco-water mixture of pulpy consistency to free the fibrous, adhesive, and humectant components of the tobacco. The resulting slurry is applied to the belt surface, dried and processed into filler-size pieces or shreds. A flow diagram for the slurry process is presented in Figure 1.

In a recent development by AMF Incorporated, a filler sheet having a density of one-half to one-sixth that of normal filler sheets has been offered. A variation of the slurry process is employed to produce this sheet, which is called Microflake* NCP Low Density Sheet. Excellent cigarette filling value is obtained. A plant layout is presented in Figure 6.

Paper Process

The paper process is widely used in the manufacture of reconstituted sheet for cigarettes. Generally, tobacco solubles are separated from tobacco insolubles after soaking in water. Chemical treatment and refining of the insoluble portion produces a fibrous mass suitable for

* Registered trademark of AMF Incorporated.

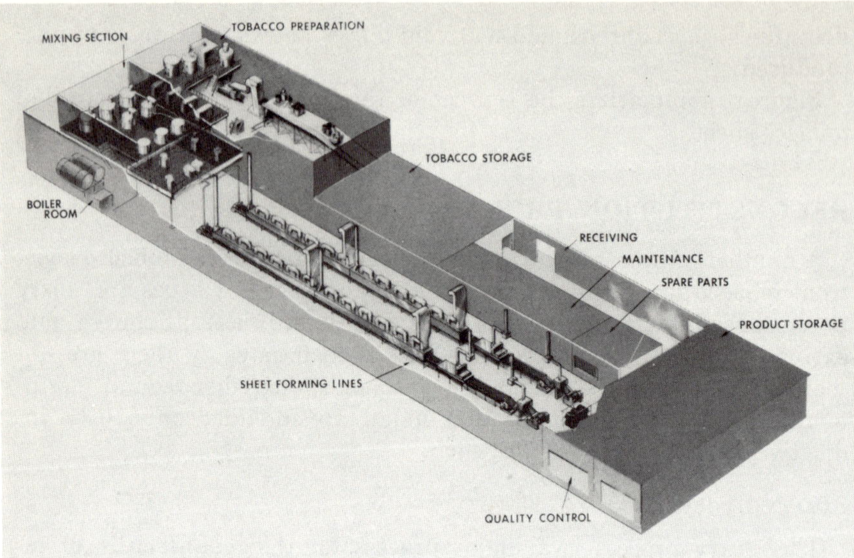

Figure 6. Microflake® NCP low-density plant.

web formation on a Fourdrinier paper machine. The paperlike dried web resulting from this process may be impregnated with the previously separated tobacco solubles, humectants and other formulation additives. A flow diagram of a paper process for tobacco sheet is presented in Figure 4.

Dust Process

One of the earliest commercial filler sheets was the so-called dust process or laminated tobacco sheet.

Dry tobacco is ground to 60–80 mesh and is bonded together by an adhesive derived from tobacco stems. The adhesive is obtained by cooking the stems in a manner similar to that used in paper processing, refining the fibers, and concentrating the natural gums and pectins in the tobacco solubles for use as an adhesive. In the sheet-forming part of the process, tobacco dust is formed in a layer on a stainless steel belt. Upon this layer of tobacco, a layer of adhesive is applied, and a second layer of dust is imposed upon it. The resulting three-component laminate is then dried and cut into pieces for blending. A schematic flow diagram is shown in Figure 5.

Extrusion or Calendaring Process

A number of processes based on extrusion or calendaring of a tobacco sheet have been proposed, but they are currently in limited use. Ground tobacco mixed with an adhesive, either a thermoplastic or solvent-liquefiable, can be extruded or calendared to form a sheet.

CHAPTER III.17

agricultural fibers in the paper industry

DWIGHT L. MILLER and IVAN A. WOLFF

Agricultural fibers were the first plant sources to serve as paper-making raw materials. As early as 500 B.C. pressed production of paper from wood began about 1800 A.D.

Wood, the basic raw material worldwide for pulp and paper, has for many decades dominated the market. Nonwood pulps account for about 3% of total production in the United States and about 5% of world production.[1] The total usage of pulp is so large that this 5% represents more than 7 million tons per year. Some 550 paper mills in more than 50 nations[2] now use nonwood agricultural fibers as raw materials.

Practically all countries have an increasing per capita demand for paper and paper products. The annual per capita consumption in the United States is about 570 pounds. As the developing countries improve their living standards, adopt modern marketing methods, and expand public education, they are consuming increasing quantities of pulp and paper.

Many areas of the world, including a large number of the developing nations, are deficient in the forest resources needed to provide sufficient pulpwood for the pulp and paper industry. Cellulosic fibers comprise a major structural component of most plants, and so nonwood sources represent a reservoir of raw materials for pulp. A sizable number of agricultural fibers (the term is used here to refer to all fibrous materials from nonwood sources) have been effectively used. Many annual plants have been investigated experimentally.[3] Pulp derived from agricultural fibers is useful either alone or in blends because of desirable properties that this pulp can give the end products.

Worldwide demand for paper and paperboard in 1985 is expected to be double that of 1965.[2] Although wood will retain its predominance, increased importance of agricultural fibers as world raw materials is anticipated.[4] Even in the United States, now that the switch from straw

Table 1 *Some Dimensional and Compositional Characteristics of Selected Pulpwoods and Nonwoody Plants Processed for Pulps and Paper*[a]

Species	Fiber length (mm)	Fiber width (μ)	Lumen width (μ)	Cell-wall thickness (μ)	Crude cellulose content (%)	Alpha-cellulose content (%)
Coniferous woods	2.0–5.7	32–44			60	39–45
Englemann spruce	3.5					
Longleaf pine	4.0				58	
Douglas fir	4.5				61	
Deciduous woods	0.6–1.9	38–50			45–62	38–49
Aspen popular	0.4–1.9	10–40			63	
Black gum	1.7–1.8				61	
White maple	1.7				61	
Paper birch	0.8–2.7	20–30				
Nonwoody plants						
Sugarcane	0.3–3.4	9–45	16	5	48–58	30–37
Depithed	—	—			56–63	37–41
Cereal-grain straws	1.1–1.5	9–13			43–54	29–39
Rice straw	0.5–2.5	4–15	3	3.5	43–47	26–36
Bamboos	1.1–3.8	12–22	2–7	4–9	44–62	30–43
Reeds	1.0–1.8	8–20			46–54	36–43
Flax	4 –8.5	10–37				
Esparto grasses	0.5–3.5	7–18	5.6–6.7	2.7	49–50	33–38
Mulberry, bast	2.3–9.9	17		5–8		

[a] Data have been adapted from the following sources:

P. R. Deshpande, *Indian Pulp and Paper,* **11,** No. 1, 51–54, July 1956; No. 3, 154–7, September 1956.

B. B. Robinson and F. L. Johnson, "Abaca—a Cordage Fiber," Bureau Plant Industries, Soils and Agricultural Engineering, USDA Monograph No. 21, Beltsville, Md., October 1956, 130 pp.

F. N. Tamolang et al., *Tappi,* **40,** No. 8, 673, August 1957; **43,** No. 6, 529, June 1960.

Forest Products Laboratory, "Density, Fiber Lengths, and Yields of Pulp for Various Species of Wood," Technical Note 191, Revised, USDA, Forest Service, Madison, Wisconsin, September 1953.

H. J. Nieschlag et al., *Tappi,* **43,** No. 3, 193–201, March 1960; No. 12, 993–998, December 1960.

E. C. Lathrop, T. R. Naffziger, and H. I. Mahon, "Methods for Separating Pith-Bearing Plants into Fiber and Pith," ARS 71-4, USDA Agriculture Research Service, Northern Utilization Research Branch, Peoria, Illinois, March 1955, 49 pp. plus 28 tables and 20 figures.

to hardwood for corrugating medium is accomplished, the usage of non-wood fibers will grow.[5] However, in some localities pulp is entirely from nonwoody fibrous plants.

In the United States usage of agricultural fibers is expected to increase, but at no greater than a 5% average annual rate—half that of the worldwide average.[5] Utilization of agricultural fibers in this country was extensive up through the early 1940s when more than 500,000 tons of straw and another 400,000 tons of other fibrous materials (exclusive of rags and wastepaper) were consumed.[6] Then economic factors caused a gradual decline in use of straw, until now no paper pulp mills in the United States rely on wheat straw as a raw material. Use of byproduct materials such as bagasse and flax straw are expected to continue; research is also being carried out on kenaf.[5, 7]

Typical dimensional and compositional characteristics of selected woods and nonwoody plants are shown in Table 1.[3] Usually, agricultural fibrous raw materials are higher in ash and hemicellulose contents and lower in lignin than either coniferous or deciduous woods. Cellulose contents are comparable. Some authors[8] subdivide agricultural fibers into the following groups: (a) straw and grasses, other than bamboo, (b) canes and reeds, (c) woody stalks with bast fibers, (d) bamboos, (e) leaf fibers, and (f) others. Comparison of chemical characteristics leads to the following conclusions:

1. Lignin content is rather low in bast and leaf fibers, but it is high in bamboos and woody portions of the stems of dicotyledons from which bast fiber has been removed. Straws and grasses have less lignin than canes and reeds, but both groups are lower in lignin content than pulpwoods, bamboos, and woody stem portions.
2. Bast fibers are low in hemicelluloses, whereas straws, grasses, and reeds are high. Deciduous pulpwoods and leaf fibers are higher than coniferous woods, approaching the straws in amount of hemicellulose. Bamboos and woody stems of dicotyledons usually contain less hemicellulose than straw.
3. Bast and leaf fibers, as well as bamboos, are higher in cellulose content than many of the other agricultural fibers. Plant stalks can vary greatly in cellulose, ranging from about 25% to 60% total cellulose and about 15–40% alpha-cellulose.[3]

Annual dicotyledonous plants need not be separated into bast and woody portions before pulping since such a procedure may be uneconomic unless a premium-grade speciality paper from the bast is required. An entire kenaf stalk, including both bast and woody fractions,

Table 2 *Conditions and Results for Rapid Continuous Pulping of Certain Annual Fibrous Materials and Bamboo*

		Pulping Conditions			Results	
Fibrous Material	Process	Na_2O [a] (%)	Time (min)	Pressure (psi)	Yield (%)	K No.
Wheat straw[b]	Soda	4.6	8	75	67	—
	Soda	10.0	8	80	50	—
Rice straw[b]	Soda	9.8	5.5	100	39	4.9
Bagasse[c]	Sulphate	12.0	10	130	52	7.5
Reeds[b]	Soda	13.8	20	130	48	15.0
Esparto	Soda	12.4	20	120	52	6.0
Napier grass[c]	Soda	15.0	30	150	44	15.0
Bamboo[d]	Sulphate	18.2	30	130	45	10.0
Cotton linters	Soda	5.6	18	100	75	2.7

[a] Chemical applied on dry, raw fiber basis.
[b] Uncleaned raw fiber used.
[c] Depithed or cleaned raw fiber used.
[d] *Bambusa arundinacea* from India.

is high in cellulose and provides good pulp yields.[9] Since blending of fibers for paper is customary anyway, both the long bast fibers of kenaf and the shorter woody fibers make individual contributions to sheet properties.

With regard to fiber dimensions, there are again wide ranges within each plant group. The longest fibers are found in bamboos, leaf fibers, and the bast fraction of dicotyledons. These fibers are of the same order of length as those of softwoods, though some bast and leaf fibers may be considerably longer. The straws, grasses, and reeds contain shorter fibers similar to the hardwoods. Shortest of all fibers are those in the woody portions of dicotyledon stems; these fibers contribute to properties of pulp blends but by themselves do not make strong papers. An extensive compilation and interpretation of fiber dimensions have been reported.[10]

Alkaline pulping reagents have been preferred for agricultural fibers; many of these raw materials require the mild treatments for their pulping and also for refining the pulps. Conditions for the continuous pulping of annual fibrous materials are listed in Table 2.[11] Continuous pulping techniques, with compaction, are preferable for agricultural fibers because their large bulk uses excess digester capacity and necessitates too high a liquid–solids ratio in stationary digesters.

USDA economists[12] have calculated that kenaf should be able to compete with corn or soybeans in many farming areas of the South, and could be developed as a papermaking raw material.

CEREAL STRAWS

Cereal straws (wheat, rice, oats, rye, barley, and flax), but particularly wheat straw, have been pulped in various parts of the world since the early part of the nineteenth century.

Before 1950 most straw was pulped by the lime process. This was later replaced by the neutral sulfite process, which gave higher yields and better quality pulp (8). Major use of the pulp was for strawboard, egg-case fillers, and corrugating paper. As late as 1950 more than 650,000 tons of pulp was produced annually from wheat straw (14). Wheat straw is no longer used for this purpose in the U.S. Although straw is available in the field in large amounts, with modern farming techniques the grain is mechanically harvested and recovered in the field and the straw is no longer collected and is not available at low cost at central locations. Collection, transportation, and long-term storage for large, modern, year-round operating pulping plants are uneconomical. Except where special and unique properties are desired, straw cannot compete with wood. Flax straw is an exception and is discussed separately.

Straw is extensively used in certain countries were wood supply is limited and modern mechanized harvesting is not extensive. Historically, as the demand for foods and papers increases, land values become expensive and such agricultural residues as straw become uneconomical.

More than 1 million tons/yr of wheat straw is a waste of cereal production and is plowed back into the soil, although formerly used as a paper raw material. Wheat straw was formerly an important pulping raw material in the United States. The first U.S. paper mill in which wheat straw was the raw material was installed about 1831 in Chambersburg, Pennsylvania. Its capacity was about 1 ton/day. At one time, there were at least 30 plants that used wheat straw as their raw material. As late as 1950, more than 650,000 tons of pulp were produced annually from the straw. Major use was for strawboard, egg-case fillers, and corrugating paper. Wheat straw has no significant market for these purposes now.

The total amount of wheat straw in the fields has not materially

Table 3 *Wheat Straw Board: Plant Size versus Board Cost (Hypothetical Plant)*

Production, tons/day	75	150	225
Capital cost, million dollars	9.04	12.3	15.6
Board cost (Dollars per Ton)			
Straw ($16 per ton)	29.9	29.9	29.9
Manufacturing costs	88.1	65.5	59.3
Fixed costs	50.7	34.8	29.8
Total factory cost[a]	168.7	130.2	119.0

[a] Exclusive of sales and profit.

changed for many years. (The newer wheat semidwarf varieties have had only minor overall effects in most regions, but could be significant in the future.) However, the location and condition of the straw have completely changed. The wheat grain is mechanically recovered in the field and the straw is left behind and distributed on the ground. Straw is no longer available at a low cost in straw stacks at central locations, as collection becomes a separate operation. The straw cost and the supply available in collected form have thus become undependable.

The original straw-based plants were located in the Eastern Midwest. Relocating plants to follow the shift in straw supply westward was not feasible. Therefore, as the pulping plants became obsolete, they were dismantled or were more economically modernized to use readily available wood. Many of these older straw plants were also too small for efficient economical operations.

Wood yields per acre are normally 1–2 tons/yr. However, trees must be grown at least 5–10 yr before they can be harvested. Thousands of acres (particularly in the South) are already being converted to the production of annual crops.

Feasibility studies on straw pulping have been made and published by several states, among them Colorado, Kansas, Nebraska, North Dakota, and South Dakota,[38-42] showing that sites are available where water, straw, labor, land, and basic requirements for large straw pulping plants could be met. Pulping plants do not exist in these regions and locations now. The technical and economic information on straw pulping is plentiful and readily adaptable to local conditions.[2-5, 9, 13-15, 18-22, 25-28, 30-32, 34-37]

There is an adequate supply of wheat straw in several U.S. locations for a pulp plant. Only a fraction of 1% of the over 100 million tons produced annually is collected for commercial use. Adaptation of the combining operation would be required to simultaneously collect the straw in a suitable form (i.e., bales) for transportation to the processing plant.

Only a large, continuous, automated plant is economically feasible. Unit costs increase rapidly below 200 tons/day, and plants with larger capacity reflect some further economic advantage but at a decreasing unit rate.

USDA research on agricultural residues (including straw) was discontinued around 1960. Based on an integrated, continuous straw-pulping and corrugating board plant located in a major wheat production region, Table 3 and Figure 1 give estimated cost in plants producing 75–225 tons/day. The cost for board of about \$168 per ton in a 75-ton/day plant is reduced to about \$130 per ton in a 200-ton/day plant. Capital cost per ton of capacity is also reduced by about 35%. Paperboard wholesale prices are in the same order-of-magnitude as in the 200-ton plant.

The effects of straw price on board price are shown in Figure 2. A change of \$1 in straw-delivered price changes the board price about \$1.86 per ton. This effect is practically the same (regardless of plant size) in a large continuous plant.

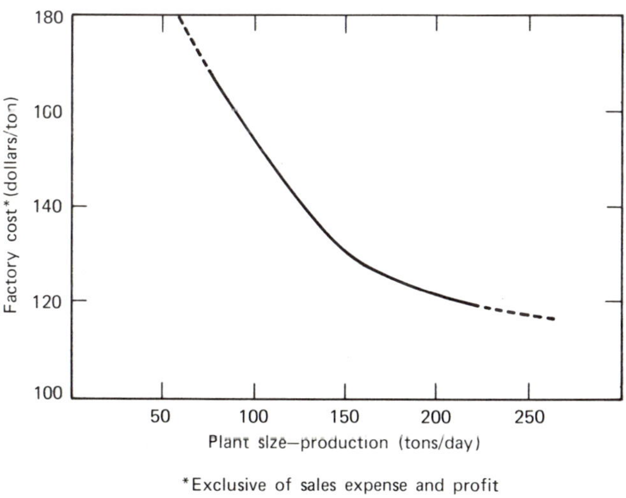

Figure 1. Wheat straw board; plant size versus board cost.

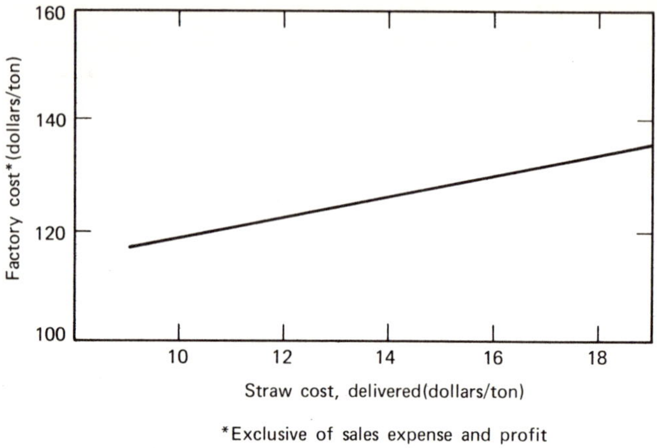

Figure 2. Wheat straw board; straw cost versus board cost (150 ton/day plant).

FLAX FIBERS

Flax straw for papermaking is a by-product of the production of flaxseed for linseed oil. Fiber flax grown for the production of linen is normally too expensive for the production of paper, although a small amount of waste fiber and textile cuttings or scraps are available. Almost all seed flax is grown in the North Central portion of the United States (Minnesota, the Dakotas, and Montana) and in Texas.

Flax straw is 12–15 in. long and relatively strong. The dry straw is baled and shipped to special pulping plants for the production of such speciality products as cigarette paper, Bible paper, carbonizing tissues, scuff-resistant papers, and currency paper. Flax fiber length averages about 25 mm but varies widely.

Only a small portion of the available flax straw is now collected. Of the approximately 500,000 tons potentially available, only about 40% is considered to be of suitable quality; one-third to one-half ton of straw per acre of flax is about normal. The price paid for baled straw delivered at a local collection point is \$15 to \$16 per ton. Approximately \$3.50 per ton is paid for usable flax straw in the field. Two companies process flax straw into paper in the United States.

The United States accounts for about 15% of the world flax acreage (approximately 15,000,000). Other large flax-growing countries are

Canada, Argentina, and India. Some flax is grown on all continents and small amounts of flax straw are utilized for paper in all world regions.

ESPARTO

Two perennial grasses are native to the arid regions of North Africa and southern Spain, *Stipa tenacissima* and *Lygeum spartum*, known as esparto. The rushlike plant has wiry stems up to 3 ft tall that are usually hand-harvested by plucking, to leave the roots for regrowth. The grass grows in clumps and is shipped in bulk or bales.

A British patent granted in 1839 describes the preparation of paper from esparto. Early usage of esparto for fine papers was mostly in Great Britain, where it was introduced by Thomas Routledge in the middle of the nineteenth century. Production was more than 300,000 tons in 1955.[43]

Caustic soda is the preferred cooking chemical. At least two modern mills have been erected for pulping esparto with 13–16% percent NaOH. One in Tunisia[44] has a capacity of 80 tons/day of bleached pulp and uses continuous cooking for 20 min at 120 psi. The other, a 55-ton/day pulp capacity mill in Spain[45] employs three stationary batch digesters with all modern controls.

Esparto fibers are round and fairly uniform; they have a moderate average length of 1.1 mm and a high ratio of length to width. Although esparto paper is bulky, it has high opacity, closed formation, resilience, and good dimensional stability with respect to moisture changes. These characteristics make the paper valued by printers for its flatness, and ink receptivity. New pulping facilities may continue to be placed in Mediterranean areas close to the centers of esparto production. Total availability of esparto is about 500,000 tons/yr.[46]

Esparto has long been the basic paper fibre for the Argentine region, which does not have trees for pulp. Cellulosa Argentina makes all the paper for Argentine money from esparto grass. Commercial stationery has the same basis.

SORGHUMS

Sorghums are high-yielding, coarse annual grasses, under wide cultivation; there are forage sorghums, grain sorghums, broomcorn varieties, and sweet sorghums. They are good crop plants and some seem to have a fibrous character for pulping; as a primary crop for

Table 4 *Comparative Chemical Characteristics of Mature Kenaf, Spruce and Maple*

Characteristics (Moisture-free basis)	Kenaf			Spruce (%)	Maple (%)
	Green		Dry		
	Florida (%)	Illinois (%)	Illinois (%)		
Cellulose					
Alpha, corrected	38.4	32.0	38.2	39.4	39.1
Crude	57.3	48.4	56.1	60.0	60.2
Lignin	14.6	13.9	17.1	27.6	23.3
Pentosans	24.6	21.4	23.1	10.6	22.6
Ash	2.5	3.7	1.9	0.5	0.5
Alcohol-benzene solubles	4.7	5.6	3.3	2.4	2.4
Solubles in 1% NaOH	27.9	39.0	27.8	13.9	15.7

pulping sorghum has been under study in both France[47, 48] and in Hungary.[49]

REEDS

Two common reeds, both in the grass family, are *Phragmites communis* and *Arundo donax.* The *A. donax* reed, a dryland plant native to the Mediterranean countries, is widely distributed through practically all of the warm-temperate and subtropical countries of the world[50]; it has been given consideration as a source of industrial cellulose in Italy and Argentina.[51] The swamp reed, *P. communis*, is a perennial that grows in river deltas and along banks, marshes, and lake shores in many areas of Eastern Europe, Africa, Asia, and South America. The *P. communis* reeds grow to 10–15 ft tall and may produce as much as 4 metric tons of dry matter per acre.[46] In colder areas of Russia, mainland China, and Rumania, mechanical harvest is carried out in winter when the ground is frozen. Elsewhere, harvest is by amphibious vehicles or by hand from small boats. Reeds provide short to medium-fibered pulps that have some of the softness and bulk characteristics of esparto.

Because of the amount of *Phragmites* reeds available as raw material, there are established pulp mills based on the natural stands. Rumania and the U.S.S.R. produce some 200,000 metric tons of reed pulp annually. Four mills have been built in Russia since 1961.[52] In one of these, a continuous digester provides 46–48% pulp yield, from a 38-min cook. Egypt has a small reed pulp mill in which the equipment is the same as used for rice straw. Alkaline reagents (caustic or kraft chemicals) are used for fully cooked pulps. Neutral sulfite may be used for semichemical pulp.

KENAF

Kenaf (*Hibiscus cannabinus*) has traditionally been grown as a textile type of fiber crop in many regions, using the bast fiber, or about 20% of the total dry weight of the stalk. The remaining portions have been discarded.

Through pulping of the whole stalk, kenaf has become a paper-making raw material. Typical comparative chemical characteristics of kenaf and wood are shown in Table 4; comparative fiber characteristics in Table 5 and chemical pulping of kenaf, in Table 6.[53]

Kenaf produces higher annual fiber yields per acre than wood in the preferred areas where it grows in the United States—usually several times higher. Yields up to 10 tons or more of dry matter per acre per year have been obtained.

JUTE

Jute is an annual plant (*Corchorus*) that is cultivated in many countries. Yields are maximum where a warm climate and heavy rainfall

Table 5 *Comparative Fiber Characteristics of Mature Kenaf, Spruce, and Maple*

	Kenaf				
	Green		Dry		
Characteristics	Florida	Illinois	Illinois	Spruce	Maple
Fiber length (mm)					
Bast	2.16	2.32	2.52		
Woody	0.45	0.48	0.51		
Combined	1.15	1.22	1.21	3.16	0.76
Fiber width (μ)					
Bast	13.6	15.7	15.9		
Woody	32.0	30.1	30.7		
Combined	—	—	—	43.8	20.2

Table 6 *Chemical Pulping of Kenaf*[a] *as Compared to Wood*

Property	Kenaf			Woodpulps	
	Green		Dry		
	Florida	Illinois	Illinois	Hardwood	Softwood
Crude yield (%)	58.0	49.1	52.7	52.0	58.0
Screened yield, 8-cut (%)	54.8	46.9	49.6	—	—
Fines (%)	3.1	2.1	3.1	—	—
Beater evaluations (unbleached)					
Characteristics at 600 ml S-R					
Burst factor $(g/cm^2)/(g/m^2)$	63.4	72.0	73.5	25	62
Breaking length (m)	11,800	11,400	11,600	4700	9300
Tear factor $(g/(g/m^2))$	80	94	93	43	120
Folds, Schopper No.	1360	1210	920	100	1600

[a] Cooking 2 hr at 170°C, 21% chemical.

prevail during the growing season; Pakistan, Thailand, and India are the major world producers. The bast fiber of jute is normally separated from the woody portion of the stem and used as a texile fiber, Jute and kenaf fibers are often used interchangeably for burlap, sacking, twine, linoleum, and rug backing. Jute bast fiber may be pulped alone or the whole stalk could be used as a papermaking raw material.

At present most of the jute fiber used for pulping comes from waste textile fibers, such as bags and wrappings, and from new burlap cuttings. Jute fiber normally averages about 2 mm in length, varying from 1.5 to 5.0 mm. Bleached fiber always retains a light yellow color because of the lignin complex.

In general, the physical data and pulping information presented for kenaf also apply to jute. Jute is particularly used for tear- and scuff-resistant papers and for heavy-duty multiwall bags.

HEMP

The hemp (*Cannabis sativa*) stalk, like other annual herbaceous dicotyledons, is composed of a central core or woody portion (hurds) and the outer bark or bast. Usually the long, flexible bast fiber is obtained by water retting the stalks; it is utilized as a textile fiber. Hemp for pulp and papermaking then becomes available to industry as secondary or waste fiber, such as used sailcloth, webbing, sacking, carpeting, cordage, twine, rope, or coarse fabrics. Hemp bast fibers are similar to those of flax. Fiber length averages about 22 mm, varying from 5 to 55 mm.

For pulping, hemp rope is chopped to about 2-in. lengths and the fibers after loosening are pulped with lime or caustic soda (about 20% based on dry fiber). In Europe, hemp fiber has been used for speciality papers for the manufacture of such products as Bibles and cigarettes.

Hemp fiber, like that from flax and jute, imparts durability and high tear resistance when added to furnishes for making heavy-duty, strong papers or light-weight papers of high strength. There is some production and usage of hemp in Europe, particularly in France. In 1964, production was about 4800 hectares, of which about 1500 was for papermaking. In the United States there is minimal hemp fiber that reaches the industry as old rope or the like.

BAMBOOS

Botanically, bamboos are woody stemmed perennial grasses. The stems (culms) are usually hollow between joints (nodes), ranging in height when mature from a foot or two to 100 ft or more. There are 700 or more known species of bamboos. These are usually subdivided into the clump types (tropical) and the running types (found in temperate zones); the two terms refer to the habit of rhizome growth. Like other perennial grasses bamboos may flower, go to seed, and die at unexpected intervals. Such potential loss of raw material from large stands for several years pending re-establishment must be considered, so as not to have a pulp mill become too dependent on a single bamboo source.

In the Orient, bamboo was first used for papermaking many centuries ago. In India it is the major papermaking raw material and accounts for almost three-fourths of that country's pulp production.[54] In 1965, the amount of bamboo pulp produced exceeded half a million tons. Commercial use of bamboo on a lesser scale, in one or several mills, also exists in Thailand, Pakistan, Indonesia, Japan, Taiwan, mainland China, the Philippines, and Brazil. Natural bamboo stands in India reportedly yield 0.5–0.75 ton/acre/yr, but under cultivation, yields of 2–4 tons/yr have been indicated. The Indian bamboo pulps are primarily bleached pulps for printing papers of various types.

Bamboo fibers typically are long, slender, and pliable. They contribute high tear resistance, as well as softness and smoothness, to paper sheets. Bamboo pulps are frequently blended with othe pulps to improve burst and tensile strengths. The softness and flexibility of the bamboo fibers makes the pulp useful for tissue as well as for bond and book papers.

The wall of the bamboo is dense, which aids digester efficiency but requires special effort for effective chipping and penetration of cooking liquor during pulping.

Bamboo is pulped by an alkaline procedure, in a single-stage kraft process. Pulped at an active alkali level of 16–17% (as Na_2O) with kraft chemicals, the yield of unbleached (bleachable grade) pulp is about 48%. The chemical requirement for bamboo is thus somewhat higher than for other agricultural fibers. Most of the Indian pulp mills still use batch digesters, though one or two new ones with continuous pulping equipment are operating.

Bamboo, grown on a limited basis in the United States for more than 50 years, has not been utilized in for pulp.

Several pertinent reviews and recent technical articles on bamboo are available.[54-59]

LEAF FIBERS

Leaf fibers, such as abaca, sisal, henequen, caroa, and phormium, are used primarily for cordage and textiles. The fibers occur in long strands, which are separated by hand operation or machine decortication. For paper the fibers, or old cordage, are first cut or chopped before pulping with kraft chemicals or caustic soda. Pulp yields for the manufacture of papers that require maximum tear strength are high. The strongest bag and wrapping papers, currency, and papers that require high scuff resistance are produced from these leaf fibers. Tea bags and stencil sheets are made from abaca (manila hemp). Only small volumes of leaf fibers are used for speciality papers.

OTHERS

Limited quantities of cotton linters and rags are chemically pulped. Normally, 170–190 lb of linters are obtained per ton of cottonseed. Most of the fiber lengths used for papers are the lowest grade and are less than 4–5 mm long. The linters contain approximately 70–85% alpha-cellulose.

Cotton rags are the basis for most rag papers. The cotton fiber itself in the rags is 10–30 mm and more in length. Primarily rags are wastes or by-products from textile mills.

The quantities of rags and linters used for papers are approximately 500,000 tons/yr, about one-half of which is rags. Chemically processed linters and rags appear mainly in fine-grade writing papers that command a premium price. Low-grade rags often go into roofing paper.

BUSHES AND BRUSH

Recent success in growing annual fiber crops for papermaking raw materials has renewed interest in the farming and pulping of wood bushes and brush. It is too early to determine the economic success of such operations. Such fast-growing woods as the silage sycamore and cottonwood are known to be under investigation by commercial companies.[32] Wood yields two to three times that of trees have been reported. However, such production would require major revisions in current tree farming and pulping practices. This development resembles the growing of nonwoody annual crops, such as kenaf.

REFERENCES

1. E. C. Lathrop and J. H. Shollenberger, "Straw for Industrial Use, Collection Problems and Quality," *Paper Trade J.*, **114,** No. 20, 46–53, May 14, 1942.

2. S. I. Aronovsky, G. H. Nelson, and E. C. Lathrop, "Agricultural Residue Pulps—Bleaching Studies on Straw Pulps," *Paper Trade J.*, **117,** No. 25, 38–48, December 16, 1943, AIC-131.

3. S. I. Aronovsky, J. O. Younger, and G. H. Nelson, "Agricultural Residue Pulps for Corrugating. Effects of Weathering and of Various Cooking Conditions," *Paper Trade J.*, **120,** No. 8, 124, 126, 128, 130, 132, February 22, 1945; *Paper Mill News,* **68,** No. 8, 102, 104, 106, 108, 110, February 24, 1945.

4. J. O. Younger and S. I. Aronovsky, "The Physical Evaluation of Strw and Other Agricultural Residue Pulps. A Proposed Method," *Paper Mill News,* **68,** No. 8, 140–142, February 24, 1945; *Paper Trade J.*, **120,** No. 19, 43–45, May 10, 1945.

5. S. I. Aronovsky, Alan Rhodes, and E. C. Lathrop, "Agricultural Residue Pulps—Comparison with Typical Wood Pulps," *Paper Trade J.*, **124,** No. 13, 49–55, March 27, 1947; *Paper Mill News,* **70,** No. 27, 12, 14, 16, 18, July 5, 1947; **70,** No. 28, 16–18, July 12, 1947; *Paper Ind. and Paper World* **29,** 277–280, May 1947; *Tech. Assoc. Papers, Series* **31,** 299–304, June 1948.

6. S. I. Aronovsky, "Collection and Quality of Straw for Strawboard," *Agr. Eng.*, **28,** No. 7, 309–310, July 1947; Collection and Quality of Straw for the Strawboard Industry, *Paper Trade J.*, **126,** No. 2, 51–52, January 8, 1948.

7. E. C. Lathrop, "Economics of Straw Utilization," *Agr. Eng., 28, No. 7, 308, 310, July 1947; Paper Trade J.*, **125,** No. 8, 59–60, August 21, 1947.

8. E. C. Lathrop, T. R. Naffziger, and R. V. Williamson, "The Small Rural Industry—A Study of the Possibility of Making Insulating Board from Straw," USDA Circ. No. 762, August 1947.

9. S. I. Aronovsky, A. J. Ernst, and H. M. Sutcliffe, "Pulping with Sodium Sulphite to Produce Strawboard," *Paper Trade J.*, **125,** No. 19, 90, 92, 94, November 6, 1947; *Paper Ind. and Paper World,* **29,** 280–281, 286, May 1947.

10. E. C. Lathrop, T. R. Naffziger, and R. V. Williamson, "The Small Rural Industry—A Study of the Possibility of Making Insulating Board from Straw. Part III. Experimental Development—Plant Design and Operation," AIC-170, November 1947.

11. S. I. Aronovsky and H. M. Sutcliffe, "Rapid Estimation of Moisture of Straw in Bales," *Paper Trade J.*, **126,** No. 3, 51–54, January 15, 1948; *Tech. Assoc. Papers, Series* **31,** 572–575, June 1948.

12. "Facts About the Industrial Utilization of Straw," AIC-189, May 1948.

13. S. I. Aronovsky, "The Manufacture of Straw Pulp," *Paper Ind. and Paper World,* Part I, **30,** No. 1, 71–78, April 1948; *Part II,* **30,** No. 2, 244–251, May 1948; *Ind. News,* Part I, **1,** No. 7, July 1948; Part II, **1** No. 8, August 1948; Part III, **1,** No. 9, September 1948.

14. S. I. Aronovsky, A. J. Ernst, H. M. Sutcliffe, and G. H. Nelson. "Agricultural Residue Pulps—Comparison of Pulping Processes," *Paper Trade J.,* **126,** No. 26, 78–83, June 1948; *Tech. Assoc. Papers, Series* **31,** 299–304, June 1948.

15. S. I. Aronovsky, G. H. Nelson, A. J. Ernst, H. M. Sutcliffe, and E. C. Lathrop, "Agricultural Residue Pulps—High-Yield Straw Pulps for Fine Papers," *Paper Trade J.,* **127,** No. 11, 154–162, September 1948; *Tech. Assoc. Papers, Series* **31,** 291–299, June 1948; *Fibre Containers,* **33,** No. 9, 34, September 1948 (summary).

16. E. C. Lathrop and T. R. Naffziger, "Evaluation of Fibrous Agricultural Residues for Structural Building Board Products. I. Methods and Equipment," *Paper Trade J.,* 127, No. 27, 55–60, December 1948; *Tech. Assoc. Papers, Series* **31,** 337–343, June 1948.

17. E. C. Lathrop, "Wheat Straw for Improved Paper and Board Products," *Pulp and Paper Mag. Can.* 50, No. 3, 167–178, February 19, 1949.

18. E. C. Lathrop and T. R. Naffziger, "Evaluation of Fibrous Agricultural Residues for Structural Building Board Products. II. Fundamental Studies on Wheat Straw Fiber," *TAPPI,* **32,** No. 2, 91–96, February 1949, *Fibre Containers,* **34,** No. 5, 60–67, May 1949.

19. S. I. Aronovsky and E. C. Lathrop, "A New Mechano-Chemical Process for Pulping Agricultural Residues," *TAPPI,* **32,** No. 4, 145–149, April 1949.

20. S. I. Aronovsky, "The Place of Straw Pulp in Fine Paper Manufacture," *Chemurgic Dig.,* **8,** No. 7, 4–6, 16, July 1949; "Straw Pulp in Fine Paper Manufacture," *Paper Ind.,* 32, No. 1, 88–90, April 1950; "Straw Paper Pulp," *Econ. Botany,* **4,** No. 1, 84, 1950 (excerpt).

21. E. C. Lathrop and T. R. Naffziger, "Evaluation of Fibrous Agricultural Residues for Structural Building Board Products. III. A Process for the Manufacture of High-Grade Products from Wheat Straw," *TAPPI,* **32,** No. 7, 319–330, July 1949.

22. S. I. Aronovsky, "Manufacture of Cellulose Pulps from Straw," AIC-280, July 1950, *Paper Ind.,* **32,** No. 6, 628, 650, 652, 654, 656, 657, September 1950. Processed: No. 6, National Farm Chemurgic Council, New York, N.Y. (1950 Series).

23. H. M. Sutcliffe, S. I. Aronovsky, R. M. Wilkinson, W. D. Burnham, W. A. Phillips, and E. R. Carpenter, "Improved Straw Pulp for Corrugating Paper," *TAPPI,* **33,** No. 8, 353–357, August 1950.

24. E. C. Lathrop, T. R. Naffziger, and E. R. Stivers, "Boxboard from Wheat Straw to Replace Wood Veneer in Wire-Bound Shipping Containers," *TAPPI,* **34,** No. 4, 145–152, April 1951; Processed No. 6, National Farm Chemurgic Council, New York, N.Y. (1951 Series); "Strawboard to Replace Wood Veneer in Wire Bound Shipping Containers," *Fibre Containers,* **36,** No. 5, 54, 59, 60, 62, 64, 66–68, 70, 71, May 1951.

25. S. I. Aronovsky and G. H. Nelson, "Mechano-Chemical Process—Rate of Delignification," *TAPPI,* **34,** No. 5, 216–222, May 1951.

26. S. I. Aronovsky, "Recent Advances in Pulping Straw in the United States," AIC-324, December 1951; Tropical Wood and Agr. Residues as Sources of Pulp, Chapter **17,** No. 3, 170–181, FAO, Rome, Italy, December 1952; *TAPPI,* **36,** No. 4, 167A–171A, April 1952.

27. E. C. Lathrop, "The Characteristics of Pulp Fibers from Agricultural Residues," AIC-323, January 1952; Tropical Woods and Agr. Residues as Sources of Pulp, Chapter **14,** No. 3, 143–154, FAO, Rome, Italy, December 1952, *TAPPI,* **35,** No. 11, 60A, 62A, 64A, 66A, 68A, November 1952.

28. E. C. Lathrop, "Pulps from Straw and Sugarcane Bagasse for Manufacture of Newsprint," AIC-327, January 1952, *Indian Pulp and Paper,* **7,** No. 1, 57–59, July 1952; *TAPPI,* **36,** No. 1, 142A-144A, January 1953.

29. S. I. Aronovsky, "The American Paper Industry Needs Straw and Bagasse," *Paper Trade J.,* **134,** No. 26, 68, 70, 72, 74, 76, 78, 80, 82, June 27, 1952; *Paper Mill News,* **75,** No. 26, 120–122, 124, 126, 128, 130, 131, 150, June 28, 1952; *Yearbook Am. Pulp Paper Mill Supts' Assoc. 33rd Ann. Ed.,* 213, 215, 217, 219, 221, 225, 227, 1952.

30. G. H. Nelson, L. E. Talley, and S. I. Aronovsky, "Bleaching Studies on Monosulphite Wheat Straw Pulp," *TAPPI,* **35,** No. 7, 301–305, July 1952.

31. S. I. Aronovsky, A. J. Ernst, R. J. Seidl, and R. M. Kingsbury, "Straw Pulp-Wood Pulp Blends for Various Types of Papers," *TAPPI,* **35,** No. 8, 315–356, August 1952.

32. S. I. Aronovsky and E. C. Lathrop, "Pulp and Paper Research at the Northern Regional Research Laboratory," *Paper Mill News,* **75,** No. 38, 92, 94, 96, 126, September 20, 1952; *Indian Pulp and Paper,* **8,** No. 1, 51–55, July 1953; *Papeterie,* **75,** No. 8, 525–527, 529, 531, 533, August 1953; *Papel,* **17,** 17–22, November 1953; *Cellulosa e carta (Rome),* **4,** No. 12, 5–9, December 1953.

33. A. J. Ernst, and T. F. Clark, "Wheat Straw Corrugating Board—An Improved Product," *TAPPI,* **40,** No. 2, 120–121, February 1957; "Research Program Pays Off. Strawboard Improving," *Fibre Containers and Paperboard Mills,* **42,** No. 3, 40, 45, March 1957.

34. T. R. Naffziger, R. S. Matuszewski, G. H. Nelson, and T. F. Clark, "Agricultural Residues as Raw Materials for Viscose-Grade Pulp," *TAPPI,* **42,** No. 7, 609–612, July 1959.

35. A. J. Ernest, T. F. Clark, M. D. Finkner, and H. J. Dutton, "A Critical Study of Short-Period Caustic Pulping of Wheat Straw," *TAPPI,* **42,** No. 3, 235–243, March 1959.

36. A. J. Ernst, T. F. Clark, and I. A. Wolff, "Corrugating Pulp from Wheat Straw by Cold Soda Process," *TAPPI,* **43,** No. 1, 34–37, January 1960.

37. T. F. Clark, J. E. Atchison, A. J. Ernst, E. R. Gremler, and I. A. Wolff, "Continuous Pulping of Wheat Straw for Corrugating Media," *TAPPI,* **43,** No. 11, 934–938, November 1960.

38. Clarence E. Grotheus, "Pulp and Paper from Kansas Straw," October 1949, Sponsored by Kansas Industrial Development Commission, Topeka, Kansas.

39. W. E. Hammond, "Survey of the Present and Potential Industrial Uses of Straw with Special Reference to Nebraska," Nebraska Experiment Station Bull. 401, December 1950.

40. Colorado Development Council, Inc., Denver, Colorado, "Paper Products from Wheat Straw," 1951.

41. R. L. Kristjanson and Winston K. Ullman, "South Dakota Sites for Straw Pulping Plants," South Dakota Experiment Station Bull. 471, March 1957.

42. Fred R. Taylor and Warren Tewksbury, "North Dakota's Potentials for a Straw Pulping Industry," North Dakota Experiment Station Bull. 406, April 1957.

43. "List of Publications and Patents on Pulping, Papers, and Related Subjects," ARS-71-9 (Retyped September 1961), USDA Agr. Research Service, Northern Regional Research Laboratory, Peoria, Illinois.

44. B. G. Watson, "The Search for Papermaking Fibers," *Paper Maker* (*U.S.*), **26,** No. 1, 1, 1957.

45. *Pulp and Paper Int.,* **6,** No. 6, 48, 1964.

46. I. M. Kasser, *Paper Trade J.,* **150,** No. 1, 48, 1966.

47. J. E. Atchison, *Paper Trade J.,* **151,** No. 9, 52, 1967.

48. René, Escourron, *La Papeterie,* **82,** No. 1, 29, 1960.

49. René, Escourron, *Afinidad,* **36,** 239, 1959.

50. Pal Lengyel and Sandor Annus, *Papiripar,* **4,** No. 3, 100, 1960.

51. R. E. Perdue, Jr., *Econ. Bot.,* **12,** 368, 1958.

52. A. W. Wilson, *Pulp and Paper Int.,* **8,** No. 7, 50, 1966.

53. D. L. Miller, *Pulp and Paper,* **42,** No. 8, 59, February 19, 1968.

54. John N. McGovern, *Paper Trade J.,* **151,** No. 5, 33, 1967.

55. R. A. Young and J. R. Haun, "Bamboo in the United States," Agr. Handbook No. 193, USDA, Washington, D.C., June 1961.

56. J. R. Haun, T. F. Clark, and G. A. White, "Fiber and Papermaking Characteristics of Bamboo," Technical Bulletin No. 1361, Agr. Research Service, USDA, August 1966.

57. T. R. Naffziger, R. S. Mutuszewski, T. F. Clark, and I. A. Wolff, *Tappi,* **43,** No. 6, 591, 1960.

58. T. R. Naffziger, T. F. Clark, and I. A. Wolff, *Tappi,* **44,** No. 2, 108, 1961.

59. T. R. Naffziger, T. F. Clark, and I. A. Wolff, *Tappi,* **44,** No. 7, 472, 1961.

ADDITIONAL REFERENCES

J. E. Atchison, *Paper Trade J.,* **151,** 52, 1967.

World Rev. Pulp and Paper, **41,** No. 29, 250, July 17, 1967.

G. H. Nelson, T. F. Clark, I. A. Wolff, and Quentin Jones, *Tappi,* **49,** 40, 1966; cf. also other earlier articles in this series.

F. T. Peterson, *Pulp & Paper,* **35,** 76, 1961.

J. E. Atchison, "Kenaf, A Potentially Important Fast Growing Nonwood for Manufacture of Pulp and Paper," presented at the Annual Meeting, Pulp and Raw Materials Group, American Paper Institute, New York, February 20, 1968.

"Pulp, Paper, and Board, Supply-Demand," House Report No. 573, 85th Congress, 1st Session, USGPO, Washington, D.C., 1957.

Tappi, **51,** No. 2, 118A, 1968.

"Raw Materials for More Paper," FAO Forestry and Forest Products Study No. 6, p. 52, Rome, 1953.

T. F. Clark, S. C. Uhr, and I. A. Wolff, *Tappi,* **50,** 52-A, 1967; cf. also other earlier articles in this series.

R. E. Perdue, Jr. and H. J. Neischlag, *Tappi,* **44,** 776, 1961.

K. W. Britt, *Handbook of Pulp and Paper Technology,* Reinhold, New York, 1964, p. 45.

W. K. Trotter and R. S. Corkern, "Economics of Kenaf for Pulp in the South," presented at the Kenaf Conference, held under auspices of the Technical Association of the Pulp and Paper Industry, Gainesville, Florida, November 1, 1967.

J. N. McGovern, *Chemical 26,* **3,** No. 6, 21, 1967.

K. W. Britt, *Handbook of Pulp and Paper Technology,* Reinhold, New York, 1964, p. 44.

CHAPTER III.18

waste statistics of food processing

H. T. Hudson, engineer in the Solid Waste Management Programs EPA, has surveyed the waste and by-product phases of the food processing industry. The industry's total yearly consumption of raw product is estimated at 33.5 million tons, with roughly 52% being vegetables, 36% fruit, 10% specialties, and 3% seafood.

In processing, the raw product becomes either primary product yield or food residual. The food residuals are essentially gross screenable solids; they do not include such things as dirt and other washings that enter the liquid waste stream without being accounted for. The industry's overall yield is about 72%, with an average residual of 28%.

Knowing the percent residual from the processing of a raw product is important to the food processor because it represents the part of his raw product that will not be marketable as a primary food product. Knowing this percentage allows engineers and planners to estimate the amount of food residual expected from a given plant's operation; They can then properly plan waste systems.

The variation in percent residuals from product to product in the processing of vegetables shows from 5–10% residuals for tomato processing to 60–65% for corn, with white potatoes producing 30–35%. The other, unlabeled vegetable products also have widely varying rates of residual generation, ranging between 5% and 65%. If, for example, the average residual of 26% for all vegetables were used to estimate the residuals from corn processing, the error would be over 100%.

The proportion of all vegetable tonnage processed in the United States varies from tomatoes at about 40% to corn at about 15%. All vegetable tonnage processed includes 14 major types of vegetables and one category of miscellaneous vegetables. The latter includes ten relatively minor products that account for about 7% of all vegetable tonnage processed.

The three major vegetable products are tomatoes, corn, and white potatoes. Overall, the range in percent residuals from fruit is 5–45%, with the average residual at 36%.

Citrus fruit (65% of the fruit tonnage processed) dominates fruit processing. Citrus residuals range between 35% and 40%; pineapple is the only fruit with a higher percent residual. The categories include ten major types of fruit plus miscellaneous fruit. The latter includes seven products that account for only 1% of all fruit tonnage processed.

The range in residuals for specialty food, baby food, TV dinners, soup, stew, spaghetti, is between 0% and 15%, compared with the average residual of 11% for all specialties.

The residuals from food processing become either by-product or waste. About 9.3 million tons of food residuals are generated each year and about 7.3 million tons, 79%, are used for by-products. The remaining 21%, about 2 milion tons, is food waste.

Of the four major product lines, fruit, vegetables, specialties, and seafood, fruit and vegetables respectively contribute 46% and 47% of all food residuals and together account for 93%. The other 7% comes from specialties and seafood, with 4% and 3%, respectively.

More than three-fourths of the residuals used as by-products comes from only three products—citrus, corn, and white potatoes, and all but 3% of the residuals used as by-products are used for animal feed. Other by-products are charcoal, alcohol, oil, vinegar, and fertilizer.

In the fruit category, 95% of the residual comes from only four products—citrus, pineapple, peach, and apple. The remaining 5% is divided among all other fruit products.

The dominance of only a few products also holds true for vegetables. Three products—corn, white potatoes, and tomatoes—generate 75% of all vegetable residuals. The remaining 25% is divided among 12 other types of vegetables.

Most food residuals are used as by-products, mainly animal feed; the remainder is disposed of as solid waste or in a liquid medium. About 84% of the total 2 million tons of food processing waste generated each year is disposed of as solid waste by filling, spreading, or burning.

Nearly all the 1,681,000 tons of solid waste are disposed of by filling or spreading, almost evenly divided between the two methods; burning is a relatively minor disposal method.

The four vegetables (tomatoes, miscellaneous vegetables, corn, and white potatoes) that contribute 40% of the total 1.7 million tons and the four fruit products (peach, apple, citrus, and pear) that account for 25%, together contribute 65% of all food residuals disposed of as solid waste.

Each of the four other vegetables, snap beans, beets, cabbage, and carrots, account for 2–4% of the solid waste. The two remaining categories, namely, all other vegetables and all other fruit, represent a total of 14 products. Eight of them produce 1–2% of the solid waste and six produce less than 1%.

Food waste not disposed of as solid waste is disposed of in a liquid medium. The terminology used here also needs explanation. Disposal in water means dumping into a stream, lake, bay, or ocean without any treatment, although some waste disposed of in this manner is ground up before dumping; in a sewer means a public treatment system; in a pond means a holding or treatment pond; and irrigation is self-explanatory. About 300,000 tons of food waste, 16% of the total 2 million tons, is disposed of in liquid.

The four major product lines are listed in the order of the amount of waste they dispose of in liquid; vegetables, seafood, fruit, and specialties. Nearly all the 319,000 tons are disposed of by water sewer, pond, and irrigation are relatively minor disposal methods. Over half of the total tons is disposed of by water.

Of the total amount of food waste disposed of in liquid, vegetables generate 50%, seafood generates 30%, fruit generates 12%, and specialties generate 8%.

PART IV
animal husbandry

Animal husbandry provides meat and protein feed to man; it generates wastes equivalent to that of two billion people. The methods employed in disposal of this waste, disposal on land directly, or after biologic oxidation, are briefly reviewed statistically in Chapter IV.1. Terminal markets or stockyards and their waste problems are discussed in Chapter IV.2 and abattoirs or meat-packing plants in Chapter IV.3. Here pouch manure of digested and undigested food is a major problem. Many products of agriculture have been discussed from the waste viewpoint and are incorporated in animal feeds. As such they have value and use and pass from the agricultural waste stage to by-products. Animal feeds, discussed in Chapter IV.4 are of the order of more than 35 million tons annually, and at approximately $100 per ton are a multibillion dollar business. Waste management in the poultry industry is handled by Ostrander in Chapter IV.5. The treatment of swine waste in apartment-type confinement areas is described in detail by Muehling in Chapter IV.6. Many lamb and sheep areas have not yet approached the confinement approximations of poultry and swine. They are grazed, with natural forces taking care of waste. Cattle feeding stations have not advanced as have poultry and swine in waste handling. Cattle feedlots are discussed in Chapter IV.1.

CHAPTER IV.1

waste quantities from animal husbandry

The number of cattle on our farms, feedlots, fattening areas, is between 48 and 50 million animals. It is also estimated that there are between 57 and 60 million hogs, more than 21 million sheep, and more than 3 billion chickens and poultry. These animals generate wastes equivalent to that of about 2 billion people; consequently, the farmer is hard pressed to find adequate means for waste disposal.

Ten thousand cattle on a feedlot will produce 260 tons of manure each day, while a poultry operation of 270,000 hens creates 35–40 tons of manure daily.

The origin, collection, and disposal of manure and excreta from animal feedlots has long been a problem for the more than 1000 cattle feedlot operators in Colorado as well as the nearly 200,000 feedlots in the 22 states where cattle feeding is of economic importance. Naturally, manure collects in corrals in the various feedlots, whence it goes back to the soil in the form of fertilizer applied to field crops. The financial burden is not inconsequential and operators are constantly seeking methods to help reduce this cost. Many systems have been tried, but nothing has been found more effective than hauling the manure to a field to be recycled. Biologic oxidation, lagoon treatment, and air oxidation have made little progress.

Animal husbandry wastes are generated in huge amounts because of the concentration of the animals at central production points. Often the fundamental problem is not the quantity but the concentration. In specialized production of poultry farming, there are thousands of birds often with insufficient land for proper application of wastes as fertilizer. Large-scale animal operations with their present means for solid waste disposal put a strain on the soil; it is questionable how much of this waste can be applied to the land without doing any damage.

There does not appear to be positive evidence that application of chemical fertilizers causes water pollution or that animal residues or fertilizers cause pollution. However, throughout the agricultural areas, the streams contain significant amounts of nitrates and phosphates, the former from animal residues and the latter from commercial fertilizers.

The animal population in the United States, with greater and greater emphasis on confinement housing and concentration, has produced a number of unsolved problems (see Table 7, Chapter I.1).

Approximately 110 million cattle and 60 million hogs are currently bred for meat consumption, with large numbers in feedlots. There are 3 billion poultry, with many operations from development of fertile eggs, hatching of chicks and breeding to fryer or broiler size, slaughtering, plucking, and preparation for market, including freezing, on a completely integrated basis. Egg production with 300,000 hens in confinement, with all services automated, are common; there are egg cities of 1 million hens in California.

There are 22 million sheep in large lammeries, complete with midwives and veterinarians. In the midwest, farm operations grow animal feeds, mix them, and feed 300 steers as an operation run only by a man–wife team on 400 acres.

Dog and cat food uses as many tin cans as all vegetable canning and is only outdistanced by beer in cans. There are approximately 21 million cats and 25 million dogs in the United States.

Statistics are unavailable on riding stables, riding horses, mules, or pet birds; It is estimated that there are a million horses in the United States.

On Long Island, in the New York megalopolis, there are six million ducks on ocean coast farms alone. These all have their own waste disposal systems.

The wastes in the form of excreta are more than 2 billion tons/yr. This is quite apart from processing the animals for food. Waste products here go into pet foods, glands into medicine, hides into leather and gelatin, wool into textiles and clothing.

Professor Tainganides of the University of Illinois has developed a table for the parameters of design of disposal systems for wastes, as shown in Table 1.

Despite the long-held beliefs that animal wastes are excellent fertilizers, they do not compete with commercial fertilizers, such as urea, phosphates, nitrates, nitrophosphates, ammonia, and mixtures of these with potassium salts, and the animal wastes need greater effort of labor to distribute them and incorporate them into the soil. The animal wastes are difficult to dry, are highly odorous, and in confined areas are dangerous as they may pollute with dangerous hydrogen sulfide.

When dry they would be roughly equivalent to 1N 1P 0K, meaning

Table 1 *Recommended Engineering Design Parameters for Wastes from Animals Reared in Confinement*

Parameter	Units	Pork Pigs	Hens	Beef Cattle	Sheep	Dairy Cattle
Raw manure per day	% live weight	5.1	6.6	4.6	3.6	9.4
Total solids (TS)	% wet basis	13.5	25.3	17.2	29.7	9.3
	% live weight	0.69	1.68	0.79	1.07	0.89
Volatile solids (VS)	% total solids	82.4	72.8	82.8	84.7	80.3
	% live weight	0.57	1.22	0.65	0.91	0.72
BOD	% total solids	31.8	21.4	16.2	8.8	20.4
	% VS	38.6	29.4	19.6	10.4	25.4
BOD/COD	%	30.7	23.2	17.4	7.8	13.8
Nitrogen	% TS	5.6	5.9	7.8	4.0	4.0
P_2O_5	% TS	2.5	4.6	1.2	1.4	1.1
K	% TS	1.4	2.1	1.8	2.9	1.7

1% nitrogen, 1% phosphorus, and 0% potassium, and would not at all be competitive with a 10N 10P 10K, selling at the same price. The prepared fertilizer is easily spread and incorporated into the soil.

In addition the animal wastes are concentrated in areas where transportation is costly. Even when desiccated, there is a market among home gardeners, florists, and organic converts for less than 1% of the amount available. It has to be returned to the land or destroyed.

A small business grows and expands until it and its production reach corporate size, and then it becomes a major polluter—a story heard all over the country. This situation now exists in the agricultural industry. In the past, farming wastes—from poultry, livestock, and the like—could be scattered over the countryside and forgotten or just neglected. With the present massive agricultural production, however, it is difficult for the farmer to find adequate means for waste disposal. Commercial poultry, beef, and pork operations are all faced with the problem of wastes that contaminate air, water, and land.

Animal wastes also create odor problems. The movement of urbanites to the country has resulted in much debate concerning the rights of the farmer and the people who live near his odoriferous farm. With little or no utilization of this material, odor will, of course, increase as the manure accumulates. Thus waste disposal has become a major problem to the modern farmer.

Large-scale animal operations and their present means for massive solid waste disposal do put a strain on the soil. How much of this waste can be applied to the land without doing damage? Thomas J. Army of the Great Western Sugar Co. states, "Because of economic restrictions in transporting manure great distances, we can expect selected soils and sites to receive inordinately heavy doses of animal wastes." Dealing with wastes in this manner leads to water pollution as well as an odor problem.

CHAPTER IV.2
stockyards

In the 1960s there were approximately 54 public terminal markets, or stockyards which handled 15 million cattle, 2 million calves, 22 million hogs, and 6 million sheep. This is a total of about 20 million animal units on the conventional equivalence of 1 cattle to 3 calves, 4 hogs, and 10 sheep. Stockyards provide a holding place for animals in the process of sale or transfer. Most are moved out within 24 hr after arrival, sometimes after 10 or 12 hr depending on the local customs. Animals sold as feeders and stockers may remain for longer periods, up to several days. Throughout the nation there are between 1700 and 1800 auction markets, where the physical handling of the animals is similar. In auction markets sales occur only once or a few times a week, compared to the continuous operations of the terminal markets. The waste production per animal is the same in both classifications. The median number of animals in the auction market is about 25,000 per year, while in the public terminal markets it is about 15 million.

The waste in these stockyards is the materials collected in the pens, composed of animal manure, plus bedding refuse and hay fed to the animal that is wasted. To this there are added truck cleanings from the trucks delivering the cattle to the yards, lumber from pen rehabilitation, plant trash from offices, hotels, and cafeterias, operated by the largest stockyards, and concrete masonry from repair of pens. The fraction of concrete masonry rubble in the total waste is very small, estimated at about 0.1%, while waste lumber is 1.5% of the total waste. The great bulk of stockyard waste is the pen waste and truck cleanings.

USDA has for many years kept and published statistics on the number of animals of each type handled by each stockyard. From these data the waste may be calculated on the basis of the daily production of waste from each animal.

Some wastes are sold for a nominal price or given away directly for fertilizer use on land. The pen waste, in some cases, after curing in a pile for six months, is hauled by merchant operators to farmers in the area. The stockyards provide handling equipment and charge $1 per load. Specialized trucks with manure spreaders have been developed to apply the material directly on the farmer's land at a charge of about $2.00 a ton for the first 20 miles of transport.

It is necessary that the stockyard be located in an area given to the type of farming that can use the fertilizer. But if the stockyard is located in a dairy region, waste is a severe problem.

It is obvious that wire, concrete, cans, and lumber have to be kept out of the pen wastes sold or taken by the farmers, and that cooperating weather must be vailable to allow the operators to get in to the piles and spread the waste on the fields.

A second method of disposal is to sell or give away the wastes for processing into fertilizer by dehydration and composting. It takes about 4 tons of pen waste to produce 1 ton of processed fertilizer. The return is only nominal since markets are specialized and limited.

A third method of disposal is to haul the pen waste to the dump located at the yards. Farmers haul away about 138 million lb of pen waste directly and about 114 million lb are processed, leaving 187 million lb a year being piled up or sent to dumps.

The problem in the stockyards is similar to that of the cattle feeding stations, and the methods of disposal at these cattle stations is of the same character as those used in the stockyards.

CHAPTER IV.3
abattoirs

In the 1960s there were approximately 1400 meat packers who slaughtered more than a thousand head of cattle or 2000 head of all livestock annually. There was an approximately equal number of meat-packing establishments that did not slaughter.

Slaughter establishments generate quantities of waste, comprising paunch manure and stomach contents. Their meat-packing plants not engaged in slaughtering produced only plant waste.

The paunch manure consists of nearly digested and undigested food, such as hay and stomach juices, obtained from the four stomachs or paunches of cattle and calves. This material is about 85% water and is also high in fat. After draining, it will burn with a greasy, persistent flame even when wet. The stomach contents of hogs consist of undigested corn. All other wastes from packing plants are either liquid or are customarily flushed with water and removed as sewage.

The paunch manure is estimated to be about 1300 million lb/yr, but the total estimated waste from the slaughtering and packaging plants is close to 2400 million lb/yr. In some plants the paunch manure is flushed to the city sewers, but there may be a point at which the large pieces are screened out and the solids dried and burned in an open pile or under burners and incinerators. In some cases where the paunch manure is used as landfill, it constitutes a problem. The solution of the problem has involved incineration, city dumps, and city sanitary landfills, as well as disposal of material through sewage systems. The amount of waste is estimated as between 2000 and 3000 million lb/yr.

Annually more than 137 million steers, cows, pigs, and sheep are slaughtered in packing plants across the country, yielding steaks, bacon, and lamb chops. The number is growing by 5–10% each year.

Meat packing—as distinguished from meat processing, which includes sausage and cold-cut manufacture, drying, and canning—gives rise to waste water with a high biological oxygen demand (BOD), offensive odor and color, and high suspended solids content.

Blood has a BOD in excess of 150,000 ppm. A gallon of blood has the same oxygen demand as the daily wastes from eight people. In all but the smallest privately owned slaughterhouses, it is an advantage to reclaim blood and scrap.

Pollution from packing operations can arise from a variety of sources. Most wastes are biological in origin and yield readily to conventional biological treatment processes with little or no modification.

Packing plants use a catch basin, a holding tank into which cleanup waters are diverted for recovery of scraps and grease. Most such holding basins have a retention time of 20–40 min; they are installed for the recovery and resale of inedible waste products.

Lagoons, air flotation, activated sludge trickling filters, oxidation ditches, anaerobic contacting, all with the sludges transported to landfills, are conventionally found in the waste waster treatment systems.

The Wilson Certified Foods plant in Cherokee, Iowa, packs both hogs and beef on a round-the-clock basis. The city agreed to build a new treatment facility and water supply for the Wilson operation.

Control starts with the unloading operations and continues until the last piece of meat is trucked away. Animal wastes accumulating before cattle or hogs are taken to the killing floor are hosed into floor drains that empty into a collection pit. Manure is screened from the water and carted away to agricultural landfill.

Inside the plant, animals are stunned, stuck, and allowed to bleed to death while they are suspended by their hind legs from an overhead conveyor. Blood is collected in a trough.

Hogs present a problem after bleeding. Hides are not removed from hogs as they are with cattle. Hogs are dipped into vats of scalding water to loosen hair. Mechanical paddles beat the hog and remove hair, which is recovered and hydrolyzed to yield an animal feed supplement.

After hide or hair removal, the animal is eviscerated. Edible organs are segregated from inedible ones and go into pet foods or other byproducts. The paunch—the portion of a cow's multichambered stomach that becomes tripe—contains large quantities of undigested or partially digested food. A 1000-lb cow may have as much as 50 lb of paunch manure. Paunch manure is flushed from the stomach with water under pressure. The waste stream from paunch removal is diverted to a separate collecting basin where paunch manure is reclaimed and sold as high-quality agricultural fill.

Other items are scraps from cutting and boning operations, inedible parts of the carcass such as ears, eyelids, or parts too bony to be of commercial value, and carcasses condemned by government inspectors. Inedible and condemned meat that does not find its way into such by-products as dog food is shredded and processed by cookers to render the grease and sterilize the residue. Inedible greases find uses in soap manufacture, and with dried processed scraps in livestock food. Edible and inedible rendering processes are kept separate as required by law and sterilized "high-grade" material is stored in glass-lined silos awaiting trucking to animal feed compounders.

CHAPTER IV.4
animal feeds

Processed feeds for animals consist largely of by-products or residues from oil seed pressing, animal processing for meat, and the left-over materials from beer, malt liquors, alcohol, whisky, and spirituous beverages, with mill feeds from grains, potatoes, sugar beets, sugar cane molasses, and alfalfa meal. It is a tremendous business with many ramifications, producing about 35 million tons/yr.

Table 1 gives the data for processed feeds and their constituents, Table 2 is a historical statistical summary of prices, and Table 3 gives production stocks and foreign trade in processed feeds. Fluctuations are extreme in the 1972–1975 period.

For each barrel of beer there are about 10 lb of spent grain and ½ lb of brewers' yeast. The bulk of these products is used as an ingredient of animal feeds. Some brewers' yeast is used for human consumption, particularly for baked foods, bread, and the like.

Brewers dried grains amount to 375,000 tons annually or 750,000,000 lb, corresponding to 75 million barrels of beer annually in the United States. Canned beer is the largest user of metal containers.

The breweries in the New York metropolitan areas deliver their spent wet grain to Farmers Feed Company, which passes it over large grates, dewaters it in centrifuges, and dehydrates the grains to very low moisture; the product is then sold as a high-protein animal feed principally to dairy farmers or to feed processing companies for beef cattle and poultry.

The dewatering operation gives a very high BOD extract that is sent to sewers and is subjected to a high sewage tax because of the difficulty in treatment in the sewage disposal plants (see Table 4). The amino acids and vitamin assay are given in Tables 5 and 6.

One study related to the utilization of urea as a protein source by ruminants (Table 7). It is known that to a limited extent proteins can be synthesized from urea in the digestive tracts of ruminants. To the extent that such utilization is possible, urea can be used as a low-cost

Table 1 *Processed Feeds: Estimated Use for Feed (in Thousands of Tons) Average **1964–68**, Annual **1964–70** Estimated **1975***

Feed	Year beginning October							
	Av. 1964–68	1965	1966	1967	1968	1969	Av. 1970	1975
High-Protein								
Oilseed meal								
Soybean meal	10,522	10,274	10,820	10,754	11,525	13,577	14,000	14,000
Cottonseed meal	2,109	2,563	1,755	1,462	2,086	1,794	1,875	2,000
Linseed meal	243	284	248	182	197	182	200	200
Peanut meal	118	108	115	133	135	122	125	150
Copra meal	106	109	90	117	111	91	100	100
Total	13,098	13,338	13,028	12,648	14,054	15,766	16,300	
Animal proteins								
Tankage and meat meal	2,008	1,961	2,068	2,059	2,021	2,012	2,100	2,000
Fish meal and solubles	800	627	827	1,083	835	560	600	600
Commercial dried milk products	250	275	255	250	235	250	655	250
Noncommercial milk products	436	454	425	400	385	400	—	400
Total	3,494	3,317	3,575	3,792	3,476	3,222	3,355	

Grain protein feeds								
Gluten feed and meal	1,487	1,481	1,486	1,514	1,550	1,574	1,600	1,600
Brewers' dried grains	318	304	324	336	333	361	375	400
Distillers' dried grains	429	426	425	447	437	429	450	450
Total	2,234	2,211	2,235	2,297	2,320	2,364	2.425	
Other								
Wheat millfeeds	4,557	4,612	4,499	4,490	4,469	4,634	4,650	4,650
Rice millfeeds	442	395	451	476	494	492	500	500
Dried and molasses beet pulp	1,243	1,153	1,129	1,130	1,523	1,600	1,525	1,600
Alfalfa meal	1,600	1,652	1,599	1,550	1,664	1,540	1,550	1,600
Fats and oils	461	427	486	496	531	545	575	600
Molasses, inedible	3,046	2,860	3,060	3,100	3,310	3,400	3,500	3,500
Miscellaneous by-products feeds	1,100	1,100	1,100	1,100	1,100	1,100	1,100	1,100
Total	12,449	12,199	12,324	12,342	13,091	13,311	13,400	
Grand total	31,275	31,065	31,162	31,079	32,941	34,663	35,480	

Table 2 *Statistical Summary of Prices, Indexes and Price Ratios*

Item	Unit	Season average		1970	1970/71			
		1968/69	1969/70	Mar.	Dec.	Jan.	Feb.	Mar.
Feed grain and hay prices								
Corn, No. 3 Yellow, Chicago	Dol. per bu.	1.17	1.25	1.22	1.52	1.59	1.57	1.56
Price received by farmers	do.	1.08	1.16	1.13	1.36	1.42	1.43	1.43
Oats, No. 1 White, Chicago	do.	0.636	0.624	0.595	—	0.820	0.840	0.778
No. 2 White, Minneapolis	do.	0.641	0.632	0.635	0.723	0.775	0.754	0.690
Price received by farmers	do.	0.598	0.586	0.581	0.650	0.668	0.675	0.658
Barley, No. 3 Minneapolis	do.	1.15	1.08	1.07	1.20	1.24	1.29	1.25
Price received by farmers	do.	0.911	0.874	0.831	0.990	1.00	1.03	1.02
Grain sorghum, No. 2 Yellow, Kansas City	Dol. per cwt.	2.00	2.10	2.01	2.28	2.38	2.36	2.33
Price received by farmers	do.	1.70	1.91	1.87	2.04	2.10	2.16	2.17
All hay, baled, price received by farmers	Dol. per ton	23.60	24.70	24.90	25.00	25.40	25.80	26.00
Alfalfa hay, baled, price received by farmers	do.	23.00	23.90	25.40	25.80	26.20	26.70	26.90
Processed feed prices, bulk								
Soybean meal, 44% solvent, Decatur	Dol. per ton	74.10	78.40	71.80	82.00	80.25	77.50	77.00
49–50%, solvent, Decatur	do.	82.50	86.60	80.10	88.20	86.10	82.40	82.00
Cottonseed meal, 41%, expeller, Memphis	do.	65.20	72.50	64.45	76.70	77.40	73.60	66.80
Linseed meal, 34%, solvent, Minneapolis	do.	68.40	67.90	73.50	64.20	63.90	63.50	59.40
Peanut meal, 50%, S.E. mills	do.	80.00	80.40	76.50	82.25	80.50	76.75	74.20
Tankage digester, 60%, Chicago	do.	94.90	109.50	114.40	105.50	105.00	98.10	102.20
[illegible]	[illegible]	91.60	107.10	109.40	102.75	101.90	92.60	97.90

Fish meal, 60%, New York (bagged)	do.	150.60	191.70	196.00	187.00	182.00	178.75	179.00
Gluten feed, 21%, Chicago	do.	43.70	47.80	45.00	57.00	59.00	50.25	50.00
Brewers' dried grains, 24%, Milwaukee	do.	45.00	48.90	42.60	55.90	60.90	56.50	49.40
Distillers' dried grains, 28%, Cincinnati	do.	55.30	57.90	58.80	66.90	68.25	68.75	64.50
Wheat bran, Minneapolis	do.	38.70	41.80	36.80	53.00	46.90	45.00	54.30
Buffalo	do.	41.60	44.50	39.50	63.90	55.75	48.10	55.70
Wheat middlings, Minneapolis	do.	39.90	42.30	36.80	53.00	47.00	44.75	54.10
Buffalo	do.	44.40	48.30	48.10	64.50	56.75	49.75	57.30
Hominy feed, Chicago	do.	41.10	43.20	34.25	53.30	52.40	43.40	46.70
Molasses blackstrap, New York	do.	23.10	27.40	28.20	25.60	26.70	27.20	27.60
Beet pulp (molasses), Los Angeles	do.	45.50	50.40	54.20	53.90	57.70	57.70	57.70
Alfalfa meal, 17%, dehydrated, Kansas City	do.	42.40	45.30	46.40	48.90	48.90	49.90	50.70
Tallow, inedible prime, Chicago	Cents per lb.	0.058	0.075	0.074	0.072	0.071	0.076	[b]
Urea, 42%, N., Ft. Worth (bagged)[a]	Dol. per ton	71.00	71.00	71.00	70.50	70.50	70.50	70.50
Prices paid by farmers								
Soybean meal, 44%	Dol. per cwt.	5.31	5.52	5.62	5.67	5.74	5.69	5.67
Cottonseed meal, 41%	do.	5.03	5.26	5.40	5.53	5.56	5.58	5.55
Wheat bran	do.	3.48	3.64	3.64	3.99	4.08	4.09	4.11
Wheat middlings	do.	3.59	3.72	3.72	4.07	4.20	4.15	4.18
Broiler grower feed	Dol. per ton	90.00	93.00	94.00	99.00	99.00	98.00	100.00
Laying feed	do.	80.00	83.00	83.00	87.00	88.00	88.00	88.00
Dairy feed, 16%	do.	71.00	73.00	73.00	79.00	81.00	80.00	80.00
Beef cattle feed, 30% and over	Dol. per cwt.	4.65	4.81	4.90	4.98	5.01	4.96	5.00
Hog feed, over 29%	do.	6.36	6.65	6.76	6.80	6.79	6.74	6.77
Alfalfa hay, baled	Dol. per ton	33.60	34.10	35.60	36.30	36.50	37.30	38.00
Price indexes (1967 = 100)								
Feed grains, prices received by farmers	Pct.	93	97	94	110	114	116	116
Eleven high-protein feeds, at principal markets	do.	96	105	100	108	109	105	103

Table 2—continued

Item	Unit	Season average 1968/69	Season average 1969/70	1970 Mar.	1970 Dec.	1970/21 Jan.	1970/21 Feb.	1970/21 Mar.
Five oilseed meals	do.	96	103	96	107	105	102	99
Tankage, meat meal, and fish meal	do.	107	128	131	121	122	116	118
Gluten feed, brewers' dried grains, and distillers' dried grains	do.	95	103	99	114	117	109	106
Feed, prices paid by farmers	do.	96	100	100	108	108	108	108
Livestock and poultry-feed price ratios[c]	Average (March)							
Hog-corn, Omaha[d]	1966–70 = 18.3	18.5	20.4	22.4	10.6	11.0	13.2	12.0
Hog-corn, U.S.	1966–70 = 18.6	18.9	20.8	22.7	11.1	10.7	13.4	11.8
Beef steer-corn, Omaha[e]	1966–70 = 22.5	24.0	23.6	25.8	18.5	19.7	21.8	21.8
Milk-feed, U.S.	1966–70 = 1.62	1.74	1.74	1.73	1.74	1.68	1.67	1.65
Egg-feed, U.S.	1966–70 = 9.1	9.5	9.9	10.2	8.8	8.2	7.4	7.2
Broiler-feed, U.S.	1966–70 = 3.3	3.3	3.0	3.1	2.4	2.6	2.8	2.7
Turkey-feed, U.S.	1966–70 = 4.9	4.7	5.2	5.6	4.6	4.4	4.4	4.3

[a] Price series at Houston discontinued.
[b] Not yet available.
[c] Units of corn or other concentrate ration equal in value to 100 lb of hog or beef steer, 1 lb of chicken, milk, or one dozen eggs.
[d] Based on packer and shipper purchases of hogs and No. 2 Yellow corn; prior to January 1971, No. 3 Yellow corn.
[e] Based on price of beef steers all grades sold out of first hands for slaughter, and No. 2 Yellow corn; replaces Chicago.

Table 3 *Production, Stocks, and Foreign Trade in Processed Feeds, Foreign Trade in Feed Grains, and Corn and Barley Processing, Specified Months*[a]

Commodity	Production					Stocks[b]				
	1970	1970–71				1970	1970–71			
	Feb.	Nov.	Dec.	Jan.	Feb.	Mar. 1	Dec. 1	Jan. 1	Feb. 1	Mar. 1
	(1000 tons)					(1000 tons)				
Processed feeds										
Soybean meal	1388.9	1550.4	1606.0	1605.4	1433.1	154.3	182.7	129.7	185.5	191.1
Cottonseed meal	213.4	219.2	218.2	215.8	203.3	108.6	82.9	85.8	103.2	123.6
Linseed meal	20.6	24.6	25.0	28.6	25.8	2.0	12.1	[c]	9.2	15.1
Peanut meal	13.2	13.5	16.9	17.1	16.2	3.2	4.1	3.8	4.2	4.7
Meat meal and tankage	154.8	168.3	179.1	167.2	157.1	—	—	—	—	—
Fish meal and solubles	5.6	15.1	9.2	5.6	4.1	—	—	—	—	—
Brewers' dried grains	25.6	25.1	27.2	26.8	24.9	4.1	6.9	5.3	5.3	5.1
Distillers' dried grains	39.0	35.7	33.5	33.7	35.6	3.2	2.2	1.7	1.6	2.4
Wheat millfeeds	356.5	361.2	361.2	361.3	345.3	—	—	—	—	—
Rice millfeeds	33.5	41.8	36.8	35.5	34.9	—	—	—	—	—
Alfalfa meal, dehydrated	12.0	26.2	11.5	12.2	12.3	300.8	682.8	518.9	385.2	263.8
sun-cured	27.0	27.2	30.1	29.4	27.2	11.6	9.2	12.2	8.7	10.6
Total	39.0	53.4	41.6	41.6	39.5	312.4	692.0	531.1	393.9	274.4
Inedible tallow	42.5	40.8	49.0	45.1	42.6	—	—	—	—	—

Table 3—continued

	Foreign trade									
	Exports					Imports				
	1970	1970–71				1970	1970–71			
	Feb.	Nov.	Dec.	Jan.	Feb.	Feb.	Nov.	Dec.	Jan.	Feb.
	(1000 tons)					(1000 tons)				
Soybean meal	287.0	434.3	353.0	438.7	310.2	—	—	—	—	—
Cottonseed meal	0.4	0.7	0.1	2.2	11.1	0	0	0	0	0
Linseed meal	1.2	9.4	8.4	0	0	0.2	0.1	0.1	0.2	0.1
Other oilseed meals	3.8	5.9	7.5	12.5	11.1	3.2	0.3	0.1	0.1	0
Fish meal	—	—	—	—	—	21.4	6.0	22.3	33.3	15.8
Molasses	7.4	0.5	0.5	0.7	1.1	154.1	142.6	241.1	187.0	107.8
Corn by-products	36.3	63.1	36.8	27.7	30.7	—	—	—	—	—
Wheat millfeeds	[d]	[d]	4.3	0.1	1.2	17.9	11.0	10.2	18.2	10.6
Alfalfa meal, dehydrated	23.1	29.0	12.3	18.9	—	—	—	—	—	—
sun-cured	8.3	24.5	13.9	12.2	—	—	—	—	—	—
Total	31.4	53.5	26.2	31.1	—	—	—	—	—	—
Mixed poultry feed	10.6	13.4	13.0	10.3	12.0	—	—	—	—	—
Mixed dairy feed	19.1	2.7	5.7	16.9	17.8	—	—	—	—	—

	(1000 bushels)					(1000 bushels)				
Feed grains (including products)										
Corn	51,562	48,854	50,269	40,241	44,426	226	214	1855	245	172
Grain sorghum	11,809	10,762	19,467	18,716	18,626	0	0	4	0	0
Oats	735	5969	1265	286	505	170	110	127	167	101
Barley	190	6681	6324	5932	8656	238	433	388	80	200
Total (1000 s. tons)	1791	1925	2125	1798	1981	15	18	63	11	11

Grain processing or use

	1970		1970–71				
	January	February	October	November	December	January	February
	(1000 bushels)						
Corn							
Distilled spirits & alcohol	3073	3023	2422	2182	2135	2587	[e]
Grain sorghum							
Distilled spirits & alcohol	214	210	427	375	353	268	[e]
Barley							
Malt liquors	7796	7648	8308	7481	7891	7709	[e]
Distilled spirits & alcohol	655	599	478	415	406	499	[e]

[a] Data compiled from reports of the Bureau of the Census; Consumer and Marketing Service; and Fish and Wildlife Service, Department of Interior.

[b] At processors' plants.

[c] Not available.

[d] Less than 50 tons.

[e] Not yet available.

Table 4 *Composite Proximate Analysis and Mineral Content*

H_2O	7.0%	P	0.55
Protein	27.9	K	0.62
Fat	7.4	Ca	0.30
Fiber	12.6	Mg	0.19
Ash	3.8	S	0.14
Nitrogen-free extract	4.3	Fe	293 ppm
		Cu	24 ppm
		Mn	40 ppm

Table 5 *Amino Acid Composition (Percent)*

Lysine	0.72
Arginine	0.79
Histidine	0.47
Ammonia	0.82
Aspartic	1.89
Threonine	0.77
Serine	2.12
Glutamic	6.51
Glycine	1.20
Alanine	1.84
Valine	1.69
Methionine	0.57
Isoleucine	1.37
Leucine	3.24
Tryosine	0.94
Phenylalanine	1.82
Cytine	0.39
Tryptophane	0.24

Table 6 *Vitamin Analysis*

Thiamine	0.34 mg/kg
Niacin	46.4 mg/kg
Riboflavin	1.27 mg/kg
B_6	1.72 mg/kg
Panthothenic acid	5.35 mg/kg
Choline	2.11 g/kg
Folic acid	0.15 mg/kg
Biotin	0.075 mg/kg
B_{12}	3.74 mEq[a]/kg
Paba	4.75 mg/kg
Carotene	0.64 IU
Vitamin E	65.1 IU

[a] Miligram equivalents.

Table 7 *Composition of Rations for Urea Utilization Study (lb/**1000** lb Feed)*

Ingredient	Dehydrated Alfalfa Meal	Distillers Grains with Solubles	Brewers Grains (95%) Brewers Yeast (5%)
Cane molasses	140	140	140
Alfalfa meal (17% dehy.)	510	—	—
Distillers grains with solubles	—	540	—
Brewers grains plus yeast	—	—	540
Urea (45% N)	200	170	170
Dicalcalcium phosphate	105	105	105
Iodized salt	35	35	35
Vitamin-T.M premix	10	10	10
Total	1000	1000	1000

Table 8 Effect of Grains on Urea Utilization

Item	Alfalfa Meal	Distillers Grains	Brewers Grains Plus Yeast
No. steers	24	24	24
Daily gain (lb per animal)	2.50	2.39	2.40
Dry feed (lb/100 lb gain)	718	724	727
Total daily N/100 lb gain	478	480	482

protein source in rations. Dehydrated alfalfa meal and distillers grains with solubles improve the efficiency of such utilization.

Feed results over a six month period are given in Table 8.

There was no significant difference in the gains (or feed efficiency) of steers fed a supplement containing either dehydrated alfalfa meal, brewers grains with yeast, or distillers dried grains with solubles. All three of these products can be used effectively in high-urea supplements for beef cattle.

CHAPTER IV.5

waste management in the poultry industry

CHARLES OSTRANDER

Waste management has become a very important part of the poultry industry, primarily in the egg production phase, it is becoming more of a problem with meat production and processing.

In the 1950s there were more laying hens in the United States than today and at that time there was no waste problem. The poultry industry has developed and used new technology faster than any other type of agriculture. For many years almost every farm had a flock of chickens that consisted of a few hens for home consumption to a few hundred for small income or to use up surplus grains and waste products from the farm. These were housed in a part of the barn or a small poultry house, with litter consisting of straw or poor quality hay to absorb the droppings. When the litter was cleaned out, once a year to several times a year, it was placed on the home vegetable garden or the "back 40." There were few people living in the country and most of these were farmers; therefore, no one complained if odors occurred occasionally. At one time there were approximately 400 million laying hens scattered throughout the country.

Because of better breeding, nutrition, housing, and management, there is now a considerably increased population of over 325 million layers. However, due to competition and low margins for eggs, most of the farm flocks have disappeared. Eggs are now produced in fewer but larger operations. The poultry farm today ranges from 5000 layers to well over 1 million and the 5000 bird farms are either getting larger or going out of business.

At one time most state and local health regulations stated that a condition had to be harmful to health to be prohibited in an area. Today this has changed and many of these now refer to enjoyment of life. The

New York State Air Pollution Control Laws, as an example[1] state as follows:

> § 1267 (4) 'Air pollution' means the presence in the outdoor atmosphere of one or more air contaminants in quantities, of characteristics and of a duration which are injurious to human, plant or animal life or to property or *which unreasonably interfere with the comfortable enjoyment of life and property* throughout the state or throughout such areas of the state as shall be affected thereby; excluding however all conditions subject to the requirements of the labor law and industrial code.
>
> (6) "Air contamination source" means any source at, from or by reason of which there is emitted into the atmosphere any air contaminant, regardless of who the person may be who owns or operates the building, premises or other property in, at or on which such source is located or the facility, equipment or other property by which the emission is caused or from which the emission comes.

One of the real problems is that odors are subjective; there is no definitive means of measuring odors. We can identify the volatile

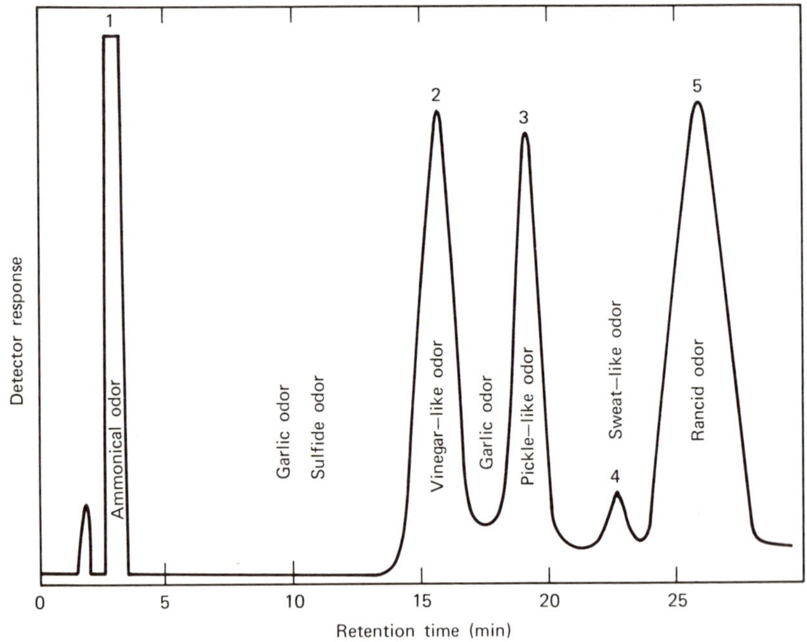

Figure 1. Liquid manure volatiles; organic components.

substances and the quantity of each with the gas chromatograph, but this tells us very little about the odor from each[2] (Figure 1). It appears that many of the malodors from poultry manure result from a combination of gases mixed together. By a "sniff tube" on the gas chromatograph we can detect a different odor for each of the peaks, and these have been given an identifying description. Many of these are not too objectionable in themselves but when mixed they are very unpleasant.

Sense of smell varies considerably between individuals; therefore, what may be satisfactory to one person may be very objectionable to another. This makes it very difficult to mediate a complaint. In fact an individual's sense of smell may vary depending on what he has been eating and drinking, and whether he has been smoking or has been exposed to an odor. However, by screening, "sniff panels" can be developed that are reasonably accurate and consistent. At Cornell University a system has been developed whereby sniff panels can rate the odors from 1 to 8 (Figure 2) depending on the similarity to strong manure odors.

QUANTITIES AND PHYSICAL PROPERTIES

A chicken (4.5–5 lb) will produce approximately 0.25 lb of 70% moisture manure each day. This is approximately 0.066 lb of total solids per day (Hart[3]). This amount appears insignificant until one considers there are some 325 million laying hens in the United States. These are replaced by young pullets every 16–24 mo with an average of every 18–20 mo. The broiler and fryer industry produces approximately 2.8 billion birds per year; the turkey industry approximately 110–115 million; the duck industry approximately 12–15 million each year, plus squabs, fancy birds, and hobby poultry.

It is estimated that the chicken industry produces approximately 5.8 million tons of manure each year, therefore a conservative estimate for the total poultry industry would be 7–8 million tons/yr.

Most broilers are still produced on litter, with the manure and litter ranging from 25% to 35% moisture. On the other hand a very large percentage of the egg-producing flock is now housed in cages with the manure, as defecated, ranging from 70 to 80% moisture.

Table 1 lists many factors to aid one in figuring quantities, nutritive value, and BOD of poultry manure. Table 2 lists physical, chemical, and biological properties of freshly excreted poultry manure.[4]

Poultry litter is variable owing to dilution, time, and concentration of birds. Cage manure is much more constant in composition and liquid manure is least constant (Table 3).

Study ____________________ Name __________________________

Day ____________________ Date __________________________

Rate the samples as to the presence of odors and the odors as to offensiveness according to the following scale using sample "0" as having an 0 rating.

Describe the odor of each sample by giving an appropriate descriptive term. *Possible* terms that might be used are given in the list below or you may use a term of your choice which you feel properly describes the odor.

Presence		Offensiveness	
No odor	— 0—	No offensive odor	Mold, musty
	1		Fish
Very faint	— 2—	Very faint offensive odor	Stagnant water
	3		Sulfide, rotten eggs
Faint	— 4—	Faint offensive odor	Petroleum
	5		Earth
Definite	— 6—	Definite offensive odor	Yeast
	7		Ammonia
Strong	— 8—	Strong offensive odor	Grain, feed
	9		Sour, fermented
Very strong	—10—	Very strong offensive odor	Rotten cabbage, mercaptans

Presence Rating	Offensiveness Rating	Sample	Odor Description
________	________	1	____________
________	________	2	____________
________	________	3	____________
________	________	4	____________
________	________	5	____________
________	________	6	____________
________	________	7	____________
________	________	8	____________
________	________	9	____________
________	________	10	____________

Figure 2. Odor rating. Adapted from Burnett.[2]

Table 1 Poultry Manure

Amount produced

$0.003–0.005\ ft^3$/day per layer

Approximately 0.4 ft^3 per 100 layers or 4 ft^3/day 1000 layers.

Approximately 25 lb/day (as defecated) per 100 layers.

Moisture content 75–80% as defecated.

Littler may range from 20% to 40% moisture.

1,000 layers produce approximately 250 lb/day of wet manure
10,000 layers produce 2500 lb/day wet manure
100,000 layers produce 25,000 lb/day wet manure

Human sewage averages about 0.18 lb BOD/capita.

Chicken manure approximates 0.015 lb BOD per hen per day.

Seven to 12 layers = one human in pollutional load.

50–80 gal of water used per person per day in sewage

100,000 layers approximately equal to a city of 10,000 people in sewage load.

Fertilizing value of poultry manure varies considerably depending on age, how it is stored, moisture content, etc.

Poultry manure loses nitrogen very rapidly.

Various figures have been given for the NPK values for poultry manure (in percent)

N	P	K
1–3	0.5–1.5	0.5–0.8

In England dried cage manure runs as high as 5.85% N, 4.02% P, and 1.47% K.

One set of figures shows the following production (in pounds) of NPK per 1000 lb live weight per year.

	N	P_2O_5	K_2O
Dairy cow	160	40	120
Steer	90	20	70
Hog	150	100	120
Sheep	120	40	125
Hens	85	40	35

With N at 10¢, P_2O_5 at 12¢, and K_2O at 5¢ per pound this is approximately 6.5–7.5¢ per hen per year.

***Table 2** Poultry Manure Analysis*

Property	% of Wet Manure	% Of Dry Matter
Moisture content	75–80	
Total solids	20–25	
Volatile solids	15–18	75–80
Fixed solds (ash)	5–7	20–25
Crude protein		36.87
Total carbohydrate		34.39
Lipids (petroleum ether extract)		1.3
Amino acids, resins, alkaloids, tanins (alcohol soluble fraction)		9.6
Pectin, hemicellulose, cellulose, polysaccarides (alcohol insoluble fraction, corrected for ash)		62–75
Calcium (as Ca)		9.3
Phosphorus (as P)		1.5
Sulfur (as S)	0.3	
Potassium (as K)		0.7
Total Nitrogen		5.9
Organic Nitrogen (% of total N)		88.2
Ammonia Nitrogen (% of total N)		11.8
Ammonia Nitrogen		0.60
Uric acid (7.8% N)		3.7–10.1
Uric acid nitrogen		0.29–0.78

Property	*Value(s)*
Particle density (gm/m^3)	1.75–1.85
pH	6.9–7.4
C/N ratio	8–12
BOD (mg/kg)	30,000–35,000
COD (mg/kg)	150,000–160,000

Poultry manure differs from other livestock manure in that the urates are part of the total mass rather than being excreted separately. It also contains considerable calcium (Table 4) and considerable fine and dense materials. This is of importance if storing or holding liquid poultry manure for a period of time.

It was once thought that the stored mass should be agitated frequently to keep solids in suspension. This was found to be undesirable because every time the mass is agitated the particle sizes are separated and the dense particles settle to the bottom. This results in a very dense mass at the bottom of the storage tank. This reaches a point at which it is practically impossible to move, thereby reducing the capacity of the storage tank. Because of this, it is advised to place the

Table 3 *Values for Various Forms of Poultry Manure*

	N	P	K
Fresh poultry manure—cages	1.88	0.48	0.48
	1.5	1.0	0.49[a]
	1.89	1.29	0.47
	1.05	0.82	0.51
Litter (Broilers)	4.4	2.69	1.95
Hens	2.91	2.94	1.36[a]
Liquid poultry manure	1.04	0.96	0.44
In tank before mixing	0.61	0.17	0.35
In tank after mixing	0.60	0.21	0.34
	0.73–1.51	0.5–.78	0.25–.50
Lagoons (Preston)	0.57	0.68	0.16
	0.43	0.04	0.16
Some solid accumulation	0.75	0.64	0.17
Dried 3–7 days old	4.8	1.87	2.44
Six months old	2.53	3.20	1.11
Cages	5.85	4.02	1.47[a]
Littler	3.93	3.97	1.84
	4.14	3.7	2.27

[a] Values from poultry farms in England.

Table 4 *Distribution of Particle Size of Fresh Poultry (White Leghorn) Manure*

Sieve No.	Particle Size (ml)	Percent of Total Solids Held on Sieve	Percent Finer By Weight
8	2.380	0.00	100.00
16	1.190	1.57	98.43
30	0.595	11.54	86.89
50	0.297	15.19	71.70
100	0.149	12.69	59.01
200	0.074	10.35	48.66
finer than 200[a]		48.66	

[a] Includes dissolved solids.

slurry in the tank and not disturb it until the tank is to be emptied. At this time thorough agitation should take place, then the mass removed before settling occurs. If this is done the tank can be cleaned completely with no difficulty.

One should never go into an empty storage tank without a respirator, thoroughly ventilating the tank and having an attendant. Gases collected in these tanks can be lethal.

Feathers in liquid poultry manure have caused difficulty in pumping, since they tend to plug impeller-type pumps unless considerable clearance is allowed. Chopper pumps have not been too successful in handling poultry manure because the feathers do not chop well and dull chopper blades very rapidly. Vacuum systems with full opening valves appear to work well with poultry manure. New types of pumps are being developed to handle poultry manure.

Broiler Manure

Broiler manure is usually defecated on litter that consists of wood shavings, peanut hulls, dried sugar cane, sawdust, chopped straw, or other types of absorbent material. This is usually pushed out of the broiler house by small tractors with blades or carried out by conveyors. The material usually has 25–35% moisture.

Very little odor occurs from broiler litter unless it is piled and allowed to heat. Usually it is hauled directly to fields to be spread on the land. It is quite common for crews to come in and clean broiler houses for the litter. At one time it was common practice for neighboring farmers to provide the litter material for the flock in return for the manure. Today many broiler growers have to haul the manure to farms to get them to take it.

Broiler litter and litter from started pullets is sometimes used to feed to other animals, particularly ruminants. At the present time it is not legal to sell poultry manure of any kind as a feed ingredient in the United States but it can be fed to one's own animals. This is discussed in more detail later.

Duck Manure

Much of the maure from ducks is produced in the form of a very dilute slurry. Many ducks are raised on wire floors over troughs or pits and the droppings are flushed out by large quantities of water. At one time much of this was flushed into streams or estuaries, but this is no longer allowed without treatment.

A common practice today is to flush the duck droppings into aerated ponds where large quantities of oxygen are entrained into the water to aid in aerobic digestion to reduce odors and BOD. The number of aerators on a pond depends on the rating of the aerators and the load being added.

Most surface aerators are rated at approximately 3.2 lb of oxygen per horsepower hour based on 70°F tap water. However, the material being aerated is often below 70°F and is contaminated to various degrees depending on the loading rate; therefore the oxygen uptake is less.

The volume of the pond should be in the range of 1 ft^3/lb of chicken or duck. Ducks are usually marketed at approximately 7 lb; therefore we can estimate that the loading be based on a 5-lb duck because much of the waste is obtained from 1- to 6-wk-old ducks and the average would be approximately 5 lb. Because of this we should allow approximately 5 ft^3 of pond water per duck.

Based on 0.02 lb BOD per duck per day this would be 20 lb BOD per 1000 ducks per day. Estimating 2 lb of oxygen per horsepower hour under existing conditions would indicate that for 1000 ducks one needs approximately

5000-gal pond capacity
$\frac{1}{2}$–$\frac{3}{4}$ hp aerator capacity

For more detailed information pertaining to aerated lagoons see "Aerobic Treatment of Livestock Wastes[6] as well as the section in this volume on Biologic Oxidation.

Duck slurry is handled in two ways. In one system the material is flushed to a settling tank where the solids settle to the bottom and are carried off by a continuous scraper. They are then conveyed to a vibrating screen where considerable water is removed and the remaining solids hauled away for disposal. This puts less load on the aerated pond but necessitates getting rid of large quantities of wet solids by other methods. The effluent then goes to the aeration pond where it is aerated. From there it goes through a chlorine treatment before going to the stream or estuary.

Another system is to flush the material directly to the aeration pond. After aeration it goes to a settling pond where the solids settle. Here it is treated to precipitate the phosphorus. After this it is chlorinated before releasing it to the stream or estuary.

Several methods are being tried to precipitate phosphorus including the use of alum, ferric chloride, or lime. Research is continuing to determine the most satisfactory and practical solution and the proper amounts of the various chemicals.

After the effluent is removed from the settling pond the solids are allowed to dry for a few days, after which they are scraped out and disposed of by field spreading or in a sanitary landfill. This is essentially an activated sludge with very little odor. While one pond is settling and being cleaned a second one is being filled to follow the same procedure.

Provided that tillable land is close by, one might dispense with the settling ponds by pumping directly from the aeration pond through an irrigation system. This would remove solids as well as phosphates. Soil ties up phosphates readily, and these could be used by a crop placed on the field.

Provided that one can irrigate the contents of the pond on tillable land, it is not necessary to be concerned about BOD. The only concern would be odor, and this may allow for heavier loading or less aeration. Much of the solids will be removed with the mixed liquor. The aerated pond may be as much as 40–50% efficient in reduction of volatile solids.

Another factor in favor of the floating aerators is that one need not be as careful about exact water depth because the aerator goes up and down with water level but continues to maintain a constant immersion depth and oxygen entrainment.

FACTORS FOR CONSIDERATION IN SELECTING A SYSTEM

Because laying hens are located in various climatic and geographical areas, management and housing vary considerably. Layers are housed in open houses in some areas and in totally enclosed, environmentally controlled houses in others. Eggs are produced in hot dry areas as well as in very cold climates. These factors have considerable effect on drying, odor production, and fly propagation in relation to waste management and therefore may influence the water handling system best suited for the situation. No one system appears to be best for all situations and locations: each situation must be evaluated and a system selected that best fits the conditions and the desire of the management.

Regardless of the system, location can be very important. In some situations this may be regulated by planning and/or zoning ordinances.

Wet, anaerobic systems, which produce odors, may not be acceptable near centers of population but may be very satisfactory in isolated areas.

Much planning and thought should be given to building sites including topography, prevailing winds, and relation to inhabited areas. Soil type should also be given consideration, including drainage, permeability, and so on.

No new poultry house should be erected less than 100 ft from a highway; some areas require 400 ft or more. No ventilation fans should face a highway unless the poultry house is 150 ft or more from the highway. It is preferable to face fans away from the highway.

A commercial poultry farm should contain adequate acreage or be assured of the use of such acreage to permit land disposal of the wastes, if this method is to be practiced, without creating a nuisance.

When selecting equipment for transporting waste products, consideration should be given to public highways and routes that must be used when transporting. Good access roads should be provided.

Good housekeeping should be carried on at all times assuring that the buildings and adjacent areas have a neat, well-kept, pleasing appearance. The removal and disposal of old equipment, supplies, and debris in areas adjacent to the buildings should be a continuing project.

Manure should not be hauled or spread on weekends or holidays during June, July, and August, except in dire emergencies. If this is necessary, it may be wise to treat with chemical maskers even though they are costly. It is not advisable to spread manure in close proximity to residences, hotels, motels, or other places of business. Plow down or soil injection are acceptable in areas such as this.

All manure, in liquid or semiliquid form, should be transported on highways in liquid-tight containers. These should be kept in a clean and neat condition.

Spreading manure over ground during cold weather does not create odors to the extent that occurs when spreading is done during warm, humid weather. However, heavy spreading on frozen ground may lead to runoff and possible stream pollution and should be avoided where pollution might occur.

A paved and preferably covered area for loading manure is desirable for control of odors and flies and as an aid to public image. This should be maintained in a clean condition.

Areas to be fertilized should be chosen carefully. Approximately 2 acres of land should be available for each 1000 layers. This allows approximately 20 tons/acre/yr and may vary up or down depending on the crop, soil type, and rainfall. Much less should be used for small grains to prevent "lodging" or sticking together.

Consideration should be given to minimize conflicts caused by air and water pollution. This will require consideration of atmospheric conditions, topography, prevailing wind direction, and other situations that may aggravate conflicts, when spreading manures.

There should be no dumping over river banks, into swamps, lakes, or streams.

SYSTEMS FOR HANDLING MANURE IN EGG PRODUCTION OPERATIONS

Litter System

For many years most poultry manure was produced in litter, a combination of litter and dropping boards or dropping pits under roosts. The manure dropped on the floor is mixed with straw, shavings, shredded sugar cane, or other absorbent material. The manure expelled during roosting time is accumulated on dropping boards or in pits. This is not mixed with litter and is clean periodically, depending on the area in which it is collected.

Manure content of litter varies considerably and often becomes very wet during cold weather, especially in uninsulated buildings. Consistency and analysis vary due to the concentration of manure and litter (Table 2). This material may be very dusty when moisture content is below 15%. With 15–30% moisture the material is crumbly and does not stick together. When moisture content gets above 30% the material becomes sticky and slippery. Litter in the range of 40% moisture is likely to produce ammonia, which may be harmful to the birds and is sometimes objectionable to attendants and perhaps to neighbors.

Litter from layers, like that from broilers, will heat if piled and release strong odors when the pile is opened. Litter removed from the poultry house fresh will give off very little odor unless it is wet and has become anaerobic. Usually few problems occur when handling and/or hauling dry litter except from blowing. It is usually wise to cover the load with a tarpaulin if transporting on the highway to prevent feathers and other material from littering the area. Many areas require this.

Producing eggs from poultry on litter has become less popular due to low margins of profit that have forced high-density housing. With high-density housing it becomes difficult to maintain dry, satisfactory litter conditions; consequently most commercial egg production is now on wire floors or in cages.

Litter continues to be used for breeding flocks due to the general belief that higher fertility is maintained under these conditions, because of ease of mating.

Slatted or Wire Floors

The slatted floor that gained some popularity in the early 1960s has practically disappeared. However, wire floors have continued in some areas. The sloping wire floor house developed at Pennsylvania State University has gained some acceptance. This system incorporates a drag rake underneath the floor that agitates and moves the manure enabling it to air dry and remain aerobic. The drag is operated four to eight times each day to mix and aid in aerating the manure. This prevents odors and fly breeding. It is not uncommon to maintain manure in the range of 30% moisture with this system. Fans underneath the wire floor are used to move air, warmed by the birds over the agitated manure to aid the drying process.

The dry (30% moisture) manure can then be spread on the land with little danger of odors or it can be further dried to 12–15% moisture by dehydration for packaging to merchandise as organic fertilizer or be recycled through livestock.

The drag rake system can also use be used under cages. The semidried manure must be removed once or twice each week to reduce stress and strain on the equipment, particularly in long houses.

Dry Collection Systems

In dry southern areas the manure can be dried quite successfully under cages by using open houses, holding density to two birds per cage, allowing for good air circulation, and not allowing water to be spilled onto it.

The droppings form a cone with the result that additional droppings tend to run down the cone, exposing larger surfaces to drying. This results in rapid drying, thereby preventing fly breeding and odor production. During most of the year coning and drying should be sufficient to prevent fly breeding and odor productiond. During the time of year when coning and drying are not occurring, fly breeding is at a minimum. For best results it is suggested that when cleaning, a 6–10-inch pad of dry manure be left under the cages. This provides heat for rapid drying of new droppings, absorbs considerable moisture from the droppings, and disperses the moisture throughout the remaining mass for more rapid drying.

When this system is managed properly, one can dig into the cones finding the material to have an "earthy" smell with complete elimination of undesirable odors. Some operators add enzymes to hasten the process but usually if managed properly, this is not necessary.

Others use various parasites that feed on fly larvae to eliminate fly production and also hasten drying. This has been shown to be reasonably successful in research trials and on the farm trials in California.[7] A Puerto Rican wasp that feeds on fly pupae appears to be very successful against the common house fly (*M. domestica*) and the lesser house fly (*fannia*). This wasp attacks the early stage of the house fly pupae. It may bore a hole in the pupa and suck out its contents or it may lay an egg in the pupa. When the egg hatches, the parasite feeds on its host. This appears to be most successful where a concrete pad is placed under the manure area, which prevents the fly pupae from boring into the ground where the parasites cannot get to them.

Other biological treatments consist of using fungi and types of beetles such as *Ophyra leucostoma.*[8] These kill and feed on house fly maggots and other fly larvae that commonly inhabit poultry droppings.

Some have found they can increase density to three birds per cage by using 3–4-inch slats (bank boards) hung under the cages to catch part of the manure. As the manure builds up on the boards it runs off, again creating more drying surface. This is not always quite as successful but appears to work for some. It is very important to prevent surface water and spilling from the water fountains. Excess water completely upsets the process, enabling flies and odors to generate.

This process does not appear to be successful in closed houses or where good air movement is not provided. Any obstructions or weed growth around the houses should be removed. Wet conditions inhibit parasite development.

The tops of the cones can be removed for field application, further composting or dehydration.

Mechanical Cleaning

Manure may be removed from under the birds by hand with a fork or shovel, by garden tractors with sidearms, or by one tractor and scraper for several houses.

Mechanical cleaners are about the only practical equipment for houses 300–400 ft or longer. The investment is higher with mechanical cleaners and they require maintenance; they also require frequent removal of manure for efficient operation. The problem here is to provide the flexibility of spreading desired. Generally it is not desirable to spread poultry manure daily. In many areas, particularly in the northern latitudes, weather conditions may make it practically impossible to spread daily. There are other times when it is not desirable. When manure is spread during hot humid weather, odors may be very objectionable because the manure does not dry quickly and the odors continue over a longer period of time.

If flexibility of spreading can be provided, fewer complaints occur. This allows one to spread when conditions are more desirable, windows are not open, people are not out of doors as much, wind is in the right direction, better drying conditions exist, or the manure may be spread and plowed under quickly.

A system that is being researched at Cornell University Animal Waste Laboratory by Dr. D. C. Ludington and associates is to make an air tunnel of the walkway between cages. Air blowing through this is released through a ⅜-inch slot, blowing over the manure in the pit.

There is a slot on each side of the pit. Air is alternated in the tunnels by time clocks controlling the fans, to blow from one side for several hours and then from the other side. The air is taken from within the house, at the end of the cage rows. This circulates the modified air (room temperature, usually a minimum of 50°F), which tends to evaporate considerable moisture from the manure mass. This moisture is then released through the ventilation system. There is very little odor present in the house. This material should be spread immediately or further dried by composting or dehydration. If it were stockpiled, it would heat and be odorous when the pile was opened. In many respects this is a modification of the Penn State System. However, here the air is directed through a confined channel and emitted through the ⅜-inch slot and no agitation occurs. This appears to be economically feasible but all results have not been compiled. The manure is removed by mechanical scrapers every 72 hr with a moisture content of approximately 40%.

Deep-Pit Houses

Deep-pit and high-rise poultry houses provide maximum flexibility for removal of manure. These consist of cage houses over basements approximately 8 ft high. The deep pit usually consists of half the basement area sunken into the ground. This reduces the total height of the building, but it sometimes presents a problem of ground water getting in through the foundation wall or the floor. It is important to first select a well-drained area. Drain tile should be placed at the footings and eave troughs installed on the house or good drainage away from the house provided.

The high-rise house differs in that it is all above ground. In fact many poultrymen build a pad of 12–18 inches of gravel, cinders, or other porous material and place the poultry house on this. Then by building a diversion ditch around the house there is very little danger of ground water problems.

Disagreement continues as to whether the basement should have a concrete floor. Some claim that if there is no floor much of the moisture will seep away and maintain a drier condition; others disagree. Usually poultry manure has a tendency to seal surface areas to prevent much penetration by moisture.

If no floor is provided and any moisture exists, a quagmire occurs at cleaning time. If this occurs the owner usually puts in a concrete floor after the first cleaning. Cleaning is very simple when a concrete floor is provided. The houses are usually cleaned with a front end loader on a tractor. The hauling equipment can be brought directly into the house for maximum efficiency.

If it is allowed to become wet, manure becomes very odorous. The odors are emitted through the ventilation fans and many be objectionable to residents in the area. To keep the manure in the pit dry, one must not only prevent ground water seepage but he must prevent fountain leaks or overflow. A solenoid valve or other shut-off device should be provided to prevent overflow during the night, when no one is around to correct it.

Many poultrymen are using an intermittent water supply to help keep droppings dry. However, one must be careful with this system, particularly during hot weather, to be sure birds are not without water for too long a period. This can be disastrous as it can throw birds into a molt or even cause heavy mortality. One cannot afford to trade one problem for a more serious one.

It is important that heat from the birds be drawn into the pit area to help evaporate the moisture. Some accomplish this with the ventilation system while others install circulating fans in the basement under the cages to draw the heat from the birds into the pit. The latter system may require a great many circulating fans because very little turbulence is created more than 25–30 ft from the fan.

In England there has been reasonable success with the deep-pit system without circulating fans in the basement. The air is brought into the house over the birds, drawn down through the birds picking up heat from the birds, and then expelled through ventilation fans located in the pit area. These are located approximately 12–16 ft from each other on both sides of the house. Artificial heat can be supplied to aid in drying the manure. However, this is not economically feasible. In northern climates it is important to provide sufficient insulation at the ceiling and side walls to ensure sufficient fan operation to aid drying.

The deep-pit system appears to be more successful with heavy breeds than with leghorns, probably because droppings of the former are lower in moisture content. However, there has been enough success with leghorns to prove that reasonably dry manure can be produced to justify the system if the house is constructed and operated properly.

There is probably no other system as efficient from the standpoint of labor efficiency for cleaning as the deep-pit house. While many poultrymen clean the house every year, it can be left for three to five years or more. However, the longer it is left the greater the job when manure is removed.

Poultrymen with crop land tend to clean enough houses each spring to provide fertilization for the crops being grown. These houses can then be left for a period of time and manure from other houses used the next spring.

Provided that the manure in the deep pits can be kept dry, this system holds a great deal of promise because of its maximum flexibility. When it is not kept reasonably dry, odors are very likely to be a problem.

Water Pits

Some poultrymen have used water pits under cages to collect the manure. These range in depth from 1 to 8 ft. The 1-ft pits do not supply enough flexibility for spreading and should be supplemented with storage tanks to be used to best advantage. A 30-in-deep pit will store manure under stair step cages for approximately 6 mo.

After each cleaning 4–6 in. of water are placed in the pit. This serves as a partial water seal to prevent emission of odors, but unfortunately many of the gases produced by anaerobic action of the manure pass through water and may cause odor problems. This system can be suggested only for isolated areas.

When cleaning, the liquid manure is pushed through a gravity draw-off or to a sump by a garden tractor with side arms for shallow pits or by a false end gate pulled by a winch for deeper pits. The material can be pumped or drawn by vacuum from the sump.

This system has lost popularity because of odors from the ventilation system and from the anaerobic material at clean-out time. Odors at spreading time can be prevented by plow down or soil injection application. This will be discussed in more detail in the section on field application.

Another disadvantage of the water-pit system is that one is handling a great deal more volume of material than with a dry system. Unless one uses an irrigation system for field application this will require a great deal more time when hauling.

The Hydraulic System

This system requires pits 10–12 in. deep under the cages. A floating dam scraper is placed into one end of the pit where water is pumped in at approximately 100 gal/min behind the dam. The water pressure pushes the dam, forcing the manure ahead of it. When resistance occurs, water flows over the dam lubricating the manure in front, decreasing the resistance and enabling the manure to flow.

The manure–water mass flows into a cross trough at the end of the cage row and then into a storage tank, lagoon, or aerated pond. Water is recycled from the effluent of the storage tank or pond to provide the flushing.

The dam is constructed of 2 × 6 in. lumber and is approximately 2.5–3 ft long and about 1 in. less in width than the pit. Leather or rubber flaps are fastened to the sides at the rear of the dam to retain the water. Enough weight is placed on top of the dam to prevent floating but not enough to keep it from being pushed forward. There is no set formula for this; one keeps adding weight just to prevent the dam from floating. Once the proper weight is found, it should be maintained.

To clean the pit one places the dam into the end of the pit and turns on the water. This should be done every second to fourth day for best results.

The operator can be doing other chores while the manure is being removed. It will usually move approximately 7–10 ft/min; however, this depends on the frequency of the cleaning.

There is no physical handling of the manure and odors are minimal in the house because the manure does not become anaerobic in the 2–4 days of collecting.

When this system is used with a storage tank, the tank should be compartmentalized with a partition to within 1 ft of the top. The storage tank is filled with water to start. Water is pumped from the effluent side of the tank. When the solids and liquid flow into the solids section, water is forced over the partition to the effluent section for recirculation and the solids will settle to the bottom.

The solids collection section will require pumping every 3–6 mo, depending on the size of the tank. This material will be anaerobic and therefore very odorous. It should be spread in isolated areas or soil injected. Tank size is discussed in the section dealing with storage.

This system can be used with an aerated pond keeping the material aerobic and nonodorous. The material from the pond can be hauled by tanker or be spread by sprinkler irrigation with no odor problem. Very little labor or equipment is required for this system. Aerated ponds are discussed in the section dealing with duck wastes.

Lagoons

Natural lagoons may be used in southern climates to handle poultry manure. These do not perform as well in northern areas because bacteria do not function well in cold temperatures.

Natural, aerobic lagoons should be over 3–4 ft deep. These depend on photosynthetic action. If the water is too deep, light will not penetrate to the bottom and anaerobic conditions will develop.

In northern climates it is suggested that 1 acre of natural lagoon be supplied for every 2000–3000 layers. In some areas in the south, lagoons are loaded much more heavily. Reports have been made of loading rates of 20,000 birds per acre without problems.

Natural lagoons are not as popular as they once were. They require considerable care to operate properly. When they do not operate properly, they emit odor constantly and in some instances they have caused considerable conflict with neighbors.

Storage

Storage of poultry manure is important to provide flexibility if one depends on land application for disposal. There are times, particularly in northern climates, when it is impossible to spread manure because of deep snow. Many areas are now prohibiting the spreading of manure on frozen ground, to prevent runoff into waterways. Where land is intensively cropped, there is often no open ground on which to spread manure during the growing season.

It is advised not to spread poultry manure near residences, camps, resort areas, or other populated areas during the summer, for reasons already detailed.

Because of this it is often necessary to provide a storage area to hold poultry manure until it is possible to apply it to the land. This may be done by using the deep pit, as described earlier, but it may require providing a specific storage area.

Several factors should be considered in making plans to store poultry manure. One needs to know the approximate holding time required and how much is being produced daily. If the material will be odorous, one must make plans to contain these odors to prevent annoyance to neighbors.

The holding tank may be constructed under part of the poultry house or it may be located outside. It is most often below ground level, but may be above ground if strong and tight. When placed below ground, filling is much easier. Often the manure can be pushed or will flow into the storage area.

Once one knows how long he must store the material, he can easily figure the size required. Referring to Table 1, it will be found that 1000 layers produce approximately 4 ft^3 of manure per day. To store the material for 5 mo requires 600 ft^3 for each 1000 birds, plus space for any water used in a liquid system.

Round storage tanks are preferred for poultry manure because they clean more easily and thoroughly than square or rectangular tanks.

Holding tanks under the poultry house should have provisions for closing the opening from the pit into the tank. These should be kept closed when not being used and particularly when agitating the material in the tank, before emptying.

Many poultry houses are ventilated by exhaust fans, creating a negative pressure in the house. This tends to draw gases from the tank. Many of these gases are toxic and even lethal. There have been cases where birds have been killed because of these gases. It is wise to increase the speed of the ventilation fans or turn on more fans and open the air intakes to their maximum during agitation of storage tanks under the house. This is particularly important during the winter when ventilation rates are usually at their minimum and air inlets are not fully open.

Again, never go into a storage tank without a respirator, gas mask, good ventilation within the tank, and one or more persons in attendance.

When the material in the tank is agitated, it should be hauled as soon as possible. Once the material is agitated particulate matter is separated (Tables 2 and 4). The dense materials such as calcium compounds tend to settle more rapidly than the less dense material. This tends to build a very dense, hard mass at the bottom of the tank. If the material is agitated several times before hauling, the buildup of this dense material may become very hard and reduce the holding capacity of the tank.

It is best to place 4–6 in. of water in the tank, unless one is dealing with a very liquid material, then dump the manure in, until the tank is approximately two-thirds full. At this time start the agitation. After 15–30 min of agitation start hauling and continue until the tank is empty.

The reason for suggesting that the storage tank be emptied when two-thirds to three-quarters full is to allow for flexibility. If one waits until it is full, flexibility is reduced and one may have to haul and/or spread when conditions are not desirable.

Storage tanks should have a reinforced concrete top, with covers for the agitator and draw-off holes. Wooden tops are not desirable because wood will eventually rot, as a result of the excess moisture. This presents a serious hazard and someone may fall into the tank.

Where a dry manure system is used, it is wise to store manure in a covered shed. This should be completely enclosed or screened to prevent fly breeding during warm weather.

Stored manure will develop anaerobic conditions during storage, unless it is very dry, and will be very odorous when the mass is opened or moved. Because of this it is wise to consider wind direction and humidity when moving or spreading this material.

Aerated Storage

Aerated storage is gaining a great deal of favor because it reduces or eliminates odor problems; there is evidence that this will reduce solids also, because the aerobic bacteria digest the organic matter. This process changes some of the organic matter to CO_2 and H_2O.

The aerated pond was discussed previously in relation to duck wastes. Odors are eliminated by this process so that there should be no problem with neighbors from this method of storage. The material may be spread without concern when aerobic conditions are maintained.

Oxidation Ditch

This is a system for collecting, storing, and treating poultry waste. As it is recommended here, the oxidation ditch would be constructed inside the poultry house under the cages where it is actually storing the material.

The oxidation ditch is a racetrack-shaped ditch usually involving two rows of cages (Figure 3). There may be one or more ditches within each house. Oxidation ditches may be placed outside the house but have been found, expecially for poultry, to be more successful when placed inside, particularly in northern climates; this provides more uniform temperatures and loading. Oxidation ditches are subject to problems with odors and foam when "shock" or "slug" loaded (large amounts of waste dumped in at one time). When the oxidation ditch is placed under the cages, loading is very constant. It has been shown that chickens defecate approximately 20 times per day; therefore, when they are directly over the ditch each dropping is mixed and aerated al-

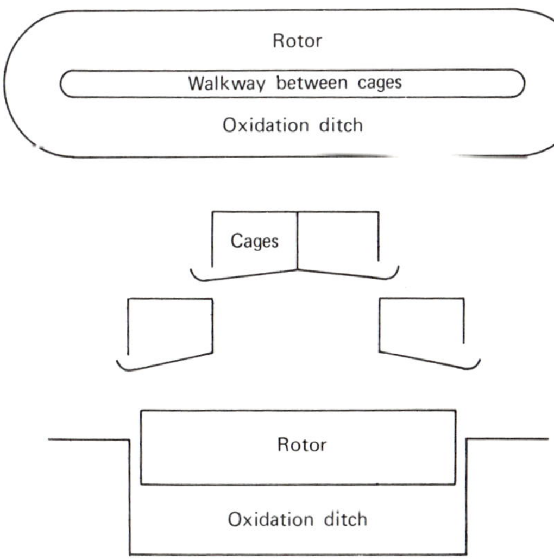

Figure 3. Schematic drawing of a batch type oxidation ditch under laying cages.

most immediately, thereby preventing large amounts being introduced at one time.

Various types of rotors have been used; the cage type, with many fingers in the shape of a drum, has been most popular.[6] At Aberdeen, Scotland, research has been performed with many types of rotors. A series of large perforated disks appears to give excellent results and requires the least power. The Alfa Laval organization has developed a vertical aerator (Carousel) that appears to hold promise and requires minimal power.

It is recommended that the oxidation ditch be constructed wide enough and long enough to catch all the droppings from the cages above it. The ideal depth is obtained by dividing the volume required by the surface area of the ditch. The volume required is approximately 0.60–0.75 ft^3 per layer.

The basic principles of the oxidation ditch have been covered in other sections of the text (Chapter IV.6) and in other publications by Jones, Day, and Dale[6] and Loehr.[9]

Oxidation ditches are basically alike whether they are used for poultry, swine, dairy or beef animals, or human sewage. However, there are slight differences in volume, BOD requirements, etc. for the various species.

One difference is that because of the lower moisture content and because the urates and solids are mixed, a batch type of ditch is often used for poultry. This is not required but because of the high evaporation rate in laying houses there is usually no overflow. In fact it is required that water be added to maintain immersion depth, as this must remain constant.

In experiments at Cornell University[9] it was found that evaporation was considerable. Make-up water can be supplied by using a float valve.

Solids content buildup tends to reduce the speed of flow, which should be maintained at 1–2 ft/sec. When the flow rate is reduced materially, liquor can be pumped or drawn off and more water added to reduce the solids content.

In the research house at Cornell University the flow of material in the ditch slowed considerably when the solids content reached 7–8%.[9] Most farms would not have adequate equipment for measuring solids content. Therefore, all one needs to do is observe the flow of the material in the ditch. When this begins to slow, usually after 4–6 mo, draw off or pump one-half to three-fourths of the material from the ditch and refill to proper depth.

It is best not to remove all the material from the ditch. The remaining material serves as "seed" to prevent ammonia release and perhaps foaming, which sometimes occurs at start-up. However, should foaming occur one can add antifoam agents such as vegetable oil, petroleum oil, or various commercial products. Sometimes this can be controlled with a water spray. However, if one is using the batch system too much water should not be added as a spray. Once the microbial population increases to approximately 2000 mg/liter of suspended solids, the foaming should subside. Every 2–4 yr it may be wise to clean the ditch completely to remove heavy insoluble material that has settled at the bottom. When this is done, if possible, save some of the material to be replaced in the ditch as seed for restarting.

It is felt that more satisfactory results are obtained if ditches are cleaned during warm weather. Microbial recovery appears to be better, and foaming is less of a problem, during warm weather.

Like the material from an aerated pond the contents from the oxidation ditch may be spread on the land almost any place at any time since there should be no odor. It should not be placed in waterways. Usually the BOD remains high enough to tie up oxygen in the water, which may cause fish kill and other problems.

The material from the oxidation ditch may be pumped into tankers. Vacuum tankers may be used or it can be pumped through an irrigation system. However, if a pump is used, one should be selected that will handle feathers. Feathers are not digested in the oxidation ditch.

Spreading Poultry Manure

Manure odors sometimes linger for a long time, particularly if the humidity is high and there is no air movement. These odors may move in air drainage pathways. If a poultry operator spreads manure on a hill during hot sultry weather, the odors emitted from the spreading operation may drift down the hill, offending everyone in their path.

If the manure is spread very thinly, the problem decreases since more rapid drying decreases the odor. Heavy spreading creates the most undesirable situation because the material does not dry rapidly, and thereby releases odors over a much longer period of time.

The ideal situation is to locate a large poultry operation in the center of several hundred acres of land with no neighbors. The land would be used to grow crops or for pasture to raise beef, dairy heifers, sheep, or some other type of livestock.

However, there are several disadvantages to this. First, many poultrymen do not want to grow crops or other types of livestock. Secondly, often large quantities of land are not available, particularly at reasonable prices. However, some poultrymen have chosen this route and seem to be getting along successfully. One such operation purchased approximately 1200 acres of very poor quality land. The manure has been placed on the land, which now grows very good pasture grasses for beef and dairy heifers. A large poultry operation in England owns 12,000 acres of land. The poultry units are located long distances from residences and the land is used to grow grains that are used in the poultry rations.

However, many poultry operations are already located where they cannot obtain vast expanses of land, so these operators must rely on other alternatives.

Manure Spreading Equipment

Many types of manure spreaders are on the market. Some of these are designed to handle only dry material, some will handle dry or wet material, and some will handle only liquid material. If one must use public highways to transport the manure, he must have equipment that is tight and does not spill manure along the highway. In some areas regulations are stringent regarding spilling. Farmers have been made to clean up the spilled manure and place limestone on the remaining material. Usually one or two such treatments causes the operator to comply very quickly. If a poultryman does not have to leave his farm, any type of equipment that will handle the product will be satisfactory.

When one must transport manure off the farm, it is a good practice to wash the spreader by hosing it off regularly. There are some situations where it is almost impossible to spread manure on the land without offending someone. In such situations if one must apply the material to the soil he can use soil injection equipment. Some soil injectors will inject in sod or untilled soil, while others must be used in light soils or soil that has been tilled.

Soil injectors are usually used with liquid or slurry. They apply the material directly behind the soil opener, letting the soil fall back in place covering the manure. One must operate the equipment at the correct speed to prevent too much material being injected to allow oozing of manure through the soil opening. Soil injection can be entirely odor free, even with anaerobic material. Research has shown that soil is a good filter. Usually 6-in. of soil will prevent all odors from being emitted. Soil injection is usually somewhat more time consuming than surface application, but again, with practice it can be applied very rapidly.

Some poultrymen have considered using soil injection entirely as a disposal method rather than as a fertilizer. They set aside a plot of ground where they go over and over again injecting manure. This may not be desirable in very light soils, particularly if there is a high water table in the area, because it could cause pollution of underground water which may get into wells or other potable water.

Some areas require that if a poultryman plans to use soil application of manure he either have land enough of his own or a firm commitment for land on which to spread before he will be given permission to build poultry houses.

Liquid poultry manure may be spread with irrigation equipment, which is very fast and requires little physical effort except to move the irrigation lines. One of the complaints against liquid manure is that one must handle two to three times the volume of material that is actually produced. This is no problem with irrigation spreading because volume is of little concern.

Raw liquid manure spread by irrigation is more odorous than normal spreading of liquid manure because of the breakup of droplet size: the odors are dispersed into the air. This is no problem when aerated material from an aerated pond or an oxidation ditch is spread by this system; such material is odor free.

When spreading by irrigation, one must be careful not to irrigate in one place too long; the irrigation heads should be moved regularly to prevent runoff, taking into account variations in soil type conditions. When irrigating on frozen ground, the sprinkler heads must be rotated more often. In some areas regulations prohibit any spreading of manure on frozen ground.

DEHYDRATION

Dehydration may be accomplished by mechanical or thermal methods or by absorption. Absorption is the principle involved with the use of litter and this, while satisfactory for low concentration of layers, is not satisfactory with high density material.

Mechanical dehydration by pressure or centrifuging can remove considerable water from manure; however, it does not reduce it to a point where the manure can be stored. Usually, there is approximately 60% moisture remaining after mechanical dewatering. This manure will heat and become odorous if stored but will improve efficiency for thermal dehydration.

Vacuum filtration is sometimes used for dewatering sewage sludge but has not generally been found to be economically feasible for poultry manure.[11]

Thermal dehydration appears to be the most practical method for dewatering poultry manure. With thermal dehydration Btu's are required to drive off the moisture from the material being dehydrated. Efficiency of the dryer is reduced with increased moisture content of the raw product. For reasonable results the raw product should not be over 65% moisture and for every percent moisture below this, the efficiency of the dehydration process increases tremendously. One of the problems, even with thermal dehydration, is that poultry manure is approximately 70–80% moisture when defecated, but there are several ways to overcome part of this disadvantageous moisture content. The moisture may be reduced from the manure if air is moved over it for a period of time before removing from the pits. Often, this may reach 60% moisture or below just by natural air movement within the house if good air movement over the manure is maintained.

The air slot system in the collection pit, being researched at Cornell University, described in the section on mechanical cleaning can reduce the moisture content to 40% or below. This increases the efficiency of dehydration considerably. It may be less costly to remove part of the water by this method than all by thermal dehydration.

Ludington[10] showed that a reduction in initial moisture content of 15% allows the plant to process 2.5 times its original capacity for a given period of time.

Centrifuging can be used to reduce the moisture content of poultry manure. This is more practical with liquid manure; centrifuging is not efficient unless the raw material is 80–90% or more in moisture content. It will reduce high-moisture material to 60% moisture, which is practical for thermal dehydration. However, in the centrifuging process a very high BOD effluent is obtained, and this must be disposed of. It cannot be placed in waterways unless treated to obtain acceptable standards, but it can be spread on land in isolated areas or soil injected in populated areas. This may have a very offensive odor, depending on how anaerobic it has become. There is a question as to the economics of centrifuging and then dehydrating: it is necessary to balance the advantage and cost of mechanical dewatering against the cost of driving the extra water off by thermal dehydration.

Moisture content of the raw product going into the dehydrator may be reduced by recirculating part of the dried product, mixing it with the material being fed into the machine. Some operators have mixed anhydrous materials with the raw product to reduce moisture content, but one must be sure the anhydrous material is not deleterious to the product.

There are many types of dehydrators on the market. However, the principle is about the same for most types. Ludington[10] (Figure 4) has shown a schematic drawing showing equipment and a flow pattern.

Dehydrated poultry manure varies considerably in analysis. In the early days of poultry waste, dehydration much of the product was produced from broiler litter. The analysis varied depending on the ratio of manure to litter and how much sand it contained. Many broiler houses had sand floors and much of this was picked up when cleaning

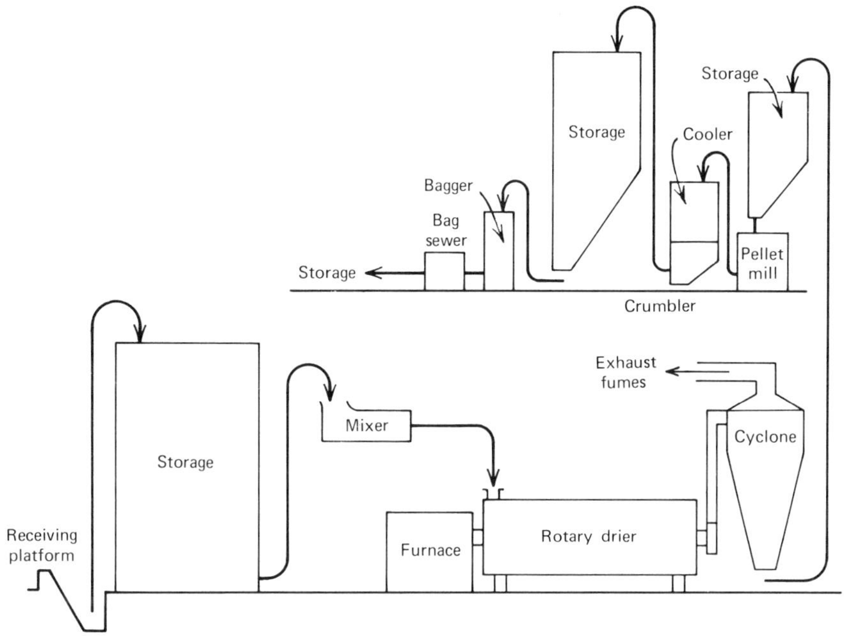

Figure 4. Flow pattern for poultry manure drying and packaging plant.

the houses. This gave dehydrated poultry manure a bad reputation because some of the merchandised product contained as much as 50% sand.

Today most dehydrated poultry manure is produced from manure collected under wire floors or cages and is not diluted with litter. Analysis continues to be somewhat variable, but the extremes are as great as in product from litter. Usually dehydrated poultry manure will be dried to 10–12% moisture, which is satisfactory for packaging and storing. The range in analysis is shown in Table 2. The age of the raw product before dehydrating will affect the analysis considerably. It tends to lose nutrients, particularly nitrogen, which is very volatile, if held for any length of time.

If the product is dried to 5% or 6% moisture, the nitrogen content may be up to 5% or above, but the total volume is reduced and the product may be excessively dusty. This has also increased the cost of drying. Should one plan to use the product immediately, without storing, drying costs can be reduced considerably by drying only to 15–20% moisture.

Until recently the main use of dehydrated poultry manure has been for organic fertilizer and as a soil conditioner. The primary market has been for home owners, to be used in flower gardens, and for the organic enthusiast in growing vegetables. From an economic standpoint this product cannot compete with inorganic fertilizer for nutrient value cost.

To produce a top-quality dehydrated poultry manure it should be dried when fresh, before anaerobic action takes place.

Interest has increased in recylcing dehydrated poultry manure through livestock. Recycling is not new. Chickens have scratched in manure piles and pigs have grazed after beef for years. The new part is that today the manure is dried, sterilized, granulated to a convenient form, sometimes even pelleted, and packaged. Dehydrated poultry manure is somewhat low in energy value but it is quite good as a protein source, particularly for ruminants, which can convert the urates to a protein equivalent substance.

There is an esthetic resistance to this but most of it is due to lack of knowledge of the material and the processing technique. The product has been sterilized, and when properly handled, preventing recontamination, it should be be acceptable. It should be comparable to tankage or meat scrap as a feed ingredient source, but may be somewhat lower in protein.

The Food and Drug Administration does not allow the sale of this product as a feed ingredient in the United States. However, in England it can legally be sold as a feed ingredient.

COMPOSTING

Composting is the aerobic thermophylic decomposition of organic waste. Decomposition is a result of the biological activity of microorganisms that exist in these wastes. These organisms are primarily bacteria, actinomycetes, and fungi. If adequate air is supplied, aerobic types of organisms predominate and they use oxygen in their respiration and metabolism, giving off CO_2, H_2O, and heat. Compost may contain up to 25% dead or living organisms and is subject to further, slower decay but should be sufficiently stable not to reheat or cause odor or fly problems. This is the way it should function and usually does if litter material is composted. However, when dealing with fresh undiluted droppings without litter, the situation is different.

The desirable moisture range for windrow composting is 35–50%. Bacterial activity is reduced if the moisture content is below this and if the moisture content is above this the material seals out the required oxygen.

To compost clear manure from laying hens it is necessary to mix it with inorganic fibrous material to provide good aeration. For good composting a carbon to nitrogen ratio of 20:1 is best; poultry manure is often in the range of 8:1 to 12:1.

For good composting and when conditions are right a temperature of 120–150°F will be reached. Many claim the ideal temperature to be 144°F. If temperatures go too high spontaneous combustion may occur and if temperatures are too low bacterial action decreases or stops.

If cage poultry manure can be mixed with domestic garbage, it is advantageous to both products. Most domestic garbage is too low in organic matter for good composting and cage poultry manure is too low in carbonaceous material. The combination of the two, in correct proportions, creates an ideal situation.

Composting can be carried out in biostabilizers for part of the process and finished off by windrow treatment, which hastens the procedure. The complete process can be carried out in windrow treatment, turning the mass each week, or when temperatures reach a critical point (150°F). This may be done by bucket loaders, snow loaders, or by a large machine with a huge auger mixer which is now available for large operations.

ANAEROBIC DIGESTION

Several people have investigated this procedure and in some areas of the world it may have application. In general it is too complicated and too costly for application in disposing of poultry wastes.

Some years ago work was done by Eby[12] at USDA, growing crops in water fertilized by poultry manure. It was shown that grasses, particularly tall fescue and rye, did quite well and produced significant growth.

PRODUCING PROTEIN

Workers at USDA[13] conceived the idea of producing true, high-quality protein by using poultry manure to produce fly pupae, harvesting and drying the pupae, and grinding them to be used as a high-protein feed ingredient.

Their procedure was to spread the fresh poultry manure in a thin layer over a ⅛-inch screen floor; the fresh manure is a very good medium in which to grow fly larvae. This was seeded with three fly eggs per gram, the eggs allowed to hatch, and the larvae to tunnel and aerate the manure. On the fifth day, 40-watt fluorescent lights were placed over the seeded manure and left for 2 days. During the first hour under the light most of the larvae migrate out of the manure, through the screen into a pan placed under the screen. The remaining time under

Table 5 *Analysis of Dried Ground Fly Pupae*

Proximate Composition[a]		Amino Acids (% Protein)		Fatty Acids (% Total Fatty Acid)	
Protein	63.1[b]	Arginine	4.2	Lauric	0.6
Fat	15.5	Glycine	3.9	Myristic	3.2
Moisture	3.9	Histidine	2.9	Palmitic	27.6
Ash	5.3	Isoleucine	3.5	Palmitoleic	20.6
Other	12.2[c]	Leucine	5.3	Stearic	2.2
		Lysine	5.2	Oleic	18.3
		Methionine	2.6	Linoleic	14.9
		Phenylalanine	4.2	Linolenic	2.1
		Threonine	3.4	Unidentified	10.5
		Valine	3.4		
		Glutamic acid	10.8		
		Alanine	4.2		
		Cystine	0.4		
		Tyrosine	4.9		
		Proline	3.1		
		Serine	3.2		
		Aspartic	8.5		
		Ammonia	2.1		

[a] Determined according to *Official Methods of the Association of Official Agricultural Chemist*, 9th ed, 1960, Washington, D. C.
[b] Nitrogen times 6.25.
[c] Primarily nitrogen-free extract and fiber.

the light drives out any remaining larvae and results in an additional loss of moisture from the thin layer of manure. Also during this period the larvae enter into the third stage of their life cycle, the pupae. The pupae are then collected for drying and grinding into pupae meal.

The remaining manure is quite dry and odorless, even after being restored to its original moisture content; it may be spread on the land or may be further dried and pelleted for packaging and sale. Some nitrogen has been lost from the manure mass, and further tests are needed to determine its fertilizing value.

The resulting pupae produce a high-protein (60%+) supplement that may be used as a feed ingredient. The pupae meal is very high in essential amino acids and contains approximately 15% fat (Table 5).

This process results in deodorizing a very obnoxious product, reducing its moisture content and volume, and produces a by-product in the form of high-protein pupae meal and a potentially useful fertilizer or soil conditioners. The unanswered questions are the quality of the processed manure, the possibility of increasing the yield of pupae meal, currently 3–3½% of the original manure weight, and the economic feasibility of adapting this process to commercial poultry operations.

DISPOSAL OF DEAD BIRDS

Another waste problem confronting poultry operations is that of disposing of birds dying from natural causes or disease outbreaks. The normal loss is approximately 1% per month but may be 2% or more. With today's commercial flocks this can amount to many birds each year.

When flocks were small the disposal pit, properly constructed, proved very satisfactory, but this is not practical with flocks of over 10, 000 layers.

Sanitary landfill operations are satisfactory if one is near enough to be practical. However, the dead birds should be disposed of daily unless refrigeration is provided to accumulate them for several days.

Rendering plants are satisfactory if one is available close by or again if one has refrigeration available to accumulate birds for periodic pick up.

Rendering plants are probably the best solution to the problem for very large operators where hundreds of birds accumulate over a short period of time. In some areas routes are set up where trucks pick up the dead birds regularly. When this procedure is used, it is advisable to accumulate and hold the dead birds some place away from the production unit to avoid possible contamination from trucks going from one farm to another.

Incineration of dead birds has become popular on many poultry farms. There are several manufacturers of poultry incinerators. One should first determine the average mortality over a period of time and purchase an appropriate incinerator.

Some incinerators are not equipped with afterburners. If not, one should be sure that enough heat is generated to completely burn any fly ash or fumes that might be emitted from the stack. Afterburners are required by some states before permission will be granted to use an incinerator. Incinerators can be a serious source of air pollution when not operated properly. The odor of burned flesh and feathers can be as annoying as that from manure. Some incinerator manufacturers have determined stack emission rates which are often required by health officials. It would be wise to check on this before purchasing an incinerator.

WASTE DISPOSAL FROM PROCESSING PLANTS

Egg Processing

Ideally the egg processing plant should be connected to a municipal sewage plant. With this system the waste is diluted with large amounts of water and can be handled satisfactorily in the municipal plant.

Attempts have been made to handle plant washing waste with septic tank systems but most of these have been failures; too often large amounts of egg albumin are contained in the wash water. Albumin is a high-protein substance that tends to plug lines and filter beds. Very often not enough dilution occurs, with the result that the system becomes inoperative in a very short time.

One possibility for the on-the-farm plant is to use an aerated pond. If sufficient area and aeration are provided, this should be satisfactory. To be successful one needs to know the BOD load and design the pond and aeration requirements accordingly.[6]

As much of the waste product should be salvaged as is possible to relieve the load on the disposal system and to supplement the income from the operation. Much of this product can be used for nonfood purposes and may supply some supplemental income.

POULTRY PROCESSING

Products to be disposed of include blood, feathers, feet and legs, heads, viscera, and tremendous quantities of water. Most of the viscera, feet and legs, and heads are processed into meat meal that is used in the feed industry. Often blood is salvaged and dried into blood meal for the same purpose. Feathers are usually dried and ground into feather meal.

Processing of this material can create offensive odors; some plants have been closed because of air pollution. Isolation is desirable when possible, but again, the low value of the product does not justify the extra cost of hauling great distances.

Wash water, usually consisting of hundreds of thousands of gallons, may be disposed of through municipal sewage plants or through aerated ponds. With proper treatment this water may be reused for certain parts of the operation.

REFERENCES

1. Air Pollution Control Laws of New York State. "Article 12-A; Air Pollution Control As Amended by Chapter 902 of Laws of 1966" § 1264–1267).

2. William E. Burnett, "Air Pollution From Animal Wastes, Determination of Malodors by Gas Chromatographic and Organoleptic Techniques," *Environ. Sci. Tech.* **3,** No. 8, 744–749, August 1969.

3. S. A. Hart, "Fowl Fecal Facts," Proceedings of the National Symposium on Poultry Industry Waste Management, May 13–15, 1963, Univ. of Nebraska, Lincoln, Nebraska, pp. 7–13.

4. W. E. Burnett, "Some Physical, Chemical and Biological Properties of Poultry Manure," Unpublished Mimeo 1968, Department of Food Science, Cornell Univ., Ithaca, New York.

5. A. T. Sobel, "Distribution of Particle Size of Fresh Manure," Unpublished Mimeo 1965, Department of Agricultural Engineering, Cornell Univ., Ithaca, New York.

6. D. D. Jones, D. L. Day, and A. C. Dale, "Aerobic Treatment of Livestock Wastes," Illinois Bulletin 737, 1970, Univ. Ilinois, Urbana, Illinois.

7. E. Fred Legner, "Parasites Control Houseflies," *Pacific Poultryman,* December 1964.

8. J. R. Anderson and J. H. Poorbaugh, "Biological Control Possibility For House Flies," *California Agriculture,* September 1964.

9. Raymond C., Loehr, D. F. Anderson, and A. C. Anthonisen, "An Oxidation Ditch For The Handling and Treatment of Poultry Wastes," Proceedings of the International Symposium On Livestock Wastes, Columbus, Ohio, 1971.

10. D. C. Ludington, "Dehydration and Incineration of Poultry Manure," Proceedings of the National Poultry Industry Waste Management Symposium, Lincoln, Nebraska, May 1963.

11. E. A. Cassell, A. F. Warner, and G. B. Jacobs, "Dewatering Chicken Manures by Vacuum Filtration," Proceedings of National Symposium of Animal Waste Management, May 1966, ASAE, St. Joseph, Michigan, pp. 85–91.

12. Harry J. Eby, "Evaluating Adaptability of Pasture Grasses to Hydroponic Culture and Their Ability To Act as Chemical Filters," Proceedings of National Symposium On Animal Waste Management, May 1966, ASAE, St. Joseph, Michigan, pp. 117–120.

13. C. C. Calvert, N. O. Morgan, R. D. Martin, and H. L. Eby, "Biodegradation of Poultry Manure From Cage Layers," 1970, USDA Beltsville, Maryland.

CHAPTER IV.6

the treatment and handling of swine wastes

ARTHUR J. MUEHLING

When hogs were fed on pasture, the handling and disposal of hog wastes were kept to a minimum. With the trend toward confinement housing, hog wastes are being concentrated on paved feeding floors and in liquid storage pits. The handling, treatment, and final disposal of the wastes have become a major problem when planning and operating a hog production unit. The magnitude of this problem was pointed out by the Secretary of Agriculture who stated that animal wastes in the United States probably exceed wastes from any other segment of our agriculture-industrial-commercial-domestic complex. The report estimated that approximately 1.7 billion tons of animal wastes is produced annually, about one-third of which is liquid. This estimate is for wastes from all animals, but the total wastes from hogs are also a sizable amount.

PROPERTIES OF SWINE WASTES

An understanding of the properties of swine wastes is necessary to develop an adequate system of waste management. When procedures used in the treatment of municipal sewage have been applied, without modification, to the treatment of swine wastes, they have seldom performed as expected.

The properties of swine wastes can be classified as physical, chemical, and biological. The physical and chemical properties may be affected by the physiology (size, sex, breed, and activity) of the animal, the feed ration (digestibility and the protein and fiber content), and the environment (temperature and humidity).

The addition of antibiotics in the ration for disease control may affect the biological properties of hog wastes. When large doses of antibiotics are administered to the hogs, a sufficient portion may pass through the digestive tract to severely limit biological treatment of the

Table 1 *Approximate Daily Manure Production (Median Values for Undiluted Manure Without Bedding)*[a]

Size of hog		Waste production			
		Liquid and solids		Wet solids only	
Age (wk)	Weight (lb)	(ft^2)	(Gal)	(ft^3)	(lb)
Pigs					
6–9	40	0.06	0.5	0.04	2.4
9–13	100	0.13	1.0	0.1	5.9
13–18	150	0.21	1.7	0.15	8.8
18–23	210	0.30	2.2	0.2	12.0
Sows, boars					
20–52	300	0.43	3	0.3	17.5
52 +	500	0.71	5	0.5	30.0
Sow + litter	—	0.55	4	0.5	30.0

[a] Adapted from Midwest Plan Service.

manure unless it is diluted sufficiently so that toxic concentrations of these antibiotics do not develop. Copper in the form of copper sulfate or copper oxide is frequently fed in the ration as a growth stimulant. Much of this copper is passed on to the waste and has also been found to limit bacterial growth.

Physical Properties

Daily production of wastes from hogs varies considerably, depending on type and size of hog, method of feeding, and what is included in the wastes, especially the amount of wastewater. Average daily manure (feces and urine) production per 100 lb of live weight are ⅛ ft^3, 1.0 gal, or 7.7 lb. The average figures in Table 1 can be used as a guide to calculate the amount of manure that must be handled from hogs of different weights. The total amount of liquid to be handled can be affected more by the amount of wastage from the many pressure waterers in a confinement system than by any other single factor. The values in Table 1 are for undiluted fresh manure without bedding, so any expected dilution must be added to these figures.

The solids in hog manure amount to about 1 lb of total dry matter per 100-lb hog per day. The volatile solids (VS) are normally about 0.75 lb/day.

Chemical Properties

The fertilizer value of livestock manure has been researched quite extensively, but the reported values vary considerably. The plant nutrients in hog manure come from the feeds consumed by the hogs. There is less urine than solids, but the urine contains about 50% of the fertilizer value of the manure. An average fertilizer value of hog manure is 0.7% nitrogen (N), 0.4% phosphoric acid (P_2O_5), and 0.4% potash (K_2O). This means that a ton of hog manure (about 265 gal) would contain 14 lb of N, 8 lb of P_2O_5, and 8 lb of K_2O.

Today, in light of the plentiful supply and low costs for commercial fertilizers, many wonder if it pays to collect and use hog wastes as a fertilizer. The wastes are often spread on the land as a means of disposal more than for the fertilizer value. When sufficient land is available, field spreading provides adequate treatment and final disposal with one operation if the associated odors are acceptable.

Liquid hog wastes spread on the land will add some nutrients to the soil. Table 2 serves as a guide to help determine how much liquid waste must be applied to furnish a given amount of plant nutrient. For example, if you want 80 lb of nitrogen per acre, you would need to spread about 3010 gal of hog wastes per acre. If considerable dilution water had been added to the wastes, larger amounts would be needed. For nitrogen, this is allowing a 50% loss before it is utilized.

Of course, by applying manure to fulfill the requirements of one of the elements, the soil will receive certain quantities of other nutrients present in the manure. For example, note in Table 2 that to get 50 lb of phosphoric acid (P_2O_5) in the soil, you must apply 1855 gal of liquid hog manure, but at the same time the soil will receive about 49 lb of nitrogen and 58 lb of potash (K_2O) per acre, not to mention trace minerals and humus value.

Biological Properties

The BOD test is used by sanitary engineers to determine the pollutional strength of organic waste in terms of the oxygen demand that the waste will exert on a body of water in which aerobic conditions must be maintained. It measures the amount of oxygen required by the aerobic bacteria for biological decomposition of the organic matter at a standard time of 5 days and a standard temperature of 20°C (68°F). The average BOD per 100-lb hog is about 0.32 lb/day.

The COD test is also used as an index of the strength of the wastes. It is a measure of the chemically oxidizable organic matter in the waste. The COD test has the advantage of taking only 3 hr while the BOD test

Table 2 *Estimated Gallons of Hog Manure per Acre Required to Provide Given Amounts of Nitrogen, Phosphoric Acid, and Potash*[a]

	Nitrogen[b] (N)			Phosphoric Acid (P_2O_5)			Potash (K_2O)		
Estimated requirements, (lb/acre)	Manure needed, (gal/acre)	P_2O_5 applied, (lb/acre)	K_2O applied, (lb/acre)	Manure needed, (gal/acre)	N applied, (lb/acre)	K_2O applied, (lb/acre)	Manure needed, (gal/acre)	N applied, (gal/acre)	P_2O_5 applied, (lb/acre)
60	2260	73	71	1855	49	58	1900	50	61
80	3010	97	95	2475	66	78	2535	67	82
100	3765	122	118	3090	82	98	3165	84	102
120	4520	146	143	3710	99	117	3800	101	123

[a] For swine wastes with no dilution. Use this table only as a guide. Additional amounts of wastes will need to be applied depending on additional amounts of dilution due to water wastage.

[b] It is assumed that 50% of the N is lost before utilized.

requires 5 days of incubation time. The average COD for a 100-lb hog is about 1.40. The COD is also a rough measure of the BOD since BOD/COD is usually about constant at 30% for swine wastes. Knowledge of these properties will help engineers design successful waste treatment facilities.

The population equivalent (PE) has been used as a method of comparing the strength of livestock wastes with domestic sewage on the basis of BOD. It is the calculated human population that would contribute the same amount of BOD as the waste in question. The population equivalent for a 100- to 125-lb hog is about 2. This means the wastes from one such hog have a BOD equivalent of waste from about two humans, and aerobically treating the wastes from 2000 hogs could be compared to aerobically treating the wastes of 4000 people. The volume to be treated would be considerably different since the hog wastes are much more concentrated.

RETURNING SWINE WASTES TO THE LAND

Where sufficient land is available and is not located too close to neighbors, the return of swine wastes directly to the soil is still the most used system and about as good as any of the disposal methods. This method has several problems: It must be done regularly to prevent excessive accumulation; it generally results in a highly odorous condition; and spreading on frozen soil may be prohibited in the future.

Spreading as a Solid

This has been the most common method for years. It is probably one of the more economical methods, but has these disadvantages: (1) The manure is highly odorous; (2) stacks of manure provide ideal places for flies and other insects to breed; (3) floors must be cleaned daily; (4) the manure is unsightly; (5) sufficient land must be available; (6) bedding is required to soak up liquids; and (7) more nutrients are lost than with liquid handling. Another disadvantage is the very undesirable working conditions, which are not conducive to attracting dependable labor. Because of the many problems, along with the wide adoption of slotted floors that almost eliminate manure handling in the solid form, this method is much less important today than it has been. For the small or beginning swine producer, it probably has a place, but it will be difficult to consider this method for swine producers in the future.

Spreading as a Liquid

Handling hog manure as a liquid has become a popular method in recent years. With early liquid manure systems for hogs, the floors were washed off and the liquid manure collected and stored in collection tanks. Various types of pumps and augers were used to convey the manure into a tank wagon, in which it was hauled and spread on the field.

When hog wastes are handled as a liquid, a storage tank must be used to accumulate the liquid wastes. The storage tank may be located under the slotted floors or it may be a separate tank into which the liquid is either drained, washed, or scraped. It is important that the tank be properly located, have proper size, and adequate strength.

The most common method of hauling liquid hog manure is to use a vacuum tank wagon that sucks the liquid manure through a rigid vacuum tube into the tank wagon and then uses the same pump to distribute the manure on the land under pressure. These tank wagons come in sizes of 800–2000 gal.

A major nuisance problem while spreading liquid manure is odors. Odor-reducing agents can be mixed with the manure prior to hauling. Odors seem to be less when the spreading is done on a cold day and they also spread less when it is not windy. Disking the manure into the soil soon after spreading also reduces odors. The plow–furrow–cover method of waste disposal was developed to reduce odors during spreading. In this method, the upper layer of the soil is used as the disposal medium for the liquid manure. These is no opportunity for flies to breed or even feed, and in a closed system there are no detectable odors. The manure slurry is deposited in a previously plowed furrow 6–8 in. deep and immediately after deposition, the manure is covered by a plow. This can be done in one or two operations. Today there are tank wagons with chisel teeth that inject a band of slurry under the surface in one operation. This can even be applied to sod without damaging the plants. Application rates of 210, 240, and 360 tons/acre have been tried.

There has been much interest in disposing of the liquid manure through an irrigation system. Some feel that this method offers a potential for a drastic reduction of labor and possibly of total system cost. Six parts of a typical liquid manure system intended for field spreading through an irrigation system are:

1. a means of collecting manure
2. a holding tank
3. a means of agitating manure before pumping
4. a means of pumping the material to the sprinkler
5. a piping network to carry the pumped material to the field; and
6. a suitable sprinkler.

Recommendations for applying hog manure with irrigation equipment are:

1. Dilute the manure as much as possible; maintain a maximum solid content of 5 to 10% if at all possible.
2. Agitate adequately before pumping.
3. Flush sprinkler lines with clear water as soon as practical after pumping manure. This washes the lines and equipment and also washes the manure from the foliage if it was sprinkled on pasture or crops.
4. Drain system in freezing weather.
5. Operate the system as often as practical to reduce the duration of each application. This allows for smaller hauling tanks and results in less runoff during cold weather.

TREATMENT OF SWINE WASTES

Early researchers considered adapting the processes used for the treatment and disposal of municipal wastes—anaerobic and aerobic decomposition. Many of the first attempts at treating swine wastes proved unsuccessful because of failure to understand their true characteristics.

Municipal sewage is composed of large quantities of water polluted relatively lightly with waste organic matter—usually about 0.1% solids (1000 mg/liter), and even strong industrial wastes are seldom more than 0.3–0.4% solids (3000–4000 mg/liter). The BOD concentration for untreated municipal sewage is generally about 300–500 mg/liter. Hog manure, in contrast, contains very large amounts of organic matter—often containing barely enough water to keep the wastes in a fluid form for handling. The BOD of liquid hog manure varies from 30,000 to 50, 000 mg/liter, approximately 100 times that of municipal wastes. With present treatment methods, effluent from hog waste treatment facilities obviously cannot be allowed to discharge directly into a stream. Even if the pollution potential of the hog wastes were reduced 90% by treatment, the waste would still have a BOD of 3000–5000 mg/liter, which is a considerably higher than untreated municipal sewage and much too high to discharge into any stream.

The development of swine confinement housing systems with slotted floors has resulted in the need for producers to store liquid manure for periods of time. Some form of treatment may be necessary to control odors and to reduce pollution to a minimum, even though complete treatment is not feasible at this time.

Anaerobic Treatment

Anaerobic decomposition is one of the most common processes used in treatment hog wastes. Anaerobic processes are those which take place in organic matter when dissolved oxygen is not available. In this environment, chemically bound oxygen is used for the final energy-producing metabolic step. The anaerobic process is usually associated with undesirable odors. The major anaerobic systems currently of importance in waste treatment are the anaerobic lagoon, the anaerobic digester, and the septic tank, which is not considered feasible for treating swine wastes.

The major benefit of anaerobic treatment is its great ability to break down organic matter. The usual purpose of anaerobic lagoons is the destruction and stabilization of organic matter, not water purification. The major attraction of the anaerobic process is its ability to decompose more organic matter per unit of volume than its aerobic counterpart.

A second characteristic of anaerobic digestion that can be beneficial under controlled conditions is the production of a useful fuel, methane, as a principal end product. The gas produced in a properly operating digester is 60–80% methane. The remainder of the gas is carbon dioxide with small quantities of various intermediate products. It is the intermediate products, less than 1% of the gas produced, that are responsible for the odor associated with anaerobic digestion.

Other uses and applications of the aerobic treatment process that may be as important as organic disposal and methane production are (1) improved dewatering characteristics, (2) reduction of solids volume, (3) odor reduction, and (4) pretreatment ahead of aerobic systems.

The anaerobic digestion process is complex, and environmental control is one way in which man can influence it. The environmental factors of primary importance are temperature, pH, and the presence of toxic materials.

Anaerobic Lagoons

Anaerobic lagoons were first adapted for the storage and treatment of hog wastes because of their low initial cost, ease of operation, and lack of serious alternatives. They have not always been successful, for sometimes they release objectionable odors, they can pollute surface and underground water, they sometimes do not decompose the manure adequately, and in time they are apt to fill with sludge.

Loading Rate and Lagoon Size

The maximum loading rate for anaerobic lagoons, which determines the lagoon size, varies within the United States because of the wide variation in temperature. The rate of digestion increases as the temperature increases, with the required detention time halved from 100 to 50 days for an increase in temperature from 50 to 70°F. Because of this relationship between temperature and rate of stabilization, anaerobic lagoons are best suited for warm climates.

The recommended design loading rate over the United States for anaerobic lagoons varies from 0.001 to 0.01 lb of volatile solids per cubic foot of lagoon per day. Since a 100-lb hog will produce about 0.75 lb of volatile solids per day, loading rate would vary between 75 and 750 ft^3 of lagoon per 100-lb hog. For continuous loading in the moderate midwestern climate, a loading rate of 0.005 lb of volatile solids per cubic foot of lagoon per day is recommended, which amounts to about 150 ft^3 of lagoon per 100-lb hog. Additional capacity is desirable in colder climates or where intermittent loading is planned.

Location

Locate a hog operation with a lagoon as far from any home as practical and where the prevailing winds will carry odors away from any residence. Some odors can be objectionable for a distance of half a mile and detectable at a distance of one mile or more.

Locate over impervious soils. Some research indicates that hog wastes serve as a sealant, but other tests indicate that seepage through the bottom can be a problem several years after operation. Do not locate over sandy soils or over limestone formations without using some special means to seal the bottom. Locate at least 150 ft from a well. The construction of a lagoon is much like building a farm pond.

Construction

The lagoon can be round, square, or rectangular. Avoid naturally shaped lagoons with long narrow appendages that are easily isolated from the remainder of the system. When possible, limit width to about 50 ft so the sludge can be removed by drag line or suction pump.

Since anaerobic lagoons do not depend on light penetration or free oxygen, they can be made as deep as the soil characteristics will allow. A deep lagoon keeps the temperature as stable as possible, which aids anaerobic decomposition. A deep lagoon also minimizes surface area, which in turn minimizes the escape of odors. Depths up to 10–12 ft have proved popular and appear to be satisfactory.

The inlet should enter the lagoon near the center or at several locations so the solids will be distributed over the lagoon and not allowed to accumulate near the edge. Both above-surface and below-surface inlets have been used.

Above-surface inlets must be provided with a tight stopper or valve at the building to prevent liquids from freezing and the pipe from clogging. During winter, always collect a quantity in a gutter or collection pit and then empty instead of having a constant overflow which could freeze.

Below-surface inlets require water pressure to work well. They seem to discharge the manure more evenly when there is ice cover. The inlet line can plug near the water level within the pipe where the wastes contain large amounts of solids. Provide access to the submerged inlet line so it may be rodded and cleaned.

Overflow from anaerobic swine lagoons is not permitted to enter the natural water courses by most water pollution control agencies. When the lagoon is full and the addition of new wastes will cause some liquid to overflow, the overflow must receive additional treatment. When the lagoon approaches the overflow stage, consider pumping the liquids out on the land. Surface runoff from rain should be diverted from the lagoon as it would help cause overflow.

All lagoons must be fenced for the protection of children, pets, and livestock. This is particularly true of anaerobic lagoons, which can be fairly deep.

Shape the area around the lagoon to prevent surface runoff from entering the lagoon. Grade the banks so they can be sealed and maintained with mechanical equipment.

Management Practices

A lagoon should not be constructed and then forgotten about. Certain management practices help ensure satisfactory operation.

Fill a new lagoon with surface run-off, roof run-off, or water from a well. After the lagoon is full, direct the run-off water away. Add water any time the lagoon level gets low, and be sure to keep all solids covered with water.

The best time to start a lagoon is in the summer to take advantage of warm weather. Do not add wastes until the lagoon is over half full of water, and then add wastes gradually at first. Seeding a new lagoon with sludge from another well-functioning anaerobic lagoon or from an operating anaerobic digester of a municipal sewage plant is not essential but will speed up bacterial action.

A lagoon functions best by adding wastes continuously rather than storing for long periods of time and then dumping excessive loads. Empty wastes from pits and storage tanks every two weeks if possible. Do not exceed the design loading rate.

The operation of the lagoon can be improved and the lagoon life extended if most of the solids are restricted from entering the lagoon. It is often possible to settle out much of the solids and let only the liquids go to the lagoon. Similarly, at certain times of the year it is often possible to spread all the manure onto cropland, using the lagoon when fields are not available.

The combination of hauling and using a lagoon has worked well. When time and weather permit, the producer hauls the wastes to the field. He hauls during most of the winter when the fields are accessible and when little stabilization occurs in the lagoon. But when the weather does not permit hauling or when he is busy in the field, he lets the storage pits fill up. The liquid wastes run out of the overflow into the lagoon, and the solids remain in the pit to be hauled away later. Some agitation may be required to remove the solids, if they are permitted to build up for a long time.

Adequate mixing of the incoming organic load and the sludge banks that have accumulated in the lagoon is necessary for satisfactory operation of a hog lagoon. Mixing occurs from wind action, gas bubbling, discharge of the wastes from the inlet, and any mechanical means.

Aerobic Treatment

Aerobic decomposition is the reduction of organic matter by aerobic microorganisms, which require free oxygen. The process may be described as one in which oxygen is supplied to the waste, making free oxygen available for the growth of aerobic bacteria. These bacteria use the waste as substrate, breaking part of the organic portion down into the more basic compounds, such as water and carbon dioxide. Some nitrogen may be released to the atmosphere, but much of it is converted into nitrites and nitrates. Unfortunately, aerobic bacteria are not 100% efficient in decomposing organic matter, and there is usually only 40–50% degradation of volatile solids.

The major benefit of aerobic decomposition is that the entire process is essentially odorless. Some form of aerobic treatment appears certain to be used in the future in most systems for animal waste treatment because of odor control alone. Other benefits of aerobic decomposition, are (1) partial decomposition of volatile (organic) solids into odorless gases, (2) destruction of most (maybe all) pathogenic organisms, (3) reduction in the pollutional characteristics of the wastes, and (4) concentrating of minerals that may be more readily applied to the land by some system.

Natually Aerobic Lagoons

The generally acceptable naturally aerated aerobic lagoon is normally a shallow basin not more than 3–4 ft deep excavated in the ground that purifies sewage or other waste water. Purification is done under climatic conditions that favor the growth of algae—warmth and sunshine. If an aerobic lagoon is properly constructed and holds the wastes for a sufficient time, a good destruction of organic matter and a satisfactory reduction in BOD occur.

Many of the early livestock lagoons were expected to function as aerobic lagoons but were usually anaerobic because of too concentrated a loading.

Mechanically Aerated Lagoons

Much interest has developed recently in the possible use of aerated lagoons for the treatment of swine wastes. The principles of operation of the aerated lagoon and the oxidation ditch are essentially the same, other than the container and the type of aerator. In an aerated lagoon, a lagoon is used instead of a ditch and a floating aerator is usually used instead of a rotor aerator. The advantages of the aerated lagoon over an indoor oxidation ditch are: It will handle a larger volume of wastes with a lower initial cost and give longer detention time and possibly greater breakdown of solids; foaming, if it occurs, is outside, giving unlimited freeboard space; in event of equipment failure, the anaerobic conditions with associated odors are outside; a combination of anaerobic, facultative, and aerobic decomposition processes break down a larger percent of the volatiles; and less aeration of free oxygen is probably required. There are four disadvantages: Manure must be carried or flushed into the lagoon; the lagoon may be unsightly; it requires more outside space than oxidation ditch; and it is subject to freezing.

Design suggestions for the mechanically aeratal lagoons are:

1. Use round or oblong shape.
2. Locate near hog house to limit piping problems. Locate in tight, waterproof clay.
3. Provide 1 ft^3 of lagoon for each pound of hog. Depths up to 15–20 ft may be used.
4. The aerator must run continuously.
5. Continuous loading is most desirable.

Oxidation Ditch

There has been considerable interest in aerobically treating hog wastes in an oxidation ditch because of the need for a low-odor method of manure storage and treatment. The oxidation ditch was developed in the Netherlands as a low-cost method of purifying sewage coming from small communities.

The oxidation ditch has a continuous open channel shaped like a racetrack that holds the liquid waste (see Figure 1). An aeration rotor churns the liquid wastes, mixing in air, which supplies the necessary oxygen for the aerobic bacteria. The action of the rotor also keeps the ditch contents circulating so that the solids will be kept in suspension. The rotor must turn continually.

Design

The most common shape of an oxidation ditch is in the form of a racetrack as shown in the figure. Probably any configuration could be used as long as a continuous loop is maintained. The channels should be such that the radius of curvature at the end of the ditch is not too sharp. A sharp bend increases the frictional resistance and retards the flow, causing solids to settle out. Where sharp bends do exist, it is suggested that deflector vanes be installed around the corners to maintain a uniform flow and to reduce settling.

Allow at least 1 ft^3 of ditch volume per 15–20 lb of animal weight to provide adequate dilution of the waste. This allows for about 50 days of detention time. Always design the ditch for the maximum pounds of pigs per building, which is probably at finish weight for the pigs. Designing for only the average weight of the pigs will not be adequate when the ditch is loaded at a maximum.

The major role of the aeration rotor is to inject air into the liquid wastes as a means of adding oxygen. The recommended oxygen that must be supplied equals 2 times the 5-day BOD of swine wastes, or

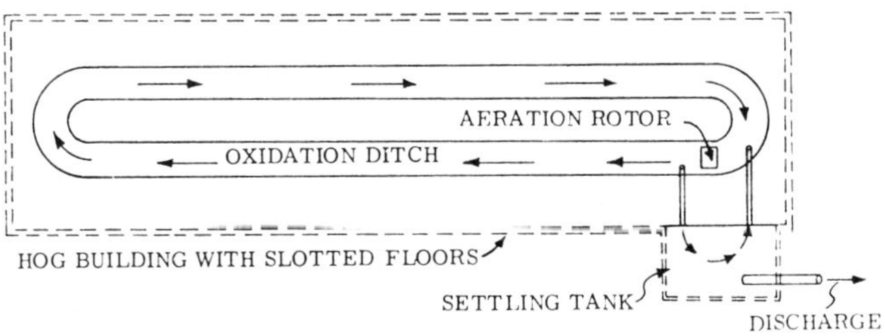

Figure 1. Oxidation ditch.

about 0.64 lb of oxygen per 100-lb pig per day, or per 100 lb of liveweight.

Each rotor must be calibrated in water at the immersion depth at which it will operate to determine its own oxygenation capacity. Capacities have been measured from 1.3 lb of oxygen per foot of rotor at 6-in. immersion and 100 rpm to 5.65 lb of oxygen per foot of rotor at 12-in. immersion and 100 rpm. A general guide for good treatment is that a foot of length of rotor be provided for each 20–25 finish-weight hogs with a 6-in. immersion and turning at 100 rpm. If only aerobic storage without much treatment is desired, it is possible to operate with a few more hogs per foot of rotor.

The depth of rotor immersion affects the rate of oxygenation, rate of liquid flow, and power consumption. The depth of liquid in the channel affects detention time and depth of immersion and rate of rotor rotation necessary to keep the solids in suspension. The simplest method of operation for a constant rotor immersion is to keep a constant liquid level in the ditch by using an overflow that discharges into a holding tank or lagoon.

As a rule of thumb for finishing hogs, the rotor immersion should equal about one-third and liquid depth of the ditch in order to prevent settling of the solids. A lesser immersion depth could possibly be used if a fine grind of feed were used. A 6-in. immersion is about maximum since too many rotor operational problems occur at greater immersion depths. This would give a recommended ditch depth of between 1 and 2 ft.

The aeration rotor must circulate the liquid wastes fast enough to keep the solid particles in suspension. Liquid transport can be the limiting factor in the design for a rotor even when adequate oxygen is being added. The liquid velocity required depends on the weight and size of waste particles in suspension circulating in the ditch. Normally, about 1–1.25 ft/sec velocity of the liquid is recommended. At these velocities, the liquids should not travel over 300–350 ft without passing another rotor since air should be added to the wastes about every 5 min for most efficient treatment.

Operation

Fill the channel with water and start the rotor at the desired blade immersion. Never start with untreated (septic) manure in the ditch. Add the waste gradually at first to minimize foaming. Initial foaming can be reduced with antifoam agents such as vegetable oil, petroleum oil, various commercial products, or by a water spray. Once the ditch establishes an adequate microflora, the start-up foaming should subside. This can take up to 2–3 wk. A slight ammonia odor may be noticed during start-up. The start-up period can be reduced by initially seeding the ditch with activated sludge from a satisfactorily operating ditch or from a municipal sewage plant but it is not necessary. Never drain the ditch completely unless absolutely necessary. Always leave at least one-third the volume for enough activated sludge to provide a seed when the ditch is refilled with water.

The major operational problem has been foaming along with incomplete treatment. Provide at least a 12-in. freeboard depth to allow a reasonable foam level without obstructing liquid circulation. Changes in appearance of foam can serve to alert the operator to imminent foaming problems. Odors can also alert the operator that something is wrong. Clinging odors are a sure sign of an improperly operating ditch, while generally an earthy smell is associated with a properly operating ditch. The dissolved oxygen level (which can be measured with a meter) also helps indicate how the ditch is operating. A reading of 3–5 ppm of dissolved oxygen about 15 ft downstream of the rotor is common. The oxygen level in the last one-fourth of the ditch normally falls below 0.5 ppm, showing that aerobic bacteria have been active. Color of the liquid can also give an indication as to the aerobic condition of the waste. A greenish-black mixture indicates that it is mostly anaerobic and more oxygen is needed, while a dark rich brown indicates a properly operating ditch. If aerobic conditions are not maintained, anaerobic bacteria will predominate causing odors and foam. Open up the house and turn on ventilation before starting a rotor that has not been operating for over 2–3 days. Pigs have been killed by gases released when a rotor was started after having been turned off for some time and the ditch contents had become septic. The rotor should never be off more than 2–3 days without draining part of the wastes and refilling with tap water before restarting.

Regular maintenance is required. Maintenance of the rotor bearings can be a major problem. Hog hair and other foreign materials seem to get into the bearings, so they should be located where they can be easily serviced and replaced. Also, belt drives seem to operate better than chain drives.

After a ditch has been in operation for several months, it will be necessary to remove some sludge to keep it from having excessive amounts of solids in the ditch. A properly operating ditch should stabilize the sludge adequately so it can be handled without causing objectionable odors.

Since the rotor creates an inch or so of hydraulic head, the inlet to the sludge trap can be placed in the ditch just after the flow passes the rotor. If the outlet is placed just behind the rotor, flow should occur just opposite the direction of flow in the ditch (see Figure 1). As the flow enters the sludge trap, the velocity will be reduced and the solids will be deposited in the trap. The sludge trap will need to be cleaned out periodically and can be pumped on to drying beds or hauled directly to the fields for fertilizer.

If no sludge trap is available, the solids content can be lowered by draining several inches of liquid in the ditch and diluting the remainder with tap water.

The liquids from the oxidation ditch will not receive adequate treatment to allow disposal into a natural watercourse. Allow overflow to discharge into a lagoon after passing through a settling tank to settle out remaining solids. Irrigate from the lagoon if surplus water is a problem.

If it is not desirable or possible to have a lagoon, build a storage tank outside the building to store the overflow from the ditch. Haul with a tank wagon and spread on the fields when necessary.

OTHER METHODS OF WASTE DISPOSAL

There are several other methods of waste disposal. Some of them show very little promise for use with swine wastes.

Dehydration

Dehydration appears to have little value for the disposal of swine manures. Swine manure when sufficiently dry may be stored, sacked, or spread without problem, but the process of drying appears to be difficult as well as expensive and the market for dried swine manure is questionable. Swine manure as it is produced generally contains 85% water and 15% dry matter. To produce dried manure containing 90 lb of dry matter and 10 lb of water, one must start with 600 lb of wet, raw manure. This means that 500 lb of water must be removed in the drying process. Not only must drying equipment dry the manure, but it must reburn the gases to prevent odors. The Michigan State Agricultural Experiment Station conducted experiments on the drying of poultry manure, but the results are not directly applicable to swine wastes since liquid swine wastes contain much more moisture than poultry manure. Ludington (1963) calculated that poultry manure must sell for at least $30 per tone before dehydration is profitable.

Incineration

Incineration (see Chapter II.13) also appears to have little to offer for the disposal of swine manure. It will reduce the solids, but very large amounts of heat must be provided. Odors and gases from the system must be controlled or they will pollute the air and become objectionable to nearby inhabitants. Hog wastes contain so much water that this method does not seem applicable to their disposal unless some new process is developed.

Composting

Composting (Chapter II.11) is a method of waste disposal not directly suited to liquid-manure disposal systems but more particularly to solid-waste disposal systems. It could be used for the wet solids from a settling tank, however, by the addition of dry material, such as chopped straw, so that the moisture content is reduced. Composting requires time, and equipment that is beyond the scope of most swine producers. The only way composting offers any possibility is if large quantities of manure are collected along with other wastes.

Use of Swine Wastes in Feed

The utilization of livestock manure as feed has been of interest to some researchers since before 1960. This is understandable since the feces of farm animals consist chiefly of undigested food that has never really been within the body proper. This undigested food is mostly cellulose and lignin. Also, a portion of the other nutrients usually escapes digestion.

Some of the first research was done with the substitution of poultry manure in the ration of beef cattle. In feeding poultry litter to hogs, a ration of 15% poultry litter in the feed of pigs up to 100 lb did not reduce their performance, and for hogs 100 lb and over, and for breeding sows, as much as 40% was unobjectionable. This could save from \$9 to \$15 per ton of feed. Others investigated the nutritional value of oxidation ditch sludge. They found the feed-value comparisons between corn and dehydrated oxidation ditch sludge from finishing pigs shown in Table 3.

Rat-feeding trials were performed to test the potential of the dried sludge as a protein supplement. It was noted that the test rats did not refuse the feed until the sludge made up nearly half of the protein requirement of the ration. Rats fed rations containing less sludge than this had appetites and weight gains comparable to those of rats fed balanced rations containing no sludge. Sludge feeding appears promising, but more testing is needed before animal performance can be predicted accurately.

Diggs et al. (1965) compared the performance of 100-lb pigs fed rations with 0%, 15%, and 30% dried pig feces. The pigs fed 15% pig feces gained faster than the control pigs receiving no feces, and the pigs

Table 3 Analysis of Sludge

	Corn	Oxidation ditch sludge
Energy (calories/g)	4000	3490
Calcium (%)	0.02	0.87
Phosphorus (%)	0.28	1.25
Protein (%)	9.0	27.0
Lysine (%)	0.21	0.34
Methionine (%)	0.20	0.25
Threonine (%)	0.31	0.69
Isoleucine (%)	0.35	0.47

receiving 30% pig feces in their ration gained as fast as the control group. The pigs receiving 15% feces were as efficient as the control group, but the pigs receiving 30% feces were less efficient. No undesirable flavor was detected in the pork as a result of feeding the various levels of feces. The digestible energy of the feces was 1.312 kcal/g, which is about 37% the value of corn and about 50% the value of oats.

GASES FROM STORED SWINE WASTES

There has been much concern about the quality of the air inside confinement swine buildings with slotted floors since the wastes may be held in the building long enough to produce gases and odors. Noxious gases and odors formed from stored wastes can be irritating to both the hogs and operator and have been the cause of numerous complaints and even lawsuits by neighbors.

The most important gases generated from stored manure are carbon dioxide, ammonia, hydrogen sulfide, methane, and a large group of trace organic compounds. In the following discussion, allowable concentrations are expressed as threshold limit values (TLV), given in parts per million (ppm). A threshold limit value is the maximum concentration to which industrial workers may be repeatedly exposed for an 8-hr working day during a working lifetime without adverse effect. Concentrations can be expressed either as percentage or ppm (1% would equal 10,000 ppm). The TLV for swine, especially small pigs, is probably lower than the TLV for man since swine are exposed 24 hr a day, 7 days a week.

Before reviewing the effects of noxious gases, it must be emphasized that *under normal conditions in an adequately ventilated confinement unit there is no evidence that any noxious gases reach lethal or even harmful concentrations for pigs or humans.*

Carbon Dioxide (CO_2)

Carbon dioxide is a colorless, odorless gas considerably heavier than air and highly soluble in water. Normal atmosphere contains about 300 ppm (0.03%), and more is released by respiration of the pigs and by manure decomposition. Most of the gas in the bubbles coming from stored liquid manure or from lagoons or oxidation ditches is carbon dioxide. Concentrations of 600–1800 ppm have been measured in a ventilated confinement hog building, compared with more than 4000 after 6 hr in an unventilated house.

Carbon dioxide is not highly toxic to humans. Its main danger is that its formation contributes to an oxygen deficiency and so results in asphyxiation. The TLV for carbon dioxide is 5000 ppm. Concentrations of 100,000 ppm (10%) cause violet panting, and higher concentrations have a narcotic effect. Death will occur after a few hours of concentrations of 250,000 ppm (25%) or more.

No significant distress is felt by the pigs from inhaling air containing 20,000 ppm (2%) carbon dioxide if normal quantities of oxygen are present. Air containing 40,000 ppm causes deeper and faster breathing. More than 100,000 ppm may produce dizziness and even unconsciousness. Even concentrations of 200,000 ppm are not necessarily dangerous to pigs of more than 150 lb for no more than an hour. Pigs tend to recover completely after they are put back into a normal atmosphere. Pigs have died after the failure of ventilation equipment caused by power failure has resulted in excessive carbon dioxide.

Ammonia (NH_3)

Ammonia is colorless with a characteristic pungent odor. It is lighter than air and highly soluble in water. It does not burn readily, but air mixtures of over 16% ammonia are considered explosive. Ammonia is released from fresh manure and during anaerobic decomposition of organic matter. The problem is less with slotted than with solid floors, particularly heated ones, because of ammonia's high solubility in water. Concentrations in confinement hog buildings have been measured as high as 35 ppm in a ventilated building and 176 ppm in an unventilated building.

A TLV for humans of 50 ppm has been set to protect against irritation to the eyes and mucous membranes of the respiratory tract. Air containing 50–100 ppm, however, can be inhaled for hours without apparent harm. A concentration of 5000 ppm may cause suffocation.

Because it is an irritant, ammonia tends to produce discomfort in pigs, but only in extremely high concentrations does it cause permanent damage. At 100–200 ppm, it induces sneezing, salivation, and loss of appetite; the pigs express their irritation by shaking their heads. Prolonged exposure may make the pigs more susceptible to respiratory diseases.

Hydrogen Sulfide (H_2S)

Hydrogen sulfide is colorless and smells like rotten eggs. It is somewhat heavier than air, is soluble in water, and burns with a bluish flame. It is produced from the decomposition of organic wastes under anaerobic conditions. A concentration of 0.09 ppm in a normal ventilated confinement building rose to 0.28 ppm after the ventilation was shut off for 6 hr. Dangerous concentrations can be released by vigorous agitation of stored liquid manure. Concentrations reaching 200–300 ppm have been reported a few minutes after pumping out a storage pit and as high as 800 ppm during vigorous agitation. You can detect the presence of hydrogen sulfide by noting the black accumulations of copper sulfide that form on copper thermostats and electrical wiring, the white deposits of zinc sulfate on galvanized steel, and the black discolorations where lead-pigmented paint was used.

Hydrogen sulfide is one of the more toxic gases associated with liquid manure storage, even at low concentrations. Even its characteristic odor does not give adequate warning because the sense of smell can be rapidly fatigued by hydrogen sulfide, and high concentrations do not give a proportionally higher odor. Its TLV has been set at 10 ppm. Concentrations of 20–150 ppm can affect the eyes after 6–8 min and cause severe irritation to the eyes and respiratory tract if inhaled for an hour. Exposure to 500 ppm for 30 min will cause severe headaches, dizziness, and excitement. High concentrations of 800–1000 ppm cause immediate unconsciousness and death through respiratory paralysis unless artificial respiration is immediately applied.

Pigs are made uncomfortable by prolonged exposure to low concentrations of hydrogen sulfide. Pigs exposed continuously to at least 20 ppm develop fear of light, loss of appetite, and nervousness; 50–200 ppm cause vomiting, nausea, and diarrhea. In acute poisoning, hydrogen sulfide acts so rapidly that there are few symptoms of imminent danger. Sudden nausea and unconsciousness are followed by death at concentrations of 800 ppm and above. The pigs may recover completely from exposure to high concentrations, but they may be susceptible to pneumonia and other respiratory diseases.

Methane (CH_4)

Methane is colorless, odorless, and much lighter than air. It is not very soluble in water and is highly flammable. Methane is generated from the anaerobic decomposition of manure. Since it is lighter than air and insoluble in water, methane tends to accumulate near the ceiling in stagnant corners or in the top corners of tightly enclosed manure pits. When the air mixture reaches 5% (50,000 ppm), as long as it does not exceed 15%, any small spark can cause a dangerous explosion.

Methane is not considered toxic, but concentrations can develop that produce an asphyxiating atmosphere. The maximum allowable concentration has been listed at 1000 ppm.

Pigs suffer no harmful effects from inhaling methane. Since it is considerably lighter than air, it will be dissipated rapidly if there is adequate ventilation. It may accumulate in stagnant corners, with considerable risk of explosion if ignited by a spark or flame. Concentrations in confinement houses are normally well below the hazardous 5% level, although they have been reported as high as 8.4%. Explosions attributed to methane have occurred around manure storage pits.

Potentially Lethal Situations

Manure gases have been fatal to both pigs and humans. Pigs have died after a ventilation failure in a tight building, when liquid manure was stirred or agitated vigorously in a manure pit usually under slotted floors, or when an animal fell into a manure storage pit. Humans have died when they entered a manure pit without first ventilating it properly.

When the ventilation system in a tightly constructed building full of pigs stops, the pigs may die from asphyxiation because of lack of oxygen and increased carbon dioxide, or from heat stress. Conditions become critical when the oxygen content drops from the normal 21% to 10% or below. A confinement unit should have some type of safety or warning device to notify the operator if the mechanical ventilation system fails.

Vigorous agitation of liquid manure that has been stored for some time can release noxious gases and create dangerous or even lethal conditions. If manure stored under slotted floors is agitated while the pigs are allowed to stay inside the building, extreme care should be taken to ventilate the building and the pigs should be watched for any signs of ill effects. Instances have been reported of pigs dying during the agitation of manure, and an operator in Scotland was overcome during the emptying of a storage pit.

The air in a manure storage pit is not safe to enter without first ventilating the pit. Both hydrogen sulfide and carbon dioxide are heavier than air and will tend to accumulate in the pit. Several human deaths have been reported of operators who entered a covered manure pit, were instantly overcome, and drowned in the remaining liquid wastes. *Never enter a manure storage pit unless it has been properly ventilated and you are equipped with a proper mask.* Always have someone stand by with a rope attached to you to pull you out at the first signs of dizziness.

Odors from Hog Wastes

Until a few years ago, odors were considered unfortunate but inevitable. With the growth of public awareness of air pollution, odors are not longer accepted without complaints.

Odors arising from the wastes associated with the confinement of large number of animals have been a persistent problem. These offensive odors are usually associated with septic or anaerobic conditions of biological decomposition. Under normal storage conditions of hog wastes, an anaerobic environment will quickly develop. If aerobic conditions were maintained, an "earthy" odor would persist, which is not usually a problem. This correlation between odor and the presence or lack of oxygen has been established by many researchers.

The odors from animal wastes are complex mixtures of malodorous gases (hydrogen sulfide and ammonia) and organic compounds. Odorants such as amines, mercaptans, sulfides, organic acids, and skatole are the major components of the odors in hog wastes. The typical characteristics of these compounds partly explain why these odors carry over great distances and often persist for long period.

There are odor-control products on the market. The masking agents and counteractants have been found to be the most effective. All are fairly expensive for treating any sizeable amount of wastes.

REFERENCES

1. A. C. Dale, "Aerobic Treatment of Animal Wastes," ASAE Paper 67-927A, 1967, 9 pp.

2. D. L. Day, E. L. Hansen, and Scott Anderson, "Gases and Odors in Confinement Swine Buildings," *ASAE Trans.* **8,** 118–121, 1965.

3. H. J. Eby, "Manure Disposal Lagoons," USDA, ARS 42–75, 1963, 12 pp.

4. T. E. Hazen and D. W. Mangold, "Functional and Basic Requirements of Swine Housing," *Agr. Engr.* **41,** 585–590, 1960.

5. D. D. Jones, D. L. Day, and A. C. Dale, "Aerobic Treatment of Livestock Wastes," Ill. Exp. Sta. Bull. 737, 1970, 55 p.

6. R. C. Loehr, "Anaerobic Lagoons: Considerations in Design and Application," *ASAE Trans.,* **11,** 320–322, 330, 1968.

7. R. C. Loehr, "Pollution Implications of Animal Wastes—a Forward Oriented Review," USDI. FWPCA, 1968, 175 pp.

8. R. C McKinney and Robert Bella, "Water Quality Changes in Confined Hog Waste Treatment," Completion Report, Kansas Water Resources Res. Institute, Univ. Kansas, 1967, 88 pp.

9. Midwest Plan Service, "Handling Swine Wastes," Midwest Plan Service AED-11, Ames, Iowa, 1969, 4 pp.

10. J. R. Miner and T. E. Hazen, "Ammonia and Amines: Components of the Swine Building Odor," ASAE paper 68–910, 1968, 13 pp.

11. A. J. Muehling, "Swine Housing and Waste Management—A Research Review"—Univ. Illinois A Eng-873, 1969, 91 pp.

12. A. J. Muehling, "Swine Confinement Buildings," 1969 Illinois Swine Seminar, Univ. Illinois Cooperative Extension Service, 1969, 14 pp.

13. A. J. Muehling and G. R. Carlisle, "Farrowing Houses for Swine," Univ. Illinois College Agricultural Circular 973, 1967, 17 pp.

14. A. J. Muehling, "Production-Line Confinement Swine Buildings," ASAE paper 67-912, 1967, 11 pp.

15. National Symposium on Animal Waste Mgt. Proc., ASAE publication SP-0366, 1966, 161 pp.

16. E. P. Taiganides and T. E. Hazen, "Properties of Farm Animal Excreta," *ASAE Trans.* **9,** 374–376, 1966.

17. E. P. Taiganides and R. K. White, "Origin, Identification, Concentration, and Control of Noxious Gases in Animal Confinement Production Units," Report Ohio State Univ. Res. Foundation, 1968, 22 pp.

PART V
particulate collection

Particulates are particles of dust and fume, or aerosols, ranging in size from 1 to 100 μm. (A micrometer or micron (μm) is 0.001 mm or about 0.00004 in.; 100 μm = about 0.004 in.)

The approach is that of basic collection theory, with reference to available equipment, its design, utility, and limitations. Specific case studies are offered of successful practice in the production of steel; in electric arc furnaces; ferroalloy furnaces, where conditions are especially difficult; and a secondary lead plant. A concluding section considers fly ash, particulates from high-intensity boilers, and utility power plants.

Particulate collection, theory and practice is discussed by Hardison in Chapter V.1, while the specific application in steel processes is written by Jensen of Lukens Steel in Chapter V.2. Lukens is a steel company with no ore treatment, no blast furnaces, no hot metal and with scrap as raw material, so that the section emphasizes particulates in steel processing. The specific case of dust collection from an electric arc furnace is given by Phillips in Chapter V.3. Details, operating, and design data collected over years are tabulated in Chapter V.4 dealing with Particulate Collection in ferroalloy plants of Union Carbide, primarily at Marietta, Ohio, and Alloy, West Virginia, and presented before the AIME. Particulate collection in a lead smelting plant is in Chapter V.5. Fly ash, the collected particulates of power plants and utilities, finds industrial application as described by Meikle in Chapter V.6. Dust collection problems have all the available technology for their solution. The economic aspects are always disheartening, for large expenditures are required for continuing success. Cost factors are continuously compared throughout the section.

If the reader wants further specific data on related equipment, reference should be made to the dust collection section of Perry, *Chemical Engineers Handbook,* McGraw-Hill, New York, the handbooks relating to heating, ventilating, and air conditioning, the data and manufacturing section of the Portland Cement Association, and the manufacturers of the specific equipment of interest as shown in the mechanical engineering catalogs (Sweet, etc.).

Collecting equipment for wood, sawdust, pulverized nuts, corncobs, and cellulose based materials are relatively simple applications since the particulates are large and settle easily. Particulates from pyrolysis operations involving removal of insulation from conductors, burning of rubber, cooking of coal, and quenching of coke, are less difficult than fly-ash collection. The pyrolysis type of operations must have afterburners to complete combustion of odorous substances.

CHAPTER V.1

particulate collection, theory and practice

L. C. HARDISON

This chapter deals with the separation of finely divided solid and liquid waste materials from gas streams in which they are suspended. Most of the discussion is limited to fairly small particles, as those over about 100 μm tend to settle quickly under the influence of gravity.

Large-particle-size solids are frequently referred to as *granular*. They are ordinarily measured by screen size for sizes about 325 mesh (44 μm, or just under 0.002 in.). Granular materials are of considerable interest in product recovery, pneumatic conveying, fluidized catalyst beds, and other applications where materials of large particle size are deliberately suspended in gas streams. They are frequently separated by passing the particle-laden gas through a large chamber in which the gas velocity is relatively low compared to the settling velocity caused by gravity, or inertial forces may be used to augment gravity settling in centrifugal collection devices. These present relatively little difficulty aside from problems involved with conveying or otherwise handling the collected solids. Most of the problems with separation of particulate matter are involved with smaller particles.

Solids in the range between 1 and 100 μm are classified as *dusts;* liquids in this size range are *mists*. Techniques for characterizing dusts and mists and for separating them from the gas streams in which they are suspended are well developed and will be described. Materials smaller than 1 μm are *fumes*. Fumes may be dispersions of either solid or liquid particles. These materials are far more difficult to categorize than are the dusts and, in many cases, more difficult to remove from gas streams.

DUSTS

The name implies not only the particle size range (1–100 μm) but also something about the origin of the particles. Dusts are ordinarily

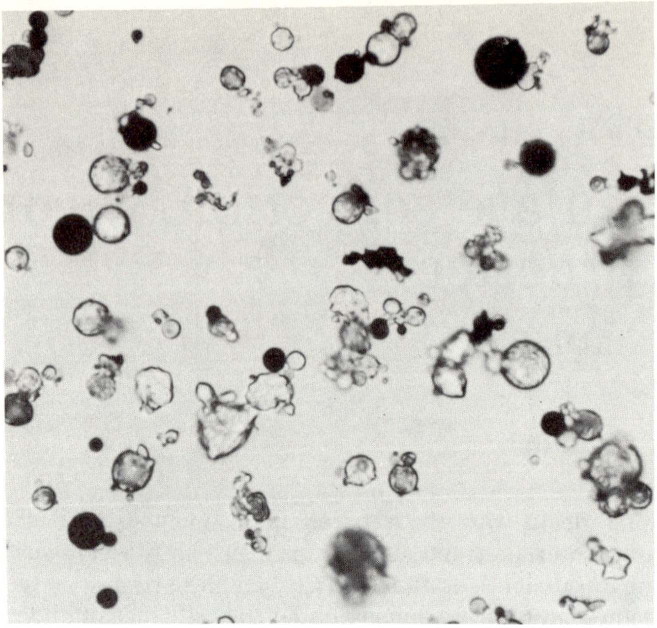

Figure 1. Photomicrograph of fume material from rockwool kiln.

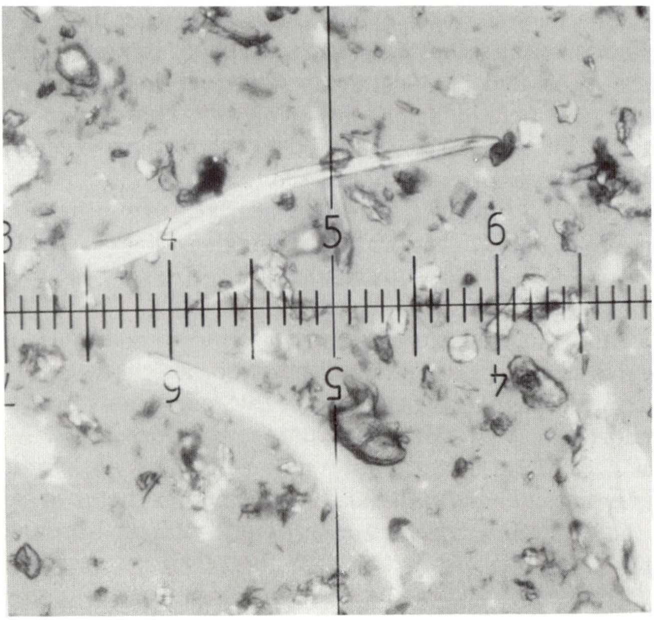

Figure 2. Photomicrograph of dust from a coal fired furnace (Courtesy Walter C. McCrone Associates, Inc.).

generated by mechanical action. They are formed by the abrasion of large particles, or by fracturing large particles to produce smaller and smaller ones. Crushing, grinding, pneumatic conveying, and other mechanical processes produce dusts. The particles are irregular in shape and have sharp edges, fracture planes, and the like, as given in Figures 1, 2, and 3.

To effectively control the emission of dusts from a process, it is necessary to characterize both the gas stream and the particulate matter. In selecting equipment for collection of dusts, the properties of the particulate matter are of prime importance in selecting the type of collector and the level of energy input for adequate separation. The properties of the gas stream are usually of importance in setting the equipment size and materials of construction.

Particle size is the most important single characteristic of dusts. Wet scrubbers and mechanical centrifugal collectors are extremely sensitive to particulate size and collect large dust particles far more efficiently than small ones. Although fabric collectors and electrostatic precipita-

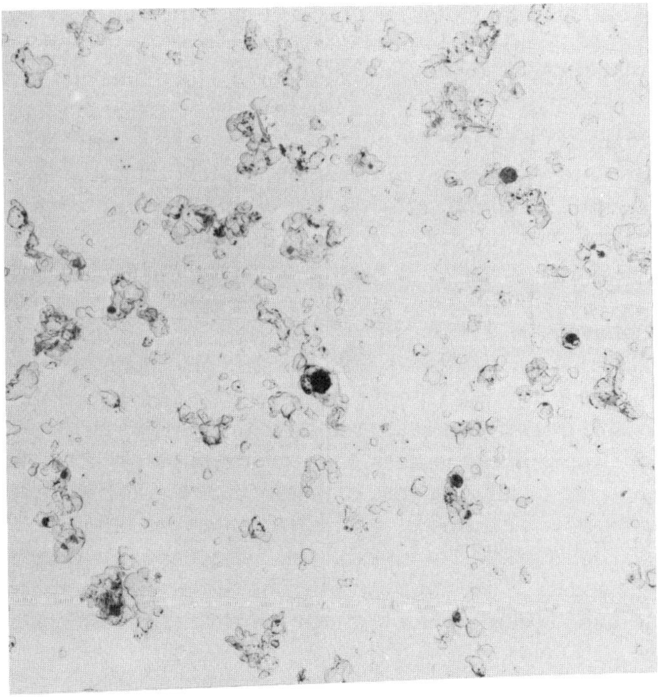

Figure 3. Photomicrograph of Dust from Grain Elevator (with ocular micrometer) (Courtesy Walter C. McCrone Associates, Inc.).

tors are less size selective, a reasonable knowledge of particle size is necessary to ensure good performance for these devices. Other variables important in the selection of dust collecting equipment are particle density, bulk density, abrasiveness, bridging and caking tendencies, and the ability to absorb moisture from the gas stream. For electrostatic precipitator applications the electrical resistivity is of particular importance.[2]

Dusts are very irregular in shape, so it is necessary to discuss "particulate size." Because many particles must be dealt with in any given sample, only one dimension can be given for each particle. "Diameter" may be substituted for particle size. The measurement of diameter in the manner used for measuring macroscopic objects—that is, by placing a calibrated measuring stick next to the object—is done by placing a very small representative sample of dust on a microscope slide and comparing the length of the particle with marks on a calibrated reticule. Measurements of this type yield "optical diameters," meaning that they have been established by visual inspection of the particles.

A microscopist may describe the particles in terms of several alternative optical diameters. He may estimate the diameter of a sphere having the same silhouette area as the particle in question by using a counting grid. He may measure the length along one direction as the diameter. The first measures the diameter that splits the particle into two approximately equal areas above and below the reference line. The second measures the longest dimension parallel to the reference line.

Microscopic analysis affords a relatively simple method for characterizing particle size, and also offers additional insights into the properties of the particulate matter which may be important in selecting collection equipment. However, the optical particle size relates to the behavior of the particles less well than some of the other measures which may be used. The particle size may be measured by complex light scattering techniques. These techniques relate both the direction and intensity of scattered light to particle size. However, the light scattering apparatus does not provide pictures of the individual particles as does a microscope. It is customary to calibrate light scattering apparatus in terms of equivalent spherical particles. Thus, irregular particles would be reported as having the same diameter as an equivalent sphere. This approximates the diameter of a sphere having the same outside surface area as the particle. Because the area of a sphere is

$$A = \tfrac{4}{3}D_t^2 \tag{1}$$

where A is the surface area and D_t is the particle diameter, it is apparent that the equivalent spherical particle diameter is given as

$$D_t = \frac{3A}{4} \tag{2}$$

In the same manner, the diameter may be reported as that of a sphere having the same mass, the same settling velocity through air or water, or on the basis of any other measurable property.

Several effective particle diameters are defined below.

Mass equivalent diameter:

$$D_t = \sqrt[3]{\frac{6W}{\rho}} \tag{3}$$

where W is the particle weight and ρ is the particle density.

Equivalent aerodynamic diameter:

$$D_t = \frac{18V\eta}{(\rho - \rho_G)G} \tag{4}$$

where V is the measured settling velocity, ρ is the particle density, ρ_G is the gas density, and η is the kinematic viscosity of gas.

For characterization of particles to be collected by inertial effects such as gravity settling or centrifugal separation, the aerodynamic equivalent particle size is used.[2] Where light scattering or visible effects are of principal concern, the surface area equivalent particle diameter is likely to be of the greatest importance.

PARTICLE SIZE DISTRIBUTION

Because dust particles are ordinarily present in great numbers in gas streams, all of the common techniques for characterizing size do so statistically. That is, any observed count of particles is smoothed and reported as though a continuous range of sizes were present.

Table 1, a tabulation of particle counts by optical microscopy, gives the number of particles in each of the size ranges. To provide meaningful data for particulate collection, the distribution is recalculated to give the fraction by volume (or weight) of the particles in each of the size categories and plotted as a histogram or bar graph. The true statistical distribution is represented better by the smoothed curve drawn through the midpoints of the bars than by the bar graph.[4]

The distribution does not have the familiar bell-shaped curve common to normal distributions. However, if the log of the frequency (or likelihood of the particles being at a specified size) is plotted, the bell shape is obtained for most dust samples. This type of distribution is log-normal. The fact that dust particles produced by mechanical processes usually conform to this shape makes it possible to use statistical methods that are very simple in describing the particle size distribution.

If the size distribution can be represented by a straight line, only two pieces of information are required in order to completely characterize the size distribution. That is, the dust may be described as having 50% by weight less than 10 μm and 84% by weight less than 25 μm. It may be described as having a median diameter (the 50% point on the log-normal plot) of 10 μm, and a geometric standard deviation of 2.5. The geometric standard deviation is the ratio of the particle size at the 84% point to the particle size at the 50% point.

The linearity of size distributions when plotted on log-normal paper is usually good in the range of above 1 μm. There is likely to be a deviation for very small particle sizes, which may have been formed by condensation of solid or liquid droplets rather than by mechanical action.

The particle size distributions for several common sources of dusts are given in Table 2.[2, 4]

Table 1 *Tabulation of Particle Counts by Optical Microscope*

Interval Number	Greater Than (μm)	but	Less Than (μm)	Count	%
1	60		—	38	2.5
2	40		60	56	3.7
3	30		40	72	4.8
4	25		30	65	4.3
5	20		25	101	6.7
6	15		20	150	10.0
7	10		15	269	18.0
8	6		10	314	21.0
9	4		6	195	13.0
10	2		4	180	12.0
11	—		2	60	4.0
				1500	100.0

Table 2 *Typical Particle Size Distribution for Several Common Sources of Dust*

Application	% < 5 μm	% < 10 μm	Ref.
Cement kilns	20	40	2
Municipal incinerator	20	33	25
Pulverized coal-fired furnace	15	30	4
Cyclone furnace	67	76	4
Spreader stoker furnace	4	10	4
Steel blast furnace	30	60	2

OTHER PARTICLE SIZING TECHNIQUES

Dust particles are frequently sized by methods based on settling through air or water. These methods involve measuring the terminal velocity of the particles and the specific gravity, so that the diameter can be calculated from Stokes' Law, by Eq. 4.

Measurements based on settling through air are ordinarily the most meaningful in selecting equipment for collection by gravity or centrifugal separation. Separation of particulate matter from liquid streams is better characterized by measuring the terminal velocity of the particles through a liquid medium. However, because many dust samples are collected in liquid-filled impingers, the sedimentation method of size measurement is frequently used.

The manufacturers of centrifugal collectors have adopted the Bahco Elutriator[5] as a test device for characterizing dust particles, of which Figure 4 is a schematic drawing. A 10-g dust sample is placed upon a spinning rotor where centrifugal force acts to drive the particles toward the outside of the rotor. The flow of gas passes radially inward and tends to carry the particles in the opposite direction. The weight fraction leaving the rotor at each measured gas velocity gives a measure of the particle size distribution when the instrument has been calibrated using standard dusts of known particle size distribution and specific gravity[6]. Other methods are shown in Figures 5–9.

Dust tends to fall into three basic ranges of density:

1. Organic materials at near specific gravity 1.0.
2. Metal oxides such as dirt, ash, rock, at 2.0–3.0.
3. Heavy metals and oxides such as iron, lead, lead oxide, etc. at specific gravity 5.0 and higher.

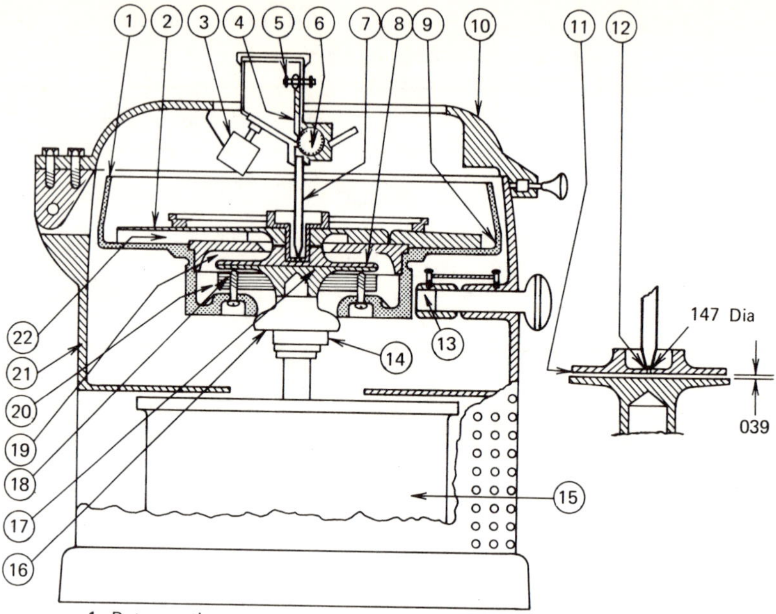

1. Rotor casting
2. Fan
3. Vibrator
4. Adjustable slide
5. Feed hopper
6. Revolving brush
7. Feed tube
8. Feed slot
9. Fan wheel outlet
10. Cover
11. Rotary duct
12. Feed hole
13. Brake
14. Throttle spacer
15. Motor – 3520 RPM
16. Grading member
17. Threaded spindle
18. Symmetrical disc
19. Sifting chamber
20. Catch basing
21. Housing
22. Radial vanes

Figure 4. Cross section of centrifugal classifier.

Most industrial dusts are in the second category, and it is cutomary to report the particle size distribution as though the specific gravity of the dust were 2.5. This allows the manufacturer of the centrifugal collector to size the collection device without making any correction for deviations of the true density of the particle matter away from the 2.5 specific gravity standard. ASME Power Test Codes 27 and 28 contain descriptions of the sampling methods that should be employed to obtain representative dust samples and the methods of analysis for particle size distributions. Power Test Code 28 describes the Bahco procedure.[6] Table 3 lists the particle size distribution for a pulverizing coal-fired flyash. Figure 10 is a fractional efficienty curve for a mechanical dust collector which gives the efficiency of collection for each particle size fraction as a function of the aerodynamic diameter.[7]

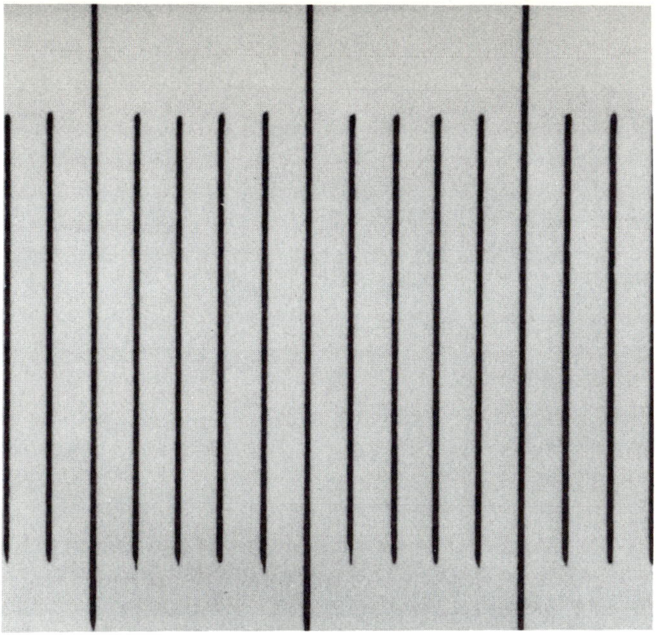

Figure 5. Stage micrometer (Courtesy Walter C. McCrone Associates, Inc.) 10 μm between lines.

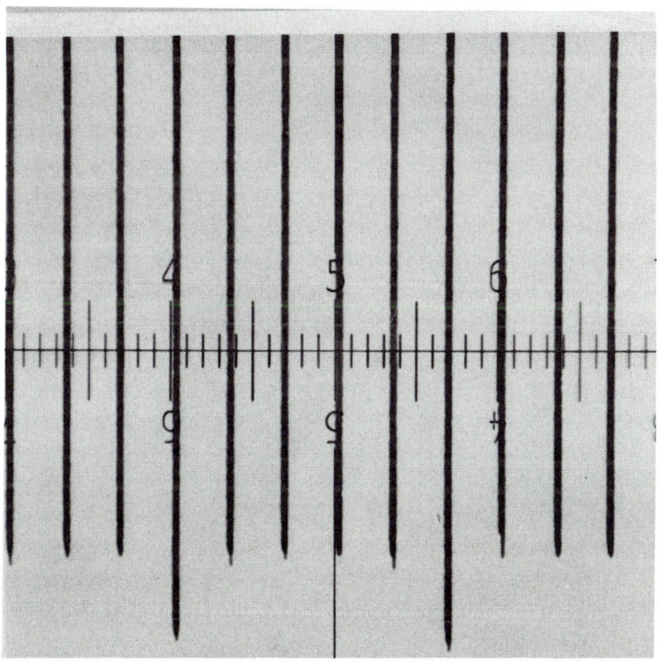

Figure 6. Stage micrometer (10 μm between divisions, with ocular micrometer) (Courtesy Walter C. McCrone Associates, Inc.).

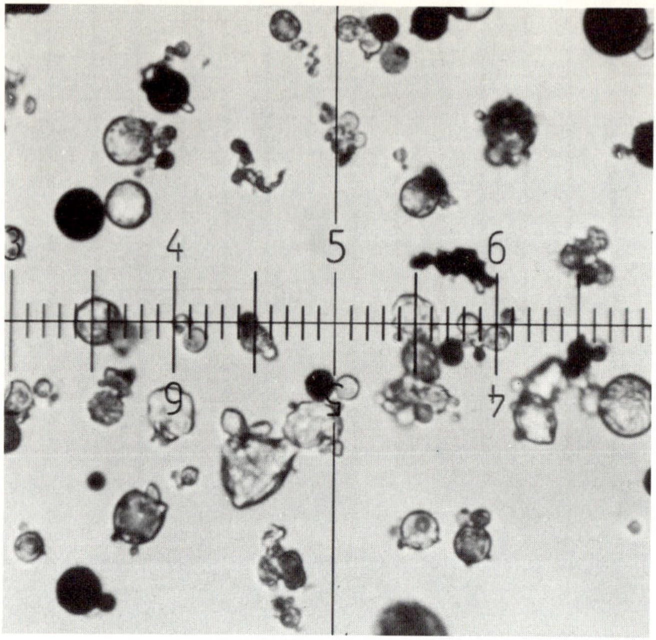

Figure 7. Ocular micrometer with power plant fly ash. (Courtesy Walter C. McCrone Associates, Inc.).

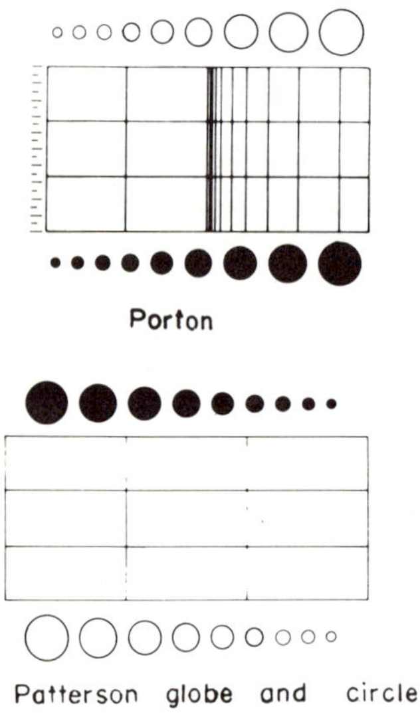

Figure 8. Counting grids for measuring particle diameters (Courtesy Walter C. McCrone Associates, Inc.).

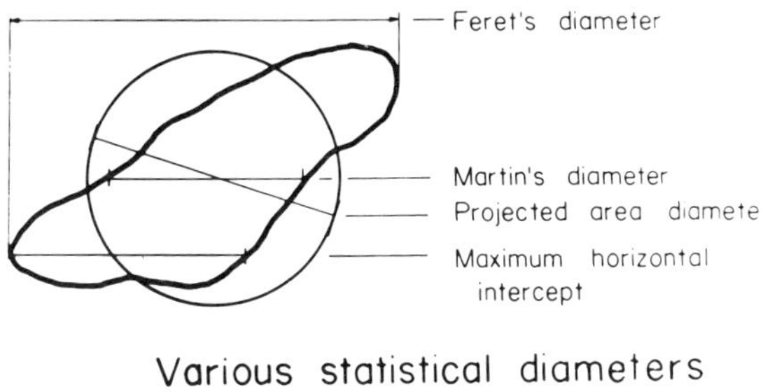

Figure 9. Two methods of classifying particle diameter, statistical diameters (Courtesy Walter C. McCrone Associates, Inc.).

For particle size distributions likely between 1 and 5 μm diameter, the Cascade Impactor offers a better approach to classifying the particles.[9] Particles are impinged upon microscope slides or filters that have been weighed accurately and that collect the largest particles in the carrier gas. A second impactor directs the remaining sample at a second slide with a higher velocity, and collects finer particles. A series of impaction tubes separates the sample into various size ranges. There is some overlap, and an optical microscope determines the efficiency of

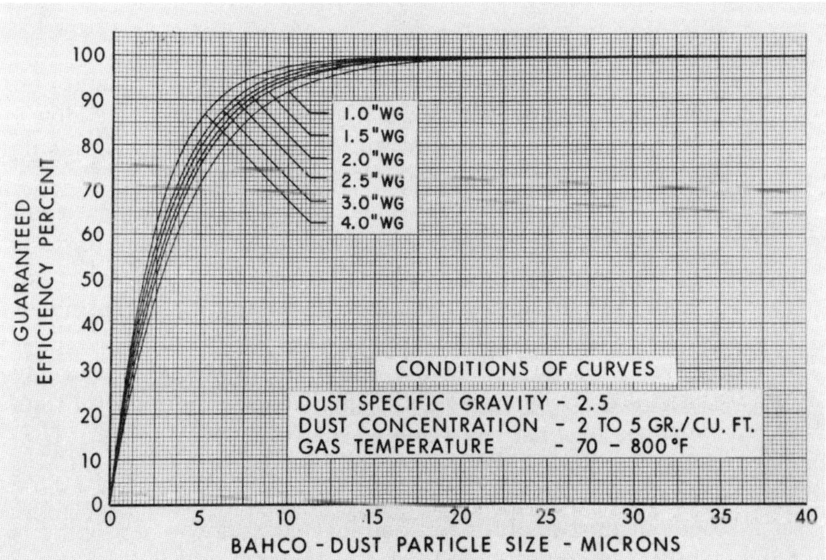

Figure 10. Fractional efficiency curve for mechanical dust collector (Courtesy U.O.P. Air Correction Div., Darien, Conn.).

Table 3 *Typical Particle Size Distribution by Bahco (Courtesy UOP Air Correction Division, Daridon Corp.)*

Pressure drop	2.5 in. WG
Relative humidity	8–15%
Grain loading	2.5 grains/ft

Test dust—Cos Cob (Bahco Analysis: Specific Gravity = 2.5)

Micron Range	Analysis
+60	3.5
+40–60	1.5
+30–40	1.9
+20–30	4.6
+15–20	5.9
+10–15	13.6
+7.5–10	12.0
+0.0–7.5	57.0

separation. While impactors are capable of providing good results for dusts, they are more frequently used for measuring materials below 1 μm in diameter, where the more conventional elutriation methods do not work well.[9]

Several other techniques are used for characterizing dry dust samples. Light-scattering devices are capable of measuring particle concentration and size distribution simultaneously in a flowing gas stream. These devices are costly but offer the possibility of continuous records of particle size distribution of operating equipment.

For larger dust sizes, dry screening is limited to particles above 44 μm or 325 mesh size. Other methods of analysis are more useful than screen size analysis for dust collection work. Table 4 is a table of U.S. Standard screen size and the equivalent micron diameter that will pass through each size.[10]

Wet methods of dust analysis are employed where wet scrubbing is to be the collection method. Samples are collected in a water-filled impinger and presented to the laboratory in suspension. Sedimentation yields an aerodynamic diameter similar to elutriation and wet sieving.

Table 4 *Specifications for U.S. Standard Sieves, Fine Series. Nominal Dimensions, Permissible Variations, and Limits for Woven Wire Cloth of Standard Sieves. (U.S. Standard Series (4th Root of 2 Ratio))*

Sieve Designation		Sieve Opening		Permissible Variations in Average Openings (%)	Permissible Variation for not more than 5% of Openings (%)	Permissible Maximum Variation in Individual Openings (%)
Standard	Alternate	μm	in. (approximate equivalents)			
5.66 mm[a]	No. $3\frac{1}{2}$	5660	0.223	±3	+5	+10
4.76 mm	No. 4	4760	0.187	±3	+5	+10
4.00 mm[a]	No. 5	4000	0.157	±3	+5	+10
3.36 mm	No. 6	3360	0.132	±3	+5	+10
2.83 mm[a]	No. 7	2830	0.111	±3	+5	+10
2.38 mm	No. 8	2380	0.0937	±3	+5	+10
2.00 mm[a]	No. 10	2000	0.0787	±3	+5	+10
1.68 mm	No. 12	1680	0.0661	±3	+5	+10
1.41 mm[a]	No. 14	1410	0.0555	±3	+5	+10
1.19 mm	No. 16	1190	0.0469	±3	+5	+10
1.00 mm[a]	No. 18	1000	0.0394	±5	$+7\frac{1}{2}$	+15
841 μ	No. 20	841	0.0331	±5	$+7\frac{1}{2}$	+15
707 μm[a]	No. 25	707	0.0278	±5	$+7\frac{1}{2}$	+15
595 μm	No. 30	595	0.0234	±5	$+7\frac{1}{2}$	+15
500 μm[a]	No. 35	500	0.0197	±5	$+7\frac{1}{2}$	+15
420 μm	No. 40	420	0.0165	±5	$+12\frac{1}{2}$	+25
354 μm[a]	No. 45	354	0.0139	±5	$+12\frac{1}{2}$	+25
297 μm	No. 50	297	0.0117	±5	$+12\frac{1}{2}$	+25
250 μm[a]	No. 60	250	0.0008	±5	$+12\frac{1}{2}$	+25
210 μ	No. 70	210	0.0083	±5	$+12\frac{1}{2}$	+25
177 μm[a]	No. 80	177	0.0070	±6	+20	+40
140 μm	No. 100	149	0.0059	±6	+20	+40
125 μm[a]	No. 120	125	0.0049	±6	+20	+40
105 μ	No. 140	105	0.0041	±6	+20	+40
88 μm[a]	No. 170	088	0.0035	±6	+20	+40
74 μm	No. 200	074	0.0029	±7	+30	+60
63 μm[a]	No. 230	063	0.0025	±7	+30	+60
53 μm	No. 270	053	0.0021	±7	+30	+60
44 μm[a]	No. 325	044	0.0017	±7	+30	+60
37 μm	No. 400	037	0.0015	±7	+30	+60

[a] These sieves correspond to those proposed as an International (ISO) Standard. It is recommended that wherever possible these sieves be included in all sieve analysis data or reports intended for international publication.

[b] These sieves are not in the fourth root of 2 series, but they have been included because they are in common usage.

NOTE—All measurements of openings and wire diameters ahall be made on the completed sieve.

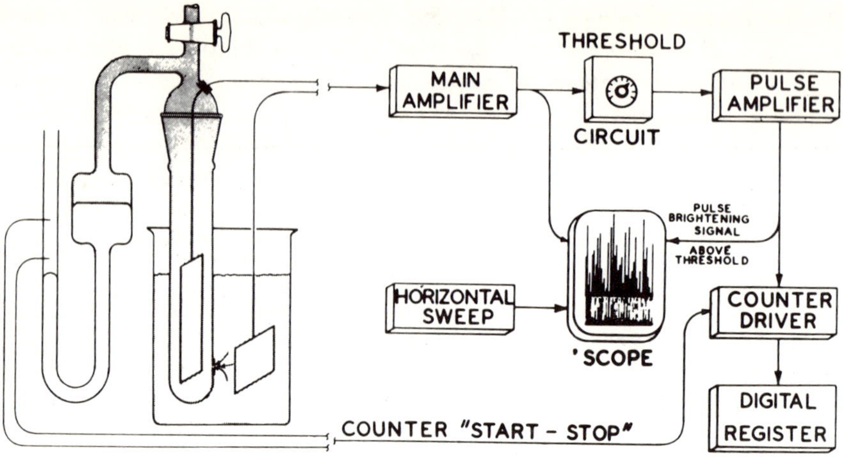

Figure 11. Schematic drawing of the Coulter Counter (Courtesy Coulter Electronics Hialeah, Fla.).

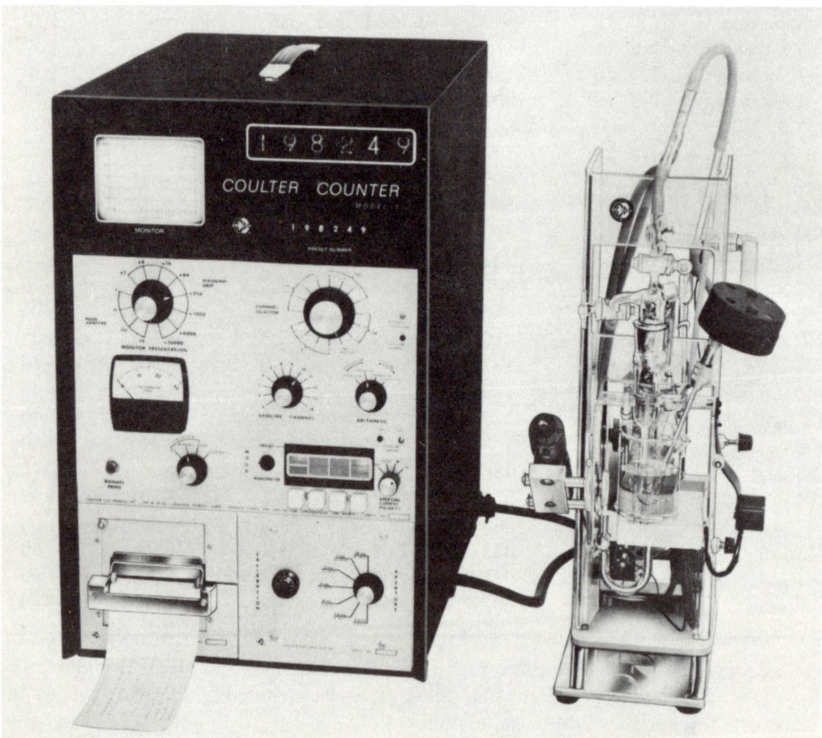

Figure 12. The Coulter Counter Assembly (Coulter Electronics, Hialeah, Fla.).

Several sedimentation methods are used. The Coulter Counter uses a separating principle illustrated schematically in Figure 11.[11] Individual particles can pass through the orifice between the liquid reservoirs and displace some of the electrically conductive liquid from the orifice. This increases the electrical resistance of the liquid circuit, and the change is related to the volume of the particles. By electrical accounting techniques, it is possible to count the electrical pulses generated by individual particles, and to segregate them according to pulse size. An electronic display of particle count versus particle size is produced over a very short sampling time. Figure 12 shows a Coulter Counter and the electronic display apparatus.[11]

PARTICLE DENSITY

Alternative measurements for particle density are:

1. The density of the individual particles themselves. The particle density is measured by weighing the sampling and then determining the volume occupied. The weighed sample is added to a measured volume of water, and the increase in volume is taken as the volume of the particulate matter. The particle density is reported in grams per cubic centimeter or specific gravity referred to water at 4°C.
2. The apparent bulk density is of interest in establishing the volume of the particulate matter that will be collected. Dusts with specific gravity of 2.5, or a density of 156 lb/ft^3 are likely to have an apparent bulk density in the range of 20–60 lb/ft^3. Measurements of bulk density are less precise than measurements of particle density because settling, compaction, and deaeration have effects upon the bulk density. Conservatively high estimates are employed for sizing structural members to support dust hoppers and conservatively low estimates of bulk density for figuring the volume of the hoppers required to hold a dust catch over a length of time.

Particle density is the same for the large particles and for the small ones in a given sample. In stoker-fired furnace applications, the larger particles may represent partially unburned coal fines of high carbon content whereas the smaller particles are likely to represent completely burned particles of ash. Such variations are difficult to detect.

OTHER PROPERTIES

No accepted methods of characterizing particles with regard to abrasiveness and caking characteristics are available. Most characteristics of this sort are based upon past experience with collection of the specific material in question. Some generalizations can be made.

Materials of very high specific gravity are likely to be more abrasive than those of low specific gravity. Similarly, large particles are more inclined toward abrasiveness than smaller ones. Microscopic examination of particles may reveal sharp edges that will abrade or erode dust collector surfaces, as opposed to smooth, rounded particles, which are likely to be less abrasive.

The tendency to absorb moisture from the atmosphere can be measured by determining the weight when a dried sample is exposed to a moist gas at the temperature of the system in which the dust collector will be placed. It is not customary to run such moisture absorption tests unless there is reason to suspect that the particles will, in fact, be hygroscopic.

Resistivity tests are frequently made where electrostatic precipitators are used. These are performed with an apparatus that measures the resistance between the two parallel plates of known surface area separated by a fixed depth of the dust. The resistivity measurement requires careful control of the test conditions, influenced not only by temperature, but by absorption of chemical constituents from the gas phase. It is necessary to control the chemical composition of the gas stream very closely to obtain useful resistivity values for the dust.

Severe dusting problems are involved in loading grains into trucks, railroad cars, and ships for transport to mills. Unloading of the grains produces similar dusting problems at storage and milling facilities. Within grain mills, the granular material is transported by conveyor or through pneumatic tubes. Conveyors must be enclosed to entrap the dust generated by mechanical action and evacuated through dust-cleaning devices. Large cyclones are effective in separating the major grain constituents, but frequently fabric collectors are required as a backup for collecting the finer dust particles. Grinding, milling, crushing, and the like generate similar dust problems, which are most frequently controlled by a combination of mechanical or cyclonic collectors and fabric filters.

Grains are screened to remove extraneous materials such as rocks and "scalped" to eliminate metallic or other inclusions. Processing of waste products, such as corn cobs, oat hulls, and bagasse, frequently involves buring of these materials with the concomitant production of fly ash. Coffee roasting also produces particulate by-product chaff that is separated from the coffee beans in the roasting process. Dusts from grain and other food handling processes are not returned to the process and must be disposed of as waste product. The waste products cannot be sold as constituents of the grain products—they must be burned or buried.

METALS PROCESSING

Mining of most ores involves transport of bulk materials with the accompanying dust problems. Ore conveying, washing, drying, crushing, grinding, and screening all involve the production of substantial concentration of dusts. In many cases, the dust can be recovered and included in the ore fed to various smelting processes. Many smelting operations produce fine particulates made by vaporization of metal components and subsequent condensation. These particulates are fumes. In addition to the dust generated by handling ores, smelting processes involve the other raw materials such as coke, limestone, and fluxes. All are handled in bulk and subject to the same conveying and processing operations as the ores. Mechanical precleaners and fabric collectors are the most satisfactory particulate control devices. Wet collectors minimize the problems in handling large quantities of very fine dry dust.[12]

ROCK PRODUCTS

The crushing of rock (usually limestone base) to produce gravel for concrete aggregate and road beds generates fine dust, controlled by fabric collectors, although wet scrubbers are also widely used. Electrostatic precipitators are more common in Europe than in the United States for these applications. Unlike the ore processing industries, the by-product produced from rock crushing cannot be returned to the product and must be disposed of by landfill methods. The by-product dust is nonreactive, however, and may be settled in lagoons or ponds.

Examples of rock production are cement, metallurgical and agricultural lime, and phosphate fertilizer. Fabric collectors and wet scrubbers are the most widely used for dust collection in these industries.

A substantial problem exists in the control of emissions from fluid catalytic cracking units in which very fine granular catalysts (50–200 μm in particle diameter) are used. Internal cyclones within the catalytic cracker limit the emission of particulate matter sufficiently to prevent losses of catalyst. However, the attrition dust of the catalyst is lost from the cyclones and ordinarily exceeds regulations imposed for particulate emissions from the end process.[4]

The fines escaping from the internal cyclones have been treated by the installation of a third, external, cyclone stage, but this is not efficient enough to meet stringent air pollution regulations. Electrostatic precipitators collect catalyst fines, substantially valueless, which are discarded as a waste product and disposed of in landfills.

Several sources of particulate matter exist in chemical plants where petroleum by-products are oxidized to form chemical building blocks. Examples are phthalic anhydride and maleic anhydride manufacture. These materials are produced by air oxidation of petroleum-derived chemicals. The products have high boiling points and form solids or liquids that are condensed out of the gas stream leaving the oxidizer. However, the condensation process is never complete and aerosols of the chemical product are discharged into the air unless air pollution control equipment is provided. Because these materials are organic, they can be destroyed by thermal incineration, which is the most common method of treatment. However, they can also be collected by wet scrubbers.[16]

Inorganic chemical manufacturing processes often produce dusts as by-products where the chemical product is ground, spray dried, tabletted, or otherwise processed.

UTILITY POWER PRODUCTION

The largest single source of particulate matter discharged into the atmosphere in the United States is from the utility power industry. Fly ash from even the smallest coal-fired utility boiler would produce a great nuisance if no control equipment were provided. Requirements for control of fly ash vary from simple multiple-tube cyclonic collectors for small feed stoker-fired units to extremely efficient electrostatic precipitators for large pulverized coal-fired units.[12] Practically all existing utility plants are equipped with either mechanical dust collectors, precipitators, or a combination of the two. A few installations of wet collectors have been made in order to obtain sulfur oxide removal along with fly-ash collection. There is a trend toward use of wet equipment for utility plants. Utility plants involve several dust problems in addition to those concerned with fly-ash discharge from the furnace. Three examples are:

1. The system for conveying fly ash collected by precipitators to storage hoppers and subsequent transfer to rail cars or trucks for disposal.
2. Coal conveying and bunkering equipment.
3. Coal pulverizing equipment.

Small industrial boilers share the particulate generation problems common to the large utility plants. In most cases, the severity of the problem is related to the size of the unit, and dust generation problems are more likely to remain untreated for the smaller industrial boilers than for large utility units. However, as air pollution control regulations are tightened, the unit size requiring treatment is becoming smaller and smaller.

MISCELLANEOUS SOURCES

Several major sources of particulate matter in the dust size range fall outside of the industrial categories described. A prime example is the generation of fly ash by municipal or other incineration equipment.

Municipal incinerators have traditionally discharged partially burned refuse and fly ash into the atmosphere. Modern incinerators are being designed so as to burn all of the carbonaceous matter in the refuse, and are being equipped with particulate collection devices to limit dust emissions. Wet scrubbers and electrostatic precipitators appear to be the favored devices for this service, although mechanical collectors have been used to a limited extent.[17] Wet collectors for municipal incinerators are subject to corrosion problems because of hydrochloric acid in the effluent. The hydrochloric acid product is proportional to the concentration of polyvinylchloride plastic materials in the refuse being incinerated. Electrostatic precipitators would not ordinarily be subject to severe corrosion by the hydrochloric acid if operated at a high enough temperature to prevent condensation. However, the control of refuse incinerators is frequently poorer than that characteristic of industrial processes, and periodic condensation may be unavoidable.

Table 5 gives the concentrations of suspended matter in industrial process gases.

FUMES

Whereas dusts are formed by mechanical action of solids on each other, fumes are the product of vaporization and subsequent condensation of solid or liquid fume materials. Fumes are particles of very small size. Average particle diameters may be as small as 0.1 or 0.2 μm and they may grow as large as 3–5 μm. The fume particles are extremely small as they are formed by condensation, and grow larger with time because of collisions with other fume particles and additional condensation of vapor on the already-formed surface of the particle. Particles tend to be spherical, as opposed to dust particles, which are

***Table 5** Concentrations of Suspended Matter in Industrial Process Gases (Research Corp.)*

Suspended matter	Concentration (g/ft^3) of gas, STP
Acid mist—sulfuric acid (contact plant)	
After roaster	
Multiple-hearth roaster—zinc and pyrites	1.08–5.8
Flash roaster—zinc blende	0.00475–0.0555
After absorbers—tail gases	0.722–2.31
Acid mist—phosphoric acid—from burning phosphorus (basis 100% H_3PO_4)	48.3–66.2
Carbon black	
From cracking natural gas	5.1–17.0
From oil cracking	19.0–40.0
Carbureted water gas	
Dry-tar basis	0.765–1.59
Wet-tar basis	1.08–2.26
Catalytic cracking units—oil	
Atmospheric-pressure units—after mechanical collector	
Natural catalyst	19.45–85.6
Synthetic catalyst	16.5–52.9
High-pressure units—after mechanical collector	
Natural catalyst	7.19–22.75
Synthetic catalyst	4.69–94.6
Coke-fired producer gas	0.03–0.06
Coke-oven gas	
Ahead of exhausters, dry-tar basis	4.51–4.88
After positive displacement exhausters, wet-tar basis	3.14–4.58
After centrifugal exhauster, wet-tar basis	1.66–3.74
Fly ash from boilers burning pulverized soft coal	1.0–5.0
Incinerators burning dry sewage sludge	3.17–4.35
Silica-rock treatment	
Oil-fired rotary dryer	6.42–23.5
Preheater gases	6.65–15.8
Ventilating system	9.37–26.2
Tin smelting	
Reverberatory furnaces	2.20–3.12
Calcining tin ores—rotary kilns	1.44–4.59
Zinc sintering machine—straight and chloridized roast	0.311–1.908
Zinc-ore roasting	
Flash roaster	3.92–45.05
Multiple-hearth roaster	3.82–7.07
Zinc-oxide—Waelz plant	12.65–28.62

characteristically irregular in shape. They tend to be more uniform in particle size in a given gas stream than dusts, although this is likely to be a function of concentration and of time available for particle growth.[9]

Some of the common materials that can be characterized as fumes are cigarette smoke, fumes from smelting furnaces, such as open hearth steel furnaces, reverberatories for smelting copper and brass, and the black carbonaceous effluent from jet aircraft engines. Fume materials have a number of properties that set them apart from dusts and require special study.[2]

OPTICAL EFFECTS

The most striking property of fumes is their high visiblity even at low concentrations; this is the result of two factors: The ability to obscure or reflect light depends upon the silhouette area of the particles rather than upon the mass. Finely divided particles have a larger silhouette area per unit weight than large particles. Figure 13 contains information concerning the specific surface area, or surface area per gram of particulate matter, as a function of particle diameter. It is apparent that particles averaging 0.5 μm diameter have 1000 times as much surface area as those averaging 50 μm. On this basis it would be reasonable to expect that a 0.5-μm fume material would exhibit about the same smoke density as a concentration of 50 μm particles 1000 times as great. Another factor that adds substantially to the visible nature of fumes is the ability of small particles to scatter light. The particles in the size range approximating the wave length of visible light have the ability to bend or scatter light rays that do not impinge upon them. Wavelengths of visible light lie between about 0.3 and 0.7 μm, and particles in this size range have an exaggerated ability to scatter and diffuse light. Fume-forming processes are characterized by extremely dense, opaque effluents and may require control to levels far below those required by dust-generating processes.[10]

Respirability

Large dust particles suspended in the air are not likely to cause health problems because they are filtered out of the air in the nose and upper respiratory tract. Complex body mechanisms are capable of cleaning these areas and rejecting particulate matter in the range above about 5 μm at adequate rates to protect the lungs.[10]

At the other extreme, very tiny particles, on the order of 0.1 μm, are inhaled into the lungs without capture by the cilia in the respiratory

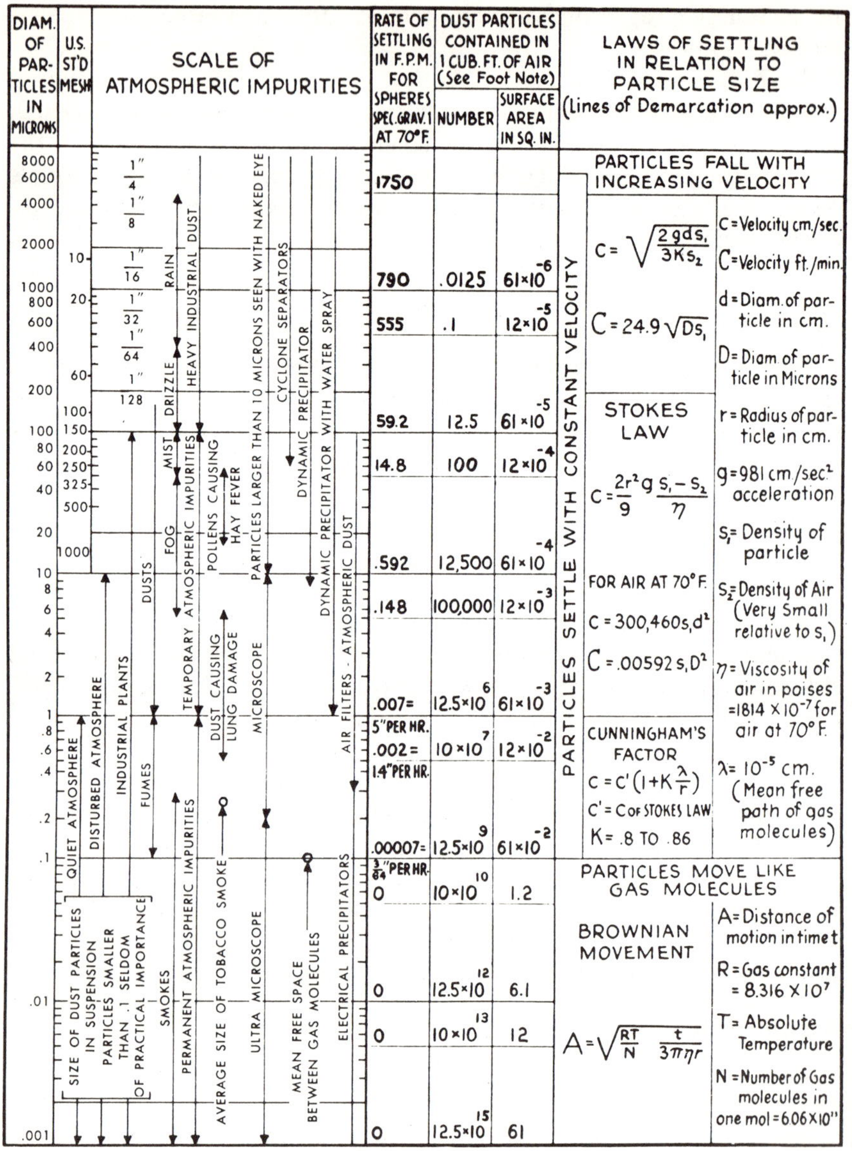

Figure 13. Size and characteristics of air borne solids (Permission American Air Filter Co., Louisville, Ky.).

tract. However, these particles are so small that they are not likely to be wetted by the moisture in the lung tissues and they are exhaled without retention on lung tissue.[10]

Not only do the particles in the submicron size range tend to enter and obstruct the lung tissue, but they serve as vehicles for carrying other air pollutants into the lungs trapping them there. Sulfur trioxide in the air is relatively harmless at low levels in otherwise clean air. The acid absorbs in moisture in the respiratory tract and does not reach the lungs. If a substantial quantity of respirable particulate matter is suspended in the air, the sulfur trioxide absorbs on the surface of these particles and is carried into the lungs and retained. There is sufficient evidence to indicate that fine particulate matter plays a significant role in the health problems associated with sulfur oxides, nitrogen oxides, and so on.[10]

Figure 3 indicates the range of settling velocities associated with a variety of particle sizes and specific gravities. It is apparent that the settling velocity of particles at the 100-μm range is sufficiently high that they are unlikely to remain in the atmosphere for very long. Particles below about 20 μm settle very slowly and are subject to transport for long distances through the atmosphere. Submicron particles remain suspended for indefinite periods and are removed from atmospheric air only by turbulent collision with solid objects or surfaces, by condensation of rain droplets, etc. The submicron fume materials discharged into the atmosphere tend to have the same sort of cumulative effect that gases discharged into the atmosphere have.[10]

Properties of Fumes

Particle size is easier to describe for fumelike materials than for dusts, because of the tendency of the fume particles to be spherical. Describing the particles in terms of diameter for distribution of weight among various diameter ranges is a more specific characterization for fumes than for dusts.

Particle diameter is the most importanct single characteristic of fumes with regard to optical effects, health effects, and collectability. However, unlike the dusts, the mechanisms for measuring fume particle size are not adequate for routine characterization of the fumes in meaningful terms for collection. There are several reasons for this.[9]

Light microscopy is inadequate for characterizing submicron fumes. A skilled microscopist can recognize particles in the 0.5–1.0-μm range, but he cannot make a good classification as to particle size. The ability to distinguish between two points in space is limited to the wavelength of the light used to view them. In the case of visible light, this is in the 0.3–0.7-μm range, so that the particles in this size range are hard to see.

Electron microscopy is suitable if good representative samples of the fume can be obtained. Electron microscopy substitutes an electron beam for the light rays used to view particles optically. The electron bean has an extremely short equivalent wavelength compared to visible light; this makes resolution of particles as small as 0.01 μm in diameter feasible. The scanning electron microscope, which uses electrons emitted from the particle after electron bombardment, provides very high resolution of fume samples.

Very fine fumes tend to be unstable in size. One of the factors that brings particle size change with time is the ability of small particles to acquire significant electric charges by random action of the electrons in the particle and in the surrounding gas. Collisions with a gas molecule are likely to cause an electron transfer to or from the particle. Large particles are not influenced significantly by the transfer of one or a few electrons. However, a tiny particle acquires a substantial electrical charge by the transfer of a single electron. Small particles, therefore, acquire substantial random positive and negative electrical charges which cause them to agglomerate by electrostatic attraction. As they grow larger, the electrical charging effects become less significant.[2] However, sampling of aerosols in the 0.05–0.2-μm diameter range is characterized by extremely rapid particle growth with time.

The stability of the small fume particles is the result of random motion imparted to them by collisions with gas molecules. The Brownian motion of very fine particles results in substantial velocities being imparted to individual particles; these velocities result in collisions that may bring about agglomeration. The particles formed by condensation are subject to growth by further condensation as the temperature is reduced. These factors tend to make the measurement of submicron particulate matter difficult.[2]

The measurement of an aerodynamic particle diameter is not as good a characterization for fume materials as it is for dusts, because mechanical settlers or centrifugal separators are of little use in the collection of fume particles. These small-diameter particles settle so slowly as to make gravity settling useless, and even the most efficient mechanical separation devices fail to remove substantial concentrations of submicron particulate matter. The collection of fumes is thus limited to filtration, electrostatic precipitation, and wet scrubbing.[9]

Many methods of characterizing fume materials for the selection of air pollution control equipment are available, or under development. It is not possible to select wet scrubbing or electrostatic precipitator equipment entirely on the basis of field sampling. It is necessary either to run pilot tests or to simulate the operation of the specific piece of equipment involved in the field if there are not sufficient background data available for establishing the design feature and operating requirements of the particulate collection system. This is not true in fabric collectors, where particle size has less effect upon the performance of the filter medium. The determination of particle size should be taken to mean "characterization for the purpose of selecting wet scrubber or electrostatic precipitator design features and operating conditions."

By far the most satisfactory basis for projecting performance of collection equipment for materials in the fume category is demonstrated satisfactory performance for other units carrying out the same process under the same conditions. Fabric filtration, shown Figure 14, is well known to be a suitable method for recovery of zinc oxide fume from lead smelting operations. Many installations have given satisfactory performance over a period of many years. Wet collection systems do not have a history of satisfactory performance in this service, nor do electrostatic precipitators. Therefore, it would ordinarily be far less expensive to select a fabric collector of conservative design based on the parameters established by past experience than to

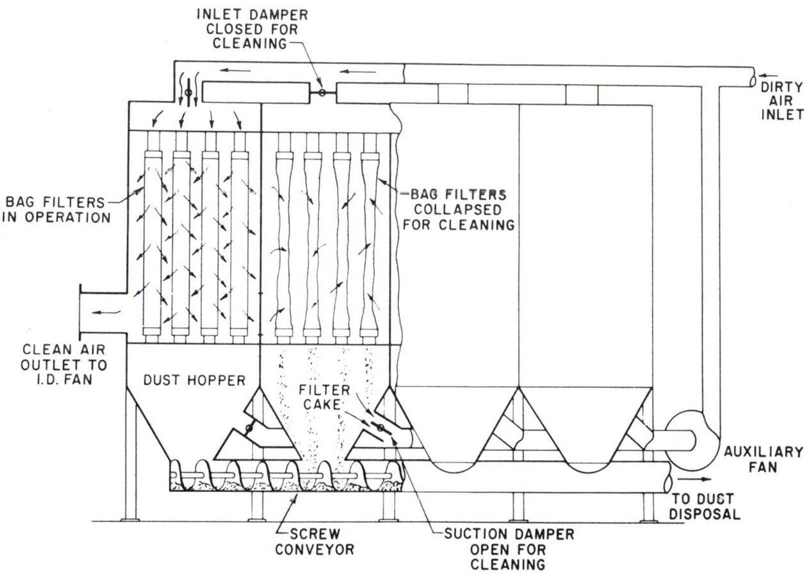

Figure 14. Fabric filtration.

undertake a program to characterize the lead oxide fume and experiment with a wet collector or electrostatic precipitator.[14]

Food and Processing

Few fume-forming situations are encountered with handling of grains, cattle feeds, and other food products. One exception is the case when animal and vegetable matter is heated. In cooking, rendering, etc., there is likely to be an evolution of vaporized organic material. This leaves the surface of the product as a vapor but recondenses to a liquid or solid fume as the temperature is reduced. Such fumes are produced by cooking (the blue-white haze exhausted from steak grills in restaurants, smoke house operations, rendering, deep frying, grain drying, etc.). The particle size is likely to be quite small, which makes collection by wet scrubbing difficult. The material is likely to be combustible, which is a substantial hazard for electrostatic precipitators since they may serve as an ignition source. Fabric collectors are not well suited for collecting oily or greasy materials since these wet the cloth fibers or other and blind the fiber material.

Most of the fumes produced by food processing operations can be controlled by thermal incineration. The principal drawback of thermal incineration is the high operating cost, for sufficient fuel must be used to heat the complete gas stream into the 1100–1200°F temperature range in order to burn the organic materials as vapors before they condense as a fume. Catalytic incineration is suitable in many cases and operates with a lower fuel cost. However, the catalytic surfaces may be fouled if condensation of the fume occurs in the ductwork leading into the catalytic incinerator.[18]

Mining and Metals Processing

No fume problem exists in the mining and preparation of ores for smelting. In any situation where molten metal is produced, the high temperatures involved are likely to result in vaporization of the product metal to some extent, and very frequently of the impurities or alloying material in the melt as well. Metal condensation results in the production of a metallic fume when the gas stream passing through the furnace is cooled. Such fume-forming situations exist in substantially all ferrous smelting processes. Blast furnaces have a mixture of iron oxide, fume, kish, coke breeze, and the like.[13] Because blast furnaces opeate at high pressures, there is a substantial amount of available energy in the gas stream leaving the blast furnace; this energy may be used to advantage in a high-pressure drop wet scrubber.

Open hearth furnaces produce substantial quantities of metallic oxide fume. However, because they operate at atmospheric pressure, there is no motive force for the operation of the high-energy Venturi scrubber. Open hearths are characterized by high gas flows, considerable variation in flow, and high carbon monoxide contents. The latter tends to introduce an explosion hazard when precipitators are used. Open hearth furnaces have been treated by electrostatic precipitation of the fume material and by fabric collection, but treatment tends to be costly and many open hearths remain untreated and are major emitters of fume materials.[19]

Open hearth furnaces are being replaced by basic oxygen process furnaces (BOF). Oxygen is used as the reagent for removing excess carbon and other impurities from the hot metal produced by the blast furnace. The gas volume from basic oxygen furnaces are smaller per ton of steel than from open hearth furnaces. The concentration of particulate matter is higher and would create an intolerable nuisance if BOFs were installed without particulate collection equipment. Consequently BOF furnaces have been equipped with either electrostatic precipitators or with high-energy Venturi scrubbing systems. In the case of electrostatic precipitators, the potential explosion hazard created by the high carbon monoxide content of the effluent gas necessitates introduction of dilution air at the top of the furnace for combustion of the CO. Wet scrubbing systems are less likely to serve as an ignition source, and it is possible to design BOF ventilating systems to exclude air and process the CO at high concentration. This substantially reduces the horsepower requirements for the fan that evacuates the effluent gases through the scrubber.

Substantial fuming problems exist in gray iron foundries in which cupolas are used for melting pig iron and scrap iron for the production of iron castings. A cupola emits not only iron oxide fume but also a mixture of dusts containing coke fines, limestone dust, iron oxide dust, and sulfur oxides and other gaseous pollutants.[4] Good treatment of cupola emissions is likely to require a capital investment greater than that for the cupola itself. This is an example of a case where both dust and fumes exist in the same gas streams, and where mechanical dust collectors can be used to remove more than of 90 weight percent of the particulate matter involved without substantially changing the appearance of the plume emitted. The physical properties of the plume relate to the fume components rather than to the mechanically produced dusts. In many cases, it is more economical to abandon the cupola and substitute electric melting in a closed crucible for production of molten iron.

Foundries have another problem in the fume category, because they contact sand in the molds with molten iron. The sand contains organic binders that are vaporized when the molten metal is poured into the molds. The resulting oil fume is difficult to control by particle capturing devices, but can be adequately handled by thermal incineration.[18]

Nonferrous smelting involves many fume production problems in roasting, converters, and reverberatory furnace melting. These involve heating the material above the melting point and give rise to substantial fume production. For some of the less common materials such as molybdenum, antimony, and mercury, the value of the particulate material emitted from the furnace is high and it is economically attractive to recover it. Even were this not the case, the production of particulate fume from each of these smelting operations would necessitate the inclusion of an air pollution control device.

For nonferrous smelting operations, fabric collectors provide the most satisfactory fume collection equipment. The temperature and humidity of the flue gas must be controlled very carefully to maintain the bag filter in good operating condition. Where sulfur oxides are also emitted, wet scrubbing has provided a satisfactory solution to both the fume and sulfur oxides problems.[14]

Secondary smelting of nonferrous alloys gives rise to the formation of lead, zinc, and other fumes that must be controlled. Fabric collectors serve very well for the collection of fumes of this sort. Secondary smelting operations are equipped with large cyclones for after cooling and drop-out of large particulate matter and fabric collectors for air pollution control.[14]

Aluminum smelting deserves a special comment because of the nature of the fume in both primary and secondary aluminum smelting. Aluminum metal is produced by electrolytic reduction of bauxite, an impure oxide of aluminum. The bauxite is dissolved in molten cryolite and electrolysis is carried out within the molten bath. Some gaseous by-products are formed, including hydrofluoric acid, aluminum fluoride, and volatile organic material from the electrodes. In addition to the fumes formed when the aluminum fluoride condenses or hydrolyzes to aluminum hydroxide, there is a substantial dust formation whenever the crust of solid cryolite on top of the bath is broken.[20]

Wet scrubbers collect the dust and the gaseous fluoride materials. HF is collected with scrubbing water. The ability of the scrubbers to collect the fume materials is very limited at the low-pressure drop levels (1–6 in. water column). Collection efficiency of 70–80% may be obtained because the dust constituents from the crust breaking are easily captured. Collection efficiency for the fume portion of the particulate emission is likely to be quite low. Scrubbers of high efficiency (usually 95%) are likely to require pressure drops of about 20 in. water column. One major processor has developed a system for removing HF by reaction with aluminum in a fluidized bed, after which fabric collectors remove the particulate matter.[21]

A similar complex fume scrubbing problem occurs in the purification of aluminum. Dissolved hydrogen gas, water, and other impurities are removed from the primary aluminum product by fluxing with chlorine. This gives rise to a complex fume consisting of chlorine, hydrogen chloride, aluminum chloride, aluminum hydroxide, and some dust from the slag layer on top of the aluminum melt. Here again, wet scrubbers are required to capture the gaseous constituents. In aluminum chlorination, the concentration of particulate fume formers is considerably higher than is the case for aluminum reduction, and it has been found necessary to equip these with high-energy scrubbers to limit the particulate emission nuisance. Scrubbers with a pressure drop of approximately 40 in. water column have been found necessary for adequate treatment of aluminum chlorination furnace effluents. Gaseous chlorine and hydrogen chloride are scrubbed out effectively by an alkaline scrubbing liquor such as caustic or lime solution.

Rock Products

Few fume formation problems are encountered in the handling of rock products. The principal exception lies in the manufacture of rock wool, glass wool, fiberglass, and plate glass, where the rock products are melted to produce spun or cast products. The melting furnaces have a problem similar to those for preparation of metal products. However, in glass manufacture the problem is often complicated by the inclusion of volatile chloride or fluoride compounds in the melt. Whenever such volatile constituents are present, they complicate the air pollution control problem; in the case of fluorides, they are very likely to attack filter fabrics. In the latter case, the equipment most suitable for capture of fume materials is unsuitable, and it becomes necessary to use wet scrubbers at very high energy levels to control the emission. Glass-melting furnaces are likely to require pressure drop levels of about 60–80 in. water column in order to adequately control the glass fume emitted concurrently with fluoride compounds from the melt. Rock wool furnaces have a lower pressure drop requirement and may be treated satisfactorily at pressure drops 30–50 in. water column.

Industrial Boilers

A widespread problem exists when boilers are operated on high-vanadium high-sulfur residual fuel oils. These oils form a complex fume of metallic oxide ash materials and sulfur trioxide. Mechanical

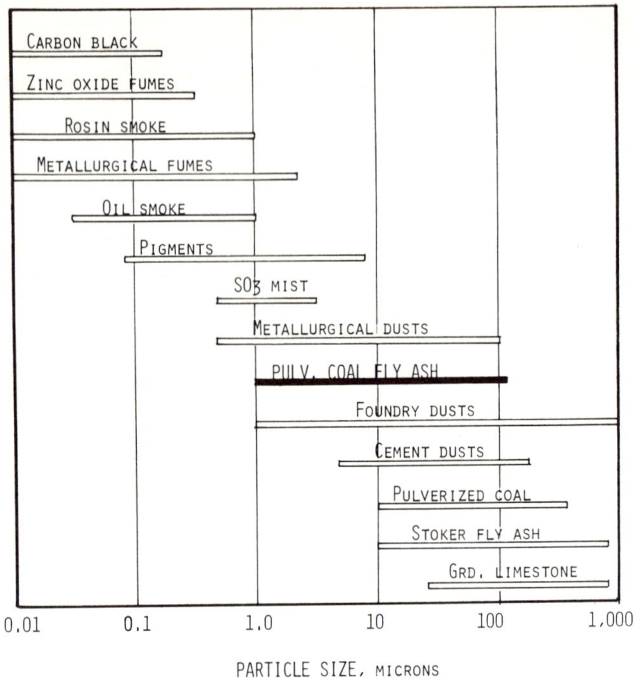

Figure 15. Size ranges of industrial aerosols. *Source:* Munger and McCabe (eds.), *The Spectrum of Particle Size and its Relation to Air Pollution,* McGraw-Hill, New York, 1952, 160 pp.

dust collectors have been used but are generally ineffective; electrostatic precipitators appear to do a satisfactory job. However, the experience has been almost entirely with precipitators designed for collecting coal-derived ash, which are on the boilers converted to oil firing.

Chemical Manufacture

The chemical manufacturing industries have few problems with the formation of fumes; only a few combustion processes are likely to form fumes. The disposal of liquid waste materials by incineration will produce a serious fume problem if there are sodium compounds (such as spent caustic) mixed with organic materials. The combustion of certain plastic wastes produces fume problems if sufficient residence time and temperatures are not provided to complete combustion. Chlorine and fluorine-containing plastics such as polyvinylchloride and Teflon® are apt to require long residence time and high temperature for complete incineration. Where such fume problems occur, the smoke product should be treated by modifying of the incinerator, rather than by installing of a particulate collection device.

For halogen-containing plastic, it is of course necessary to use a wet scrubber for collection of the resulting acid. However, a low-energy-level scrubber will suffice to capture hydrochloric or hydrofluoric acid at high efficiency, whereas fume problems would require a much higher-energy input to the scrubber. Plastic molding and extrusion frequently produce fumes by vaporization of plasticizers at elevated temperatures. These are difficult to control, but satisfactory performance can be obtained with incinerators and with high pressure drop filters if the effluent is free of solids. Size ranges of fumes and dust are given in Figure 15.

PARTICULATE COLLECTION EQUIPMENT

Four types of equipment are commonly used for the collection of particulate material, namely (1) mechanical separators, (2) fabric collectors, (3) electrostatic precipitators, and (4) wet scrubbers. (Combustible particulate matter is treated by incineration.) Some applications are particularly suited to one of these equipment types, while other applications may use all of the methods as alternatives. Some particularly difficult problems are hard to handle by any of the conventional methods. The principles of operation of each of the particulate collection devices are described below from the standpoint of the application engineer.

Mechanical Dust Collectors

Mechanical dust collectors fall into two basic categories. Some rely almost entirely on gravity settling of the particulate matter. The settling velocity is high enough that large particles, say 50 μm and above, will settle out of the gas stream under the influence of gravity provided sufficient time is allowed, and the gas velocity is reduced so that there are no turbulent eddies with velocity higher than the settling velocity.[23] Settling devices are not efficient enough to be of great value as the primary dust collection device for an industrial problem. For particles too small to settle by gravity, the application of centrifugal force may provide enough additional thrust on the particles to move them to the collector wall or hopper. Devices employing centrifugal force for separation of particulate matter are centrifugal collectors; they operate with a minimum of cost, generally have no moving parts, and require little attention. They are capable of operating at any temperature at

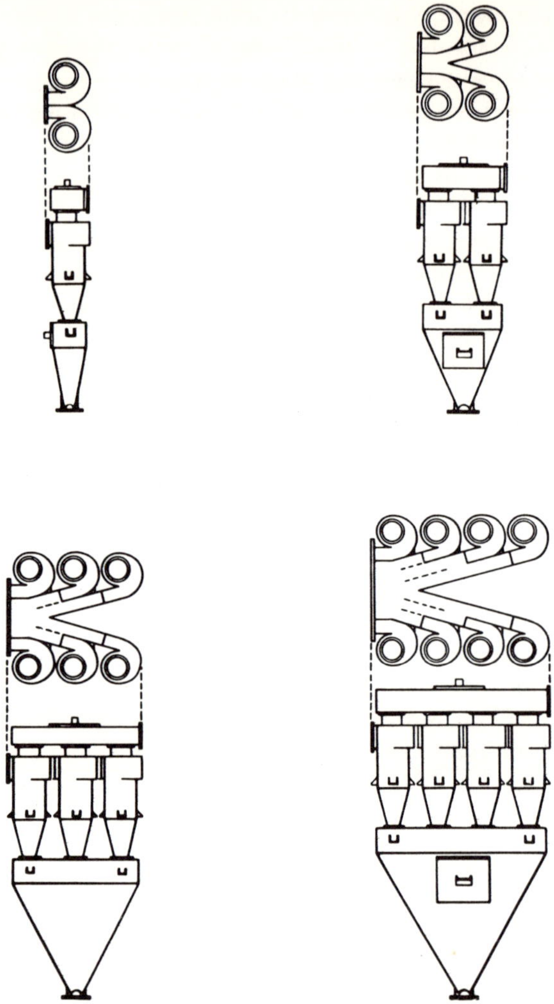

Figure 16. Multitube cyclone collectors.

which the collected material can be kept flowing freely, subject only to limitations in the materials of construction.

It is not practical to make these collectors efficient on small particles. Whereas settling chambers are of limited use for particles below about 50 μm, large cyclones (with diameters of greater than 12 in. are of limited use for particles below about 20 μm, and small cyclones, which are usually arranged in multiple-tube banks inside a common inlet box, are of little use below 5 μm in diameter.[24] Most industrial dust contains particles of sizes too small to be collected at high efficiency by mechanical collectors. Figures 16–18 are cutaway views of a multitube collector of the type used for collecting fly ash from stoker-fired boilers.

Principle of Operation

Figure 17 is a schematic drawing of a large cyclone. The gas entering the cyclone passes through a convergent section, which accelerates it to a high velocity. The dust particles suspended in the gas stream also ac-

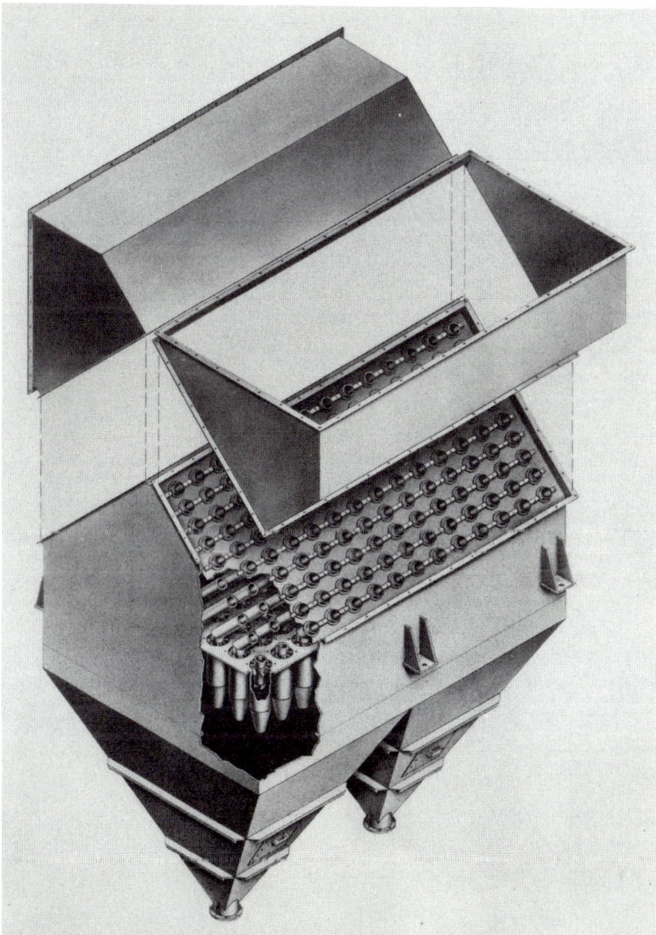

Figure 17. Large cyclone dust collector.

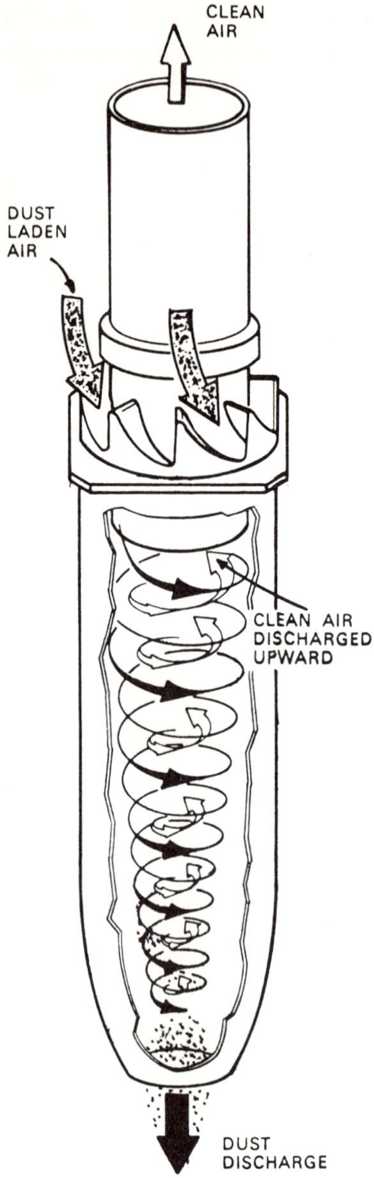

Figure 18. Cyclone dust collector.

celerate although there is some time required for the particles to reach the gas velocity. The gas is introduced into the cyclone tangentially and spins around the central outlet pipe or dust tube at high radial velocity. The dust particles suspended in the gas stream are subjected to a centrifugal force that tends to move them toward the outer wall of the cyclone, and to viscous drag forces exerted by the gas that tend to keep the particle moving along with the rotating gas. The particle moves under the influence of these forces and accelerates toward the outer wall. The effect of gravity on the particle can be neglected for even the largest particles in a centrifugal collector, and the time required to accelerate to "terminal velocity" toward the outer wall of the collector can be neglected.

On this basis, the centrifugal or "collecting" force can be written as

$$F_c = \frac{1}{g}\left(\frac{\pi\, d_p^3}{6}\right)(P_p - P_g)\,\frac{W^2}{R}, \qquad (5)$$

where

F_c = collecting force, lb
d_p = particle diameter, ft.
P_p = particle density, lb/ft^3
P_g = gas density, lb/ft^3
R = mean radius of path, ft
W = angular velocity $(= V_t/2\pi R)$
V_t = tangential velocity, ft/sec
g = acceleration due to gravity = 32 ft/sec^2

The viscous drag of the carrier gas is given by

$$F_v = 3\eta\pi\, d_p V_R \qquad (6)$$

where

F_v = viscous drag, lb
η = gas viscosity lb-sec/ft^2
V_R = radial velocity, ft/sec

and the solution of these two equations for the velocity of the particle yields[25]

$$V_R = \left(\frac{2V_T^2}{9}\right)(P_p - P_g)\, d_p^2 \qquad (7)$$

where

V_T = gaseous viscosity
V_R = radial velocity, ft/sec

The time required for a particle entering towards the center of the collector to move to the outside wall and be collected should be the prime measure of the efficiency for collecting particles of a given size. If the particles were collected with a high degree of certainty upon contact with the wall, and the gas moved in laminar flow around the cyclone, the efficiency for a given particle would be given by

$$E = 1 - \frac{HV_R}{D_c} \tag{8}$$

where

E = fractional efficiency
H = treatment time, sec
D_c = distance of travel, ft

However, the collecting velocity is imposed on a turbulent gas velocity so that a particle does not move with certainty toward the outer collecting surface during its passage through the collector. Also, there is a significant change of reentrainment of particles at the outer wall of the cyclone. The equation of the cyclone efficiency versus particle size curve may be represented by

$$E = 1 - \exp \frac{-HV_R}{D_c} \tag{9}$$

where the collecting velocity V is not precisely the velocity calculated in the foregoing equation, but a lower number which takes into account the possibility of reentrainment.

Statistical equations of this type would lead one to the conclusion that the efficiency of cyclones is unlimited because a radical velocity and diameter exist for which any calculated efficiency desired could be obtained. This is, however, not the case because of mechanical limits. If the velocity is increased the probability of reentrainment also increases. In addition, there is a practical limit imposed upon the inlet velocity to cyclones by the abrasiveness of the dust being collected. Cyclones of ordinary metals such as cast iron and steel are seldom used with pressure drops above 6 in. water column. Large cyclones with specially prepared abrasion-resistant linings are frequently used at pressure drops of 20 in. or more water column, but these are relatively short-lived.

Small cyclones are more efficient than large cyclones. When high efficiency is desired it is common to manifold several smaller cyclones together rather than install one single large cyclone. The ultimate in manifolding cyclones involves multitube collectors, which have radial inlets connected into a common inlet header. This type of collector has been used almost universally on stoker-fired and cyclone-fired utility boilers. Many industrial boilers have also used multitube dust collectors. A typical cutaway view of such a collector is shown in Figure 16.

Size has been a limiting feature in the manifolding of mechanical collectors. Multitube collectors compromise collection efficiency in order to obtain a higher gas volume throughput. Collectors of this type are capable of handling a higher gas velocity per unit of size than are large high-efficiency cyclones. Typically, collectors of this type are sized in accordance with a rated gas flow per tube, which varies with the pressure drop available for the particular application.

The size of the overall collector is obtained by counting the number of tubes required to handle the total gas flow.

Mechanical collectors are less efficient when the gas flow and velocity through them is low than when the gas flow is high. It is necessary that mechanical collectors be sized as accurately as possible for the gas flow, rather than oversized. Collectors that have been inadvertently oversized can be made to function effectively by blanking off some of the individual tubes. In addition, gas flow into the collector inlet may be increased by installing a barometric damper or simply by cutting holes in the ductwork.

Collectors are usually sold on the basis of a fractionation efficiency curve, which relates the efficiency for each particle size to the collector pressure drop.

Based on the fractionation efficiency for 4-in. water column pressure drop and the particle size distribution, the efficiencies were calculated as follows:

Spreader stoker	93
Underfeed stoker	91
Chain grate	90
Cyclone fired furnace	88
Pulverized coal-fired furnace	85

A rule of thumb for estimating the efficiency of multitube collectors involves presuming that they are 100% efficient on all particles above 10 μm and that they are 50% efficient on all particles below 10 μ. For example, the pulverized coal-fired furnace example had 53% less than 10 μm. The estimated efficiency would have been $(100-53/2)\%$ or 73.5%.

Limited efficiency of mechanical collectors prevents their use as primary dust collectors on pulverized coal-fired boiler installations, lime kiln, grain processing plants, and so on. They are capable of

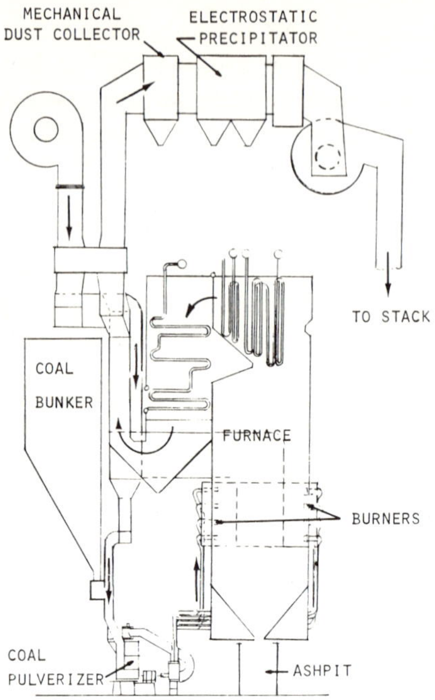

Figure 19. Corner-fired dry bottom boiler showing cyclone collector and electrostatic precipitator in series.

collecting the product in dry form with a minimum of first cost and operating expense, and may be very valuable as precollectors ahead of wet scrubbers or in some cases fabric collectors,[14] or precipitators as shown in Figure 19.

Fabric Collectors

Fabric collectors function to collect particles of all sizes at high efficiency, just as the household vacuum cleaner does. The fabric bag serves as a support upon which a layer of dust is built. The dust layer is capable of separating the tiniest particulate materials. A wide variety of fabrics and mechanical designs for support of the fabric bags or tubes are available commercially.

Fabric collectors are suited for operations in which dust or fumes are released into ambient air streams. Cotton sateen bags serve to collect almost any kind of dust released at ambient temperature. Mechanical designs have been developed along many lines; common are the tubular shaker-type baghouse and the reverse jet collector.[27] Figure 14 (p. 657) illustration of the conventional shaker type baghouse and illustrates some of the principal features of baghouse design.

Air enters the top of the baghouse and flows downward through tubular bags or sleeves. The filter cake builds up on the inside of the tubes as the air passes through them flowing radially outward. The filter cake is periodically loosened by mechanically shaking the bags; it then falls into a dust hopper. The pressure drop across the filter cake is several inches of water. It is necessary to interrupt the flow of gas and relieve this pressure difference to free the dust from the fabric. Baghouses are compartmented and arranged so that the flow of gas can be blocked off from the compartment while the bags are being cleaned.

Reverse jet cleaning allows design of fabric collectors without compartmentalization. The gas flow is radially inward with filter cake depositing on the outside of the tubular sleeve. Cleaning is accomplished by blowing a pulse of compressed air down the tube from a jet at the outlet. This momentary reversal of the gas flow flexes the fabric and causes the filter cake to break loose and fall into the collection hopper. Reverse jet collectors can be operated at higher velocities than the mechanically cleaned version; this results in a lower first cost. However, fabric life is shorter for the reverse jet filters.

A variety of filter fabrics is available for specific services. Cotton fiber is limited to approximately 160°F, whereas wool can be used up to 180 or 190°F; the synthetics, Orlon (polyester fiber) and Dacron (acrylic fiber), are operable at temperatures up to 270°F. These are widely used in the smelting industry where furnace gases contain fume materials.

Nomex (nylon fiber) can be used as high as 450°F and graphite lubricated fiberglass up to 550°F; fiberglass bags have the highest temperature limits of any material in wide use. The fibers tend to abrade each other and, in spite of the inclusion of the high-temperature lubricant to reduce friction, they will wear very quickly unless handled in an extremely gentle way. The most satisfactory method of cleaning fiberglass bags involves a slow reverse flow of gas, which necessitates compartmentalization and the inclusion of a blower for the cleaning cycle.[27]

Synthetic fibers such as Orlon and Dacron cost two to four times as much as natural fibers. Glass increases the price by a factor of two to four over synthetics and Nomex is about twice as expensive as Orlon or Dacron. A very high cost of treatment of elevated temperatures suggests that a careful economic study of cooling equipment versus high-temperature fabric filtration is in order when hot gas streams are to be treated.

The basic sizing variable is the *air-to-cloth ratio,* the number of cubic feet of gas at flowing conditions per square feet of fabric area. The air-to-cloth ratio for a coarse granular material at ambient temperature is likely to run around 5 $ft^3/min\ ft^2$ for shaker-type collectors.

After the bags are used for a while, the pressure drop level never returns to the "clean" level after shaking; some residual filter cake stays on the fabric after cleaning. If this were not the case, the efficiency of collection would be considerably lower, because clean bags do not capture small particles effectively.

Several factors tend to cause filter cake to build rapidly, bring about a high-pressure drop, and require the use of lower air-to-cloth ratios, or larger filters, for a given gas flow volume. High grain loadings (that is above 1 or 2 grains per cubic foot), very small particle sizes, and a particularly narrow particle size distribution range are factors that require reductions in air-to-cloth ratios below the 5 $ft^3/min\ ft^2$ level. Very light grain loadings, large particles, and wide particle size distribution sometimes allow selection of collector and high air-to-cloth ratios. For example, lead reverberatory furnaces produce high grain loadings with fine particle sizes falling within a narrow size range. Collectors for this application are frequently sized for 1–1½ $ft^3/min\ ft^2$. At the other extreme, grain elevators produce relatively large particles with a broad range of sized particles when they are preceded by mechanical dust collectors. Here ratios of 10 or 12 $ft^3/min\ ft^2$ are often employed.

Reverse jet filters are often used at much higher ratios than the shaker-type collectors. It is not uncommon to design them for 20 or 30 $ft^3/min\ ft^2$, depending upon the service involved.

Two specific problems relating to fabric collectors should be considered thoroughly in any installation. The first is the sensitivity of bags to heat. No matter how much safety margin is allowed between the normal operating temperature and the temperature limit of the fabric, if there is a source of heat in the process there is some possibility of bag failure owing to loss of temperature control. An emergency dilution air system is probably the most foolproof way of protecting bags against excessive temperature. Emergency water sprays and high-temperature fan shut-down are less reliable alternatives.[14]

The other principal drawback of fabric collectors is the sensitivity to condensation of liquids. The filter-cake pressure drop goes up sharply if the slightest condensation of moisture takes place on the cake. This condition, known as blinding, has the effect of shutting off the gas flow as effectively as if a damper within the baghouse were closed. If the bags blind, the fan or process may put up enough pressure to rupture bags. The operating temperature should always be substantially above the calculated dew point of the gas stream.

Even when this condition is met, condensation may take place in the walls of the collector, or in hoppers not subjected to through-flow of hot gas. This may lead to buildup of heavy deposits of solid or to plugging of the hoppers and dust removal equipment. Steam or electric heating of the hopper bottoms is sometimes the only way to prevent this problem in cold climates.

Placement of fans is of considerable importance. If the fan can be located on the upstream or dirty side of the collector, the dirty gas is confined within the tubular collector and the housing contains only clean gas. The housing construction may be simplified in this case because leakage of the clean gas through joints is not important. Occasionally, the structure or "housing" around the bags can be omitted altogether if the air being cleaned can be recirculated within the building. When the fan is located on the discharge or clean gas side of a fabric collector, it is usually necessary to make sure the housing is gas tight, in order to prevent ambient air from being drawn into it and providing moisture for condensation.

Electrostatic Precipitators

Electrostatic precipitators function by imparting an electrical charge to the dust particles and then passing the dust-laden gas through a strong electrostatic field. The field reacts with the charged particles and imparts a drift velocity, or collecting velocity, to them just as centrifugal force provided a collecting velocity in the mechanical dust collectors. Figure 20 is a sketch of one configuration of a plate electrostatic precipitator, while in Figure 21 the charging electrode runs vertically through the center of a metal pipe. Gas flow is ordinarily downward through the pipe. Particulate matter is charged in the "corona discharge" around the charging electrode, and the strong field between the central electrode and the grounded outlet pipe produces the collecting field. The earliest precipitator designs were tubular. However, in order to make the most efficient use of the limited space available for precipitator installations, plate-type units have become much more widely used. The plate-type precipitators charging electrodes are located between rows of parallel plates. The gas flow is horizontal, and the gas passes perpendicularly to the length of the wires. The depth of the plate-type precipitator in the direction of gas flow is sufficient to allow the gas to contact many charging electrodes so that

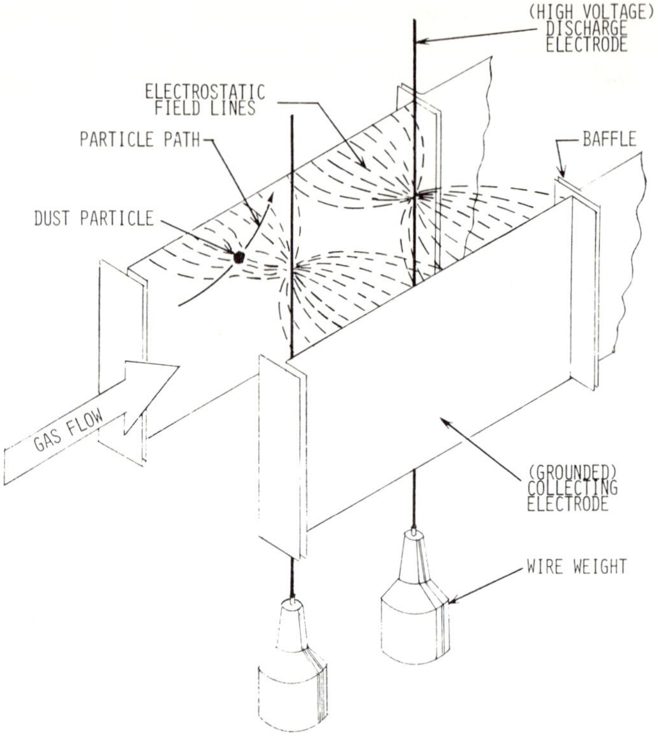

Figure 20. Electrostatic precipitator.

any dust not collected as the gas passes the first wire may be collected at the second or third.[2] A large industrial unit is shown in Figure 22.

Plate-type precipitators ordinarily operate with inlet gas velocity between 1.5 and 10 ft/sec, and with "treatment time" or time in contact with the plates of 2–10 sec. Operating voltages are in the range of 55,000 RMS volts from the charging wires to the grounded plates. Temperatures up to 800°F are handled routinely by precipitators consisting almost entirely of carbon steel. There is no upper temperature limit, but special design techniques are required for temperatures above 1000°F. Precipitators of this type are ideally suited to large hot gas streams and are difficult to build economically for small flows.

A second basic precipitator type is sometimes used for smaller gas flows. Here the gas passes over rows of charging electrodes and is subjected to a positive corona discharge with voltages of 12,000–15,000 volts. After charging, the gas flows between parallel plates that are alternately negatively charged and grounded. The type is "two-stage" because of the separation of the charging stage and the collecting shape and is well adapted to air cleaning services, where the particulate loading is small. This is because the plates cannot be rapped to remove dust as easily as in the single-stage precipitator.[4]

The two-stage precipitation was developed for air cleaning service to deodorize and remove dust from ventilating air going into the building. The high-voltage two-stage precipitator creates a sufficient amount of ozone in the negative corona discharge to deodorize. Because of the lower voltage level, it is sometimes assumed that two-stage precipitators are not hazardous for the collection of combustible materials from air streams. Nevertheless, caution should be exercised in applying precipitators to the collection of combustibles from gas streams with an oxygen content high enough to support combustion.

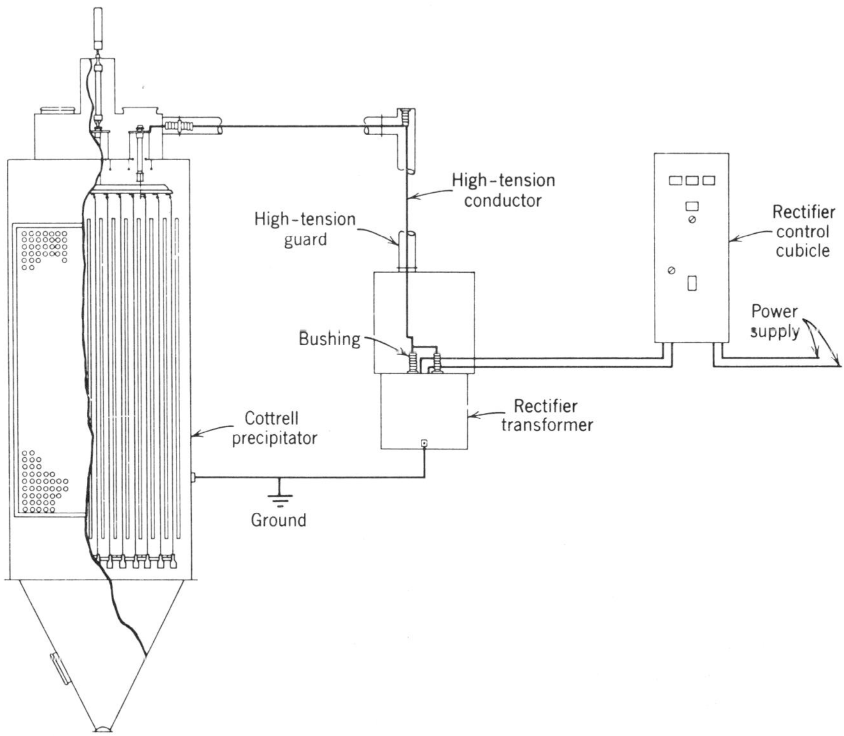

Figure 21. Schematic of an electrostatic precipitator.

Figure 22. Large industrial electrostatic precipitator.

Principle of Operation

In a single-stage precipitator, charging of the particles is accomplished in a negative corona discharge surrounding the discharge electrodes. Corona discharge occurs in the air whenever the voltage gradient becomes sufficiently high to tear electrons away from the surface of the charged conductor. The voltage gradient at a surface is proportional to the voltage, and it requires a relatively high voltage difference between two surfaces to produce a corona. Even with voltages in the 15,000 range, corona discharge does not occur between flat surfaces separated by distances great enough to prevent arcing. If one of the surfaces is curved, such as the outside of a small-diameter wire, the voltage gradient around that surface is higher than when there is a flat surface. Small-diameter wires are used to produce the corona discharge.

The precipitator must operate between the voltage levels indicated as the critical voltage, or voltage in which corona discharge starts, and the arc-over voltage at which point the direct arc between the charging electrode and the charging plate occurs. This range is much broader for small-diameter wires than for large ones; the lower limit of wire diameter is set by mechanical considerations. Ordinarily, these wires are about $\frac{3}{32}$ in. diameter placed between parallel plates 9 in. apart.[28]

The corona is an irregular visible glow with a crackling sound; the gas molecules become ionized and are driven towards the collecting plates. Particles passing through the corona are charged by collision with negative ions. Conducting particles acquire a greater charge than nonconducting particles, since the gradient lines converging upon a conducting particle, and separating to pass around a nonconducting particle. Both types of particles are readily charged, however, and theoretically acquire charge as follows:

$$n \cdot e = E_c r^2 \left(1 + 2\,\frac{k-1}{k+2}\right) \tag{10}$$

where

n = number of electron charges
e = electronic charge 1.6×10^{-19} coloumb
E_c = electrostatic field strength
r = radius of particle (assumed spherical)
k = dielectric constant of particle

In addition to charging by collision with negative ions, there is some random charging of particles that takes place independently of the electric field imposed by the precipitator. This random transfer of electrons is one of the principal reasons for agglomeration of small particles when they are present in high concentration. Small particles acquire a charge spontaneously when random transfers of electrons from particles to gas molecules or vice versa take place. These charges are significant for very small particles in the submicron size range, but do not produce significant voltage for particles above a few microns in

Table 6 Particle Charges

Diameter (μm)	Dielectric Constant	(A) Max. Field Charge	(B) Max. Heat Motion Charge	Ratio (A)/(B)
20	4	21,000	21,000	10
2	4	210	210	1
0.2	4	2	20	0.1

diameter. The ratio of this "thermal" charging effect to the electrostatic charging produced by the corona for a number of particles sizes are in Table 6.

Particles that have acquired a charge in the corona which are subjected to a voltage gradient are acted upon by a calculable electrostatic force. This force is given by

$$F = \frac{E_1 E_2}{R^2} \tag{11}$$

where

F = attractive force
E_1 = charge on particle
E_2 = field strength
R = distance between particles

Using the charge imparted to a nonconducting particle and the voltage difference between the charging and collecting electrodes to give the voltage gradient, the force is

$$F = \frac{r^2[1 + 2(K - 1)/(K + 2)]}{R^2} \cdot E_2{}^2 \tag{12}$$

When this force is equated with the viscous drag imposed on the particle to resist motion toward the collecting electrodes, the equation may be solved for the migration velocity:

$$V = \frac{r^2[1 + 2(K - 1)/(K + 2)]}{6\pi\mu} \cdot E_2{}^2 \tag{13}$$

where

V = migration velocity
μ = viscosity of gas
E_2 = field strength

The drift velocity is equivalent to the collecting velocity calculated for mechanical dust collectors. This may be used in the Deutch equation[28] that relates the size of the precipitator in terms of square feet of plate area, the gas volumetric flow, and the velocity to the efficiency:

$$E = 1 - e^{-AV/Q} \tag{14}$$

where

A = plate area, ft^2
Q = gas flow, actual ft^3/sec
V = drift velocity, ft/sec

The theoretically derived equation for drift velocity does not provide accurate efficiencies when used in the Deutch equation. A number of factors enter into the deviation from theory in real precipitators. Significant among these are:

1. The particles may become reentrained after collection, and require collection more than one time.
2. The gas flow through a precipitator is turbulent rather than laminar. The particle does not progress toward the collecting electrode with a uniform drift velocity, but rather moves randomly from side to side with only a slight velocity bias towards the plate.
3. For an electrostatic capture to take place at the collecting electrode, the particle must come within a very short distance of the plate. At any other point, it is subject to being moved back into the bulk gas stream by random velocity components.
4. The high voltage gradient does not exist uniformly in the precipitator.

Precipitators are sized on the basis of drift velocity values obtained by experience. The efficiency of an operating precipitator is measured, and the drift velocity that would produce that efficiency is calculated. Practical drift velocities range between about 0.1 and 0.55 ft/sec. The most significant factor in establishing the drift velocity seems to be the electrical conductivity of the layer of dust formed on the collecting electrode surface.

To obtain good performance, electrical resistivities of 10^6 to 10^{10} ohm-cm are required. The resistivity is a measure of the electrical resistance of a dust sample 1 cm^2 in cross-sectional area and 1 cm thick. The resistivity of most metal oxides falls within this range when the temperature is above about 450°F. It is necessary to collect particles at much lower temperatures than this, and the resistivity of the particles themselves is likely to be quite high. The resistivity of particles may be decreased by adsorption from the gas strea of materials

Table 7 *Total Utilization of Electrostatic Precipitators in the United States, 1912–1972*

Industry and Application	Total 1965–1969			
	No. Pptrs.	Gas Volume ACFM-Millions[a]	No. Pptrs.	Gas Volume ACFM-Millions
Electric utility fly ash	1331	531	419	257.61
Pulp and paper recovery Blr	229	34.4	62	12.4
Iron and steel				
Coke oven detarring	600	5.0	0	0
Iron cupola	15	0.5	0	0
Blast furnace	196	11.36	1	0.06
Open hearth furnace	62	12.06	8	3.79
Basic oxygen furnace	24	6.3	12	3.39
Scarfing machine	10	0.78	2	0.16
Sintering machine	29	7.5	9	2.36
Electric furnace	N.A.	N.A.	*c*N.A.	N.A.
Total	936	43.50	32	9.76
Rock products				
Cement	300	41.5	36	5.19
Gypsum	70	1.8	2	0.80
Total	370	43.3	38	5.99

that serve as conductors. In utility power plants, precipitators have served very well owing to the fortunate presence of traces of sulfur trioxide in the flue gas. As little as 5–10 ppm of sulfur trioxide at 300°F is capable of reducing the resistivity of fly ash from 10^{13} ohm-cm to 10^9 ohm-cm, which makes the difference between poor and good performance. Other materials that work as "conducting agents" for fly ash are water, ammonia, ammonium sulfate, and ammonium chloride.[29]

Rapping

To maintain a continuous process of precipitation, it is necessary to continuously or periodically remove the precipitated material from the collecting electrodes. In the case of dry dusts, this is usually done by vibrating or rapping the collecting electrodes to cause the material to be dislodged and fall into the hopper. Most large single-stage precipitators are similar in basic dimensions. Plates or collecting electrodes are spaced about 9 in. apart. Wires of about $\frac{3}{32}$ in. diameter used for

Table 7 Continued

Industry and Application	Total 1965–1969 No. Pptrs.	Total 1965–1969 Gas Volume ACFM-Millions[a]	No. Pptrs.	Gas Volume ACFM-Millions
Chemical process				
Sulfuric acid mist	120	2.33	36	1.00
Elec. furn. phosphorous	22	0.16	4	0.06
Phosphoric acid mist	21	0.13	0	0
Carbon black	100	4.0	0	0
Total	263	6.62	40	1.06
Nonferrous metallurgy				
Roasters and reverb	58	1.44	12	0.38
Sintering machine	26	1.02	0	0
Misc. smelting	21	1.76	N.A.	N.A.
Aluminum—prebake potline	59	4.88	0	0
Aluminum—carbon vent and misc.	5	0.17	N.A.	N.A.
Aluminum—calciners and misc.	13	0.71	N.A.	N.A.
Total	182	9.98	12	0.38
Petroleum				
Catalytic cracking	42	3.86	0	0
Carbureted water gas	57	0.32	0	0
Misc. detarring	64	0.40	2	0.26
Total	163	4.58	2	.26
Miscellaneous				
Refuse incineration	17	1.91	17	1.91
Pipeline demisting (proto)	1	Proto	1	Proto
Fluid bed coal (pilot)	2	Pilots	2	Pilots
Fluid bed refuse (pilot)	1	Pilot	1	Pilot
Total	21	2.0	21	2.0
Grand total	3495	675.38	626	289.46

[a] ACFM, actual ft³/min.

collecting electrodes are hung between the plates. Plate material is usually 18-gage hot rolled carbon steel, and some sort of stiffening arrangement is required to maintain the plates fast. The stiffeners are designed to contribute to the collector performance as much as possible by creating eddies of gas adjacent to the plates in which the linear velocity is very low. A number of designs accomplish this purpose, without any one design showing any significant advantage.

Table 8 *Operating Characteristics of Electrostatic Precipitators in Major Application Industries*

Industry and Application	Typical Precipitator Capacity (ft^3/min, Thousands)	Temp. (°F)	Gas Properties Press—psia	Moisture (% Vol)	Aerosol Properties at Precipitator Inlet: Type	Aerosol Properties at Precipitator Inlet: Concentration (Grains/scf)	Precipitator Properties: Type[a]	Precipitator Properties: Input Power (Watts/1000 ft^3)	Efficiency (%)
Electric utility, fly ash	25–2000 (400)	225–900 (295)	14.7 ± 0.5	6.0	Dust	1.5–7.5 (3.1)	DHP	25–175 (80)	90–99.6
Pulp and paper, Sulfate recovery boiler	95–200 (150)	225–375 (310)	14.7 ± 0.5	20–40	Dust and fume	1.0–9.0 (3.8)	DVP-DHP	100–700 (150)	90–99.5
Iron and Steel									
Coke oven detarring	5–10	(113)	14.7 ± 0.5	50	Tar and mist	(0.3)	WVT	(100)	95–98
Iron cupola		300–400	14.7 ± 0.5		Dust and fume	0.3–0.7	DHP		90–96.6
Blast furnace	30–100 (50)	70–130 (97)	14.7 ± 0.5		Dry fume	0.0–0.7 (0.18)	WVT-WHP	50–320 (120)	90–98
Open hearth furnace	50–700 (200)	500–600 (510)	14.7 ± 0.5	7–16	Dust and fume	0.01–2.5 (0.45)	DHP	(53)	95.5–99.2
Basic oxygen furnace	200–330 (250)	(425)	14.7 ± 0.5		Dust and fume	(4.7)	DHP	(29)	99.2–99.8
Scarfing machine	50–100 (77)	(97)	14.7 ± 0.5		Dust and fume	0.0–0.8 (0.06)	DHP	(268)	85–97.5
Sintering machine	150–400 (250)	150–400 (245)	14.7 ± 0.5	5–15	Dust and fume	(1.5)	DHP	(71)	90–98.5
Electric furnace		300–500	14.7 ± 0.5		Dust and fume	1.0–3.0	DHP	N.A.	92–97
Rock Products									
Cement									
wet process kiln		353–650	14.7 ± 0.5	30	Dust and fume	4.1–48.4	DHP	60–333	92.8–99.94
dry process kiln	50–300 (140)	500–700	14.7 ± 0.5	1–8	Dust and fume	3.7–53.6	DHP	61–210	98.8–99.98
mill and dryer		93–160	14.7 ± 0.5		Dust	7.5–40.6	DHP	95–277	96.9–99.91
Gypsum									
kettles	(8)	220–342	14.7 ± 0.5	19–47.6	Dust and fume	4.8–48.0	DVP-DHP	124–928	94.4–99.90
rock dryers	(25)	125–240	14.7 ± 0.5	3.6–17.4	Dust	4.6–156	DVP-DHP	69–282	97.2–99.85
rotary calciner	(47)	N.A.	14.7 ± 0.5	14.2–35.2	Dust	32.9–46.3	DVP-DHP	62–174	99.1–99.82
holdflite calciner	(25)	200–339	14.7 ± 0.5	17.6–54.0	Dust	30.8–62.6	DVP-DHP	576–983	98.2–99.90
comb. kettle/ dryer	(36)	140–250	14.7 ± 0.5	6.7–22.2	Dust and fume	7.7–63.9	DVP-DHP	65–390	93.7–99.96

[a] Precipitator nomenclature: W—wet; D—dry; T—tube type; P—plate type; V—vertical gas flow; H—horizontal gas flow;Parentheses denote mean or average.

Table 9 *Operating Characteristics of Electrostatic Precipitators in Minor Application Industries*

Industry and Application	Typical Precipitator Capacity (ft^3/min, Thousands)	Temp. (°F)	Gas Properties Press (psia)	Aerosol Properties at Precipitator Inlet: Type[a]	Aerosol Properties at Precipitator Inlet: Concentration (Grains/scf)	Precipitator Properties: Type	Precipitator Properties: Input Power (Watts/1000 ft^3)	Efficiency (%)
Chemical process								
Sulfuric acid mist	10–50 (20)	80–180 (130)	14.7 ± 0.5	Mist	0.25–4.5 (0.75)	WVT	100–1000 (380)	95–99
Elec. furnace phosphorus	0.2–50 (8)	550–900 (620)	14.7 ± 0.5	Dust and fume	6.0–16.0 (12.0)	DVT	N.A.	96–99.9
phosphoric acid mist	3–13 (6)	140–285 (203)	14.7 ± 0.5	Mist	4–33 (20.3)	WVT	N.A.	85–98
Carbon black	20–75 (38)	410–500 (454)	14.7 ± 0.5	Fume	6–50 (22.3)	DHP	N.A.	90–97
Mining & metallurgical (nonferrous)								
Roasters & Reverb. furnace	20–31 (25)	450–1100 (674)	14.7 ± 0.5	Dust and fume	1.9–40 (11.4)	DHP	N.A.	94.9–98.0
Sintering machines-zinc	22–80 (39)	131–250 (213)	14.7 ± 0.5	Dust and fume	0.38–5.0 (2.2)	DHP	N.A.	91.7–97.4
Miscellaneous smelting	5–240 (88)	N.A.	14.7 ± 0.5	Dust and fume	N.A.	DVT-WVT	N.A.	90–98
Aluminum—prebake potline	50–115 (83)	90–250 (190)	14.7 ± 0.5	Dust	.01–0.10	DHP	N.A.	90–95
Aluminum—carbon vent and misc.	35	120–175 (150)	14.7 ± 0.5	Carbon dust and tar	7.0–13.4	DVT-WVT	N.A.	98–99
Aluminum—calciners and misc.	20–200 (55)	70–565	14.7 ± 0.5	Dust	0.1–45	DHP-DVP-WVT	N.A.	95–99.9
Petroleum								
Catalytic cracking	60–240 (90)	450–850 (610)	14.7 ± 22.2	Dust	5–40	DVP-DHP	N.A.	90–99.7
Carbureted water gas	3–10 (5.7)	70–110	14.7 ± 164	Tar and mist	0.8–2.5	WVT	N.A.	95–99
Miscellaneous detarring	1–11 (6)	90–200	7.5–115	Tar and mist	N.A.	WVT	N.A.	98–99
Miscellaneous								
Refuse incineration	30–150	450–550	14.7 ± 0.5	Dust and fume	.5–4.0	DHP	50–200	95–99.2
Pipeline demisting (proto)	5.5	90–130	780–825	Mist	$7–34 \times 10^{-3}$	WHP	1000	99+
Fluid bed coal comb. (pilot)	1	1200–1400	65–115	Dust	1–3	DVT	N.A.	98–99.5
Fluid bed refuse comb. (pilot)	0.1	1500–1700	65–115	Dust and fume	N.A.	DVT	N.A.	N.A.

[a] Precipitator nomenclature: W—wet; D—dry; T—tube type; P—plate type; V—vertical gas flow; H—horizontal gas flow; Parentheses denote mean *or* average.

Table 10 *Precipitator Erected Costs (1959–1970) Miscellaneous and Minor Applications*

Application	Installation Capacity (ACFM, Thousands)	Erected Cost Pptrs. ($/ACFM)
Refuse incineration	64–569	0.66–0.44
Pipeline demisting	5.5	10.0
Catalytic cracking	150–254	1.16 (F.O.B.)–0.48 (F.O.B.)
Miscellaneous detarring	11.4–55.6	2.50–1.10
Nonferrous roasters	20–47	6.0–4.25
Aluminum calciners	50–70	0.72–1.10
potlines	115–1380	0.68–0.66
Acid mist sulfuric	10–100	6.70–4.40
phosphoric	10–40	12.40–4.90
Electric furnace phosphorus	10–20	16.7–14

Walker of Research Cottrell in a paper before the December 1970, meeting of the International Union of Air Pollution Prevention Association at Washington, D. C., gave the data in Table 7, Overall Utilization of Electrostatic precipitatators in Major Application Industries; Tables 8 and 9, Operating Characteristics in Minor Application Industries; Table 10, Erected Costs; Table 11, Total Use; and Table 12, Operating Conditions.

Table 11 *Total Use of Electrostatic Precipitators*

	1912–1969		1965–1969	
	Number	Capacity (Millions ACFM)	Number	Capacity (Millions ACFM)
Electric utility	1331	531	419	257.6
Pulp and paper	229	34.4	62	12.4
Iron and steel	936	43.5	32	9.76
Rock products	370	43.3	38	5.99
Chemical process	263	6.6	40	1.06
Nonferrous metals	182	9.98	12	0.38
Petroleum	163	4.58	2	0.26
Refuse combustion	17	1.91	17	1.91
Misc.	4	0.09	4	0.09
Total	3495	675.38	626	289.46

Table 12 Operating Conditions of Electrostatic Precipitators

	Temp. (°F)	Pressure (psia)	Concentration (Grains/SCF)	Efficiency (%)
Electric utility	225–900	14.7	1.5–7.5	90–99.6
Pulp and paper	225–375	14.7	1.0–9.0	90–99.5
Iron and steel	70–600	14.7	0.01–3.0	85–99.8
Rock products	350–700	14.7	3.7–156	92–99.9
Chemical process	80–800	14.7	0.2–50	85–99.9
Nonferrous metals	70–1100	14.7	0.01–45	90–99.9
Petroleum	70–850	7.5–164	0.8–40	80–99.7
Refuse combustion	450–550	14.7	0.5–4.0	95–99.2
Misc.	90–1700	65–825	10^{-5}–3.0	95–99.5

Scrubbers

The wet scrubber is unique among air pollution control devices in that it is used for abatement of either particulate or gaseous emissions. Scrubbers are frequently chosen for services where both forms of pollutants are present in the gas stream. Scrubbers are also widely used where the dirty gas stream has a high moisture content.

Scrubbers collect particulate material above several microns in diameter with efficiencies approaching 100%. They are, however, selective with regard to particle size, and diameters below 1 μm present challenging problems in design and application of scrubbing equipment. Scrubber designs are much more varied than those of mechanical, fabric, or electrostatic collectors.[9]

Scrubber Types

The steps taking place within wet scrubbers involve saturation of the incoming gas stream, wetting or capture of the particulate matter in a liquid phase, and separation of the liquid phase from the gas stream.

The initial contact of effluent gas with water may take place within the scrubber or in a separate presaturation chamber designed to prevent evaporation of liquid within the scrubber, or to protect the scrubber from exposure to high-temperature gases. Figure 23 is a drawing of a scrubber with a presaturation chamber. Saturation of the gas is incidental to particulate removal, and for practical purposes the scrubber may be presumed to receive a saturated, particulate-laden gas stream.

The principal mechanisms for wetting the particulate material are inertial impaction, agglomeration, and diffusion.[9]

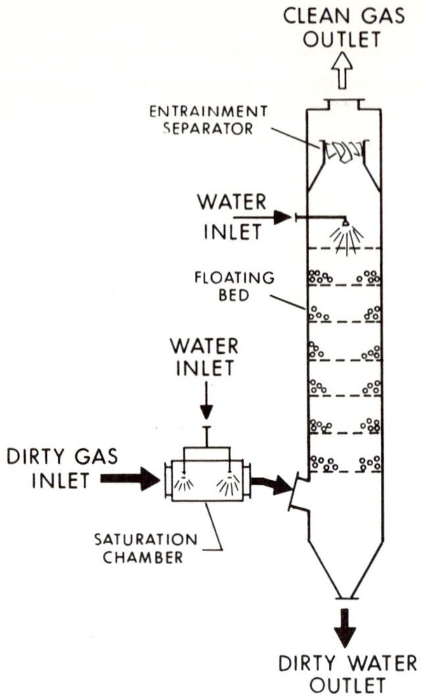

Figure 23. Scrubber with presaturator.

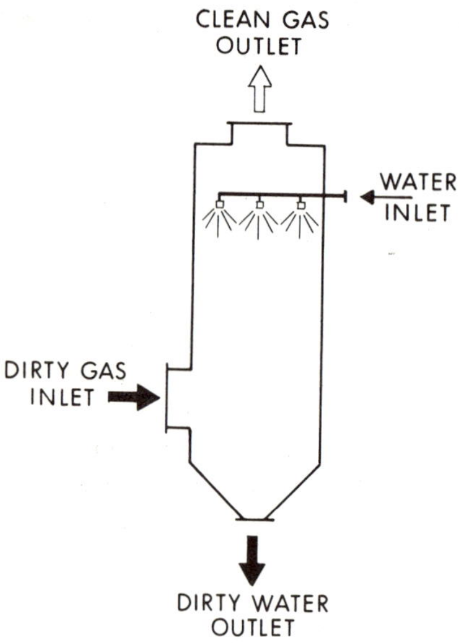

Figure 24. Spray tower.

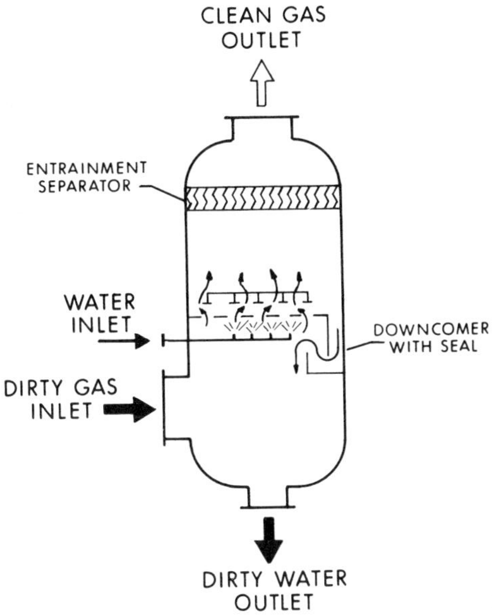

Figure 25. Impingment scrubber.

In the simplest scrubbers, the inertial impaction mechanism simply involves passing the particulate laden gas stream through a contacting zone in which water is dispersed either by spray nozzles, or a packing material. Figure 24 is a drawing of a spray tower such as these, for the collection of cryolite particles in the aluminum reduction industry. The energy imparted to the particulate matter is limited by the velocity of the gas passing through the scrubber, and only particles of several microns and over in diameter are likely to have sufficient mass to impinge upon the water droplets and be wetted by them. Particulate matter in the submicron size range may pass through such a scrubber and leave as a dry solid.

One mechanism for producing higher gas velocities, and consequently higher particle energies, involves passing the gas through orifice plates as shown in Figure 25. Other mechanisms are illustrated in Figure 26 (packed counter-current scrubber), and Figure 27 (mobile packed counter-current scrubber). Each of the scrubbers illustrated here is effective in removing particles above 5 μm in diameter and may show good performance in the 1–5 μm range. None is very effective when treating particles between 0.1 and 1 μm in diameter. These very small particles require gas velocities of one or two orders of magnitude higher than those which can be obtained in counter-current or cross-

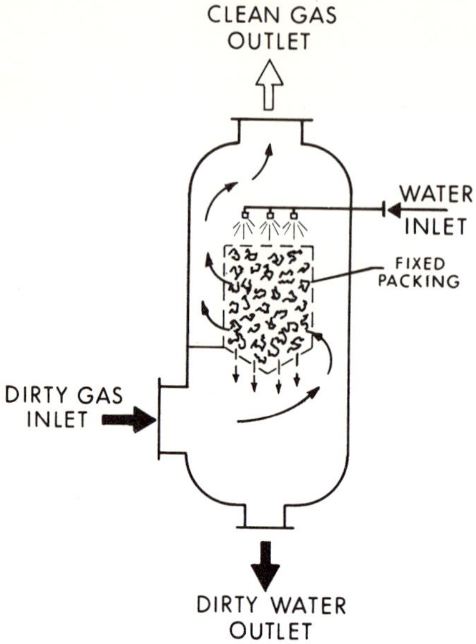

Figure 26. Cross-flow packed scrubber.

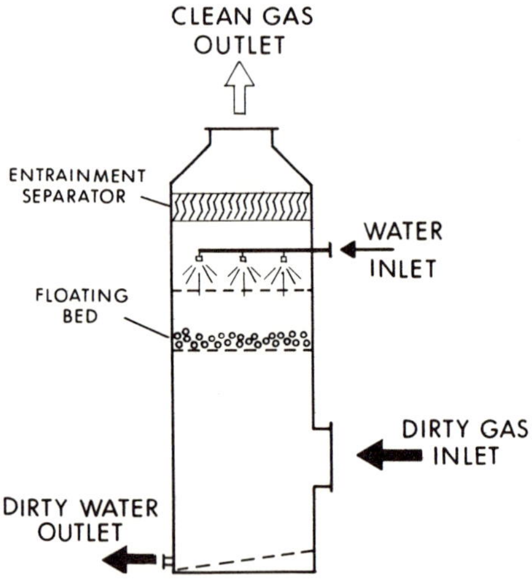

Figure 27. Mobile packing scrubber.

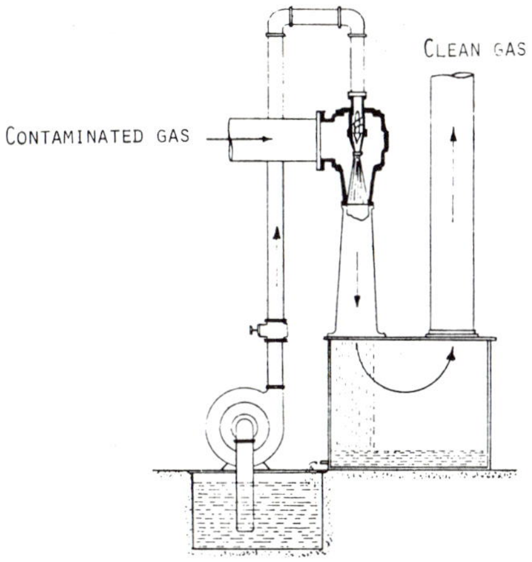

Figure 28. Venturi scrubber.

flow scrubbers. These are collected in cocurrent Venturi scrubbers that operate with velocities between 200 and 400 ft/sec at the throat.

Figures 28–31 are illustrations of Venturi scrubbers. Here the gas is accelerated in a convergent nozzle and water is introduced at the throat or is allowed to flow into the convergent section with the gas. The water may be atomized mechanically at the throat, or it may be atomized by the shear forces imposed by the gas.

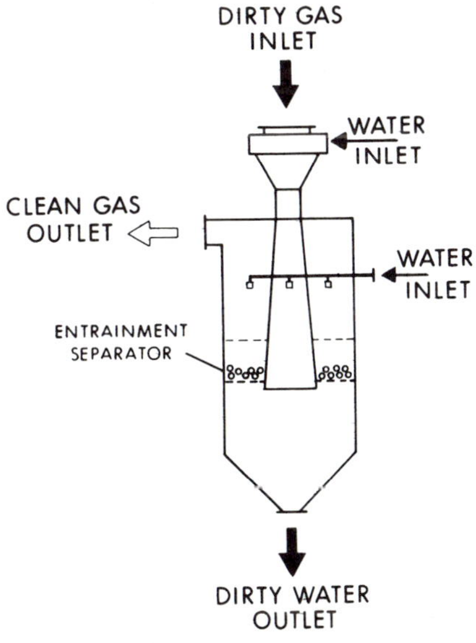

Figure 29. Venturi scrubber with entrainment separator.

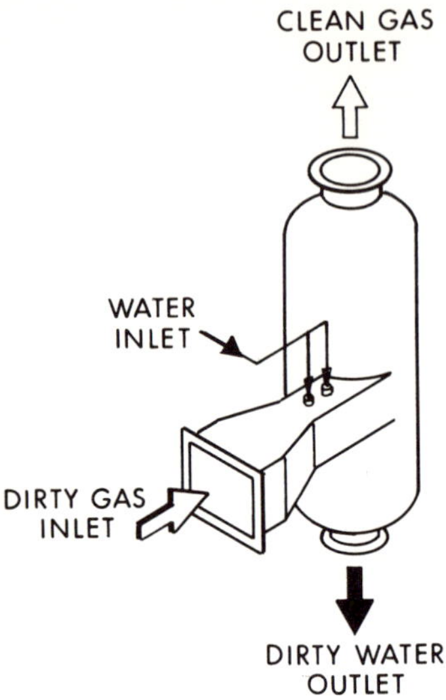

Figure 30. Venturi cyclonic scrubber.

In the Venturi, it is important that the water droplets be dispersed in a fine mist or fog. The energy required to impact a dust particle upon a water droplet is smalllest when the two are of substantially the same diameter. The water droplets are not broken down to sizes as small as the aerosol particles being collected. Atomization of the water does contribute significantly to the performance of the scrubber.

Other mechanisms to enhance contacting of the particulate material and the water are (1) condensation of water droplets around the

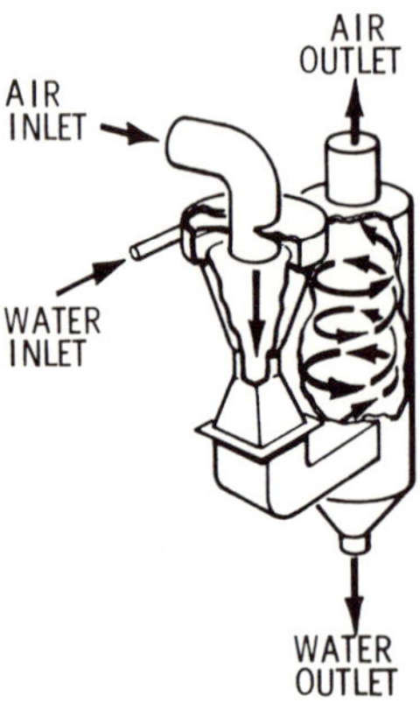

Figure 31. Venturi cyclonic scrubber.

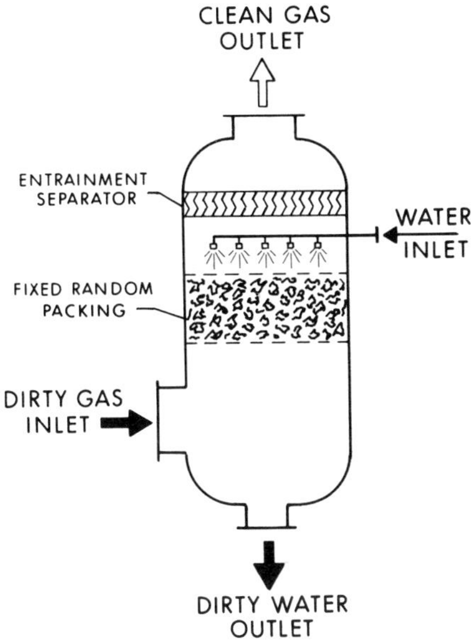

Figure 32. Packed scrubber.

particulate matter as nuclei, and (2) electrostatic agglomeration of the particulate matter.

Below abour 0.1 μm diameter, diffusion of the small particles through the gas stream is probably an important mechanism. The kinetic energy required to reach the surface of the liquid is provided by molecular bombardment of the particle by the gas molecules, rather than by the velocity of the gas.

The final step in the collection of particulate matter in wet scrubbers involves the recapture of the dispersed liquid and its separation from the gas stream. The liquid is not dispersed into droplets below about 50 μm, and presents a much easier collection problem than was the case

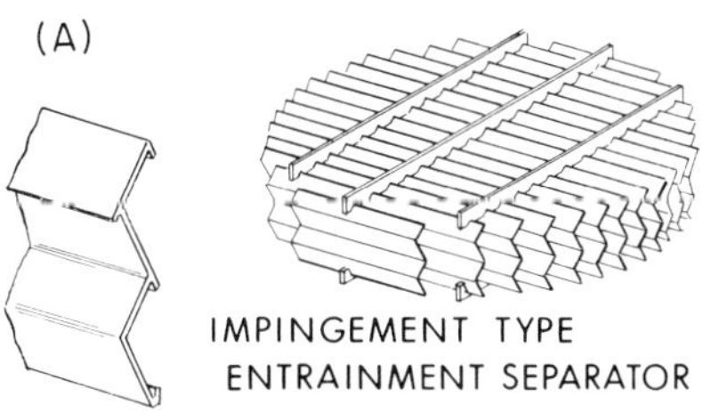

Figure 33. Impingment separator.

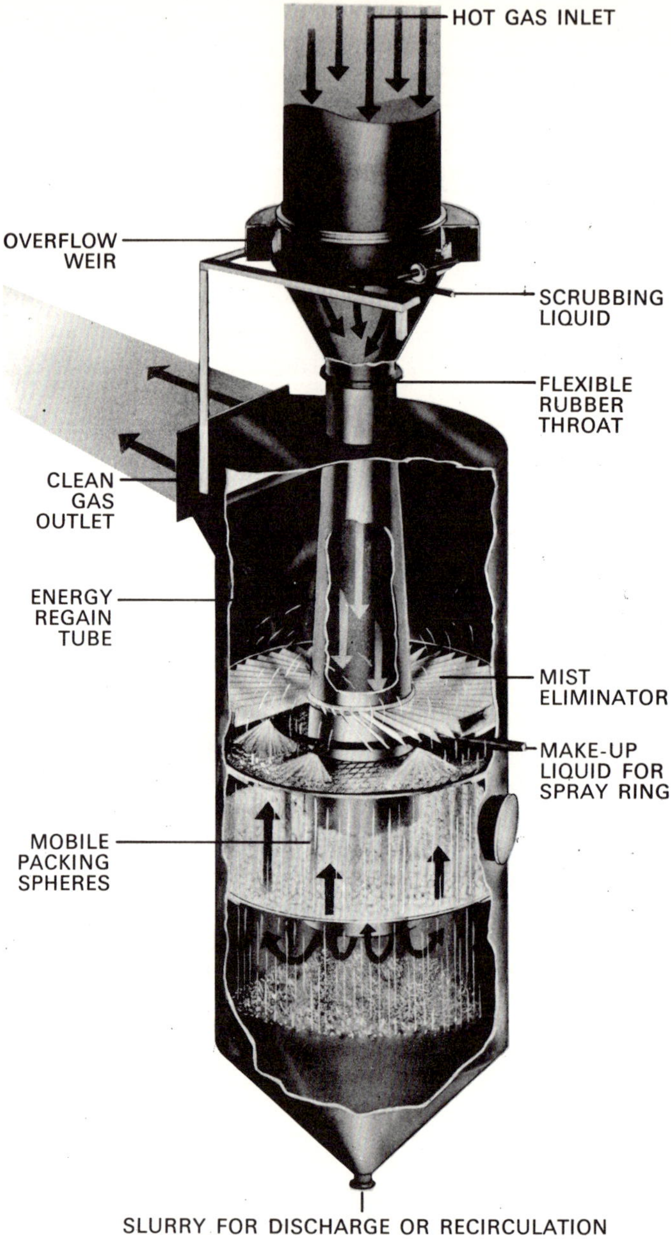

Figure 34. Venturi sphere wet scrubber with mist eliminator.

(B)

SPINNER TYPE ENTRAINMENT SEPARATOR

Figure 35. Spinner type entrainment separator B.

with the original particulate material. Figure 32 illustrates a centrifugal collection system employed in the "Venturi-Cyclonic" scrubber. Figure 33 illustrates a baffle-type impingement separator wherein liquid droplets are impinged upon wetable surfaces and drained back into the scrubber. Figures 34 and 35 show spinner arrangements used as a centrifugal entrainment separating device. The separation of the liquid and gas phases after particulate collection is accomplished presents no unusual technical problem. Other engineering arrangements are given in Figure 36.

The Small Particle Problem

The scrubbing systems described in the foregoing paragraphs are used to collect particulates of all sizes. All scrubbers are selective for larger particles; they the larger particles at higher efficiencies than small ones. The particles below 1 μm in diameter are the major cause of the visible stack emissions.

The Venturi scrubber has the capability to collect very fine particles and is frequently employed for particulates which lie within the 0.1–1 μm diameter range. However, these applications present difficulties. Sampling small particulates for total concentration in gas streams is not difficult. The principal mechanism involves collection of the particulates from a gas sample stream, with fabric filters and low-velocity electrostatic precipitators. The fabric filter is the most used. Adequate field testing can be done with portable instrumentation to provide a design basis for an air pollution control system.

In the absence of good data on particle size distribution, or good background expeience with similar applications, it is necessary either to design for pressure drops well in excess of the expected values or to perform pilot studies with small scale equipment in the field.

Several companies have developed miniature scrubbing devices that can be used as analytical instruments. These are patterned after commercial scrubbers, and have been designed to produce the same rela-

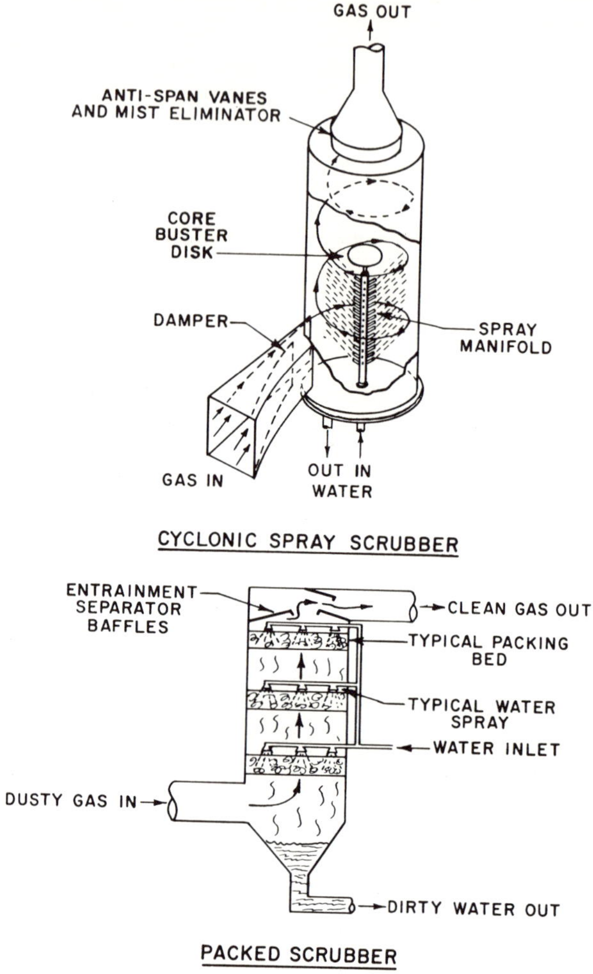

Figure 36. Other scrubber designs.

tionships between pressure drop, gas velocity, and collection efficiency. Initial tests in the field indicate that such instruments can be used to select design condition for commercial equipment with good reliability.

Venturi Scrubbers

At temperatures up to about 400°F saturation is allowed to take place in the Venturi throat. There is, however, a significant pressure drop associated with the change in gas volume, and for higher temperatures it is preferable to saturate the gas before accelerating it to throat velocity.

One of the most significant problems in all types of scrubbing equipment is the area in which liquid and gas are first brought into contact. A liquid that has solids dispersed in it must be kept away from hot dry surfaces, where it will evaporate and leave the solids behind as deposits. Significant corrosion problems are associated with variable liquid interfaces. The gas–liquid interface should be made as clear-cut and stable as possible. Those surfaces that are wetted should be wetted continuously; those that are dry should be kept free of splashes.

The throat may be restricted by a variable mechanism to provide for changes in gas throughput without reducing the efficiency. A flexible throat of variable cross section may be used, or a disk or vane placed in the throat.

The diffuser section following the throat is of concern in ensuring good utilization of the energy requirements. It is here that a portion of the energy in contacting the gas with the liquid can be recovered to reduce the total energy requirements of the scrubber. For the vertical Venturi-sphere scrubber with a straight diffuser section, as much as 80% of the energy required to accelerate the gas and liquid at the throat can be recovered in the diffuser section.

Costs of Particulate Collection Equipment

Costs of air pollution control equipment involve three separate items: equipment cost, installation cost, and operating cost.

There is a tendency to place more emphasis on the equipment cost than on the other two items, although both installation and operating costs over the life of the equipment are likely to exceed the equipment first cost. The principal reasons for the overemphasis on first cost are that this is the first expenditure made in a program to control emissions and the cost of equipment can be obtained rather easily and explicitly by soliciting quotations from manufacturers.

Accurate assessment of the installed cost of the system can only be approximated after detailed plans for the installation have been made. Unknown costs are more often underestimated (or entirely disregarded) than overestimated. Operating costs are often disregarded during equipment selection. Equipment manufacturers often fail to call attention to high maintenance costs during their efforts to sell their products. Figure 37, gives relative costs of various dust collectors and Table 13 gives the relative performance on which the costs are based.

This section contains some general guidelines to costs of pollution control systems for a typical application. They should be used only for

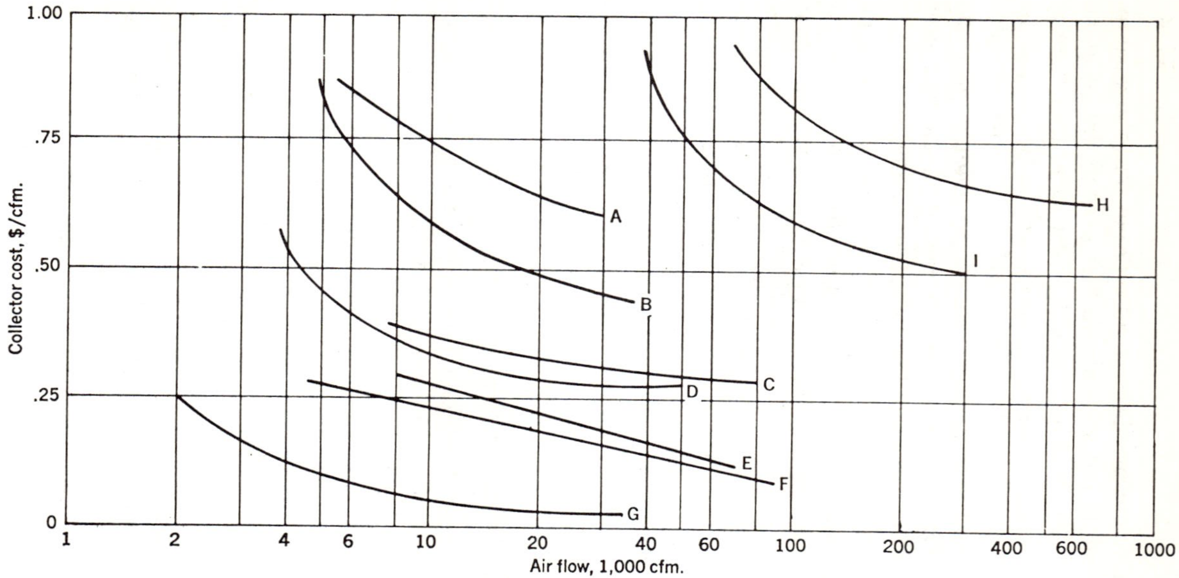

A — High temperature fabric collector (continuous duty)
B — Reverse jet fabric collector (continuous duty)
C — Wet collector (maximum cost range)
D — Intermittent duty fabric collector
E — High efficiency centrifugal collector
F — Wet collector (minimum cost range)
G — Low pressure drop cyclone (maximum cost range)
H — High voltage precipitators (fly ash installations)
I — High voltage precipitators (minimum cost range)

Figure 37. Scrubber capabilities.

Table 13 Scrubber Capabilities

Type of Scrubber	Usual Range of Particle Sizes (μm)	Normal Draft Loss, (in. Water)	Maximum Efficiency (%)	Typical Applications
Open spray tower	>10.0	$\frac{3}{4}$–2	85	Iron pyrite roasting
Packed tower	>10.0	1–6	85	Absorption of fluorine compounds
Wet centrifugal	>2.5	2–6	95	Phosphate dust from fertilizer plants
Flooded bed	>2.5	2–8	95	Primary washer for blast furnace gas
Orifice	>2.5	To 6	95	Rock and cement dust
Wet dynamic	>2.0	None	98	Heavy concentration of fine particles
Venturi	>0.5	6–80	99+	Oxygen converter and ferrous cupola gas, acid mist, lime dust, coal dust, incinerators
Flooded disk	>0.5	6–70	99+	Same as for Venturi

the most preliminary estimates, however, because of the wide variability in cost for treating even similar applications. Wide job-to-job variation is particularly characteristic of retrofit, or add-on, installations where there may be a great deal of process rearrangement involved in the air pollution control installation.

The most reasonable approach to obtaining good estimates of overall cost of equipment is to hire a competent engineer or contractor to rough out a design and make cost estimates based on the specific job. If more than one approach is applicable, it is usually good practice to estimate costs of each on a total annual cost basis and select the least expensive alternative consistent with good performance over the expected life of the installation.

EQUIPMENT COST

The basic apparatus is likely to amount to at least 25% of the installed system cost or as much as 75%, depending upon the complexity of the installation. It is customary to estimate equipment prices based on generalized cost data or quotations from equipment manufacturers, and apply a factor defined as the ratio of installed cost to equipment cost in order to get an estimate of the overall capital cost. These factors may be as high as 4.0/1 or as low as 1.33/1.

Generalized equipment costs are often expressed on the basis of cost per cubic feet per minute (cfm). Most collectors are designed on the basis of actual Fefm (acfm) at the following temperature, which is the best index of cost versus size. This means that anything that tends to reduce the flowing gas volume, including reduction of gas temperature, is likely to bring about a reduction in collector cost. Wet scrubbers are sized by the following volume after saturation, which is almost always lower than the inlet gas volume. Calculated saturated gas acfm is used here.

Incineration equipment, such as that used for burning fumes from chemical processes or food processing, is usually sized on the basis of standard cfm (scfm), because the temperature, and hence the acfm, changes in the unit.

Some notes of caution are in order with respect to the basis for quotation used by equipment manufacturers. Some items of equipment, such as small wet scrubbers, are shipped as completely assembled units. Others, such as multitube cyclonic collectors, fabric collectors, and precipitators, are customarily quoted on a "knocked-down" basis; that is, the price includes the pieces and parts from which the unit can be assembled in the field, but not the actual assembly.

INSTALLATION COST

The cost of installation ordinarily must include engineering design; foundations; structural support steel, accessways, etc.; a fan or blower; pumps, piping, etc.; site electrical supply work; dust disposal equipment, etc.; a stack, breeching, and related equipment.

It is ordinarily unwise to obtain quotations on overall installation costs or "turnkey" installations without a considerable amount of preliminary engineering design. Contractors who have made a number of sucessful installations of equipment in similar installations may be able to make good estimates of installed costs without plans and specifications. Both customers and suppliers should approach contractors of this type with caution if they do not have an outstanding record of sucesses in nearly identical circumstances.

OPERATING COSTS

For electrostatic pricipitators, the fan horsepower is nominal, as a pressure drop less than 1 in. water column is the rule. Fabric collectors ordinarily run between 2 ½ and in., as do multitube centrifugal collectors.

Wet scrubbers are frequently used in the 2–6-in. water column pressure drop range. Only for relatively severe scrubbing jobs do the horsepower requirements become very large. The first cost of the fan is likely to be a substantial part of the total installation cost at high pressures.

Water and fuel costs, where they are applicable, are easy to estimate. They are usually smaller than the fan cost, and are too often neglected without a rough check to establish whether they are negligible or not. Fuel costs are ordinarily significant only in cases where thermal incineration is used for combustible particulates, or where reheat is used to prevent steam plume formation in very large wet scrubber installations.

Finally, the cost of disposal of the collected particulate matter should be taken into account. Costs are likely to run high if particulates are collected as a fine, dry powder that is unsuitable for storage. On the other hand, solids collected by wet scrubbers are likely to be expensive to handle, for they cannot ordinarily be discharged into sewage systems or natural bodies of water.

REFERENCES

1. A. C. Stern, *Air Pollution,* Academic Press, New York, 1968.
2. Harry J. White, *Industrial Electrostatic Precipitation,* Addison-Wesley, Reading, Mass., 1963.
3. Walter C. McCrone, R. G. Draftz, and J. G. Delly, *The Particle Atlas,* Ann Arbor Science Publishers, Ann Arbor, Mich., 1967.
4. John A. Danielson, *The Air Pollution Engineering Manual,* USHEW PHS Cincinatti, Ohio, 1967.
5. "Cyclonic Mechanical Dust Collector Criteria," Publication M-2, Industrial Gas Cleaning Institute, P.O. Box 448, Rye, N.Y., June 25, 1969.
6. "ASME Power Test Code #28," ASME, New York.
7. J. L. Burdock, "Present Applications of Mechanical Collectors to Boilers," 62nd Annual Meeting of the APCA, New York, June, 1969.
8. "ASME Power Test Code #27," ASME New York.
9. L. C. Hardison, *Wet Cleaning of Particulates,* American Ceramic Society, Washington, D.C., 1969.
10. "Air Quality Criteria Document for Particulate Matter," USHEW, PHS, 1969.
11. Coulter Counter Reference.
12. "Control Technology for Particulate Air Pollutants," USHEW, PHS, 1969.

13. Harold E. McGannon, *The Making, Shaping and Treating of Steel,* 8th Ed., U.S. Steel Corp., Pittsburgh, Pa., 1964.

14. L. C. Hardison, "Cost and Effectiveness of Air Pollution Control Equipment in the Lime and Secondary Smelting Industries," USPHS Contract No. CPA-70-150.

15. A. B. Walker and R. F. Brown, "Statistics on Utilization, Performance and Economics of Electrostatic Precipitators for Control of Particulate Air Pollution," 2nd International Clean Air Conference, Washington, D.C., December 6–11, 1970.

16. Allied Chem. paper on incineration.

17. Walter R. Niessen, "The Emission and Control of Air Pollutants from the Incineration of Municipal Solid Wastes," Second International Clean Air Conference, Washington, D.C., December 6–11, 1970.

18. L. C. Hardison, "Industrial Waste Disposal," First Executive Symposium on Pollution, St. Louis, Mo.

CHAPTER V.2

particulates in steel processes

HAROLD B. JENSEN

The process of converting iron ore, which is a mixture of oxides (Fe_2O_3 and Fe_3O_4) in nature to pig iron produces very large quantities of dust. Early blast furnace facilities were characterized by the light coating of red dust that was normally found on almost everything in the vicinity. Modern installations collect almost all of the dust that is produced. Approximately 58 lb of dust is produced and collected for each ton of pig iron produced in the United States. Approximately 92 million tons of pig iron were produced in the United States in 1970 in blast furnaces. The total flue dust from the production of this pig iron in one year is therefore approximately 5,336,000,000 lb or 2,668,000 tons of dust.

The molten pig iron produced in the blast furnace then goes either to an open-hearth furnace or to the more modern basic oxygen furnace for conversion into steel. Both of these processes also produce large quantities of dust. The older open hearths do not have dust collectors on them, and these furnaces are being rapidly phased out as they are obsolete and normally do not warrant having extensive dust collection and smoke abatement equipment installed on them.

In the basic oxygen-furnace method of making steel, approximately 50 lb of dust per net ton of steel produced is collected. In 1970, 63,330,000 tons of steel were made by the basic oxygen process in the United States. From this production, 3,166,500,000 lb or 1,583,250 tons of dust was collected.

The chemical composition of the dust varies between the blast furnace and the basic oxygen furnace from one furnace to another of the same type. In addition, the types of materials being charged to the blast furnace and to the BOF vary and therefore the dust collected will vary. A typical analysis of the dust produced by these two processes is given in Table 1.

Virtually all of the 4,250,000 tons of dust collected from the blast

Table 1 Properties of Furnace Dusts

	Blast Furnace	Basic Oxygen Furnace	Electric Furnace
Production			
(millions of tons/yr)	2.7	1.6	0.5
Chemical analysis (% iron oxides)	46.0	56.0	39.0
Sulfur (S)	0.2	0.6	0.35
Zinc oxide (ZnO)	0.0	2.00	2.50
Lead oxide (PbO)	0.0	0.09	0.50
Carbon (C)	18.0	1.45	2.00
Silica (SiO_2)	4.5	1.70	3.00
Alumina (Al_2O_3)	1.8	0.25	2.50
Lime (CaO)	5.0	9.00	16.00
Magnesia (MgO)	1.2	1.57	13.80
Particle properties			
Fluff density (g/cm^3)			0.62
Packed density (g/cm^3)			1.37
Particle size distribution			
(% retained on 325–43 micron sieve, by weight)			3.3
20–30 micron (%)			Scattered
10–20 micron (%)			Scattered
2–10 micron (%)			1–2
1–2 micron (%)			20–30
Less than 1 micron (%)			70

furnaces and the basic oxygen furnaces in this country is processed through a sintering plant in order to agglomerate the flue dust to put it into a product acceptable for recycling into the blast furnace. If this dust were not recycled to the blast furnace, it would mean that over 2, 000,000 tons more iron ore would have been required to produce the steel that was made in 1970 in the United States. This is roughly equivalent to two additional shiploads of ore every week of the year.

The electric-furnace method of making steel has come into increasing prominence in recent years and it appears that the percentage of steel made by electric furnaces will continue to increase in the coming years. The electric arc furnace uses only scrap in making steel, and it therefore depends 100% on recyclable ferrous scraps and wastes for its system. This process also generates large quantities of dust. Early electric furnaces simply vented this dust to the atmosphere and it settled over the landscape and in some respects enriched the soil because of its iron content. As electric furnaces became more sophisticated, dust collecting systems were installed on them. At the present time, all new electric furnaces are being built with elaborate collecting systems that will meet the high standards for smoke abatement in the United States. In addition, existing electric furnaces that do not meet current standards are being required to spend large sums of money in installing equipment that will meet very high standards. While the amount of dust collected in an electric furnace process will vary with the types of steel being made and the types of scrap being charged, an efficient dust collection system will probably collect around 50 lb of dust per ton of steel produced. In 1970 20,162,000 tons of steel were produced in electric furnaces in the United States. From this production, a total of 501,620 tons of dust was collected. This dust has a relatively low density and is extremely fine. Physical properties and chemical analysis of typical electric furnace dust are given in Table 1.

The particles in the electric furnace flue dust are not much larger than that found in cigarette smoke. The material flows like water; it is magnetic and difficult to handle. The chemical composition of this material varies greatly from one furnace to another and from one location to another, depending upon the types of material being charged to the furnaces (see Table 1).

Considerable efforts have been and are being made to find a use for this material. It cannot be recharged to any type of steel making in the form in which it is collected. Even if it were agglomerated by a sintering process, it would be an unsatisfactory material for recharging to the electric furnaces: it has an unacceptably high level of undesirable residual elements and the electric furnace is a poor way to reduce ferrous oxide. It is also an unsatisfactory product for the blast furnace without considerable and costly beneficiation. Extensive efforts are under way to develop economic ways to beneficiate this material so that is is satisfactory for recycling to the blast furnace. To do this, the material must be agglomerated, and various cements have been tried to make the dust into a pellet that would have the strength necessary for charging to a blast furnace. While methods to date have not been totally successful, it is anticipated that means will be found to recycle this by-product of the electric furnace steelmaking process. It has been tried as a filler material for plastic, asphalt, rubber, and concrete block. Experiments have been conducted for its use in molding machinery parts. It has been tried as fertilizer. Continuing efforts are in process of finding ways to dissolve the dust in acid and precipitate recoverable mineral elements from it.

In the meantime, the collection and disposal of the dust is costly for the steel maker. The consistency of the dust is such that it cannot be hauled in an open truck on the highways, since it will be blown off the truck in the process of hauling it a short way. A number of mills pelletize the material into a small but not very strong pellet, simply in order to hold it in a truck as it is hauled to a landfill area.

CHAPTER V.3

dust collection from an electric arc furnace

N. D. PHILLIPS

One of the many industrial applications of cloth filter air pollution control systems is collecting the off-gases from electric arc furnaces. Electric arc furnaces are used to melt iron and steel for castings. Capture of the dust generated from these furnaces during their various processing steps cleans the atmosphere in the plant for the working personnel and prevents emission of the dust to the outside atmosphere to pollute the surrounding area.

The problem of providing adequate equipment to solve these problems includes many engineering facets.

1. In order to capture the fumes being emitted from the furnace at the doors, inspection ports, pour spouts and electrode openings, proper hooding must be supplied. These hoods can consist of any of four types[1] as shown in Figure 1: (a) canopy hood above the crane rail to capture all of the dust that is emitted from the furnace during charge, melt, and tapping; (b) side-draft hoods, which provide horizontal openings at the electrode ports with small hoods over any door opening and spout; (c) the "pancake" hood, which is a refinement of a side-draft hood providing more complete capture of the gases being emitted at the electrode ports; (d) roof tap for evacuating the fumes directly from the furnace and maintaining a nearly balanced or slight negative pressure within the furnace.

The choice of any one of these systems depends on the type of electric furnace used, the interfering equipment and building structure, and the particular melting cycle and metal product being produced.

2. The ducts conducting the hot gases away from the furnace must be designed to withstand the particular pressure and temperature of the gases. In many cases the gases are hot enough to require either a heat-resistant alloy construction of the pipes or water cooling of certain zones to protect the metal. Little fume and gas cooling occurs so that later treatment is required. The gas ducts must be very accurately

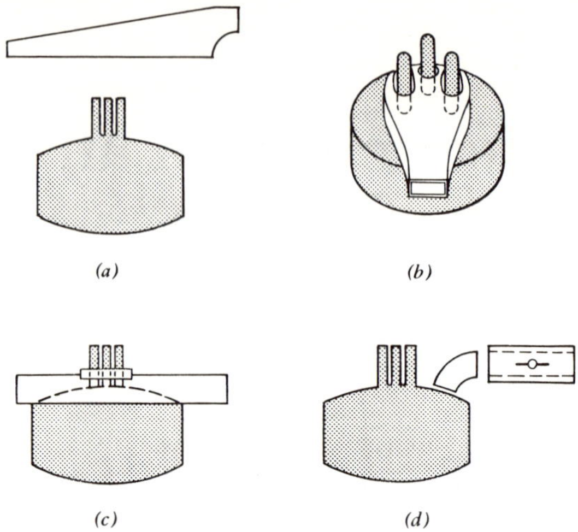

Figure 1. (*a*) Canopy hood; (*b*) side draft hood; (*c*) full roof hood; (*d*) roof (or shell) tap.

engineered to fit within the building structure to take the fumes outside the building and transfer them to the dust collector. Additional gas cooling by dilution or radiation/convection usually is required as shown in Figure 2.

3. The dust collector details depend on the temperature of the gases, their chemistry, and the particle size of the dust being emitted.

The following description is a case history of a particular dust collection installation for electric arc furnaces. The National Castings Division of Midland-Ross Corporation (1971) had a Fuller-Dracco dust collection system[2] installed in their Sharon plant. This dust collection system serves three Lectro-melt electric arc furnaces. These nose tilt furnaces presently are melting 60,000–70,000 tons of steel per year. Each furnace produces about 7 ton/hr of plain carbon, low alloy steel. The furnaces are top charged, utilizing a 15,000-lb-capacity charging bucket. The melt time for the furnaces is approximately 3 hr.

About 85% of the castings produced at the Sharon works are side frames, bolsters, and couplings for railroad cars. In producing castings, the company has found it economical to use about 25% steel turnings in the furnace charge. However, the high oil content of the turnings created a smoke and fume problem. Concerned with improving conditions for its employees and with contributing to cleaner air in the community, National Castings called upon Fuller Company to engineer and install a dust collecting system at the Sharon works.

The dust collecting system, consists of dust collectors and all necessary equipment and controls. The system is unusual in that each furnace—the two new 11-ft units and a renovated 9-ft furnace—has its own collector. The particulate material is collected from the furnaces by adjustable-volume side-draft hoods with auxiliary hoods over the slagging door tapping spout. The gases are drawn into 42-in.-diameter conducting pipes. The ducts on each furnace are interconnected by dampers, so that the air volume can be withdrawn from any or all furnaces. This allows one collector to be shut down, if necessary, without affecting the emission control volume of the rest of the system. The collectors are programmed to be shut down for cleaning on off-fume intervals—or about every 3 hr—again, so overall efficiency is not affected.

Each collector has 384 synthetic fabric filtering bags, giving a total of

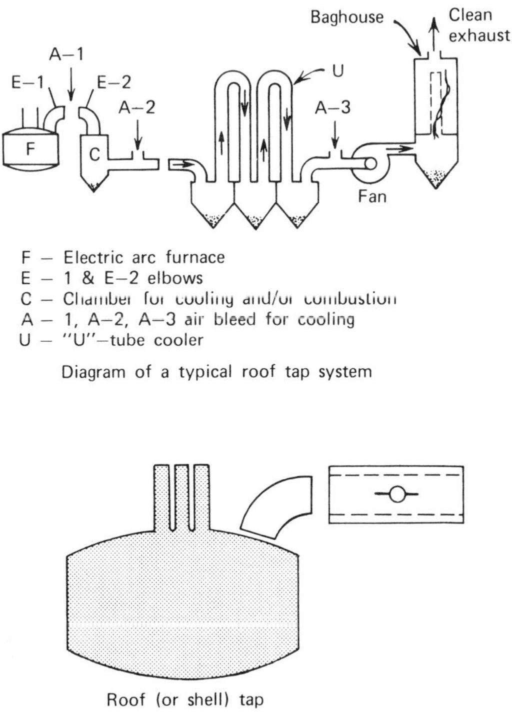

Figure 2. Roof top system.

17,280 ft^2 of cloth per collector; the air-to-cloth ratio is 2.5 to 1, and each collector has a capacity of 42,500 ft^3/min. Design operating temperature limit of the bags is 260°F at steady state. Normally, however, the gas coming into the bag area has a temperature of less than 200°F. If the temperature should reach 250°F or more, tempering dampers automatically open and keep the temperature at acceptable levels. In the baghouse, the bags hang from heavy shaker bars and are fastened to the inlet bag collars with a quick-release clamp at the bottom of the compartment. This arrangement permits quick adjustment or removal of the bags.

A timer controls the cleaning operation of the unit. Periodically the timer closes a damper to one of the compartments and initiates the shaking cycle. As the dust is dislodged, it falls into a trough hopper that runs the full length of each collector. From the hopper, the dust drops to a screw conveyor that conveys the dust to a dust-tight container for disposal.

Exhaust fans on the clean air side of the filter provide pressure differential for the process and expel the clean air to the atmosphere. These fans are driven by 150-hp 1200-rpm motors.

For every ton of metal melted at the National Castings Foundry, the units collect 14–20 lb of dust. Considering that National Casting melts from 60,000 to 70,000 tons of steel per year at its Sharon works, this means that a million pounds or more of dust would have escaped into the atmosphere since the units were installed. No visible effluent comes from this plant.

COSTS

Equipment

The original equipment only cost for units of this type is approximately \$0.90 per actual ft^3/min of gas volume. This value is lower than generally quoted values because this particular operation permits the use of intermittent type collectors which can be cleaned between furnace heats. Applications requiring continuous on-line duty and on-line inspection for maintenance require different controls and design features at increased costs.

Installation

Erection cost for the necessary hooding and ducting for a unit as described requiring only dilution air cooling is approximately \$0.75 per actual ft^3/min. It must be understood that this item can vary widely depending on the plant layout and specific application. Therefore, installation and equipment variations must be carefully considered when comparing different installations.

Maintenance

Maintenance costs can vary widely depending upon operating conditions, that is, variation in temperature, gas flow, moisture content of gases, and frequency of start-up and shut-down cycles, and upon frequency of inspections to correct and replace failed parts. The last factor is especially critical when holes in filter bags can cause failure of adjacent bags.

The largest factor in the maintenance cost of fabric filter units is replacement of the bags. Maintenance costs including cost of bags and labor to replace them for electric arc furnace installations are about \$0.028 per/ton of steel produced. Total operating and maintenance costs including power and dust disposal have been reported[3] at \$0.46 per/ton steel produced.

REFERENCES

1. R. J. Wright, "Concepts of Electric Arc Furnace Fume Control," *JAPCA,* **18,** No. 3, 175–178 March 1968.
2. *The Fuller Engineer,* **19,** No. 2.
3. "J. L. Venturini, Operating Experience with a Large Baghouse in the Electric Arc Furnace Steelmaking Shop at Bethlehem Steel Corporation's Los Angeles Plant," APCA 6/18/70 Paper 70–28.

CHAPTER V.4

dust collection in cement plants

N. D. PHILLIPS

Cement plants have a major dust collection and waste disposal problem. The sources of dust from these plants are kiln exhaust gases, clinker cooler vent gases, grinding mill vents, and material handling ventilation.

In most cases only the first source presents a waste disposal problem. The other dust streams usually consist of a product which can be returned to the total system. Kiln exhaust gases will be discussed in detail later in this section; at this time a brief review of the other three source problems can be summarized.

SOURCES OF GAS

Clinker Cooler Vent Gases

Many systems for cooling the clinker material discharged from the kiln utilize large volumes of air. Some of the heated air after cooling the clinker is induced into the kiln to be used for secondary combustion air in the kiln system while the excess hot air and entrained dust is vented to the atmosphere. At present many vent air streams are cleaned of clinker dust by means of cyclone collectors. This material can be added to the clinker discharged from the cooler. With more stringent air pollution codes being enforced, it is becoming necessary to apply fabric filters on the discharge of the clinker coolers to remove all the dust from the hot vent air. This vent air volume is approximately 11 scf/min per daily barrel capacity of the kiln.

Grinding Mill Vents

The grinding mill circuits used to produce the finish grind cement of the desired fineness are generally connected to the silos and mill vents for collection in fabric collectors. All of the collected material is usable product.

Material Handling

Since most of the material handling in cement plants is accomplished by pneumatic conveying, the silo and receiver vents are usually all protected by fabric collectors. Again all the material collected in this manner is usable as a plant product or as a return feed stream.

Kilns

The gas stream in the cement plant operation that produces the most difficult dust collection and waste disposal problem is the exhaust gas from the calcining kiln. The hot kiln gases contain the mechanically entrained dust particles as well as some volatilized components from the calcining process. Depending on the chemistry of the stone being calcined, some plants can return the kiln dust with the raw feed for more complete calcination. Many plants cannot do this because the alkali content in the raw stone is too high to be acceptable and a recycle of the collected dust would increase the alkali content of the finished product. In plants were the alkali content is a problem, the material collected in the dust collector is usually treated as waste to be dumped into old quarries.

The kiln discharge gases are usually treated for temperature control and dust removal as follows:

1. Heat removal can be achieved by a material preheater to heat the raw feed and cool the gases. A material preheater works on the principal of direct contact between the raw feed and the exhaust kiln gases to recover some of the heat from the gas by heating the raw feed before it enters the kiln. This improves the heat balance of the calcining operation.
2. A very common method of cooling the kiln gases from the kiln exit temperatures of approximately 1000°F is by using evaporative spray coolers to reduce the gas temperature by evaporation of water into the gas stream. Special atomizing nozzles are available that can utilize high-solids-contents pond water as an inexpensive source of water for the purpose. Final temperature control in almost every case is obtained by air dilution to the desired end temperature, usually about 550°F.
3. Dust removal is usually accomplished in two stages. Cyclone collectors are utilized to remove the coarse material, which can be recycled to the kiln effectively. The final stage consists of many fiberglass bags assembled in enclosures called baghouses to filter the fine particulate from the gas stream before the clean gases are emitted to the atmosphere.

The baghouses for this application can be either pressure or suction flow systems. Generally the bags are nominal 12 in. diameter and 30 ft long. The air-to-cloth ratio, or face velocity, is about 2 $ft^3/min/ft^2$. The dust is removed from the cloth by reverse air cleaning, often assisted by the use of sonic generators. The dust is carried away from the hoppers by screw conveyors or air-assisted gravity conveyors.

Design of the baghouse enclosure must include consideration for the type of feed to the kiln. Kilns have either wet or dry feed systems, depending on the method of grinding utilized by the plant. The selection of the grinding method sometimes depends on the type of stone used as raw feed to the process, or upon the availability of raw water for the process.

A dry feed usually results in higher discharge gas temperatures requiring controlled cooling of the gases. The wet feed system introduces large amounts of moisture, which produces a high dew point temperature for the gases. Proper insulation design must be utilized to prevent condensation of the water vapor, which can cause corrosion within the enclosure. Condensation can also produce a wet cake of the collected dust, causing a handling problem as well as providing the potential for blinding the filter fabric surface to cause high pressure drop and inefficient venting of the kiln exhaust.

One such wet kiln installation[1] at the Universal Atlas Cement Company, Northampton, Pennsylvania, treats 15,000 acfm at 490°F for each of two kilns. The coarse dust is removed from the gas by 6 Mark V Fuller-Dracco cyclone units. The coarse dust collected by these cyclones is returned to the kiln system. A 12-compartment glass baghouse of the Fuller-Dracco design consists of 12 compartments each containing 78 filter bags, 12 in. diameter by 30 ft long. This system is totally automated and designed to operate at temperatures up to 550°F. The baghouse is a pressure unit located after an induced draft fan that pulls the kiln gas through the cyclone collectors and blows it into the baghouse with the clean air being discharged from the unit through gravity ventilators. The fans are driven by 800-hp motors. The mechanical collectors are designed to achieve approximately 75% efficiency to reduce the dust loading of the kiln gases from approximately 5 g/scf to about 1 g/scf entering the baghouse.

This particular baghouse is fully insulated with a pressure type of structural panel design. The heavy duty prefabricated "sandwich" insulated structural panels used are designed to withstand temperature and load requirements within the baghouse. Each panel is 45 ft high, weighs 9 tons and consists of a 2-in. layer of Spintex mineral wool insulation between two sheets of No. 10 gage USS Corten steel. A critical factor in controlling emissions from a wet process kiln is the need to prevent gases within the baghouse from reaching the dew point. This was achieved by applying an outer layer of Spintex insulation with a water proof coating on all of the cyclones, duct work, and hoppers. Consequently, the water vapor in the gas stream exits through the vents without any condensation in the system, thereby eliminating the problems of clogging and plugging.

This unit is designed for an air-to-cloth ratio of 1.72 $ft^3/min/ft^2$ of cloth when all compartments are in service and 1.91 $ft^3/min/ft^2$ of cloth when one compartment is out for cleaning. The fiberglass cloth used in this unit is a crow foot weave with a thread count of 54/52, a weight of approximately 8.5 oz/yd^2 and permeability of 16–20 $ft^3/min/ft^2$ at 0.5 in. w.g. The cloth is treated with silicone, graphite, and teflon for maximum flex life.

Cleaning of the filter bags is accomplished by reverse air flow. At hourly intervals, each compartment is cleaned automatically for approximately 2 min under the direction of a program timer. The time between cleaning of the compartment can be varied depending upon the static pressure drop across the collector, which under normal operating conditions is 5–7 in. w.g. A small cyclone collector ahead of the reverse air fan collects the loosened particles from the reverse flow air stream before it passes through the fan and completes the circuit back to the baghouse collector through the supply duct. To release collected dust into the 12-in. screw conveyor each of the twelve hoppers is equipped with a Fuller 10-in. cast iron rotary lock. Two identical locks also serve the reverse air cyclone and a precleaner cyclone. The dust is fed into screw conveyors and delivered to a bucket elevator that discharges by gravity into a 50-ton storage tank adjacent to the collectors. A truck hauls the dust from this point to the plant's dumping area.

The control room under the baghouse contains baghouse and kiln controls, including U-tube manometers to indicate inlet pressures and pressure drop across each compartment, fan static pressure readings, and an alarm system in the event of dust collector malfunction. In addition, the control room contains compressors that supply dried instrument air. To provide ultrafiltration of ambient air for ventilating the control room, a Dracco Atmos-filter collector is in continuous operation.

COSTS

The initial cost of the equipment for a system such as described is approximately \$2.30 per ft^3/min for dust collection equipment, \$0.60

per ft^3/min for auxiliary equipment including fans, motors, etc., \$0.20 per ft^3/min for controls, and \$1.30 per ft^3/min for erection costs. Since this particular unit was made of structural panel sides with integral insulation, the erection costs were substantially less than a typical on-site fabricated baghouse construction.

Maintenance Costs

Since the primary maintenance costs for fabric filters consists of replacement of the filter media, the anticipated maintenance costs for a unit such as this is approximately \$7500 per year, which equals an average cost of \$0.05 per year per ft^3/min capacity. This maintenance cost is anticipating a 2-yr life of the glass bags, which is typical for most installations.

REFERENCE

1. "Glass Bag Collector at Universal Atlas' Northhampton Plant," *Pit and Quarry,* **63,** No. 8, 113–115, February 1971.

CHAPTER V.5

particulates from electric furnaces, ferroalloys

R. A. Person[1] of Union Carbide, Metallurgical Division, summarized the problems and available data in control of emissions from ferroalloy processing, including calcium carbide units. The furnaces are large, 25–45 MW or 25,000–45,000 kW.

Table 1 gives fume characteristics and properties. Table 2 gives gas generation and flows. Tables 3 and 4 give allowable discharge from a 20 MW open 75% ferrosilicon furnace, based on the New York State regulations for allowable emissions for environmental rating C. Table 5 gives data on application of wet scrubbers on collection of different types and sizes of furnace, and Table 6 gives data on furnace bag collectors. Table 7 gives estimated costs in dollas per actual ft^3/min on open furnaces, including furnace hood and interior duct work, but not including liquid waste disposal. Table 8 gives operating cost.

Analysis shows no preeminent advantage for any of the three methods. It should be pointed out that maintenance and operating costs for the precipitator are more speculative. It should also be emphasized that individual plant conditions and power costs may tend to favor a particular system.

The Industrial Gas Cleaning Institute found a wide variation in the cost to repair and replace precipitators. After surveying users of the precipitators, the institute concludes that the big cause of variations is the length of time the equipment has been in service. It found the least expense in coal-fired electric utility boiler applications, where an

Table 1 *Typical Furnace Fume Characterizations*

Furnace Product	50% FeSi	SMZ[a]	SiMn[c]	SiMn[c]	FeMn	HC FeCr[b]	Chrome Ore Lime Melt	Mn Ore[c] Lime Melt
Furnace type	Open	Open	Covered	Covered	Open	Covered	Open	Open
Fume shape	Spherical, sometimes in chains	Spherical, sometimes in chains	Spherical	Spherical	Spherical	Spherical	Spherical and irregular	Spherical and irregular
Fume size characteristics (microns)								
Maximum	0.75	0.8	0.75	0.75	0.75	1.0	0.50	2.0
Most particles	0.05–0.3	0.05–0.3	0.2–0.4	0.2–0.4	0.05–0.4	0.1–0.4	0.05–0.2	0.2–0.5
X-ray diffraction								
Primary	All fumes were primarily amorphous							
Trace constituents	FeSi $FeSi_2$	Fe_3O_4 Fe_2O_3 Quartz SiC	Mn_3O_1 MnO Quartz	Quartz SiMn Spinel	Mn_3O_1 MnO Quartz	Spinel Quartz	Spinel	CaO
Chemical analysis (%)								
SiO_2	63–88	61.12	15.68	24.60	25.48	20.96	10.86	3.28
FeO		14.08	6.75	4.60	5.96	10.92	7.48	1.22
MgO		1.08	1.12	3.78	1.03	15.41	7.43	0.96
CaO		1.01	—	1.58	2.24	—	15.06	34.24
MnO		6.12	31.35	31.92	33.60	2.84	—	12.34
Al_2O_3		2.10	5.55	4.48	8.38	7.12	4.88	1.36
LOI[d]		—	23.25	12.04	—	—	13.86	11.92
TCr as Cr_2O_3		—	—	—	—	29.27	14.69	—
SiC		1.82	—	—	—	—	—	—
ZrO_2		1.26	—	—	—	—	—	—
PbO		—	0.47	—	—	—	—	0.98
Na_2O		—	—	2.12	—	—	1.70	2.05
BaO		—	—	—	—	—	—	1.13
K_2O		—	—	—	—	—	—	13.08

[a] Si, 60–65%; Mn, 5–7%; Zr, 5–7%. [b] HC, high carbon. [c] Manganese fume analyses in particular are subject to wide variations, depending on the ores used. [d] LOI, loss on ignition.

Table 2 *Furnace Gas Generation (Without Combustion, Sat. at 100°F, scf at 30 in. Hg, 60°F)*

Silicon Metal	140–150 scfm/MW	
50% ferrosilicon	130–140	
Standard ferromanganese	160–170	
Silico manganese	120–130	
Ferrochrome-silicon	110–120	
High carbon ferrochrome	80–90	
Calcium carbide	70–80	
Gas flows, acfm at temp.	Closed fce	Open fce (low hood)
Ferrochrome silicon	8700, 1100°F	230,000 at 430°F
scfm	2900	135,000
50% ferrocilicon	20,000 at 1100°F	410,000 at 760°F
scfm	6600	175,000

Table 3 *Allowable Discharge from 20-mW Open 75% Ferrosilicon Furnace*

	A (lb/hr)	B (lb/hr)
Mix	21,300	21,300
Electrodes	300	300
Air stoichiometrically required to oxide CO	24,400	
Air stoichiometrically required to oxide SiO	1,800	
Free moisture	−1,500	−1,500
Total	46,300	20,000
Allowable discharge (from formula)	31	18

Table 4 *Collector Characteristics to Satisfy Conditions of Table 3*

	A	B
Avg. inlet loading	1.9 grains/scf	1.9 grains/scf
Avg. outlet loading	0.031 grains/sef	0.018 grains/sef
Min. collection efficiency required	98.4%	99.1%

Table 5 *Examples of Furnace Wet Scrubbers*

Collector	Open Furnace—Low Hood			Semiclosed Furnace
	I	II	III	IV
Original completion date	1967	1968	1968	1965
Furnace rating for collector design (Mw)	25	30	30	45
Furnace product	FeCrSi	SiMn	H.C. FeCr	50% FeSi
Measured collection efficiency, avg. (%)	92.6	98.6	98.2	98.4 (particulates) 79% (organics)
Inlet loading (grains/scf)	1.43	1.31	1.07	4.93
Outlet loading (grains/scf)	0.106	0.017	0.019	0.08
Design volume at furnace hood (acfm)	230,000	255,000	210,000	—
Design temp. (°F)	430	620	590	—
Actual duct or offtake temp. (°F)	500–570	490–550	480	1100–1200
Design volume handled by fans or blowers	194,000 acfm	196,000 acfm	196,000 acfm	6500 scfm
Operating pressure drop across collector (in. water)	55	57	57	72 to 80
Installed hp—fans	2600	2800	2800	600
—auxiliaries	50	50	50	—
Tons/day collected dust	11–13	14–17	11–17	5–8
Tons/day uncollected emission	0.75–1	0.20	0.26	None
Water circulation (g/min)	1800	1800	1800	75–100
Water usage (g/min)	310	350	350	75–100
Problems	(1) System falls 25% short of design flow. (2) High fan outage. (3) Necessity to add lime to neutralize scrubbing liquid.	(1) High fan outage.	(1) High fan outage.	(1) Continuous kerosene injection necessary for blowers.

[a] acfm, actual cubic feet per minute.

Original completion date	1963	1967	1969	1969
Revisions completed	1965	—	—	—
Furnace rating for collector design (Mw)	12	18	20	10.5
Furnace product	Si and FeSi	Lime-manganese ore melt	FeMn-Si	SiMn
Area of radiant tube cooler (ft^2)	None	57,100	26,000	None
Design volume at furnace (acfm)[a]	180,000	630,000 (peak)	254,000 (peak)	100,000
Design temp. at furnace (°F)	450	1700 (peak)	860 (peak)	400
Design volume at collector (acfm)	180,000	250,000	185,000	125,000
Design temp. at collector (°F)	450	400 (peak)	500 (peak)	325
High temp. bypass set at (°F)	500	500	500	450 (tempering valve)
Average operating temp. (°F)	350–450	250–350	—	—
Operating pressure drop across collector (in. water)	12–15	4–8	—	5–9
Tons/day collected dust	5–10	15	7 (estimate)	4–6 (estimate)
Bag cleaning	Shaking	Shaking	Shaking	Shaking
Cycle	Continuous	Intermittent	Continuous	Continuous
Cleaning cycle	8 min every 30 min	4 min every 100 min	—	6 min every 90 min
Number of isolatable compartments	1	1	8	6
Installed motor power—fans (hp)	900	800	900	500
—auxiliaries (hp)	100	100	100	45
Bags, graphite and silicon-treated				
Number	1024	1152	1152	864
Size	11.5 in. diam × 30.5 ft long			
Bag failures	7 per week	None		
Problems	(1) Electrostatic charge. (2) Continuing nature of bag failures. (3) Lack of compartment isolation.	(1) Bypass maintenance. (2) Flow of dust from storage bin.		

[a] acfm, actual cubic feet per minute.

Table 7 *Installation Costs per Actual Cubic Feet per Minute*

Venturi scrubber	$2.40–3.60
Bag filter	2.90–3.60
Electrostatic precipitator (with conditioning)	3.80–4.2

Table 8 *Dollars per Actual Cubic Feet per Minute*

Venturi scrubber	$0.40–0.60
Precipitator	0.20–0.30
Bag filter	0.30–0.40

average of $6 per 1000 ft^2/min was spent annually on labor and $5 per 1000 ft^3/min for labor annually, and $70 per 100 ft^3/min for parts.

REFERENCE

1. R. A. Person, "Particulates from Ferro Alloy Electric Furnaces," *J. Metals,* 17–29, April 1971, Prize Winning Paper of the Metallurgical Society of the AIME.

CHAPTER V.6

particulate collection in a lead smelting plant

General Battery Corp. (GBC) was faced with a court order to abate polluting emissions resulting from their manufacturing process. As a result the company opened its new secondary lead smelting plant in Reading, Pennsylvania, at an investment of about \$3.5 million, of which more than \$1.5 million was for particulate controls.

GBC reclaims batteries, recovering the lead from old batteries, resmelting it, and using it to make new ones. The company had two smelting plants: an old one at the Reading site, and another in Hamburg, Pennsylvania. The Reading smelter started operations in 1933. Each of the plants had a capacity of 6500 batteries a day.

GBC batteries are produced for major oil companies and auto accessory stores. The new Reading plant has 28,000 ft^2 of floor space and can accommodate the smelting operations of the two earlier plants.

About 80% of the lead used in batteries is recycled. In the recycle operation the tops of old batteries are sheared off, the contents spilled, and the lead resmelted. The lead is then melted in one of two blast furnaces or in a reverberatory furnace; the three furnaces have a daily capacity of 100 tons of alloyed lead. Antimony content is critical.

The battery cases are crushed and used as landfill. They have not been successfully employed as fuel in the blast furnace. All equipment in the shearing operation must be acid-proof. All water and acid from this shearing operation are collected, retained, and subsequently neutralized to eliminate acid before discharge to a water course.

In 1963, all battery cases were made of rubber. Today, many cases are all plastic; some are made of polystyrene, others of high-shock polypropylene. The lead from the old batteries is contaminated with solid particles from the casing, which adheres to the lead. So, when the lead is resmelted, odors resulting from the plastic are produced.

The air pollution control systems satisfy not only the complaints but also the requirements of the Pennsylvania Department of Public

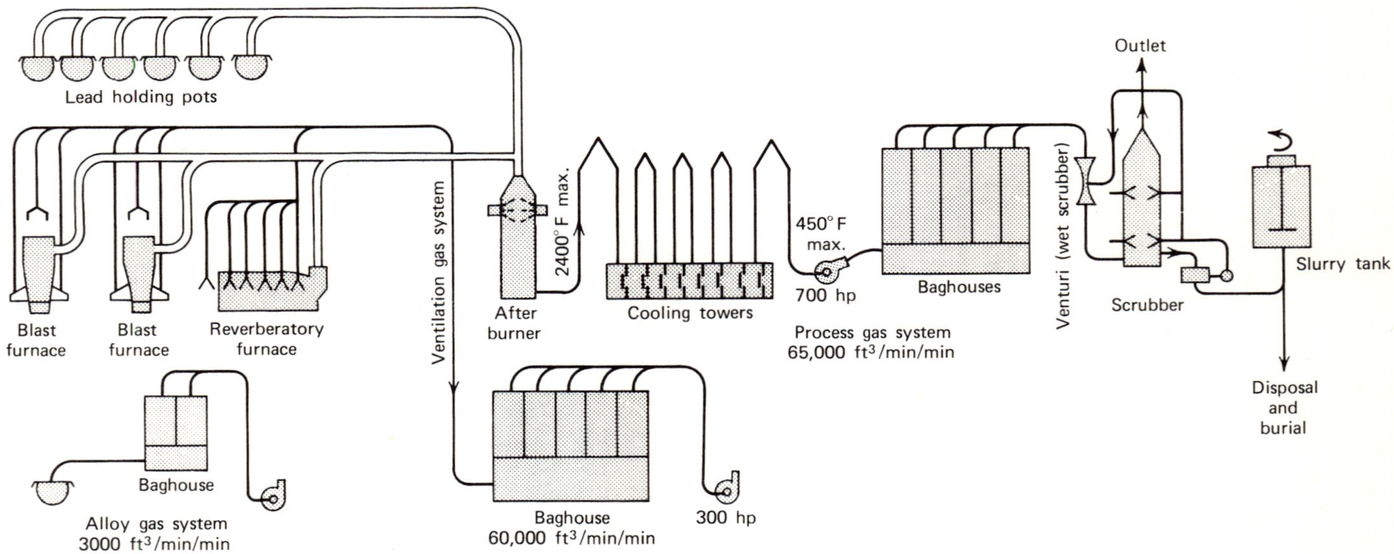

Figure 1. Control of lead particulates.

Health. Odors are one type of complaint, but other emissions include SO_x, lead oxide, and lead aerosols from the smelting operation.

Odors are disposed of by incineration at 2400°F; lead oxide emissions and lead aerosols are trapped by baghouses. Before the hot gases are passed through the baghouses, they are cooled in 14 cooling towers and the temperature of the gas is reduced from 2400°F to less than 450°F. Lead oxide emissions from the ventilation and alloy systems also are removed by baghouses. SO_2 emissions are scrubbed out with a Venturi scrubber and mist eliminator. See Figure 1.

CHAPTER V.7

fly ash

PHILIP G. MEIKLE

By-product ash, the residue from the combustion of coal at steam-electric generating plants, amounts to over 30 million tons annually and is the ninth most abundant solid mineral in the United States (see Table 1). Efforts to use it are of growing importance.

Since 1967 a concerted effort on the part of the Edison Electric Institute (EEI), in cooperation with the Economic Commission for Europe, has resulted in the gathering of accurate annual worldwide data on ash produced and utilized. Figures developed by EEI for the United States with projections for coal consumption and ash production through 1980 are shown in Figure 1. It is estimated that the 1980 consumption of coal will approach 480 million tons, with ash production of 52 million tons, consisting of 36 million tons of fly ash and 16 million tons of bottom ash and boiler slag.[1]

The physical and chemical characteristics of fly ash emitted from any particular boiler furnace depend on the type of combustion equipment and its design, the combustion or firing rate, the chemical content, burning characteristics and particle size of the coal, variations in steam load, draft and combustion controls, and operator skill.

Most modern coal-fired generating stations burn pulverized coal, with boiler capacities in excess of 200,000 lb of steam per hour. Coal is pulverized in grinding mills so that about 80% passes through 180 mesh and is burned with additional air. Because the temperatures may range to over 3000°F, most of the mineral matter present in the coal is fused and is altered physically and chemically. A portion of the ash (20–50%, dependent on the type of equipment) drops to the bottom of the furnace as bottom ash or slag. Most of the ash, however, is carried through the boiler in the flue gases and is extracted from these gases by electrostatic and/or mechanical collectors.

Pulverized coal-fired units are "wet" and "dry" bottom, and the amount of fly ash and bottom residue in each is related to furnace design and the pulverization level, volatile matter, and ash content of

Table 1 *Estimated Annual Production of Minerals and Solid Mineral Fuels (Short Tons)*[a]

Sand/gravel	917,739,000
Stone	819,403,000
Coal (all types)	557,000,000
Iron Ore	91,766,000
Portland cement	73,042,700
Clays	57,233,000
Salt	41,274,000
Phosphate rock	41,251,000
Coal Ash	31,661,000
Slag (all types)	28,742,000
Lime	18,637,000
Gypsum	10,018,000

[a] Source: Minerals Yearbook, Volume I–II, *Metals, Minerals, and Fuels,* USDI. Estimated accuracy, ±10%.

the coal. In the "wet" bottom, up to half of the total ash can be trapped within the furnace as ash or slag; however, bottom ash in "dry" bottom units amounts to only 20–25% of the total ash. Dust loadings in the stack gases can run as high as 12 lb per 1000 lb of gas and the fly ash usually is 40–50% less than 10 μ in particle size and occasionally as fine as 65%, 10 μ.

Another type of modern coal firing—though less prevalent in the United States—is the cyclone method. Minus 4 mesh or ⅛-in. coal is introduced tangentially under high pressure into a cylindrical combustion chamber. From 80% to 85% of the ash of coals suitable for this type of firing never reaches the combustion gas because burning is so rapid that ash melting temperatures are quickly exceeded; the ash is tapped from the furnace bottom as a slag. The small amount of ash that reports to the gas stream, that is, fly ash, is very fine, up to 90% below 10 μ.

Typical fly ash particle size distributions from pulverized, cyclone, and stoker type furnaces are shown in Figure 2.

Fly ash particulates control and the devices for collection are covered in the particulate section (pp. 669).

When markets have been established for fly ash, it is handled dry by a pneumatic system, stored in silos or bins on the plant site, and then transported in closed tank trucks or railcars (Figure 3). Slag to be sold is pumped from quench tanks below the boiler to draining bins and then to market or open storage. Bottom ash and slag are granular materials and lend themselves to mechanical handling schemes.

Ash that is not marketed is disposed of by dumping in impounding ponds or lagoons or in landfill areas. Fly ash and bottom ash are frequently combined at the plant and transported to the disposal site by truck or pipeline.

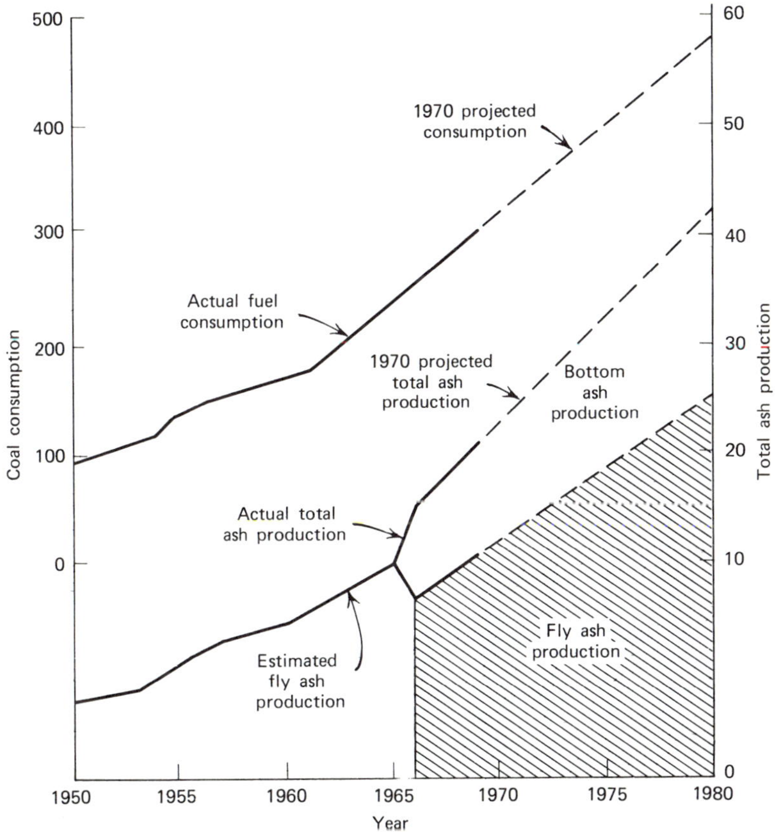

Figure 1. Coal consumption and total ash production by U.S. electric utilities (in millions of tons). (Source: Ref. 1.)

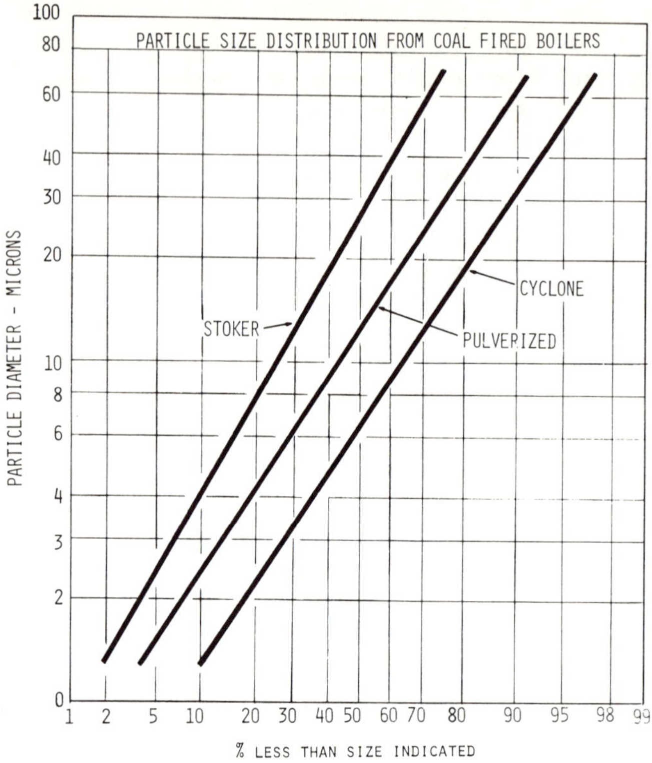

Figure 2. Typical fly ash particle size distribution. (Source: Air Preheater Co., Inc.)

For transport distances of 1–2 miles, pumping to ponds is the most popular disposal method. One typical hydraulic system handles about 200,000 tons of fly ash per year through a 10-in.-diameter, 1½-mile-long pipeline at a rate of 1750 gal/min against a total head of 370 ft. The material is discharged into an impounding basin formed by constructing an earthen dam across a valley.

Trucking of ash is cheaper than hydraulic handling even within short distances and is the common disposal method in urban areas. Disposal at sea has been practiced to a limited extent along the East Coast.

Fly ash hauled to landfill disposal sites must be conditioned with up to 20% water to prevent dusting and is transported in covered trucks. Dust conditioners of the type shown in Figure 4 are in wide use. Fly ash disposal ponds stabilize in time and a variety of flora will cover the

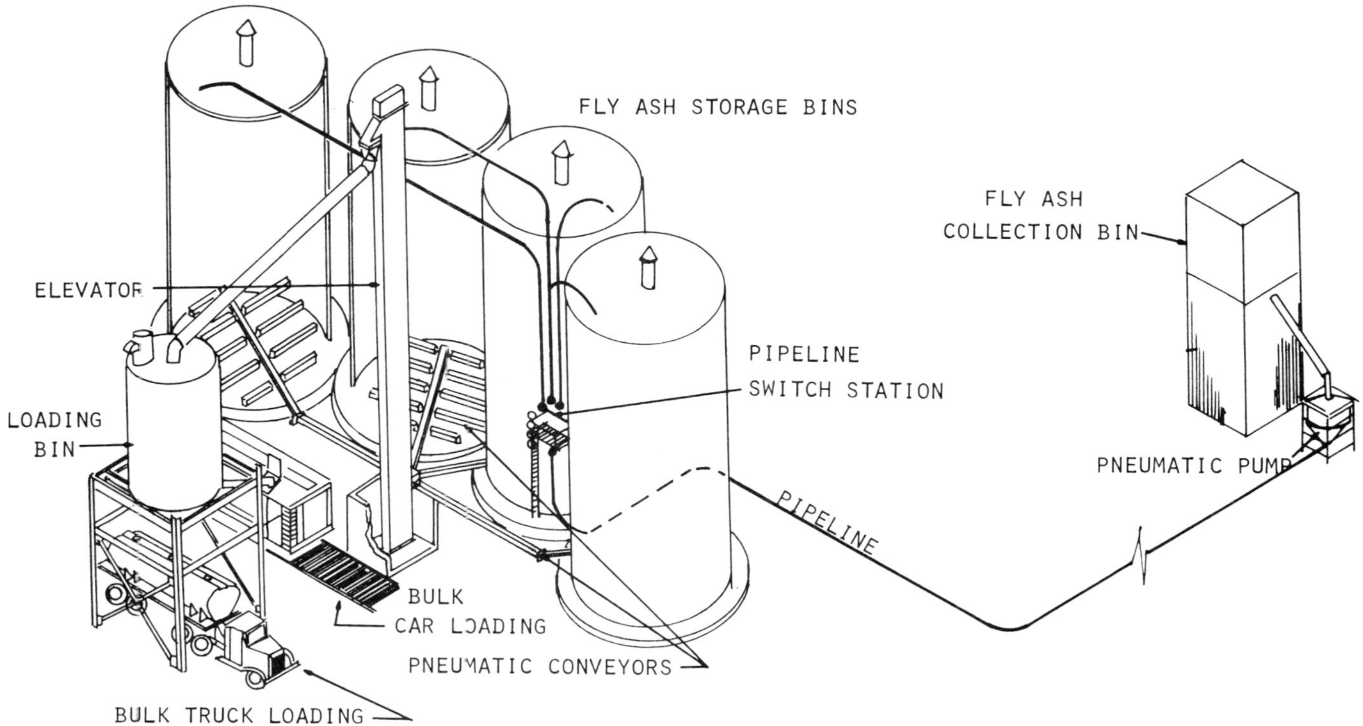

Figure 3. Pneumatic system for storage and handling of fly ash. (Source: Concrete Industries Yearbook 1969, Pit and Quarry Publications, Inc.)

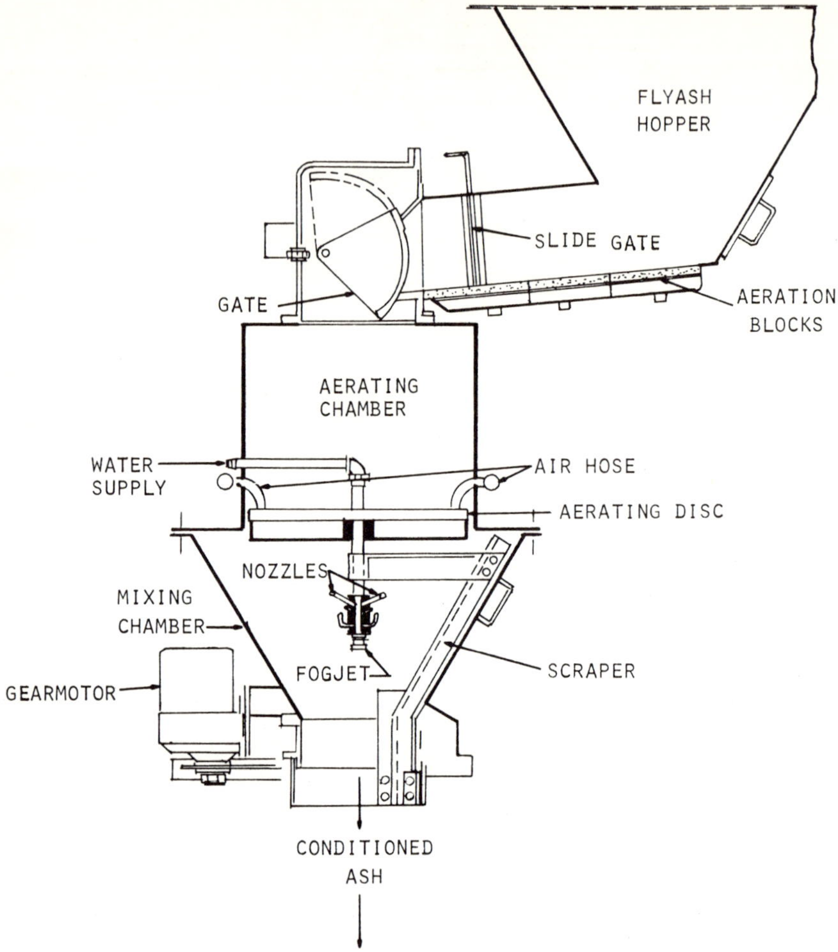

Figure 4. Dust conditioner.

area, preventing dusting. Additions of soil and suitable fertilizers are frequently made to encourage plant growth. To prevent dusting during prior to overgrowth, the surfaces of ponds are periodically flooded.

Disposal of ash in an abandoned deep coal mine adjacent to a power plant has been demonstrated.[6] This method eliminates air pollution problems and renders acid drainage from the mine neutral or alkaline. Continuous monitoring of the mine water discharge is necessary, and a water treatment plant may be required to meet water quality standards.

The cost of transporting ash, dyking and maintaining disposal areas, and preventing stream and air pollution from ponds and landfill areas is $3.00 per ton in metropolitan areas and will probably average $1.00 countrywide by 1975.[7] Local ordinances limiting the dumping are becoming widespread.

Ash residues are utilized in the manufacture of cement, concrete and concrete products, construction fill, roadbase and soil stabilization, lightweight aggregate, and mineral filler for asphaltic pavement. Minor uses include oil-well grouting, foundry core sand, filler in plastics and chemicals, blasting grit, ice control, and mine fire and subsidence control. Table 2 outlines the production and utilization pattern of coal ash of all types in the United States. Table 3 is a survery of end use in 1969 for each of the major utilization categories. Total ash collected in 1969 amounted to 31.6 million tons, consisting of 21.1 million tons fly ash, 7.6 million tons bottom ash, and 2.9 million tons boiler slag. Of that amount about 1.9 million tons of fly ash, 2 million tons of bottom ash, and almost 1 million tons of slag were utilized. The major uses for bottom ash and boiler slag are in construction or road fill and ice control on roads.

Fly ash consists mainly of glass spheres together with crystalline matter and varying amounts of carbon. It is composed principally of silica (SiO_2), alumina (Al_2O_3) and iron, carbon, potassium, calcium, magnesium, sodium, sulfur, and traces of titanium, phosphorus, and germanium. Chemical analyses are given in Table 4.

Silica is present as (1) quartz, (2) with alumina as mullite ($2\ SiO_2 \cdot 3\ Al_2O_3$), and (3) in a glassy phase with calcium, alumina, and iron.

Table 2 *U.S. Production and Utilization of Ash, 1955–1970*

	Millions of Tons		%
	Production	Utilization	Utilization
1955	15	0.3	2.0
1965	25	1.3	5.2
1966	25.2	3.1	12.0
1967	27.5	3.8	13.8
1968	29.6	5.2	17.5
1969	31.7	4.9	15.3
1970	32	4.9	20
1980 (est.)	52	5.0	—

Table 3 Ash Collection and Utilization Survey Year 1969

	Fly Ash tons	Bottom Ash tons	Boiler Slag (If separated from Bottom Ash) tons
1. Total ash collected	21,091,406	7,648,567	2,921,226
2. Ash utilized			
a. Mixed with cement clinker or cement (pozzolan cement)	40,701	—	—
b. Mixed with raw material before forming cement clinker	132,226	5,000	—
c. Partial replacement of cement in:			
1. Concrete products	170,450	5,686	—
2. Structural concrete	176,788	—	—
3. Dams and other mass concrete	161,874	—	—
d. Stabilizer for road bases, parking areas, etc.	148,193	19,417	73,240
e. Lightweight aggregate	263,564	19,288	—
f. Fill material for roads, construction sites, etc.	203,100	1,101,642	548,606
g. Filler in asphalt mix	110,829	10,084	89,970
h. Miscellaneous (see footnotes)	235,732[a]	522,883[b]	252,262[c]
Total, item #2	1,643,457	1,684,000	964,078
3. Ash removed from plant sites at no cost to utility but not covered in item 2 above	284,036	275,900	500
Total utilized, Items #2 and #3	1,927,493	1,959,900	964,578

[a] Oil well cementing, grouting, mine fire and subsidence control, etc.
[b] Ice control, asphalt shingle mfr., blasting grit, etc.
[c] Ice control, blasting grit, aggregate.
Source: Edison Electric Institute.

Alumina is usually combined with silica but may occur in a crystalline form of clay. Iron appears as ferrous oxide (FeO), hematite (Fe_2O_3) and magnetite (Fe_3O_4), and in the glass. Fly ash is, for the most part, glass; the $SiO_2 + Al_2O_3$ content generally exceeds 70%.

Fly ash ranges in size from less than a micron to greater than 100 microns in diameter. Associated carbon, in the form of irregularly shaped coke, exceeds 10 microns in size and in extreme cases, exceeds 300 microns.

Fly ash varies in color from tan through various shades of gray to almost black, but is usually a light gray. Specific gravity ranges from just under 2.0 up to 3.0; compacted dry bulk density varies from 70 to 90 lb/ft.[3]

Fly ash in concrete, concrete blocks, precast concrete, and mortar has been the largest market amounting to about ½ million tons annually. Fly ash in concrete acts as a mechanical filler, supplementing or replacing fine (sand) aggregate. Its spherical shape aids in the workability of plastic concrete and in the ease with which it is finished.

Fly ash is a *pozzolan,*[10] a siliceous or silico-aluminous material that has no cementing value itself but which, in finely divided form and in the presence of moisture, reacts chemically with calcium hydroxide (lime) to form cementitious compounds. The pozzolanic property of fly ash reduces permeability, dry shrinkage and heat of hydration, reducing cracking.

Fly ash, to be suitable as a pozzolan should be fine and low in carbon. A substantial portion of the silica should be in the amorphous (glassy) phase rather than the crystalline form. Tables 5 and 6 list the chemical and physical requirements for fly ash as an admixture in concrete.[11-13]

The fineness requirements for pozzolan limit fly ash in concrete to that collected from electrostatic precipitators. Mechanical collectors usually furnish fly ash having a surface area of 2200–4500 cm^2/cm^3 whereas electrostatic precipitators will collect materials having finenesses greater than 6600 cm^2/cm^3.[14] Note that these values are average only. Mechanical collected ash as fine as 7300 cm^2/cm^3 and electrostatically precipitated material as coarse as 4500 cm^2/cm^3 have been reported.

Table 4 *Range of Chemical Analyses of U.S. Fly Ashes*

	Bituminous	Lignite
	(%)	(%)
SiO_2	30–55	20–41
Al_2O_3	10–35	10–18
Iron oxides	10–40	7–16
CaO	0–15	20–40
MgO	0–5	2–11
TiO_2	1–3	1–3
Na_2O & K_2O	1–5	½–9
SO_3	0–3	½–9
Moisture	½–¾	—
Loss on ignition	½–20	0.1–1.4

Table 5 *Chemical Requirements for Fly Ash as an Admixture in Concrete (PFA)*

	Bureau of Reclamation	Corps of Engineers	ASTM C-618
	(%)	(%)	(%)
Silicon dioxide (SiO_2) + aluminum oxide (Al_2O_3) + ferric oxide (Fe_2O_3), minimum	75.0	70.0	70.0
Magnesium oxide (MgO), maximum	5.0	5.0	—
Sulfur trioxide (SO_3), maximum	4.0	4.0	5.0
Loss on ignition, maximum	5.0	6.0	12.0
Moisture content, maximum	3.0	3.0	3.0
Exchangeable alkalies as Na_2O, maximum	2.0	1.5	1.5

Table 6 *Physical Requirements for Fly Ash as an Admixture in Concrete*

	Bureau of Reclamation	Corps of Engineers	ASTM C-618
Fineness			
Surface area, Blaine air permeability apparatus, cm^2/cm^3 minimum	7000	6500	6500
Material retained on No. 325 mesh sieve, percent, maximum	12	—	—
Compressive strength			
With portland cement, percent of control, 28 days, minimum	80	75	85
With lime, 7 days, min. psi	800	900	800
Increase in drying shrinkage of mortar bar, percent shrinkage of pozzolan bar minus percent shrinkage of control bar, maximum	0.04	—	0.03
Water requirement percent, maximum	103	0[a]	105
Reduction of reactive expansion, at 14 days, percent, minimum	60	75	—
Mortar expansion at 14 days, percent, maximum	—	—	0.020
Soundness, autoclave expansion, percent, maximum	—	0.50	0.5

[a] This specification requires that a mortar with a natural pozzolan and 115% of the water content of the control mortar or with a fly ash pozzolan and 103% of the water content of the control shall have a flow equal to or greater than that of the control.

The pneumatic collection and mechanical classification system in Figure 5 is typical of installations capable of producing a specification ash from a marginal material. Air velocities in the mechanical separator are controlled so as to separate coarse fly ash particles from fine material, with the former being sent to disposal and the latter (approximately 85% of the total) being stored in silos for subsequent sale.[15]

Low carbon content is required and uniformity from day to day is necessary. Ash from base-loaded power stations where steam loads are relatively constant is uniform in size and the carbon content is often under 5%.

Fly ash transportation from utility plants is similar to that for Portland cement; the material, once aerated, flows readily. Several rows of low-pressure air pads around the bottom of the bins are useful in getting the material to move once it is deaerated. The delivered cost of fly ash in most cases is less than one-half that of Portland cement, depending on transportation distances.[16]

The amount of fly ash as percent of total cementing medium for concretes is: ready mix, 10–30%; block, 25–35% (low-pressure steam curing) and up to 50% (high-pressure steam curing); and mass concrete, up to 35%.

Fly ash can be used to manufacture raw material (portion of the kiln feed) and may be interground with clinker to form Portland-pozzolan cement. Less than 200,000 tons/yr are so utilized. The French cement industry uses over 1,000,000 tons of fly ash yearly in the manufacture of cement.[17]

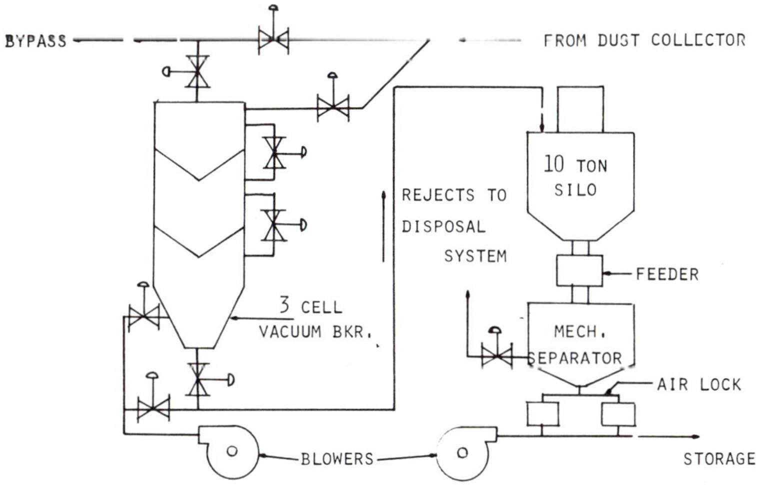

Figure 5. Fly ash collection and beneficiation system.

Table 7 *Typical Strength Retention values of Asphalts Made with Traprock Dust, Limestone Dust, or Fly Ash*[a]

	Strength Retained (%)
Traprock dust:	
Massachusetts	90
New Jersey:	
Sample 1	83
Sample 2	86
Virginia	93
Limestone dust	87
Flyash:	
Illinois	97
New Jersey	
Sample 1	96
Sample 2	91
Sample 3	91

[a] 4-day immersion in water at 120°F; asphalt content $5\frac{1}{2}$ parts/100 parts coarse aggregate; filler utilized as-received.

Source: J. P. Capp and J. D. Spencer, "Fly ash Utilization—A Summary of Application and Technology", USBM 1C 8483, USDI, 1970.

As early as 1931 it was found that fly ash could be used to replace limestone dust as filler in bituminous concrete.[18] Lower cost, proper size gradation, and resistance to water in dense-type bituminous concrete are some of the advantages.

Several states approve fly ash as mineral filler in asphalt paving and highway construction, but current consumption in this market is about 110,000 tons/yr.

In studies by the Bureau of Public Roads wherein fillers were rated as to resistance of compacted mixtures to loss of strength after immersion-compression tests, fly ash was found to be equal to other fillers.[19] Strength retention of asphalts containing eight different ashes ranged from 85 to 100%; 75% is the critical minimum. The results for four of the fly ashes, one limestone dust, and four traprock dusts are shown in Table 7.

Specifications stipulate that the material be dry and free of lumps consisting of aggregations of fine particles, a maximum carbon of 12%, by loss on ignition as AASHO M17 and ASTM D242, given in Table 8.

Lightweight aggregate for structural concrete, insulating concrete, and masonry is produced in the United States, employing over a quarter million tons of coal fly ash annually.

Lightweight aggregate gained 45% to a total of 15 million tons from 1960 to 1960.[20] Fly ash in aggregate is available in large quantities where lightweight aggregates are in demand.

Fly ash lightweight aggregate is dry ash, mixed with 20–25% water, a binder material such as lignin, and agglomerated by extrusion or balling. The formation into pellets is required so as to form a porous bed on the grates of the sintering machine, allowing air for combustion to enter the bed. The agglomerates are spread on a travelling grate and ignited by a gas flame. The sintering is maintained during the travel of the grate through its length by inducing a draft through the grate and the bed of burning pellets. The hot, sintered cake is discharged from the end of the sinter strand as clinker and crushed to marketable sizes. Unsintered fines are screened and returned to the process. A schematic diagram of a fly ash lightweight aggregate sintering plant is given in Figure 6.

The requirements for fly ash suitable for sintering are carbon content, fineness, and iron oxide content.[21] If residual carbon varies widely, combustion control is difficult to maintain; 4–8% loss on ignition is the preferred range. Fineness is related to porosity, absorption, and strength of the fired aggregate.

Table 8 *Size Gradation Requirements for Mineral Fillers for Bituminous Paving Mixtures*

Sieve No. (U.S.S. Mesh)	Percent Passing	
	AASHO M17	ASTM D242
30	100	100
50		95–100
80	Not less than 95	
100		95–100
200	Not less than 95	70–100

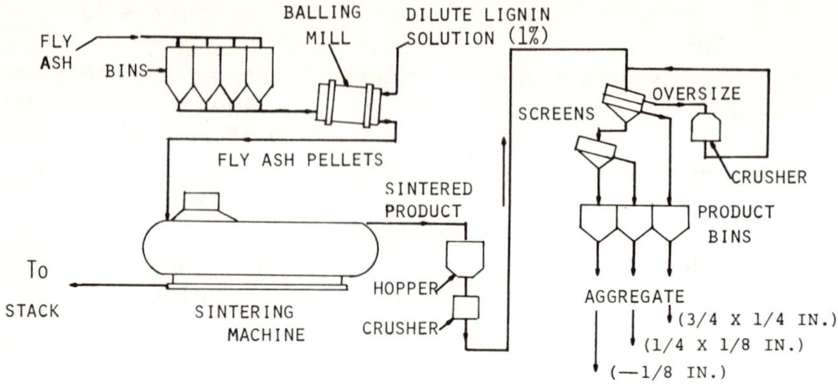

Figure 6. Sintering process for manufacture of fly ash light weight aggregate.

Satisfactory aggregate may be produced from raw fly ash; however, beneficiation is sometimes required to make the material more amenable to processing.[22]

Considerable field and laboratory experience has been accumulated on fly ash with lime and aggregates in the construction of base courses for highways, airports, parking areas, and the like. The load-carrying member is a hardened composition of lime-pozzolan (generally fly ash)-aggregate; a wearing surface—bituminous or Portland cement concrete—is placed over this layer. Specifications for lime- fly ash-aggregate base courses exist in Illinois, Maryland, New Jersey, Ohio, Pennsylvania, Virginia, and West Virginia.[23]

New York Port Authority's Newark Airport Redevelopment Project[24] has received considerable notice. Port Authority research, initiated in 1965, developed a mixture of hydrated lime, Portland cement, fly ash, and the in-situ sand at the airport as material for the base of the new runways. The mixture is hydrated lime (2.8–3.6%), Portland cement (0.7–0.9%) fly ash (12–14%). The remainder is hydraulic fill sand and where high base strength is required, 30% crushed stone is added. The material LCF (lime-cement-fly ash), is expected to develop five-year strengths of between 2000 and 2400 psi in compression; A cubic yard costs about $3.80 (1970). In the New York area, crushed stone is $5.00 to $6.00 per ton and a yard of lean concrete containing three sacks of cement costs about $12.00. It is reported that a 30-in.-thick LCF base has a load performance equivalent to 16 in. of Portland cement concrete or 60 in., or more, of aggregate-base asphaltic concrete. Cost estimates for placing the material vary between $7.00 and $7.50 per square yard, compared to about $12.00 for an equivalent conventional asphaltic concrete and up to $20.00 for a concrete pavement capable of sustaining equivalent loads. The redevelopment of Newark Airport calls for placing 2 million yd^2 of the new pavement.

Specifications for the quality of fly ash to be used with lime are given in ASTM C593. This specification sets no limit on carbon content. It has been suggested that any available fly ash containing less than 15% carbon is satisfactory if it meets the other criteria specified in C-592.[25] Around 150,000 tons/yr of fly ash are being used in pozzolanic pavement construction.

Fly ash is also used with lime to stabilize soils where lower strengths are satisfactory but where absorption and freezing-thawing are more important. These include subgrade and subbase preparation where upgrading of fine-grained soils (silts and clays) is necessary.

Several other uses are oil well cementing and structural grouting, mine void filling for fire and subsidence control, and sewage sludge dewatering. Total tonnage annually is about 175,000 tons.

Oil well cementing pumps a cement-water slurry down a steel casing to critical points in the annulus around the casing, in the open hole, or in a permeable rock formation. The process was introduced to the oil industry in the early 1920s.[26] Cementing provides protection against corrosion of well casings from subsurface minerals and electrolysis and conserves the producing value of the formation by eliminating salt water flow from other possible producing horizons behind the well casing. Fly ash has been found to be a suitable admixture for oil well cements; in 1969 over 80,000 tons were reported used.

Abandoned coal mine fires and surface subsidence have been controlled by using fly ash to fill voids left by underground mining. Ash is introduced both in dry and water slurry forms through boreholes drilled into mined areas. A number of projects have been completed, several of them funded partially through the Appalachian Redevelopment Act. In 1969, approximately 90,000 tons of fly ash went into this application. Patents covering various aspects of the techniques involved in mine fire and subsidence control with fly ash have been awarded.[27, 28]

Figure 7 is a diagram of a method and equipment employed to pneumatically inject fly ash into underground voids.

Studies at Notre Dame University[29] have demonstrated that fly ash may be used in the treatment of polluted waters and sewage, through adsorption, which promotes the removal of organics; water extracts of lime and gypsum in some ashes, which cause precipitation of inorganic phosphorus and neutralize acid; and a conditioner of sludge for dewatering on vacuum filters. Ash in the 10 to 30 micron size range and of

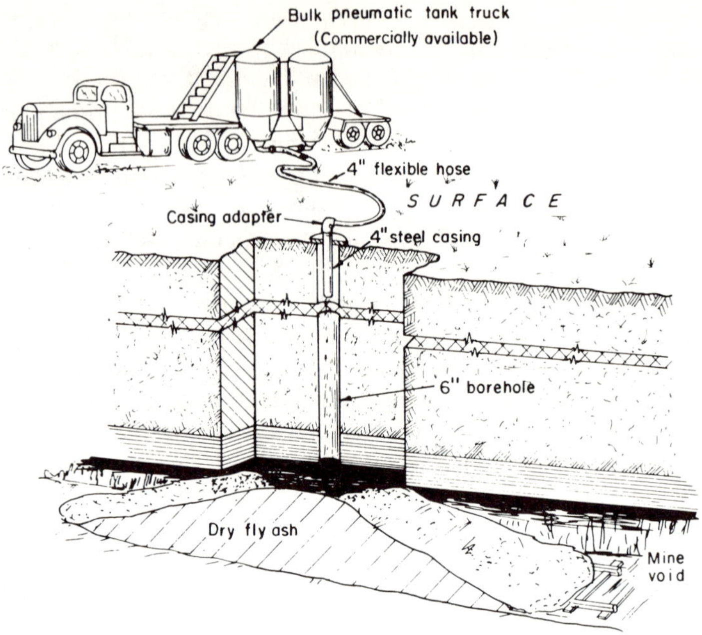

Figure 7. System for pneumatic injection of dry fly ash into mines (Source: USBM)

high carbon content was studied through a grant from the FWPCA using municipal sewage.[30]

The British, principally through the concerted efforts of the Central Electricity Generating Board, use large quantities of fly ash in structural or load-bearing fill and as a main constitutent in lightweight aerated concrete.

Work in Great Britain has shown that fly ash has self-hardening properties to make it suitable as a lightweight load-bearing fill for road projects.[31] Emphasis has been placed on highway embankments, where low bulk density and high shear strengths are advantageous. The compacted dry density is about 70–90 lb/ft^3, which is less than most conventional fill materials. Fly ash is conditioned with 25–30% water, laid down in 6–12-in. layers, and immediately compacted with rollers between lifts. Two million tons of fly ash are being used in this manner in England yearly in highways and railway embankments, bridge abutments, and airport runways. Several projects in the United States have been completed.[32, 33]

Aerated concrete is a homogenous fine-grained silicate enclosing small air cells. Cellular concrete, gas concrete, and foamed concrete also describe the material made by introducing gas into a paste composed of cement and/or lime plus a siliceous filler. The filler is often fly ash, and it is commonly employed in amounts of as much as 70% by weight of the total.[34]

Aerated concrete is produced in two forms: precast units such as blocks and cast-in-place or in situ. Precast blocks and slabs are steam-cured at high pressure. In situ material is air-cured, but there are methods for speeding up curing by heat. It is only through autoclaving that it is possible to produce a material that is both lightweight and has strength and drying shrinkage properties that make it suitable for structural purposes.

Aerated concrete has good thermal and acoustic properties, is dimensionally stable, fire resistant and termite and rot proof; it can be drilled, nailed, screwed, and sawed, and its weight allows it to be shipped and handled in large economical sizes. Although most of the material is made into masonry block, the larger reinforced units are felt to have the greatest potential in the construction industry. The prefabrication of large sections of buildings with structural aerated concrete containing fly ash is now being practiced in some European countries.

Application in the United States has been limited to insulating roof decking. Restoration and reclamation of ash disposal sites to make them suitable for agricultural purposes has been studied.[35] The most tolerant plants grown in ash soils are grasses, particularly rye, and legumes; therefore, these make starter crops on ash.

Other efforts have demonstrated the feasibility of reclaiming surface coal mine spoil with large amounts of fly ash.[36] Some fly ashes have a high pH (for neutralization of acid spoil), a high content of some macroplant and most microplant nutrients, moisture retention, and a diluting effect when mixed with heavy soils. Phosphorus and nitrogen, generally absent in most fly ashes, must be provided through normal fertilization. Because of the large amounts of the silt-sized fly ash that are required (up to 800 tons/ace) the soil texture is improved. Table 9 shows amounts of the more common elements that could be expected in large quantities of fly ash. The percentages listed are about the same as or a little higher than those found in many natural soils.

There is continued pressure from a number of sources to develop improved methods of recovering minerals while reducing damage to surface and streams. Over 30% of this nation's coal is obtained by open-cast methods, and this technique will continue to be an important way of obtaining that fuel. The application of fly ash as a means of reclaiming mined areas will continue to receive attention.

Table 9 *Elements in Fly Ash*

Element	Percent	Approximate Amount per 600 tons ash (lb)
Calcium	1.00	12,000
Magnesium	0.36	4,000
Potassium	1.74	21,000
Phosphorus	0.13	1,600
Cobalt	0.01	60
Molybdenum	0.01	80
Boron	0.01	100
Manganese	0.02	250
Aluminum	14.01	170,000
Iron	9.94	120,000

Source: J. P. Capp and J. D. Spencer, Fly ash Utilization—A Summary of Application and Technology, USBM 1C 8483, USDI, 1970.

Materials such as peat, calcined clay, expanded shales, sewage sludge, sand, and sawdust are used to modify soils or turf grasses. Benefits include increased aggregation, water infiltration and percolation, reduced runoff, improved hydraulic conductivity, and decreased soil compaction. Field work by the USBM[37] has indicated that sintered fly ash lightweight aggregate is a potentially valuable soil modifier. Applications of 30% by volume to the top 6 in. of soil supporting turf increased the water infiltration rate, pore space, root density, and grass yield, thus indicating the advantages of the material for soils underlying sod on golf courses, athletic fields, public parks, and similar areas.

Fly ash lightweight aggregate in ¼-in.-diameter pellets is commercially marketed in Great Britain for horticulture.[38] The British Ministry of Agriculture and Fisheries has approved the material for land drainage improvement programs. A finer material is mixed with heavy clay soils to improve drainage and with lighter soils to improve moisture retention.

Fly ash has been found to be both chemically and physically suitable for brick making. Investigations have been carried out on the manufacture of brick with fly ash, boiler slag, sand, and clay.

Bricks were produced with mixtures of fly ash, boiler slag, and 1% high-swelling bentonite clay binder in 1950 by both extrusion and dry-press methods.[39] More recent pilot plant work[40-42] has produced dry-pressed face brick using approximately a 3-to-1 proportion of bituminous coal fly ash to boiler slag with 3% sodium silicate as binder. A schematic diagram of this plant is shown in Figure 8 and physical properties of the product are compared with those of clay brick in Table 10.

Laboratory investigations with lignite fly ash have indicated that sodium silicate is not required to make dry-pressed brick from that material.[43] Clay-fly ash brick with up to 74% by weight of ash have been made[39]; nearly 100,000 bricks were produced in beehive kilns. The use of ash with clay for certain types of brickmaking has been practiced in the United Kingdom since the 17th century, to reduce shrinkage of the clay and make use of unburned carbon in the ash.

Steam-cured (unfired) brick composed of sand and fly ash have been produced.[44] Optimum fly ash percentage was 30–50.

Critical to the establishment of a market for fly ash brick are:[45]

1. Quality fly ash and slag must be available.
2. The fired brick color must be in the acceptable range for the market area.

Table 10 *Physical Properties of Clay Brick and Fly Ash Brick*

ASTM Standard Specifications	ASTM Grade NW clay brick Commercial	ASTM Grade NW clay brick Prime	ASTM Grade SW clay-face Brick	Fly ash brick[a]
Compressive breaking strength (psi)	1500	2500	3000	3000–20,000
Saturation coefficient			0.88	0.78
Water absorption (%)	[b]	22	17	14–22
Shrinkage	[b]	[c]	[c]	0.5
Initial absorption rate	—	—	—	75–300
Bulk density	130–160	130–160	130–160	90–120

[a] Varied depending on fly ash source, slag, sodium silicate, and manufacturing conditions.
[b] No limit.
[c] Standard not established; shrinkage probably higher than that for fly ash brick.
Source: J. P. Capp and J. D. Spencer, Fly ash Utilization—A Summary of Application and Technology, USBM 1C 8483, USDI, 1970.

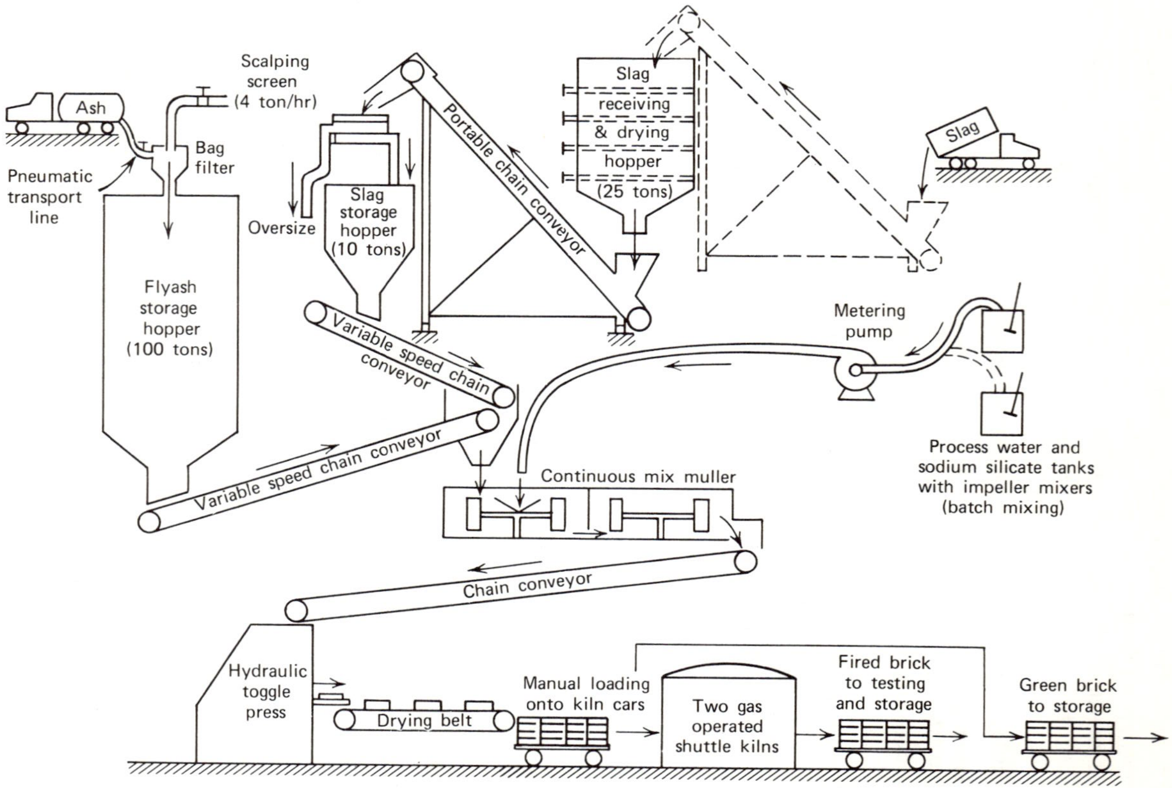

Figure 8. Fly ash brick pilot plant.

3. The quality of clay brick in the area must be met or exceeded by the fly ash product.
4. The specifying sector must endorse the product, as must the building departments through the various codes.
5. Cost considerations must indicate competitiveness.

Uses for fly ash is an active current effort in solid waste utilization and recycling. Some of the technology has been employed for 40 years.

The electric and utility and coal industries of the United States, realizing their stake, in 1968, organized the National Ash Association to act as the catalyst by which communication in this area will improve and information be coordinated and disseminated.

REFERENCES

1. C. E. Brackett, "Production and Utilization of Ash in the United States," Proceedings Second Ash Utilization Symposium, Pittsburgh, Pa. March 10–11, 1970.

2. W. C. Helt, "Fundamentals of Dust Collection," Proceedings of the 1968 Industrial Coal Conference, Purdue Univ., Lafayette, Ind., October 9–10, 1968.

3. S. T. Cuffe and R. W. Gerstle, "Emissions from Coal-Fired Power Plants: A Comprehensive Summary," USHEW, PHS 1967.

4. S. T. Slome, "Capabilities of Equipment for the Removal of Particulates from Coal Fired Boilers," paper presented at Industrial Coal Conference, Univ Kentucky, Lexington, Ky., April 9, 1969.

5. Private communication.

6. L. R. Love and S. F. Whirl, "Fly Ash Problem—Buried in Deep Mine," Proceedings 82nd Annual Meeting Coal Mining Institute of America, December 12–13, 1968, Pittsburgh, Pa.

7. R. E. Morrison, "Ash Utilization—A Liability to an Asset," *Combustion,* **42,** No. 4, 29–32, October 1970.

8. R. W. Styron, "Fly Ash—A Commercial Product," *COAL-Today and Tomorrow,* 31–33, April 1965.

9. E. G. Barber, "The Production and Properties of PFA," PFA Congress, CEGB, London, England, October 1968.

10. E. J. Hyland, "Practical Use of Flyash in Concrete," *Construction Specifier,* March 1970.

11. ASTM Designation C618-68T, "Fly Ash and Raw or Calcined Natural Pozzolans for use in Portland Cement Concrete."

12. Corps of Engineers Specifications for Pozzolan for use in Portland-Cement Concrete, CRD-C 262-63.

13. Bureau of Reclamation Specifications for Pozzolan for use in Portland Cement Concrete.

14. M. J. Snyder et al., "Properties and utilization of Fly Ash," Battelle Memorial Institute, Research Report, June 30, 1966.

15. C. E. Brackett, "Availability, quality, and present utilization of fly ash," USBM 1C 8348; "Fly Ash Utilization," Proceedings: EEI, National Coal Assn.; USBM Symposium, Pittsburgh, Pa., March 14–16, 1967.

16. G. H. Roman, "You can save money with Fly Ash," *Coal Age,* 60–63, August 1968.

17. A. Jarrige, private communication.

18. F. V. Zimmer, "Fly Ash As a Bituminous Filler," Proceedings 2nd Ash Utilization Symposium, Pittsburgh, Pa, March 10–11, 1970.

19. C. A. Carpenter, "A Cooperative Study of Fillers in Asphaltic Concrete," *Public Roads,* **27,** No. 5, 101–110, December 1952.

20. R. A. Grancher, "The Lightweight Aggregate Boom," *Rock Products,* 59–62, 94, December 1969.

21. L. J. Minnick, "Lightweight Concrete Aggregate from Sintered Fly Ash," *Highway Research Record,* **307,** 21–32, (1970).

22. "Fly Ash Goes Commercial," *Coal Age,* 64–69, April 1969.

23. E. J. Barenberg, "Life-fly ash-aggregate mixture," USBM 1C 8348; "Fly Ash Utilization," Proceedings: EEI National Coal Assn., USBM Symposium, Pittsburgh, Pa., March 14–16, 1967.

24. "Ancient Concrete Mix Makes Jet-Age Runway," *Eng. News-Record,* May 22, 1969.

25. G. & W. H. Corson Company, private communication.

26. D. K. Smith, "Utilization of Fly Ash in the Cementing of Wells, USBM 1C 8348; "Fly Ash Utilization," Proceedings: EEI, National Coal Assn. USBM Symposium, Pittsburgh, Pa., March 14–16, 1967.

27. J. L. Heavilon, et al., "Method for Mine Fire Control," USP 3,421,587, January 14, 1969.

28. M. O. Magnuson, "Fly Ash Injection Method and Apparatus," USP 3, 500,934, March 17, 1970.

29. M. E. Tenney and W. F. Echelberger, "Fly Ash Utilization in the Treatment of Polluted Waters," Proceedings 2nd Ash Utilization Symposium, Pittsburgh, Pa., March 10–11, 1970.

30. "Pressure Filtration of Secondary Digested Sludge with Fly Ash Filter Aid," Cedar Rapids, Iowa, Water Pollution Control Plant, 1968, report by Howard R. Green Company, Consulting Engineers, August 1968.

31. "Load Bearing Fill," *PFA Data Book,* Central Electricity Generating Board, England, September 1967.

32. A. M. DiGioia and W. L. Nuzzo, "Fly Ash for Structural Fill," paper presented at Joint Power Conference, Pittsburgh, Pa., September 29, 1970.

33. "Fly Ash Used As Common Borrow in Railroad Bridge Embankments," *Construction Bull.*, 18–21, July 16, 1970.

34. "Aerated Concrete," *PFA Data Book,* Central Electricity Generating Board, England, September 1967.

35. E. G. Barber, "Win Back the Acres: Ash and Agriculture," Central Electricity Generating Board, England, January 1966.

36. J. P. Capp, "Fly Ash In Agriculture," USBM 1C 8348, "Fly Ash Utilization," Proceedings: EEI, National Coal Assn., USMB Symposium, Pittsburgh, Pa., March 14–16, 1967.

37. J. D. Spencer, "Review of Bureau of Mines Coal Program, 1969", USBM IC 8479, June 1970, pp. 13–15.

38. Lytag Ltd., private communication.

39. L. J. Minnick, and W. H. Bauer, "Utilization of Waste Boiler Fly Ash and Slags in the Structural Clay Industry," *Am. Ceramic Soc. Bull.,* **29,** No. 5, 177–180, May 15, 1950.

40. H. E. Shafer et al., "Status Report on Bricks from Flyash," USBM 1C 8348; "Fly Ash Utilization," Proceedings: EEI, National Coal Assn., USBM Symposium, Pittsburgh, Pa., March 14–16, 1967.

41. K. K. Humphreys, "An Economic Evaluation of the WVV-OCR Process for Producing Fly Ash—Based Structural Materials," Report No. 36, Coal Research Bureau, West Virginia Univ., Morgantown, W.Va., 1968.

42. C. F. Cockrell and H. E. Shafer, "A Technical Evaluation of the West Virginia Univ.-OCR Process for Producing Fly ash-Based Structural Materials," Report No. 40, Coal Research Bureau, West Virginia Univ., Morgantown, W.Va., 1969.

43. O. E., Manz, "Ash From Lignite," Proceedings 2nd Ash Utilization Symposium, Pittsburgh, Pa., March 10–11, 1970.

44. M. Mateos, "Sand-Fly Ash Bricks," *Mater. Res. Std.,* **4,** No. 8, 427–430, August 1964.

45. J. A. Reidelback, Jr., "The Clay Brick Industry and The Fly Ash Brick Potential," National Ash Association, Inc., Report No. 3-69, July 1969.

PART VI recycling

Metallurgically, recycling is decades old, and in precious metals goes back to the days of Elizabeth I and Henry VIII of England and earlier on the continent of Europe.

Chapter VI.1 discusses recycling and obsolescence, secondary materials, by-products, and the effect of advances in technology, eliminating plants, markets, and entire industries. Chapter VI.2 enlarges on the tremendous number of specifications for engineering materials, emphasizing that scrap and waste must be processed to meet these performance standards. Chapter VI.2 also emphasizes the very high percentage of reuse of the wide variety of metals available, in the major engineering alloys; recovery in many cases is equal to primary production. Chapter VI.3 treats iron and steel, the scrap grades and usage, the tremendous scope of the industry, with its scrap yards and their huge investments, sale, shipment, and delivery, and world markets.

Chapter VI.4 describes the dream of many environmentalists—mining automotive graveyards and returning metal to the scrap cycle. Chapter VI.5 deals with nonferrous scrap and its preparation, collection, sorting, and markets. Chapter VI.6, by Kirkpatrick, is concerned with aluminum recycling, and Chapter VI.7 with the recent attention to aluminum cans. Chapter VI.8 deals with copper-bearing scrap; Chapter VI.9, by O'Reilly, with communications scrap; Chapter VI.10 is concerned with zinc and zinc alloys, and Chapter VI.11 with tin and its scrap.

Chapter VI.12, by Forbes, discusses the recovery of silver from photographic paper and film waste at Eastman Kodak, where one of the largest silver refineries in the world is the result of the program. Secondary gold is treated in Chapter VI.13. Edwards discusses the entire paper industry in Chapter VI.14, under the title of "Waste Paper Recycling."

Hayes, in Chapter VI.15, discusses recovery of pine stumps from lumbering of years ago, collecting, processing, and recovery of rosin, and use of extracted chips as a raw material for paper, on a profitable basis. Chapter VI.16 discusses experimental plants for recovery of values from urban wastes. The subject of Chapter VI.17 is forest products wastes, an area where recoveries are approaching, except for pine roots, 100%. Forest products and paper recovery are approaching that of the meat-packing industry, where all is recovered save the squeal. Chapter VI.18 reviews the problem of nonreturnables and developments adapting them to recycle and the related human problems.

CHAPTER VI.1
recycling and obsolescence

Separation of newspapers from home trash and garbage is a first step in sorting trash. Collected separately, this paper is readily marketed and can be recycled without excessive collection or disposal cost. Similarly, separation of aluminum and tin cans, and keeping the different metals separate at the source encourages recycling—which, after all, has to be economic—but it also reduces the load on the local incinerators, most of which are operated over their effective capacity.

Brewing is an example of recycling. Grains are malted and fermented and the extract made into beer. The spent grains from the breweries in our large cities are transported to central plants where they are dewatered, dried, and prepared for animal feed, particularly for milk farms not too far distant, serving the same central population. The grains after drying are high-protein feed.

However, ultimately the waste problem arises. Milk farms may have several hundred head on a small farm, but, for example, the 2500 cows with rotary milking tables at Walker Gorden Farms in New Jersey, constitute a problem. A cow produces about 9–10% of its live weight in manure per day and a cow weighs 1000–1300 lb. That is about 100 lb/day; for 2500 cows, 250,000 lb or 150 tons/day. The product is processed, dehydrated, and has no market. So it goes back to the land where the grain started, narrowing the profit margin on milk.

It is especially with metals that recycling plays a major role. About half of our steel production starts as scrap. We have outstanding speciality steel companies with no hot metal or blast furnace capacity, which are entirely on a scrap iron and steel basis, in addition to consuming their own scrap.

Secondary copper for decades has been as great a contributor to our copper supply as have been our mines. One large Eastern copper refinery works entirely on scrap, and is a smelter with by-products. Secondary lead has been with us for more than half a century—it is estimated that a lead molecule recycles as a useful component 100 times before it is dispersed so thin we cannot collect it.

Secondary aluminum is represented by many dealers, brokers, smelters, and refiners and is expected to grow in recycling so that the production equals primary metal. There are active secondary magnesium, solder, nickel, tungsten, silver, gold, platinum, and precious metal businesses, all highly competitive. In a similar fashion stainless steel scrap, in a wide variety of analyses, is collected, sorted, and reused to make more articles.

In various areas of the country, recycled glass or cullet has been the business of companies in operation for fifty years. Recycled glass has been with us for many years.

Large secondary iron and steel scrap companies such as Luria, United, Proler, and other processors, have found substantial profits—cyclical though they are, depending on the fluctuating value of scrap steel—in auto hulks, refrigerators, stoves, and other steel scrap. An auto shredder, which reduces steel scrap into fist-sized pieces, of clean ferrous metal has been developed. Another process recovers precipitation iron from tin cans and light-gauge steel trimmings.

Copper ore is leached in a solvent that produces a solution, which in turn is mixed with precipitation iron to produce copper. Profits on an output roughly half that of the steel-shredding business are superior, partly because this product is sold to the copper industry on long-term contracts that protect from variations in scrap prices. Zinc, copper, and aluminum pieces are removed from the ferrous scrap. These metals are sold.

One company, Proler Steel, has plants in Houston, Kansas City, Los Angeles, Chicago, Everett, Boston, London, Jersey City, and Brooklyn, but finds that it may run out of raw material, wrecked and discarded cars, if it expands too rapidly. Proler iron precipitates copper in El Paso, Texas, Copperton, Utah, and Chicago and has tie-ins with modern efficient copper facilities, which are the least likely to be shut down during slow business periods. Proler has several active competitors.

Twenty percent of the paper industry's raw materials are recovered wastes, saving each year the equivalent of 200 million trees. Similarly, over half the nation's output of steel is reclaimed scrap. The same is true of copper, while twice as much lead is recovered as is mined. Over 10,000 tons of food wastes, collected each year by private contractors from restaurants, institutions, markets, and to some extent private homes, are recycled by being fed to 850,000 hogs. The National Association of Secondary Material Industries lists 725 corporations involved in recycling; it estimates the value of the resulting products at $8 billion per year.

Where there is a potential or actual profit, material is recycled and has been for many years. The old-fashioned junk man is not dead, but has been replaced by large-scale, highly financed, competitive, and efficient secondary dealers of practically anything. One of the most completely integrated operations is that of Western Electric, Nassau Smelting Works, and the American Telephone and Telegraph Company, respectively the manufacturing arm, the scrap and recovery arm, of the parent A.T. & T. and the communications industry.

The Association of American Railroads wrecks freight cars for metal recovery. Other companies deeply involved in recycling are Alcoa, American Can, Continental Can, Engelhard Minerals—the list is almost endless. Catalysts used by chemical companies are recovered by scrap producers for reuse.

One resource we value in the United States is platinum and its related metals. Our annual consumption is about equivalent to a cube 3 × 3 × 3 ft. Laboratories never throw away platinum utensils; they constantly return them for credit. The metal is used by physicians, dentists, and the like, to repair our bodies, bones, and teeth. When we are cremated these pieces of platinum (and gold) are recovered by sifting the ashes and returning the metal to the cycle.

SECONDARY MATERIALS

The output of goods and services in the United States has grown as much since 1950 as it grew in the entire period from the landing of the Pilgrims up to 1950. A similar growth period is predicted between 1970 and 1980.

Urban refuse generates about 60 billion cans, 360 million bottles, 58 million tons of paper and paper products, 4 million tons of plastics, over a million abandoned automobiles, mountains of demolition debris, 180 million tires, and countless millions of tons of refrigerators, stoves, television sets, communications scrap, outmoded telephone equipment, and similar items.

Secondary sources and the recovery therefrom provide 50% of the lead of our annual production, 40% of the copper, 45% of the iron and steel, and 25% of the zinc and aluminum. These are from in-use channels that embrace nearly 40 million tons of copper, 3½ million tons of lead, and about 4 million tons of zinc, a growing man-made mine of future raw materials.

Annually discarded in municipal dumps are approximately 11

million tons of ferrous metals and over a million tons of nonferrous metals including copper, aluminum, tin, lead and zinc. Millions of junk cars remain to be reclaimed from auto graveyards.

Nearly 400,000 tons of aluminum were used for manufacturing cans, lids, and caps in 1970. Only now are reclamation procedures being started. It is estimated that in city dumps there are 25,000 tons of tin in tin-coated steel cans, equivalent to the amount of tin salvaged from all other secondary sources.

The USBM, USDI, has operated pilot plants that separate and recover, in the same way that minerals are separated from their ores, the major metal and mineral values contained in municipal incinerator residues. The process employs conventional and well-known equipment, such as magnetic separation, screening, grinding, and shredding. Each ton of residue processed is estimated to yield 700 lb of iron, 40 lb of nonferrous metals, including aluminum, copper, lead, tin, and zinc, small amounts of silver, and a half ton of glass.

The 180 million tires, having a scrap value of about 11¢ each, contain about 2 million tons of rubber. Disposal by burning results in a billowing, acrid, black smoke. Tires are difficult to compact and are nonbiodegradeable, so they add little to landfills. Pyrolysis has been estimated to give 1500 ft^3 of 800–1200 Btu/ft gas, and 140 gal of hydrocarbon liquid per ton of tires. The tire producers have constructed pilot plants.

In the Minneapolis area, urban refuse has been employed as a reductant for converting nonmagnetic taconite, an iron ore, but currently a waste product of iron ore mining, into magnetic iron ore by reduction roasting process. The nonmagnetic taconite, containing about 30% iron, is roasted with refuse in a rotary kiln at 1000°C. The recovered product analyzed 70% iron, which can be pelletized for use in a blast furnace.

OBSOLESCENCE

Obsolescence produces a great amount of waste; it is largely a function of changing technology.

Business establishments consign their obsolete records and forms to shredders, to join the waste pile of paper to be reworked; the U.S. Treasury carefully burns shredded and outworn currency.

Fires, wrecks, and explosions, add to the waste pile. These are unforeseen and unplanned events; to them are added the entirely planned obsolescence of machines, appliances, automobiles, and similar products.

Railroads contribute obsolete or outworn rails to the scrap market, which also disassembles the trains, and freight cars, after cannibalizing reusable parts; metal is salvaged but wood, ornamentation, seats, and the like are burned. Some trolley-car systems, after being displaced by buses, find a home in South America, but they often enter the scrap pile.

Obsolete or noncompetitive ships and ferries are stored for a time and if possible sold to less affluent nations; but they eventually find their way to ship breakers or salvage dealers, who convert what they can for reuse or scrap and discard the rest.

The propeller-driven airplane was rendered obsolete by the prop jet, then by the pure jet; and smaller planes were replaced by larger. Often they find new homes, but sooner or later they reach the scrap market.

Chemical and process industry plants amortize their equipment, but changing technology, or corrosion, may make equipment obsolete before it is completely amortized. There is an active second-hand equipment market, but a fair proportion becomes scrap.

Changing technology may present the sight of a newer, bigger, and better petroleum catalytic cracker being built alongside of one still in operation until the new one is ready. The twenty-year life expectancy is shortened to five years, on the road to the junk pile.

Home appliances of all kinds join the discard piles because repair facilities are increasingly difficult to find. Recycle of such mechanisms is truly complex; much is buried.

Pleasure boats pass from hand to hand; shipbreakers are usually not interested, and they are eventually beached or sunk "accidentally."

When wartime petroleum transmission pipelines were sold for transmission of natural gas, the industry expanded remarkably and town gas and water gas plants were shut down. Now we need more gas and are again building manufactured gas plants; but the old ones were destroyed and added to the waste stream.

Only about a decade ago the development of a bleachable ferrotannate ink for telephone directories, which become obsolete in a year or two, was developed; a new copy required the surrender of an old one, which was recycled. Collection and transportation cost made the process uneconomical; but now, as the amount of paper strains incinerators, paper recycling is once more a lively topic.

Liquor bottles, by law, cannot be reused and should be destroyed when empty. Recycling of glass bottles is only beginning in a major way, as a source of cullet for glass plants. Plastic bottles for liquor are entering the market.

Worked-out mines become a source of abandoned machinery, as well as the site of a pile of waste tailings and a ghost town. A tourist attraction?

DEMOLITION WASTE

Demolition waste may be divided into two classes: combustible and noncombustible. Frame houses produce a waste comprising wood, stucco, metal lath, wiring BX, and linoleum-type materials. These are generally regarded as combustible although not incineratable in a typical incinerator. The bulk density of the material is 350 or 400 lb/yd^3.

The noncombustible materials consist of bricks, masonry, rocks, concrete, and rubble, from the demolition of masonry buildings and the foundations and slab floors of frame buildings. The bulk density of this material is as high as 1800 lb/yd^3.

Some typical figures for demolition waste are: Milwaukee, 230 lb/yr per capita; New Orleans, 60–80; Forth Worth, 12–30; Dallas, 22–30; Houston, 175–200; and New York, 200–250. During the 1960–1970 period it is estimated that the waste from demolition, slum clearance, rebuilding, utilization of air rights, and reconstruction was about 40, 000 million lb/yr, or 20 million tons, with an expected rise in 1975 to 25 million tons.

Disposal is by the demolition contractor, conventionally in landfills or, where the waste conforms, in building construction as clean fills. The large trucks hold 4½ to 52 yd^3, and larger ones are planned. Truck mileages from the demolition site to the disposal areas are 10–40 miles.

With higher labor costs and scarcity of semiskilled labor, salvage is little employed in highly industrialized regions. Houses are now demolished in one day, using a small clamshell derrick. Formerly it was the practice to clamshell a building depositing the combustible wastes on an adjacent site where they were burned. Safety and related ordinances are forcing an abandonment of this practice. One coast city handles demolition wastes in the same way as dredging wastes, by barging out to sea.

Disposal of masonry rubble is usually not difficult particularly in a metropolitan area where clean fill is always needed. The combustible wastes are not welcomed at sanitary landfills because of the odd sizes of the pieces or the items. Segregation is usually impossible because of labor costs. Open burning is prohibited and incinerators of special design offer the only real solution.

Careful drafting of building codes and close attention by builders to engineering data result in longer lasting buildings, and thus less waste.

CHAPTER VI.2

metals and compounds: use and recycling

ALUMINUM

Al. Light-weight casting alloys used throughout the automotive, electronic, aircraft, and aerospace industries. Also used for agricultural machinery, appliances, and furniture. Strongly recycled, secondary metal expected in future to approach annual tonnage of primary.

ANTIMONY

Sb. Alloyed with lead for batteries, with lead and tin in type metal. Compounds used in paints, ceramic opacifiers, flameproofing agents. No recycle for compounds, comlete recycle in metallic alloy uses.

ARSENIC

As. Little use as metal. Compounds used as rust inhibitors, oxidizing and refining agent in glass, insecticides, wood preservatives. Little recycle.

BERYLLIUM

Be. X-ray windows in x-ray tubes, Be-Cu alloys in springs, components in aircraft, electrical, and mechanical systems. Expensive. All scrap carefully conserved and recycled.

BISMUTH

Bi. Additive for malleable iron, free-machining steel, and aluminum. Used for fire protective devices requiring lowmelting points. Also, compounds are an important ingredient of proprietary medicines. Low-melting alloys recycled and reused.

CADMIUM

Cd. A key component in battery cells. A coating agent in electroplating. An important ingredient of antifriction alloys for lining heavy-duty bearings. Strongly recycled from residues, metal scrap, electroplating, and bearings.

COBALT

Co. Magnetic and superstrength high-temperature alloys, dental and surgical uses. High price, strong market, recycled to maximum amount alone or as alloys.

COLUMBIUM

Cb. Also called niobium. Carbon stabilizer in stainless steels, atomic reactor equipment. Widely dispersed as alloy, little recycling.

COPPER

Cu. The most widely used electrical conductor in everything from computer circuits to pop-up toasters. Used for roofing, plumbing, and piping systems, air conditioning and hearing units. Strongly recycled, large secondary business and large secondary plants, including those of the communication industry.

GOLD

Au. Known through its wide use in dentistry, gold has a broad range of industrial applications: base pins of electron tubes, transistors, and resistors, fuse wires, high-frequency conductors for printed circuits. Because of value, recycling is complete, even to shop sweepings, and return to market sometimes over a fifty-year cycle in jewelry, watch cases, and the like.

INDIUM

In. Its low melting point makes it important in fusible alloys for fire door links and sprinkler heads. Also used for preforms for semiconductors, III-V compounds. Tonnage is small, but fusible alloys and electrical components recycled with other metals.

LEAD

Pb. Batteries, chemical apparatus, sound proofing, protection against x-rays, solder, collapsible tubes. There is about as much secondary lead production as virgin; heavily recycled time and again.

MAGNESIUM

Mg. Galvanic anodes for protection of underground or underwater steel structures. Also in space-age structural alloys for high strength-to-weight ratio. Galvanic uses dissipate metal but secondary magnesium recovery from discarded aircraft, machinery, and the like is vigorous.

MANGANESE

Mn. Iron- and carbon-free metal for stainless steels, aluminum, and magnesium alloys, copper, and instrument alloys. Used as a ferroalloy, Mn is vital for control of S in steel making. The greater part of it ends in steel slags.

MOLYBDENUM

Mo. Adds toughness and high heat resistance to alloys. Used in electrical and electronic parts, missiles and aircraft, high-temperature furnaces, nuclear applications. Recycled as component of strong steels, electrical, and aircraft alloys; at times Mo is the most valuable component of the scrap.

NICKEL

Ni. Reinforces alloys against wear and corrosion. Alloy ingredient for stainless steel products. Also used for food processing, chemical industry, pharmaceuticals, electroplating. Specialty companies for nickel-bearing material, strong export market for scrap. Largely reused and reprocessed, even as constituents of catalysts.

PALLADIUM

Pd. Will not tarnish, so it makes electrical relay contacts noise-free in voice circuits. Also a valuable catalyst in many processes. Vital element in dental alloys. Carefully recovered from electrical contact and communication industry scrap whose recovery plants have sizable palladium sections.

PLATINUM

Pt. Used as an impressed current anode for cathodic protection of ship hulls and as a crucible liner for making high-purity optical glass. Catalyst for making high-octane gasolines. Also valuable for jewelry. Similar to other precious metals, strongly recycled even from concentrations as small as 1 oz per million pounds of copper. Intense scrap market.

SILVER

Ag. For photographic film and photosensitive papers used in many office copying machines. Found in complex printed circuits, commemorative coins, silverware jewelry. Metal strongly recycled, even sweepings. Such heavy recovery from used x-ray films and waste photographic film that Eastman Kodak operates the largest silver refinery in the world at Rochester, N.Y.

TANTALUM

Ta. Corrosion resistant, chemical heat transfer equipment, electrical capacitors rectifiers; high priced and strongly recycled.

TELLURIUM

Te. Helps make free-machining steels. As copper additive increases machinability with very little loss in conductivity. Also a major ingredient in thermoelectric materials for electronics. Recycling difficult and little done except as alloys.

THALLIUM

Tl. Lowers the freezing point of alloys used in thermometers and instruments for the arctic. Highly photosensitive. Little recycling; small quantity.

TIN

Sn. Half the primary tin goes into coating other metals. It is also used with recycled tin in brasses and bronzes. Combined with lead for soft solders. Alloyed for bearing linings. Has been extensively recycled by special processes for over 50 years.

TITANIUM

Ti. Titanium dioxide pigments are the outstanding whitener and opacifier for paints, ceramics, many different types of paper, inks, and plastics. As metal, developing market for corrosion-resistent and strong alloys, jet engine frames in airplanes. Strongly recycled.

TUNGSTEN

W. The heaviest of the metals. Wire in incandescent lamps, contacts, ignition points, x-ray tubes, carbide cutting tools, wire dies, high-speed steels. Very active scrap market, strongly recycled.

VANADIUM

V. Alloying agent for improved elastic properties of steels. Strongly recycled as residues and alloys.

ZINC

Zn. Electroplating, cathodic protection, galvanizing, brass, bronze, die casting. Alloys heavily recycled as are galvanizing residues.

ZIRCONIUM

Zr. Atomic energy applications, corrosion-resistant equipment. High price and scrap market competitively alert. Heavily recycled.

CHAPTER VI.3

iron and steel processes

A blast furnace may use air at the rate of more than 100,000 ft^3/min and produce slag in excess of 1000 tons/day. Blast furnaces are operated to produce hot metal for steelmaking processes as well as pig iron, silvery pig iron, and ferroalloys. Sinter plants are operated as parts of blast furnace departments to agglomerate fine ores, flue dust, and mill scale as part of the blast furnace burden. Slag quenching operations produce granulated or expanded slag.

These operations can produce particulate emissions in blast furnace burden materials including furnace charging, from opening blast furnace pressure release valves and agglomeration operations at the sinter plant. Venturi scrubbers or electrostatic precipitators clean blast furnace flue gas to less than 0.01 g/ft^3. Hooding, Venturi scrubbers, and baghouses can effectively control particulate emissions from sinter plants. Control of slips can greatly reduce blast furnace venting, and hooding and chemical additives are solutions to controlling particulate emissions and sulfur compound emissions from slag quenching. The major solid waste problem in these operations is producing more slag than can be sold for road building.

The major steelmaking processes are the basic oxygen process (BOP), electric furnaces, and open-hearth furnaces, which are now oxygen lanced. The potential is in the fume that is generated from the furnaces and during molten metal transfer operations. Control facilities on BOF and open-hearth furnaces are Venturi scrubbers or electrostatic precipitators. New electric furnace installations use bag houses. Emissions of kish, generated during molten metal transfer operations, can be controlled by evacuation hoods and gas-cleaning equipment.

Recovered materials from BOF emissions represent one of the industry's solid waste problems. High concentrations of zinc, originating in the scrap metal charge, are generally regarded as making recovered material unusable for reuse. Acid leaching recovers zinc.

The principal solid waste problem is in sludge resulting from lime neutralization of spent pickle liquor.

Insofar as coke oven gas, blast furnace gas, flue dust, slag, soluble oils, mill scale, coke oven chemicals, and scrap are concerned, the industry has practiced conservation and reuse because of economic incentives.

IRON AND STEEL SCRAP

Iron and steel scrap in the United States is a fundamental part of the foundry and steel business; see Figure 1. In-plant scrap, recycled in the plant of a steel producer, amounts to 25–30% of the raw steel produced, depending on the product such as structurals, plates, sheets, rods, wire, nails, or coated sheets. Producers report production of raw steel, of which approximately 70–75% is shipped as steel products.

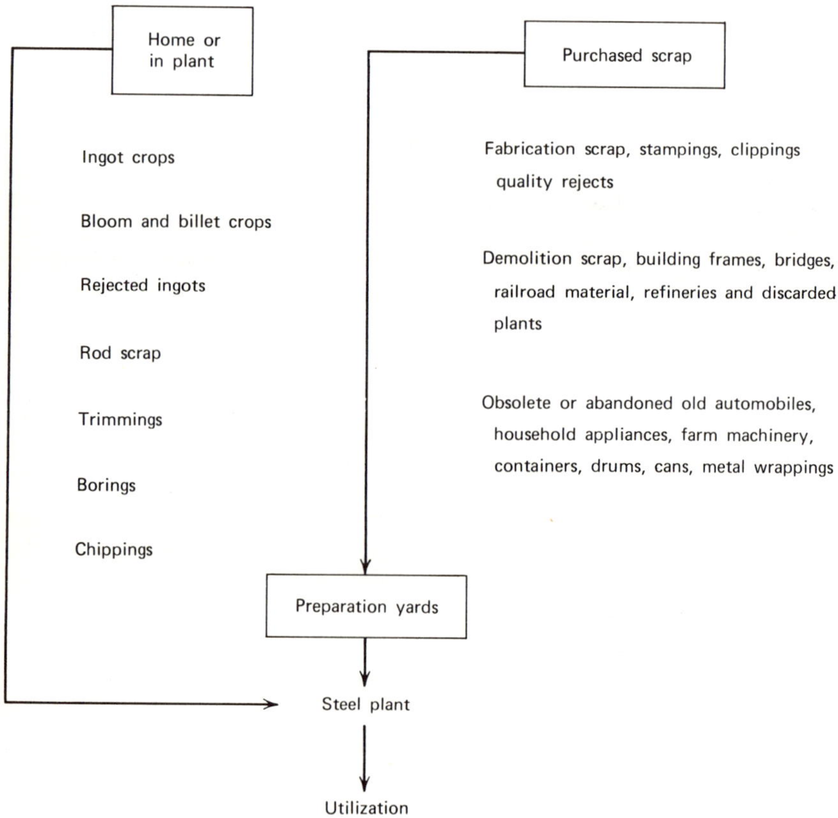

Figure 1. Ferrous scrap.

Over the industry it is estimated that open hearths can use 50% of the charge as scrap, the balance hot metal; BOP furnaces can employ 40%; and electric furnaces can operate on 100% scrap and no hot metal. There are specialty steel producers who have no blast furnaces, no coke ovens and no hot metal facilities, relying entirely on carefully selected scrap, prepared by processors, to make heavy and wide plates, tankage materials, military steel and laminated or clad materials, all meeting exacting specifications.

Others, located in areas with large consumption of reinforcing bars for concrete, are on a 100% scrap basis, not so carefully collected or bought, in that the product encounters easily met simple specifications. These minimills can use scrap that the high-specification integrated producer would be wary of and avoid.

In the 1960s in a typical year the U.S. consumption and export of iron and steel scrap was about 95 to 100 million short tons, of which 5–7% was exported, and 93–95% went to the domestic steel foundry and ferroalloys industry.

Grades of iron and steel scrap are given in Table 1.

The scrap industry is organized to collect scrap, to process it into the physical forms and grades required by consumers, and to handle its purchase and sale.

In 1965 the industry processed or otherwise handled more than 40 million tons of ferrous scrap having a market value approximating $1.2 billion. The Bureau of the Census indicated about 3500 ferrous scrap dealers in the nation. The Institute of Scrap Iron and Steel reported a membership of 1200 ferrous scrap dealers.

Processors sort scrap into grades, cut it into sizes, and press or bale the light-gage material into bundles. The smaller processors frequently do not have the facilities needed to handle certain types of scrap; this is sold to larger yards having the necessary equipment. The processed scrap is sold in carload lots to consuming mills, foundries, or to brokers.

Scrap processors prepare the scrap for sale, and brokers usually handle the sales. Brokers locate and supply quantities of scrap. There are about 200 brokers in the nation, of which about three-fourths are also processors.

Brokers purchase some scrap from railroads, demolition companies, shipbreakers, auto wreckers, and others, but scrap processors are their principal source of supply. Brokers represent processors in discussions for adjustments proposed by mills and are responsible to the mill for supplying scrap that is "suitable and acceptable." The mill determines whether the scrap is suitable and acceptable. Brokers expect a profit of $1 per ton of material handled.

Some auto scrap bypasses both the processors and brokers and is shipped by the wreckers to consumers. Most of these sales are to foundries and consist of cast metal scrap salvaged from motor blocks and the housings of differentials and transmissions. Small foundries and steel companies deal with auto wreckers for No. 2 heavy melting grade automotive scrap, particularly wheels, springs, axles, and gears. However, most of the No. 2 melting scrap and virtually all the automotive sheet was shipped to processors.

Of the total inventory of junk vehicles 6% is in scrap processor's yards. Automotive scrap tonnage was about 15% of total ferrous scrap. Inventories in scrap processors' hands are small in comparison. The inventory turnover was quite rapid; for every vehicle in inventory at the year end, 16 had been processed.

Table 1 *Iron and Steel Scrap Grades*

#1 Heavy melting steel	Wrought iron and steel, at least *4* in. thick, dimensions not to exceed charging box size 60 × 24 in. Grade most preferred
#2 Heavy melting steel	Similar to #1 but at least $\frac{1}{8}$ in. thick, black and galvanized wrought iron. Prepared to ensure compact charging May include prepared automotive scrap Box size 60 × 24 in. or 36 × 18 in.
#1 Busheling	Light steel scrap, not exceeding 12 in. in any dimension May not include old auto body and fender stock or electrical sheet over 0.5% silicon Must be free of metal coated, limed, or vitreous enamel material May include new factory busheling, sheet clippings, stampings, bunchings, etc.
#1 Bundles	New black steel scrap, chemically detinned material, clippings or skeleton scrap from metal working and fabrication, compressed to charging box size to at least 75 lb/cu ft. May not contain old auto body and fender stock, vitreous enamelled sheet, and electrical sheet containing over 0.5% silicon

Table 1 *Continued*

#2 Bundles	Old black and galvanized steel sheet scrap, compressed to charging box size and compressed to not less than 75 lb/ft³. May not contain tin-coated, lead-coated or vitreous enamelled material May contain auto body and fender stock burnt or hand stripped
Bundled #2	Variant of #2 Bundle May consist entirely of auto body, chassis, drive shafts, and bumpers
Scrap automobiles	
Body and fender	#2 bundle
Frame, axles, springs, shock absorbers, wheels, transmission, differential, drive shaft, brake shoes, and steering linkage	#2 heavy melting steel
Motor block, transmission and differential housing, and other parts	Foundry grade iron and steel scrap
Automotive slab	Sheared compressed automobile hulk into slabs of desired dimensions
Steel and iron borings and turnings	Product of machining, drilling, and fabrication operations
Shredded or fragmentation scrap	Small pieces of fragmented auto bodies, and similar materials. Pieces range from ½ to 12 in. in length and width
Foundry grade steel, Foundry grade cast iron	Includes varieties of cupola cast, charging box cast heavy breakable cast, broken motor blocks Consists of crop ends from billets, blooms, bars, forgings, broken car wheels, auto rear ends, crank shafts, front axels, springs, gears, structural shapes and plates

The independent collector has been a mainstay of the scrap industry. In some districts independent collectors obtain automobiles from owners, municipal pounds, and municipal disposal facilities, and deliver them to scrap processors.

In other instances the collector serves as a channel between the auto wrecker and scrap processor. He will take stripped hulks from the wrecker, and deliver them to the processor. In metropolitan areas where open burning is prohibited, the collector often will transport the hulk out of the restricted area and burn it before delivery to the processor. Where long-distance transportation is required, the collector often will flatten hulks.

Scrap prices are negotiated between processors and mills or brokers and steel mills. Prices reflect prevailing demand. Increased hot-metal capacity through the costly development of high-grade iron ore pellets and technological advancements, such as the basic oxygen furnace, which reduced scrap requirements, were reported to be the factors in reducing the steel companies' dependence on scrap. In many cases home scrap and hot metal were supplying the bulk of the metal requirements for the large integrated steel companies.

Prices for the principal grades of ferrous scrap in selected pricing cities are published weekly in *Iron Age* and daily in the *American Metal Market.*

Scrap Yards

Mechanized processing yards contain equipment valued from several hundred thousand to several million dollars. A scrap processor with capacity of 50,000 tons of scrap annually would require: An overhead crane, a press (baler), three shears, a scale, four trucks, a truck crane, plus assorted cutting tools. Handling, cutting, or compressing is arranged so that scrap can be processed with a minimum of handling. The larger yards have a network of buried pipes providing natural gas, acetylene, oxygen, and water. Electric power cables parallel the pipes.

Figure 2. Auto Collector (Courtesy AISI).

Figure 3. Vertical flattener—mobile or stationary (Courtesy AISI).

Underground placement of pipes and electrical wires is practiced to provide unobstructed movement of trucks, cranes, and other material handling equipment. Since the prepared scrap is shipped by rail, railroad sidings are part of all large yards for conveniently spotting cars for loading and unloading scrap.

Incoming scrap is weighed (to determine payment) and then is unloaded by the processors' equipment at storage areas. The processors' objective is to prepare, segregate, and move it out. Mixed loads are segregated by hand.

Some yards maintain laboratories to determine the composition of grades. Some scrap is moved to an area where it is torch cut to manageable size or where undesirable material is removed. Heavier grades such as No. 1 and No. 2 heavy melting scrap are cut by acetylene torches directly for shipment. Other scrap is moved to shears, balers, presses, and the like, so that a continuous flow is maintained. The same crane will move a load of scrap through the processing operations up to loading the scrap into railroad cars or onto stockpiles.

The auto collector is self-powered, equipped with cranes, high-sided and low-slung. They can grapple an old car aloft and deposit it in the huge truck body (see Figure 2). They deliver their cargo to a salvage yard or a mobile flattener, which crushes an auto into a long, flat parcel about a foot thick (see Figure 3). The flattener has an assistant, a front-end loader that lifts the hulk into the machine and removes it after the operation (see Figure 4). They hold the chassis firm with their lifting forks while their hydraulic boom pulls out the engine block from its mounts. A gravity pile driver (see Figure 5) drops its ball or hydraulic ram to break up the block for foundry scrap.

The flatteners compact hulks by bearing down on one after another with a screw press motion or open their heavy hinged doors to swallow the hulk head on, rolling or pressing it into tombstone shape.

Another type of flattener or crusher draws wheelless hulks into its jaws and nibbles them into smaller flat slabs. Twenty of these are a

Figure 4. Front-end loader with engine puller (Courtesy AISI).

payload for a flat-bed truck and one hundred, secured by chains, can ride a railroad car. The scrap processor may receive an auto without tires, wheels, battery, radiator, glass, and upholstery, which he picks up by a yard magnet and compacts in a baling press.

Cranes, such as mobile diesel-electric cranes and overhead electric cranes, are essential in conjunction with balers or presses.

Hydraulic balers or presses process light-gage ferrous scrap, including auto bodies and are designed to reduce the mass to a compact package. Balers are long bins equipped with a removable lid and hydraulic rams. Scrap is charged to the bin, the lid is closed, and the

Figure 5. Block buster—scrap breaker (Courtesy AISI).

Figure 6. Yard baler (Courtesy AISI).

scrap is compressed. Depending on size, a baler can cost from $40,000 to several hundred thousand dollars. Portable balers have been developed to process light-gage scrap, primarily auto bodies, in areas distant from a scrap market. The dense compressed bundle could be economically shipped longer distances than the bulky, light-weight unprocessed sheet metal. Portable car flatteners serve a somewhat similar purpose. Yard balers are shown in Figure 6.

One of the principal equipment types is the shear. Alligator shears consist of a mobile sharp-bladed upper jaw cutting against a stationary sharp-bladed lower jaw. The guillotine shears are equipped with a sharp blade under great pressure and cut scrap forced by ram from a charging box.

The shear tackles the slabs from the flattener. One variety can open and snap with 8 ft wide and 3 ft high jaws, while another can sweep its 9-ft upper jaw with a cutting force of 400 tons, seven times a minute. They make chunks of scrap (see Figures 7 and 8).

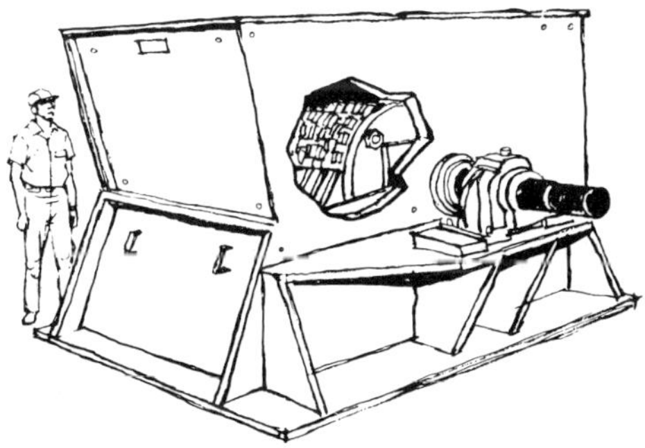

Figure 7. Hammermill shredder (Courtesy AISI).

Figure 8. Guillotine shear (Courtesy AISI).

The newest adaptations of machinery from other industries, coupled with intensive metallurgical development of hard tough surfaces, are the shredders, fragmentizers, and nodularizers, which slash, rend, batter, and pummel the hulk into fragments. They must be continuously fed by crane or conveyor (see Figures 9, 10, 11).

One type has rolls of sharp teeth followed by slowly revolving sets of teeth. The carcass is ripped to shreds. Another type is a huge heavy version of the swing hammer mill, where the hammers batter against cutting edges and shred the metal. Another type is the fragmentizer with its rows and ranks of cutting bars. Fist-sized fragments are produced.

The large shredders are reported to have capacities to process in excess of 150,000 tons/yr and cost upwards of $2 million. Smaller shredders having capacities of 20,000 tons or more per year are reported to cost upwards of $300,000. In general, the flow sheet of the

Figure 9. Alligator shear (Courtesy AISI).

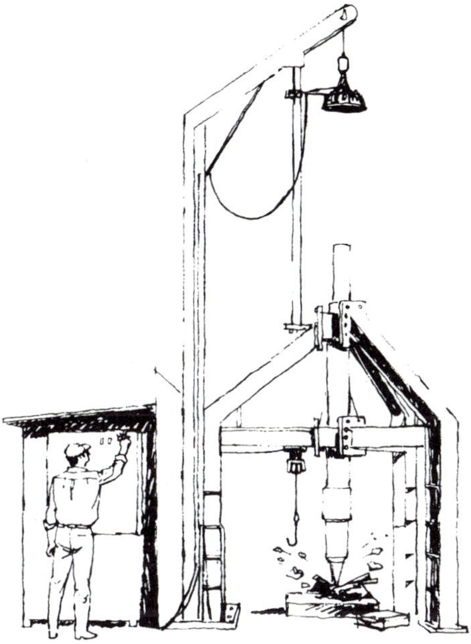

Figure 10. Hammer-cutter, combination shredder (Courtesy AISI).

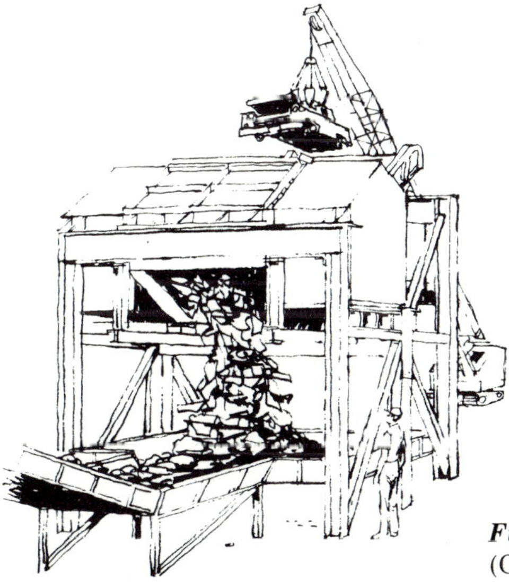

Figure 11. Ripper shredder (Courtesy AISI).

shredding operation entails the movement of vehicle hulks (stripped of motor blocks, transmissions, tires, gas tank, and radiator) up a conveyor into a modified hammer mill where the metal is hammered and torn into small fragments (½–12 in.). The ferrous fragments then are magnetically separated from the nonferrous and nonmetallic material. In one process, the fragments go through a double magnetic separation, are heated, and individually compacted. Several companies have developed shredding machines. The processes vary somewhat but involve cutting, tearing, or hammermilling auto hulks into fragments with separation of undesirable elements.

The greatest promise for shredded scrap is in electric furnaces and cupolas. The small size and low bulk density is an advantage in electric furnaces reducing electrode breakage, and shielding refractory linings. The size makes it suitable for small electric furnaces and foundry cupolas where No. 2 bundles could not be used. The automobile manufacturers made a thorough appraisal of shredded scrap for use in their foundries.

Shipment of Prepared Scrap

Most processed scrap is transported by rail, barge, or ship. The transportation costs of materials competing with scrap, such as pig iron, iron ore, and iron pellets also have an effect on the movement of scrap.

Most rail shipments are in open gondolas. Borings and turnings are shipped in hopper cars from which the scrap is unloaded from beneath and moved to the mill or foundry by conveyor belt. There has been a reduction in numbers of railroad cars in service, because worn-out rolling stock has been junked faster than new cars have been built to replace it.

Freight rates for shipments of large tonnages of scrap iron and steel are usually less than for small shipments. For example, the railroad rate from Detroit, Mich. to East St. Louis, Ill., is about 20% lower for a 112,000-lb car than for an 80,000-lb car. If the shipper agrees to a 500-gross-ton single shipment, the rate would be lower by another 20%. Likewise, the rates per gross ton for single-car shipments of 75,000 lb or even 112,000 lb, from Kansas City to Chicago, can be markedly lowered for a single shipment of 500 gross tons.

A considerable percentage of scrap is moved by barge throughout the Midwestern, Eastern, Gulf Coast, and Southern States. Barge movement affords economies and provides for a lowered delivered cost. It is often necessary to unload barges at the site of delivery and transfer the scrap to in-plant rail systems for final placement.

Barges are common carriers; hence barge rates on interstate shipments are subject to approval. For intrastate barge rates, the public utilities commission (or equivalent) takes jurisdiction. Upon approval by the appropriate commission, barge companies publish tariffs, at which time the rates become effective.

Even brief study of automotive scrap quickly leads one to realization that many of the crucial problems are in the markets. The automobile is only one of many sources of ferrous scrap. Ferrous scrap consumers purchase from any source, and at the lowest price possible, any scrap of a quality that will meet their specifications. Often, for quality and price reasons, automotive scrap fails to compete successfully.

A number of excellent reports have been published in recent years describing, in considerable detail, technical and economic aspects of the processes and industries involved in ferrous scrap generation, processing, marketing, and consumption. For detailed information on the technical and economic features of the markets for scrap iron and steel the reader is referred to the publication "Iron and Steel Scrap Consumption Problems," published by the USDI, and other publications listed in the references.

In 1965 the domestic consumption and exports of iron and steel scrap totaled 96.6 million short tons.

Ferrous Scrap Markets

The principal markets for ferrous scrap are in the United States. In 1965, consumption of ferrous scrap in domestic markets totaled 90,359, 335 tons. Of this, 55,213,000 short tons was home scrap and 41,239,000 short tons was purchased scrap.

Largest of all consumers of ferrous scrap is the steel industry of the United States. Steelmaking furnaces consume more than 68,000,000 tons of scrap annually. The American Iron and Steel Institute reported production of the order of 129,000,000 tons of steel and total revenue of $18,000,000,000.

The amount of scrap used depends on hot-metal production capacity. Integrated companies producing hot metal in blast furnaces prefer hot metal for quality and cost stability. This preference has been strengthened by improvements in blast furnace operating methods, and the use of high-grade ores and pellets. These have reduced hot metal costs and increased the volume of output.

After hot metal, integrated steel plants prefer the home scrap generated in their own operations. Typically, integrated plants have relatively little need for purchased scrap except in times of peak production when their hot metal and home scrap capacities are exceeded.

The nonintegrated steel plants consume a higher proportion of scrap (in open hearths, 80% compared with about 39% for integrated plants).[5]

Home scrap and prompt industrial scrap, both of known composition, are preferred by mills for quality steel. Obsolete scrap varies widely in composition, and some is contaminated with copper and other nonferrous metals. Some contaminants can be removed by oxidation, vaporization, or a suitable slag. Copper and nickel are very difficult to remove.

One of the first considerations for any steel producer when purchasing scrap is the nature of his finished products and specifications that must be met. When a finished product must be low in copper or other residual elements, the scrap used must be comparably low in these contaminants.

All purchased scrap, however, need not be of the highest quality because it is possible to blend lower grade scrap with pig iron or hot metal and high grades of scrap and still meet specifications. The extent to which combination of hot metal and low or high-quality scrap is used becomes a matter of economics and the attitudes of the operating personnel who are responsible for the steelmaking operation. When it is technically possible to use No. 2 bundles or other low grades of scrap, price and quality considered, many companies do so because it is more economical. The proportions of the different grades of scrap used in many instances are based on the copper content of the scrap, copper being one of the most objectionable residuals for high-quality steel.

Steel ingot producers in the United States indicated that their average metal yield from No. 2 bundles was between 85 and 90% and the copper content between 0.2 and 0.1% when the bundles were purchased under firm contract and inspection system. In contrast, ordinary No. 2 bundles had a copper content of 0.4–0.5% and a yield of 80–87%. Prices paid for bundles ranged from $22 to $30 per net ton, depending on the individual markets. At this yield, residual, and price, the better quality No. 2 bundle could be used extensively and was competitive with other grades of scrap.

The shredding of scrap is another solution to the problem of contaminants. Consumers of shredded scrap indicated that in some instances the scrap averaged between 0.20 and 0.25% copper, but that with careful processing, including hand-picking of contaminants, an average of 0.15% can be maintained. In an electric furnace operation, a steel yield of almost 97% has been reported. The bulk density of shredded scrap was lower than that of other competitive grades.

The physical form in which scrap is used in iron and steel furnaces varies according to the size of the charging box and the opening through which the charge is fed into the furnace. Blast furnaces can accommodate only small pieces of about a foot in length because of the small opening between the charge bell and the hopper at the top of the furnace. The open-hearth and large top-charge electric furnace and oxygen converters have large enough openings to permit charging boxes which can handle 5-ft-long bundles or individual pieces of scrap. In the case of the cupola, individual pieces of iron and steel scrap must be sized so that no single dimension exceeds one-third of the cupola throat diameter.

The foundry industry is an attractive market for automotive scrap. Ferrous foundries are much more numerous and widely distributed than are steel plants. The American Foundrymen's Society has compiled a list of 2240 ferrous foundries in the nation. The Society has estimated that the total output of molten metal for all foundries approximates 35 million tons and that foundries might use an additional 5–7 million tons of automotive scrap per year.

The foundryman who may be selling his castings from $500 to $1000 per ton must have confidence in the quality of the ferrous raw materials he is melting. He uses home scrap suitable for recharging, pig iron from blast furnace operators, cast iron scrap of specified quality, and selected steel scrap.[9]

The products of the foundry industry have a wide variety of compositional and physical properties. To meet these requirements the individual foundry operator obtains the necessary blend of raw materials from the most economical sources available.

As a contribution to this study, the American Foundrymen's Society surveyed several branches of the foundry industry to establish the limits of various elements that can be tolerated. The information obtained for each of the ferrous branches of the industry are shown in Table 2.

Foreign markets served by export are sizeable markets for scrap. They do not recycle as far as the United States is concerned except that some of the scrap comes back in the form of imported foreign cars.

Automotive Scrap—A Special Problem

In a half-century of growth and adaptation, the automobile has transformed many aspects of life in the United States, bringing many benefits and a number of problems, not the least of which is the problem of disposal when the automobile is no longer useful as a transportation vehicle.

Table 2 *Composition Tolerances for Tramp Elements in Ferrous Foundry Metal (Maximum Percentage)*

			Gray Iron		Ductile Iron	
Element	Steel	Malleable	½-in. and up	½-in. and down	Ferritic	Pearlitic
Aluminum	—	0.01	0.01	0.01	0.02	0.02
Antimony	0.02	0.01	0.03	0.03	0.002	0.03
Arsenic	—	0.01	0.10	0.05	0.02	0.10
Bismuth	0.02	0.02	0.001	0.002	0.001	0.03
Boron	0.10	0.003	0.05	0.10	0.005	0.05
Cadmium	—	—	0.005	0.005	0.005	0.005
Chromium	1.00	0.05	0.50	0.20	0.05	0.10
Cobalt	—	—	3.00	3.00	0.50	3.00
Copper	0.50	0.15	1.50	1.50	0.15	1.50
Lead	0.020	0.02	0.001	0.004	0.002	0.004
Manganese	No limit	0.40	1.00	1.00	0.25	0.50
Molybdenum	No limit[a]	0.02	1.00	0.25	0.05	0.50
Nickel	No limit[a]	0.15	3.00	3.00	0.50	3.00
Phosphorus	0.05	0.05	0.20	0.20	0.05	0.08
Selenium	0.10	0.001	0.06	0.04	0.03	0.03
Sulfur	0.05	0.05	0.12	0.10	0.04	0.04
Tellurium	0.10	0.001	0.001	0.001	0.01	0.01
Tin	0.020	0.01	0.15	0.10	0.02	0.10
Titanium	—	—	0.05	0.05	0.05	0.05
Tungsten	0.10	—	—	—	—	—
Vanadium	No limit	0.01	0.50	0.10	0.02	0.10
Zinc	—	0.01	0.20	0.10	0.10	0.20

[a] But range should be shown.

Ultimately, nearly all automobiles reach a point at which they become disposal problems. Auto hulks, because of their durability, are difficult to conceal. Abandoned or neglected, they soon become eyesores detracting from the appearance of urban neighborhoods or rural scenery. Accumulated in public dumps or auto graveyards, they are less widely dispersed.

To the individual owner the inoperative automobile presents a sometimes annoying disposal problem. The cars junked annually provide a source of millions of tons of metals for remelting. The scrap metal industry has been processing and selling automotive scrap for many years.

In the 1960s the rate at which cars were being junked had become so great that the esthetic and utilization problems became matters of public concern and revealed a need for additional basic information. The Department of Commerce concluded that (1) "There is no basic problem with respect to the movement of originated, acceptable iron and steel scrap through trade channels into consumption by steel plants and foundries in this country and abroad," and (2) "There is, however, a problem with respect to using the full amount of originated scrap of limited acceptability; that is, No. 2 bundles and similar scrap containing undesirable quantities of nonferrous materials."

To provide needed information, the Bureau of Mines was authorized to make a fact-finding survey of the auto wrecking industry, the ferrous scrap processing industry, and other elements in the automotive scrap situation. Consequently, a sample survey was designed. A total of 54 districts was selected to represent a variety of urban, suburban, and rural conditions throughout the United States.

A junk automobile ordinarily goes through some degree of dismantling in being reduced to scrap. Some parts may be salvaged for their value as used parts. The battery may be sold for its lead content. The radiator, electric motors, and other parts containing substantial quantities of copper are removed and marketed for their copper content. Some copper, such as in wire and small electrical parts, is often left in, however, when the labor cost of removal exceeds the copper value. The motor block and other cast iron parts commonly are removed and sold as foundry cast. The heavier steel components, including the frame, may be removed and sold as No. 2 heavy melting scrap. After the stripping procedure the remaining hulk may weigh less than 1000 lb and seldom more than 2000 lb. Therefore, between one-third and two-thirds of the metal content of the automobile usually finds a market as relatively high-grade scrap. The principal problems are encountered in marketing or disposing of the stripped hulk.

When the combustible and other nonmetallic materials have been burned or stripped, the hulk may be baled, sheared, or shredded to produce scrap in the physical form specified by the consumer.

Consumers of No. 2 bundles had difficulty with them in the past and felt the quality hazards did not justify their use. Other consumers have found that careful control of stripping and baling assured them of a copper content of 0.2% or less, and a metal yield of 85–90% at prices that were very advantageous. No. 2 bundles, which may contain 0.4–0.5% or more copper, can be used in making reinforcing bars.

Shredders have been successful in processing automotive scrap in several areas. Consumers have been willing to pay premium prices.

Two ceilings that serve as limiting factors in the scrap price structure are (1) the cost of producing hot metal from iron ore in integrated steel operations and (2) the cost of pig iron to foundries. Automotive scrap must be offered at competitive prices and qualities.

Auto Wreckers

There are at least 10,000 auto wreckers operating in the United States and the gross income of the industry is estimated to exceed $500 million. Because of the low cost of entering the auto wrecking business, there are a large number of small operators, including those who engage in auto wrecking as a sideline of service station or repair garage operation, and those in intermittent employment.

The auto wrecking industry is the primary market for junk cars. When the industry is operating effectively, junk vehicles flow in and out of the wrecking yards at virtually the same rate; parts are removed and the residue, the stripped vehicle hulk, moves forward to the scrap processor.

Auto wreckers usually operate by one of three methods: (1) Park the vehicles in yards and strip parts as they are required for sale; (2) strip the vehicles to the bare hulk, place parts in storage or immediately sell them to rebuilders or wholesale outlets, then dispose of the stripped hulks; and (3) park the vehicles in yards and let the customer remove the part he desires (the self-service yard). Economic factors, such as land values, local demand for parts, inventory tax, storage space, and community pressures, influence the choice of operating method. Location and size of the yard are major factors governing operating procedures. The cost of land usually is dependent on land utilization in the surrounding area. Zoning ordinances often control the establishment of new yards and even the expansion of existing yards. Land in rural areas is inexpensive and easily acquired and has few restrictions pertaining to use.

Individual owners sell, give, or sometimes pay an auto wrecker to take a junk car. The transaction depends on the auto wrecker's appraisal of the value of the car for parts and on the prevailing prices of automotive scrap. Many wreckers are reluctant to take old vehicles that have no parts value. Unclaimed impounded cars sold at auctions usually are purchased by auto wreckers, but in some areas they refuse to bid if the cars have little parts value. When vehicles have value as scrap, but open burning is restricted in an area, wreckers often are not interested in acquiring them. Wreckers report that the cost of hand-stripping vehicles to prepare them for scrap processors often is prohibitive. New and used car agencies accumulate vehicles in trade that are of little resale value. These cars are usually sold individually or in groups to auto wreckers.

Most wrecked vehicles are acquired directly from insurance companies or from salvage pools operated cooperatively by a group of insurance companies.

Because parts from late models are in great demand by repair and body shops throughout the country, high prices are paid for wrecked late-model vehicles.

Auto wreckers almost always remove the motor from a junk vehicle for use as a second-hand engine or for sale as scrap. Ferrous motor blocks find a market as foundry scrap. The housings for the transmission and differential also are made of cast iron. The total cast iron weighs about 360 lb for a small car and over 450 lb for larger autos. Because of the strong market for cast metal scrap, this material sometimes is marketed directly to the consumer by the auto wrecker. Metal yields are given in Table 3.

In some areas processors will accept nothing but completely stripped and burned hulks; in others they accept whole vehicles with all the parts intact.

To meet the quality standards of the steel industry for No. 2 bundles, wreckers and processors must remove glass and other nonmetallics and nonferrous metals as completely as possible. The least expensive method of removing most of the combustible material from vehicle hulks is by burning the upholstery fabric, plastics, rubber, grease, undercoating, fiber board, felt, and insulation on wiring. Enclosed burning in incinerators or handstripping the combustible material are practiced.

Transportation is an important element in vehicle disposition costs.

Table 3 *Metal Yield—Auto Wrecking*[a]

	Buick Hulk (lb)	Chevrolet Hulk (lb)
Ferrous metal	3705	2726
Copper and copper alloys	52	35
Lead	24	21
Zinc alloys	123	41
Aluminum alloys	48	8

[a] Bureau of Mines, Salt Lake City Metallurgy Research Center.

The first stage in the movement usually is to an auto wrecker. Junk vehicles often are transported by the auto wrecker by tow truck, one at a time. Some large operators that acquire junk vehicles in groups use auto transport trailers and haul loads of six or seven vehicles.

Hulks are delivered from auto wreckers to processors by truck. Not flattened, a truckload might consist of one to seven hulks. If flattened, trucks can accommodate as many as 20 or even 30 vehicles. Transportation costs range between 4 and 10 cents per ton-mile, when truckloads consist of a large number of hulks shipped long distances. When hulks are trucked in smaller numbers for shorter distances delivery costs of 40 cents to $1 per ton-mile are characteristic.

The distance hulks are transported depends on such factors as the prevailing prices of scrap, availability of flatteners, transportation rates, and the existence of price allowances for long-distance shipments. Long-distance (200 miles or more) hulk shipments are made to processors in the Pueblo, Colorado, Houston, Texas, and Los Angeles areas. On the other hand, in Boise, Idaho, and Coos County, New Hampshire, there was little or no movement of stripped hulks to market.

The scrap industry collects ferrous scrap, processes it into the grades required by consumers, and handles its purchase and sale. In 1965 the industry processed or otherwise handled more than 40 million tons of ferrous scrap having a market value approximating $1.2 billion. The American Iron and Steel Institute estimates that an unstripped auto weighing 3600 lb may yield 2500 lb of steel, 500 lb cast iron, 54 lb zinc, 51 lb aluminum, 32 lb copper, 20 lb lead, and 400 lb of nonmetallics.

CHAPTER VI.4

mining the automotive graveyards

Auto graveyards are tracts of land containing accumulations of junk vehicles that are not apparently destined for use as sources of parts or for scrap. These vehicles are not in the industrial flow—that is, the flow from an auto wrecker to a scrap processor and finally to a steel plant.

Auto wreckers were a contributory factor in the creation of many graveyards. Some of the large graveyards were the sites of auto wreckers who had gone out of business. These graveyards were found most frequently in rural areas. The owners reported that there were insufficient sales for either body or mechanical parts. Some attributed low sales volume to small unlicensed operators underselling them, or to the theft of salable parts from their unfenced inventory. Some auto wreckers having small yards in urban areas with limited space acquired land in rural areas for storing vehicles. The vehicles were to be used in their auto wrecking yards when space became available. The urban auto wreckers used the rural land to store stripped hulks arising from their auto wrecking. There was no local demand for vehicles hulks, and the auto wrecker was forced to remove stripped hulks from his yard to make room for vehicles with parts value.

Graveyards were maintained by new or used car dealers to store vehicles accepted in trade. Usually auto wreckers in those areas would not accept vehicles that were badly rusted or which had limited value for parts, and processors would not accept unstripped vehicle hulks. The dealers had no alternative except to either lease or purchase land in a rural area to store the vehicles, or to get into the auto wrecking business. Generally these cars were destined to sit and rust. Car dealers in such areas revealed increasing reluctance to accept old rusted vehicles as trades. This gave rise to another graveyard. This type was started by individuals having old cars in poor condition who could not get a car dealer to accept it in trade. Since auto wreckers would not take these cars, the car owner in many instances would drive the vehicle to a rural area and abandon it, most likely on a farm property. Before abandoning the vehicle, the owner would remove license plates and other identifying numbers. This vehicle sometimes served as the nucleus for an accumulation of similarly abandoned vehicles, and the eventual result would be a large graveyard.

After they had served a purpose, the partially stripped inoperative vehicles were permitted to accumulate. The location and size of the accumulation depend on land availability, ordinances, and community action. Similar accumulation of vehicles occurred in the vicinity of service stations and repair garages.

Inoperative vehicles were most commonly found in small groups. Private residences were characterized by inoperative vehicles standing in their yard. These were frequently in pairs or in groups of three or four. These were old-model vehicles that had been purchased second hand and after a major breakdown, and were permitted to stand on the owners' property. The owners rationalized that some day they might repair these cars, or if they purchased a similar model they would have a source of parts. Nothing was done with these cars. Auto wreckers who would handle derelict cars for their scrap value said the public had no concept of the scrap value of a vehicle and would not part with their vehicles except for a price higher than an auto wrecker would pay.

Auto graveyards served the purpose of localizing derelict vehicles that ordinarily would have been scattered over the countryside. Those graveyards in rural areas with large accumulations of vehicles were attractive to portable bailing operators.

Graveyards were not characterized by orderly storage of vehicles. The premises of a graveyard were unguarded and unfenced. Through vandalism and theft, the stored vehicles would be cannibalized of parts and have most of their window glass broken. Usually the cannibalized vehicle hulks were scattered haphazardly, some turned upside down, some on their sides. The graveyards maintained by car dealers usually were the best organized.

The AISI states than annually almost nine million cars and tank trucks begin an average of 5½–6 yr of service; about an equal number change hands out of used car lots. Meanwhile, eight million vehicles became too disabled for further service. The dropouts became the treasure trove of the auto wrecker or are gathered up by salvage dealers. Finally after being cannabalized of everything serviceable, the unsold crumpled residues are candidates for recycling or end up in a graveyard or dump.

Dragged to a remote roadside, the hulks escape and the annual crop of derelicts joins an older population of its fellows. The castoffs cluster behind outbuildings, barns, side yards, ditches, and bottoms of hillsides as a myriad of mini junk yards. They present a problem of collection for recycling.

Estimates run as high as 20 million wrecks in auto wreck shops, scrap processing plants, and in graveyards across the nation.

Wrestling hulks out of their hiding places has become a problem of logistics and economics, spawning a new generation of machinery.

There are road blocks that retard the movement of abandoned cars into the scrap flow. There is little understanding of the high cost of correcting abandonment. In cities, there is time-consuming towing, title searching and investigation, impounding, and final transfer to the scrapyard. All this must be paid for by tax dollars.

In a General Motors Corporation report, "Progress in Areas of Public Concern" (Milford, Mich., February 1971), Philip West discussed abandoned cars. He presented a table of the distribution of junk car accumulation, given here as Table 1.

It thus appears that abandoned and derelict vehicles as well as those in graveyards, according to General Motors, amount to 4 million abandoned automobiles, not 20 million junk cars.

General Motors wanted to find out what kind of problems a community would encounter if it initiated an abandoned car cleanup campaign. Traverse City, a resort area in Michigan, was far enough away from major scrap producing areas so that economics became important.

Quoting West, after General Motors offered free removal of all derelict and abandoned cars in a campaign which had strong public support:

> When our test was concluded, we had removed almost 2,500 cars from an area within a 20-mile radius around Traverse City. Actually only 101 of these cars were truly abandoned—that is, discarded on public or private property without the consent of the land owner. The remainder were what we technically call derelict vehicles.

Table 1 *Distribution of Junk Car Accumulation*

	Urban (%)	Rural (%)	Total (%)
Auto wrecker	28	45	73
Abandoned and derelict	6	12	18
Scrap processors	3.5	2.5	6
Graveyard	0.7	2.3	3

Although we did not plan it this way, we found we had done a better job than we realized in helping the cleanup of Traverse City. Not only did we take out 6,000 tons of junk cars and 57 tons of engines and parts, but because they were there and we couldn't ignore them, we also removed 100 tons of abandoned refrigerators, stoves, and other appliances, as well as 148 tons of miscellaneous metals ranging from barbed wire to old bedsprings. We even took out 25 tons of tin cans.

Of course, what we really acquired in Traverse City was a thorough education in how to harvest abandoned cars. For example, we learned that to many people the old derelict rusting in the back yard is not junk—it's valuable to them and they want to keep it. Legally, they have every right to do so.

If the community really wants to get rid of these privately owned derelicts, it may have to consider legislation that would penalize car owners who don't get rid of these old vehicles in a reasonable length of time. Our conclusion: Anyone contemplating an abandoned car harvest should first learn all they can about the titling laws and local ordinances affecting the ownership and disposal of vehicles. We were fortunate. Michigan is one of the states that has modern titling laws affecting abandoned car disposal. We briefed ourselves thoroughly on the law before we started the Traverse City Project. I'm sure that spared us some embarrassing incidents.

We also learned that collecting abandoned cars isn't a simple task. You don't just go out with some volunteer help, and over a few weekends, drag in all the derelicts you can find. Sound planning and organization are needed. We are convinced that with good planning, and with the cooperation of its citizens, a community can indeed collect all of its legally abandoned cars and it can do so effectively.

An ingenious approach was employed in Columbia County, near Albany, New York. The county health department declared that abandoned and derelict cars were breeding grounds for rats. Funds from a rodent control program administered by the New York State Department of Health were used for a junk and abandoned car drive. A portable crusher was used to smash the hulks hauled in by volunteer help, netting 12,000 derelicts at a cost of $20,000 net, or $1.66 per car collected, after the scrap was sold.

Schoharie County, New York used funds and trained some unemployed in the stripping of derelict vehicles, and ran a scrap cleanup program.

Maryland encourages the movement of old cars by paying a bounty of $10.00 to the scrap processor for each car recycled. State law also discourages an owner from abandoning his old car by fining him $200.00.

Only by involvement of the cities, the states, the wreckers, scrap processors, steel industry, and the individual auto owners will the chain of recycling be securely forged.

CHAPTER VI.5

nonferrous scrap

The reclamation and utilization of nonferrous scrap metals is an old and well-established industry engaged in a multimillion-dollar business annually. Thousands of dealers, smelters, and refiners have been strategically located throughout the United States to collect and process scrap metals and metal-bearing products. A list of such companies is presented in editions of *Metal Statistics,* and illustrates the geographical distribution of firms and the kinds of scrap materials with which those firms deal. Several of the companies listed also operated mines, smelters, and refineries internationally, and some had more than one plant in the United States for the treatment of nonferrous scrap metals. Export of scrap is an important part of the business. The *Waste Trade Directory* contains lists of scrap-metal processors arranged geographically and according to kinds of metals they are equipped to process.

The large number of dealers and processors for gathering, preparing, smelting, refining, and marketing scrap metals ensures competition and promotes efficiency in reclamation and utilization.

Large tonnages of new scrap metals generated in manufacturing operations are transferred to secondary processors to produce marketable metals.

The operations to produce marketable metals are complex, and they require metallurgical skills, expensive facilities, and knowledge of marketing. They must be conducted on a large scale to be competitive.

The recovery and refining operations are much simpler and hence less expensive when different types of materials are treated separately instead of as mixtures.

The USBM compiles statistics on the recovery in the United States of the major nonferrous scrap metals. These data may be found in *Minerals Yearbook*. The industry has differentiated between "primary" production, deriving metals from ores and concentrates, and

Table 1 *Nonferrous Scrap Recovered in the United States*

Metal	Approx. Tonnage per Year	Percent of Total Consumption	Percent of Mine and Smelter Production
Aluminum	700,000	30–45	33
Antimonial lead	22,300		
Secondary	20,800		
Primary	800		
Copper and brass	1,500,000	45	100
Lead	700,000	52	115
Magnesium	10,000	10	10
Mercury	900	Highly variable	
Nickel	25,000		131
Tin and solder	23,500	30	
Zinc	225,000	20	50

"secondary" production, deriving them from scrap metals and industrial by-products. "New scrap" designates cuttings, turnings, and materials generated during the fabrication of equipment and merchandise. "Old scrap" refers to obsolete equipment, piping, wire, and other materials reclaimed in dismantling buildings, ships, and the like. Typical of these are the listings in Tables 1 and 2.

The referenced data on the quantities of industrial scrap metals showed a number of classifications for the materials processed and the products obtained. Actually, there are many more classifications of nonferrous scrap metals. The National Association of Secondary Material Industries, Inc., has issued Circular NF-66, "Standard Classification for Nonferrous Scrap Metals," which contains standard classifications for 119 types of nonferrous scrap metals.

Precious metals have been recovered from a variety of materials, including photographic film and solutions, no longer used coinage, discarded articles and sweepings; see Table 5.

Four categories of nonferrous scrap materials have been particularly important because of the large tonnages involved. Scrap aluminum, copper, lead, and zinc are sufficiently important industrially to warrant daily publication of prices.

Most of the nonferrous scrap metals returned to industry have been obtained from industrial plants that generally have not processed their scrap metals for reuse. Plants that generate large amounts of scrap metals may have contracts with wholesale dealers for the collection of those metals. The remainder of scrap metals has been brought to dealers by collectors from sources such as small shops and municipal refuse collection. The dismantling of obsolete machinery and buildings has also provided a source.

Scrap metals have been sold by dealers to wholesalers and to smelters. The value of the scrap and whether a specific kind of scrap will be sent to a primary or a secondary smelter depend on the proximity of the smelters, together with other considerations. The most important factor is the composition of the scrap in desirable metals and undesirable metals or foreign materials. The composition influences the expenses for transporting, sorting, sampling, preliminary treating, smelting, and refining.

Nonferrous scrap materials frequently require treatments to prepare

Table 2 *Partial List of Precious Metal Scrap*

Turnings, chips, shavings.
Silver on steel bearings.
Silver steel turnings.
Grindings.
Blanking scrap, stampings, strip, and wire.
Powder mixtures.
Screen scrap, catalysts
Solder scrap (silver solder)
Brazing alloy scrap (silver alloy)
Electric wiring material containing precious metals
Bimetal scrap.
Silver-paint waste, wipe rags, paper, and cans.
Plating solutions, gold & platinum, and plated materials.
Filter pads.
Anode ends.
Tank scrapings.
Hypo solutions.
X-ray film.
Coated plastics, ceramics, glass, mica, quartz, etc.
Chemicals.
Mirror solutions ($NaNO_3$).
Platinum-bearing material.
Gold on molybdenum or tungsten wire.
Jewelry and dental scrap.

them for melting, smelting, and refining operations. Turnings, borings, and punchings of new metals may be subjected to magnetic separation to remove iron or steel fragments. In some instances, degreasing has been practiced to overcome the adherence of the ferrous contaminants to the turnings and borings. Stripping machines or manual methods have been used to remove lead sheathing and insulation from electrical cables. Materials that are not entirely metallic, such as sweepings, drosses, ashes, and slags, may be ground to liberate metallic prills, drippings, and grindings. The liberated metallic particles can then be recovered.

If the source of a particular lot of scrap metal is known, the lot can be sold without analysis. Mixed materials of uncertain origin must be sampled and analyzed to provide a basis for payment. These operations add to the expense of the transactions, but they can be minimized or avoided by careful segregation or sorting of scrap metals.

Scrap metals acquired by dealers have been sorted piece by piece by workman who have attained skill in identifying the types of metals and alloys encountered. The sorting process has frequently been repeated at the smelting plant. The surest and simplest means for identifying scrap is by knowledge of its use or origin. Thus, plumbing fixtures, copper wire, automobile radiators, and lead battery plates are easily distinguished, even by only slightly experienced people. Less obvious metallic articles may be subjected to tests depending on specific physical and chemical properties of metals and alloys.

Bulky materials (e.g., wire, turnings, thin plates, sheets, and radiators) have been baled by compression in hydraulic presses to facilitate handling and charging the materials to furnaces. Compaction of small pieces also promotes heat transfer in the furnaces and minimizes losses of metals by oxidation and by entrainment of metal particles in slags. In contrast to the requirements for massive charge materials in furnacing processes, chemical methods require the large surface areas of particulate materials for the most efficient reaction. Therefore, in contrast to baling or briquetting of fine particles, it may be advantageous to grind or shred materials when chemical processes are employed.

NONFERROUS SCRAP, GENERAL

The data on percentage of each of the metals produced that are recovered and recycled are harder to obtain but generally they are in the same order showing that recovery of the more noble metals, lead and copper, is easier than that of the more reactive and readily oxidized metals, zinc and aluminum.

In copper reclamation there are two grades of consumer-generated secondaries that are metallurgical challenges. These are "insulated wire" and "iron-bearing" material. The first consists of a copper conductor coated with insulation such as paper, rubber, polypropylene, polyethylene, neoprene, polyvinyl chloride, Teflon, or Tygon. The copper in these two-phase conductors can be separated from the insulation by automated mechanical means. The more complex form would be all the electrical components of a wrecked building including panel boards, complex multicomponent cables, B-X cables, and fuse boxes all tied together in a tangled mess or a demolished telephone central exchange, or all forms of complex electronic scrap such as telephones, television sets, radios, and even computer components.

The basic problem is the organic insulation. Mechanical means, such as chopping, grinding, or hot pressing to separate the plastic from the copper conductor, have been devised for stripping a plastic coating from a simple conductor. These procedures become less effective as a cable becomes more complex or if a simple conductor is a part of a scrapped electrical or electronic apparatus. Organic solvents have been proposed but have failed to be a universal solvent for the many types of rubber and plastic.

Burning to remove insulation brings problems as to how to burn completely and how to scrub the products of combustion to free them of pollutants such as hydrochloric and hydrofluoric acids. Plastic coatings break down on heating to form gaseous hydrocarbons and HCl or HF. The burning is characteristic of a dried Christmas tree. The flame and heat is all well above the pile as it burns. Combustion is not complete, so that soot forms.

Burning the organics from a pile of scrap in an open field or a crude incinerator was legislated out of existence. Large consumers of secondaries design and operate a smelting unit that uses as much of the heat of combustion as possible to melt the copper, slag off some impurities, burn the resulting gases to completion in an after-burner, and scrub the products of combustion with a basic solution to neutralize the hydrochloric and hydrofluoric acids. Such a system is expensive and sophisticated. Treatment charges have to be higher than they were when such material could be burned in a smelter yard or smelted in a furnace that expelled unburned carbonaceous material and acid mist to the atmosphere.

The "iron-bearing" secondaries are a mixture of many different types of consumer-generated scrap including iron-backed bearings, motors, generators, fuse boxes, and similar electrical components. The electrical gear all have insulation or enamels as well as a high percentage of steel.

The problem is to expose the copper by finding some means of removing the organic coatings. At one time, scrap dealers could afford to disassemble a motor and remove the armature and other copper containing components from the iron shell. With rising labor costs, this operation is uneconomic. Burning off the organics presents the same pollution problems as with insulated wire. Fused salt treatment of copper coated steel has been suggested. In this procedure, the copper is melted and drops to the bottom of the fused salt container, if no volatile coatings are present, such as varnish, plastic, or rubber insulation. But anyone who drops an unassembled scrap motor or generator with its contained lubricants and insulation into a fused salt bath is in a hazardous job.

Smelting is done in a unit with fume collection, after burning, and gas scrubbing equipment. The high iron content produces large quantities of slag. Decreasing slag losses result when hot slag from a secondary copper blast furnace is sent to an electric arc furnace to fume off volatile impurities such as zinc and lead, and collect copper and nickel as a black copper.

The fumes which come off the blast furnaces and converters treating the smelter grade copper scraps are essentially zinc oxides contaminated with large quantities of lead, tin, iron, copper chlorides, and fluorides. These are caught in baghouses.

A similar type of impure zinc oxide fume results from smelting of galvanized steel scrap. Both of these by-products are too impure to be acceptable feed for conventional zinc smelters either because of impurities such as chlorides and iron or because tin, copper, and precious metals are not recoverable.

CHAPTER VI.6

aluminum recycling

JAMES E. KIRKPATRICK

The public has been aware of aluminum and its many products and applications for many years. The names of the large companies who make primary aluminum from bauxite ore are household words. The recycling industry has been quietly manufacturing secondary aluminum from scrap since the turn of the century. The recyclers have had a tremendous impact on the economy and ecology. By understanding the technology of the industry and the value of recycled aluminum, the problem of solid waste becomes a little less difficult.

ECOLOGICAL IMPACT OF RECYCLED ALUMINUM

For thousands of years nature and man have been polluting the air and waters of the earth. The natural processes of nature have taken care of most of the polluting until very recently, when huge concentrations of people in urban areas began to create more pollution than nature could handle. Pollution has become an obvious problem and people are demanding that something be done.

The problem of diminishing natural resources has also become of great interest to the citizen. Government agencies were established to monitor new programs for the conservation of resources. Money was allocated for research to determine how man could more efficiently and less wastefully use the known supply of materials.

By 1965 the USDI, and especially its Bureau of Mines, had found that more intelligent reuse of solid waste materials was part of the answer to less pollution and better conservation. The word "recycled" was coined to cover the general processes of salvage and reuse of materials. "Recycling our resources" was a phrase that caught the fancy of the people.

Aluminum has been recycled since it became a commercial metal at the beginning of the 20th century. Like many other valuable and precious metals, aluminum, when it becomes obsolete, has enough value to attract collectors, dealers, and smelters to reprocess it into new products for which there is a ready market.

An efficient collecting and manufacturing industry has grown right along with the ever-increasing use of aluminum; recycled aluminum has taken its place along with the many other valuable materials that are never totally lost to mankind.

THE ALUMINUM INDUSTRY

Primary (virgin) aluminum is produced from an ore, bauxite. High-grade deposits of bauxite exist all over the world. The ore is a reddish brown material and looks much like red dirt. Four tons of bauxite when processed, make two tons of alumina (white powder). The alumina is fed into electrolytic cells (pot line) where electrical current breaks down the alumina into molten aluminum and other residues and gasses.

Primary aluminum is usually carried through the manufacturing process to as close to a consumer product as possible. The large aluminum companies that do this are called "integrated" producers. By 1971 there were 13 companies operating 31 plants in the United States

Table 1 *The U. S. Aluminum Industry*

	Type of Product	Companies	Plants
Primary (*virgin*) producers	Primary Ingot	13	31
	Sheet and Plate	39	61
	Foil	18	27
	Extrusions	156	209
	Rod and Bar	17	31
	Wire and Cable	122	242
	Tube	36	46
	Powder and Paste	16	21
	Forgings	63	70
	Impacts	26	28
	Total	506	766
Aluminum recyclers	Secondary Ingot	42	52
	Sweaters Ingot	35	35
	Total	77	87
	Total industry	583	853

Table 2 *World's Production of aluminum (in tons of 2000 lb)*

1962	5,493,777
1963	5,968,119
1964	6,648,423
1965	7,139,597
1966	7,767,928
1967	8,519,723
1968	9,087,209
1969	10,133,885
1970	10,968,317
1971	11,375,000
1972	12,103,000
1973	12,400,000(E)

producing primary aluminum from bauxite. Many were integrated and produced both consumer products (foil) and semifabricated products (sheet, plate, extrusion, wire, etc.).

Aluminum has proved such a versatile metal that specialized manufacturing has grown rapidly. By 1971 there were 766 plants in the United States (including the large integrated plants) producing one or more aluminum products (see Table 1).

The primary industry has rapidly become a worldwide industry. In the ten-year period from 1961 to 1970 worldwide primary aluminum production doubled (see Table 2)—a unique achievement in a highly competitive market.

When a ton of primary aluminum is processed into consumer products, approximately 500 lb of scrap aluminum is generated. This scrap is the life blood of the recycling (secondary) industry that has grown up to process the scrap. By 1971 there were 42 companies operating 52 plants to produce specification grade recycled ingot. There were also 35 companies that produced sweated ingot, which is consumed by the smelters to increase the supply of recycled aluminum (see Table 1).

Many of the technologies for producing various fabricated products were well advanced when aluminum became available in commercial quantities and at competitive prices. The older metalworking industries had already established most of the necessary methods for making

Table 3 *Recycled Nonferrous Metals (Short Tons)*

Year	Copper	Aluminum	Lead Zinc and Tin	Total
		Worldwide		
1960	2,367,537	831,459	1,877,477	5,076,473
1961	2,394,095	877,743	1,845,520	5,117,358
1962	2,465,505	1,045,137	1,840,671	5,351,313
1963	2,579,672	1,169,663	1,908,113	5,657,448
1964	2,914,900	1,297,715	2,111,101	6,323,716
1965	3,140,920	1,445,163	2,273,976	6,860,059
1966	3,190,621	1,591,949	2,229,456	7,012,026
1967	3,010,995	1,645,837	2,177,883	6,834,715
1968	3,245,170	1,905,799	2,286,099	7,437,068
1969	3,494,001	2,242,570	2,311,115	8,047,686
Total	28,803,416	14,053,035	20,861,411	63,717,862
		United States		
1960	998,081	329,388	614,695	1,942,164
1961	965,903	340,187	591,884	1,897,974
1962	1,042,382	461,628	635,524	2,139,534
1963	1,115,224	505,487	671,448	2,292,159
1964	1,262,231	505,543	732,610	2,500,384
1965	1,405,932	640,593	796,195	2,842,720
1966	1,497,287	692,827	819,998	3,010,112
1967	1,303,556	697,566	764,567	2,765,689
1968	1,381,798	816,692	807,105	3,005,595
1969	1,542,139	952,679	818,896	3,313,714
Total	12,514,533	5,942,590	7,252,922	25,710,045

sheet, plate, bar, rod, wire, forgings, and castings. Metal fabricators soon found their expertise could be directly applied to aluminum. Aluminum was very quickly absorbed as a standard material by many existing plants. Warehouses and service centers were soon established. The small-quantity buyer could have his aluminum delivered very quickly at little or no premium.

By 1971 mill products were being produced at a rate of nearly 3.8 million tons per year.

The average shop generates about 20–25% scrap. This new (prompt) aluminum scrap is vital to the recycling industry. The continued growth of aluminum fabrication assures the recycler of his raw material.

VALUE OF RECYCLED METALS

Recycling of metal is not a new concept. Many metals have been recycled for thousands of years. Notable examples are gold, silver, copper, and iron. In the period from 1960 to 1970 nearly 64 million short tons of nonferrous metals were recycled in the free world. This metal had a value of 35 billion dollars. Of this total nearly 26 million tons (worth about 14.5 billion dollars) were recycled in the United States (see Table 3).

During this same 1960–1970 period 14 million short tons of aluminum were recycled in the free world with a value of 7 billion dollars. In the United States, 6 million tons of aluminum worth about 3 billion dollars were recycled (see Table 3).

The steady dollar value growth of recycled aluminum is best shown by its percentage of all nonferrous metals recycled in this period. Aluminum was only 17% of the total in 1960 but by 1970 had risen to 23% (see Table 4).

Even more interesting is the actual tonnage growth of aluminum. In 1960 aluminum was only 17% of the total tonnage of all recycled nonferrous metals in the U.S. By 1970 this total was 29% (see Table 5).

Some of the nonferrous metals (copper is an outstanding example) have a much higher price per pound, therefore fewer tons can be of much higher dollar value. Because 1 lb of aluminum has about 2½

Table 4 *Dollar Value of Aluminum Recovered from Scrap*

Year	U. S. Value, All Nonferrous Metals ($)	Aluminum Value ($)	Aluminum (% of U.S. Total)
1960	1,007,485,705	171,281,760	17
1961	932,717,971	173,209,613	19
1962	1,040,879,041	220,427,370	21
1963	1,118,215,353	228,712,648	20
1964	1,321,756,596	240,041,927	18
1965	1,633,895,288	313,980,253	19
1966	1,750,767,975	339,485,230	19
1967	1,628,824,567	348,476,071	21
1968	1,857,829,481	417,868,628	22
1969	2,240,762,672	517,800,090	23

Table 5 *Aluminum's Relationship to Total U. S. Scrap Recovery*

Year	Total Nonferrous Recovery (short tons)	Aluminum Recovery (short tons)	Aluminum (%)
1960	1,942,164	329,388	17
1961	1,897,974	340,187	18
1962	2,139,534	461,628	22
1963	2,292,159	505,487	22
1964	2,500,384	505,543	20
1965	2,842,720	640,593	23
1966	3,010,112	692,827	23
1967	2,765,689	697,566	25
1968	3,005,595	816,692	27
1969	3,313,714	952,679	29

times more volume than copper, it is sometimes difficult to make direct comparisons. Another example: on a *volume* basis the 11 million tons of aluminum produced in the free world would equal about 31 million tons of iron.

HISTORY OF RECYCLED ALUMINUM

The first aluminum recycling company was established in 1904, a scant 16 years after the first commercial production of primary aluminum in 1888. They pioneered a new industry on the premise that the scrap from metal with such outstanding qualities was too valuable a part of solid waste to be allowed to go into discard.

The aluminum recycling industry has always used scrap as its raw material. In its early days, following the pattern of other metal industries, it was called the *secondary* aluminum industry, to differentiate it from the primary industry which uses virgin aluminum. The term "secondary" is unfortunate because for many people it carried a connotation of less important, but it is so deeply ingrained in the nonferrous industry that it is difficult to eliminate.

The early recycling industry was quick to recognize that the new metal had excellent properties that made it attractive to product manufacturers and design engineers. Some of these properties were:

1. Exceptional strength-to-weight ratio
2. Simplicity of fabrication with conventional equipment
3. Excellent corrosion resistance
4. Electrical and heat conductivity
5. Excellent reflectivity
6. Nonmagnetic

The new aluminum industry grew slowly (because of the limited scrap available) until the early 1940s. During the World War II period, very large tonnages of aluminum were used by the aircraft industry, and this created large volumes of new processing scrap. As obsolete

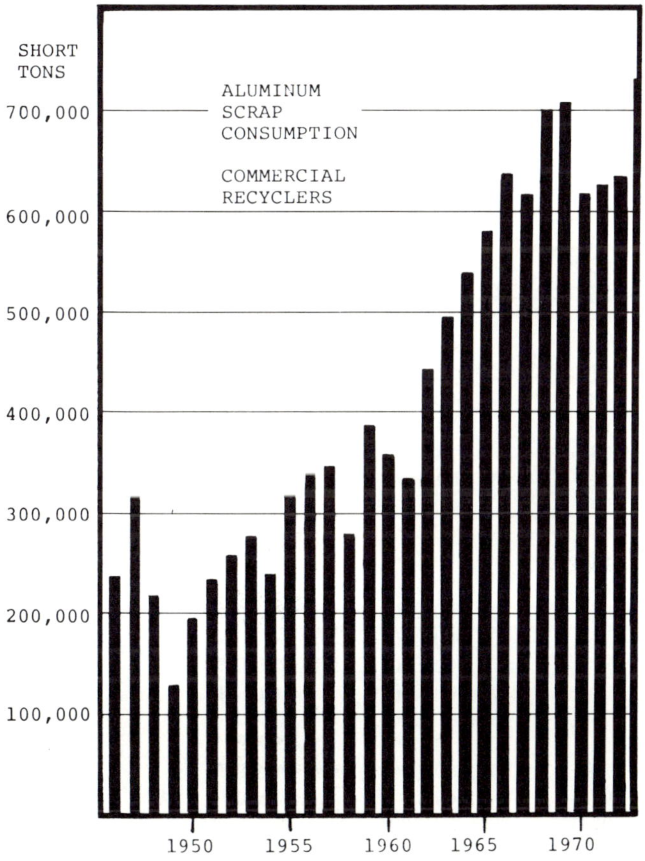

Figure 1. Aluminum scrap consumption from commercial recyclers.

Table 6 *Recycled Aluminum (from Scrap)*

Year	Pounds	Short Tons	Accum. Tons
1913	9,308,000	4,654	4,654
1914	9,044,000	4,522	9,176
1915	17,000,000	8,500	17,676
1916	38,600,000	19,300	36,976
1917	32,200,000	16,100	53,076
1918	30,100,000	15,050	68,126
1919	37,382,000	18,691	86,817
1920	31,000,000	15,500	102,317
1921	17,800,000	8,900	111,217
1922	32,580,000	16,290	127,507
1923	42,600,000	21,300	148,807
1924	54,000,000	27,000	175,807
1925	88,000,000	44,000	219,807
1926	88,400,000	44,200	264,007
1927	92,400,000	46,200	310,207
1928	95,600,000	47,800	358,007
1929	98,800,000	49,400	407,407
1930	77,200,000	38,600	446,007
1931	60,600,000	30,300	476,307
1932	48,000,000	24,000	500,307
1933	67,000,000	33,500	533,807
1934	92,800,000	46,000	579,807
1935	102,800,000	51,400	631,207
1936	103,000,000	51,500	682,707
1937	125,120,000	62,560	745,267
1938	77,600,000	38,800	784,067
1939	107,894,000	53,947	838,414
1940	160,724,000	80,362	918,376
1941	213,714,000	106,857	1,025,233
1942	392,000,000	196,000	1,221,233
1943	628,000,000	314,000	1,535,233
1944	650,000,000	325,000	1,860,233
1945	596,000,000	298,000	2,158,233
1946	556,000,000	278,000	2,436,233
1947	690,000,000	345,000	2,781,233
1948	574,000,000	287,000	3,068,233
1949	362,000,000	181,000	3,249,233
1950	486,000,000	243,000	3,492,233
1951	594,000,000	297,000	3,789,233
1952	608,000,000	304,000	4,093,233
1953	736,000,000	368,000	4,461,233
1954	626,000,000	313,000	4,774,233
1955	828,000,000	414,000	5,188,233
1956	856,000,000	428,000	5,616,233

Table 6—continued

Year	Pounds	Short Tons	Accum. Tons
1957	880,000,000	445,000	6,061,233
1958	708,000,000	354,000	6,415,233
1959	898,000,000	449,000	6,864,233
1960	876,000,000	438,000	7,302,233
1961	970,000,000	485,000	7,787,233
1962	1,164,000,000	582,000	8,369,233
1963	1,308,000,000	654,000	9,023,233
1964	1,414,000,000	707,000	9,730,233
1965	1,658,000,000	829,000	10,559,233
1966	1,774,000,000	887,000	11,446,233
1967	1,756,000,000	878,000	12,324,233
1968	1,878,000,000	939,000	13,263,233
1969	2,060,000,000	1,030,000	14,293,233
1970	1,790,000,000	895,000	15,188,233
1971	2,100,000,000	1,050,000	16,238,233
1972	2,252,000,000	1,126,000	17,364,233
1973	2,296,000,000	1,148,000	18,512,233

[a] Source: R. J. Anderson, Secondary Aluminum, Bureau of Mines, Aluminum Association, BDC, USDC.

aircraft became available, the recycling industry entered the real growth period. New technologies had to be developed to cope with the new quantities and qualities of aluminum scrap available.

In 1969 over a million tons of recycled aluminum were produced. In the period from 1913 to 1973 the industry had produced 15 million tons of aluminum—the equivalent of 3 years of primary production at the 1973 rate (see Table 6).

The growth curve for recycled aluminum has been relatively steady since World War II (see Figure 1). The relationship of recycled aluminum to primary aluminum has been somewhat more spotty on a percentage basis. During the first 40 years of the commercial recycling aluminum industry (1913–1953), there were great fluctuations in the supply relationship of primary and recycled aluminum. Both industries were relatively small. From 1953 until 1973 the recycled metal remained about 18–23% of the total production (see Table 7).

When the primary producers cannot make enough metal for their plants, they will purchase scrap metal. When the primary industry has more metal than they can consume, primaries will sell the surplus in the same ingot markets which are serviced by the recycling industry.

Table 7 Aluminum Statistics (Short Tons)

Year	Primary Production	Recycled Commercial Aluminum	Total	Recycled as a Percent of Total
1913	23,639	4,654	28,293	16.4
1914	28,986	4,522	33,508	13.5
1915	45,252	8,500	53,752	15.8
1916	57,553	19,300	76,853	25.1
1917	64,930	16,100	81,030	19.9
1918	62,362	15,050	77,412	19.4
1919	64,238	18,691	82,929	22.5
1920	69,021	15,500	84,521	18.3
1921	27,266	8,900	36,166	24.6
1922	36,816	16,290	53,106	30.7
1923	64,329	21,300	85,629	24.9
1924	75,282	27,000	102,282	26.4
1925	70,058	44,000	114,058	38.6
1926	73,693	44,200	117,893	37.5
1927	81,803	46,200	128,003	36.1
1928	105,272	47,800	153,072	31.2
1929	113,986	49,400	163,386	30.2
1930	114,518	38,600	153,118	25.2
1931	88,772	30,300	119,072	25.4
1932	52,444	24,000	76,444	31.4
1933	42,562	33,500	76,062	44.0
1934	37,088	46,000	83,088	55.4
1935	59,647	51,400	111,047	46.3
1936	112,464	51,500	163,964	31.4
1937	146,340	62,560	208,900	30.0
1938	143,441	38,800	182,241	21.3
1939	163,545	53,947	217,492	24.8
1940	206,280	80,362	286,642	28.0
1941	309,067	106,857	415,924	25.7
1942	521,106	196,000	717,160	27.3
1943	920,179	314,000	1,234,179	25.4
1944	776,446	325,000	1,101,446	29.5
1945	495,060	298,000	793,060	37.6
1946	409,630	278,000	687,630	40.4
1947	571,750	345,000	916,750	37.6
1948	623,456	287,000	910,456	31.5
1949	603,462	181,000	784,462	23.1
1950	718,622	243,000	961,622	25.3
1951	836,881	297,000	1,133,881	26.2
1952	937,330	304,000	1,241,330	24.5
1953	1,252,013	368,000	1,620,013	22.7
1954	1,460,565	313,000	1,773,565	17.7

Table 7—continued

Year	Primary Production	Recycled Commercial Aluminum	Total	Recycled as a Percent of Total
1955	1,565,721	414,000	1,979,721	20.9
1956	1,678,954	428,000	2,106,954	20.3
1957	1,647,709	445,000	2,092,709	21.3
1958	1,565,557	354,000	1,919,557	18.4
1959	1,954,112	449,000	2,403,112	18.7
1960	2,014,498	438,000	2,452,498	17.9
1961	1,903,711	485,000	2,388,711	20.3
1962	2,117,929	582,000	2,699,929	21.6
1963	2,312,528	654,000	2,966,528	22.0
1964	2,552,747	707,000	3,259,747	21.7
1965	2,754,478	829,000	3,583,478	23.1
1966	2,968,366	887,000	3,855,366	23.0
1967	3,269,259	878,000	4,147,259	21.2
1968	3,255,042	939,000	4,194,042	22.4
1969	3,793,062	1,030,000	4,823,062	21.4
1970	3,976,148	895,000	4,871,148	18.4
1971	3,925,200	1,050,000	4,975,200	21.1
1972	4,122,400	1,126,000	5,248,400	21.4
1973	4,450,000	1,200,000	5,448,000	22.0 (E)

In 1950 the first "hot-metal" contract was consummated in the United States. Molten metal was transported by truck from the primary smelter to his customer. The haul was about one mile. In 1964 the first recycled hot metal was delivered over the highway. By 1971 the delivery of molten metal from the recycler's plant to the customer was relatively common. Hauls of over 200 miles were successful.

COLLECTION: THE FIRST STEP IN RECYCLING

New or prompt scrap is generated when new aluminum metal is processed into consumer and industrial products. Industrial operators will never generate scrap if they can avoid doing so. When aluminum scrap is generated, it has value and is often segregated and packaged in order to receive the highest bids from scrap dealers.

In the United States there are thousands of scrap collectors. These collectors usually accumulate truckloads of solid waste, including paper, rags, wood, discarded durables, obsolete appliances, and both ferrous and nonferrous metals. The collector then sells these materials to a dealer or scrap processor. The dealer must prepare the various materials for resale. He usually has the capability for sorting, magnetic separation, cleaning, cutting, shearing, bailing, and identification.

The aluminum is then offered to the aluminum smelting and recycling industry. The flow of scrap is shown in Figure 2.

The dealers must have standard classifications of aluminum scrap to deal intelligently with the smelting and refining industry. The industry has a standard classification reference sheet that is distributed to the dealers (Table 8). All visible contaminants have to be removed and specific methods of packaging and shipment met. The more carefully the classification sheet is followed the better the position of the dealer to negotiate the best price for his material.

Some aluminum has iron and steel mechanically attached to it in such a manner that normal sorting cannot remove it. In the trade this is called "high irony" scrap. The dealer can ship this to the smelter for processing on a recovery basis. Some dealers prefer to remove the aluminum by operating their own "sweat" furnaces. The "high irony" scrap is placed in a special furnace and the temperature is raised to about 1350°F. The aluminum liquifies and slowly runs off the furnace hearth into molds of standard ingot size, or very large molds called "sows."

Unfortunately other metals such as zinc, lead, magnesium, etc., melt at the same or lower temperatures than aluminum. It is not economically feasible to remove lead and zinc in the smelting plant, so the sweated material must be slowly dissipated by small additions to con-

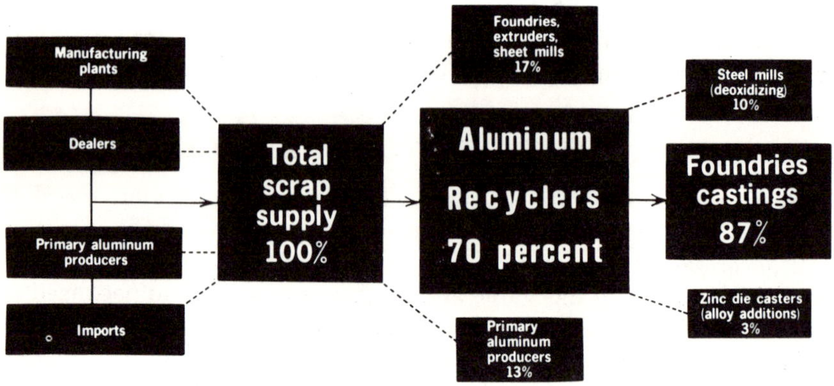

Figure 2. Aluminum recycle.

Table 8 *Aluminum Scrap Classifications*

1. New pure aluminum clippings: Shall consist of new, clean, unalloyed sheet clippings and/or aluminum sheet cuttings.
2. New pure aluminum wire and cable: Shall consist of new, clean, unalloyed aluminum wire or cable.
3. Old pure aluminum wire and cable: Shall consist of old, unalloyed aluminum wire or cable.
4. Segregated new aluminum alloy clippings: Shall consist of new, clean, *uncoated* aluminum clippings of one specified aluminum alloy.
5. Mixed new aluminum alloy clippings: Shall consist of new, clean, *uncoated* aluminum clippings of two or more alloys.
6. Mixed low copper aluminum alloy clippings: Shall consist of new, clean, *uncoated* aluminum clippings of two or more alloys.
7. Segregated old aluminum alloy sheet: Shall consist of clean, *uncoated,* old aluminum sheet of one specified alloy only.
8. Mixed old alloy sheet: Shall consist of clean, *uncoated,* old alloy sheet aluminum of two or more alloys.
9. Scrap sheet and sheet utensil aluminum: Shall consist of clean, *uncoated* manufactured sheet aluminum.
10. Segregated new aluminum castings, forgings, and extrusions: Shall consist of new, clean, *uncoated* aluminum castings, forgings and extrusions of one specified alloy.
11. Mixed new aluminum forgings and extrusions: Shall consist of clean, new, *uncoated* aluminum forgings and extrusions of two or more alloys.
12. Mixed new aluminum castings: Shall consist of clean, new, *uncoated* aluminum castings of two or more alloys.
13. Aluminum auto castings: Shall consist of all clean automobile aluminum castings of sufficient size to be readily identified.
14. Aluminum airplane castings: Shall consist of clean aluminum castings from airplanes.
15. Mixed aluminum castings: Shall consist of all clean aluminum castings which may or may not contain auto and airplane castings.
16. Aluminum pistons:
    (a) Clean aluminum pistons: Shall consist of clean aluminum pistons to be free from struts, bushings, shafts, iron rings and any other foreign materials.
    (b) Aluminum pistons with struts: Shall consist of clean whole aluminum pistons with struts to be free from bushings, shafts, iron rings and any other foreign materials.
    (c) Irony aluminum pistons.
17. Wrecked airplane sheet and/or breakage aluminum.
18. New aluminum foil: Shall consist of clean, new, pure, *uncoated,* unalloyed aluminum foil.
19. Old aluminum foil: Shall consist of clean, old, *uncoated,* pure, unalloyed aluminum foil.
20. All other aluminum base foils including etched foil, radar foil and chaff.
21. Segregated aluminum borings and turnings: Shall consist of clean, uncorroded aluminum borings and turnings of one specified alloy.
22. Mixed aluminum borings and turnings: Shall consist of clean, uncorroded aluminum borings and turnings of two or more alloys.

Table 8—continued

23. Sweated aluminum: Shall consist of aluminum scrap that has been sweated or melted into a form or shape such as an ingot, pig, or slab for convenience in shipping.
24. Aluminum grindings.
25. Aluminum drosses, spatters, spillings, skimmings and sweepings.
26. Aluminum hair wire.
27. Aluminum wire screen.
28. Coated aluminum (painted or plastic coated, etc.)
29. Containers or all types (oil, food, beverage, aerosol)
30. Items not covered specifically by above classifications: Any new item that might appear and which is not covered specifically by above classifications.

trolled melts to dilute it to specification limits. Magnesium can be removed during the refining process.

Aluminum scrap is received from the dealer in railroad cars or over-the-road trailers. The smelter stores the scrap by alloy until he is ready to process the material. The seven basic categories for the scrap are (1) clips and solids, (2) borings and turnings, (3) sheet and castings, (4) residues and dross, (5) irony scrap, (6) sweated ingot, and (7) cans and containers. The basic operations are shown in Figure 3.

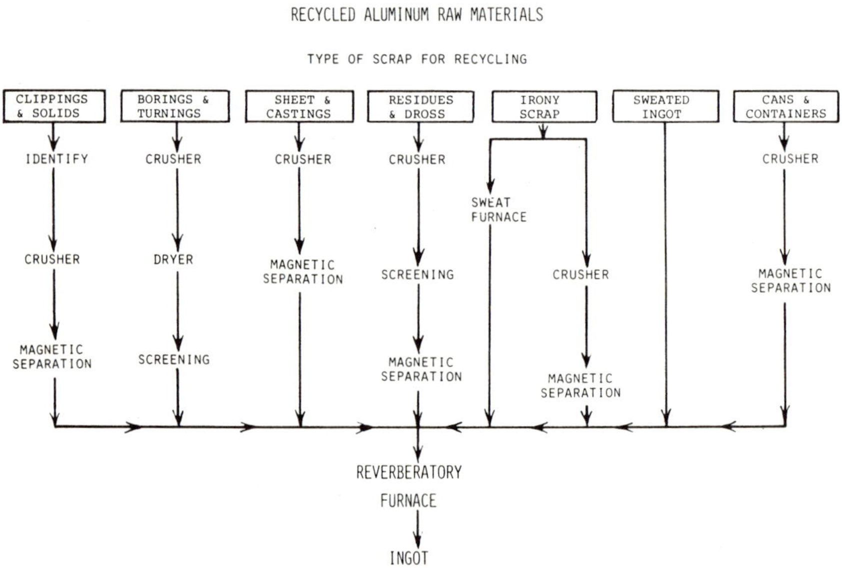

Figure 3. Recycled aluminum materials.

Identification is usually accomplished by trained sorters. These men can do an excellent job of separating contaminants visually and by weights. Chemical kits are also available to give positive readings—usually by the color of the chemical reaction of the kit materials. The metal laboratory technicians can give positive identification by spectrographic analysis when required.

The dealer or broker receives both new (prompt) scrap and old scrap. The new scrap is usually well classified and often delivered to the dealer in good segregated form.

Old scrap becomes available when consumer products have reached the end of their useful life and are discarded. Old aluminum scrap accounts for about one-fourth of the total domestic scrap consumed in the United States. The reason for the much greater amount of new scrap is that much more aluminum is being fabricated. Old scrap depends on obsolescence of products and some have extremely long lives, such as curtain walls and siding for architectural applications, home cookware, and commercial entrances.

From 1950 to 1970 total aluminum consumption grew from about 1 million to 5 million tons/yr. The ratio of scrap used remained fairly constant during this period. New and old scrap stayed at relatively the same level. Projections for the next ten years (to 1980) indicate that the consumption pattern of primary metal and old and new scrap will grow at about the same levels as in the past (see Figure 4).

Every foundry, die casting shop, and metalcaster generates aluminum scrap. This scrap is usually in the form of gates, runners, and sprues. Large integrated primary plants generate their own type of scrap such as slab chips (scalpings), saw chip, heads of slabs, logs and billets, etc. These items are usually put back into the melting furnaces in the plant where they are generated and are called "home" or "run around" scrap.

Sometimes these in-house scrap figures get mixed with the true commercial scrap which is defined as aluminum scrap that is bought, sold and traded in the commercial market place. Anyone interested in recycled metal should be aware that different statistics are available depending on the definition of the scrap involved. The aluminum scrap, and the attending statistical data in this section deal only with the truly *commercial scrap*.

CANS, CONTAINERS, AND PACKAGING

One of the most rapidly growing uses of aluminum is in the field of cans, containers, and packaging. The total aluminum for these items

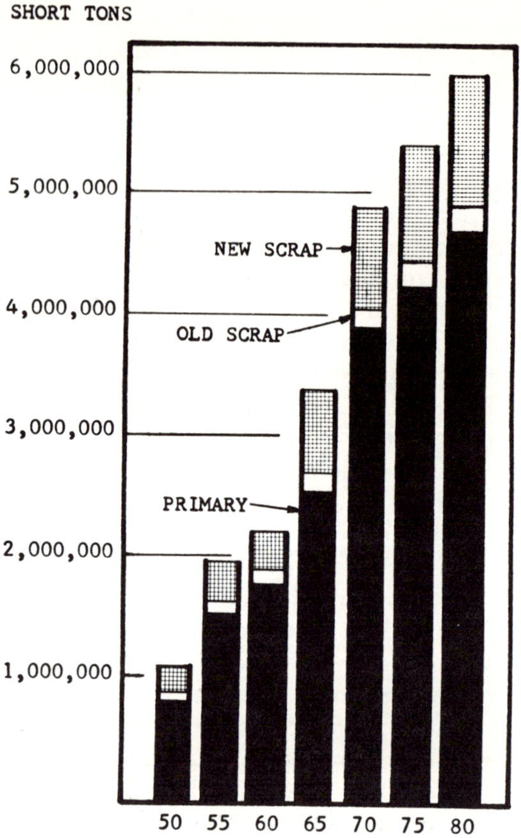

Figure 4. Ratios of primary, old and new scrap, 1950–1980.

amounted to over one million tons in 1973 and should reach over 1,100, 000 tons in 1975 (see Figure 5).

Citizens are demanding a cleaner environment, yet when they act as consumers they have not been sufficiently motivated to recycle their own trash. Marketing experts refer to this as a lack of understanding of reverse distribution. Most of the aluminum in these cans and containers find their way into city collection systems. As more efficient methods of segregation and collection are devised, containers will become a vast new mine of aluminum to be recycled.

More and more aluminum is moving through America's six million vending machines. Institutional food tray packaging is growing. Thaw-and-serve foods develop more recyclable aluminum in quantities that invite saving for resale to solid waste collectors.

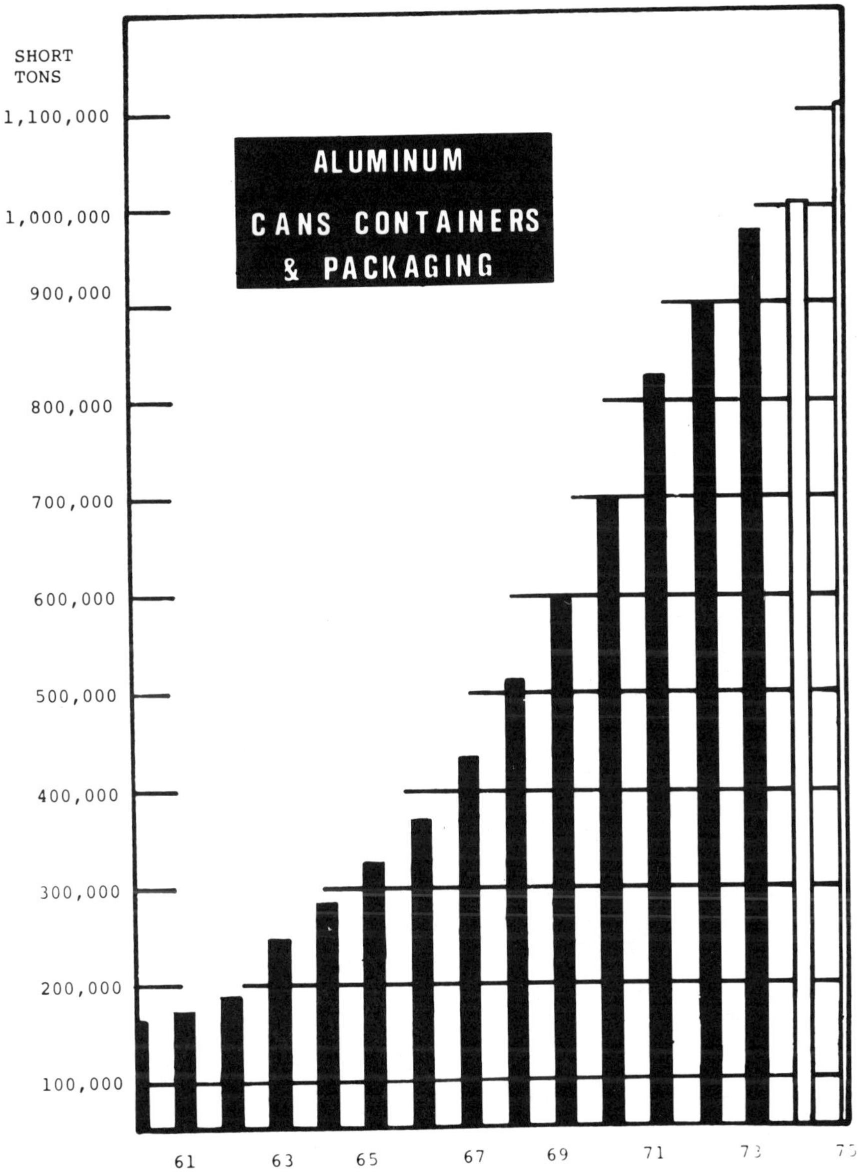

Figure 5. Aluminum cans and containers, 1961–1975.

Institutional food services, and others, are using large amounts of aluminum food containers. Where volume warrants, "washer-crusher" compactors reduce large volumes of light weight cans and containers into high density bundles. These have good value and are salable to collectors. The aluminum soon reaches a smelter who manufactures recycled aluminum products.

A new item of great market potential is hot canned food from vending machines. This market jumped nearly 40% in 1970; 205 million hot canned servings worth 75 million dollars were sold. The ease of "peel top" openings and the attendant cleanliness is attractive to the consumer.

Food distribution and food packaging using aluminum is destined to generate large quantities of aluminum for recycling in the next 10 years.

RECYCLED ALUMINUM METALLURGY

Aluminum, like all other metals, is never used in its pure state. Other metals are added to give the aluminum the desired physical and mechanical properties and the end product is called an alloy.

The recycler uses dilution chemistry in manufacturing recycling aluminum alloy ingot. Because the alloying ingredients are already in the scrap delivered to the smelter, he must adjust that alloy by adding pure scrap, such as electrical conductor scrap, thereby diluting the melt until it comes into specification. When the scrap does not contain enough alloying materials, these are added by the smelter.

Some of the more common alloying additives are:

1. Silicon: increases fluidity, thereby helping the metalcaster manufacture a better casting at a lower cost.
2. Zinc: helps produce strength and toughness in aluminum alloys.
3. Copper: adds hardness, strength and machinability. It also helps elevate property characteristics. Too much copper can cause hot-cracking in a casting and will reduce corrosion resistance.
4. Manganese: can improve corrosion resistance and change the iron constituent to improve ductility and resistance to impact.
5. Iron: small percentages increase strength and hardness and can reduce hot-cracking of castings.
6. Magnesium: can improve the properties of corrosion resistance and machinability. Increases dross problems and in higher amounts, can impair elongation and impact values. For these reasons most metalcasting alloys often limit the amount of magnesium.

7. Titanium: is often added to foundry alloy as a grain refiner.
8. Nickel: used in some special casting alloys to improve properties at elevated temperatures.

Specification grade alloys manufactured from recycled aluminum by the dilution chemistry process are sold on an equal basis with the traditional alloys manufactured from primary (virgin) aluminum.

In the case of primary alloys, all constituents must be added by the smelter.

SUPPLY AND DEMAND

Unlike primary aluminum, which is produced from ore, scrap is inelastic in that increased price or demand will not increase the volume of scrap. The aluminum recycling industry is probably as close to pure competition as any industry in our economy. There are a large number of firms; no one firm controls the market, and it is easy to enter the market.

The price offered for scrap is essentially the same within any market area. The selling price of recycled ingot is extremely sensitive to the general economy. The index of the production level for the automobile and heavy durable markets develops a similar level of demand for metalcasting aluminum alloys.

Excess capacity or lack of demand for integrated producers' products will cause virgin aluminum to be offered in the foundry market.

Based on the projected growth of the total aluminum industry, there will be an adequate supply of scrap for recycling in the years ahead. By 1971 there was over a million tons of old and new scrap being generated every year; by 1975 this should reach 1.6 million tons and if the economy follows the projected pattern the total should reach 2 million tons a year by 1980 (see Table 9).

Selling prices of recycled aluminum are published daily in the *American Metal Market*. During periods of a slow economy actual prices are somewhat below the published prices due to discounting. When the economy rises and production approaches capacity, prices move up to or near published prices. A recap of these published prices is shown in Table 10.

MANUFACTURING

Aluminum scrap is not ready for processing as it is normally received at the recycling plant. It must be sorted, crushed, dried, have foreign

Table 9 *Aluminum Scrap Available (New and Obsolete)*

Year	Short tons
1948	300,000
1949	200,000
1950	250,000
1951	300,000
1952	310,000
1953	380,000
1954	280,000
1955	370,000
1956	375,000
1957	360,000
1958	300,000
1959	400,000
1960	460,000
1961	505,000
1962	550,000
1963	610,000
1964	660,000
1965	800,000
1966	850,000
1967	900,000
1968	1,000,000
1969	1,050,000
1970	1,100,000
1971	1,200,000
1972	1,300,000
1973	1,400,000
1974	1,500,000
1975	1,600,000

materials removed, melted, refined, blended, cleaned and brought into specification for the alloy being produced. The recycler is totally dependent on the scrap dealer for his raw material.

Recyclers tend to locate near large industrial operations where aluminum scrap is being generated. The recycler is not restricted to geographic locations near raw materials and low-cost electricity, as are the primary producers.

Not all recyclers can process all the types of scrap shown in Figure 3. The more sophisticated his equipment, the more types of scrap he can process.

Table 10 *Average published Prices (Cents Per Pound)*

Year	Primary Aluminum	Recycled Aluminum
1936	20.50	18.81
1937	20.08	19.45
1938	20.00	18.40
1939	20.00	19.38
1940	18.69	18.74
1941	16.50	17.36
1942	14.00	15.00
1943	14.00	14.15
1944	14.00	11.73
1945	14.00	12.55
1946	14.00	14.59
1947	14.00	14.92
1948	14.74	21.36
1949	16.00	18.15
1950	16.69	21.69
1951	18.00	25.84
1952	18.40	20.48
1953	19.70	22.19
1954	20.20	20.61
1955	21.88	28.73
1956	24.03	27.01
1957	25.42	22.70
1958	24.79	22.06
1959	24.75	23.44
1960	27.23	24.67
1961	25.46	22.56
1962	23.88	21.20
1963	22.62	21.10
1964	23.72	22.05
1965	24.50	24.21
1966	24.98	24.74
1967	25.57	24.75
1968	27.18	25.00
1969	27.18	26.82
1970	29.00	27.72
1971	29.00	27.92
1972	26.45	27.00

Many recyclers have large crushers. These heavy machines tear, chop, or hammer the scrap into small fist-size chunks, and remove much of the contaminating materials. The fist-size chunks are passed under a magnet that lifts out the magnetic materials, such as iron and steel. The chunks then pass over more screens where smaller dust and dirt are removed.

At this point some recyclers put all the crushed material through dryers that remove moisture, oil, and any remaining organic materials. The scrap is now ready to be either stored or melted. Many smelters are crushing and drying borings and turnings to achieve a uniform "feed material."

The processing of residues and drosses requires special equipment and technical knowledge. Many smelters do not process their own residue. They ship them to recycling plants who are equipped to process them. These materials are purchased on the basis of metal recovered.

The recycling plants that are equipped to manufacture specification grade alloys buy large tonnages of sweated ingot and sweated sow from the smaller dealers who operate sweat furnaces only. This material is commonly known as high irony ingot.

The workhorse of the recycling industry is the reverberatory furnace. These furnaces are rectangular in shape. They are lined with refractory brick and are direct-fired, usually with gas, but in some cases oil. Heat is radiated into the melt from the ceiling and walls as well as from the direct flame. Reverberatory furnaces (known as "reverbs" in the industry) vary in capacity from 20,000 to 180,000 lb of aluminum.

Reverb furnaces are equipped with side doors for cleaning and initial charging. The initial charge is usually heavy pieces, sweated ingot or sweated sows (very large ingot). This initial charge is called the heel and varies from 20% to 25% of the furnace capacity.

All reverb furnaces have an opening at one end (and often at each end), which is called a well. When the heel material is melted, the molten aluminum rises in the well and after an analysis is made, the furnace operator is ready to charge the furnace. From this point on the side door is closed and selected and prepared scrap is charged into the well.

If the reverb furnace is in continuous operation, the operator will leave about 20% of the previous melt in the furnace to act as a heel. This will shorten the furnace cycle time and improve furnace efficiency. It requires 15–35 hr to completely charge a reverb furnace, depending on its size. The molten metal must be constantly mixed so that the whole melt is a homogeneous mixture. At many stages of charging, analyses must be made and the next charge chosen for proper blending of the mixture.

Molten aluminum reacts with air to form oxides. The liquid aluminum in the well of the furnace must be constantly covered with a molten (wet) flux to protect the molten metal. The flux also helps clean the aluminum as it is carried under the furnace by new charges, or is puddled into the metal by the operator.

If the melt contains more magnesium than is allowed in the specification, it must be removed. This is done by fluxing with aluminum fluoride. Elemental chlorine gas can be injected into the molten aluminum, which will also eliminate the magnesium.

During the charging and melting aluminum absorbs hydrogen from oil, moisture, etc. All gasses must be removed before the metal is allowed to solidify. Degassing is accomplished by bubbling dry nitrogen, chlorine, or a mixture of the two through the molten metal. Degassing is the last operation before the furnace is tapped and ingot poured or hot metal loaded for delivery.

Aluminum recycling plants in areas such as California became aware of air pollution problems before they were generally recognized by the public. The equipment and technology to control gaseous effluents from dryers, reverb furnaces, wells, and sweat furnaces did not exist. During the 1960s aluminum recyclers spent millions of dollars in engineering, research, and testing equipment to combat pollution. Baghouses, afterburners, scrubbers, and precipitation equipment were installed. Although some of this equipment did a fair job, much of it would not meet the more stringent codes that were being enacted.

Most equipment meeting present codes was developed by cooperative efforts of the recycling industry and pollution control equipment companies. The cost of such equipment adds to manufacturing cost but does not add to profitability.

QUALITY CONTROL IN MANUFACTURING

To produce specification grade recycled alloys the smelters requires a metallurgical laboratory. Most plants maintain a quantometer or direct-reading spectrograph. Aluminum can be analyzed in a few minutes.

Rapid analysis of the melt in the reverb furnace is a necessity. Dilution chemistry is possible only if the operator knows the exact composition of the melt at all times. He must also know the composition of the material he will charge into the furnace. Final samples are taken at the time of tapping and pouring. Analyses are a part of the permanent record for each heat and are available for reference.

Some recyclers' laboratories are equipped to test mechanical properties such as elongation, tensile strength, and yield strength. Impact and hardness tests are sometimes required. Microscopic studies are made to determine that there are no inclusions or voids in the alloy. X-ray diffraction tests are used to identify any internal defects or nonmetallic inclusions.

The recycled ingot is packaged in protection envelopes. Many ingot storage rooms are maintained dust free, and controlled atmosphere is now common in the industry.

The smelting industry is aware of the customers' demand for clean, high-quality casting alloys. Any metal made from scrap must meet the same standards of quality as the original metal or the market would be lost.

PRODUCTS FROM RECYCLED ALUMINUM

The bulk of products developed by the recyclers of aluminum are used in manufacturing aluminum castings. Castings vary in size from less than an ounce to hundreds of pounds. Alluminum castings have become a ubiquitous product in manufacturing, and a wide product mix of casting alloys is required to satisfy its many uses. Like all the older metals, aluminum has developed certain workhorse alloys that do a fine job in thousands of applications. The industry developed hundreds more alloys as they were needed and many are still in use in specialized uses.

The aluminum casting alloys were given new designations and registered with the Aluminum Association in 1969. The new system standardized all alloys. The alloys and their chemical composition are shown in Table 11.

The largest buyer of recycled aluminum is the metalcasting industry, but the sale of recycled aluminum to the steel industry is increasing. The consumption of aluminum to remove oxygen from steel (deox) reached 130,000 tons by 1969 and more than half of this tonnage was recycled aluminum. It is estimated that free world use of deox aluminum is now about 300,000 tons/yr.

Recycled aluminum is used in alloying steel. Any steel that has to be nitrided must contain about 1% aluminum. Zinc alloys also contain aluminum as an alloying ingredient. In 1965 zinc alloys consumed about 40,000 tons of aluminum of which 45% were recycled alloys. Some recycled alloys are used to alloy magnesium, and a small amount for copper and brass alloys.

Table 11 *Aluminum Alloy Ingot Chemical Composition Limits*

New Designation (A.A.)	Former Designation	Cu	Fe	Si	Mn	Mg	Zn	Cr	Ti	Ni	Sn	O.E.	O.T.
208.1	108	3.5–4.5	1.0	2.5–3.5	0.50	0.10	1.0	—	0.25	—	—	—	0.50
212.1	112	6.5–7.5	1.2	2.0–3.0	0.6	0.10	2.5	—	0.25	0.35	—	—	0.50
213.1	113	6.5–7.5	1.2	3.0–4.0	0.6	0.10	2.5	—	0.25	0.35	—	—	0.50
222.1	122	9.2–10.8	1.2	2.0	0.50	0.20–0.35	0.8	—	0.25	0.50	—	—	0.35
238.1	138	9.0–11.0	1.2	3.5–4.5	0.6	0.20–0.35	1.5	—	0.25	1.0	—	—	0.50
242.1	142	3.5–4.5	0.8	0.7	0.35	1.3–1.8	0.35	0.25	0.25	1.7–2.3	—	0.05	0.15
A242.1	A142	3.7–4.5	0.8	0.6	0.10	1.2–1.7	0.10	0.15–0.25	0.07–0.18	1.8–2.3	—	0.05	0.15
295.1	195	4.0–5.0	0.8	0.7–1.2	0.30	0.03	0.30	—	0.20	—	—	0.05	0.15
B295.1	B195	3.5–4.5	1.0	2.0–3.0	0.35	0.05	0.50	—	0.25	0.35	—	—	0.50
308.1	A108	3.0–4.0	0.8	5.5–6.5	0.50	0.10	1.0	—	0.25	0.35	—	—	0.50
319.1	319	3.0–4.0	1.0	5.5–6.5	0.50	0.10	1.0	—	0.25	0.35	—	—	0.50
328.1	RedX-8	1.0–2.0	0.8	7.5–8.5	0.20–0.6	0.25–0.6	1.5	0.35	0.25	0.25	—	—	0.50
A332.1	A132	0.50–1.5	1.1	11.0–12.5	0.35	0.8–1.3	0.35	—	0.25	2.0–3.0	—	0.05	—
B332.1	B132	1.0–2.0	0.7	11.0–12.5	0.50–0.9	0.45–1.0	1.0	—	0.25	0.05	—	—	0.50
F332.1	F132	2.0–4.0	1.0	8.5–10.5	0.50	0.6–1.5	1.0	—	0.25	0.50	—	—	0.50
333.1	333	3.0–4.0	0.9	8.5–10.0	0.50	0.10–0.50	1.0	—	0.25	0.50	—	—	0.50

Table 11—continued

New Designation (A.A.)	Former Designation	Cu	Fe	Si	Mn	Mg	Zn	Cr	Ti	Ni	Sn	O.E.	O.T.
355.1	355	1.0–1.5	0.6	4.5–5.5	0.50	0.45–0.6	0.35	—	0.25	—	—	0.05	0.15
C355.1	C355	1.0–1.5	0.15	4.5–5.5	0.05	0.45–0.6	0.05	—	0.20	—	—	0.05	0.15
356.1	356	0.25	0.50	6.5–7.5	0.35	0.25–0.40	0.35	—	0.25	—	—	0.05	0.15
A356.1	A356	0.10	0.12	6.5–7.5	0.05	0.25–0.40	0.05	—	0.20	—	—	0.05	0.15
357.1	357	0.05	0.15	6.5–7.5	0.03	0.45–0.6	0.05	—	0.20	—	—	0.05	0.15
A357.1	A357	0.20	0.20	69–7.5	0.10	0.40–0.7	0.10	—	0.20	—	0.02–0.4	0.05	0.15
360.1	360	0.6	1.0	9.0–10.0	0.35	0.45–0.6	0.35	—	—	0.50	0.15	—	0.25
A360.1	A360	0.6	1.0	9.0–10.0	0.35	0.45–0.6	0.35	—	—	0.50	0.15	—	0.25
380.1	380	3.0–4.0	1.0	7.5–9.5	0.50	0.10	2.9	—	—	0.50	0.20	—	0.50
A380.1	A380	3.0–4.0	1.0	7.5–9.5	0.50	0.10	2.9	—	—	0.50	0.20	—	0.50
B380.1	B380	3.0–4.0	1.0	7.5–9.5	0.50	0.10	0.9	—	—	0.50	0.20	—	0.50
384.1	384	3.0–4.0	1.0	10.5–12.0	0.50	0.10	2.9	—	—	0.50	0.20	—	0.50
A384.1	384	3.0–4.0	1.0	10.5–12 0	0.50	0.10	0.9	—	—	0.50	0.20	—	0.50
390.2	390	4.0–5.0	0.6–1.0	16.0–18.0	0.10	0.50–0.65	0.10	—	0.20	—	—	0.10	0.20

A413.1	A13	0.10	0.6	11.0–12.5	0.05	0.05	0.05	—	—	0.05	0.05	—	0.10
443.1	43	0.6	0.6	4.5–6.0	0.50	0.05	0.50	0.25	0.25	—	—	—	0.35
A443.1	A43	0.30	0.6	4.5–6.0	0.50	0.05	0.50	0.25	0.25	—	—	—	0.35
B443.1	B43	0.10	0.6	4.5–6.0	0.10	0.05	0.10	—	0.20	—	—	0.05	0.15
C443.1	C43	0.6	1.0	4.5–6.0	0.35	0.10	0.35	—	—	0.50	0.15	—	0.25
444.1	344	0.05	0.12	6.5–7.5	0.05	0.05	0.05	—	0.20	—	—	0.05	0.15
514.1	214	0.15	0.40	0.35	0.35	3.6–4.5	0.15	—	0.25	—	—	0.05	0.15
A514.1	A214	0.35	0.50	0.35	0.8	3.6–4.5	1.4–2.2	—	0.25	—	—	0.05	0.15
B514.1	B214	0.35	0.50	1.4–2.2	0.8	3.6–4.5	0.35	0.25	0.25	—	—	0.05	0.15
F514.1	F214	0.10	0.40	0.30–0.7	0.30	3.6–4.5	0.10	—	0.20	—	—	0.05	0.15
518.1	218	0.25	1.0	0.35	0.35	7.6–8.5	0.15	—	—	0.15	0.15	—	0.25
B518.1	B218	0.05	0.12	0.10	0.04	6.5–7.5	—	—	0.10–0.25	—	—	0.05	0.15
520.1	220	0.20	0.20	0.20	0.10	9.6–10.6	0.10	—	0.20	—	—	0.05	0.15
535.1	Almag 35	0.05	0.15	0.10	0.10–0.25	6.3–7.5	—	—	0.10–0.25	—	—	0.05	0.15
705.1	Ternalloy 5	0.20	0.6	0.20	0.40–0.6	1.5–1.8	2.7–3.3	0.20–0.40	0.25	—	—	0.05	—
707.1	Ternalloy 7	0.20	0.6	0.20	0.40–0.6	1.9–2.4	4.0–4.5	0.20–0.40	0.25	—	—	0.05	—
D712.1	D612	0.25	0.40	0.25	0.10	0.50–0.65	5.0–6.5	0.40–0.6	0.15–0.25	—	—	0.05	0.20
713.1	613	0.40–1.0	0.8	0.25	0.6	0.25–0.50	7.0–8.0	0.35	0.25	0.15	—	0.10	0.25
771.1	Precedent 71	0.05	0.10	0.10	0.10	0.8–1.0	6.5–7.5	0.06–0.20	0.10–0.20	—	—	0.03	0.10
850.1	750	0.6–1.3	0.7	0.7	0.10	—	—	—	0.20	0.7–1.3	5.5–7.0	—	0.30
A850.1	A750	0.7–1.3	0.7	2.0–3.0	0.10	—	—	—	0.20	0.30–0.7	5.0–7.0	—	0.30
B850.1	B750	1.7–2.3	0.7	0.40	0.10	0.6–0.9	—	—	0.20	0.9–1.5	5.0–7.0	—	0.20

Table* 12 *Where Cast Aluminum Is Used

Application	Alloy	Application	Alloy	Application	Alloy
Air compressor block	380.0	Gasoline tanks	319.0, 380.0	Radio components	380.0, A380.0
Air compressor pistons	355.0		413.0	Railroad car frames	520.0
	380.0	Gear housing	355.0, 380.0	Rear axle housing	356.0
Aircraft structural	355.0	Generator end plates	360.0	Reflectors	308.0, 380.0
	356.0, 520.0		380.0		518.0
Air junction block	384.0	Generator housing	A242.0	Rifle parts	360.0, 380.0
Archery bow	360.0		380.0		A380.0, 518.0
Architectural trim & fittings		Golf club cart	380.0	Rocker arm covers	355.0, 380.0
	380.0, A514.0, 518.0	Grenade launcher, rifle	A380.0	Roller mill bearings	A850.0
Automotive components	355.0	Gun control, aircraft	B295.0	Rubber molds	380.0, C443.0
	356.0, 360.0, 380.0	Hand truck	356.0, 380.0	Seat frames, railroad	B295.0
Axle housing	295.0	Hangars, spring	295.0	Sewage plant fittings	514.0
Bearings	850.0	Hardware	514.0, 518.0	Sewing machine housing	380.0
Blower rotors	355.0, 380.0	Hi-fi system horn	384.0	Sewing machine parts	413.0
Bobbins (textile)	A332.0	Impeller	380.0	Sign letters	514.0, F514.0
	360.0, 380.0, 518.0	Impeller blades	380.0, 518.0		518.0
Bomb components	380.0		C712.0	Sole plates, irons	238.0
Brackets, hardware	380.0	Instrument cases	360.0		380.0, 413.0, C443.0
	L514.0, 518.0	Intake manifolds	355.0	Spindles, adaptors	413.0
Brake shoes, aircraft	518.0	Jacks, mine	356.0	Stators	319.0, 355.0, 360.0
Bridge railing	A444.0	Jacks, railroad	356.0		380.0
Bus bar	A360.0	Jet engine case	355.0	Street lamp housings	360.0
Bushings	850.0, B850.0	Juicer, fruit	356.0, 380.0		413.0
Calculator cover	380.0	Kitchen ware	360.0, C443.0	Supercharger covers	355.0
Cam shaft cover	380.0		514.0, 518.0	Table leg	360.0, 380.0
Cam shaft timing gears	F332.0	Knife handle	380.0, A514.0	Tape recorder components	380.0
	355.0, 380.0		518.0, 520.0		A380.0

Application	Alloy
Carburetor components	380.0, 384.0, C443.0
Card rolls (textile)	A332.0, 380.0
Casters	356.0, 380.0
Clutch housing	319.0, 356.0, 360.0, 380.0
Computer equipment	380.0
Conductor, bus	EC cast
Connecting rods	B295.0, 380.0, 413.0
Cooking utensils	360.0, C443.0, A514.0
Conveyor buckets	514.0, 518.0
Crankcase	319.0, 355.0, 380.0
Cylinder block	355.0, 356.0, 360.0, 380.0
Cylinder heads	242.0, 380.0
Dairy equipment	380.0, 514.0
Directional sign boxes	380.0
Drafting eraser housing	380.0
Electrical appliances	A360.0, 380.0, A380.0, 413.0, 443.0
Electrical conductor	514.0
Electrical equipment	360.0, A360.0
Electronic equipment	A360.0, A380.0, 413.0, 518.0
Escalators	518.0
Extension housings, auto	360.0, 380.0
Large intricate parts	413.0
Lawn mower components	380.0
Light hood	384.0
Machine tool door	A380.0
Machine tool parts	356.0, 380.0
Manifolds	208.0
Master cylinder pistons	360.0
Meter parts	355.0, 360.0, 380.0
Missile electrical system	413.0
Muffler	380.0
Nameplates	360.0, 364.0, C443.0
Office machine housing	380.0
Oil filter bases	360.0
Oil pans	319.0, 356.0, 380.0
Oil pumps	360.0, 380.0
Optical measuring equipment	380.0
Ornamental grilles	308.0, 380.0, 518.0
Ornamental hardware	F514.0, 518.0
Outboard motor parts	360.0, 413.0
Outdoor light housing	380.0, A380.0
Photographic equipment	380.0, A380.0
Temperature controller	380.0
Therapeutic equipment	380.0
Timing chain covers	360.0, 380.0
Timing gears	355.0, 380.0
Tire molds	443.0, 514.0, 518.0
Torpedo propellers	360.0
Traffic signal closures	413.0
Transmission bell	355.0, 356.0, 360.0, 380.0
Transmission cases	356.0
Transmission housing	355.0, 356.0, 360.0, 380.0
Transmission plates	360.0, 380.0
Transmission valve bodies	380.0, 413.0
Transportation components	380.0
Tricot beams (textile)	A332.0, 380.0
Truck chassis	A356.0
Truss joint	535.0
Turbo-charger shroud	380.0
Typewriter frames	380.0, 413.0
Typewriter housing	380.0
Union, pipe	356.0, 514.0
Unloader cap and body	380.0
Utensils, cooking	443.0, A514.0, 518.0
Vacuum cleaner	238.0, 380.0

Table 12—continued

Application	Alloy
Fan blades	356.0
Filer housing cover	380.0
Fittings, aircraft	B295.0 518.0
Flywheel	295.0, 380.0, 384.0
Flywheel housing	319.0, 356.0 380.0, 443.0
Food handling	360.0, 443.0 514.0, 518.0
Fuel pump bodies	B295.0, 380.0
Gas barbecue grille	380.0
Gasoline fittings	319.0 D332.0, F332.0, 356.0, 380.0

Application	Alloy
Piano plates and frames	319.0
Pipe fittings	C443.0
Pistons	242.0, F332.0, 380.0
Postage meter housings	380.0
Power brake housing	360.0 380.0
Pressure-tight parts	360.0 380.0, 413.0
Process equipment	514.0, 518.0
Pump parts	356.0, 360.0, 380.0 413.0
Radar components	356.0, 360.0 380.0

Application	Alloy
Valve housings	208.0, 380.0
Valve parts	208.0, 222.0 413.0
Valve rocker arms	360.0, 380.0
Valve seat	413.0
Washing machine agitators	213.0, 360.0, 380.0
Water outlet connections	360.0
Water pump bodies	360.0
Water pump covers	360.0
Wheels, aircraft and bus	295.0
Wheels, auto, racing	380.0 A380.0
Wrench handles	514.0, 518.0

Some recycled aluminum is also sold for explosives, exothermic applications, and miscellaneous chemical use.

New products often require special properties peculiar to their use. For example, automobile sports car type wheels are largely aluminum castings and new alloys are being developed for them. In today's weight conscious world, aluminum castings are being considered for more and more uses. The recycling industry can usually develop the alloys to do the job.

The widespread application of aluminum castings is illustrated by end use (Table 12).

MARKETS FOR RECYCLED ALUMINUM

Any recycled material is of value only if there is a market for it at a price that ensures profitability. It is possible to recycle many of the materials in solid waste, but only those that have markets have been successful over the long haul.

The metal casting industry has been a buyer of recycled aluminum since the inception of the industry. Total aluminum casting use has grown from 236,000 tons in 1948 to over 900,000 tons in 1972. Over one-half of the castings produced were manufactured from recycled aluminum.

The largest single user of recycled aluminum is the diecasting industry. There are over 3500 metalcasting plants in the United States; about 1500 of these have die casting facilities. Some companies produce castings for their own use (captive) and some produce castings for sale to others (custom).

Aluminum die castings are important as cost saving components in nearly every form of consumer and industrial manufacturing. They can be produced in high volume with consistent size. They eliminate assembly since they are practically ready to use as cast.. They can be used without surface finish in many cases because of the excellent corrosion resistance. Aluminum die casting production has risen from 150, 000 tons in 1958 to 630,000 tons in 1973 (see Figure 6).

The use of aluminum in passenger cars has been rising. In 1974 it reached more than 82 pounds per car. Approximately 80% of this was die cast aluminum (see Figure 7).

Other foundries produce 40% of the aluminum castings. Other processes used are sand, permanent mold, plaster, investment and centrifugal. These methods yield castings at lower tooling costs and need less lead time.

Other markets for aluminum die castings are heavy durables, home

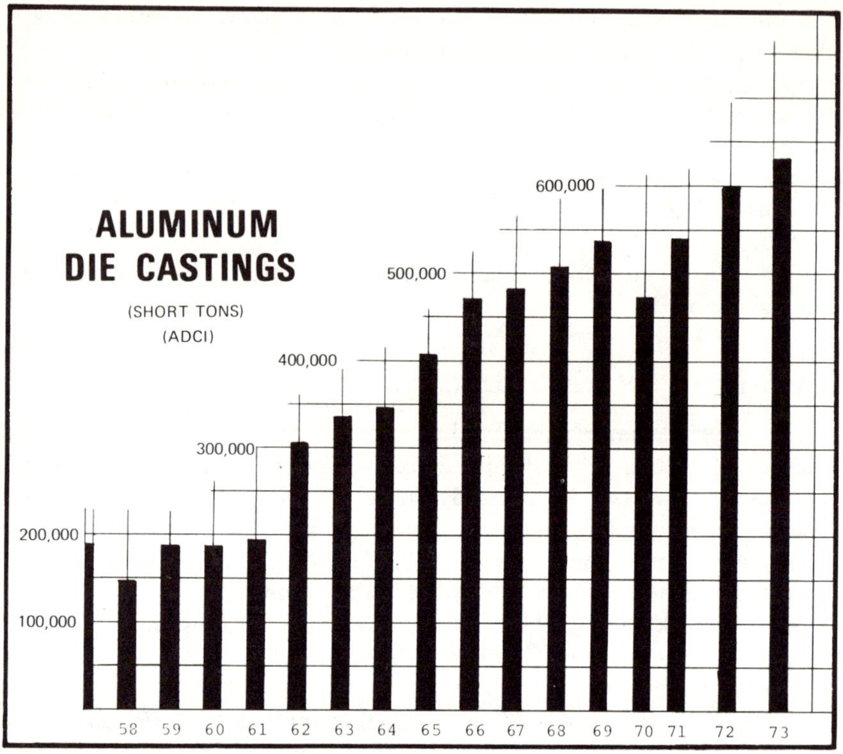

Figure 6. Aluminum die castings.

appliances, tools, machinery, office machines and equipment, photo and optical equipment, and electrical devices.

Another market for recycled aluminum is in destructive uses. It is used to remove oxygen from molten steel (deox). It is used as an alloying material to improve other metals. It is also used in explosives, pyrotechnics and exothermics.

The value added to the economy by aluminum castings in 1971 was nearly two billion dollars. The use of recycled aluminum in all areas is growing. These markets are a constant outlet which allows the recycling industry to dispose of the aluminum it is constantly processing. Research indicates that there are more markets ahead, especially for die cast aluminum.

TECHNOLOGY

In the early 1930s a group of aluminum recyclers met to establish working committees of company metallurgists. These men clearly

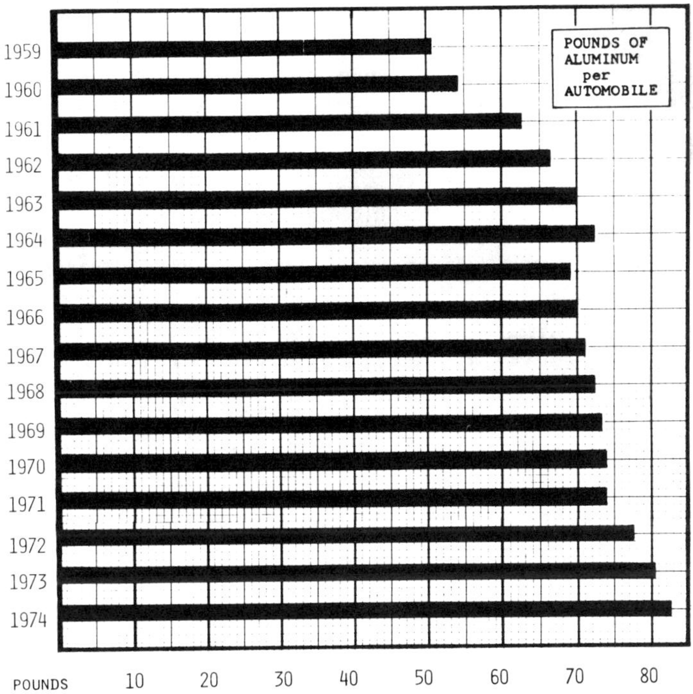

Figure 7. Pounds of aluminum per automobile.

recognized that, for the new industry to be successful in the marketplace, they would have to show the consumers of recycled aluminum that consistently high-quality, specification grade ingot would be available at all times and at competitive prices.

The alloys developed during this period from recycled aluminum met the requirements of the metalcasting industry. During World War II the scrap mix changed and many new alloys were developed to add specialized materials for the increasing applications of the foundries and die casters.

Manufacturing aluminum casting by the sand and permanent mold processes was well established commercially by 1950. Mass production technologies required consistently reliable recycled alloys. The productivity advantages of aluminum die casting was well established, and new alloys were formulated to supply this new, rapidly growing, industrial technique.

Spectrographic analysis became available to the aluminum recycler, and a few years later quantometers made possible the analysis of a melt in a few minutes instead of the hours or days required by the older wet chemical techniques. As the demand for recycled alloys grew, the size of reverberatory furnace kept pace. By 1973 furnaces with a melting capacity of 180,000 lb of aluminum were built.

In the early 1960s the technology for delivering molten aluminum had been developed, and the hot metal systems were operating. Molten aluminum is poured into preheated crucibles. Two crucibles are hauled on a flat bed trailer. Each crucible usually carries 15,000 lb of liquid aluminum. The metal is transported directly to the user's plant and discharged into his furnace.

Many related technologies were required to understand and develop alloys for specialized applications. Alloys had to be improved to meet the higher rates of casting production, castings had to have special finishes, some alloys had to withstand high temperatures (piston alloys). Alloys were developed for castings that came in contact with food, and items such as tools and appliances required castings that could be buffed and polished and retain their appearance in these consumer products.

By 1970 nearly 60% of the aluminum castings used by American industry were being manufactured from alloys produced from recycled aluminum. Specialized processes were developed to reclaim aluminum from dross and residues that would otherwise end in dumps as unused solid waste.

Recycling companies trained engineers who could offer the metalcaster technical services for any metallurgical problems that came up in his plant. Representatives of the industry are members of national committees who write specifications, test materials, and assure the consumer a consistent quality.

IMAGE IMPROVEMENT

Technical personnel who specify what material will be used in their product have not always understood or looked kindly on materials which have been reprocessed. The recyclers of aluminum have published technical literature and specification brochures for customers and potential customers.

Aluminum castings in products have often been identified by the Mark of Aluminum. The aluminum industry has spent millions of dollars to make users aware of the product improvements made possible by the use of aluminum. In 1970 the Recycle Aluminum Mark was developed to help make the consumer more aware of the value of *recycled* aluminum (see Figure 8).

By 1971 there were nearly 500 aluminum can recycling centers in operation in the United States. The public had become aware that alu-

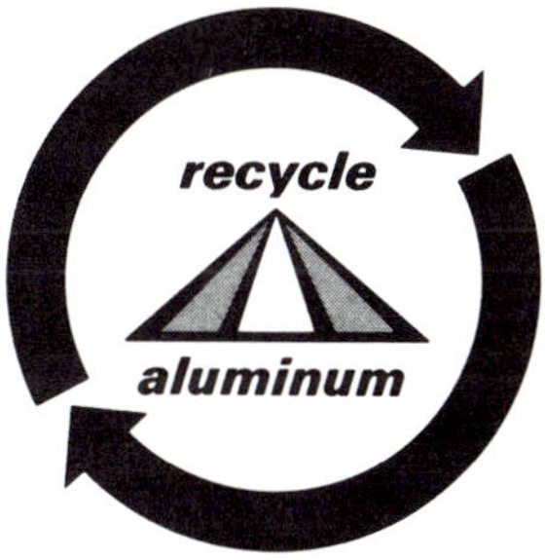

Figure 8. Logo of recycled aluminum.

minum as scrap still had monetary value and its collection eliminated litter from the landscape.

The aluminum in solid waste can be salvaged. The public is aware that if more metal is saved we are really recycling our resources.

ALUMINUM RECYCLING—WORLDWIDE

In the past decade aluminum has become a major worldwide industry. There are now 69 companies producing primary aluminum. It was inevitable that recycling industries would follow closely this world growth.

In 1970 the free world produced 1,668,000 metric tons of recycled aluminum (1,839,000 short tons) (Table 13). While exact data on how this aluminum was used are not available, it can be assumed that it was used to manufacture products in much the same pattern as has been shown in the U.S. figures. The technology required to recycle aluminum has also become sophisticated on a worldwide basis.

During 1970 few if any recyclers were producing at capacity. The aluminum recycling industry will have to help create new markets through research and development, technical services for the customer, better market planning, and aggressive selling.

Very recently recycled aluminum has been internationalized. In the past years some metal had been traded on a world basis. By 1970 metal was being transported all over the world in spite of high freight costs.

One area of important consumption for recycled aluminum ingot is in the production of automotive die castings. In many countries the cost of gasoline has been so high that manufacturers have been forced to build smaller and lighter weight cars. Aluminum components are almost a must for accomplishing this weight reduction.

World industry has quickly learned that aluminum die castings can often reduce manufacturing costs. Less secondary operations are

Table 13 Recycled Aluminum, Worldwide (Metric Tons), 1970

United States	560,200
Japan	310,000
Germany	258,500
England	214,400
Italy	154,000
France	87,400
Spain	29,000
Sweden	20,000
Switzerland	15,400
Austria	7,000
Norway	4,200
Finland	3,400
Holland	2,500
Belgium	2,000
Total	1,668,000

needed. The cost of machine operations are much less than with traditional materials.

The use of aluminum to deoxidize steel is growing on a world basis. The world consumption of aluminum for this purpose is estimated at 300,000 metric tons in 1970. Much of this is recycled aluminum.

FUTURE RECYCLING

The primary aluminum industry, through its trade association and working with government and other industry, have proposed a total recycling concept for all the solid waste collected by a city.

Although aluminum is only a small part of all solid waste, its use is growing. The proposed plan would reduce all trash to fist size chunks, pass these over magnetic separators to remove iron and steel, then incinerate or use pyrolysis for all combustibles. Exhaust gas would be cleaned and returned to the air. All noncombustibles would be processed over a Bureau of Mines type separating unit that would give five categories of material: (1) magnetic glass, (2) nonmagnetic glass, (3) nonferrous metal, (4) ferrous metals, and (5) sand. Such a system should yield approximately 6 lb of aluminim per ton of collected solid waste.

While such sophisticated installations require large capital investments, they will be tested during the 1970s. Engineering studies indi-

Table 14 U. S. Exports of Aluminum Scrap

Year	Short Tons	Year	Short Tons
1946	640	1962	65,534
1947	788	1963	71,040
1948	438	1964	68,615
1949	397	1965	38,547
1950	800	1966	48,666
1951	1,460	1967	54,532
1952	1,027	1968	49,427
1953	4,582	1969	86,256
1954	39,339	1970	57,155
1955	18,290	1971	30,675
1956	19,329	1972	66,040
1957	18,166	1973	115,120
1958	18,906		
1959	32,164		
1960	79,513		
1961	82,005		

Source: Metal Statistics (American Metal Market).

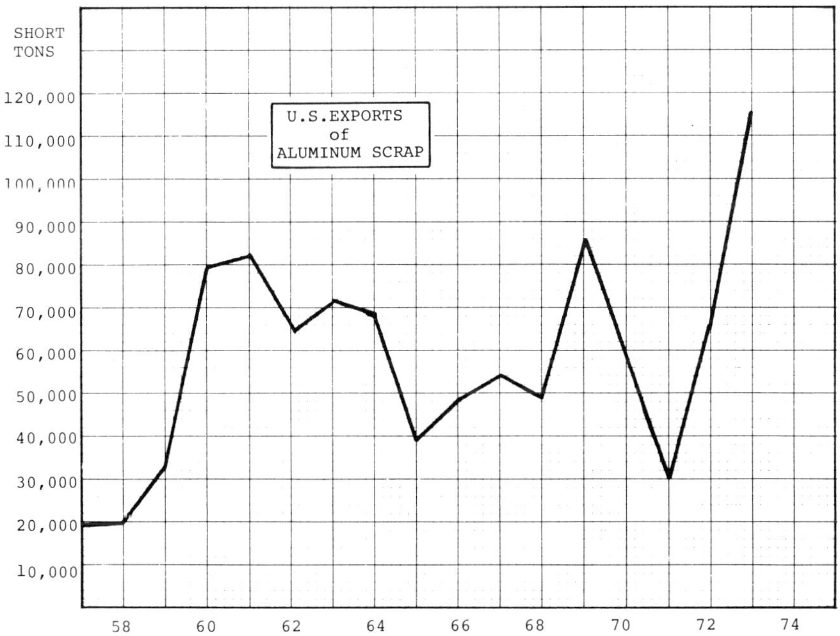

cate that by properly recycling solid waste and using generated heat and all by-products, a city might be able to develop a break-even point rather than take the losses of present less efficient systems. The recycled aluminum would have a resale value to contribute to this system.

EXPORTS OF ALUMINUM SCRAP

Aluminum scrap has become a very important commodity in the world market (see Table 14). Each ton of new industrial aluminum production generally produces about 25% aluminum scrap. During periods of very high industrial activity as in 1961, 1969, and 1973, the worldwide demand for aluminum scrap has caused exports of the scrap to increase drastically. This demand, in turn, has caused the price of scrap in the United States to almost double during these periods.

During the 1970 recession the aluminum scrap exported from the United States was only 30,000 short tons, but during the industrial boom of 1973 this rose to 115,000 tons. The United States is the largest user of aluminum in all forms, and of course the generation of scrap is the highest in the world. Bidding for the scrap in the United States by foreign industry has caused the shortages and higher prices that have made the supply and price problems very real to the aluminum casting industry, which depends on secondary ingot for a large part of its raw material.

As the industrial-intensive countries of the world fabricate more aluminum they will generate scrap on the same basis as the United States. Until the scrap generated outside the United States increases, there will continue to be periods of export and higher prices on the world market that will be felt in the United States.

Aluminum, like all valuable metals, is worth being recycled. Extensive programs for energy conservation will force more and more reclaiming of aluminum. The recycled material is in high demand for casting and other uses, so exporting will in order whenever the industrial activity is high.

CHAPTER VI.7

aluminum can reclamation and recycling program

ALAN C. KLEIN and R. J. LOGAN

Grass-roots can reclamation and recycling is a relatively new concept, but recycling of aluminum can scrap generated by manufacturers has been in effect since initial production of the all-aluminum can. A complete reclamation and recycling program involving the "closed-loop" concept is shown in Figure 1.

The production processes for can sheet and recycled cans are the same; therefore, the closed-loop concept neither eliminates jobs nor affects the economy by changing the value of aluminum. Recycling enables the producer to conserve metal reserves and power requirements. In addition to conserving aluminum, recycling saves other products alloyed with aluminum, such as manganese and magnesium.

Alcoa recognized early in the development of the all-aluminum can that market potential for such a product could exceed the aluminum industry's ability to produce enough base metal. It was thus decided to approach the production of aluminum can sheet in much the same manner as other aluminum products, by ensuring that scrap generated in can production and the can itself could be recycled. The challenge was to find a method for economically remelting aluminum containing not only protective coatings and appropriate packaging decoration but also a mixture of two different aluminum alloys. Alloy mixtures occur according to the different properties required by the body and lid of the can.

The task was to determine how best to process bales of scrap for proper segregation and separation of foreign material and elimination of potential explosion from entrapped moisture. The decision was made to shred scrap material into silver dollar size. This process used an apron feeder, machine similar to those used to reduce automobile bodies to pellets, to convey each bale into a shredding. Aluminum fines and dust generated during this cycle are collected and removed by cy-

Figure 1. The closed-loop concept.

clones to eliminate in-plant pollution and explosion hazards. After shredding, magnetic pulleys placed on the conveyor remove the remains of steel straps and any magnetic material. The shredded scrap is pneumatically processed to remove traces of paper and other nonmagnetic contaminants and cleaned to control pollution and reduce melt loss in the remelting cycle.

The usual practice in remelting aluminum is to charge the material into the main hearth of a reverberatory type of remelt furnace. Shredded aluminum can scrap, however, is not heavy enough to submerge in the molten bath. Rather than compress the material for direct charging, it was decided to use a dipwell furnace designed for remelting light scrap. This furnace has two outside wells or blisters, connected to the conventional main hearth of the remelt furnace by a submerged arch. Molten metal is transferred by a pump from one well to the other. The shredded can scrap is fed into the stream of molten metal at the second well and returns via a submerged arch to the main hearth of the furnace. The can scrap is remelted by excess sensible heat of the molten aluminum and not by radiant heat from the refractory walls of the main hearth. This method oxidizes less aluminum. The molten metal is gravity-tapped into refractory lined crucibles, transferred to holding furnaces where composition is adjusted, and poured into ingot molds. Ingots are transferred to the rolling mill for production into can sheet (see Figure 2).

Commercial research determined that nationally only 13% of all litter is cans, and that aluminum represented only 10% of the cans, just 1.3% of all litter. In 1970, most aluminum cans were in use in the West, Southwest, in Florida, and Metropolitan New York.

San Diego, California was chosen as the first test market. A relatively high volume of all-aluminum cans, an environmental concern of the citizens, and Alcoa's desire to evaluate a program under conditions of high costs, such as long freight shipments, were the basic reasons for the selection.

As critical as location, perhaps more so, was determining how best to handle the collected cans. Alcoa turned to the scrap processing industry, the National Association of Secondary Materials Industries, Inc., whose members to perform scrap handling services. To make the program effective, these services had to include the availability of large bins at all locations where cans were to be collected, routing pick-up and replacement of those bins, segregation of nonaluminum items from the collected material, weighing and paying for the material collected, packing for shipment, storage to accumulate 60,000-lb quantities, and loading of railroad cars for shipment. In San Diego, Industrial Metals and Salvage Company was able to perform these services. It also was necessary to stimulate local business, government and civic leaders to become a part of this program.

Eleven locations were established to serve as the collection centers

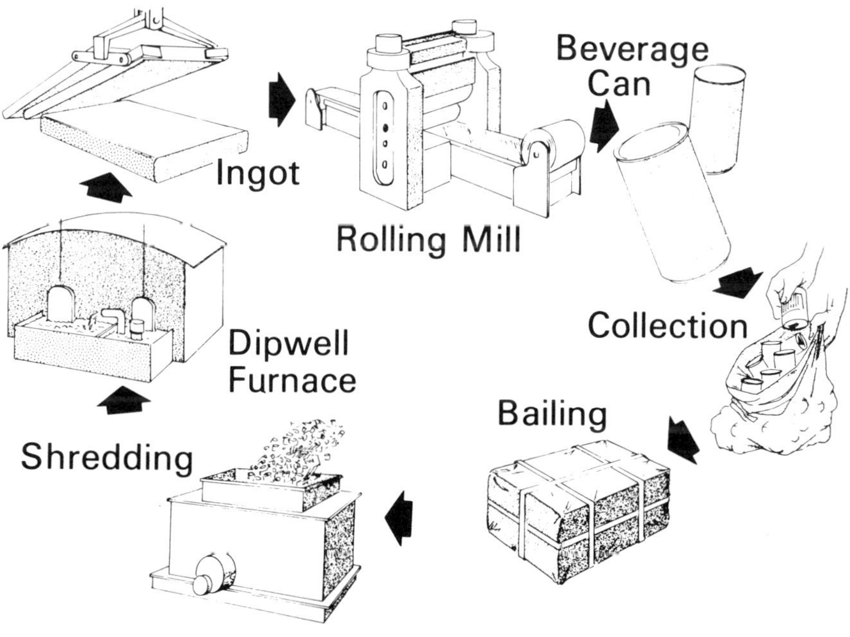

Figure 2. Manufacturing in the closed loop.

where citizens could redeem used all-aluminum cans for cash. Six shopping center sites manned by local Explorer Scouts, three Coors distributors, one Anheuser-Busch distributor, and the secondary metals processor made up the 11 collection locations. Later, private citizens, clubs, and other businesses and governmental agencies began collections, all of which were incorporated into the program.

A public announcement explained that the all-aluminum can had a rounded bottom, had no side seam, and would not be attracted to a magnet. It detailed the materials available to collectors, such as bags and magnets. Citizens were urged to collect all-aluminum cans, crush them, and bring them to any one of the collection locations. There, the cans were weighed and the collectors were paid ten cents per pound. The full bins were picked up by the secondary processor and after weighing and inspection, the cans were hydraulically bundled into 30 × 40 × 60-in. bales weighing 600 lb. The response was over 500,000 all-aluminum cans and subsequently averaged more than 600,000 cans per week, equal to 30,000 lb of recyclable aluminum.

When the secondary producer had baled 60,000 lb of all-aluminum cans, the bales were shipped to Alcoa's Warrick, Indiana plant. At that plant the recycling sequence is put in motion: unloading, shredding, steel or iron removal, cleaning, resmelting, casting, and rolling, to produce the sheet used to fabricate cans. Coils of aluminum are shipped to can manufacturers for production into more all-aluminum cans. The scrap generated at the can producer's plant is returned to Warrick, added to Warrick's own scrap generation along with the all-aluminum cans purchased in the litter program, and the "closed-loop" cycle continues.

The San Diego program was paralleled almost immediately by similar programs in Dallas and in Fort Worth, Texas, and after nearly 3 years has been enlarged to include many other localities, which return 2,000,000–3,000,000 lbs each month.

As more and more all-aluminum can scrap is removed from the solid waste stream, this material has become like other types of aluminum scrap: a commodity that is bought and sold daily as part of the operation of the secondary metals industry. This has established a market value for aluminum cans as a package with a scrap value.

CHAPTER VI.8

copper-bearing scrap

Copper-bearing waste is a heterogenuous mixture of plastics, iron articles, corrosion products, paper and insulating materials, sweepings, innumerable varieties of trash, repair waste, household appliances, etc.

Its treatment is that of a complex ore; Figure 1 gives major steps. The copper-bearing waste is charged into a blast furnace with fuel (coke or coal), producing gases whose particulates are largely zinc and lead oxides. These are collected in baghouses and processed for sale as concentrates to a zinc smelter, which will separate the zinc, lead, and cadmium.

The refinery slag and fluxes, black from iron, contain tin values. This material is processed for sale to secondary tin smelters, usually outside of the United States.

The black copper follows conventional smelting practice, starting with ores, and is purified in converters, the gases containing particulates of oxides of other metals being collected in baghouses or in combination with electrostatic precipitators.

The blister, with plant scrap, clean copper scrap, and heavy copper sections, is purified in the anode furnace and cast into anodes for electrolytic refining. A case study of U.S. Metal Refining plant of Amax follows.

The Carteret, New Jersey, plant of American Metal Climax accounts for about 10% of refined copper in the United States and 35% of all refined copper from secondary sources. Annually the unit produces more than 200,000 tons/yr of refined copper, 1 million ounces of gold (largest U.S. refiner) 20 million ounces of silver, platinum group metals, selenium, tellurium, and lead, zinc, nickel, and tin by-products (usually sold to other treaters).

The processing circuit is given in Figure 2. Technical and operating details are given by Mantell.[1]

Raw materials enter the processing at a number of points. Miscellaneous copper-bearing scrap, the residue of other preparation operations (telephone, communication and electronic scrap, containing fibres,

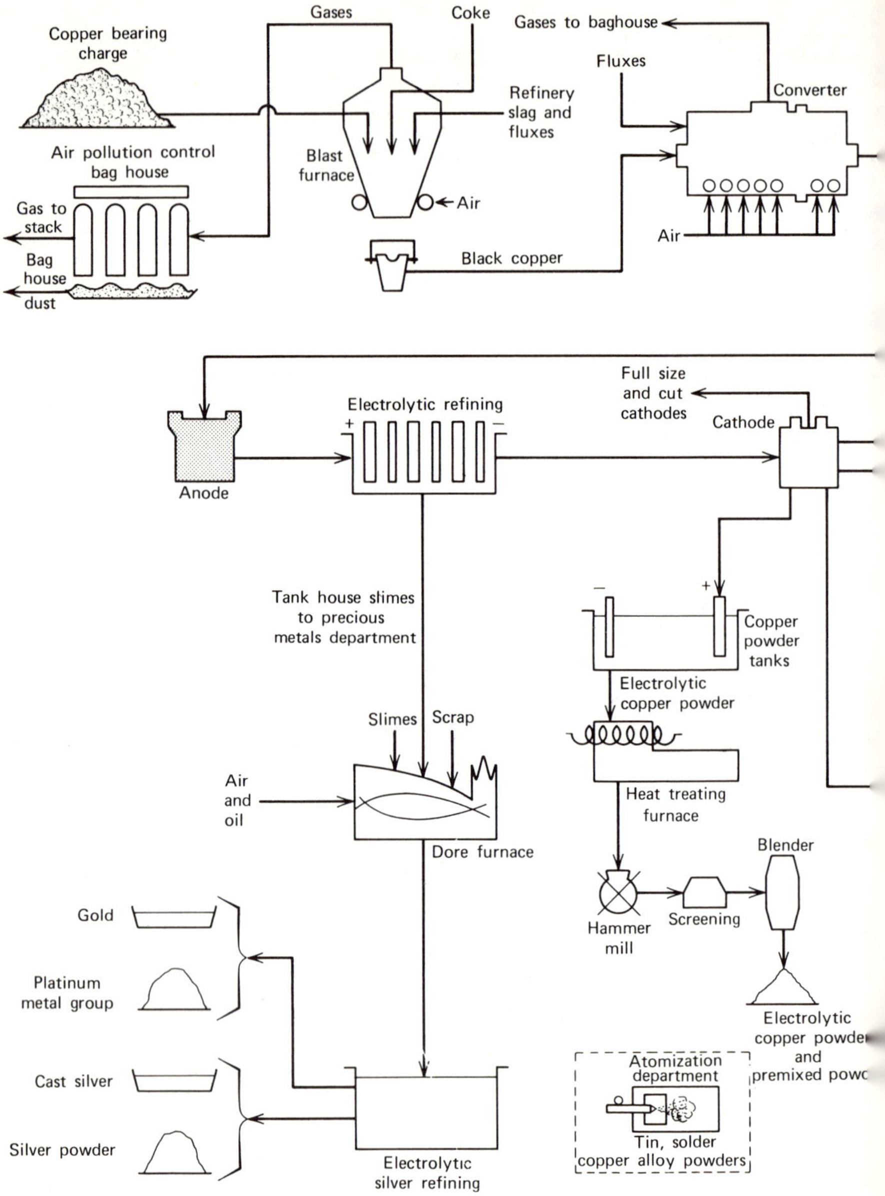

Figure 1. Refining of copper bearing scrap.

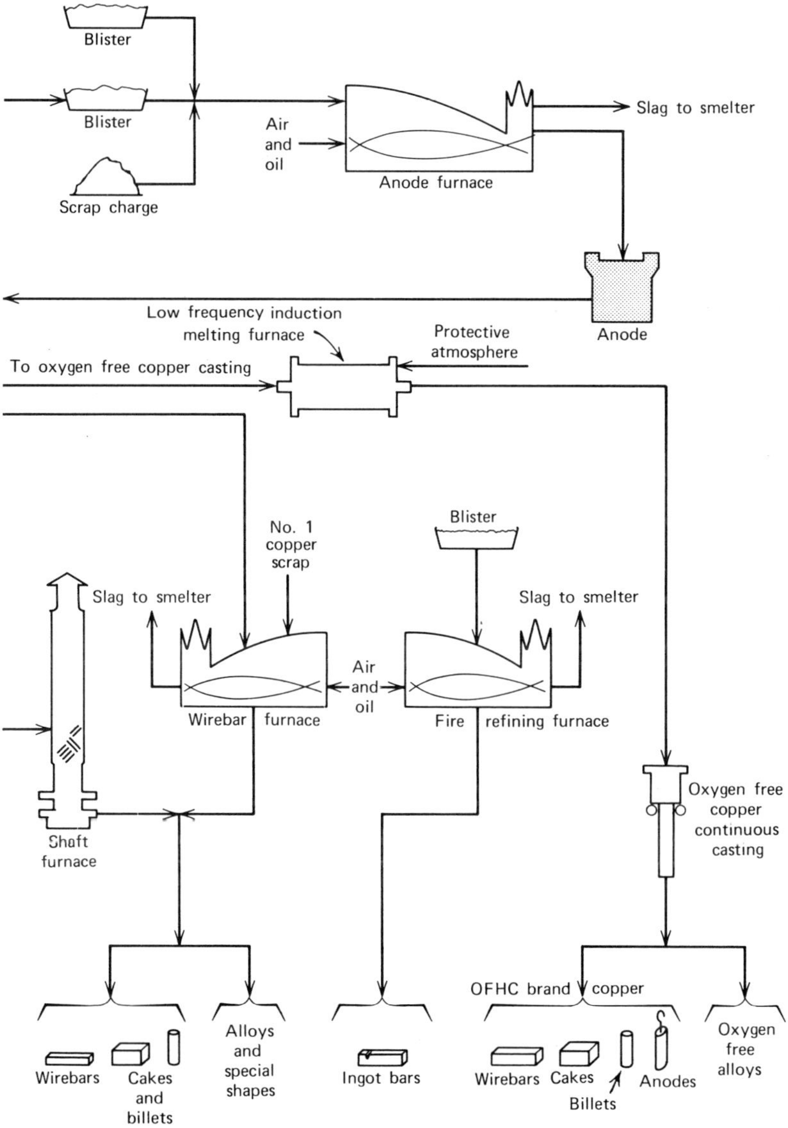

Blister
Blister
Scrap charge
Air
and
oil
Anode furnace
Slag to smelter
Anode
Low frequency induction
melting furnace
Protective
atmosphere
To oxygen free copper casting
No. 1
copper
scrap
Blister
Slag to smelter
Slag to smelter
Air
and
oil
Wirebar furnace
Fire refining furnace
Shaft
furnace
Oxygen free
copper
continuous
casting
OFHC brand copper
Wirebars
Cakes
and
billets
Alloys
and
special
shapes
Ingot bars
Wirebars
Cakes
Billets
Anodes
Oxygen
free
alloys

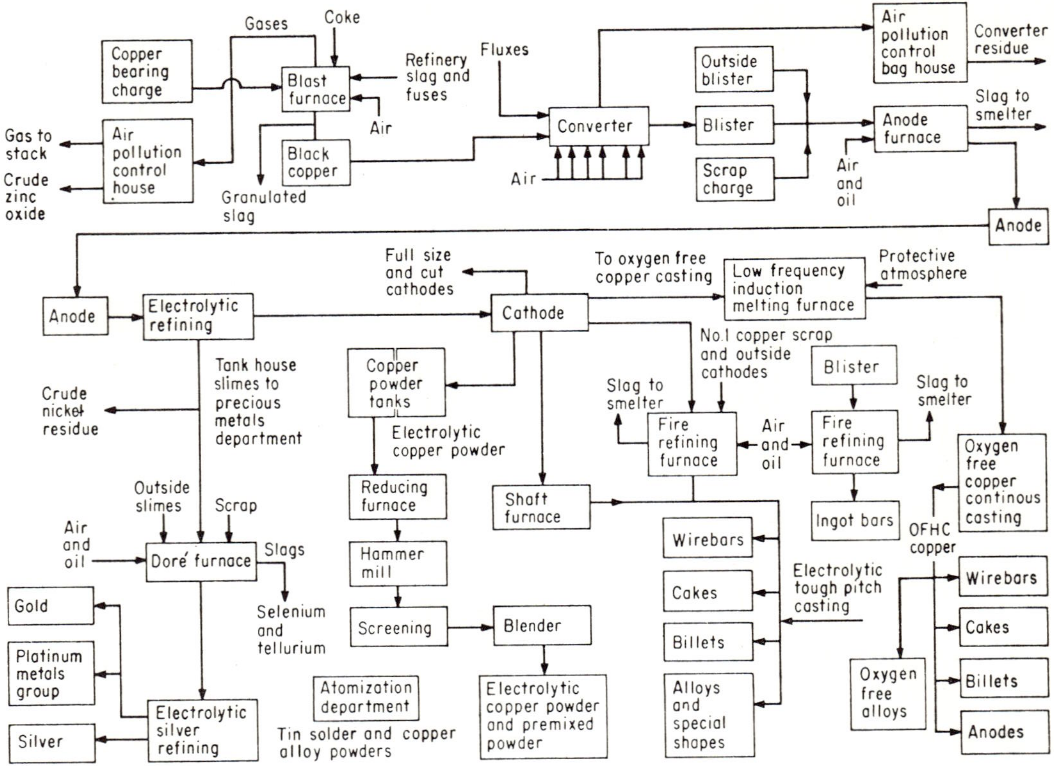

The flowsheet for the Carteret plant

Figure 2. Refining at Carteret.

insulation, plastics, iron materials), enters at the first step, the cupola blast furnace, along with other residuals. Better scrap and blister go to the anode furnace for preparation for electrolytic refining. Number 1 scrap, prepared wire and cable (free of insulation), paper, plastics, lead and iron, segregated manufacturing scrap, and outside cathodes from other plants, by-pass electrolytic refining, are remelted for casting into commercial shapes.

The plant buys scrap from hundreds of dealers, from copper and alloy fabricators, electronic manufacturers, utilities, brass mills, wire mills, and several dozen industrial plants for whom much of the scrap is handled on a toll basis, the refined copper being returned to the customer.

Buying and selling scrap is a complicated business, involving sorting into grades, at different prices; recognition of admixtures, plated, and coated materials; checking and applying highly sophisticated analytical procedures; and highly automated sampling and accounting methods.

Major production units at the smelter and refinery include a cupola blast furnace, a Peirce-Smith converter, an anode furnace, and the tank house. Precious metals extraction equipment includes doré furnaces, electrolytic extraction facilities for silver and gold, and chemical extraction facilities for platinum, palladium, iridium, rhodium, ruthenium, selenium, and tellurium.

Blast furnace charge contains the lowest grades of scrap, described as copper-bearing junk. This material averages about 33% copper—a figure that has declined by about 10% over the past 10 years. The blast furnace charge also contains some copper-bearing refinery slags.

Because of the nature of the feed, blast furnace gases are rich in metallic dusts. Metal values in this product average about 55% zinc and 10% lead, with smaller amounts of tin, copper, and precious metals. These are recovered in baghouses. Product sales net little above the cost of operating the baghouses used to recover them.

The main blast furnace product is black copper, averaging about 80% copper and containing additional percentages of iron, lead, tin, zinc, and nickel. Molten black copper is moved to the Peirce-Smith converter, which also accepts some grades of brass and copper scrap. The converter product assays 97% copper, 40–50 oz of silver per ton, and 0.5–1 oz of gold per ton.

Converter gases are routed to a baghouse, which traps by-products to be packaged and sold in the open market.

Blister, outside blister, and No. 2 copper scrap are melted down in the anode furnace and fire refined to upgrade quality to 99% copper,

60–120 oz of silver per ton, and 1 oz of gold per ton. This product is cast as 525-lb anodes for use in the tank house, where electrolysis produces cathodes that average 99.95% copper and tank house slimes containing 6000 oz of silver per ton and 200–300 oz of gold per ton.

Cathodes are routed for melting and tough pitch casting either to the shaft furnace or to an older reverberatory furnace, which also accepts No. 1 scrap. Melt from these furnaces is cast into a variety of shapes having a minimum of 99.9% copper, with silver counting as copper. Alternatively, cathode may be melted in a low-frequency induction heating furnace that excludes oxygen to produce 99.99% plus copper. This copper, Amax's proprietary OFHC brand, is cast or alloyed into a variety of products.

Electrolytic copper and blended powders are also produced, with output totaling about 7.2 million lb/yr. For powder production, the cathode is electrolyzed a second time to a powder, dried, agglomerated in a reducing furnace, ground down in a hammer mill, screened, classified, and blended, for application in a variety of sintered parts, including bearings, bushings, and brake linings.

Precious metals production started up as a natural adjunct to copper production, with tank house slimes being treated to extract valuable traces of gold, silver, and other metals. Eventually, this precious metals processing capability attracted precious metals scrap from industrial and other sources, so scrap sources now account for more of the plant's precious metals output than do slimes from the electrolytic plant.

The precious metals treatment plant is a closely guarded unit located inside the larger copper production complex. Slimes and scrap are smelted in Doré furnaces. The principal furnace product is cast as silver anode and then electrolyzed in a nitrate electrolyte to produce silver crystals. The crystals are cast for marketing.

Slimes from the silver electrolysis cells contain gold, platinum, and palladium. The gold is separated chemically, cast as gold anode, and treated electrolytically for final purification. The gold cathodes are cast as gold bars.

Platinum and palladium muds remaining after the gold separation are parted chemically and purified to produce the two metals as valuable end products.

Other end products are extracted from slags produced at the doré furnaces. These slags are divided three ways: (1) iridium metals slag, which is treated to extract iridium, rhodium, and ruthenium; (2) selenium-tellurium slag, which is treated chemically to produce selenium and tellurium; and (3) the final slag, which, defying further individual separation, is recirculated to the smelter for another run through the process.

The Carteret plant, after nearly 70 years in operation, remains an important and growing contributor to U.S. copper and precious metals supply.

REFERENCE

1. C. L. Mantell, *Electrochemical Engineering,* 4th ed., McGraw-Hill, New York, 1960.

CHAPTER VI.9

communications scrap

BERNARD O'REILLY

The Nassau Smelting and Refining Co., a branch of Western Electric, is a scrap processor for the Bell Systems communication industry, A.T.&T. The plant at Tottenville, Staten Island, New York, is one of the largest palladium metal producers because of the use of the metal in electrical contracts that come in as scrap. The plant also is a large producer of fire-refined copper in rod form, originating in scrap cable conductors, contacts, obsolete switching sets, component parts of apparatus, repair parts, and portions of communication equipment.

Early in the history of the Bell System, the telephone companies disposed of their scrap for the best obtainable prices. After the turn of the century, Western Electric Co. acted on behalf of the operating telephone companies by disposing of scrap materials. By 1909 it had become standard operating procedure for Western Electric to dispose of materials of no further value except as scrap.

The scrap volume had reached the point, in 1916, where it became necessary to organize a facility for processing scrap in Chicago; by 1920 another similar unit was set up in Kearney, New Jersey. By the late 1920s these facilities were up to capacity and consideration was being given to establishing more scrap handling units throughout the telephone system's operating areas.

In 1923 fire destroyed the Tottenville Copper Works, a fire-refined copper producer. The plant was rebuilt and expanded. Western Electric Co. was looking for a place to process its growing volume of scrap. At this time the Tottenville Copper Works joined the Bell System.

In 1931, the premises became the Nassau Smelting and Refining Co. Supervisory and engineering personnel, experienced in processing waste materials, were transferred from the parent company to Nassau and the plant was quickly equipped to process scrap peculiar to the communications industry. The advent of the telephone cable strippers and the wire bar furnace had the greatest immediate impact on Nassau by providing for the immediate return of copper and lead and alloy products.

The volume of scrap materials increased to the point that today Nassau has about 800 employees. The operating site takes up 42 acres and while some of the work is done outdoors, Nassau has about 15 acres under cover. The outdoor work areas are paved and are accessible by various types of industrial handling equipment.

The plant receives over a million pounds of scrap every day in the year. The shipping of this material to the plant requires 3000 railroad cars and 6000 trucks on an annual basis or 12 rail cars and 24 trucks daily, based on a 5-day week. Almost half of this tonnage returns as suitable raw material for manufacturing telephone equipment. The balance is composed of waste, refining losses, and materials sold on the open scrap market.

As the material is received into staging areas, it is sorted and processed into classifications. This is accomplished by hand sorting, stripping, shearing, pulverizing, magneting, and sweating. These operations upgrade the scrap by providing for better material classification and densification. This allows for more efficient refining, recycling, or sale of the material.

Basically there are three objectives:

1. Purchase Bell System scrap at a price that will net the telephone companies or Western Electric at least as much as and generally more than could be obtained by outright sale on the open scrap market.
2. Assure the Bell System a dependable supply of secondary nonferrous and precious metals and alloys such as copper, bronze, lead, solder, gold, and palladium. Secondary metal products of the type processed at Nassau are identical in quality to primary metals. Nassau annually satisfies about one-third of Bell System copper requirements.
3. Obtain the maximum scrap value for junked materials by sorting, classification and processing, or sale to other processors. Certain types of telephone equipment scrap has very low nonferrous and precious metal content and do not fit into Nassau's present refining operation. This scrapped equipment is mutilated, upgraded by the removal of iron and nonmetals, and subsequently sold on the open scrap market.

In view of the reusable condition of a portion of this category of scrap, programs are under way within the manufacturing parent company to recycle usable equipment which is not obsolete.

The magnitude of this scrap recycling operation is the result in part of the rapid obsolescence of communication equipment, which has forced the development of an efficient scrap system. The manufacturing specifications in the communications industry are the most exacting in the world's industries.

TYPES OF SCRAP

There are two types of scrap in the Bell System. The scrap is classified according to its source, that is, manufacturing scrap or distributing house scrap, and amounts to about 1 million lb/day. Scrap with which we are not concerned, such as iron and steel, plastics, textiles, and paper, amounts to about 500,000 lb/day, but is sold directly to purchasers at the collection point.

Manufacturing Scrap

Scrap generated by the Western Electric manufacturing locations represents 40% of the receipts at Nassau, while the rest represents communication equipment, repair scrap, obsolescent equipment, exchanges, wiring scrap, and cable. As this scrap material is generated, it is packaged, weighed, classified, and shipped to Nassau. Since 76 of the 92 naturally occurring elements are used in communication apparatus, care in classification is of extreme importance.

There are 22 basic product classifications of manufacturing scrap, each of which is further subdivided into grades of the secondary metal industry:

1. Aluminum and magnesium
2. Brass
3. Bronze
4. Copper
5. Iron and steel including alloys and stainless steel
6. Lead
7. Miscellaneous materials
8. Miscellaneous mixed metals and miscellaneous metals
9. Nickel and nickel silver
10. Paper
11. Rubber
12. Textiles
13. Tin
14. Zinc
15. Mercury
16. Gold
17. No. 1 contact metal, gold, silver, and platinum
18. No. 2 contact metals, pure palladium

19. Silver and silver bearing
20. Platinum and platinum metals
21. Palladium, gold, silver mixed
22. Refined precious metals

Distributing House Scrap

Distributing house scrap is old scrap in Nassau. This is material that has been used and taken out of service; it represents about 60% of the total receipts and in general is not as easily handled as manufacturing scrap because of conditions at the point of origin.

Materials are separated into major categories as follows, while omitted numbers represent highly specialized materials of small tonnage:

1. Copper-bearing scrap material containing precious metals
2. Textile and rubber covered power cable
3. Cable, alpeth or stalpeth with paper or polyethylene insulated
4. Cable, lead-covered telephone
5. Unsorted material; telephone apparatus
6. Reclaimed mercury
7. Points, contacts, precious metal, clean
8. Cable, alpeth or stalpeth telephone
9. Scrap sold after sorting
10. Cable splicing scrap
12. Solder pot skulls
13. Mercury relay switches
16. Clean sheet lead
17. Clean copper line wire and bus bars
18. Wire, underground, gopher protected
20. Insulated wire, twisted with solid copper conductors
23. Aluminum covered telephone cable with polyethylene insulated copper conductors
25. Copper line on reels
27. Tangled copper line wire
36. Telephone and switchboard cords
49. Lead-covered cable with tape and armor protection
53. Miscellaneous textile and plastic-covered switchboard wire (flameproof)
68. Metal tools, discarded
80. Copper-clad steel wire

It is apparent that items 3, 4, 18, 20, and 27 need mechanical processing, removal of paper, insulators, and the like before further processing can be done for copper recovery. It is also apparent that

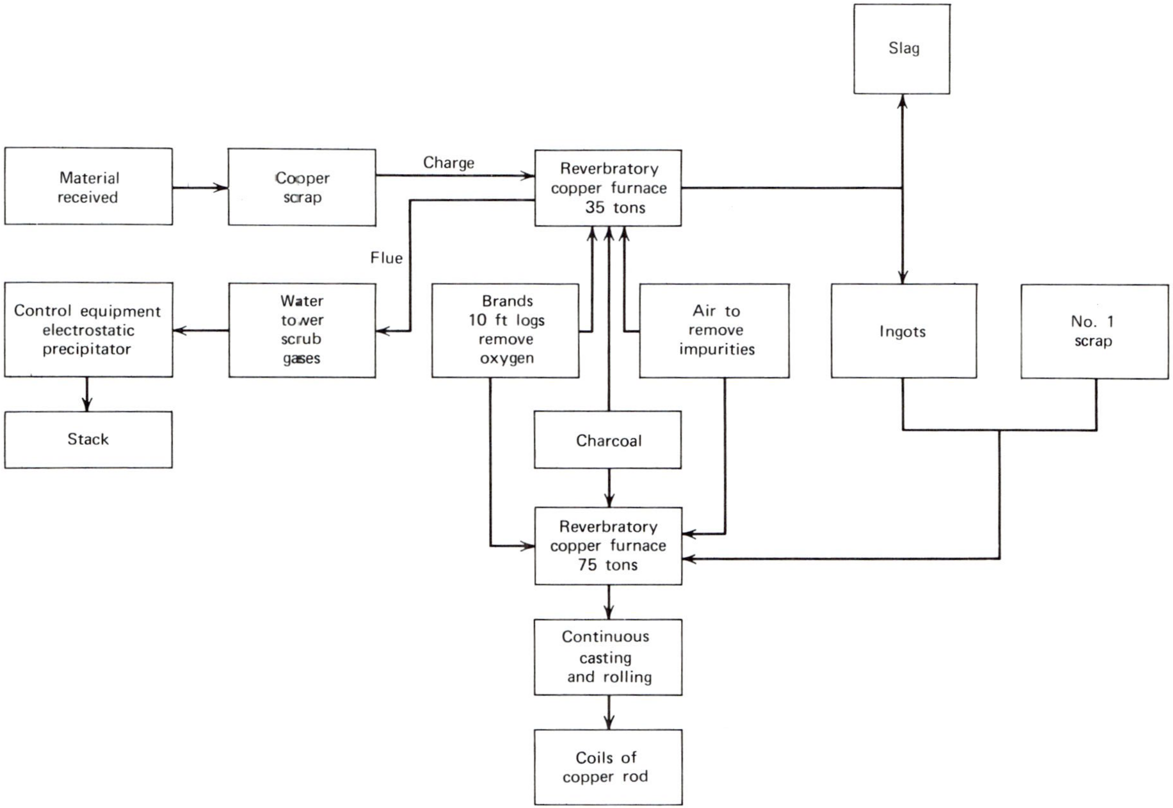

Figure 3. Flow diagram prerefining and final refining of copper from scrap.

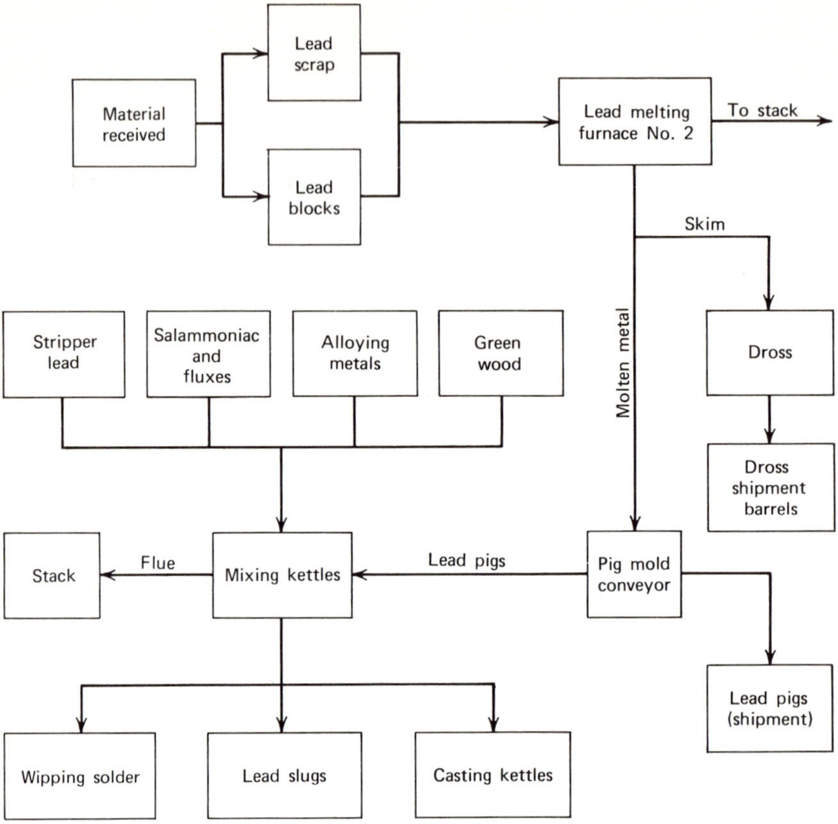

Figure 4. Flow diagram, lead refining.

there must be further sorting and classification into grades on which the market sets prices, extent, and quality of treatment or conversion into residual copper-bearing scrap for toll treatment by a secondary copper producer such as U.S. Metals Refining Co. at Carteret, New Jersey across New York Harbor.

The flow diagrams, Figures 3–6, illustrate broadly the spectrum of activity at Nassau. The ingredients common to all processes are heat, pollution control, blending of metal scrap, and refining of molten metal.

Pollution control in the refining of metal is an absolute necessity, and Nassau has top quality pollution control on all its operating equipment.

The scrubbing tower in conjunction with the electrostatic precipitator cleans smoke from the copper refinery. Scrubbing and neutralization are applied to gases in precious metal processing and in recovering copper from insulated wire. Subsequently in insulated wire processing hydrocarbons are filtered in a baghouse type of collector. The lead operations are basically for melt down with blending and do not require sophisticated pollution control apparatus.

The low-temperature pyrolysis treatment of insulated wire scrap volatilizes the hydrogen chloride, plasticizers, and a broad spectrum of hydrocarbons from the insulation. The copper itself is not significantly changed at the low temperatures used. As the scrap travels through the

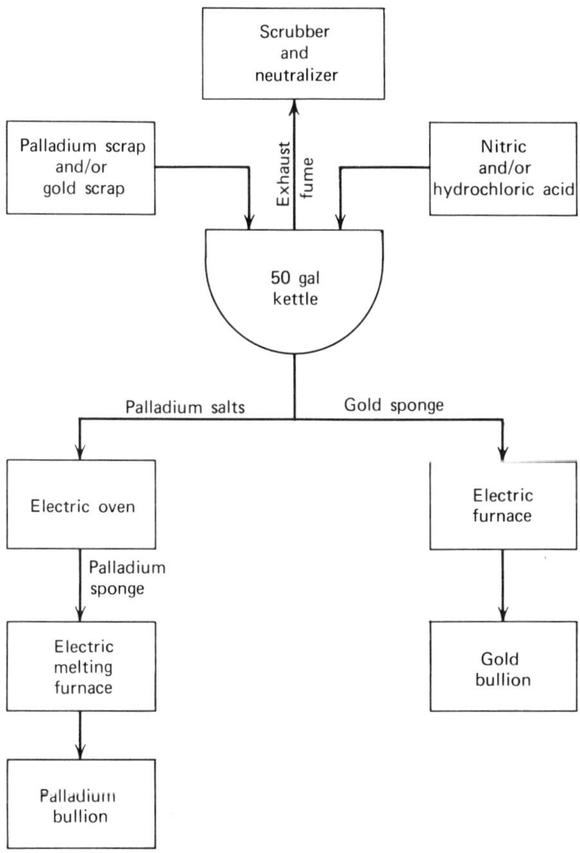

Figure 5. Flow diagram, precious metal refining.

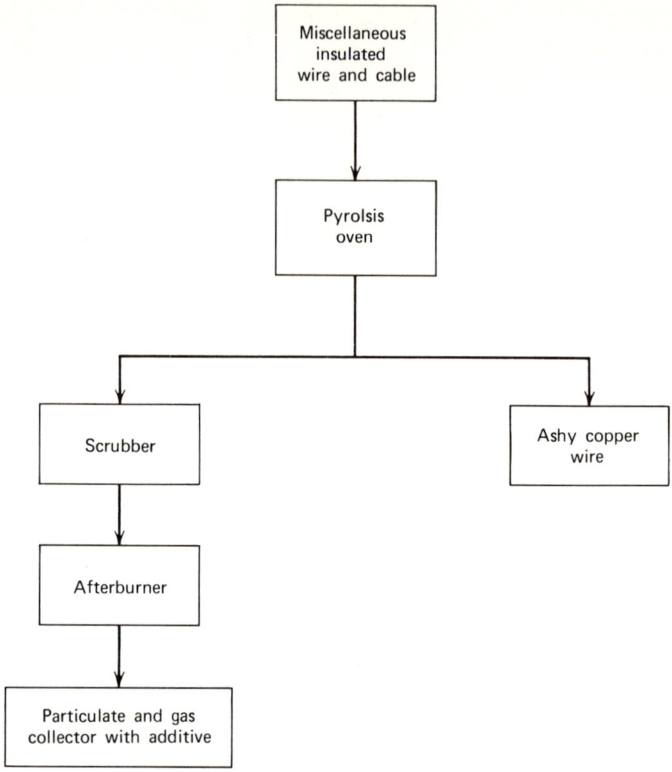

Figure 6. Flow diagram, pyrolysis of miscellaneous insulted wire.

system, the coating loses its plasticity and upon leaving the oven is subjected to vibration and impact with subsequent baling. A slight residue remains on the copper wire which causes no further problems in refining operations.

The products derived are competitive in cost and quality essential to the communications industry.

CHAPTER VI.10

zinc-bearing scrap

The marketing of zinc metal is and has long been a highly competitive trade. Ranking third in volume of world production among the nonferrous metals, zinc is distinctive among them in that a higher proportion of its total output is marketed by custom smelters than is the case for copper, aluminum, or lead. For example, Belgium ranks third in the Free World in the volume of its zinc metal production and yet only one of the three large Belgian zinc smelting companies has any directly owned mining properties and even in this case, the mines it owns (located in countries other than Belgium) produce only a fraction of its metal output. This emphasis on custom smelting leaves its mark on the ways in which zinc metal reaches the market.

Another respect in which zinc differs from its sister metals is the relatively low proportion of zinc metal production that is derived from scrap. Considerably less than 10% of the output of the world's zinc smelters is of secondary origin. It is true that there is also a substantial quantity of secondary zinc circulating in the form of the zinc content of brass scrap; but since the principal element of value in brass is the copper rather than the zinc content, this secondary trade is rarely considered in calculations of either zinc production or consumption. Further, the volume of zinc in brass scrap has only a minimal influence on zinc prices.

Throughout the world the largest use of zinc is in the coating of steel. With improvements in galvanizing techniques that have been developed in recent years, the steel industry has found that zinc-protected steel is a commodity for which markets are expanding more rapidly than for other of its products. Particularly in the developing countries, galvanized steel is an ideal construction material.

In the United States and Canada, die casting ranks about equally with galvanizing as a zinc outlet. Die-cast products enter largely into automobiles and all other forms of durable consumer goods, electrical appliances, business equipment, etc. In other parts of the world, however, brass or zinc-rolled products are relatively larger users of zinc than die casting. This pattern may change as manufacturers in Europe

Table 1 *Slab Zinc Specifications*

	Lead (max. %)	Iron (max. %)	Cadmium (max. %)	Zinc (min. %)
Special high grade (SHG)	0.003	0.003	0.003	99.99
High grade (HG)	0.07	0.02	0.03	99.90
Intermediate	0.20	0.03	0.40	99.5
Brass special (BS)	0.6	0.03	0.50	99.0
Prime Western (PW)	1.6	0.05	0.50	98.0

Shapes—all grades, the standard is a slab weighing approx. 55 lb and measuring 8½ × 17 × 1¾ in. Prime Western and special high grade are supplied by most producers in 2400-lb ingots measuring approximately 19 × 20 × 31 in.

and Japan adopt production methods comparable to those that have found favor in the United States. Another major outlet for zinc is in the production of zinc oxides, but a large proportion of the U.S. oxide output is obtained through the direct conversion of ores and concentrates.

Table 1 shows the specifications for the different marketable grades. The following discussion breaks down the uses of these grades by the four principal markets:

Galvanizing

The two major areas of galvanizing activity are galvanizing of steel shapes prior to fabrication and galvanizing of fabricated products. In the first instance, the galvanized coating is applied, as a rule, by the basic steel producer, although there are some exceptions to this on the part of companies that purchase sheet or pipe from the basic producer and then galvanize the steel purchased. Before World War II, steel companies normally bought straight PW (prime Western) zinc for their galvanizing operations. With the development of the new continuous lines for sheet and strip operations, however, this practice has changed.

Zinc metal meeting standard PW specifications is no longer satisfactory. Special qualities are specified by the major steel companies; the zinc producers now refer to this as "tailor-made" zinc. Particular attention is paid to the lead and aluminum contents. Lead content affects "spangle"—that is the pattern or glitter on the surface of the coating. Highly prized on galvanized sheets used for exterior construction, particularly in the agricultural trade, spangle is now felt to be undesirable in applications where the surface is to be painted, as in automobile production. Control of the aluminum content is considered essential in connection with adherence of the zinc surface to the steel substrate. The high rate of growth in the production of continuously galvanized steel sheets is illustrated by the fact that at the beginning of 1966 the major U.S. steel producers had in operation 45 lines for such products, five more were under construction, six more were being designed, and at least another six were under consideration.

The bulk of the galvanizing of products already fabricated is carried on by a large number of job shops scattered throughout the country. The products galvanized vary in size from nails, bolts, nuts, scrub pails, and garbage cans—to name a few consumer products—to utility towers, highway guard rails, and other major items of capital goods. Between one-third and one-quarter of the zinc used in galvanizing is applied to products already fabricated; the balance is applied at the time the basic steel shape is produced.

The die-casting trade insists on and can use only SHG (special high grade) zinc. This is true even though the zinc is alloyed with varying amounts of aluminum, magnesium, and sometimes copper before being fed to the die-casting machines. Early disappointments in the drive to expand production of die-cast products were in large measure attributed to the use of zinc with more than 0.01% in impurities, lead being the chief offender. Once the die casters learned to limit their zinc supplies to metal of 99.99% quality, the enormous growth in die-casting production began and has continued uninterruptedly since—except for occasional pauses caused by the economic cycle. The most recent growth has also been stimulated by improvements in plating methods and casting techniques.

Brass Mills

The zinc content of the various copper alloys produced by the brass mill industry varies from 40% down to as little as 7% or 8%.

Rolled Zinc

Less than 5% of the total consumption of slab zinc is in the form of rolled zinc sheet and strip.

The marketing of zinc and zinc base scrap is not a complex procedure. The principal buyers in order of importance are (1) the primary zinc smelters engaged in secondary smelting, (2) strictly secondary zinc smelters, (3) chemical plants, and (4) remelters.

This lineup is natural because primary smelters are the largest suppliers of slab zinc to the galvanizing industry, which is, in turn, the main source of scrap zinc by-products. These by-products of galvanizing are generated domestically at the rate of approximately 11, 500 tons monthly. Zinc dross comprises nearly one-half of the total, with zinc ashes and steel-strip mill-top skimmings making up most of the balance, allowing for 600–800 tons monthly for salammoniac skimmings, which usually are consumed by chemical plants. Secondary smelters are also anxious buyers, purchasing tonnages in competition with the primary smelters, particularly when they have a freight advantage.

Buying of this class of zinc scrap is generally on a continuing basis, with the slab zinc supplier usually being the purchaser either on a contractual basis or a mutual understanding arrangement. However, some primary smelters who do not process scrap, or cannot quote competitively because of a freight disadvantage, refrain from buying the galvanizing scrap, and it will flow either to another primary plant or a secondary smelter.

Pricing is usually based on a percentage of the new slab zinc market and fluctuates with it. Zinc dross and steel-strip mill-top skimmings, which are metallic in form, carry the highest value, ranging from approximately two-thirds to three-quarters of the new slab zinc market. Zinc ashes are bought mainly based on the zinc contained, with deductions for smelting charges and deleterious impurities. Some secondary smelters will purchase zinc ashes of known quality on a flat price basis. Salammoniac skimmings are purchased both on a flat price basis and on a percentage of the slab zinc market. These salammoniac skimmings are valued at relatively low prices of approximately 5–10% of the slab zinc price because of lower zinc content and impurities not readily conducive to smelting. Thus, salammoniac skimmings are sold to chemical plants that consume nearly all of the accumulations generated by galvanizers.

Scrap metal dealers, or other agents, may purchase some galvanizing residues, the amounts are not large and the arrangements are of convenience to the galvanizer. In the past, some large galvanizers would have their residues "tolled" by a primary or secondary smelter. In such a transaction, the scrap was shipped and a formulated portion of slab zinc returned. This type of disposal has given way to a less complex purchase and sale arrangement. Primary smelters who do not produce galvanizing grades of zinc do not purchase zinc scrap.

The small metal dealer who sells as fast as he buys usually sells his small lots to the wholesale dealer. These purchases are completed on an "as is" price with payment rendered immediately in cash. The wholesale dealer will sell mainly to the secondary smelter or remelter. According to government reports, the sale of this source of scrap amounts to approximately 3,600 tons/mo.

The main item of zinc scrap in this group is scrap die cast which makes up over 80% of the total. The automobile is the principal source of scrap die cast, supplying approximately two-thirds of the total. Scrap printers plates, scrap zinc castings, formed sheet zinc, and waste zinc clips make the balance. There are many small items not shown on the governmental reports, and these represent additional available tonnage for the smelter and remelter.

The secondary zinc smelter is a custom smelter. He buys from more, and usually smaller, suppliers than a primary smelter. Simple buying methods and generally more prompt payments are his chief assets in purchasing. The remelter of die cast has a somewhat different function. He attempts to purchase a more selective and less contaminated scrap die cast. By careful sorting of this choice scrap, he can produce a salable product equal to, or nearly equal to, the standard grades used by the die caster. The remelter, under normal conditions, must sell his product below the price of virgin ingot. However, he still is able to buy his selective range of die cast scrap by paying more than the smelters who use it for production by distillation.

Buying by a remelter is not contractual, but primarily on a spot sale basis, although very often from recurring sources due to close personal relationships. The remelter must carefully inspect and manually sort much of the scrap. Consequently, with increased labor costs his role as a die-cast scrap buyer becomes more difficult. He must not only compete with the secondary smelter, but he is expected to pay a somewhat higher price because he requires a selective and narrow range of scrap.

The secondary smelter, who refines by distillation, primarily needs only zinc units and is not so concerned with contamination. The smelter, therefore, is able to purchase all grades of scrap zinc and die cast from the accumulator, who usually prefers to ship all his scrap to one consumer.

In addition to the galvanizer and metal dealer, there are a few other sources of usable scrap. Aircraft builders occasionally sell zinc-base forming dies. These are always desirable for the secondary smelter and frequently attractive to the remelter. If the tonnage offer is sizable, the remelter normally will have less interest, since his tonnage requirements are usually less than those of a secondary smelter. These dies are valued at the approximate price of zinc dross or pig die cast. Sometimes the dies are unusually large, weighing many tons each, and then the price is lowered to compensate for extra melting or cutting costs.

Chemical residues are another source of zinc scrap. These consist mainly of spent textile bleaches in the form of zinc sludge. A small source that will probably be increased is flue dust from foundries and others melting zinc or zinc base scrap. Air pollution controls now being required by city, county, and state agencies will force removal of zinciferous smoke. This dust will find a market at either secondary or primary smelters. The value of this material is small because it is relatively low in zinc content, high in contaminants, and very bulky.

Most scrap zinc and zinc base alloy scrap has been carefully enumerated and specified by the National Association of Secondary Material Industries Standard Classification of Nonferrous Scrap Metals. Very little zinc scrap is either imported or exported, so the foreign market is of small consequence.

Some new trends are developing in the zinc scrap market, particularly die-cast scrap originating from automobile wrecking. In the past, all cars were wrecked by hand and the scrap iron portion cut up by an acetylene torch. Then, when the costs became too great, the car bodies were hydraulically pressed into large cubical bundles. In recent years, a large guillotine shear was developed, and this shear cut iron scrap into the sizes needed by steel mills. In each of these operations, some scrap die cast was lost because of excessive cost of removal by the scrap iron processors.

Now, a new and more efficient method has been developed that fragments the auto body by the use of a hammermill. After breaking the car body to small pieces, the iron is magnetically removed. By this method, it is expected that more of the formerly lost die cast will be salvaged.

CHAPTER VI.11
tin-bearing scrap

Bureau of Mines statistics available from 1921 show that U.S. consumption of virgin tin varied over the 50-year period from 60,000 to 70,000 long tons annually, while 23,000–30,000 long tons of secondary tin were recovered each year from scrap and reused. The sources of tin scrap are clean tinplate cuttings, trimmings and waste accumulated during manufacture of the tin-plated steel and containers; tin-lead solder as dross or sweepings; dross from tinning pots, sludges from electrolytic tinning lines, bronze rejects and used parts; babbitt from discarded bearings and type metal scrap.

The recycling of tin alloy scrap is handled by secondary melters; refineries accept solder, tin drosses, babbitt, and type metal. Automobile radiator solder has been in a closed cycle for years. The secondary operator has developed fumeless processes and produces solder to specifications by the addition of virgin metals to satisfy desired analysis. Type metal, a ternary alloy of lead, tin, and antimony, is sold with the agreement that dross, residues, spills, and sweepings resulting from his processing will be returned to the seller for credit against new deliveries. This is remelted, impurities (iron, copper, antimony, and zinc) are removed, and the scrap is returned to the market as an alloy. Copper refineries accept scrap brass and bronze articles. Copper-bearing waste is processed in blast furnaces and the tin-bearing slags are sold to the secondary smelters. Tin-bronze scrap, free from objectionable impurities, is recovered by remelting to new metal. Bronze scrap if contaminated with iron, nickel, zinc, lead, or precious metals is channeled through copper smelting and electrorefining processes.

Clean tinplate scrap generated by the tinplate mills, can and bottlecap makers, etc., may amount to 10–15% of production. The scrap is bundled, transported to detinning plants, the tin removed and recovered, and the steel returned to the mills. Tin-plated steel scrap has little economic value without treatment since the tin contaminates the remelted steel. Detinning processes are nearly 100 years old. The alkaline process is now employed; the tin is recovered as metal or as tin chemicals. Since 1941, continuous electrotinning processes have reduced the tin coating thickness. Tin recovered per long ton of tinplate scrap is about 9 lb, whereas before 1940 it was 36 lb. The return in tin values is low. Thinner steel is used in manufacture and tin-free cans have been developed. Resistant lacquer coatings must be removed, and the economics of recycling are less attractive.

Tin for tinplate is primary tin. Tinplate production produces 10–15% scrap so only 3000 tons of tin for tinplate returns to the market as chemicals or as secondary metal. Tinplate nonreturnable containers for the packaging of foods and beverages, motor oil, aerosol products, and the like are lost.

During World War II, when the United States suffered from a shortage of tin, not only did we build our own tin smelter (now a secondary recovery plant) at Texas City, Texas, but we collected cans from waste. These were referred to as "garbage cans" and were treated by detinning processes for tin recovery. The added cost of cleaning, washing, and shredding was justified by the tin shortage. After World War II, detinning plants for garbage cans ceased operation or were converted to other uses.

Steel cans from incinerated waste were not processed for tin as this constituent was effectively lost during incineration.

Mantell[1] has described and analyzed the factors in tin usage and recovery as well as the effect of replacement of hot dipped tinplate by electrolytic tinplate brought on by World War II shortages. The United States now makes four times as much tinplate per ton of tin employed as it did in the 1930–1940 period.

REFERENCE

1. C. L. Mantell, *Tin, Its Mining, Production, Technology and Applications,* 1st ed., 1930, Reinhold, New York; 2nd ed., 1949, Reinhold, New York; 3rd ed., 1970, Hafner, New York. See also W. E. Hoare, E. S. Hedges, and B. T. K. Barry, *The Technology of Tinplate,* St. Martin's Press, New York, 1965.

CHAPTER VI.12

recovery and disposal of waste photographic papers and films

RICHARD H. FORBES

Over 200,000,000 lb of photographic films and papers are manufactured annually in the United States. These consist of a light-sensitive emulsion coated on a photographically inert, flexible substrate or base that is a carrier for the emulsion. The emulsion is normally a gelatin matrix in which light-sensitive silver chloride, bromide, and iodide grains, together with special organic dyes and sensitizers, are dispersed. The inert base in photographic films is a flexible, transparent plastic sheet made from either cellulose esters or synthetic polyesters. Photographic papers have a paper base. The value of the principal raw materials used in making these films and papers exceeds $150,000,000. The gross silver usage alone is more than 50,000,000 troy ounces, in 1969 worth over $90,000,000. Only a small fraction of this material is retained in permanent archival records; the balance becomes solid waste within a relatively few years.

The waste handling challenge can be considered in two parts. Because of the nature of the manufacturing process, as much as one-fourth of the raw material used may be discarded in the production plant. Fortunately, this manufacturing waste is continuously generated at a single location and, like many industrial wastes, is an excellent source of valuable raw materials for reprocessing and recovery operations. Finished photographic products become waste when the user has no further need for the information which the exposed and processed film or paper contains. This may vary from within a few minutes of use, as in the case of picture-in-a-minute type negatives, to several years, as in the case of diagnostic x-rays. The waste from any one user is likely to be variable in quantity and type. Even though it contains the same basic raw material values as manufacturing wastes, it requires a different handling system to realize these values. Since both manufacturing wastes and trade wastes contain significant quantities of valuable materials, waste handling involves salvage and recovery wherever possible. It becomes waste disposal only when recovery is not practical. The handling of manufacturing wastes and trade wastes will be discussed separately.

In addition to the photographic films and papers themselves, packaging wastes and processing wastes are associated with the use of photographic materials. These packaging materials may be handled and disposed of in the same manner as other packaging materials. Generally they become part of general municipal or commercial trash; in some cases the volume may be sufficient to warrant segregation and salvage of the fiberboard used in boxes and cartons. Processing wastes are liquid effluents containing developers, bleaches, and fixers used in converting the latent image formed in the emulsion by exposure to a useful permanent image. Information on the disposal of these water-borne wastes is available to photographic processors.[1] Since waste water treatment is a subject in itself, the recovery of valuable materials and ultimate disposal of these water-borne wastes will not be discussed here.

MANUFACTURING WASTES

In the manufacturing process material is discarded for a number of reasons at a number of different stages in the process. In order to show how one manufacturer handles this problem, the operations of the Eastman Kodak Company at their Kodak Park Division in Rochester, New York will be used as an example. This is the largest of the seven photographic manufacturing plants that Kodak has throughout the world. It is a fully integrated manufacturing operation involving the production of chemicals and gelatin for emulsion and film base manufacture, paper base for photographic papers, all types of sensitized photographic goods, and a myriad of other supporting products such as reels, spools, and boxes that are required to deliver films and paper to the users of photographic products in the form and size that best meet their needs.

The Kodak Park plant covers over 1900 acres in the northwest section of Rochester. This discussion will concentrate on the handling of those solids wastes generated directly from the manufacture of sensitized films and papers. Because of the interrelationships in the waste handling systems at Kodak Park this will, of necessity, touch on other aspects of the waste handling, recovery, and disposal operations.

One of the best ways to handle waste is not to generate it. All raw materials entering the process are tested thoroughly to ensure that they meet the high quality standards required for the manufacture of sensitized goods. These tests include measurements for contaminants such as silicones, mercury, and radioactive elements that can damage photographic goods when present in concentrations of only a few parts per billion. Photographic emulsions are one of the most sensitive detectors that man has invented. In addition to infrared, ultraviolet, and visible light, x-rays, and gamma rays, a great number of chemical compounds affect the properties of the emulsion. These can either desensitize the emulsion so that no image is formed on exposure to light, or sensitize the emulsion so that it is totally developed. Either way, the information intended to be recorded is lost. Extensive testing of raw materials is essential to prevent waste from being produced further along in the manufacturing process.

At the Kodak Park Division, the manufacturing process for both sensitized films and papers begins with the production of the flexible base. The base is made in long, wide rolls, which must be of uniform width to be properly handled in the ensuing sensitizing operations. Consequently, waste is generated from edge trimmings. Waste is also made during machine startup, shutdown, and change over from one product type to another. These same sources of solid waste occur again in the emulsion coating or sensitizing operation in which the light-sensitive emulsion is applied to the base to make photographic film or paper. In the emulsion manufacturing operation itself, on the other hand, the wastes are primarily water-borne wastes that contain significant quantities of silver. This waste is processed in the silver recovery operations, which will be described later, to recover the silver.

Another source of waste in the manufacturing process is from the necessary testing that must be conducted at each stage in the operation to ensure that the product has the proper physical and sensitometric characteristics. Unfortunately, there is no way to determine the sensitometric characteristics, that is, the way film will react to exposure and development, other than to expose and develop the material under controlled conditions. Consequently, these are destructive tests and the material must be discarded after testing, adding to the total waste.

Packaging products for sale results in another source of waste. Because of the many needs of different users of photographic products, many different sizes of sheets and rolls of films and papers must be provided to meet the needs of the photographic industry. These range from the tiny $\frac{7}{8} \times 1\frac{3}{8}$-in. sheets of x-ray film used for dental x-rays to rolls of film and paper up to 52 in. wide and hundres of feet long, which are used for special purposes in drawing reproduction and the graphic arts. To produce all of these many sizes from the master stock rolls coming from the sensitizing process involves a high degree of planning to minimize the unusable portion of the rolls. Despite the best attempts at planning and diagraming, customers rarely demand a product mix that is the optimum for use of the material, particularly with sheet films. As a result, a certain amount of good product must be discarded because it is the wrong size.

In addition to the above sources of waste that are an inherent part of the manufacturing process, another cause of waste is manufacturing defects rejected by quality control. In addition to the normal mechanical defects resulting from cutting and handling that all manufacturers of sheet and roll products encounter, the photographic industry has a large number of specialized problems relating to the extreme sensitivity of the light-sensitive emulsion. All defects affecting the emulsion must be removed to ensure that the user of the final product will obtain the picture he desires without extraneous marks, lines, spots, or blotches.

As can be visualized from the many sources, manufacturing waste appears in a number of forms. The sensitized waste all contains silver in various amounts and is too valuable to be disposed of as trash. Unsensitized film and paper waste represents a potential source of raw materials. Handling of this material is a result of an economic balance between the costs of a recovery system that will convert the scrap to a useful raw material and the costs of the most efficient waste disposal method. Factors that influence the route chosen are the specifications that the recovered material must meet to be suitable for use and any special problems that may be involved in disposal. Film base is both more valuable and more difficult to dispose of than paper. Many photographic papers contain fillers, additives, and coatings used to increase light-reflecting power and decrease the permeability to water, all of which make reuse of recovered pulp difficult or raise the cost of processing above the economic value of the recovered material.

Polyester Scrap Recovery

Polyester film base is manufactured by reacting terephthalic acid or dimethyl terephthalate with ethylene glycol to produce a polyethylene terephthalate polymer. Additives may be used in the polymer to impart color or other properties to the finished base. The polymer melt is extruded into a thin, tough, transparent, flexible sheet having extremely good dimensional stability. Several thin coatings, known as subbing layers, are applied to one or both sides of the sheet to ensure that the final film product will have the desired properties. This base is used primarily for products such as x-ray films, portrait films, films for the graphic arts industries, and aerial films. When they have served their purpose they are returned to Kodak as waste.

Dimethyl terephthalate sells for about 20¢/lb and terephthalic acid for only a little less, so there is considerable economic incentive to process waste into a usable raw material. Several processes can be used to remove the applied subbing layers from the base so that it can be reused in manufacture. This must be effectively done before the base can be reused because traces of foreign materials can give rise to additional waste. Recovered materials must meet the same high quality standards established for all materials used in photographic products.

Cellulose Ester Scrap Recovery

The principal cellulose ester film base made at Kodak Park is cellulose triacetate, which contains various plasticizers to make the material flexible so that it can be used for film manufacture. Some cellulose acetate propionate and cellulose acetate butyrate esters are used for certain types of films because of the special characteristics these mixed esters impart to the finished products.

Cellulose acetate films are used principally for motion picture films and other roll films where the ease of splicing and simplicity of handling are more important characteristics than the toughness and dimensional stability offered by polyester films.

The film base, commonly referred to as acetate safety base, is made by dissolving the cellulose esters together with plasticizers in suitable solvent mixtures, principally methylene chloride and methanol. Plasticizers are very high boiling organic compounds that are necessary to give the base the desired flexibility. Other solvents and additives are often used in the formula to provide desired product and manufacturing characteristics. The resulting solution is a thick, viscous liquid resembling honey and known as dope.

The dope is cast onto a highly polished metal surface. Sufficient solvent is evaporated so that the material can be stripped from the metal surface and handled as a sheet. The sheet is dried to remove the remaining solvents, and the various thin surface coatings, or subbing layers, that may be required for the final product are applied. The evaporated solvents are condensed in extremely efficient refrigerated condensers. The condensate is processed by fractional distillation, which separates and purifies the individual solvents so that they can be reused many times. Scrap is generated in these operations for the reasons previously discussed. Once again the surface coatings must be removed before the scrap can satisfactorily be used in place of new esters in making film base. Aqueous washing can be used to remove the emulsion from sensitized film scrap for recovery.

Film scrap is segregated by type, chopped, and transferred to large screen-bottomed tanks equipped with high-intensity agitators. The scrap is first treated with an aqueous alkaline wash to remove the emulsion and some of the subbing layers from the base. The material is drained and rinsed several times. Water that contains more than three parts per million of silver is diverted to the silver recovery water treating system. All other water used in washing is drained to the industrial sewer and then to the main industrial waste water treatment plant. The wash water is salvaged from cooling water used elsewhere in the plant. The water used for chopping is largely recycled, so that the net usage of fresh water makeup to the washing process is quite small. This not only conserves water and heat, it also reduces the load on the main waste water treatment plant, thereby increasing its efficiency.

Following the initial treatment, the acetate film can be subjected to an oxidizing treatment to remove the remaining surface layers. Following this second treatment, the material is again thoroughly rinsed with water and finally pumped as a water slurry to a perforated holding bin where excess water is allowed to drain from the film. It is then fed to continuous rotary dryers where the film is dried by direct exposure to hot, clean, filtered air.

An alternative emulsion removal treatment, which can be used for some types of scrap, employs enzymes in place of the alkaline wash. This is a less aggressive treatment and does not affect the base to the same extent.[2] Certain types of surface coatings may not require an oxidation treatment for removal. For these, it is possible to use a simpler treatment employing a detergent to remove the coatings. The aim in processing film scrap is to use the simplest treatment required to meet the stringent quality requirements of the film manufacturing process. The dry processed scrap is sampled and tested to ensure that it meets the necessary specifications for material to be used in film base manufacture.

Many thousands of tons of manufacturing waste are processed through scrap recovery operations at Kodak Park each year. A major disposal problem is avoided by the ability to economically process these scrap materials to yield usable raw materials that meet the required specifications for the manufacture of film base.

Disposal of Unrecoverable Base Material

Film scrap that is mixed and is unsuitable for use is burned. All sensitized paper wastes generated in the manufacturing process are burned because of the difficulties in removing various materials used in manufacturing paper products from paper pulp so the pulp can be reused in manufacturing photographic paper.

For a number of years a paper recovery process was operated that used solutions of sodium thiosulfate to remove silver from sensitized paper waste. The desilvered paper was then pulped and reused in manufacturing certain grades of paper. Unfortunately, the changing technology demanded by the needs of the product resulted in a different kind of waste that no longer could be satisfactorily recycled.

Sensitized paper and film waste is segregated from general plant trash in the production departments and is delivered to a separate sensitized waste incinerators.[3] This unit has a capacity of 55 tons/day of waste having an average heat content of 7500 Btu/lb. This incinerator has an automatic ram-type feed, reciprocating grates, and a vibrating conveyor ash discharge that discharges the ash directly into storage and handling bins. The primary combustion chamber is equipped with atomizing water sprays designed to permit close temperature control and reduce silver losses.[4] The burning temperature is maintained at 1400°F by controlling the rate of feed and the rate of feed and the rate of flow of both the primary combustion air and the cooling water to the spray nozzles. The hot flue gases from the primary combustion chamber are mixed with additional air in a secondary combustion chamber. This chamber is large enough and has sufficient turbulence and mixing to ensure complete burnout of carbon containing compounds to carbon dioxide. Following the secondary chamber, the gases are drawn by the induced draft fan through a spray quench chamber where they are cooled from 1400°F to 600°F. From the quench chamber the gases pass through an electrostatic precipitator that removes over 99% of the particulates from the gas and discharges a clean, particulate-free gas stream to the atmosphere. In normal operation there is no visible plume discharging from the stack. The dust collected by the electrostatic precipitator is discharged through screw feeders into the same bins that receive ash from the incinerator. The ashes from the incinerators are transported to the silver recovery smelting operations. The cost of this incinerator and its highly efficient electrostatic precipitator was approximately $1,000,000. The value of the silver recovered in a single year is several times the cost of the incinerator.

Unsensitized paper waste generated in the paper manufacturing operations is chopped and fed through an air conveyor directly to the power boilers in the Utilities Division.[5] This is a simple, clean way to handle and dispose of this material in such a way that the heat value of the waste is recoverd through the steam generated in the power plant boiler.

Any paper waste that cannot be handled in this manner and any unsensitized film waste that cannot be recovered is transported to the general plant incinerator where it is burned along with other combustible trash from all the Rochester divisions of the Eastman Kodak Company.[6] This unit has a capacity of 89 tons of refuse per day.

Silver Recovery

The waste waters containing silver from the scrap washing process are mixed with the waste waters containing silver generated as a by-product of emulsion making. These waters amount to several hundred million gallons per year. The water also contains substantial amounts of gelatin, which tends to keep the finely divided silver and silver halide particles in suspension. The gelatin is degraded by proteolytic enzymes in the first step of the recovery process. The enzymes break down the gelatin into soluble fractions so that it no longer interferes with settling. This pretreated water is then pumped continuously through sedimentation basins at rates of up to 75,000 gal/hr. Two basins in series are used. The first is a 60-ft-diameter Dorr thickener with peripheral feed and center discharge. The second is a 70 × 70-ft cross-flow tank. The settled sludge is raked to the center and pumped from both tanks. Sufficient caustic soda and aluminum sulfate are added prior to each settling step to form an aluminum hydroxide floc, which assists in coagulating and settling the silver-bearing particles. The addition of the coagulant is important to ensure effective silver recovery in a reasonable residence time of several hours in the settling basins.

Following the settling process the waste water, which is now practically free of silver, goes to the main industrial waste water purification plant. The sediment is pumped from the settling basins and treated with hot sulfuric acid. This regenerates the aluminum sulfate and further degrades any remaining gelatin. This acid-treated mixture is allowed to settle so that the silver-bearing compounds can be removed from the liquor. The sludge collected in this settling step is dewatered by centrifuging to yield a wet centrifuge cake containing about 25% silver by weight. The silver is present primarily as silver halides along with some elemental silver and some silver sulfide. The silver is recovered by roasting, smelting, and electrolytic refining.

Roasting is done in a gas-fired Nichols-Herreschoff furnace. The centrifuge cake is dried and the organic matter burned off to leave silver-rich ash that can be conveniently handled in the smelting operations. The roasting furnace operates at temperatures of 1200–1400°F. Flue gases from the roaster are cleaned by an electrostatic precipitator that removes over 99% of the particulates from the smoke.

Smelting consists of melting the ash together with oxidizing chemicals such as sodium carbonate and calcium carbonate and fluxes such as borax and silica. Most metallic compounds are converted to oxides. Iron may be used in order to aid in reducing silver sulfide. Silver oxide is unstable and spontaneously decomposes to yield metallic silver at temperatures above 900°F. The smelting operation is carrried out at temperatures between 2100 and 2600°F, where the material forms a nonviscous molten pool or liquid in the smelting furnace. The silver metal is not miscible with the oxide slag, so it separates in a liquid layer beneath the lighter slag layer. The molten materials are drawn from the furnace and the silver separated from the slag for further refining. Gases from the smelting furnaces are also cleaned by the same electrostatic precipitator used for the roaster before discharge to the atmosphere. The slag contains enough retained silver to make it economically possible to sell it to a primary copper refiner where it is handled together with copper concentrates that contain silver.

Crude silver from the smelters contains several undesirable impurities. This must be purified to 99.95% silver before it is suitable for use in making silver nitrate for photographic emulsions. The principal impurities are copper and lead. These, together with the other contaminants, are removed in an electrolytic refining process. The crude silver from the smelter is melted in an induction furnace and cast into bars that are used as the anode of the electrolytic cell. These anode bars are placed in a cloth bag and immersed in the electrolyte, which is an acidic solution of silver nitrate. The cathode is a polished stainless steel plate. The anode is dissolved and the silver is preferentially plated at the cathode as silver crystals. These crystals are scraped from the cathode, drained, washed to remove residual electrolyte, and dried. They are then tested and used in making silver nitrate. Over 10,000,000 troy ounces of silver are refined annually at Kodak Park, all recovered from either manufacturing wastes, from photographic processing operations, or discarded films. The electrolytic cell used at Kodak Park is a modified Balbach-Thum cell in which all of the anode bars are suspended in a single tank and the electrolyte is continuously circulated through the tank. The stainless steel cathode forms the bottom of the tank.

The impurities in the crude silver are removed in two ways. Some impurities, such as gold and silica, are not dissolved by the electrolyte. These collect in the cloth anode bag and are concentrated by washing them free of all soluble portions. Soluble impurities must be removed from the electrolyte. Two processes can be used to do this. One, called cementing, uses copper to displace the silver from the electroyte solution before the solution is discarded. The silver metal is collected and returned to the refining process. The other method of impurity removal uses controlled precipitation of metal oxides from the electrolyte.[7] By proper choice of precipitating conditions, an oxide sludge containing most of the contaminants and a relatively small amount of silver can be produced. This concentrated oxide slude and the anode sludge are dried and sold to other refiners of precious metal wastes for additional recovery. Eventually, most of the metals in these wastes are recovered and returned to use through a chain of recovery operations.

The sludges from the waste water purification plant are collected and dewatered on rotary vacuum filters. The filter cake is burned in a Bartlett-Snow rotary kiln.[6] The ashes from this incinerator contain enough silver to warrant shipping them to primary metals smelters who handle them in the same way as silver-bearing ore. This is also true for the ashes from the general plant incinerator.

TRADE WASTES

Photographic products are used by many different kinds of customers to record many different kinds of things. Amateur picture-taking is familiar to everyone. But this is only one of the many ways in which photography is used today.

A major use of photographic film today is for medical and industrial x-rays. X-ray pictures, or radiographs, are essential to the modern practice of medicine and dentistry. The ability to reveal the invisible internal structure of materials as well as of people makes radiographs a vital part of inspection procedures for castings and welds used for space vehicles, atomic reactors, and similar critical applications where one defect can mean a disaster.

The motion picture industry is another big consumer of photographic products, not only for the feature films shown in theaters, but for films used on television, for educational movies, and a wide variety of commercial advertising and promotional films.

Data recording is an important use for photographic films and papers that is not so familar to the general public as previously mentioned uses. Flight test programs require the simultaneous recording of vast amounts of data during test flights so that it will be possible to reconstruct the events leading up to any kind of failure or unsatisfactory performance. Although not so dramatic, the high capacity of photographic data recording systems is important in the geologic explorations and drilling activities that accompany the quest for oil throughout the world.

The graphic arts industry is another big and growing user of photographic goods because of the speed and flexibility provided by photographic systems. Printing parallels photography in that both are techniques to process information and produce images, so that many relationships can develop between these image and information-handling technologies.

Document copying, drawing reproduction, and microfilming are related areas where photographic processes find many specialized applications, even though other systems may also be used for these purposes.

In addition to the above somewhat specialized uses, large quantities of photographic goods are used throughout industry and commerce in much the same way as those used by the amateur photographer. Aerial films are widely used in crop surveys, engineering studies, map making, urban planning, and other applications in addition to military reconnaissance.

All of these uses result in a final image on film or paper which is disposed of when it is no longer of value to the user. Like so many other wastes that are potential raw materials, the value of photographic wastes depends not only on the unit value of the material itself, but also on the amount available in any one location or group of locations. The major cost in waste recovery is in the collecting and sorting of the material so that a sufficient volume is available to allow economic operation of a reprocessing plant. The unit value of photographic waste can vary from less than \$0.01 to over \$1.25 per pound. This is principally a function of the silver content. The silver content of typical waste films varies from 0.05 to 0.10 troy ounces of silver per pound for typical commercial and professional negatives up to 0.15–0.30 troy ounces of silver per pound of graphic arts films and radiographs. Most conventional photographic paper prints have less than 0.05 troy ounces of silver per pound. Special photographic data recording papers and masters used in certain document copying and printing systems retain substantially all of the silver originally present after processing. These papers can contain 0.1–0.7 troy ounces of silver per pound of discarded paper.[8] Processed color products that have been conventionally processed through developers, fixers, and washes do not contain any silver. The image is formed by dyes. The silver is removed in processing and is reclaimed from the solutions. A recovery efficiency of over 90% is possible.

In black-and-white products the image is formed by tiny grains of metallic silver that have been developed from exposed silver halide grains. Conventionally processed black-and-white products may retain 10–90% of the silver originally present. Of course, this varies with the type of product and kind of image. Specially processed products, such as recording or document copying papers that do not go through the normal developing sequence, can retain all of the silver originally present.

The amount of waste at any given location is characteristic of the way in which photographic goods are used and whether the images are retained for long periods or discarded regularly. For example, most hospitals file radiographs and retain the files for from four to ten years. An entire year's accumulation may be discarded at one time. Since x-ray film has an emulsion on both sides, the pictures range from 8 × 10 in. to 14 × 17 in., and as many as 100,000 individual sheets of film may be discarded at one time, collecting used radiographs is attractive economically. Indeed, there are many firms that will purchase discarded x-ray film for anywhere from $0.25 to $0.50 per pound, depending on the location, the amount of material available, and the current price of silver.

Used motion picture film is also a valuable waste, again because of the amount available at locations such as film rental exchanges where prints are scrapped after they have been used. Other concentrations of material occur at graphic arts houses, printing plants, professional photographic studios, photofinishing plants, and military installations. Because of the problems involved in segregating, storing, and handling photographic wastes, a reasonable economic incentive is needed before establishing such a program. This usually means that at least 50–100 lb of waste must be available on a fairly regular basis. Of course, if accumulation takes place over a long period of time as part of a filing system, then several hundred pounds may be available at one time and recovery becomes very practical.

A number of firms are engaged in the recovery of values from photographic wastes.[8] Most used photographic materials are processed to recover the silver value only, generally by burning the waste and reclaiming the silver from the ashes by conventional metallurgical processes. One reason for this is the variety of base materials used and the mixing of bases. For many years the Eastman Kodak Company purchased used motion picture film as a service to the industry to reclaim the silver and the film base by the same processes used in reclaiming manufacturing waste that have been previously described. Increasing amounts of color products, changes in base materials, difficulties and higher costs in sorting products into appropriate categories for recovery, and the availability to the trade of alternative outlets for discarded motion picture films finally resulted in a decision to discontinue this service to the motion picture industry.

Burning of processed photographic wastes for silver recovery can be done in a variety of equipment.[9] These many be classified into two general types, depending on whether the burning rate is primarily controlled by the rate of fuel addition or the rate of air addition to the burning zone. Incinerators using fuel rate for control use an excess of air in relation to the theoretical quantity required for combustion. These are similar to the incinerator used at Kodak Park for the recovery of silver from sensitized manufacturing waste, although usually not as large or elaborate. Both dry systems, such as electrostatic precipitators, and wet scrubbers are used to clean particulates from the flue gas before it is discharged to the atmosphere. Instantaneous control of burning temperatures may be achieved by the use of water sprays or air volume, but the primary control on temperature is the amount of fuel available.

Incinerators operating with an excess of fuel and a minimum of air are batch-type units similar to coke ovens or charcoal furnaces. A batch of waste is loaded into the furnace, the furnance is sealed, and the charge ignited. When the desired temperature is reached, the air volume is reduced to control the temperature. The amount of air is below the theoretical amount required for complete combustion, hence the off-gases contain some flammable compounds. The low gas velocities in the burning chamber minimize the amount of particulates in the gas. Since the volume is small, the flue gases can be cleaned by either a water scrubber, an afterburner, or both if necessary.

The end product of incineration in all cases is an ash containing the silver. Depending on the starting material, the silver content of the ash can vary from less than 10% to over 60%. The silver can be recovered from the ash by relatively simple treatments and subsequently refined to the desired purity.

Recently, higher costs for burning brought about by the need to regulate air pollution have resulted in renewed interest in various chemical treatments that lend themselves to recovery of the base as well as the silver.

Chemical treatments, such as those used at Kodak Park for handling sensitized film scrap from the manufacturing process, minimize air pollution but may result in a water pollution potential that must be dealt with. They do permit recovery of the base material as well as the silver. In general, the process chosen is tailored to the requirements that exist for the reclaimed base. For example, it may not be necessary to remove the transparent subbing layers from film base in order to use it for some applications. In this case, only the emulsion needs to be removed.

Three major chemical processes are used. One employs hot, dilute alkaline solutions to strip the emulsion. The alkalinity can come from potash, caustic, lime, or soda ash, all of which have been used at various times. The second uses sodium hypochlorite solutions at moderate temperatures to strip the emulsion. Concentrations range from a few percent to nearly 20%. Depending on the concentration and temperature, emulsion removal can be accomplished in seconds. This technique is particularly adapted to the recovery of motion picture films or aerial films. These can be continuously treated in a series of tanks in almost the same manner as is done in film processing machines. The third treatment involves the use of proteolytic enzymes to soften and degrade the emulsion so that it can be easily removed.[2] This is similar to the technique previously described for handling some manufacturing wastes, although more time is required to remove the hardened gelatin from processed film.

The silver in these processes is collected from the waste water as a sludge by settling, filtering, centrifuging, or combinations of these techniques. Additives may be used to promote coagulation and ensure effective separation of the solids from the supernatant liquor. The silver-rich sludge is dried and processed to recover the silver in the same manner as other rich silver-bearing wastes.

Other techniques that have been proposed or used in special cases involve dissolving the silver from the film with subsequent recovery of the silver from solution. Such a process can be modified to reclaim the gelatin separately, if desired, and if a market exists for it. One simple method uses sodium thiosulfate, or fixer, to remove the silver. This is particularly adapted to unprocessed materials. Processed silver can be converted back into silver compounds by using photographic bleaching solutions such as are used in color processing. These silver compounds can then be removed by photographic fixers. The use of free chlorine to convert silver to silver chloride to permit the use of fixing baths to remove the silver has also been proposed.[10] Other agents, such as nitric acid and sodium cyanide, will also dissolve the silver from processed film and paper but are not widely used because of the hazards of handling them.

The preceding discussion has attempted to summarize the present status of solid waste handling of sensitized products in the photographic industry. Methods of recovery of waste products are described. Essentially all of the silver contained in manufacturing wastes is recovered, and a substantial portion of the base value is also reclaimed. Recovery of photographic base materials is hampered by the variety of materials used. This factor has affected the recovery and reuse of manufacturing wastes in the recent past, and is a much greater factor in the recovery of trade wastes. It is anticipated that collection and recovery systems for photographic scrap will be expanded to reclaim the silver content of the wastes.

It is estimated that about one-third of the silver on photographic goods sold throughout the world is being currently recovered, and that an equal amount is potentially recoverable with existing technology and systems at silver prices in the range of $2.00 to $3.00 per troy ounce." Since processed films and papers retain about 20,000,000 troy ounces of silver annually, they represent a significant potential source of silver. A major barrier to complete recovery exists in the need to segregate and concentrate the widely dispersed wastes so that recovery can be done economically.

REFERENCES

1. Disposal of Photographic Processing Waste," Kodak Publication J 28, EK Co 1969.
2. E. Gansel, "The use of Proteolytic Enzymes for the Removal of Gelatin from Film Base," *PSAJ.*, Sec B., **10,** 82–86, 1952.
3. T. W. Ewell and J. Piper, "Design of a New Silver Recovery Incinerator," Proc. 1970 Nat Incinerator Conf, ASME, New York, 1972.
4. W. E. Fisher and J. E. O'Brian to EK Co, USP 2944886, July 12, 1960.
5. E. K. Tanzer, "Pneumatic Conveying for Incineration of Paper Trim," Proc. 1968 Nat Incinerator Conf, ASME, New York.
6. R. L. Merle, "Kodak Park Waste Disposal Facilities," Proc. 1966 Nat Incinerator Conf. ASME, New York.
7. G. R. Green to EK Co, USP Application, Ser No 734,848, filed 6/6/68.
8. "Recovering Silver from Photographic Materials," Kodak Pub J 10, EK Co, 1969.
9. "Recovery of Silver from Discarded Photographic Films and Paper," Kodak Pub J 32 EK Co.
10. B. Rychly, "Recovery of Silver from Used Light Sensitive Materials," Rudy Metale Niezelazne 10, #4, pp 213–215, 1965.
11. J. T. Hurd and R. H. Fobes, "Profit Center: Silver Recovery," *Photo Marketing,* **45,** 31–33, May 1970.

CHAPTER VI.13
secondary gold

Gold, formerly a monetary standard and a backing for paper currency, now is an important industrial metal because of its malleability into very thin sheet, its ductility, its resistance to corrosion, its electrical conductivity and heat reflectivity, and its ability to be electroplated, rolled, brazed, joined, and alloyed.

In the United States the annual consumption of gold is more than six million troy ounces, a basic unit equal to twenty heavy weights, 480 grains, or 1.097 avoir. ounces. Approximately 60% is consumed in jewelry, about 12% in dentistry, 28% in industrial uses, primarily in electronics, switch gear, decorative applications, book binders gold, and electroplating of articles. Gold is a precious metal whose price from 1837 to 1934 was $20.67 an ounce, set by the Treasury. It was raised to $35.00 an ounce by presidential proclamation from 1934 to March 1968, when the Treasury Department sales to the domestic market ceased and a free gold market came into being, and the price was above $35.00 an ounce. Regulations governing the use and holding of gold by U.S. citizens were set under the Gold Reserve Act of 1934, which gave the Secretary of the Treasury authority to regulate the acquisition, retention, transportation, treatment, disposal, and use of gold with the requirement that specific licenses were necessary to anyone handling gold with the exception of specialized small transactions.

A major supplier of gold before March 1968 was the Department of the Treasury, while others were commodity firms licensed to buy and sell gold. Since then the Treasury has issued additional licenses to banks and commodity firms, manufacturers, refiners, and fabricators.

In 1967, the total mint issue was about 8,618,000 oz, of which 28% was from secondary sources, and 72% from primary sources. The value of the metal dictates as high a cycling rate as possible. The purity is measured in fineness, the proportion of pure gold in parts per thousand, so that fine gold, 100% pure, is 1000 fine; or in karats, in which pure gold is 24 karats, and lower numbers express the weight proportion of gold in an alloy. A 10-karat or 416 fine, therefore, contains

41.6% gold. Gold bullion is metal that has been smelted or refined; its value depends on the gold content, not its shape or form. Electroplate is a coating of gold after manufacture, with a minimum thickness of 7 millionths of an inch, while gold-filled material is covered on one or more surfaces with a layer of gold alloy to form a clad metal; with a gold content one-twentieth or more of total weight. Gold flashings are thin coatings of gold, less than 7 millionths of an inch, while heavy gold electroplate has a minimum thickness of 100 millionths of an inch.

In scrap the gold content is measured and reported as a percentage of gold. Usually 60–70% is reported in this class. It consists of all jewelry, brazing alloys, chemical ware, electronic contacts, relays, dental alloys and fillings, ion-exchange units, high-grade bench grindings, and fillings. Low-grade scrap consists of plated or brazed gold on base metal, electronic circuits, such as printed components associated with plastics, ceramic and glass decorations, gold-plated jewelry, gold polishings and sweepings, rags from cleaning or polishing of which the gold content is measured in ounces per ton. Residues are the gold-bearing product in a form other than bullion, being the results of burning, crushing, and leaching of sweeps and miscellaneous scrap. Secondary gold is the metal produced from scrap or metallurgical wastes, such as fillings, clippings, polishing sweepings, and semiprocessed material whose value depends on its gold content. Semiprocessed scrap gold is produced by melting, burning, crushing, grinding, blending, leaching, or other procedures, but is not refined. Sweeps are collected along with sweeping compound, the cleaning of benches, drawers, and floors where gold is produced or manufactured articles are made.

The several hundred million tons of municipal solid wastes handled in incinerators have residues of precious metals, silver and gold, from 0.02 to 0.05 oz/ton. To date this has not warranted recovery on an economic basis.

The secondary gold industry is primarily composed of the Bureau of Mines, precious metals semirefiners and refiners, precious metal refiner fabricators, secondary nonferrous smelters, and primary base metal smelters and refiners. The latter two are the major sources of gold that are by-products of base metal processing and refining.

Scrap flows to the secondary gold refining industry from numerous enterprises throughout the nation. The cyclic period, from product to scrap to fine gold, can range from a few days to several decades. Methods of collecting vary from collectors soliciting from door-to-door, to refiners who collect, refine, and market fine gold. Flow back can involve several intermediate steps, or it can follow a direct line of return

Table 1 *Cycle of Secondary Gold; Estimated Gold Use and Secondary Gold Returned to Supply in Thousands of Troy Ounces*

Issue of the mint, refineries, mines and scrap returned		8618 oz; 50% of issue
Jewelry	4877 oz	
New scrap	1122 oz; 23% of jewelry	
Used scrap	3755 oz; 77% of jewelry	
Dentistry	796 oz	9.2% of issue
New scrap	24 oz; 3% of dentistry	
Used scrap	772 oz; 97% of dentistry	
Industry	2945	34.2% of issue
New scrap	1178; 40% of industry	
Used scrap	1767; 60% of industry	
Consumption	6294 oz	
Old scrap	85 oz	
Domestic refineries	2296 oz	
return to mint	884 oz	
Export for refining	113 oz	
Return to supply	2409 oz	

to fine gold. Old scrap normally flows through intermediate steps and has the longest cyclic period, while new scrap follows a more direct path and has a shorter cyclic period.

The cycle of secondary gold is shown in Table 1. Precious metal scrap consists of new scrap and old scrap. New scrap is generated in the manufacturing of jewelry, dental, and industrial products. Old scrap comes through the scrap collector channel, but at times is sold directly to a unit in the recovery sector if a large volume is generated, such as old scrap from the Department of Defense.

Precious metal scrap can go to any of the precious metals semi-refiners, refiners, and refiner-fabricators, or if bullion, to the Bureau of the Mint. Associated materials limit the number of firms to which certain types of scrap can be sent, although some refineries can handle nearly any type. Nonferrous scrap, with some precious metal content, is sent to secondary nonferrous smelters, primary base metal smelters and refiners, and to selected precious metal refiners. Scrap sold for base metal content normally flows through a scrap dealer who sorts out the gold and sells it to a refiner, or sells all the scrap to a secondary nonferrous smelter or primary base metal smelter and refiner.

Primary base metal smelters and refiners are more important because (1) more industrial low-grade scrap is being generated, and (2) they can accept large volumes of scrap with a minor amount of gold, and usually recover both base metals and gold. The primary base metal smelter and refiner is an important receiver of both electronic components and other low-grade precious metal scrap. Precious metal refiners handle input measured in percent of gold and base metal refiners in traces of gold. Sweeps and obsolete electronic components may be exported for recovery of precious metals, chiefly because of the cost advantage.

Most precious metal scrap is recovered on a toll basis because of the fluctuating price for gold. Recovery charges are negotiable and are usually higher if fine gold return rather than a product is returned.

In general, all scrap ends as bullion or sweeps, which are amenable to recovery by smelting, fire refining, and chemical and/or electrolytic refining. Physical methods may be used to prepare scrap for smelting to bullion.

Scrap may be crushed or burned, leached with aqua regia, dried, screened, and shipped to a refiner. Scrap can also be hand sorted, trimmed, tumbled with potassium cyanide, and the dissolved gold precipitated with zinc dust. The zinc in the precipitate may be dissolved with hydrochloric acid or later volatilized. The dry precipitate is fluxed and smelted.

Solutions are treated in different ways. Gold can be precipitated with zinc or sodium hydrosulfite, and formed into a bullion. Solutions can also be evaporated, with the resulting residue containing as much as 90% gold.

Bullion high in precious metals content goes to a silver electrolytic cell, and the insoluble gold residue is treated and sent to the gold electrolytic cell. Residues received from consumers or semirefiners are converted to the metallic state chemically and refined electrolytically or chemically.

Chemical methods of refining can yield up to 999.5 fine gold. Some refiners claim a yield of 999.9 fine, using chemical methods. Silver can be removed by treating with hot sulfuric acid or warm nitric acid, resulting in up to 998 fineness if several treatments are made. Residues can be treated by dissolving in aqua regia, adding hydrochloric acid, and precipitating the gold by addition of ferrous sulfate or sulfur dioxide. Further washing and melting produces a gold of 999.5 fineness.

Tin-lead solders are often used, in the electronics industry, for joining gold-plated surfaces, and the gold dissolves in the solder. This gold has been recovered on occasion; electrolytic separation of solder results in the accumulation of gold in the sludge. This sludge is chlorinated and converted to the metallic state for recovery in gold electrolytic cell. Some solder has been sent to lead smelters where tin and gold are recovered. Solder can also be melted, sampled, and exported to smelters for recovery of tin, lead, and gold with a resulting recovery of metals of about 98%.

The three refineries operated by the Bureau of the Mint have processed unrefined gold and silver by electrolytic methods. Three types of cells are commonly employed, depending upon the quality of the bullion to be processed. The electrolytic silver cells are the vertical and flat cells, and the electrolytic gold cell is the Wohlwill cell.

The Wohlwill cell is used exclusively for the electrolytic refining of gold. It utilizes the gold recovered from silver cells and/or high-grade gold, 875 fine or better, for anode makeup. The anodes are suspended vertically in an electrolyte of gold chloride and hydrochloric acid. By electrolytic action, pure gold is deposited upon the fine gold cathode sheet, silver is converted to silver chloride and falls to the bottom of the cell, and base metals along with platinum and palladium go into solution in the electrolyte. When the cathodes have grown to a sufficient size, they are removed, washed, dried, melted, and cast into gold bars, 999.7–999.9 fine. Electrolytic methods are usually more economical and result in finer gold. Parting and refining charges are given in Table 2 and 3.

Table 2 *Parting and Refining Charge in Cents, Rate per Gross Troy Ounce to the Nearest Hundredth; Bullion Containing Gold*

Base metal content (thousandths)	Gold content (thousandths): Up to 250	$250\frac{1}{4}$ to 500	$500\frac{1}{4}$ to $949\frac{1}{4}$
Up to 100[a]	7	9	17
Over 100 to 250	9	13	21
Over 250 to 500	21	23	30
Over 500 to 750	27	30	—
Over 750	30	—	—

[a] Gold content 950–999.5: Refining charge 8 cents; gold content 999.6 and over, none.

Table 3 *Refiners Charges (Minimum $25.00 to $200.00)*

	$/oz Gold	$/gal or lb + $/oz RECOVERED
Solutions	1.75–$3.00	$1.00–$1.50/gal + 0.15
Sludges and Sweeps	$3.00	$0.5–$2.25/lb + 0.15–1.75
Electronic Components		$0.15–$0.50/lb + $3.00/oz
Resins		$0.70–$1.00/lb or $2.00/oz
		$1.00/lb + $0.15/oz
Alloys		$0.35–$4.00
		$0.45–$0.60/lb $0.15

Treatment charges made by the Bureau of the Mint for treating and processing gold or silver are as follows:

A melting charge of $5.00 for the first 1000 gross troy ounces of each deposit of bullion and 50 cents for each additional 100 gross troy ounces or fraction thereof, applied to the after melting gross weight.

When there is a melting loss in excess of 15% there is an additional melting charge of $3.00 for the first 100 gross troy ounces plus $1.00 for each additional 100 gross troy ounces or fraction thereof.

When the assayer determines the presence of white gold alloys in a deposit of bullion, an additional melting charge of $3.00 for each 100 gross troy ounces or fraction thereof is made. This charge is applied to the before melting gross weight of the deposit.

At the discretion of the officer in charge of the Bureau of the Mint or Assay Office, deposits of bullion which require abnormal treatment are subjected to additional charges equal to the extra cost, including remelting and retreatment if necessary. When charges for abnormal treatment are assessed, a charge is not made for an excess melting loss.

CHAPTER VI.14

recycling waste paper

J. RODNEY EDWARDS

The paper industry has recycled waste paper and rags since its very beginning in the United States. The industry was started some 280 years ago making new paper from old rags. As the production of the paper industry increased, greater quantities of waste paper have been recycled. In 1969, almost 20% of the paper consumed was recycled.

The highest rates of recycling were achieved by the industry during World War II with a high of 35.3% in 1944. In that year, Bureau of Census data show 6.86 million tons recycled. For 1969, Census data show 10.44 million tons recycled. This tonnage figure does not include "broke," the trim and off-quality paper recycled in the mill, nor does it include a portion of the tonnage returned to the mill from converting plants. Much of this tonnage is conveyed from converting plant to the mill without a financial transaction taking place; therefore, it is not likely to be reported to Census. Thus, 10.44 million tons represents only the tonnage sold to the consuming mills.

However, while the tonnage of recycled waste paper has increased, the percentage of total paper consumption recycled has decreased. This means that more paper products are entering the solid waste stream each year. Of the 58.5 million tons of paper and paperboard consumed in the United States in 1969, over 7 million tons remained in permanent use in such forms as books and building materials, or were disposed of in home fireplaces or in the case of such products as tissues, in sewer systems. Some 11.4 million tons or 20% of the total were reclaimed and reused. Most of the remaining 40 million tons entered the solid waste stream.[1]

DISPOSABILITY OF PAPER IN SOLID WASTE SYSTEMS

Recycling offers the greatest potential for reducing the volume of material that enters the solid waste stream. Recycling extends natural resources and lets their use be concentrated in the products that require virgin materials. It should be pointed out that if the paper industry was not consuming the increasing tonnage of recycled paper, the tonnage entering the solid waste stream would be further increased each year by that amount. In fact, if the amount of waste paper recycled in 1969 was disposed of as solid waste, it would have increased the tonnage collected by municipalities some 5%.

However, the percentage of the total consumption of paper being recycled as waste paper is going down, and industry is being urged to explore plans by which this trend can be reversed. To do this will require new methods of collection, transportation, sorting, and utilization to make existing and new products economically, and will require a commercial demand for these products.

Collecting and Sorting

Paper that is collected for recycling is separated from other solid waste at the point where it is discarded. Commingling of waste paper with other refuse contaminates the waste paper so that it cannot be reused to make most of the products that are made from waste paper.

The primary grades of waste paper are as follows.

Used Corrugated Boxes

The largest single source of waste paper is used corrugated boxes. In 1969, 14.5 million tons of paperboard were produced for conversion into this type of box, and about 25% of this total production was recycled. About 8–10% of this tonnage or over 1 million tons is generated at corrugated box plants in the form of box cuttings. Of these, almost all are baled in the plant and shipped directly to mills for recycling. A small percentage have various coatings and wet-strength additives than render them difficult to repulp using conventional equipment. However, where the quantity is substantial, specialized equipment is available to separate such materials as plastics from paper fibers and sometimes reclaim the plastic as well.

Corrugated boxes used to ship merchandise to retail stores, supermarkets, department stores, and factories are discarded after the contents are removed. In and around large cities where recycling mills are located, these used corrugated boxes are separated from other solid waste and shipped to mills. Some are picked up loose or in string-tied bundles. There are small in-store balers that operate on 110 V and can be wheeled down the aisles of the store. After the contents of the box are placed on the shelf, the empty box is fed into the baler. The small bales are placed where trucks can pick them up easily. Some stores uti-

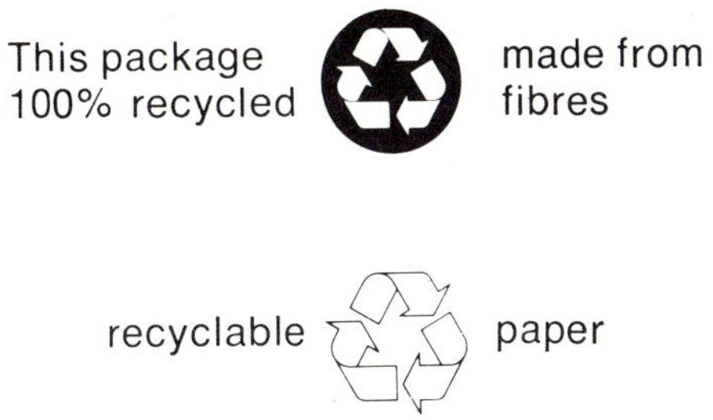

Figure 1. Logo for recycled paper.

lize very large balers installed in the warehouse area of the store. Some supermarket chains take the used boxes back to a central stock processing center where they are baled for shipment directly to the mills.

In 1969, almost 2.5 million tons were separately collected for recycling. It is expected that in the future additional quantities of used corrugated boxes will be separated for pickup by paper stock dealers because:

1. The greater awareness of the solid waste problem, along with the potential in recycling, is alerting more businesses to separate these boxes from other waste and to seek a dealer or mill that will ensure the recycling of these fibres.
2. Many stores and factories no longer incinerate solid waste, particularly paper, because of local ordinances. In those cities where municipal collection is not available to commercial establishments, a private hauler must be contracted with to remove wastes. As charges to dump this waste increase, there will be increased incentive to separate the waste for which there is a salvage market. In fact, corrugated boxes are being imprinted with a symbol (Figure 1) that immediately identifies the box as free from contaminants and recyclable.
3. Production of corrugated boxes is forecast to increase over 5% per year during the 1970s,[2] which means both clippings and used boxes will increase.

Old Newspapers

The second largest source of waste paper is old newspapers. In 1969, domestic consumption of newsprint was 9,741,000 tons, which includes about 6.5 million tons imported from Canada. In 1969 over 2.2 million tons were collected and recycled, about 23% of the total. Old newspaper is the only paper that is collected from homes in any quantity for recycling. While residents sometimes include other papers such as magazines and books with their newspapers, these items are not as easily repulped for various reasons and are sorted out by dealers packing old newspapers in bales.

Newspapers are collected from homes for the most part by Boy Scouts, by schools, various civic and religious organizations, and in a few instances by municipal sanitation departments. A pilot program conducted in Madison, Wisconsin, has produced most interesting results and a review of this project is available from API upon request.

Some newspapers are collected from office buildings, hotels, and terminals.

To increase collection of old newspapers will require one or more of the following:

1. Greater efforts on the part of community groups.
2. Greater use of receptacles that can be placed in school yards, church parking lots, train and bus stations.
3. Expanded collection by municipal or private refuse haulers such as in San Francisco, California, or Madison, Wisconsin.

Residents of many areas and particularly in Madison have shown a willingness to separate newspapers for recycling, and many more will be doing this as emphasis is placed on recycling. However, new collection systems must be provided; otherwise these papers will be commingled with other refuse and their chances of being recycled greatly reduced, if not eliminated. In England much of the recycled waste paper is collected in small trailers that are towed behind sanitation trucks.

Waste Paper Collected from Office Buildings

A waste paper grade known as "mixed papers" is collected from office buildings by paper stock dealers. The collected paper is sorted and baled for shipment to mills. In 1969, about 2.8 million tons were collected for recycling.

Included in mixed papers is a significant percentage of high-quality white paper made from virgin wood pulp. This, if it could be separated, could be recycled into the high-quality products made from waste paper. However, labor costs are too high for manual sorting and mechanical sorting systems have not been developed as yet for this purpose.

Office waste is being increasingly contaminated with other materials that cannot be removed in the repulping process. New techniques are needed to clean the repulped fibers of these contaminants.

Waste Paper from Printing and Converting Plants

The trim or cuttings from plants producing folding cartons, envelopes, business forms, ledgers, and general printing are almost 100% recycled. Most of these plants have conveyers removing these cuttings from the factory to a baling station. Here they are baled for shipment directly to the mill.

The clippings from the paper industry's converting plants amount to about 2.2 million tons in 1969 (and this does not include the clippings from corrugated box plants covered in that section of this chapter).

In the future, the recycling of these plant wastes will increase only as the tonnage of paper converted by these plants increases. The waste is easy to separate, is generated in large quantity, and full carloads or truckloads can be shipped from the generator to the consuming mill.

Another grade included in this category, but not generated in converting plants, is used tabulating (IBM) cards. These cards are discarded from data processing centers. Usually they are collected by dealers from several sources, packed in bales and shipped to mills. They are of very high quality and are used as a substitute for woodpulp. If increasing quantities of white cuttings from printing plants and IBM cards could be collected, the paper industry could use them. If ever an oversupply of these high grades developed, they could be exported.

Products Made from Waste Paper

In 1969, the paper industry in the United States produced 53.8 million tons. Of this, 26.4 million tons was classified as paperboard and 23.5 million tons as paper; the remainder includes construction grades.

Paperboard production is divided into four categories (Table 1). The paperboard definitions allow for up to 15% recycled fibers in unbleached kraft and solid bleached packaging paperboard and up to

***Table 1** Paperboard*

	1969 Tons
Unbleached kraft paperboard	11,830,300
Semichemical paperboard	3,694,800
Bleached packaging paperboard	3,529,900
Combination paperboard	7,322,800
Total	26,377,800

25% waste paper in semichemical paperboard. Combination paperboard is made from almost all recycled fibers. The definition of this grade is "manufactured from a combination of recycled fibers from various grades of paper stock with the predominant portion of its furnish being recycled fibers."

Production of paperboard has experienced a dramatic growth trend. Since 1950 production has increased from 11.1 million tons to 26.4 million tons in 1969 or almost 2½ times. However, during this period, combination paperboard has accounted for a decreasing percentage of the total.

Of the 3.7 million tons of linerboard produced in 1950 to make corrugated boxes, about 25% or 921,000 tons was made from combination paperboard. In 1969, only 347,100 tons or 3.6% of the total of 9.7 million tons of linerboard for domestic use was combination paperboard. In 1950, 23% of corrugating medium (the fluted center) was made from waste paper and in 1969 the percentage was 17.5%. However, there has been an increase in the tonnage of waste paper used to make semichemical paperboard. It is estimated that 10% of the total fiber used to make this grade is waste paper.

In boxboard, all set-up boxes have been made from recycled fibers, both in 1950 and 1969. However, production has declined from 729,000 tons to 497,300 tons during this period. For folding cartons, as late as 1960, combination paperboard supplied 75% of folding carton requirements. However, by 1969 this market share had declined to less than 62%.

Therefore, in looking at the paperboard packaging markets, while total production of combination paperboard has increased slightly, it has decreased measurably as a percentage of the total production of paperboard. The reversal of these long-established trends will not come about without considerable pressure and energy. As combination paperboard consumes 75% of the waste paper recycled, and the demand for this product is growing at a very modest rate, it is not likely that the mills producing this grade will increase their demands for waste paper. More waste paper might be used to make paperboard in future years if mills using solid wood pulp utilize increased percentages of waste paper.

Paper production in 1969 was 23.5 million tons; in the various grade categories some products are made from a furnish which consists of a portion or even as much as 100% waste paper (Table 2).

A new deinking process was developed during the early 1960s to make new newsprint out of old newspapers. In 1969 one company operated three mills which consumed 380,000 tons of old newspapers to

Table 2 Paper Production

Grade	1969 Tons
Newsprint	3,232,000
Uncoated groundwood	1,123,000
Coated papers	3,314,000
Uncoated book and writing	5,486,000
Bleached bristols	1,073,000
Unbleached kraft packaging and industrial converting	3,866,000
Other packaging and industrial converting	1,274,000
Special industrial	442,000
Tissue	3,606,000
Total	23,416,000

make new newsprint. This is one of the few examples where a waste paper grade can be recycled into the same grade. This company has announced plans to increase the capacity of their California mill 30% and plans further expansions to capacity in present mills, as well as building new mills. Increases in capacity depend upon a constant flow of good, clean, collected newspapers. That flow must be at least 350 tons/day or almost 130,000 tons/yr to support the capital investment in an entirely new paper machine.

Several grades of printing, writing, coated, and tissue papers are made using varying percentages of waste paper grades. However, these grades are usually made from the high grades generated from printing and converting plants that are 100% recycled today. Their production will increase gradually, as fast as printing plant cuttings increase, but to meet vastly increased fiber demands to produce these grades will require new methods of sorting high-grade waste paper from the mixed paper collected from office buildings. There is also a possibility that these high grades of waste paper may be available in reduced quantities. Tabulating (IBM) cards may phase out because of the increasing use of magnetic tapes. As the contaminants such as inks that are difficult to repulp, coatings, laminates, adhesives, and plastics, become more stubborn to remove in the repulping, deinking, and bleaching process, more waste will be unacceptable for reuse.

Building Products

A major user of waste paper is the building products industry. Roofing papers, flooring papers, and insulation board are manufactured using waste paper fibers. Manufacturers of gypsum wallboard are important users of waste paper, but gypsum wallboard facing is included as a combination paperboard grade. Also, it should be pointed out that the waste paper consumed to make building products does not reenter the solid waste stream until the structure is scheduled for demolition, and when this occurs the rubble is handled separately from other solid wastes and does not tax regular municipal systems.

Conclusion

Potentially, waste paper is certainly a source from which further substantial additions to the nation's wood fiber supply may be forthcoming, but some of the projections made for the future reuse of waste paper will require further critical review. For example, in a report prepared for the USHEW, the National Academy of Sciences has recommended that by 1985 the nation reuse, in effect, 35% of the paper annually consumed, as against the present 20%. Since it is estimated that by 1985 paper consumption will have more than doubled, this proposed goal implies an expansion of enormous proportions in the industry's recycling facilities. Achievement of a goal of this magnitude would almost certainly require new recycling technology, new systems of collection, new marketing programs for new products, greater market acceptance of present products, and generally the solution of far-reaching economic, engineering and environmental problems, many of which have not so far been adequately elucidated.

If the production of paper increases at the forecast rate in the next 15 years, it is inevitable that the volume of paper discarded after use and entering the solid waste stream will increase substantially, even if recycling could achieve the optimistic percentages cited by nonindustry sources.

Forecasts also show the total volume of solid waste increasing in future years at a rate comparable with that of paper, indicating clearly the need for new processing plants engineered to reduce volume prior to disposal. Such systems are most necessary for major cities where landfill space is rapidly disappearing. Modern incineration is the most promising system developed to date for reducing the volume of solid waste.

Modern incineration, using paper as an essential fuel, can reduce the volume of other solid wastes to one quarter of the original charge. Table 3 lists the Btu per pound yielded from paper items.

The view is held in some branches of government that solid waste pollution can be controlled by new legislation. This legislation may be in the form of new taxes, banning certain products, changing certain

Table 3 *Colorific Values*

Product	Moisture Content (%)	Dry Matter (%)	Dry Basis Btu/lb
Corrugated boxes	5	95	7429
Newspaper	6	94	8480
Magazines	4	96	6710
All other paper	15	85	7793

specifications and designing new types of incentives for industry. As various solutions to the solid waste problem are explored there will be an excellent opportunity for government officials and business executives to work jointly toward solving the solid waste problem.

At a time of rapidly changing attitudes and technology in the field of environmental protection, when no one can predict the developments that lie ahead, there is reason to believe that a considerable degree of flexibility in government policy on waste paper will prove essential in working out effective solutions. The entire waste paper situation continues to be under careful analytical study by the industry, which will share its findings with the responsible government agencies.

REFERENCES

1. "The Paper Industry's Part in Protecting the Environment," API 1970, p. 21.

2. "Long Range Demand for Fiber Boxes," Fiber Box Association, Chicago, Ill.

CHAPTER VI.15

waste pine stumps: basis for a modern industry

JOHN T. HAYS

A vast area in southeastern United States, from the Savannah River to the Mississippi and from southern South Carolina to eastern Louisiana, was once a 60-million-acre virgin forest of great value. By far the largest part was covered by stands of the majestic longleaf pine mixed with slash pine in the low areas.[1, 2] The widely recognized commercial value of the longleaf pine led to destructive lumbering operations that reached a peak in 1907 and continued until about 1930. By 1932 the U.S. Forest Service estimated that only about one-tenth of the virgin longleaf forest remained. Drastic cutting had led to devastation, and the trees that remained were the poorest in the original forests and insufficient to reseed the land. By 1935, the forested land had been reduced to about half its original area and consisted of only about 60% longleaf as reforestation was mostly in slash pine and other species. This shrinkage of forest land left large areas of waste land, too encumbered by stumps for agricultural use and not easily reforested without special care.[2]

RECOVERY AND LAND RECLAMATION

This legacy of a more wasteful era comprised millions of acres, now predominantly of virgin longleaf stumps. The stumps have been developed into a natural resource of striking proportions. They currently yield about half the rosin produced in the United States and nearly one-fourth of the turpentine. Since 1910, when the separation of rosin and turpentine from stumps was begun, over 10 million tons of wood or steam-distilled rosin and nearly 7 million barrels of steam-distilled turpentine have been produced.[3, 4] Impressive as these tremendous volumes of basic chemical raw materials are, the achievement of clearing millions of acres of cutover land for agricultural use and reforestation is most certainly of great long-range importance. Until relatively recently, the land reclaimed in this fashion corresponded approximately to the shrinkage in the forests. Currently, the more than one-half million acres cleared annually by stump removal is about the same in total area as the acreage being replanted by large pulp and paper companies on their land holdings in the Southeast.[5] Utilization of stumpwood as a source of naval stores chemicals, chiefly wood rosin, turpentine, and pine oil, and the accompanying reclamation of the cutover land is a continuing process. Production of wood naval stores increased steadily from 1910 to 1941, dropped during World War II, and peaked in 1950 and again in 1954–1956. Since the 1950s, there has been a downward trend, from about 1,300,000 drums to about 1,000,000 drums in 1966 and in 1967, and dropping more sharply in 1969–1970 to 816,000 drums.[6]

The continuation of the wood naval stores industry depends solely on the availability of virgin longleaf pine stumpwood. Second-growth stands are cut when the trees are about 9 in. in diameter (breast height). Consequently their seasoned stumps have essentially rotted away as the trees are too young to have an appreciable heartwood. The uprooting of old stumps is thus a mining operation with no replacement from second-growth. According to a USDA estimate, in 1961, the remaining supply of old longleaf and slash pine stumpwood totaled about 73 million tons distributed over some 27 million acres from the Carolinas to Texas. Depending on assumptions on future conditions and rates of processing, estimates of 20–30 years of continued production of wood naval stores were made.[3] Recent estimates suggest that the stump supply will allow production to continue for another 25 years.[7] Therefore, the opportunity for recovering valuable chemical raw materials still exists and the accompanying need for reclaiming and reforesting the still very large areas of cutover land appears as important as ever.

COMPETITIVE SOURCES

To get the proper perspective on the chemical raw materials from pine stumps, other sources of these materials must be considered. For centuries, exudations from living trees have been the source of naval stores products; up until the early 1900s all rosin and turpentine came from this source. About 1910 gum products began to be supplemented by wood rosin and turpentine. As the wood naval stores industry struggled to solve problems of collection, separation, and quality, its products amounted to only a small fraction of gum rosin and turpentine. From about 1922, growth of wood naval stores was steady; in 1941, production of wood rosin first exceeded that of gum, and since 1945, wood rosin production has been higher than gum with the difference increasing. Wood turpentine production has also been larger than gum but the difference is not so great since the yield of turpentine per unit of rosin is much less in wood naval stores than in gum. Thus, wood naval stores, a by-product of lumbering operations, has surpassed the earlier basic gum industry. Now the pattern is being repeated as the sulfate naval stores industry, based on by-products of the sulfate process for paper, increases in relative importance. Sulfate turpentine production became important in the 1930's and since 1954 has been the largest source of turpentine; since 1957 it has exceeded the total gum and wood turpentine. Production of tall oil rosin from sulfate skimmings began about 1950; in 1968 it was about 75% that of wood rosin and in 1969 reached 94% (6). The proportion of naval stores products from stumpwood is thus affected by supplies from the alternative sources of these products. Such factors affect fields of application and price, but the continued removal and processing of stumps and reclamation of the cutover land will depend on stump supply and economics of the total operation.[3, 4]

RECOVERY OF WASTES FROM GUM NAVAL STORES AND THE SULFATE PROCESS INDUSTRIES

The utilization of waste stumpwood has been emphasized in the preceding discussion. The other sources of naval stores materials, the gum industry and the sulfate paper process, have also reached their present status by recovery of waste products. The area covered by wounds and streaks made in the living tree to stimulate the flow of the exudate, which yields gum turpentine and rosin, was originally too impregnated with resin for use in pulping. These areas or "catfaces" were sawed off and left in the woods or used as fuel. Use of improved methods of turpentining,[8] developed and promoted by the U.S. Forest Service, have reduced the resin in the catface area to the level of that in the whole tree. This has made it feasible to use trees for production of gum rosin and turpentine and still to use the entire tree for paper pulp.

In the sulfate pulping process, cellulose fibers are freed from other constituents, including lignin and resin and fatty acids. These acids are separated as sodium salts in the "skimmings" from the black liquor and may be acidified to produce tall oil.[9] Until the late 1940s, most skimmings were burned, but expansion of the sulfate industry was accompanied by their collection and utilization for tall oil production.

This former waste material now is comparable in importance to stumpwood as a source of rosin.

COLLECTING STUMPS

Removal of stumps from cutover timber land and transporting the stumpwood to central processing plants obviously presents many practical problems. The prospect of using this operation to provide a raw material for low-cost chemical products such as rosin and turpentine would not seem immediately attractive. Nevertheless, pine stumps have been the basis for the successful development of the wood naval stores industry.

The methods for stump removal have evolved and adapted to changes over the years. In the 1920s, dynamite was used to break up the stumps. This use of explosives was in fact the initial basis for the interest of the Hercules Powder Company (now Hercules Incorporated) in the naval stores industry. (Hercules has pioneered all phases of development in the wood naval stores industry over a period of 50 years and has been the major producer.) Mechanical pulling of the stumps soon largely displaced explosives and field loading also became mechanized. The critical stump collection operation was a major effort, with companies maintaining company-owned woods camps in which employees lived, and which were relocated periodically. In recent times the importance of harvesting operations by the large naval stores producers has decreased; a major portion of the stumpwood is now collected by independent contractors and delivered to the processing plant at a flat price. This situation fluctuates with time and location.

The caterpillar tractor was a major breakthrough for stump removal. Large tractors were found to be more effective in pushing rather than pulling the stumps out. Harvesting equipment became larger and more efficient. Then in the 1960s a complete reversal took place. The multithousand acre tracts of stumps began to disappear and harvesting of small, scattered, low-yield plots became more common. The latest development is a three-man mobile unit capable of harvesting 500 tons of stumps per month at the lowest man hour per ton rate yet achieved.

Currently a splitter yoke is used to break up stumps before pushing. A field loader tractor with a hydraulic boom loads the stumps onto a steel pallet. The pallets are delivered to a huge, tandem-axle, diesel truck designed for both off- and on-highway use. The pallets are drawn onto the chassis and hauled to a rail siding where another boom loader transfers the stumps to rail cars that carry them to the processing plant.

Stumps are delivered to the plant currently at a harvesting and collection cost of $10–15 per ton. This does not include payment to the land owner, which is a smaller but generally significant cost under present conditions. The maintenance of low stumpwood costs over the years by ever-increasing efficiency in the face of increasing distances from the processing plant is a major achievement and one that has been essential for successful operation in the wood naval stores industry.

Not only does the owner receive payment for stumpwood (under various types of contracts), but the cleared land is suitable for agriculture. It is immediately suitable for grazing or for cultivation; an increasing portion is being reforested since much of the stump land is owned by pulp and paper companies. Reclamation of the land is a most attractive bonus for the owner and for the nation. For the past 30 years, land has been reclaimed at rates of 500,000–750,000 acres/yr.

PRODUCTION OF ROSIN AND TURPENTINE

A typical analysis of the wood of an aged virgin pine stump, as it is received at the processing plant, shows 19% rosin, 4% turpentine, 4% of a gasoline-insoluble resin, 23% water, and 50% of cellulose and lignin type of materials.[10] The harvested stump at the processing plant is first reduced by means of hoggers and shredders to small splinter-type chips suitable for extraction. Chips may first be steam-distilled to remove the oils (hence the designation steam-distilled rosin for wood rosin). The preferred method, however, is to extract both the oils (turpentine and pine oil) and resinous materials (rosin and other resins) in a single step under pressure. The solvent used determines the composition of the extract. Aliphatic solvents (e.g., gasoline) extract only oils and rosin; aromatic solvents extract additional more polar resinous materials that have commercial uses.[10, 11]

As indicated by the stump analysis, rosin is the major component of the extract. As first obtained it is of limited utility because of its dark color. The crude rosin may be refined to remove color by an adsorbent, such as fuller's earth, or by selective solvents, such as furfural.[11, 12] The latter process consists of a liquid-liquid extraction in which an aliphatic hydrocarbon is used as the solvent for the refined rosin and furfural for the highly colored components. The operation is flexible in that a range of color grades of pale rosin can be obtained by varying the solvent ratio. In this way, refined wood rosins are obtained that have uniformity superior to gum rosin and are equivalent to the corresponding color grades of gum rosin. This degree of refining was

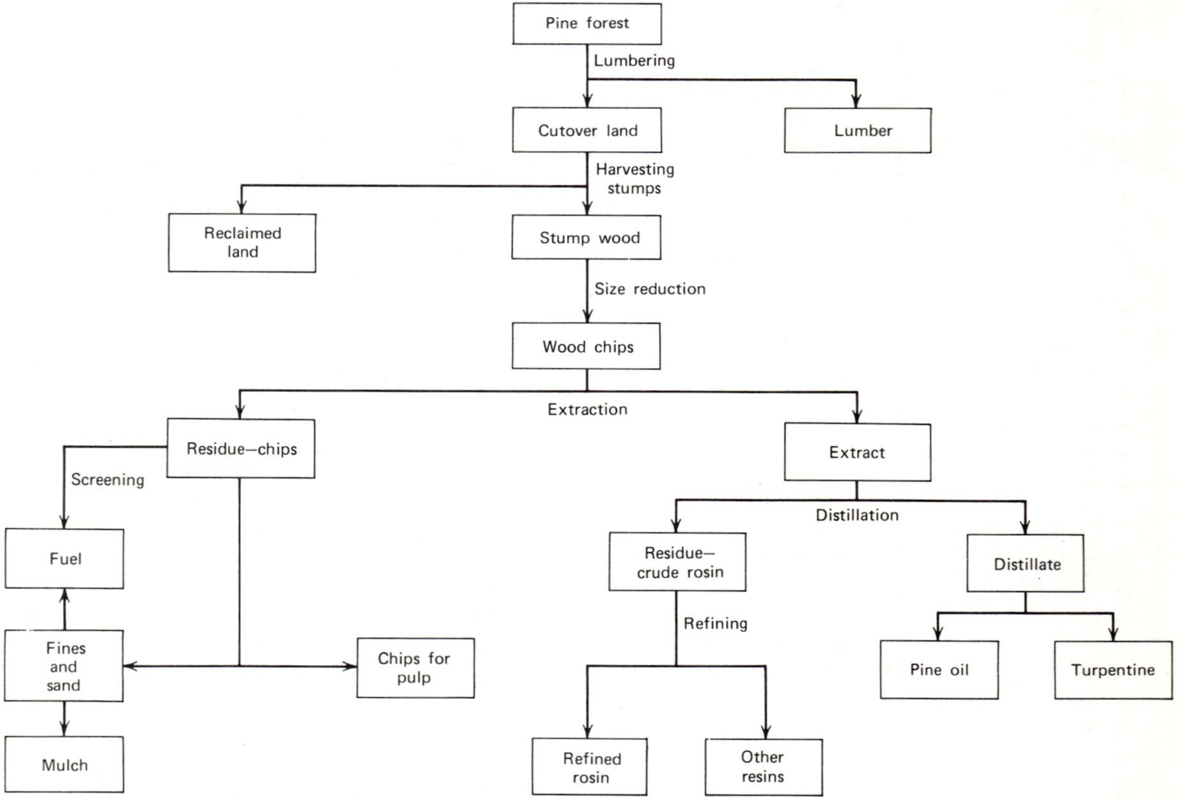

Figure 1. Flow sheet for rosin.

necessary to ensure the commercial profitability of the wood naval stores industry. Figure 1 is a generalized flow sheet showing the steps in the production of rosin and other products.

ROSIN MODIFICATIONS

Once the color problem in wood rosin was controlled, additional steps were required to maintain and improve its economic status. It was necessary to overcome deficiencies, such as oxidation by air, crystallization from solvents, and reaction with heavy metal salts, as in varnishes. Such difficulties have essentially been eliminated by chemical modifications. These reactions involve the two chemically active centers in the resin acids that make up rosin, the double bonds and the carboxyl group.[10, 11]

Double-bond reactions (Figure 2) involve reactions with maleic

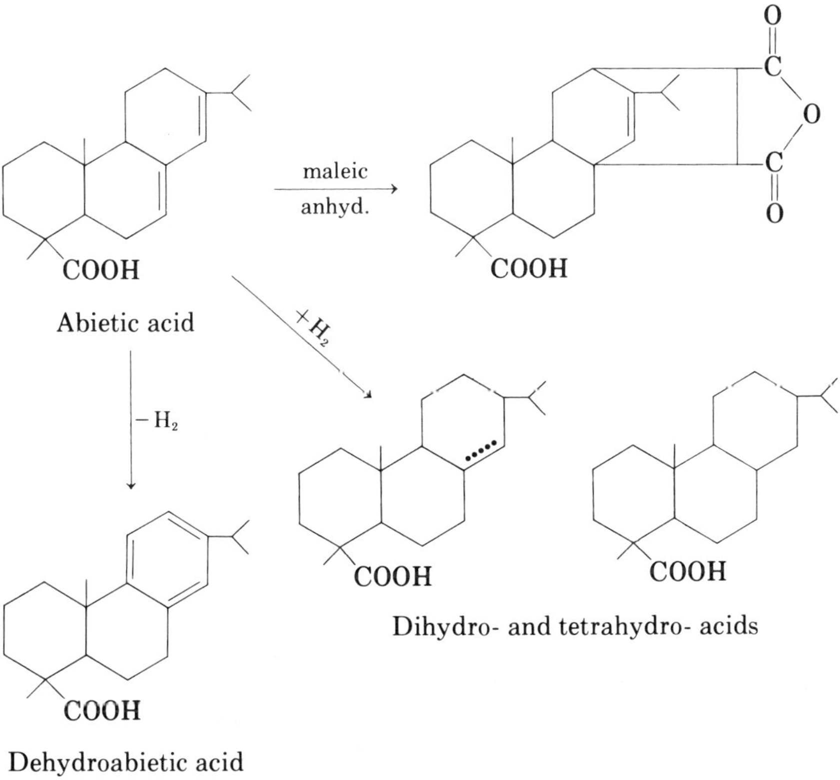

Figure 2. Chemical modification of rosin.

anhydride (or fumaric acid), hydrogenation, dehydrogenation, polymerization, and addition of formaldehyde. These reactions have led to commercial modified products used in synthetic rubber, paper size, adhesives, coatings, and inks in a manner not possible for unmodified rosins. Reactions at the carboxyl group include salt formation (the basis for production of rosin size), esterification, hydrogenolysis, and ammonolysis. Esterification, frequently in combination with double-bond reactions, gives products used in lacquers, printing inks, coatings, and adhesives. Hydrogenolysis gives a primary alcohol and ammonolysis a primary amine of commercial interest.

The net result of efforts on chemical modification of rosin has been to divert almost all of the rosin into production of upgraded modified products with improved markets and profitability.

TURPENTINE UTILIZATION

In the early stages of development of the wood naval stores industry, turpentine brought a high price as a solvent and paint thinner. This is a minor use today with most of the turpentine serving as raw material for chemical processes,[8] once more to a considerable extent due to the continuing effort to upgrade the initial products from waste pine stumps.

Steam-distilled wood turpentine from southeastern stumps consists chiefly of α-pinene, less than 5% β-pinene, and 10–15% camphene.[13] α-Pinene has two major uses: hydration to synthetic pine oil and isomerization to camphene. Pine oil is used in flotation, as a bactericide, as a dispersing agent, and as an odorant. The major use of camphene has been in chlorination to toxaphene, a large-volume commercial insecticide. The hydration and isomerization reactions give rise to by-product *p*-menthadienes, which may be dehydrogenated to *p*-cymene, is the raw material for production of *p*-cresol by an oxidation reaction[13] (Figure 3.)

These developments provide a much more stable and diversified base for marketing wood turpentine than the original solvent uses.

EXTRACTED WOOD CHIPS

On a dry basis, about 60% of the original stumpwood is made up of nonextractable material consisting largely of cellulose and lignin type materials,[10] and this corresponds approximately to the dry weight of extracted wood chips obtained after the rosin and turpentine are removed. The "spent" chips after extraction generally contain about 35% moisture along with about 5% sand. For years it has been cus-

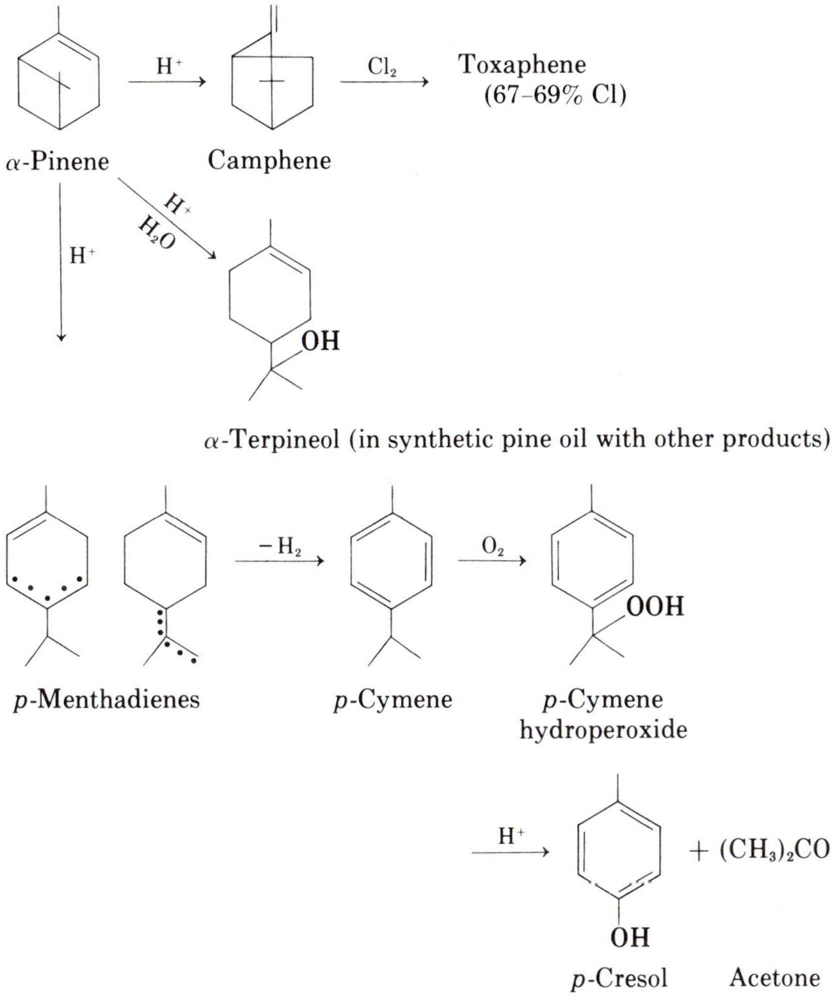

Figure 3. Uses of alpha pinene.

tomary to burn these chips for fuel. A relatively large ash is obtained owing to the sand present, and this is separated for use in landfill operations. The chips have a direct fuel value as high as $5 per ton in some locations. Use as a fuel thus recovers a significant portion of the cost of the stumpwood and helps to balance the economics of the entire operation.

More profitable use of the extracted wood chips has been the subject of continual investigation since 1920. The considerable value as fuel coupled with the volume limitations and requirements due the by-product status of the chips have in the past always combined to prevent new projects from being economically attractive enough to justify large capital investment.

The main technological problem in the utilization of wood chips has been to remove contaminants, particularly sand, along with carbon, clay, and bark. It was established many years ago that a satisfactory dissolving pulp could be made from the spent chips after removal of contaminants. This was accomplished, however, only after screening the chips, with a large loss in weight, riffling and washing the chips, and riffling the unbleached pulp. To make such operations profitable appeared to require that the pulp be priced higher than competitive materials.

Another approach was developed in Hercules sponsored research at the Forest Products Laboratory in Madison, Wisconsin, in 1948, which showed that the spent chips could be converted to ethyl alcohol by a combination of hydrolysis and fermentation. Yields of about 50 gallons of ethyl alcohol were obtained per ton of spentwood (dry basis, after removal of fines and sand). This was satisfactorily in line with predictions but still not competitive with large volume production of ethanol from ethylene.

Efforts in pulping have continued, and in 1970, breakthrough was announced in utilization of the extracted chips in paper products.[14] Interstate Paper Corporation used 125,000 tons/yr of spent stumpwood for production of linerboard at its Riceboro, Georgia, mill in 1970. Hercules supplies the spent stumpwood from its Brunswick, Georgia, plant. The processed chips are mixed with green pine chips for conversion into pulp and provide up to 30% of Interstate's pulpwood requirements. The spent stumpwood used is equivalent to annual production of pulpwood from 175,000 acres of forest land and will expand the total forest resources to that extent.

Processing the chips involves screening operations which reduce sand to about 1%. The fraction containing fines and most of the sand may have an attractive potential as a mulch for horticultural uses, but, for the present, it will be burned as previously. The degree of purification attained, blending with green chips, and selection of an appropriate use appear to be the factors that have combined to make commercial use of the main fraction possible. Perhaps all the spent chips will be diverted from use as fuel to upgrade this major fraction of stumpwood and improve the overall economics of the wood naval stores industry.

ACKNOWLEDGMENT

It is a pleasure to acknowledge the assistance of Mr. Stanley Fenelon in supplying material on collection and processing of stumpwood and the contributions of earlier Hercules authors (Ref. 8 and 10–13).

REFERENCES

1. C. E. Ostrom and J. W. Squires, "Naval Stores: The Forests," in *Trees, The Yearbook of Agriculture 1949,* USDA, pp. 291–298.
2. W. G. Wahlenberg, "Longleaf Pine—Its Use, Ecology, Regeneration, Protection, Growth, and Management," Charles Lathrop Pack Forestry Foundation, Washington, D.C. and Forest Service, USDA, 1946, pp. 8–14.
3. D. K. King, H. B. Wagner, and G. H. Goldsborough, "The Outlook for Naval Stores," USDA, 1962, pp. 1, 17, 18, 22, 63.
4. "Agricultural Statistics 1969," USDA, p. 575.
5. W. L. Pritchett and H. Hanna, *Forest Ind.,* **96** (11), 26, 1969.
6. Naval Stores 1969–70 Annual Report, USDA Statistical Reporting Service and Crop Reporting Board, Washington, D.C., May 1970.
7. T. H. Risch, Jr., "Naval Stores Review," *International Yearbook,* Wiley, N.Y., 1966, p. 15.
8. H. Skolnik and A. G. Snow, Jr., *Paper Trade J.,* **122,** 52–53, 1946.
9. R. H. Stevens, *Encyclopedia of Chemical Technology,* Vol. 13, Wiley, N.Y., 1954, pp. 572–577.
10. H. I. Enos, Jr., G. C. Harris, and G. W. Hedrick, *Encyclopedia of Chemical Technology,* 2nd ed., Vol. 17, 1968, pp. 475–508.
11. H. Skolnik, H. I. Enos, Jr., F. H. Gardner, Jr., "The Literature of Wood Naval Stores," in *Literature of Chemical Technology,* Advances in Chemistry Series No. 78, ACS, Washington, D.C., 1968, pp. 349–361.
12. I. W. Humphrey, *Ind. Eng. Chem.,* **35,** 1062, 1943.
13. A. B. Booth, *Kirk-Othmer Encyclopedia of Chemical Technology,* 2nd Ed., 1969, Vol. 19, Wiley, N.Y., pp. 803–833.
14. "Joint Venture of Two Industry Firms Finds New Use for Spent Stumpwood," *Naval Stores Rev.,* July 1970, pp. 4–6.

CHAPTER VI.16

paper recovery from urban waste

Franklin, Ohio, population 10,000, just south of Dayton, has a $2.1-million plant that recovers paper and metal from the city's waste. It disposes of trash and garbage, and in the process recovers usable materials—metal, paper, and eventually glass—that will be sold. The sales could cover up to half the plant's operating costs, including depreciation. The federal government's EPA has paid two-thirds the cost of the plant as a demonstration experiment. The EPA and many cities are looking at a variety of processes for recovering by-products such as energy, materials, or compost from municipal waste.

Franklin's recycling plant was designed, built, and partly funded ($200,000) by Black Clawson Co., a manufacturer of papermaking machinery. Much of its technology resembles the process to produce paper from pulp. The refuse goes onto a conveyor belt, which dumps it into a large tub of water with rotating blades at the bottom. Black Clawson calls the tub an "overgrown Dispose-all." Large and heavy materials come out from the bottom of the tub and pass under a magnet that collects the iron. The water slurry containing the small and light materials goes through a battery of screens and centrifuges, which, in the end, extract cellulose fiber (for paper) and a separate mixture of glass and aluminum and other nonferrous metals. Franklin hopes to build a plant addition to recover the aluminum and the glass from the gritty mixture.

The plant is expected to recover 18% of the garbage as paper pulp, 6% as iron, 4% as glass, and 0.5% as aluminum. Most of the rest will be burned. The city, which was running out of room for landfill, will have to bury only one-twentieth the amount of solid waste it did before. If it sells all the material recycled, the city will recover about $95,000 annually. This would put Franklin's cost of solid waste disposal, including plant amortization, at about $6.00 per ton, or about half the usual cost of incineration. If the city can persuade some neighboring towns to dump their refuse at the new facility, the cost could be reduced even more.

Franklin's plant is a municipal recycling facility, and is also unique because it is integrated with the city's sewage treatment system. It will dump its waste water into the adjacent sewage plant, and get back the huge volume of clean water needed for its slurrying operations. The recycling plant will also accept industrial liquid wastes and "burn" them in a reactor where their heat value will help dispose of sludge from the sewage plant. The ash from the reactor will return to the sewage plant as a settling agent in the water treatment.

The projected economics of the Franklin plant, however, are not necessarily applicable to all other installations. Paper is a major item in city refuse. Franklin, itself an old papermaking town, is located near a string of papermaking plants in the Miami River Valley. A maker of roofing paper in Franklin has signed up to buy the recycling plant's product. Armco Steel Corp., in nearby Middletown, will take the recovered iron.

Indications are that a city should have customers already lined up before it goes ahead with waste recycling. If every city now had a plant like Franklin's, they would have to burn much of the paper they produced.

CHAPTER VI.17

forest product waste

When a typical tree in the forest is cut down for lumber, the resulting log constitutes about 40% of the felled tree. The leaves, branches, waste trees, stumps, and cutting wastes left on the forest floor to decompose constitute the remaining 60%. When the log reaches a lumber or paper mill, the bark is removed, causing a further loss, depending on species and size, of 9–15%: 225 lb of bark per cord, or 500 lb per thousand board feet of lumber. Sawing into lumber produces 10–16% waste as sawdust, slabs, and discard, or 600 lb per thousand board feet; the final result is that about 30% of the cellulose or wood in the tree is actually used as lumber.

Sawmills and planing mills, hardwood dimension and flooring mills, and special product plants, do not include logging except where the logging is conducted in direct combination with the sawmill. Waste generated excludes the logging waste and covers only that generated between the delivery of the log to the mill and the shipping of the dressed, finished lumber. Hardwood dimension is hardwood cut to prespecified dimensions for remanufacture into other wood products. Hardwood flooring mills are sawmills that proceed further in the remanufacturing in milling typical flooring board shape.

All these operations are represented on the flow sheet Figure 1. A typical sawmill starts with the log, debarking it, producing bark waste and a debarked log. This is made into a cant, that is, a squared-up log, by sawing the rounded slabs off four sides. The waste is the slabs and edges. There is a market for wood chips in the pulp industry and it is common to chip up the slabs for sale.

The rough lumber is dried, either in a kiln or the air and becomes the feed to the planing mill, producing dry shavings and finished lumber. The planing mill trims the finished lumber to length and may occasionally do resawing, producing dry sawdust, dry shavings and mill trim and ends.

A plywood plant starts either with debarked or undebarked logs, usually 8 ft long. These are placed in a lathe and turned to become cylindrical. The wood bark removed in this operation is roundup, cor-

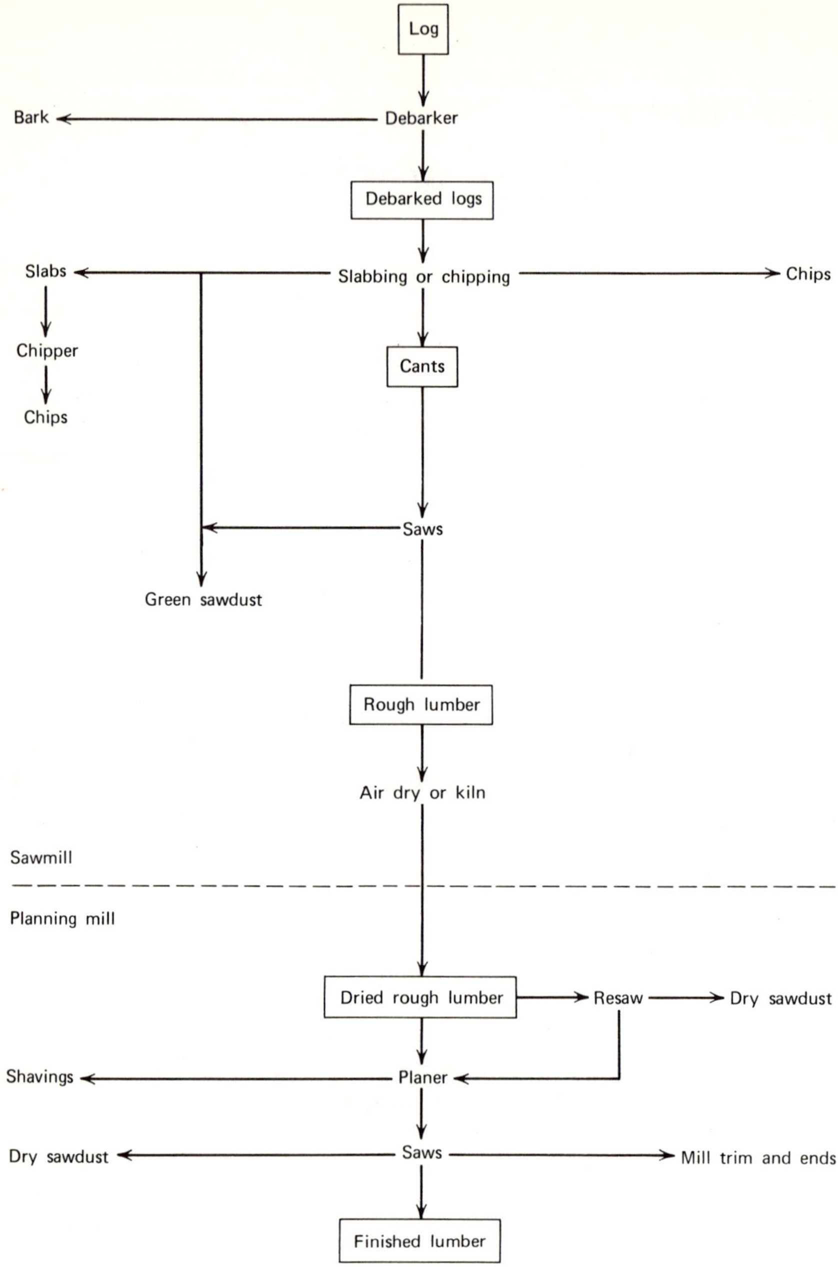

Figure 1. Flow sheet for sawmills.

Table 1 *Billion BF International ¼ in. Board Rule*

	Roundwood Products Total	Sawlogs	Veneer	Pulpwood	Misc. Ind.	Fuel Wood
North	5.732	3.393	0.240	1.335	0.423	0.341
South	41.608	9.396	0.751	3.244	0.835	0.328
Rocky Mt.	3.714	3.438	0.131	0.096	0.048	0.001
Pacific Coast	21.521	14.790	3.898	2.491	0.257	0.084
Total	72.575	31.017	5.020	7.166	1.563	0.754

responding to the bark and slabs produced in the sawmills. The log is then peeled down in the lathe, producing veneer, which corresponds to the rough lumber in the sawmill. There is no sawdust, but there is a residue comprising a "core" 4–8 in. in diameter, too small to peel further. Plywood mills make 8-ft studs from these "cores" or chip them.

The wet veneer is then dried, fabricated into plywood by glueing and the plywood sheets squared up by sawing, producing dry sawdust and trim, but no shavings.

An approximation of the nation's forest industries is given in Table 1. Round wood products refers to the logs taken from the forest and delivered to the plants. Round wood is also used for veneer, pulp wood, fuel, and industrial wood products. Pulp wood amounts to 16% of the total. Veneer manufacture probably has a waste product ratio not very much different from that of sawmills. Statistically, the figures may vary appreciably depending upon the economic conditions.

The uses of the lumber depend upon the linear dimensions of the piece. The unit of measurement used in the sawmill is associated with linear dimension and neglects product and raw material weights.

Units of measurement used in the lumber industry may cause confusion to the uninitiated, and lead to the false conclusion that little waste is produced. The unit of measurement is the board foot, the equivalent of a board 12 in. wide, 1 in. thick, and 1 ft long. By this definition a 1 in. × 8 in. × 18 ft board contains 12 board feet, the equivalent in volume of 1 ft^3. In reality such a board never actually contains 1 ft^3 of lumber: as finished dressed lumber this piece would contain only 0.875 ft^3; as rough lumber it contains about 1.2 ft^3, and in the original Douglas fir log where it originated, it was about 2.2 ft^3 of log.

These log scales seem to indicate that the industry has no waste, since from a thousand board feet of log is produces a thousand board feet of rough lumber and this in turn yields a thousand board feet of dressed lumber. Measured by the log scale, the log is claimed to contain a thousand board feet but the operator may be able to produce 1150 board feet out of it, with a yield of 115%.

In the South, sawmills sell chips to paper mills by weight and moisture content. When slabs and edgings, instead of being sold as chips, were burned for fuel, there was no record kept. Some mills purchase logs by weight. This is referred to as a weight-scale, in distinction to the log-scale. Some establishments are able to provide information on the pounds per nominal board feet, for the log, the bark, the chips, the rough lumber, the shavings, and the finished lumber. Since the national statistical figures are in terms of nominal board feet of production, which is approximately equal to the nominal board feet in the log, a unit of pounds per board foot of measure becomes directly useful in waste studies. From the many literature sources, the pounds per nominal board foot in the log was calculated at 8.1 + 10% and could be used to calculate the green waste streams. From many literature references Table 2 has been tabulated. Here the Pacific

Table 2 *Waste/Product and Other Ratios, Western and Eastern Saw Mills*

	Volume %		Pound/Board Foot		Other Ratios	
	East	West	East	West	East	West
Bark	20.0	12.2	2.40	0.99		
Sawdust	18.0	13.0	2.13	1.05		
Slabs or chips	20.0	16.2	2.44	1.31		
Shavings	8.5	6.2	0.78	0.39		
Mill trim	0	3.1	0	0.19		
Total waste	66.5	50.7	7.75	3.93		
Original log	100.0	100.0	12.00	8.10		
Nominal board foot/cubic foot in log						5.50
Pound/cubic foot of log					53.00	41.00
Nominal board foot/cubic foot in finished lumber						10.70
Bark and sawdust (volume) %					1.11	0.94
Shavings/sawdust (volume) %					0.47	0.48
Slabs or chips sawdust (volume) %					1.11	1.25

Table 3 *Percentage of Each Waste of Given End Use*[a]

	Sold	Fuel	Give Away	Teepee	Self-Dump	Not Dump	Open Burn	Total
Bark	22/0	27/10	0/0	51/48	0/25	0/0	0/17	100/100
Chips or slabs	80/100	10/0	0/0	10/0	0/0	0/0	0/0	100/100
Sawdust	18/10	45/35	0/12	37/16	0/17	0/0	0/10	100/100
Shavings	25/10	42/35	0/12	33/16	0/17	0/0	0/10	100/100
Cuttings	67/0	0/0	0/0	33/0	0/0	0/0	0/0	100/100
Total waste including chips	41/31	28/30	0/1.9	31/26	0/8	0/0.5	0/1.6	100/100

[a] Given as West/East.

Coast and Rocky Mountain regions are called West, and the southern include both the northern and southern regions, called East in the table.

The original log has 12.0 board feet in the East and only 8.1 in the West. The pound per cubic foot density in southern pine wood is greater than that of Douglas fir and Ponderosa pine. The diameter is smaller in the South than in the West. The fractional waste is greater in the South than in the West, and accordingly the cubic feet required for 1 nominal board foot is greater in the South than in the West.

During the 1960s, saw lot production was about 35 billion board feet; of this 15 billion was in the East and 20 billion in the West, these regions being as defined above. The waste from the two regions is given in Table 3. The nonsold, nonfuel, nongiveaway waste was 41 billion lb/yr for the East and 24½ billion lb for the West. In the East about ¾ of this was burned in teepee burners. Before the 1960s, pulp mills did not accept sawdust and other residues because of their short fiber length, but improved pulping processes have allowed the retention of a longer fiber length from the conventional materials, round wood and chips, and this allows the incorporation of sawdust and shavings in the mix. In 1966 the southern mills began to dispose of sawdust as well as chips to the pulp mills and there developed increasing opportunities to sell all shavings and sawdust, as well as the chips for pulp. When this is done, the shavings and sawdust now used for fuel will be replaced by bark. The result will be that the only waste left will be bark equivalent to the 1.19 lb/board foot.

Western mills succeed in using a greater percentage of their waste for fuel and in selling it. The integrated operations use all of it. Particle board is expected to take all the sawmill waste and logged-down wood. Recent development also project the inclusion of bark in particle board.

In the East, particularly in the South, natural gas is available, so that wood waste has little value for fuel. In the West, however, gas is not cheap and the replacement of wood waste by gas for fuel would be accompanied by increased costs. Therefore there must be a balance between purchased and available wood fuel. Oregon by law now outlaws teepee burners for wood waste because of air pollution. Existing units are allowed to continue operations but no new ones may be installed.

It is of interest to see the trend in a case study.

In 1962 Georgia Pacific, an international diversified producer of forest products, bought the Crossett Company and in 1963 the Fordyce Lumber Company of Arkansas, acquiring 500,000 acres of middle South timberlands. Little over a year later, Crossett's lumber production was discontinued and in its place rose two pine plywood facilities, each with a rated capacity near 150 million ft^2 of plywood, equivalent of approximately 9,375,000 panels of 4 × 8-ft construction plywood.

Georgia-Pacific has 14 such facilities in operation in the South.

In 1962, Crossett was known for its lumber and paper products. Eight years later the product mix increased substantially to include plywood, particleboard, lumber, pulp, bleached and unbleached Kraft, bleached paperboard, bags and sacks, household tissues, milk cartons, and chemical products.

Crossett today is an example of the full utilization concept. Virtually every part of the tree is used—even the bark, as an auxiliary fuel source.

Large pine sawlogs are converted into plywood; peeler cores from plywood logs are cut into 2 × 4 × 8-ft construction lumber studs; waste from plywood is transferred to paper operations for making pulp; dry waste is transferred to particle board as raw material for underlayment and furniture core stock; hardwood and smaller pine logs are used in the pulp and paper making, including kraft, paperboard and tissue; Kraft paper is converted into bags and sacks; paperboard into milk cartons; tissue converted into household tissues; and paper-making by-products are converted into industrial chemicals.

Further projects include a new $2 million lumber and chip sawmill at Columbia, Mississippi, utilizing small diameter timber in the area; salvaging the 1.8 billion board feet of damaged timber areas affected by Hurricane Camille; contracts with woodland owners to make a permanent operation.

A $6 million Southern pine plywood plant and a $6 million particleboard plant at Taylorsville, Mississippi, further reduces waste. Plywood production is scheduled with particleboard completion in 1970.

To emphasize G-P's concept for full utilization of the tree, a $1.2 million "stud" lumber mill at the Taylorsville complex will have an annual capacity of 24 million board feet, utilizing log cores from the G-P plywood operations at Taylorsville, Gloster, and Louisville, Mississippi. A $2 million expansion at the Gloster, Mississippi plywood plant increased production from 86 to 140 million ft^2/yr.

Similar case studies of Weyerhauser, Boise Cascade, Crown Zellenbach, and others show the adherence to the total concept of integrated lumber and paper mills, producing particleboard, plywood, paper articles, chemicals, and by products with bark as fuel. But even this is changing with utilization of bark. By 1975 the forest industries will have no wastes in the commonly used sense.

The Forest Products Laboratory publication "Uses for Sawdust, Shavings and Waste Chips," lists dozens of paper mills employing these materials for Kraft and groundwood paper. Bark now finds uses for charcoal briquettes and water treatment.

The forest products industry now vies with meat packing where nothing is wasted save the pig's squeal.

Mills not utilizing the chipping headrig (a rotating bit or vertical lathe) lose upwards of 25% of the total log volume in sawdust, as compared with about 5% for those utilizing the headrig.

What once was burned as sawdust is now shipped to pulp mills as a crescent-shaped chip to be converted into kraft paper. Thus, the chips are a by-product for the sawmill.

In the tree-processing centers, the lower portions of the tree stem, if 12 in. in diameter or larger, are set aside for conversion into veneer (from which plywood is made): the tops of the stem are converted into pulp chips; while the center portion is converted into lumber with the headrigs. The core left over after peeling the veneer from a log could be converted into a pair of two-by-four studs via the chipping headrig.

Southern pine forests grow about 3600 lb of overdry stemwood per acre annually. The annual needle fall per acre does not differ much from the annual wood production. Largely unutilized is southern pine bark, which in the average tree exceeds 10% of the dry weight of stemwood. It accumulates in enormous quantities and is generally burned.

Studies are required on the properties and cellular structure of pine roots to determine what use might be made of this material. The recoverable root system of a pulpwood size southern pine comprises 10% or more of the tree's total weight.

CHAPTER VI.18
nonreturnable packages

Annual consumption of cans in America has reached 308 per capita, or 1032 for the typical household; see Table 1. The hermetically sealed can prevents food spoilage and provides a convenient, and by now almost essential, source of foods and beverages for the average American family. The tinplate container also provides strength to withstand rough handling in shipment and economies in storage space.

Packaging beer and soft drinks consumes 25 billion cans a year. Cans are popular because they are convenient, easy to open, easy to ship and to store in the refrigerator. Alternative packaging materials for beverages—glass, plastic, paper—do not alter the problems arising from discarding the empties, which in any case end up as solid waste, unless they are returnable.

The returnable container, whether glass or plastic, could be an answer to the solid waste problem, but attempts to reinstate this former method of distribution have met with little success. There is, in any event, considerable controversy concerning the desirability of a return to the multiple-trip, or returnable, container. One result would surely be an increase in the costs of distribution, collection, and storage of returned bottles, dislocation of industry, and obsolescence of can plants.

When surveyed, people say they would return bottles, but the facts do not support their intentions, even when a substantial deposit value is placed on the bottles. In New York City the typical bottle makes only two trips to and from the store before it becomes solid waste. The national average is less than 20 round trips.

The world-wide trend to distribution of food through supermarkets also argues against the returnable container: the additional storage space necessary for the returns, the requirement that bottles be sorted by brand, and the high cost of labor for hauling, washing, and refilling.

In any case, most packaging materials are of necessity used in one-way containers: the typical can, once opened, cannot be reused.

The contribution of the food and beverage container to the solid waste problem and to the unsightly litter in our parks and streets is

Table 1 *Annual Consumption of Metal Cans (Approximate)*

		Billions
Food (fruit, vegetable, meat, etc.)	26,781,245,000	27
Beer	15,342,000,000	15
Soft drinks	10,027,500,000	10
Coffee	035,681,000	1
Pet food	3,100,000,000	3
Pressure packs (aerosol)	2,065,000,000	2
Oil	855,675,000	0.8
Paint and varnish	756,379,000	0.7
Other nonfood	2,451,923,000	2.5
Total	62,455,903,000	62.5

sometimes exaggerated. Such packaging materials account for only 13% of all garbage; cans and bottles to only 1.3%. Litter is highly visible, but it appears to be essentially a "people" problem rather than a technical one.

The scrap value of the metals in waste cans approaches $200 million. Less than 4% is recovered, yet technology is fully capable of recovering and recycling more than 90% of the metals. The deterrent to recovery in municipal waste is that it represents less than 5% in the typical community. Such a small fraction of the total waste, even though it is valuable, cannot by itself support the cost of processing of all waste material if the other 95% is discarded in an open dump or sanitary landfill.

Metal cans typically represent 3–5% of the total household waste. Of this approximately 95% are steel cans and the other 5% are aluminum cans. The ideal system for subsequent processing would be the segregation of the cans by the householder into separate garbage cans before pickup. Various communities throughout the United States during the last 30 years have required this separation by the householder and have maintained separate pickup service. As proved in Los Angeles in the 1960s, the reclamation of metal cans so segregated can be profitable.

The technique of separation should not be neglected as an interim measure by communities that are not otherwise reclaiming metals from their municipal waste. Even if only 10% of the community segregated their cans in this manner, the cans might have a scrap value of $25,000 a year for a city of 200,000 population. For the long term, however, separate can segregation by the householder is probably self-defeating, since it will never represent the bulk of the discarded cans. Removal of a portion of the cans from the municipal waste will only discourage separation of the remaining cans at central collection centers. The ecological response to our solid waste disposal problem must be centralized processing, if we can spend more than $4 billion per year spent on disposal of that waste.

The simplest recovery is magnetic separation of the incoming rubbish after coarse shredding. Magnetic separation would remove most of the steel cans, but would not recover the aluminum cans, which, per pound, are the most valuable element in municipal solid waste. Cans so segregated could be water washed to remove most contamination with garbage and preferably given a low-temperature incineration to remove any residual garbage and burn off the organic coating on the cans. After baling or compacting, they would be shipped to the various reprocessing facilities.

Since many of today's steel cans have aluminum easy-open ends, 3–5% of the metal recovered in magnetic separation is aluminum. This aluminum is lost in further processing and is not an asset in steel recycling. It is desirable to insert an additional shredding operation after the initial magnetic separation. Shredding to silver dollar size pieces is probably close to the best if metal cans make up the bulk of the magnetically separated fraction. The second magnetic separation after shredding results in a small aluminum-rich residue, which can be separately compacted and returned for reprocessing.

Waste reprocessing systems result in a fairly clean, finely divided metallic fraction; in preparing the finely shredded material for composting; or in total processing where most of the metal and glass are separated by magnetic techniques and/or air classification; or the metal fraction recovered from the paper pulping operation. All of these fractions are relatively clean.

However, the trend within the can industry is towards TFS or tin free steel cans, which contain no tin (see the Annual Reports of American Can Co. and Continental Can Co. of the United States, who are the leading consumers of tin plate, 1972). The projections are that within the next 5 years, TFS and aluminum will supplant tinplate as the major canmaking material. In the most rapidly growing segment of the market, that of the beer and soft drink can, it is projected that by the end of 1972, tinplate will make up less than 50% of the total beverage cans produced.

In many communities solid waste is incinerated before any attempt is made to separate out ferrous fractions. The maximum temperature attained during incineration varies depending upon the design of the incinerator, and has an important effect on the characteristic of any metal subsequently separated. At lower temperatures, that is, 1000–1400°F, not all organic material and fibers are eliminated. Any aluminum present will tend to be oxidized but generally may not have melted and become separated from the ferrous fraction. At higher incinerator operating temperatures, 1400–2000°F, practically all of the aluminum is melted off and/or oxidized. The steel cans may also have absorbed up to 0.5% of copper from copper-bearing sources within the solid waste, such as electric cable and appliances. Depending on the operating conditions of incineration, the steel in the cans will be oxidized to a certain degree, which as will be seen later, can impede further processing of the steel scrap in various remelting operations.

High-temperature incineration causes a much more complete reduction in bulk of solid waste. But, at the furthest extreme where temperatures of 3000°F are reached, complete alloying of all the metals present occurrs and any glass present completely fuses to form a slaglike mixture over the molten metal. Subsequent metallurgical processing to recover individual metals is quite difficult; consequently, the high-temperature incineration product either must be used as is or it has value only where maximum reduction of bulk is desired and the residue is not to be processed further but rather dumped in landfill space. Therefore, from the standpoint of processing to recover metal values, intermediate temperature incineration produces the product most easily handled, even though some metal values are irrevocably lost in comparison to techniques of metal separation before incineration.

While the recycling of the ferrous fraction of solid waste presents problems to the steel and foundry industries, they all seem to be solvable by present technology. In a large-scale test at National Steel Corporation at Weirton, West Virginia it was demonstrated that up to 10% of the scrap charged to a BOF could be unincinerated beverage can scrap containing almost 10% aluminum. The AISI indicated that the major steel companies might accept any properly prepared scrap cans, without the removal of tin or solder for recycling in the steelmaking operation. The cans should be cleaned, magnetically separated at least twice, shredded, and compacted. Material so prepared would be purchased by the steel industry at the #2 bundled scrap prices ($25 per ton). Foundries would prefer to use incinerated scrap since it would better eliminate the aluminum and lead impurities that can be detrimental. However, oxide buildup on the steel surface during incineration can cause refractory erosion problems during melt down and may limit the use of incinerated scrap.

Technology exists today for the separation and recovery of various elements in solid waste, but it has been cheaper and easier to throw away all solid waste material rather than to try to recover the more valuable elements of it. Federal and state governments must change some of their laws favoring virgin materials over reclaimed, and eliminate indirect subsidies that have been given to the virgin materials in depletion allowances and preferential shipping rates for virgin over reclaimed materials. Iron ore can be shipped by rail at $1.64 per ton, while steel scrap, which basically is a form of iron ore, may pay as much as $4.12 per ton for rail shipment. These differentials should be removed and if anything changed to subsidize the scrap industries indirectly. Another area that should receive attention is possible revision of impurity specifications for the various grades of steel purchased by the government and industry. In some cases the specifications are unduly restrictive and have been set 30 or 40 years ago to minimize processing problems that have since disappeared with improved technology.

Citizens and local governments must appreciate the need to conserve our resources and to accept the small economic penalty required to more fully recover valuable elements from solid waste. If the citizenry is willing to pay $4 billion for scrap and rubbish collection, they should understand that an investment of up to 50% of this figure may be necessary in order to fully recover the valuable elements in the solid waste. The charge for performing this extra separation can be paid by the householder directly in terms of an increased rubbish pickup charge, or indirectly by taxing the manufacturer or the packager. Either way the consumer must realize that he ultimately will pay for the convenience and other assets of one-way packaging.

PART VII
process industry wastes

This part deals with a few selected industries and their wastes. Petroleum from the well to the final customer, discussed in Chapter VII.1, has had wastes over its entire history. First the major waste was gasoline; then with the advent of the combustion engine, it was kerosine; then jet engines consumed that. The refiner found markets for his end product for electrodes for aluminum and electric furnaces, but high-sulfur coke was a detriment. Now the major wastes are those of housekeeping of the service station.

Chapter VII.2 by Barber deals with phosphorus wastes, largely dusts and clayey, well under control and satisfactorily treated. The same is the case for "phossy" water. Radioactive wastes, the by-products of nuclear reactors, atomic power plants and propulsion devices, are treated by Buckham in Chapter VII.3. Here are all the usual problems of waste treatment plus the extreme hazard to unprotected human beings. The reader will find, along with special terminology, a most complex problem; solved by some unorthodox disposal methods. Chapter VII.4 by Novak describes the solid wastes collection, treatment, and disposal of chemical plant waste of one of the largest diversified producers. Chapter VII.5 by Slover and Hainley deals with the selection, coordination, and preparation of a landfill for chemical wastes of wide variety, adjacent to a group of the largest plants in the country, making a wide range of organic materials. Many are the problems of organization, management, control and operation.

CHAPTER VII.1

petroleum

The petroleum industry is an excellent example of wastes turned into useful products by advancing technology, investment of tremendous sums of money in development, patient capital, and development of markets.

About 2500 individual commercial products are manufactured in petroleum refineries or allied petrochemical plants, and their number is increasing. Table 1 lists their type and number.

Only a small percentage of deep wells, despite all preparatory geology structure studies, return more than their investment cost. Until the metallurgical development of oil country piping and steel frames, derricks were of wood. In areas of reduced production or cessation of flow, they dot the countryside as ghosts of foregotten hopes. In producing wells there are sludges, brine emulsions, wild gas, which is flared or burned, sour crudes with limited markets, or carbon dioxide wells with no markets. Wildcat wells, often failures, have no money left to pay for removal of the derrick, cleaning up, and removing the casing, all of which are abandoned to let nature cover the failure.

A large share of the waste of the petroleum industry is the waste of material, equipment, controls and storage containers, caused by changing markets, process and product obsolescence, aided in a small measure by fires, explosions, floods, and hurricanes. Petroleum-based ethyl alcohol made molasses and grain fermentation processes and plants obsolete. Synthetic methanol wiped out wood distillation plants. Cracking processes, catalyst developments, and extraction processes in the industry are often wiped out in a decade by new technology or more efficient units; it is common to see a new piece of equipment, complex and huge, being erected alongside of a still operating structure it is replacing, the latter a marvel only a decade before. Plants in one portion of the country are replaced by more efficient and more profitable operations elsewhere. The industry has to be persistently dynamic in the fierce competition. There are ghost refineries just as there are ghost producing areas and ghost markets.

Table 1 Products Made by the U. S. Petroleum Industry

Class of Product	Number of Products
Fuel gas	1
Liquefied gases	13
Gasolines	40
Motor	19
Aviation	9
Other (tractor, marine, etc.)	12
Gas turbine (jet) fuels	5
Kerosines	10
Distillates (diesel fuels and light heating oils)	27
Lubricating oils (light and heavy, compounded and uncompounded	1156
White oils	100
U. S. pharmacopoeia	78
Technical	22
Rust preventatives	65
Transformer and cable oils	12
Greases	271
Waxes (crystalline and microcrystalline)	113
Residual fuel oils	16
Asphalts (including road oils, tars, etc.)	209
Cokes	4
Carbon blacks	5
Chemicals, solvents, miscellaneous	300
Total	2347

There are wastes in transportation in that ever larger tankers make smaller ones inefficient; tankers are, after all, ships navigated by man, and ships run aground, are wrecked, burned, and spill their cargoes. Despite all the development of the largest research and development group of the largest petroleum company in the world, some of its tankers have cracked apart in the open sea from brittle failure of steel. The steel specifications are the most detailed as to testing, manufacturing, and fabrication our industries have ever developed.

In the Arctic the industry is plagued by mechanical failures of materials under severe cold, materials which give excellent performance in temperate zones.

Manufacturing processes produce their share of wastes. Solid adsorbents reach the stage where they can no longer be regenerated or recycled. Bleaching clays in large tonnage are discarded containing oil that cannot be economically recovered. These must be buried, and they persist without decomposition for decades. This necessitates their removal as mineral wastes. Catalysts lose their efficiency. They must be discarded and replaced. If they contain precious metals, these are recovered but the major remaining residues must be buried. Drying agents are a form of adsorbent.

Caustic soda solutions, used in large quantities in refineries, lose their neutralizing ability and are not convertible to useful materials. They are often pumped to sea for disposal or are buried or discharged to sewage systems.

Residue oils, not attractive as fuels because of location, are often destructively burned, under restrictive conditions, to varieties of carbon black by furnace processes. These processes economically eliminated channel black processes, which came into being using very cheap waste natural gas, now no longer cheap, or wasted, and an expensive superior fuel.

Rapidly increasing tonnages of sulfur are removed from high-sulfur crudes and from natural gas to give low-sulfur fuels acceptable to large cities for power generation, with compatible sulfur dioxide content of stack gases. The sulfur can be sold.

Often asphalt containing oil bottoms, not meeting coal tar specification and having many tonnage competitors, must be disposed of by burial in available gulleys or hillsides.

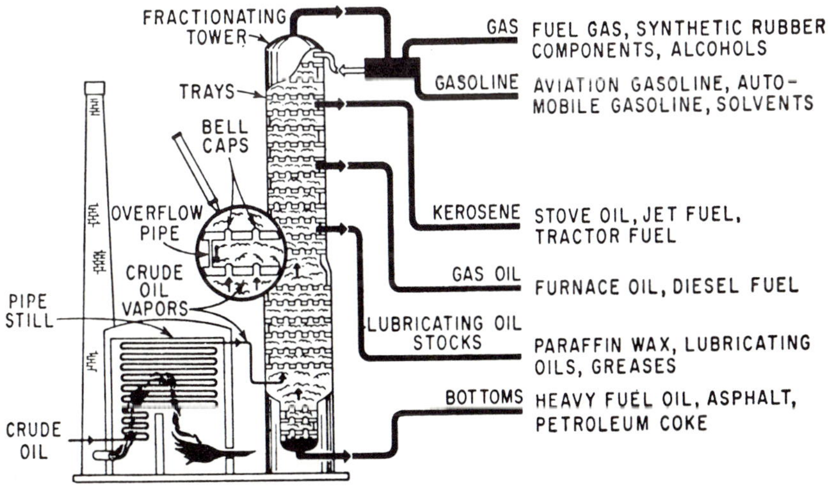

Figure 1. Simplified diagram of the distillation of crude oil in a refinery.

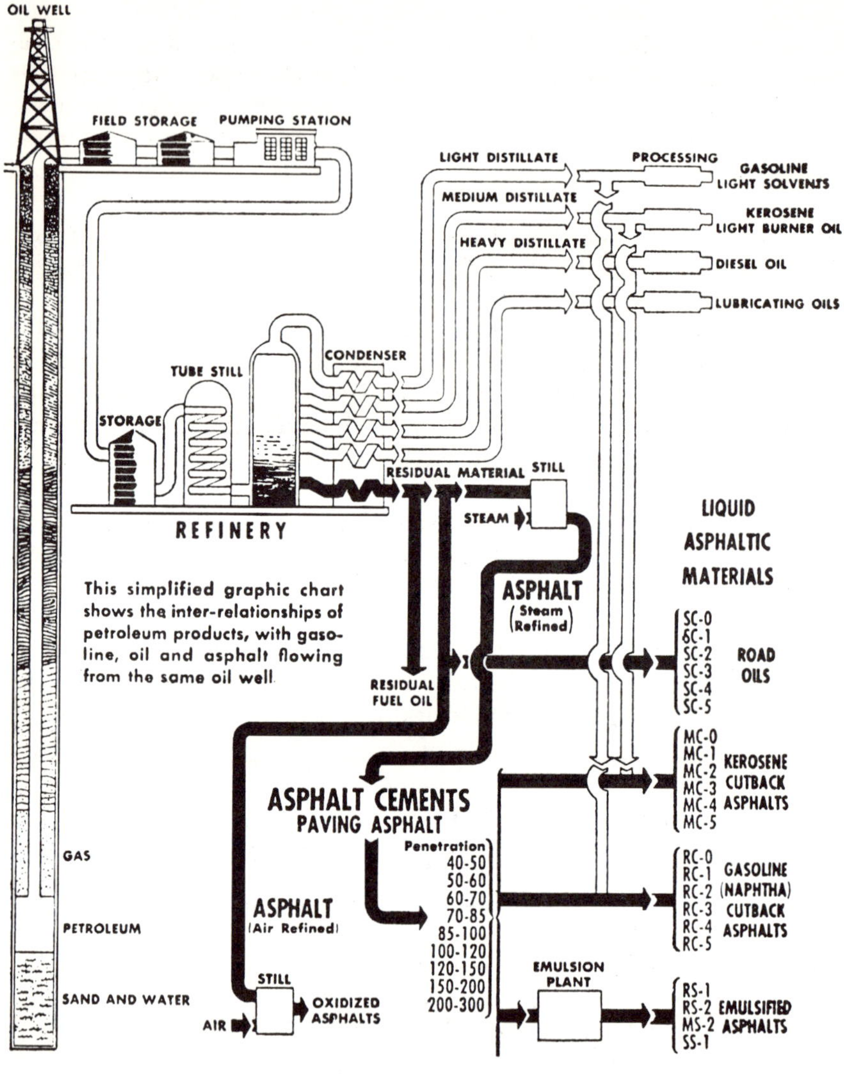

Figure 2. Petroleum flow chart.

Asphalt is prepared at refineries from the heavy residues obtained from the distillation of asphalt-base crude oils. The quality of asphalt depends on the nature of the crude oil. Crude oils containing both asphalt and wax are not suitable for the manufacture of asphalt because wax has an unfavorable effect on the properties of asphalt and it cannot be removed from asphalt commercially.

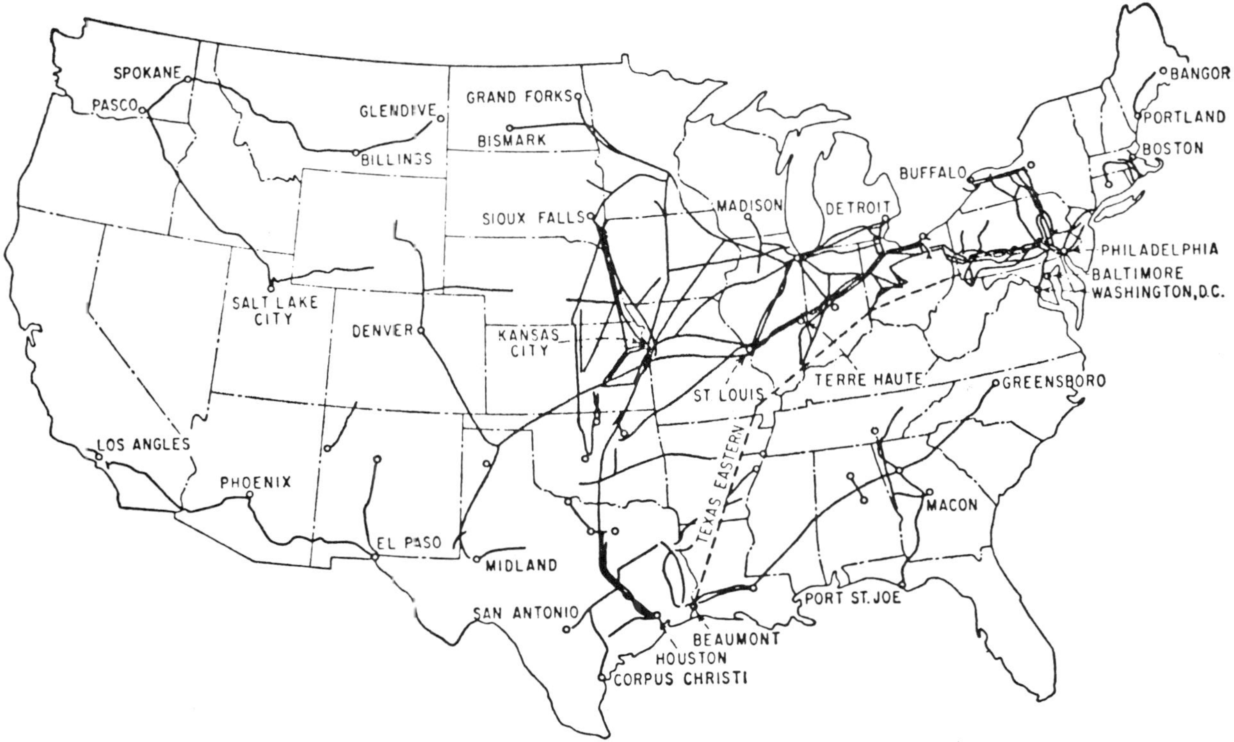

Figure 3. Principal petroleum products pipelines in operation in the United States in 1957.

Removal of the light petroleum distillates and a substantial portion of the heavier, oily distillates provides an asphaltic product of semisolid to solid consistency. Allowing more of the oily fraction to remain, blending back with petroleum distillates, or emulsifying with water, provides materials that have a greater degree of fluidity at normal temperatures. These types are called liquid asphalts. A variety of asphaltic products are prepared by blowing air to produce "blown" asphalts. The petroleum flow chart (Figures 1 and 2) illustrates in a general manner the flow of crude petroleum through a refinery, with emphasis on that portion of the process related to the recovery and production of asphaltic materials.

Now the waste problems are those connected with distribution and the service station. Empty containers pile up, along with wiping rags, cleaning paper, discarded tires, tubes, valves, engine parts, oil cleaners and filters, and housekeeping residues and tank residues. The rubbish carter is paid to take these away. He disposes of them at a dump or a landfill, which does not want them because the oil content prevents decay. Preferably he takes them to an incinerator where the oil content provides fuel.

Worn-out batteries are collected by the agency that delivers new batteries to the service station. Worn out tires, containing large quantities of petroleum-based synthetic rubber, are largely a waste and are disposed of as such. When recycled, used auto tires have an economic value of about 11 cents. Of course some are recapped or converted into snow tires for winter, but these applications are small compared with the tons that end up in dumps or incinerators. Pyrolysis of tires is only experimental but is a hope for the future. Reclaiming companies (desired for temporary adhesives, masking tape, and the like) have abundant supplies.

There are more than 12,000 companies in petroleum in the United States, including those in natural gasoline. Figure 3 shows the pipeline network. There are several hundred in refining, about 1500 in transportation, nearly 30,000 in marketing, with a total of nearly 50,000 companies employing about 2,000,000 people.

CHAPTER VII.2

solid wastes from phosphorus production

J. C. BARBER

The phosphate industries embrace mining for phosphate rock, which is beneficiated and reacted with sulfuric acid to produce superphosphate fertilizers for agriculture; or the phosphate rock is reacted in electric furnaces to produce phosphorus, which may be shipped as such, and converted to phosphoric acid and its salts, triphosphates, polyphosphates, food additives, detergents, medicinals, and chemicals.

The sulfuric acid product, other than recovered fluorides and fluorosilicates, produces no solid wastes. The furnace method gives slags, dusts, ferrophosphorus and phosphorus sludges that are difficult to dispose of.

Elemental phosphorus is produced by smelting a mixture of phosphate rock, silica, and coke in an electric furnace. Coke supplies carbon, which combines with oxygen in the phosphate material and releases phosphorus vapor. The SiO_2 combines with CaO to form calcium silicate slag. The Al_2O_3 in the charge also goes into the slag and combines with the other slag constituents. Calcium aluminum silicate is one aluminium compound identified in the slag.

Large quantities of solids have to be handled and processed. For the production of a ton of phosphorus, about 12.7 tons of solid material is required—typically, 9.5 tons of phosphate rock, 1.5 tons of coke, and 1.7 tons of silica rock.

Most of the solid material entering the furnace is tapped out as molten slag, which solidifies as solid wastes or by-products. However, some solid materials come off in the furnace gas with the phosphorus vapor and carbon monoxide. These solids are separated and treated as solid wastes.

Nearly all the phosphate ores occur in deposits formed in prehistoric seas. In these sedimentary deposits the phosphate occurs as microcrystalline apatites that differ considerably in composition from the pure mineral fluorapatite, $Ca_{10}(PO_4)_6F_2$, because of extensive substitution of carbonate for phosphate and of other metals for calcium.[15] Some commercial phosphate ores, such as Kola phosphate, are ingneous in origin and their composition will be close to fluorapatite.

Impurities present with the phosphate mineral, result from weathering or precipitation of dissolved materials from the sea. The sedimentary deposits are a mixture of phosphate particles, partly phosphatized limestone, sand, and clay. The soft, clay-rich ore is called matrix. In Florida a typical matrix contains about one-third recoverable phosphate, one-third quartz sand, and one-third slime (the minus 150-mesh material).[8]

Most commercial phosphates are readily reduced in furnaces to form elemental phosphorus. Iron compounds are reduced in the phosphorus furnace to elemental iron, which combines with phosphorus to form a metallic by-product, called ferrophosphorus. Phosphorus recovery is decreased when ferrophosphorus forms, and for this reason iron compounds in the furnace charge materials are considered undesirable. Silica occurs as an impurity but SiO_2 is needed in the phosphorus furnace to combine with CaO; therefore, SiO_2 is generally considered a desirable impurity. Other metal compounds are present as impurities in the ores, or occur in the silica rock or coke ash. Some of these are reduced and the metals mix with the ferrophosphorus. Unreduced compounds come out in the slag.

Phosphate ores usually consist of small particles, and the raw material must be agglomerated and hardened before smelting. Figure 1 illustrates a charge preparation process of drying, agglomerating, and calcining to harden the material. Volatiles are driven out during calcination. High-grade lump phosphate may be used without agglomeration and hardening and screened pebble is used in some furnaces without charge preparation.[4] Figure 1 represents TVA operation where a mixture of Tennessee phosphate matrix and Florida phosphate concentrate is raw material. The mixture is normally about three-fourths Tennessee phosphate matrix and one-fourth Florida phosphate concentrate obtained by beneficiating.

In beneficiating the phosphate ore is screened, crushed, washed, and a separation is made at 16 mesh. Slime is removed from the minus 16-mesh fraction by desliming cones and hydroseparators. The slimes are discarded as a solid waste into collecting ponds. The slimes generally contain about 9.1% P_2O_5 on a dry basis, representing about 30% of the P_2O_5 in the ore. Hydroseparator underflow can be upgraded by flotation to separate out sand tailings and produce phosphate concentrate. The tailings (minus 35 mesh) may be used for dam construction and land reclamation.[8]

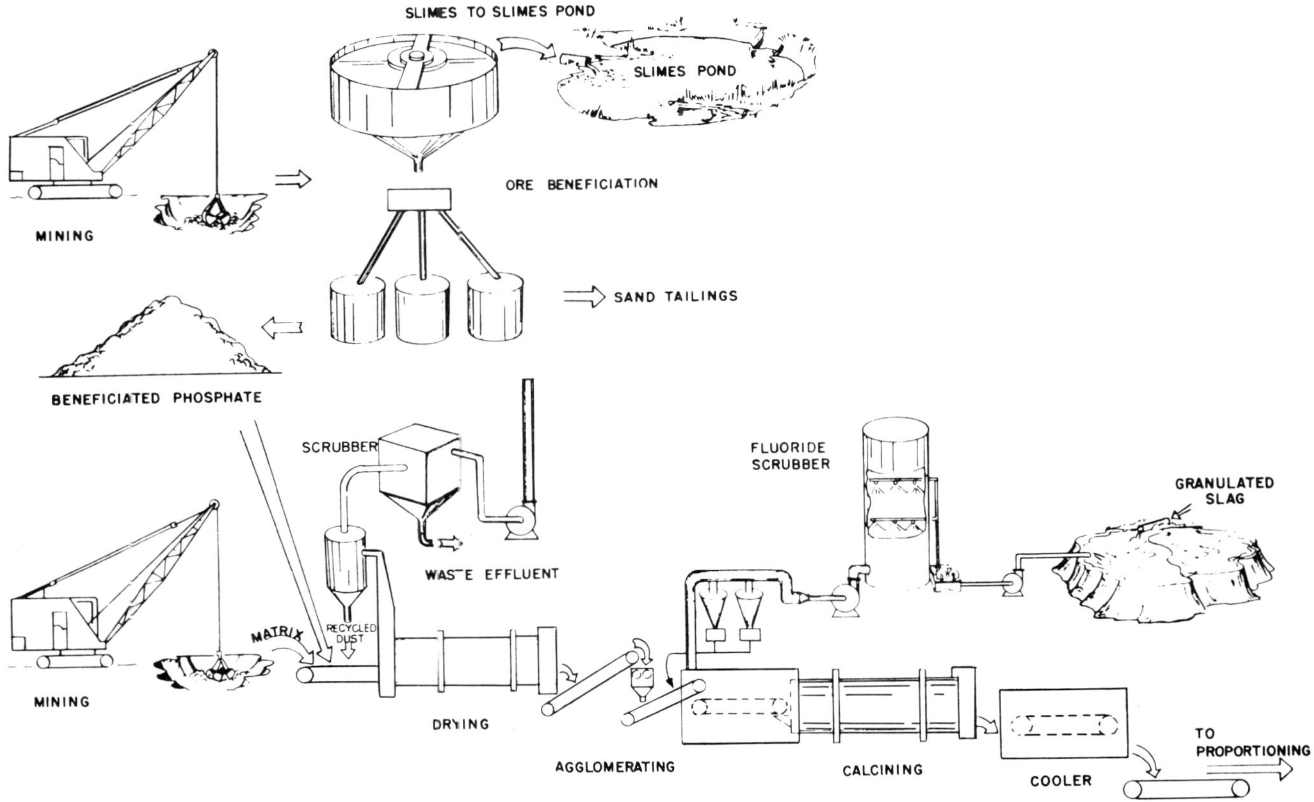

Figure 1. Phosphate charge preparation.

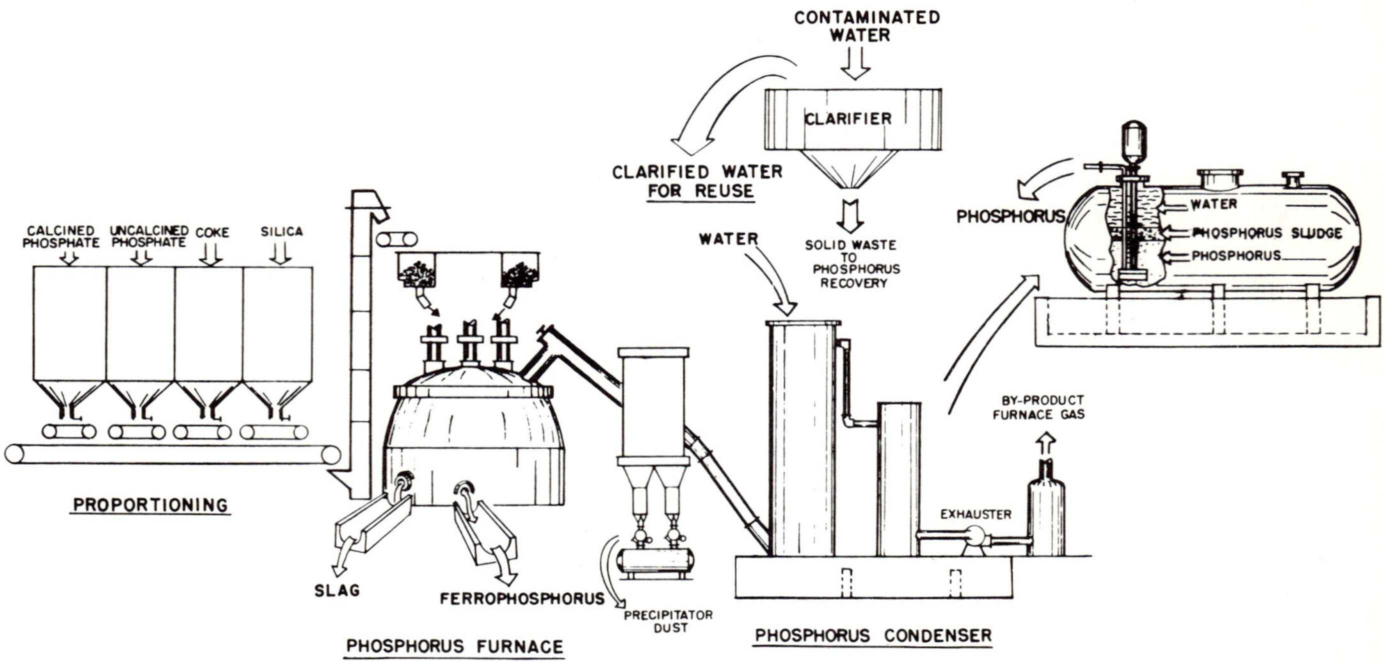

Figure 2. Phosphorus smelting.

Figure 2 shows the remainder of the phosphorus production process at the TVA plant. Coke and silica rock are added to the agglomerated and calcined phosphate in proportions needed to reduce the phosphate and combine with the CaO. The materials are furnace smelted. Some lump phosphate is removed from the Tennessee phosphate matrix ahead of the matrix dryer, and the crushed lumps can be put in the process at the proportioning operation without any agglomeration or calcination. Detailed descriptions of the TVA facilities have been published.[9, 14]

SOLID WASTE FROM CHARGE PREPARATION

Phosphates are agglomerated by compacting, briquetting, pelletizing, or nodulizing. When compacted, briquetted, or pelletized the moisture content must be adjusted by drying or moistening the material, or both. When the phosphates are agglomerated by nodulizing, the raw phosphate can be fed into the kilns without any moisture adjustment. Ores containing clay are conveyed and fed more easily when they are partially dried.

Solid wastes are discharged from the charge preparation process as dusts and as suspended solids in scrubber effluents. Fluorides are volatilized from the calcining operation and are removed from the calciner exhaust gases by water scrubbing. Precipitation of the dissolved fluorides results in additional wastes.

At the TVA operation the phosphate mixture is partially dried in a rotary dryer, shown in Figure 3. Dusts entrained in the dryer gases are collected by cyclones and returned to the process. The partially cleaned gases are scrubbed with water. Effluent from the water scrubber, containing suspended solids, is discarded.

The partially dried phosphate (moisture content about 15%) is agglomerated in compactors and calcined in a grate-kiln. There is dust collecting equipment and a fluoride scrubber at the grate-kiln, Figure 4. Cleaning of the exhaust gases by cyclones reduces the dust content to about 1.0 grain per cubic foot or less. The collected dust goes back to the compactors for agglomeration. The exhaust gases are scrubbed with water to remove fluorides and small-sized dust remaining in the gas. Exhaust gases from both the dryer and calciner contain less than 0.05 grain per cubic foot.

The amount of suspended solids in the fluoride scrubber effluent is about 50 lb/ton of phosphorus produced, and the amount in dryer scrubber effluent is about 150 lb/ton of phosphorus. These are largely clay, and the P_2O_5 content is too low to justify recovery. Both effluents contain P_2O_5 in solution. About 5 lb of P_2O_5 per ton of phosphorus

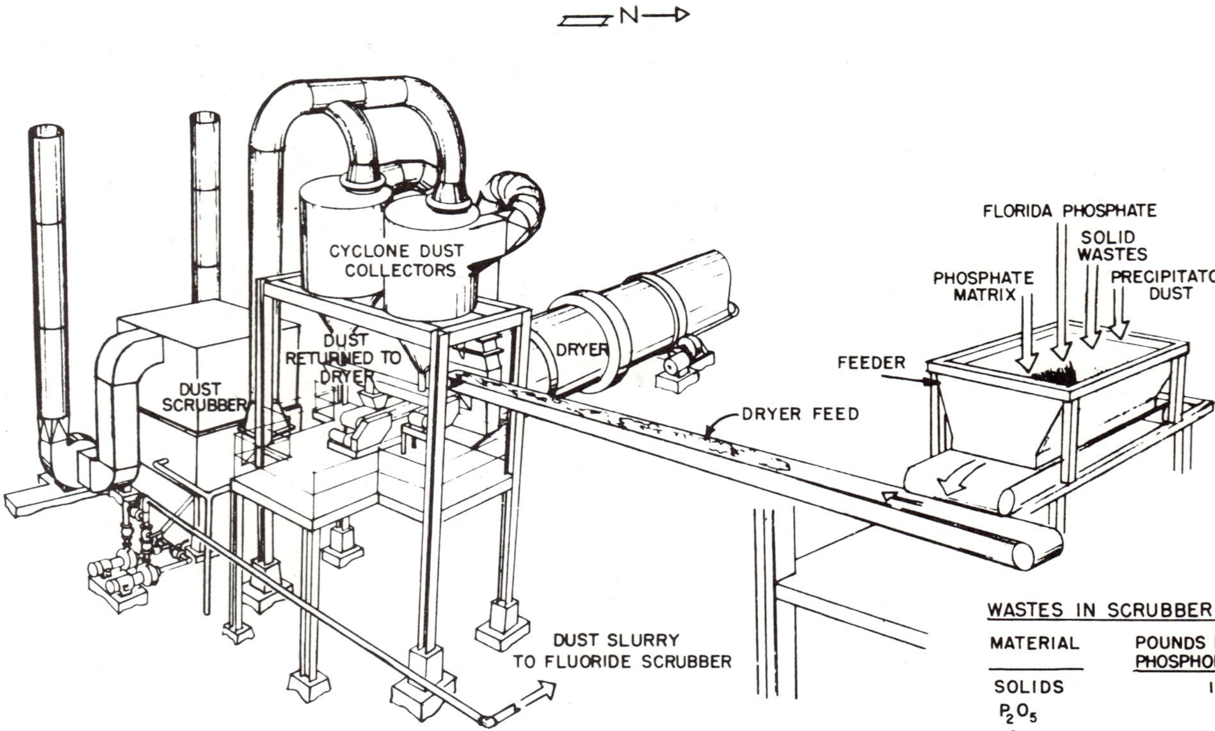

WASTES IN SCRUBBER EFFLUENT

MATERIAL	POUNDS PER TON OF PHOSPHORUS PRODUCED
SOLIDS	150
P_2O_5	2.2
S	0.5
F	0.1

Figure 3. Solid wastes from phosphorus drier.

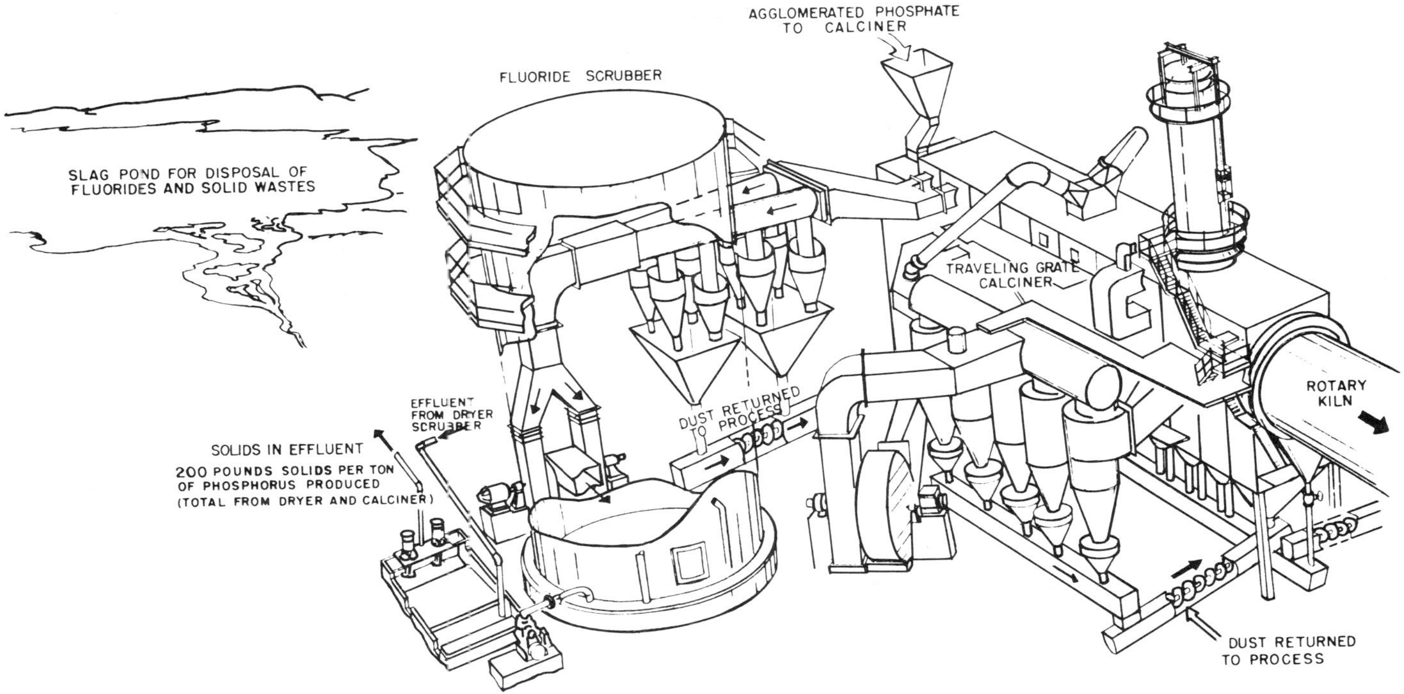

Figure 4. Solid waste from charge preparation.

produced is in the kiln scrubber effluent and about 2 lb per ton is in the dryer effluent. By-product carbon monoxide gas is burned as fuel in both the dryer and kiln, and this gas contains some unrecovered phosphorus. Some of the solid wastes returned to the process may contain small percentages of phosphorus, which burns and releases P_2O_5 to the gas stream.

The combined scrubber effluents from the dryer and calciner are pumped to a pile of granulated slag for treatment.[3, 12] Water flows through the slag and suspended solids, soluble P_2O_5, and soluble fluorides are retained in the slag.

At some plants scrubber effluents are collected in a pond, neutralized with lime or limestone, and the clarified water is reused in the scrubber. Solid wastes, consisting of phosphate dust, calcium fluoride, calcium phosphate, and excess lime or limestone, collect in the treating pond. The solids are periodically removed from the pond and discarded and buried.

HANDLING AND UTILIZATION OF SLAG, FERROPHOSPHORUS, AND PRECIPITATOR DUST

The proportions of furnace charge materials—phosphate, coke, and silica—are adjusted to provide the desired reaction mixture. Sufficient coke is added to supply carbon for reduction of all P_2O_5 and Fe_2O_3 in the furnace charge materials, including Fe_2O_3 present in the coke ash, plus an excess of 3 or 4%.

Usually, the SiO_2-to-CaO weight ratio in the furnace charge is adjusted to the range of 0.80–1.0 by the addition of silica rock. the SiO_2-to-CaO weight ratio at the TVA furnaces is 0.90–0.95.

Phosphates can be reduced to form elemental phosphorus without addition of any silica rock, or in the absence of SiO_2. However, the addition of silica rock makes the melt in the furnace more fluid and the phosphates are more readily reduced. A fluid melt is required to permit slag to be tapped and to permit separation of slag and ferrophosphorus. Other constituents in the furnace charge, such as alumina and potassium silicate, may affect the properties of the melt. The effects of these constituents are not normally considered in the operation of the furnaces. The operators adjust the SiO_2-to-CaO ratio to a value which, in their opinion, results in the easiest tapping and the best separation of slag and ferrophosphorus. Extra silica rock is sometimes added to improve furnace operation.

Molten slag (temperature 2600–2700°F) is tapped from the furnace by the arrangement of equipment illustrated in Figure 5. Streams of

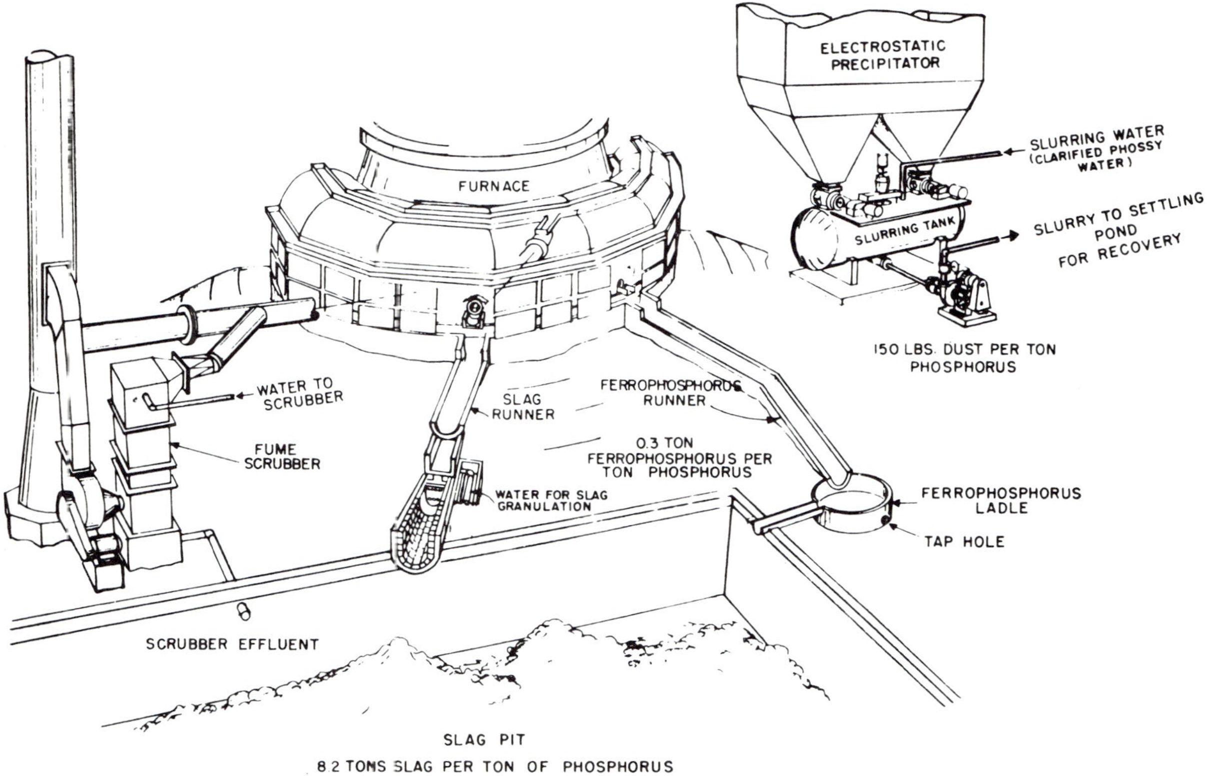

Figure 5. Equipment for handling slag, ferrophosphorus, and precipitator dust.

water at high velocity, coming from nozzles, contact the molten slag and convert it into granules less than 10 mesh in size. The granulated slag has a bulk density of about 70 lb/ft^3. Granulated slag and water flow into a pit and part of the water is reused for granulation. Most of the slag is pumped to a storage pile where it reacts with fluoride scrubber water. From 30 to 40% of the granular slag is sold for aggregate in concrete blocks. The granulated slag is used in combination with limestone—20–25% of the aggregate mixture is granulated slag. The texture of blocks is improved when part of the aggregate is granulated slag, but large amounts give too much fines and cement requirements are increased.

At most phosphorus plants molten slag is tapped into a pit, air cooled or cooled by applying water on the slag surface, and the material is crushed into lumps for use as ballast or aggregate. The specific density of the material is about 180 lb/ft^3. Air-cooled slag crushed to a size suitable for making concrete blocks (all minus ⅜ in. and about 10% through 100 mesh) has bulk densities in the range of 85–90 lb/ft^3.

A typical analysis of granulated slag is given in Table 1. In addition to the constituents shown, the slag contains 0.1–0.2% each of Na_2O, MgO, S, and MnO. Most phosphate rocks contain between 30 and 200 ppm uranium, and radium content is in radioactive equilibrium with the uranium (about 3–75 ppm). The thorium content of phosphate rock, based on a few reported analyses, is less than 10 ppm.[16] Most of the uranium comes out in the phosphorus furnace slag.[19]

Barber and Striplin[6] described utilization of the phosphorus furnace slag. The slag can be converted satisfactorily into slag wool, but the demand for this product has not been sufficient to justify its production.

Table 1 *Typical Analysis of Phosphorus Furnace Slag*

Constituent	Percent, Dry Basis
CaO	44.1
SiO_2	41.3
Al_2O_3	8.8
P_2O_5	1.3
F	2.8
K_2O	1.2

Table 2 *Composition of Ferrophosphorus*

Constituent	Percent
Fe	68.8
P	24.0
Si	1.7
Mn	1.1
Ca	0.2
V, Ti, Mo, Cu, Cr	Traces

Experiments were made to investigate water-quenched slag as a raw material for the portland cement; it appeared that satisfactory cements could be made from mixtures of slag and limestone. However, slag in cement manufacture has not been economic.

Slag is alkaline, and granulated slag has been used in place of limestone to lime acid soils. When the Al_2O_3 content of phosphates is low, the granulated slag has a liming value nearly equivalent to limestone, but high Al_2O_3 slags (8–9%) have an equivalent liming value about 70–80% as much. Plentiful supplies of agricultural limestone and low effectiveness of high Al_2O_3 slag limit the utilization of the material. Liming agents are not needed on soil where phosphorus is produced.

A process was developed to expand furnace slag for lightweight aggregate[6, 17], as shown in Figure 6. The expanded slag had a bulk density in the range of 55–62 lb/ft^3. Tests of slag produced at the TVA plant and of concrete blocks made from the TVA slag showed that radiation was negligible and presented no radiation hazard.

Considerable quantities of expanded slag were sold for lightweight building blocks, but the cost of making expanded slag was more than the extra value received for the expanded product.

Figure 5 shows equipment for ferrophosphorus, a mixture of iron-phosphorus compounds—Fe_3P, Fe_2P, FeP, and FeP_2. Ferrophosphorus contains 20–26% phosphorus, so it consists mainly of Fe_2P and a phosphide richer in phosphorus than Fe_2P.[9] Typically, about 0.3 ton of ferrophosphorus is obtained per ton of phosphorus, but the quantity of ferrophosphorus may vary from 0.1 to 0.4 ton depending on the iron content of the phosphate smelted. A typical analysis is given in Table 2.[9]

The silicon in the ferrophosphorus comes from reduction of some of

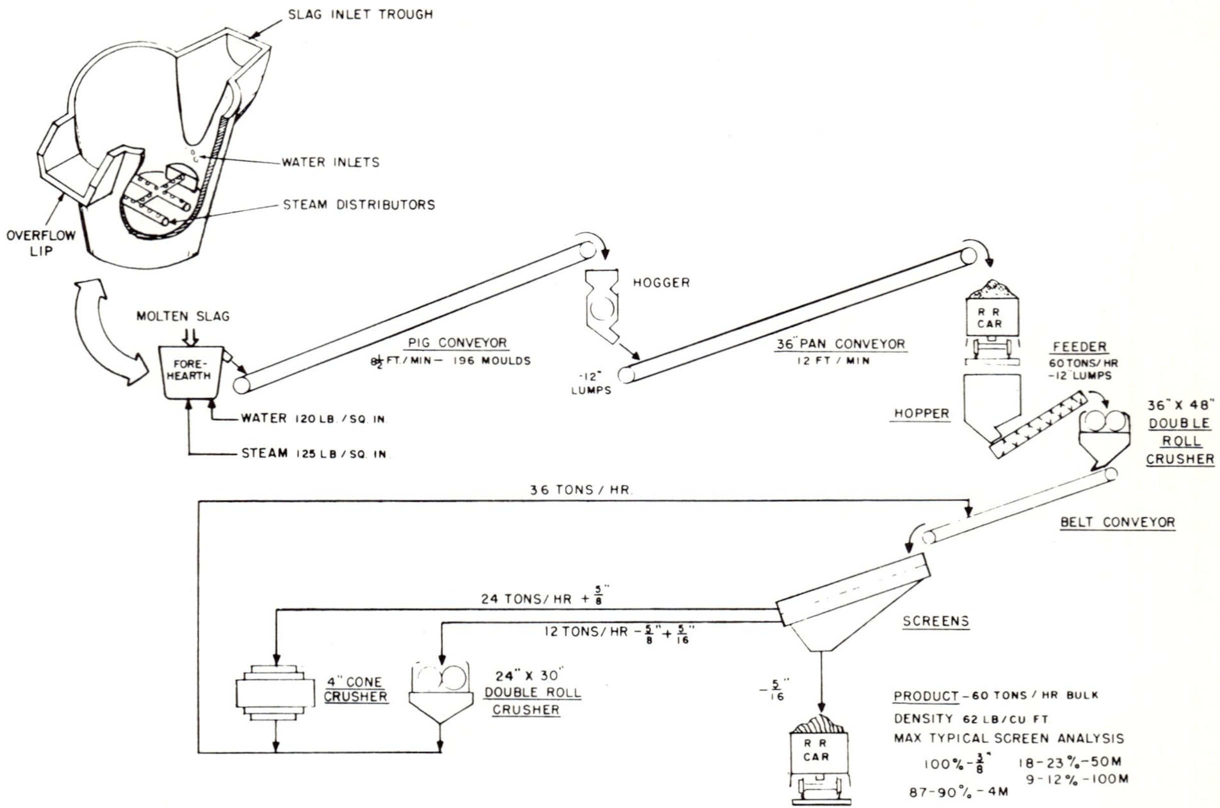

Figure 6. Process for making expanded slag.

the SiO_2 in the charge, and contamination by silicon can be controlled to some extent by changing the proportion of coke in the furnace charge. Smaller proportions of coke decrease the silicon content.

Ferrophosphorus has a variety of uses in the iron and steel industry. Material containing less than 2.0% silicon is preferred. Western phosphates contain more vanadium and chromium than the Tennessee or Florida ores, and the ferrophosphorus contains 3–6% vanadium and the same range of chromium. The higher vanadium and chromium contents are undesirable in iron and steel, but the ferrophosphorus is a potential source for vanadium and chromium. Some patents have been obtained on recovery of vanadium and chromium from ferrophosphorus produced from the western phosphates.[1, 10]

Phosphorus vapor and other gases come off the furnace in a gas mixture along with dust in suspension. The gases are cleaned by an electrostatic precipitator to remove dust before condensation of the phosphorus. The dust collected by the electrostatic precipitator, is discharged through rotary valves, collected in a receiving tank, and slurried with water from the phosphorus treating system (Figure 5).

The dry dust has the approximate analysis: 28.5% P_2O_5, 16.2% K_2O, 18.1% SiO_2, 10.7% CaO, 5.9% F, smaller percentages of Al_2O_3, Na_2O, Fe_2O_3, and MgO, and up to 3.0% elemental phosphorus; typically the phosphorus content is less than 1.0%. About 150 lbs of the precipitator dust is collected per ton of phosphorus.

Precipitator dust has been used as a fertilizer, but problems are encountered in handling a material consisting almost entirely of very small particles (1 micron or smaller), and elemental phosphorus in the material caused handling difficulties. The best way to treat the dust is to slurry it, collect the wet solids in ponds, and add these solids to the incoming raw materials. Improved methods for treating and handling the dust are needed to eliminate fuming problems.

The coke contains germanium and gallium, normally present in coal. Germanium and gallium concentrate in the dust; 0.053% Ge and 0.010% Ga have been found. The dust offers some potential for the recovery of these elements. The Bureau of Mines investigated extraction of these rare elements by a volatilization method.[18]

SOLID RESIDUES CONTAINING PHOSPHORUS

Electrostatic precipitators at the phosphorus furnaces remove 60–90% of the dust from the furnace gases. The gases are then cooled adiabatically by water sprays in a spray chember, and the phosphorus condenses.

Liquid phosphorus collects as the bottom layer in a sump under the condenser. About 10% of the phosphorus condenses as a sludge and accumulates in a layer above the liquid phosphorus. Sludge has a specific gravity of about 1.30, compared with a specific gravity of 1.73 for liquid phosphorus. Water sprayed into the condenser collects above the phosphorus and phosphorus sludge layers, is separated from the phosphorus and sludge, and is reused. Some of the recirculating water is taken off and treated to remove suspended phosphorus particles and solid particles.

Sludge is a mixture of phosphorus droplets, water and solid particles of dust and precipitated fluorides. The proportions of phosphorus, solids, and water in the sludge vary widely, but the three constituents are generally present in about equal percentages.

About 85% of the phosphorus may be extracted from sludge by centrifuging[5]; about 15% comes out of the overflow with the solids and water.

The streams of water contaminated with elemental phosphorus—centrifuge overflow, condenser water bleedoff, and water in phosphorus storage tanks—must be contained in the production process to avoid water pollution. Water contaminated with elemental phosphorus is called phossy water. Studies of the Tennessee Department of Public Health[13] and by the Fisheries Research Board of Canada[11, 20] have shown that marine animals are killed when they are exposed to small concentrations of elemental phosphorus. Pollution by elemental phosphorus may be confined to a small area, but the irreversible biological consequences will be spread over a larger area.

Solid wastes containing elemental phosphorus are also potential pollution sources, and such solids must be handled and treated in ways that will ensure that the wastes do not enter watercourses. A method was developed at TVA for collecting, clarifying, and reusing phossy water.[2] Solids are removed from the water by flocculation and clarification, and the wet solids obtained are centrifuged to recover phosphorus.

Figure 7 is a diagram of the phossy water treating system, and Figure 8 shows the clarifier used to remove solids from the phossy water. After centrifuging, the wet solids go to a settling tank; water overflows the tank and goes back into the phossy water treating system. The settled solids contain about 18% phosphorus and have a specific gravity of 1.135. The solids are incinerated along with phosphorus sludge. By-product carbon monoxide gas serves as a fuel to promote combustion. The phosphorus burns and most of the resulting P_2O_5 goes into the combustion gases and

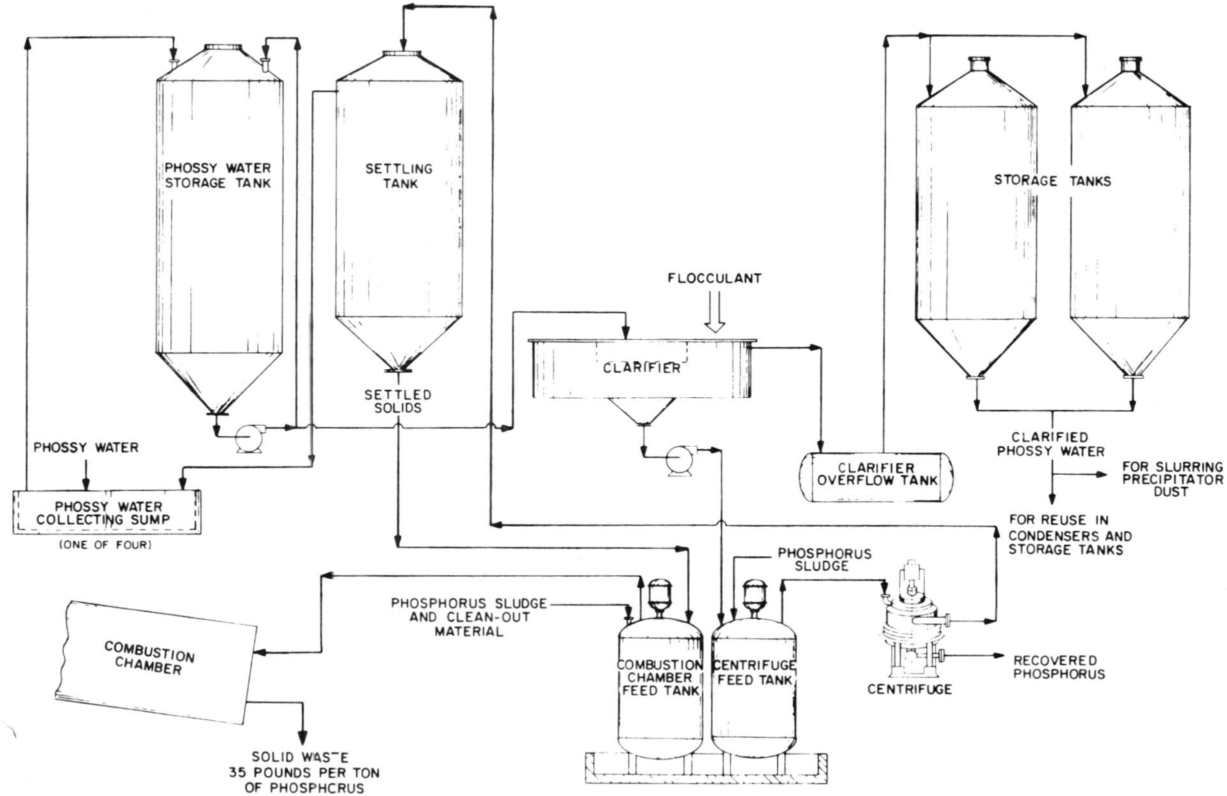

Figure 7. Equipment for treating solid wastes containing elemental phosphorus.

is recovered as impure phosphoric acid. A slag-like ash, free of elemental phosphorus, is periodically removed from the combustion chamber. The ash also contains P_2O_5. It is combined with other solid wastes containing P_2O_5, such as phosphate dusts, and returned to the process at the phosphate dryer.

Elemental phosphorus may be recovered from the solid residues by heating the materials to drive off the phosphorus and condense the vapor; TVA uses batchwise distillation of the residues. The solids were free of elemental phosphorus and could be returned to the process. Figure 9 is a diagram of a pilot plant developed for treating the residues to recover the elemental phosphorus.

Distillation is an improvement over incineration because high-grade phosphorus is recovered, instead of the impure phosphoric acid recovered by incineration. Distillation has been used in large-scale plants to recover phosphorus from residues.

Figure 8. Phossy water clarifier.

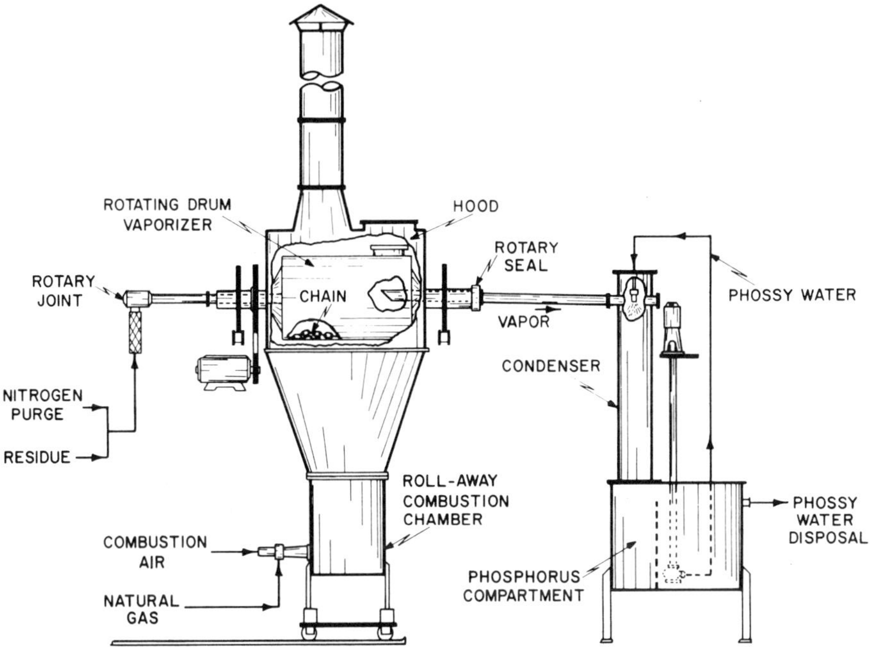

Figure 9. Pilot plant batch still for phosphorus residue.

Conclusions

Most of the dusts obtained when the phosphate charge is prepared for smelting are collected as dry materials and returned to the process. Some small-sized dusts, containing clay particles, are discarded as solid wastes because their P_2O_5 content is low. The discarded solid wastes could be collected and returned to the phosphorus furnace process without much economic loss if disposal of these materials became a problem. With beneficiated phosphate as raw material, the P_2O_5 content of the small-sized dust would probably be high enough to justify reuse of the material in the process.

Slag and ferrophosphorus are the principal solids obtained from phosphate smelting, and both are by-products. At some places it is not possible to sell all the slag produced, and this material becomes a solid waste. The slag can be converted into several useful products, and some method of utilizing the slag can probably be put into effect if its disposal as a solid waste became a problem. Utilization of the slag as a by-product is profitable at some locations, unprofitable at others.

Some of the solid wastes at the phosphorus furnaces contain elemental phosphorus, and these wastes cause particularly difficult handling and disposal problems. The wastes are toxic to marine animals, and special care must be exercised to see that the wastes are not discharged in streams. Considerable work has gone into the development of methods to extract elemental phosphorus from the solids and obtain solids that can be returned to the plant process. However, further development of methods for handling and treating solids containing elemental phosphorus is needed.

A large amount of solid waste is obtained when phosphate ores are beneficiated to remove clayey constituents. The wastes go to slimes ponds. The ponds are potential sources of water pollution, and they may occupy large land areas—about 1.25 acres per acre of ore mined.[8] Furthermore, phosphate ore beneficiation requires large amounts of water and some of the phosphate is wasted. Some western phosphate shales are smelted in phosphorus furnaces without beneficiation.[7] At the TVA operation, 70–80% of the phosphate raw material is unbeneficiated matrix, and 20–30% is purchased phosphate concentrate. Solid waste associated with the use of phosphate matrix is considerably less than it would be from use of concentrate entirely as phosphate raw material. The use of unbeneficiated phosphate increases the amount of slag and ferrophosphorus produced, but these materials can usually be marketed as by-products or they are accumulated without significant solid waste problems.

The development of procedures for using unbeneficiated phosphate from Tennessee and Florida as raw material for the production of phosphorus would nearly eliminate any solid waste discharges associated with phosphorus production. This development would also contribute to conservation of phosphate and water resources.

REFERENCES

1. Lloyd H. Banning and Wallace E. Anable, USP 2,654,655, October 6, 1953.
2. J. C. Barber, *Chem. Eng. Progr.*, **65,** No. 6, 70–73, 1969.
3. J. C. Barber and Thad D. Farr, *Chem. Eng. Progr.*, **66,** No. 11 56–62, 1970.
4. J. C. Barber and E. C. Marks, *J. Metals,* **14,** 903–6, December 1962.
5. J. C. Barber, G. H. Megar, and T. S. Sloan, USP 3,084,029, April 2, 1963.
6. J. C. Barber and M. M. Striplin, Jr., *Electric Furnace Proc., AIME,* **18,** 320–333, 1960.

7. Gordon H. Bixler, Josiah Work, and R. M. Lattig, *Ind. Eng. Chem.* **48,** No. 1, 2–15, 1956.
8. J. R. Boyle, Information Circular 8404, USDI Bureau of Mines, 1969.
9. R. B. Burt and J. C. Barber, Chemical Engineering Report No. 3, TVA, 1952.
10. Wendell L. Darrow, Chao Hsiao, A. G. Robiette, and Blair Burwell, USP 3,154,410, October 27, 1964.
11. W. J. Dyer, D. F. Hiltz, R. G. Ackman, J. Hingley, and G. L. Fletcher, *J. Fisheries Res. Board Canada,* **27,** No. 6, 1131–9, 1970.
12. F. E. Gartrell and J. C. Barber, *Chem. Eng. Progr.,* **62,** No. 10 44–7, 1966.
13. Billy G. Isom, *J. Water Poll. Cont. Fed.* **32,** No. 12, 1312–16, 1960.
14. E. C. Marks and L. S. Wilson, *J. Metals,* **17,** 306–12, March 1965.
15. Guerry H. McClellan and James R. Lehr, *Am. Mineralogist,* **54,** 1374–91, 1969.
16. Ronald G. Menzel, *J. Ag. and Food Chem.,* **16,** No. 2, 231–4, 1968.
17. W. C. Scott, Jr., and W. D. Sandberg, *Chem. Eng.,* **60,** 182–3, July 1953.
18. R. F. Waters and H. Kenworthy, Bureau of Mines Report Investigation 6940, 1967.
19. Edward Young and Zalman S. Altschuler, *Ind. Eng. Chem.,* **50,** 793–6, 1958.
20. V. Zitko, D. E. Aiken, S. N. Tibbo, K. W. T. Besch, and J. M. Anderson, *J. Fisheries Res. Board Canada,* **27,** No. 1, 21–29, 1970.

CHAPTER VII.3

management of radioactive wastes

JAMES A. BUCKHAM

NUCLEAR INDUSTRY TERMINOLOGY

Activation product—a radionuclide formed as a result of bombardment with neutrons from an external source.

Curie, C.—the principal unit of radioactivity; defined as 3.7×10^{10} disintegrations/second.

Decontamination—the removal of fission products (radioactive contaminants) from irradiated nuclear fuel in fuel reprocessing.

Decontamination factor—a measure of the efficiency of removal of radio-active contaminants, e.g., a ratio of the activity present before to that after the decontamination process.

Fissionable material—a material that will undergo nuclear fission, usually restricted to materials that readily undergo fission when bombarded with thermal neutrons (e.g., U-235, U-233, Pu-239).

Fission products—any of the primary fragments resulting from the fission (splitting) of heavy nuclei, together with their radioactive decay products.

Fission yield—the amount of energy released by fission.

Fuel reprocessing—the chemical or metallurgical treatment of irradiated reactor fuel for the purpose of recovering and decontaminating unburned fissionable material.

Half-life—the characteristic rate at which a radionuclide decays i.e., the time required for one-half the atoms in a sample of the radionuclide to disintegrate.

Meat (fuel element)—that portion of the fuel element containing the fissionable component, as contrasted to the cladding, or protective covering.

Radioactive decay—the spontaneous disintegration of unstable atomic nuclei accompanied by the emission of radiation, e.g., alpha particles, beta particles, gamma rays.

Radioactive waste levels

Unrestricted—concentration of radioactivity does not exceed the maximum permissible concentration (MPC).

Low level—MPC to 10 microcuries/gallon

Intermediate level—10 microcuries to 0.4 curies/gallon

High level—above 0.4 curies/gallon; may be hundreds or thousands of curies.

Radioisotope—a radioactive isotope of a chemical element; may be naturally occurring or, more generally, artificially produced.

Radionuclide—a species of radioactive atom (includes radioisotopes).

Transuranic elements—elements beyond uranium in the periodic table, i.e., with atomic number greater than 92; all are radioactive.

INTRODUCTION

Radioactive wastes are the necessary by-products of operation of nuclear reactors—for production of electric energy, for propulsion of ships, for production of nuclear weapons materials, for production of other special nuclear materials, or for research and testing purposes. By far the largest quantity of radioisotopes in radioactive wastes are the fission products themselves—those isotopes that are the two large mass fragments formed from fission of uranium-235, uranium-233, or plutonium-239. Other radioactive isotopes in waste materials are those formed by neutron activation of components in the reactor fuels or in structural parts of the reactor itself. Finally, a minor but important portion of the active isotopes in radioactive wastes are the small quantities of fissionable materials that remain unrecovered during various operations in the nuclear fuel cycle. For as long as mankind needs and wants to enjoy the many benefits of the energy produced by nuclear fission, it will be faced with the necessity for proper management of radioactive wastes. Furthermore, since radioactive wastes remain hazardous during the period of their radioactive decay (hundreds of years in many cases), the responsibility for proper management of these wastes represents a situation unique in man's technological development. Each generation of technologists must provide for man's safety from its actions for many future generations.

During operation of a nuclear reactor, most of the activation products and virtually all of the fission products remain inside the cladding on the fuel elements and are eventually sent to a chemical reprocessing plant as an integral part of the spent irradiated fuel. The reactors, are, however, all designed to cope with any fission products that may escape from the fuel either during normal operation or during accidental failure of a fuel cladding. During fuel reprocessing, the reusable fissionable material is chemically separated from fission products and other fuel components. This, then, is the major source of high-level radioactive wastes, whereas low-level radioactive wastes are produced at nuclear reactors as well as fuel reprocessing plants and, in fact, to some extent at all facilities employing radioactive materials.

Many of the radioactive wastes originate as solutions, sometimes very dilute, and they all contain materials other than the radioactive isotopes themselves. Usually, the mass of the nonradioactive material greatly exceeds that of the radioisotopes, and often the various components are virtually inseparable. Thus, a radioactive waste has two major characteristics, its total mass and the relative concentration of various radioisotopes.

While most radioactive wastes are at one time in the form of solutions or slurries, it is a steadily increasing trend—and in some cases a legal requirement—to convert them to solid forms. This chapter treats the overall management of these wastes and emphasizes their solidification and the storage or disposal of the solidified wastes.

MAGNITUDE AND NATURE OF THE RADIOACTIVE WASTE PROBLEM

During the first two decades of the nuclear industry, from 1942 to 1962, virtually all nuclear activities in the United States were government sponsored, and nearly all the radioactive wastes were generated, treated, and stored at government-owned facilities. During this period, the largest quantities of radioactive wastes were created and stored at the principal Atomic Energy Commission (AEC) production sites at the Hanford Plant in the state of Washington and at the Savannah River Plant in South Carolina. Lesser quantities were created and stored at the National Reactor Testing Station in Idaho and at each of the several National Laboratories. Over 80 million gallons of high-level radioactive waste solutions have been created in these plants, along with much larger volumes of lower level gaseous, liquid, and solid wastes. In these AEC operations the first effort to face the radioactive waste management problem was made, and responsibility for creation, development, and demonstration testing of new techniques for satisfactory resolution of the problem was undertaken.

Table 1 *The Volume, Weight, and Heat Generation of the Accumulated Fission Products from the Nuclear Power Program in the United States*

	Year			
	1970	1980	1990	2000
Est. 10^3 MW, installed	12	128	335	675
Vol. liquid, 10^6 gal	0.2	3.5	20.5	55
Vol. solid, 1000 ft^3	2	35	205	550
Weight, metric tons	8	485	2990	7950
Heat Generation, 10^6 Btu/hr	11	152	640	1350

Since the early 1960s, a significant portion of the effort of the United States toward development of satisfactory waste management practices has been devoted not only to management of wastes from AEC operations, but in a larger measure to management of the more significant quantities of wastes that will be created by the emerging nuclear electric power industry. The growth rate for this industry has been predicted by the AEC in its 1967 report to the President.[1] Using this prediction, the total production of fission products from this industry will be as shown in Table 1.

Fission Products

As shown in Table 1 it is anticipated that by the end of the twentieth century some 8000 metric tons of accumulated fission products will be created by the nuclear power program in the United States. Not all of this mass will be equally hazardous, however. It is instructive to examine the mass of fission products from the standpoint of the relative biological significance of each. This can best be done by comparing the yield of each fission product and the maximum permissible concentration of each in drinking water. To facilitate this comparison, the maximum permissible concentration[2] of strontium-90 of 3×10^{-7} μCi/ml can be taken as a base for defining relative biological significance factors for each isotope as follows:

$$\begin{matrix}\text{Relative biological} \\ \text{significance factor}\end{matrix} = \text{RBS} = \frac{(3 \times 10^{-7}\ \mu\text{Ci/ml})}{\left(\begin{matrix}\text{Max. permissible} \\ \text{concentration of isotope, } \mu\text{Ci/ml}\end{matrix}\right)}$$

Then the initial relative biological significance of each isotope would be the product of its fission yield and its relative biological significance factor. Correcting each of these terms for decay over the period since leaving the reactor gives a picture of the relative biological significance of each isotope as a function of time. A picture of this relationship for the most significant fission products produced by the fission of uranium-235 is given in Figure 1, and related data are given in Table 2.

The gross fission products clearly fall into several distinct categories as given in Table 2. First, there are the relatively short half-life isotopes (cerium-144, ruthenium-106, europium-155, promethium-147, antimony-125, and europium-154), which are extremely hazardous during the first 5 years but virtually harmless after 50 years. Second are the isotopes having high yields and intermediate half-lives (cesium-137, strontium-90, and samarium-151); these three isotopes are the most significant and represent by far the greatest hazard after 50 years. (While samarium-151 is only of marginal significance from uranium-235 fission, it is an order of magnitude more abundant from reactors using plutonium-239 fuel.) It is interesting to note that at the 100-year period, the biologically weighted activity of these isotopes is divided as follows:

Strontium-90	98.3%
Cesium-137	1.6%
Samarium-151	0.001%

Table 2 *Yields and Biological Significance of Significant Nonvolatile Fission Products from Uranium-235 Fission*

Isotope		Half-Life (yr)	Relative Biological Significance	Yield From ^{235}U Fission (Ci/kg)	Biologically Weighted Activity	
					20 yr	100 yr
I.	^{144}Ce	0.8	3×10^{-2}	105,000	5.7×10^{-4}	$\ll 10^{-6}$
	^{106}Ru	1.0	3×10^{-2}	5,820	1.7×10^{-4}	$\ll 10^{-6}$
	^{155}Eu	1.8	1.5×10^{-3}	25	1.5×10^{-4}	$\ll 10^{-6}$
	^{147}Pm	2.6	1.5×10^{-3}	12,260	9.0×10^{-2}	$\ll 10^{-6}$
	^{125}Sb	2.7	3×10^{-3}	163	2.9×10^{-3}	$\ll 10^{-6}$
	^{154}Eu	16	1.5×10^{-2}	15	1.6×10^{-2}	5.2×10^{-4}
II.	^{137}Cs	28.8	1.5×10^{-2}	3,138	30	4.7
	^{90}Sr	30	1.0	3,190	1910	280
	^{151}Sm	90	7.5×10^{-4}	8	5.1×10^{-3}	2.8×10^{-3}
III.	^{79}Se	6.5×10^{4}	1.0×10^{-3}	0.013	1×10^{-5}	1×10^{-5}
	^{99}Tc	2.1×10^{5}	1.0×10^{-3}	0.43	4×10^{-4}	4×10^{-4}
	^{93}Zr	9.5×10^{5}	3.8×10^{-4}	0.1	3.8×10^{-5}	3.8×10^{-5}
	^{107}Pd	4.5×10^{6}	3.0×10^{-2}	0.0006	1.8×10^{-5}	1.8×10^{-5}

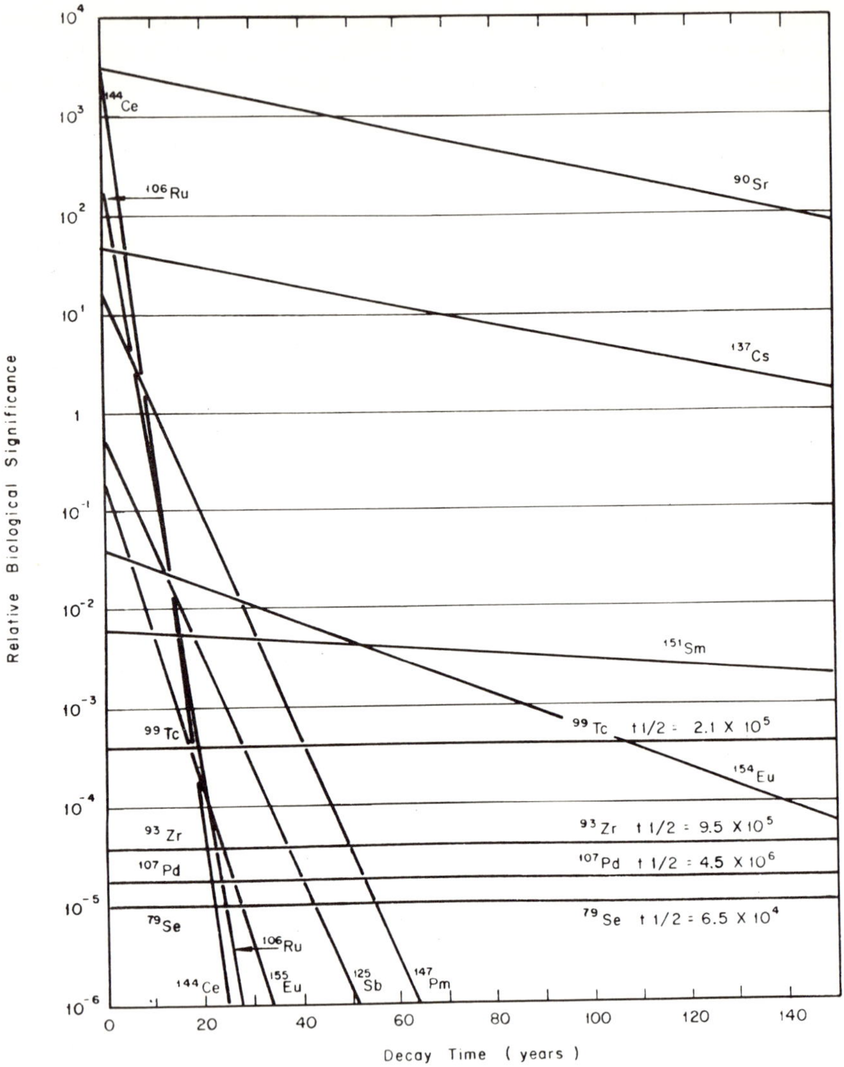

Figure 1. Relative biologic significance of uranium decay isotopes.

Thus strontium-90 stands out clearly as the most important radioisotope in the total radioactive waste picture.

Still a third group of fission products is shown in Table 2; these are the very long-lived isotopes selenium-79, technetium-99, zirconium-93, and palladium-107. Cumulatively, these isotopes have a relative biologically weighted activity of 4.6×10^{-4}, which is essentially independent of time. This activity—about 0.0002% of the strontium-90 activity at 100 years—is not too significant, but does represent a residual activity that will be present in decayed radioactive wastes essentially forever and would still be present in wastes from which the more significant radioisotopes had been quantitatively separated.

Activation Products

Many activation products are formed in nuclear reactors from the reactions between neutrons and structural components, coolants, fuel components, and minor impurities in these materials. Fortunately, most of these activation products are short-lived, and thus represent only interim problems from the waste management viewpoint.

Two of the more significant activation products are cobalt-60 (half-life = 5.2 yr) and sodium-22 (half-life = 2.6 yr). Cobalt-60 is formed in stainless steel components, such as cladding and structural members, and is often the limiting isotope in handling and storage of these solid wastes. Sodium-22 is formed in the sodium collant used in fast neutron spectrum reactors. It represents the limiting radiation level of such a reactor coolant after complete removal of fission products (from fuel leakage) and activated corrosion products.

Unfortunately, very few generalizations can be made about total quantities of activation products because the amounts are highly dependent on design details and material purities for each reactor. These materials do, however, have a significant impact on the total solid radioactive waste management picture.

Fissionable Material Contamination

Many radioactive wastes are contaminated with fissionable materials or other transuranic elements—isotopes of uranium, neptunium, plutonium, amerecium, or curium. Many of these isotopes have very long half-lives, and they are usually alpha emitters. Thus, the radioactivity represents a biological hazard primarily from ingestion. Adding to this problem in many cases is a chemical toxicity which often creates a greater hazard than does the radioactivity per se.

Many radioactive wastes are contaminated principally by these isotopes, for example, wastes from fuel fabrication plants. Many other wastes have residual quantities of uranium or transuranic isotopes mixed with fission and activation products. This is the usual situation with high-level radioactive wastes from fuel reprocessing plants in which the small amounts of unrecovered fissionable material are contained in the same waste solutions as the fission products.

U.S. POLICY ON RADIOACTIVE WASTES

The U.S. government, through its Atomic Energy Commission, has long required construction and operating licenses for all facilities handling radioactive materials. To date, operating licenses have been issued or have been requested for over 100 nuclear reactors. Only one licensed commercial fuel reprocessing plant is now in operation in the United States (the 1-ton/day West Valley, New York plant, is operated by Nuclear Fuels Services, Inc.[3], but another 1-ton/day plant is being constructed at Morris, Illinois by the General Electric Company.[4] A construction permit for a 5-ton/day plant in Barnwell, South Carolina has been applied for by Allied-Gulf Nuclear Services, Inc.[5])

In obtaining an operating license, each nuclear facility is required to submit a comprehensive safety analysis covering both normal operation and any conceivable accident. The consequences of both normal operation and all accidents must be proven to be acceptable in terms of public exposure to radiation and total environmental impact. This requires an applicant to design many features into these plants to ensure the containment of all significant radioactive isotopes at all times, including adequate facilities for containing and disposing of all radioactive wastes.

During the past several years, the AEC has extensively studied the safety aspects of siting of fuel reprocessing plants and of the attendant high-level waste management facilities.[6] These studies have culminated in publication in the *Federal Register*[7] of a U.S. federal policy on these matters. This Federal policy has been incorporated as Appendix F in 10CFR Part 50 of the Federal regulations and became effective in February 1971. In this policy the AEC concluded that public health and safety considerations relating to fuel reprocessing plants do not require that such facilities be located on land controlled by the U.S. government. Such plants, including the facilities for the temporary storage of high-level radioactive wastes, may be located on privately owned property. However, this policy does require solidification of all high-level wastes within 5 years and shipment of these wastes to a Federal repository within 10 years. The text of the portion of these regulations dealing with high-level wastes is[7]:

> A fuel reprocessing plant's inventory of high-level liquid radioactive wastes will be limited to that produced in the prior 5 years. (For the purpose of this statement of policy, "high-level liquid radioactive wastes" means those aqueous wastes resulting from the operation of the first cycle solvent extraction system, or equivalent, and the concentrated wastes from subsequent ex-

traction cycles, or equivalent, in a facility for reprocessing irradiated reactor fuels.) High-level liquid radioactive wastes shall be converted to a dry solid as required to comply with this inventory limitation, and placed in a sealed container prior to transfer to a Federal repository in a shipping cask meeting the requirements of 10 CFR Part 71. The dry solid shall be chemically, thermally, and radiolytically stable to the extent that the equilibrium pressure in the sealed container will not exceed the safe operating pressure for that container during the period from canning through a minimum of 90 days after receipt (transfer of physical custody) at the Federal repository. All of these high-level radioactive wastes shall be transferred to a Federal repository no later than 10 years following separation of fission products from the irradiated fuel. Upon receipt, the Federal repository will assume permanent custody of these radioactive waste materials although industry will pay the Federal Government a charge which together with interest on unexpended balances will be designed to defray all costs of disposal and perpetual surveillance. AEC will take title to the radioactive waste material upon transfer to a Federal repository. Before retirement of the reprocessing plant from operational status and before termination of licensing pursuant to ¶50.82, transfer of all such wastes to a Federal repository shall be completed. Federal repositories, which will be limited in number, will be designated later by the Commission.

Disposal of high-level radioactive fission product waste material will not be permitted on any land other than that owned and controlled by the Federal Government.

A design objective for fuel reprocessing plants shall be to facilitate decontamination and removal of all significant radioactive wastes at the time the facility is permanently decommissioned. Criteria for the extent of decontamination to be required upon decommissioning and license termination will be developed in consultation with competent groups. Opportunity will be afforded for public comment before such criteria are made effective.

NUCLEAR FUEL REPROCESSING AND TYPES OF WASTE GENERATED

A variety of processes have been developed for processing nuclear fuels for recovery of the reusable uranium and plutonium values. In all of these processes, a high-level radioactive waste stream containing virtually all of the fission products is created. Although pyrometallurgical and halide volatility processes have been developed and may be used for certain operations, aqueous processes are used in all existing and planned nuclear fuel reprocessing plants. This discussion is limited to aqueous fuel reprocessing.

Fuel elements often have sizable end fittings that contain no fuel but usually contain radioactive materials from activation by neutron absorption. Reprocessing economics generally dictate that these end fittings be removed before reprocessing. This is usually done by an underwater sawing operation. These end fittings constitute a solid radioactive waste of sizable volume but low, and generally short-lived, radioactivity level.

In aqueous reprocessing of nuclear fuels, the fuel is dissolved in an inorganic acid such as nitric acid, sulfuric acid, or hydrofluoric acid to create an aqueous solution of uranyl, plutonyl, and fission product ions, ions of other fuel components, and excess acid. In many cases, the cladding and meat of the fuel are dissolved integrally, which puts the components of the cladding into the same aqueous solution. This is the case at the AEC's Idaho Chemical Processing Plant[8] and Savannah River Plant,[9] for example. In other cases, such as at the AEC's Hanford Purex plant,[10] the cladding is dissolved first in a separate reagent that does not attack the fuel meat. This cladding dissolution step does dissolve some fission products, however, thus creating a separate medium-level waste. The three initial commercial fuel reprocessing plants in the United States,[3-5] however, all employ a chop-leach process in which the fuel is sheared into small pieces that are fed to a dissolver. In the dissolution process, the nitric acid reagent dissolves the exposed fuel meat but does not attack the Zircaloy or stainless steel cladding. This process leaves the cladding hulls in the form of a low- or medium-level solid waste (with most of the radioactivity coming from activation products in the cladding) and has the advantage of not incorporating components of the cladding in the high-level fission product waste.

In all these reprocessing plants, solvent extraction is employed to separate the uranium and plutonium from fission products. A typical solvent extraction cycle is shown in Figure 2. In such a cycle, dissolver product solutions are fed, sometimes through an evaporator, to an extraction column in which the primary separation occurs. In the extraction column, the aqueous solution is contacted in multiple stages by a suitable organic solution, usually tributyl phosphate dissolved in a high-grade kerosene type of solvent. The organic solution preferentially dissolves the uranyl and plutonyl ions but not the fission products, dissolved fuel cladding or structural components, or excess acids. Thus,

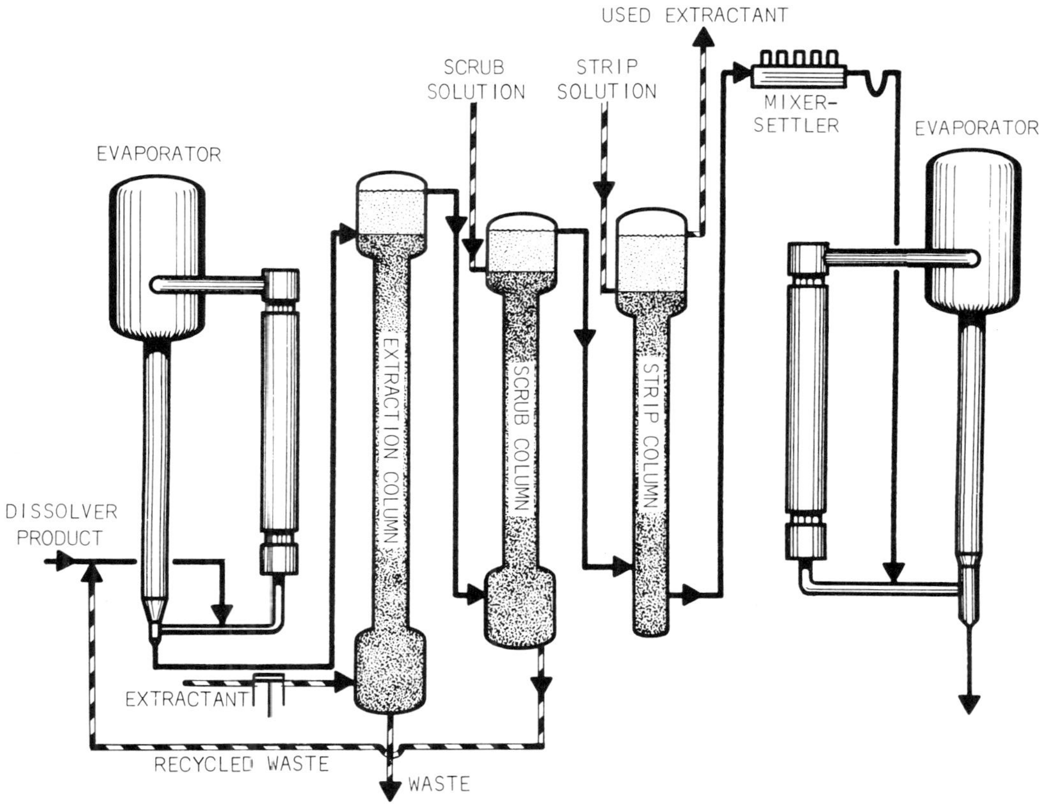

Figure 2. Typical solvent extraction cycle in reprocessing.

the aqueous solution leaves the bottom of the extraction column essentially free from uranium and plutonium and carries the balance of the constituents in the dissolver product stream. This extraction column raffinate is the high-level waste stream.

The uranium and plutonium are contained in the organic stream leaving the top of the extraction column. This stream is washed to remove additional fission products in a second column, the scrub column. Finally, the organic stream is contacted by an aqueous solution of low ionic strength in a third or strip column. Here, the equilibrium constant for uranium and plutonium favors the aqueous phase, so these components are stripped from the organic stream into the aqueous phase. This aqueous stream, which is sometimes evaporated, forms the feed solution for a second cycle of solvent extraction. Generally, three cycles of solvent extraction are required for complete decontamination of uranium and plutonium from fission products, although ion-exchange systems are sometimes used for the third decontamination step.

In plants that recover both uranium and plutonium from fuels—the usual case—the uranium and plutonium are usually separated from each other by a special extraction partition step at the beginning of the second cycle, and the two streams undergo final decontamination separately.

Columns used for these solvent extraction operations are usually pulsed sieve-plate columns because of their high separation efficiency per unit height. However, packed columns are also used to some extent. Mechanically operated mixer-settler units are also sometimes used in lieu of columns of these solvent extraction operations, but the waste generation aspects of these units are the same as for columns.

While over 99% of the fission products entering a fuel reprocessing plant are contained in the high-level waste solution from the first extraction unit, medium-level wastes are also created from the second and third extraction operations. In addition, decontamination operations, chemical analysis, and several auxiliary operations create other aqueous radioactive waste streams.

Solid wastes are created in fuel reprocessing plants not only from removal of end fittings but also from the paper, rags, metals, and so on, used in various plant operations. Of course, the largest amount of radioactivity is contained in the high-level aqueous wastes, which are generally stored in underground tanks pending solidification.

The chemical and radiochemical composition of radioactive waste solutions varies over a very wide range owing to the wide variety of nuclear fuel compositions and chemical processing flowsheets employed.

Table 3 *Chemical Composition of Typical High-Level Liquid Wastes (All Compositions Expressed as Molarity)*

Component	Nuclear Fuels Services Plant[3]	Neutralized Savannah River Plant Waste[9]	Idaho AEC Plant Aluminum Waste[8]
Acid	3.7	−0.01	1.0
Iron	0.9	0.44	—
Chromium	0.01	0.02	—
Nickel	0.005	0.01	—
Aluminum	0.001	0.001	1.6
Sodium	0.14	3.6	0.05
Mercury	<0.001	<0.001	0.01
Nitrate	7.5	2.0	6.0
Sulfate	—	0.9	—

In most high-level wastes, fission products (which amount to 1.1 kg per 1000 MW days of thermal exposure when expressed as oxides) are exceeded in concentration by other metals contained in the fuel or used in processing such as sodium, iron, aluminum, and zirconium. These are generally present as nitrate salts in acidic solutions, although some wastes are neutralized with caustic before storage and solidification. In chop-leach processing of power reactor fuels, every effort is made to minimize the quantity of materials other than fission products entering the high-level waste stream. Under such conditions, as little as 10 gal of high-level waste can be created per ton of fuel processed. With other flowsheets, and with addition of decontamination solutions and concentrates from boiling down lower-level wastes, this volume can increase to over 200 gal per ton of fuel. Some typical high-level waste compositions are given in Table 3.

INTERIM STORAGE OF LIQUID WASTES

Until recently, all high-level radioactive wastes were stored as solutions in large underground tanks. As previously indicated, current policy in the United States has deemed this method of storage permissible only on an interim basis; but it is likely to be continued for some time as an interim method to provide a reasonable working hold up between fuel reprocessing and waste solidification.

Extensive technology has been developed for underground tank storage of liquid wastes. In nearly every case, some form of protection has been provided to ensure recovery in the event of tank leakage. This precaution is now a requirement for all interim tank storage, and true double containment will be provided for all future interim tank storage.

In the older U.S. installations at Hanford and Savannah River, wastes are neutralized with caustic before storage in underground tanks having capacities of up to a little over 1 million gallons. The Hanford tanks are constructed of reinforced concrete and are lined with carbon steel. The heat from fission product decay is removed in these tanks by condensation of the vapors from the boiling liquid wastes. The Savannah River tanks are constructed of carbon steel, contained in individual reinforced concrete vaults, and cooled by cooling coils. The bottom of each vault is lined with a saucer of carbon steel, which provides a means of detecting leaks and returning liquid to the tank. Interim tank storage at the Idaho Chemical Processing Plant has employed acidic solutions in stainless steel tanks. Each tank is cooled with cooling coils and is contained in a concrete vault fitted with a stainless steel sump and jet to recover any leaked liquid. In general, new commercial fuel reprocessing plants will also use acidic storage in stainless steel tanks.

Prevention of corrosion is very important with interim tank storage of high-level radioactive waste solutions. Fifteen carbon steel tank failures have been reported,[11] but no failures of stainless steel waste storage tanks have yet occurred. The corrosion rate of stainless steel is approximately 0.005 mil/yr with acidic wastes at temperatures under 60°C, but the corrosion rate increases to over 1 mil/yr at the boiling point. Therefore, it is important that cooling coils be provided to maintain the temperature of the stainless steel tanks below 60°C. Although the total corrosion rate of carbon steel is neutralized wastes is generally less than that of stainless steel in acidic wastes, such systems are highly subject to pitting. In many cases the actual leaks that occurred have been attributed to pitting, which has often been enhanced by sludges in the bottom of the tanks.

Storage of acidic wastes in stainless steel tanks is now generally regarded as preferable to neutralization and alkaline storage for the following reasons: (1) lower volumes, (2) less tendency to leak, (3) smaller volumes of precipitated solids, (4) smaller total solids content and thus creation of less volume of solidified waste, and (5) comparable costs.

In setting up criteria for liquid radioactive waste storage tank farms, a realistic policy of providing spare tankage is always required. For example, at the Idaho Chemical Processing Plant there are a total of eleven 300,000-gal stainless steel tanks for interim storage of radioactive liquid wastes. One of these tanks is always maintained empty so that any liquid leaking from another tank can be immediately transferred from the vault sump around the leading tank into the empty spare tank. In addition, sufficient freeboard is provided in each tank at all times to allow for the increase in volume on transfer of any leaking solution due to dilution from the steam jet of approximately 5%.

In addition to continuing the practice of interim storage of radioactive waste solutions in suitable tankage, the AEC is continuing to evaluate the possibility of longer-term storage of high-level radioactive waste solutions in underground caverns at its major production sites. A statement from the AEC regarding these studies is[7]:

> The Commission is continuing to evaluate the feasibility of storage of high-level liquid radioactive wastes from its production facilities at Savannah River, S.C., and Hanford, Wash., in underground caverns beneath these sites. The over 80 million gallons of high-level wastes now stored in tanks at these installations constitute over 95 per cent by volume of all high-level wastes in this country. However, such wastes differ materially in radioactivity level, heat output and chemical composition from wastes produced by licensed fuel reprocessing plants planned or under construction. For example, Savannah River and Hanford wastes have been chemically neutralized, contain large volumes of nonfission product materials, and have heat and radioactivity outputs many times lower than the licensed plant wastes. Most of the Savannah River and Hanford wastes were generated as the result of operations to meet defense needs.
>
> There are a number of unique considerations and incentives for the Commission's decision to support research and development on long-term storage in underground caverns beneath these Savannah River and Hanford sites. These include: The large volume of the Savannah River and Hanford wastes; the relatively low heat output of these wastes as compared with those from licensed reprocessing plants; and the tremendous estimated cost and related problems of solidifying and shipping these wastes offsite. This research and development program is being performed in phases. In

the interim both Savannah River and Hanford are immobilizing their tank-stored liquid wastes by evaporating them to salt cakes in the existing tanks. Before any decision is made to store wastes in underground caverns a substantial amount of exploration work will be necessary to assure that the wastes will be isolated from the biosphere for centuries. The research and development program at Savannah River to date, and further exploratory drilling and mining to determine the suitability of the bedrock under the site, are expected to cost about $14 million. Such a research and development program is a necessary prerequisite to a decision to construct a facility solely for Savannah River wastes. Nevertheless, because of the large volume of wastes already in existence at that site, long-term storage of the waste in caverns mined in the bedrock offers the potential of significant cost savings as compared to the alternative of solidification and storage offsite. However, it is doubtful that bedrock storage of liquid wastes would be attractive for licensed reprocessing plants. Specifically, the high cost of the research and development program that would be required to prove out the bedrock at each location, the uncertainty and associated delay until the site is accepted, and the high estimated costs of constructing each such disposal facility, support this conclusion.

SOLIDIFICATION AND TREATMENT OF LOW-LEVEL WASTES

Low-level radioactive wastes occur most often as slightly contaminated solutions, generally much too voluminous to store. These solutions are treated by distillation or chemical methods to remove the radioisotopes, and then purified liquid is then discharged. Other low-level radioactive wastes occur initially as solids, and these are generally stored in burial grounds, sometimes preceded by compaction or incineration operations. These two radioactive waste disposal aspects are discussed separately.

Low-Level Waste Solutions

In the United States, evaporation is the principal method used to purify low-level radioactive waste solutions. At the Savannah River Plant,[9] aged alkaline supernates from high-level wastes are drawn off, mixed with lower-activity wastes, and then concentrated by evaporation. The concentrate is sent to a high-level waste storage tank. This form of evaporation gives a decontamination factor of from 10^4 to

10^5. Other radioactive solutions at this plant are passed through synthetic ion-exchange beds before release to effluent settling basins. At Hanford,[10] the second- and third-cycle raffinates from solvent extraction are combined and evaporated with the high-level wastes. To reduce the rate of heat production, cesium and strontium are recovered both from the supernate and sludge of aged alkaline wastes and from the acidic new waste concentrates. The residual new wastes are neutralized, interim-stored for 5–10 years, and combined with older wastes for final in-tank evaporation to salt cakes. Lower-activity waste solutions at the Hanford plant are sent to settling ponds or underground distribution cribs, depending upon their radioactivity level.

At the Idaho Chemical Processing Plant,[8] all low-level radioactive wastes are evaporated, and the acidic concentrates are combined with high-level wastes in interim tank storage awaiting calcination. Decontamination factors of about 10^4 are obtained in the evaporation step. The evaporator overhead stream is further treated by an ion-exchange system to achieve an additional decontamination factor in excess of 100. Thus, at this plant virtually no fission products are released to the ground.

The Nuclear Fuels Service commercial reprocessing plant in New York[3] employs evaporation for virtually all of its low-level wastes. The condensate from the high-level waste evaporator is mixed with other low-level wastes and reevaporated. The overhead vapor passes through a rectifier for acid recovery and is then condensed and reevaporated before discharge. All of the concentrates from these evaporation steps are stored as high-level waste. Future commercial reprocessing plants will employ similar systems.

Chemical treatment of low-level radioactive waste can be accomplished at a far lower cost than for distillation, but it generally produces only a 60–98% removal of the gross activity. Consequently, chemical treatment either can be used for only very low-level wastes or requires multiple steps to achieve the necessary decontamination. Chemical treatment steps are frequently used in Europe for treatment of low-level radioactive waste. Typical of these chemical treatments is one employed at Marcoule, France,[12] which employs successive precipitation with ferric hydroxide and nickel ferrocyanide, calcium carbonate, lead periodate, and manganese dioxide. This treatment produces a total volume reduction factor of about 200.

Treatment of Low-Level Solid Waste

Low-level radioactive solid wastes consist of fuel element end pieces, cut segments of metallic fuel claddings from chop-leach processing, contaminated equipment, and contaminated paper, rags, and clothing from decontamination operations. Of these, paper is normally the largest constituent. In the early period of nuclear industry operations, virtually all solid wastes were collected and buried in suitably located burial grounds. This practice has been adequate with limited volumes of low-level wastes and with carefully selected burial grounds. However, it is becoming increasingly evident that the large volume of low-level radioactive wastes requires improved treatment processes. One approach given serious consideration in the United States is to employ incineration of combustible solid wastes so that the large volume of these materials can be significantly reduced. At the same time, compaction of any combustible material to be buried, together with compaction of noncombustible radioactive solid wastes, is being increasingly practiced.

Wastes to be buried underground are sometimes buried in temporary storage containers such as barrels and boxes. This method is satisfactory for short-term burial or for permanent burial of short-lived radioisotopes. However, permanent underground burial of longer-lived isotopes requires additional protection. A number of processes have been developed and put into practice for incorporating such wastes in water-resistant materials such as concrete or asphalt. Typical of these processes is one developed at the Oak Ridge National Laboratory for incorporating radioactive aqueous or organic waste residues in asphalt or thermoplastics.[13] This process, shown schematically in Figure 3, has been applied to a wide variety of solutions and slurries containing high concentrations of inorganic or organic compounds, together with trace amounts of radioisotopes. Wastes that have been successfully incorporated in asphalt or polyethylene include aqueous solutions containing caustic-aluminate or mixtures of iron and other metal salts of sulfate, chloride, or nitrate; aqueous slurries of hydrous iron and other metal oxides or calcium phosphate; and organic compounds such as tributyl phosphate, carbon tetrachloride, and dodecane. Wastes that are to be incorporated in asphalt should be either neutral or alkaline, as acids degrade asphalt and the reactions of oxidants with asphalt could be hazardous.

A wide variety of process equipment for incorporation of low-level nuclear reactor wastes with concrete has been developed and is being put to use in the United States today. In general, concrete blocks containing the incorporated low-level radioactive wastes are taken to a Federally licensed burial ground.

Where incineration of low-level combustible waste is employed, the ashes must be prepared into a safe form for storage. In many current

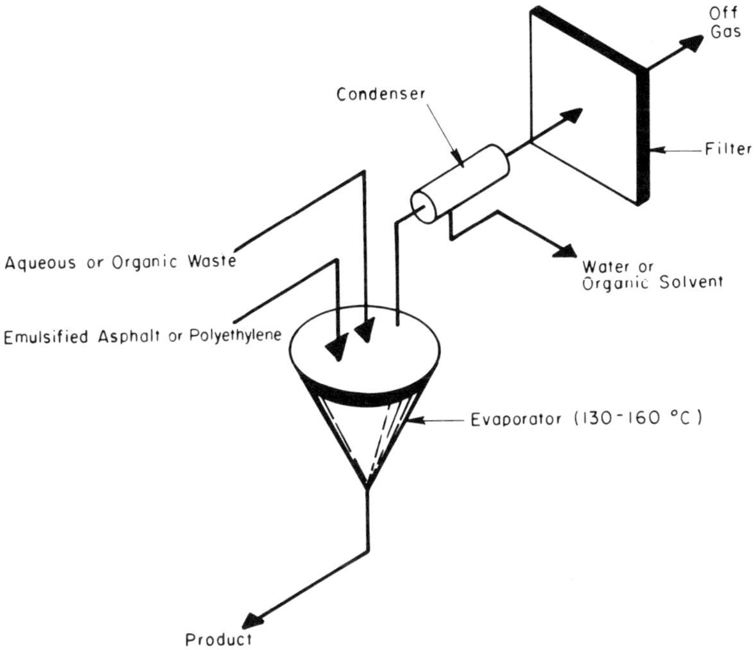

Figure 3. Incorporation of asphalt or polyethylene.

European operations, ashes from incineration operations are incorporated in concrete or asphalt before burial.

A typical low-level solid radioactive waste incineration operation is the pilot plant operation at the Karlsruhe Nuclear Research Center in Germany. There, a pilot plant incinerator was installed in 1963 and has been operating successfully since that time, achieving a reduction in volume of combustible wastes of approximately 80-fold.[14] The resulting ash from this incineration operation is incorporated with cement and stored in suitable burial sites. A larger incineration unit is currently being designed for that installation. This incineration plant is designed to burn all combustible wastes, which occur in a wide variety of forms. The inside of the combustion chamber of this incinerator was designed for maximum ease of decontamination by minimizing the surface area and number of internal parts. Off-gas cleanup from radioactive incineration operations is a vital requirement. During the first 4 years of operation of the Karlsruhe pilot incinerator, off-gas cleanup was accomplished with bag filters and final high-efficiency filters. A number of operating problems were encountered with this system; high-temperature ceramic filters have been installed during the more recent operation. Decontamination of the off-gas cleanup system has been greatly simplified through the use of the ceramic filters.

SOLIDIFICATION OF HIGH-LEVEL RADIOACTIVE WASTES

High-level radioactive wastes have been stored historically on an interim basis as concentrated solutions in underground tanks, but it is the policy in the United States that these wastes be solidified, for which a number of processes have been developed. One of them, fluidized bed calcination, has been employed in the United States since 1963 for this purpose, and in-tank solidification of some wastes at the AECs Hanford plant has been practiced for several years. Other processes have been developed through the radioactive demonstration stage and are available for further application. In addition, several other processes have been developed in European countries for solidifying high-level radioactive wastes. Each of these processes is described separately.

Fluidized Bed Calcination

Fluidized bed calcination has been employed at the Idaho Chemical Processing Plant since 1963 for solidifying high-level radioactive wastes resulting from reprocessing of highly enriched aluminum-uranium and zirconium-uranium alloy fuels at that plant. Through 1970, over 2,000, 000 gal of these high-level radioactive waste solutions have been converted to approximately a tenth that volume of bulk-dried granular oxide solids, which are stored in underground vaults at the plant site. In addition, fluidized bed calcination has been selected for solidification of the high-level wastes from the General Electric commercial fuel reprocessing plant at Morris, Illinois,[4] and similar equipment has been installed at that plant.

A cutaway sketch of the Idaho Waste Calcining Facility is shown in Figure 4. This figure shows the centralized operating corridor for remotely operating the equipment containing the highly radioactive material behind concrete shielded walls 5 ft thick. The heart of this facility is the 4-ft-diameter fluidized bed calciner shown in the flowsheet in Figure 5. Radioactive waste solutions are injected into the calciner through pneumatic atomizing nozzles where they impinge upon a heated (400–500°C) bed of granular solids that is fluidized with air. The high-level radioactive waste feed is divided equally among three

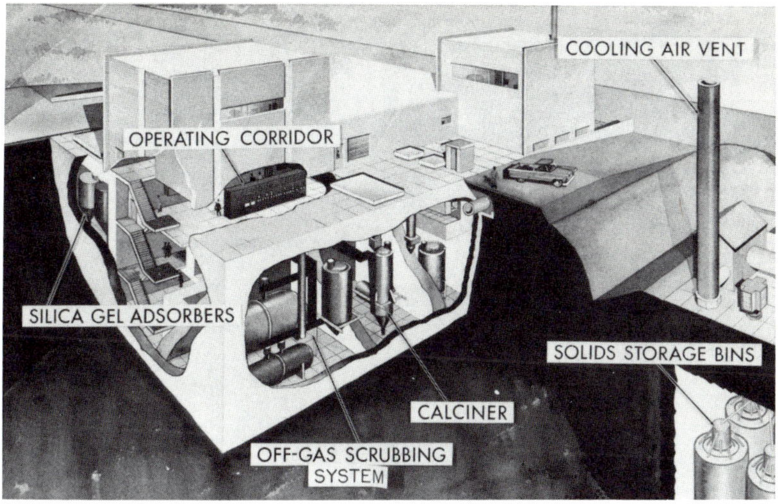

Figure 4. Cutaway sketch of the waste calcining facility.

nozzles and flows at a nominal total rate of 100–130 gal/hr. This feed rate includes batch recycle of off-gas scrubbing solution at an equivalent rate of 15–30 gal/hr.

The volumetric ratio of atomizing air to liquid feed is varied as necessary to cause sufficient break-up of bed particles and thereby control the average bed particle size at any desired value. This volumetric ratio varies from 200 to 1000 for various waste solutions. The bed level is maintained above the feed spray zone at all times so that the spray contacts bed particles in a fluidized state. Initially, heat was supplied to the fluidized bed by an internal heating bundle through which sodium-potasium eutectic alloy was circulated at temperatures up to 760°C. The capacity of the calcination process at that time was limited by the rate at which heat could be introduced into the bed. Consequently, in 1970, the liquid metal heating system was replaced with an in-bed combustion process in which kerosene atomized by oxygen is introduced into the bed through three separate nozzles located approximately 12 in. below each of the feed nozzles. With the adoption of the in-bed combustion process, the throughput rate of the facility is no longer limited by the rate at which heat can be introduced but rather by the pressure drop requirements of the off-gas cleanup system.

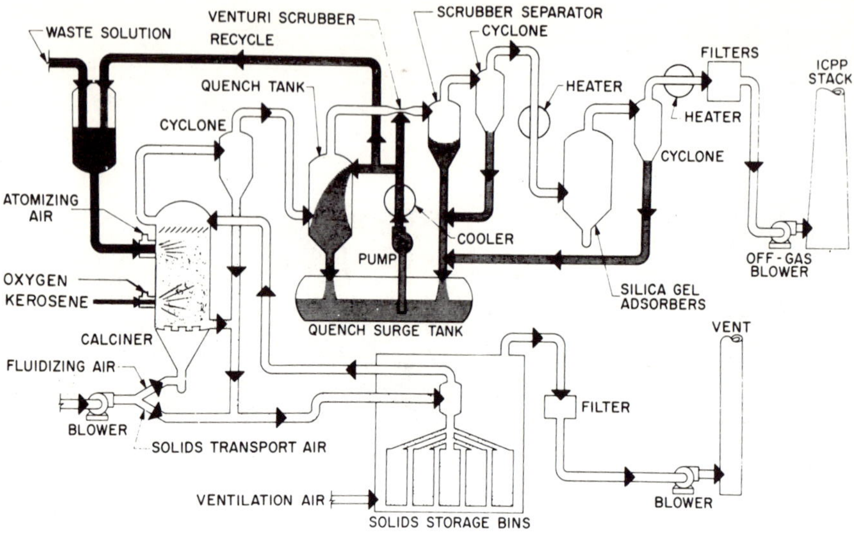

Figure 5. Fluidized bed calciner.

Preheated fluidizing air is introduced through 14 orifices equally spaced in the air distributor plate at the bottom of the fluidized bed. Each orifice is surmounted by a cap that helps distribute the air more uniformly in the bed and to minimize back flow of solids when the flow of fluidizing air is stopped. Operation at a superficial fluidizing velocity between 1.0 and 1.5 ft/sec has generally been satisfactory over the usual average bed particle size range of 0.3–0.8 mm.

Thermal decomposition of the nitrate and fluoride salt solutions to metallic oxides and fluorides, water vapor, and nitrogen oxides is the principal reaction occurring during calcination. Solid product builds up in layers on the bed particles, while the gaseous products are swept from the vessel with the fluidizing gases. A stable bed level is maintained by adjusting the rate of product removal.

Before being released to the atmosphere, the off-gas stream is decontaminated in the series of devices shown in Figure 5, including a cyclone, a spray quench tower, a venturi scrubber and associated entrainment separators, silica gel adsorbers (primarily for removal of volatile ruthenium compounds), and high-efficiency filters. About 70% of the solids carried from the calciner with the off-gas are separated from this gas in the primary cyclone. These solids are transferred by air jets into the solids transport line where they are combined with solids withdrawn at a controlled rate from the bottom of the bed. This combined product is transported pneumatically to a second cyclone separator, from which it falls by gravity into underground solids storage bins. The air that accompanies the product through the transport lines is returned to the overhead space of the calciner to be decontaminated along with the process off-gas. The driving force for this pneumatic transfer system is the pressure drop across the calciner, and no other motive force is necessary.

Most of the solids remaining in the gas leaving the primary cyclone are removed in a gas scrubbing system consisting of a spray quench tower followed by a venturi scrubber and two cyclone separators in series. The solids collected in the scrubbing solution are recycled intermittently to the calciner feed tank. Downstream from the scrubbing system the gas is heated above its dew point and then passes through three adsorber beds in parallel, each containing 6000 lb of silica gel for adsorption of gaseous ruthenium compounds. A fourth adsorber is maintained in standby condition. Following the adsorbers is a cyclone separator for removal of any water or silica gel accidentally carried over from the adsorbers during regeneration operations. This cyclone is followed by another heater and finally by three parallel high-efficiency filter units. An off-gas blower, located downstream from the final filters, maintains the entire system at subatmospheric pressure and transfers the decontaminated off-gas to the 250-ft-high plant stack.

Excellent performance has been obtained with the Waste Calcining Facility off-gas cleanup system. Overall releases of strontium-90, the limiting radioisotope, have been well under 1% of the established safe release values. The decontamination factors for strontium-90 have ranged from 1000 to 2000 in the scrubbing system, from 8 to 15 in the adsorbers, and from 1000 to 2000 in the final filters. With the liquid metal heating system, the release of ruthenium was larger than it is with the in-bed combustion heating system, primarily because with the latter system the nonvolatile dioxide of ruthenium is preferentially formed during calcination. Release of radioactive ruthenium with both systems has been well below the safe release limits; however, the much lower release with the in-bed combustion system would permit the use of this process with much wastes cooled a shorter time, which would contain much higher quantities of radioactive ruthenium—relatively short lived isotopes.

Two principal radioactive waste solutions have been processed in the Idaho Waste Calcining Facility. One is the waste resulting from reprocessing aluminum-uranium alloy fuels; this solution has the composition listed in Column 3 of Table 3. The other waste is essentially 0.3*M* zirconium fluoride and 0.7*M* aluminum nitrate along with fission

Figure 6. Calciner product.

product nitrates and other minor constituents. The first of these is converted largely to aluminum oxide, whereas the second is precomplexed with calcium nitrate before calcination (to prevent volatilization of corrosive and toxic fluorine compounds) so that the solids resulting from processing this waste are largely calcium fluoride, zirconium oxide, and aluminum oxide together with fission product oxides.

Product from the Waste Calcining Facility is granular and thermally stable up to temperatures exceeding 800°C. A photograph of some of the calciner product is shown in Figure 6 along with several other forms into which this product has been converted in bench-scale tests; these other forms, from left to right, are incorporation in a metal matrix, incorporation into a homogenous phosphate glass, and incorporation in a fused salt matrix. These other forms would be much more resistant to leaching of fission products if the calcine solids were to be contacted by water, but they are far less easily retrieved or transported. Currently the product from the Idaho Waste Calcining Facility is stored as granular solids in stainless steel bins in underground vaults. Each bin is fitted with a pipe to permit retrieval by vacuum aspiration should further processing or relocation of these solid wastes be desired at some future date.

In-Tank Solidification

At Hanford,[11] cesium and strontium are recovered from the supernate and sludge of aged alkaline wastes and from acidic new wastes. The residual new wastes are neutralized, stored for 5–7 years, and combined with older waste concentrates and alkaline aluminum decladding wastes for final in-tank evaporation to form salt cakes.

Three methods are used to solidify the contents of the tanks after the bulk of the strontium and cesium have been removed. One method uses heated air passed through the liquid in the tank. The air becomes saturated with water vapor and is discharged to the atmosphere after passing through an off-gas cleanup system and a condenser. Electrical heaters are used to heat the incoming air. A second method employs electrical heaters inserted directly into the liquid being evaporated. A small amount of air is used to sparge the contents of the tank so that a pattern of circulation can be established in the tank. Again, the off-gas is cleaned and the water is condensed. In the third method of in-tank solidification, the contents of the tank are pumped continuously into an evaporator placed over it. The concentrated solution is returned to the tank, and after an appropriate time the contents become sufficiently concentrated to solidify.

The preremoval of much (about 90%) of the strontium and cesium not only reduces the long-lived radioactivity hazard from the solidified wastes in the tank but it eliminates a significant amount of the heat generation from these aged wastes so that the final solidified wastes in the tanks do not overheat.

Spray Solidification Process

Spray solidification is a continuous process developed by Battelle-Northwest;[18, 19] it has been demonstrated in the Waste Solidification Engineering Prototype unit at Hanford with full-level radioactive wastes. This process is adaptable to a wide variety of wastes, but testing to date has emphasized its use with the type of high-level radioactive waste to be generated by the majority of the commercial fuel reprocessing plants in the United States. (See Columns 1 and 2 of Table 3.) In the spray calciner, shown in Figure 7, liquid waste (containing any necessary additives required for melt making) is fed through a pneumatic atomizing nozzle into the top of a heated cylindrical tower. The atomized waste is sequentially evaporated, dried, and calcined to a powder as it falls into a continuous melter located below the calciner. The solids are melted in this melter at temperatures of 800–1200°C. Process gases from the calciner flow into an adjacent filter chamber, carrying much of the calcined powder. The powder collects on the porous metal filters as the gas passes through. The powder is periodically blown off the filters and into the melter by sudden pulses of high-pressure steam or air directed backward through the filters by small nozzles. The molten calcine flows through an overflow

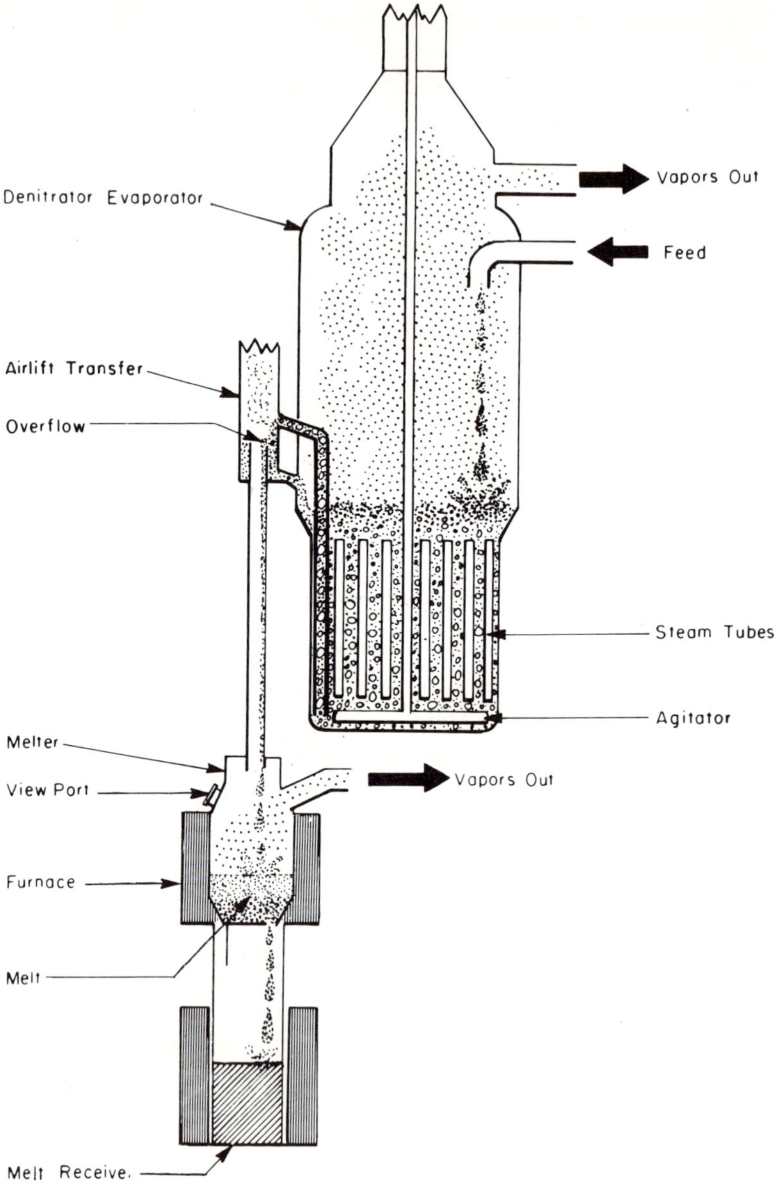

Figure 7. Spray calciner.

weir or a freeze valve from the melter into a receiver storage pot located below. After the pot is filled, it is cooled in the furnace, sealed, and taken to storage. Stainless steel Type 310 has been found to be a suitable material of construction for the spray calciner. For the continuous melter, platinum is the only reliable metallic construction material that has been found to withstand the corrosive, high-temperature environment. The pot for receiving the molten waste may be made of mild steel if its temperature is kept below 700°C.[20] Solids most frequently produced by the spray solidification process are ceramic orthophosphate solids of alkali metals and iron. Calcium is added to retain any sulfates that may be present.

Pot Calcination

Pot calcination was developed at the Oak Ridge National Laboratory, and is a second waste solidification process that has been demonstrated at the Waste Solidification Engineering Prototype at Hanford.[19, 21] In the pot calcination process, shown schematically in Figure 8, liquid waste is added to a pot heated in a multiple-zone heating and cooling furnace. The waste is concentrated at a constant volume up to the point that salt cakes form on the walls of the pot. As calcination continues, the salt cake grows in thickness and reduces the capacity for boiling the remaining sludge, therefore requiring a proportionate reduction in the feed rate. Average feed rates of 11 liters/hr are possible with this process, and when the feed rate drops to approximately 5 liters/hr the feed is shut off. At this point, the salt cake has grown inward from the pot wall and upward from the bottom to fill the pot except for a thin cone-shaped void in the upper 2–3 ft of the salt cake. Heating is continued until the liquid in this void has been boiled to dryness and all of the waste in the pot has been calcined and has reached a temperature of approximately 900°C. The pot is then cooled in the furnace, removed, sealed, and taken to storage. Total time cycles for filling, cooling, and change out of an 8-in. diameter pot with a 6-ft height of calcine have been demonstrated in the Waste Solidification Engineering Prototype to be approximately 45 hr. Increasing the diameter or length of the pot has little effect on the processing rate.

Phosphate Glass Solidification

Phosphate glass solidification is another continuous solidification process providing a good quality glass product. This process was developed by the Brookhaven National Laboratory[22] and has been demonstrated with full-level radioactive wastes at the Waste Solidification Engineering Prototype at Hanford.[19] In the phosphate glass solidification process, shown schematically in Figure 9, liquid waste

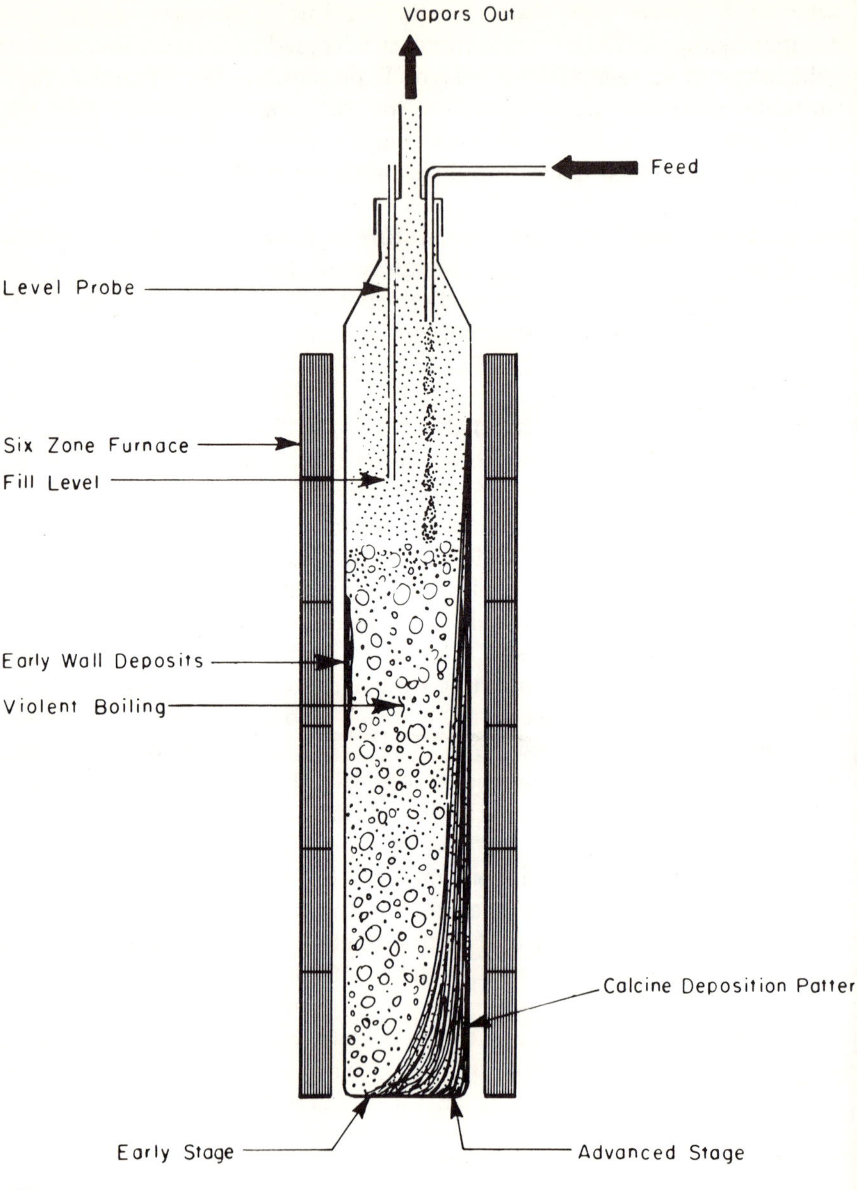

Figure 8. Pot calciner.

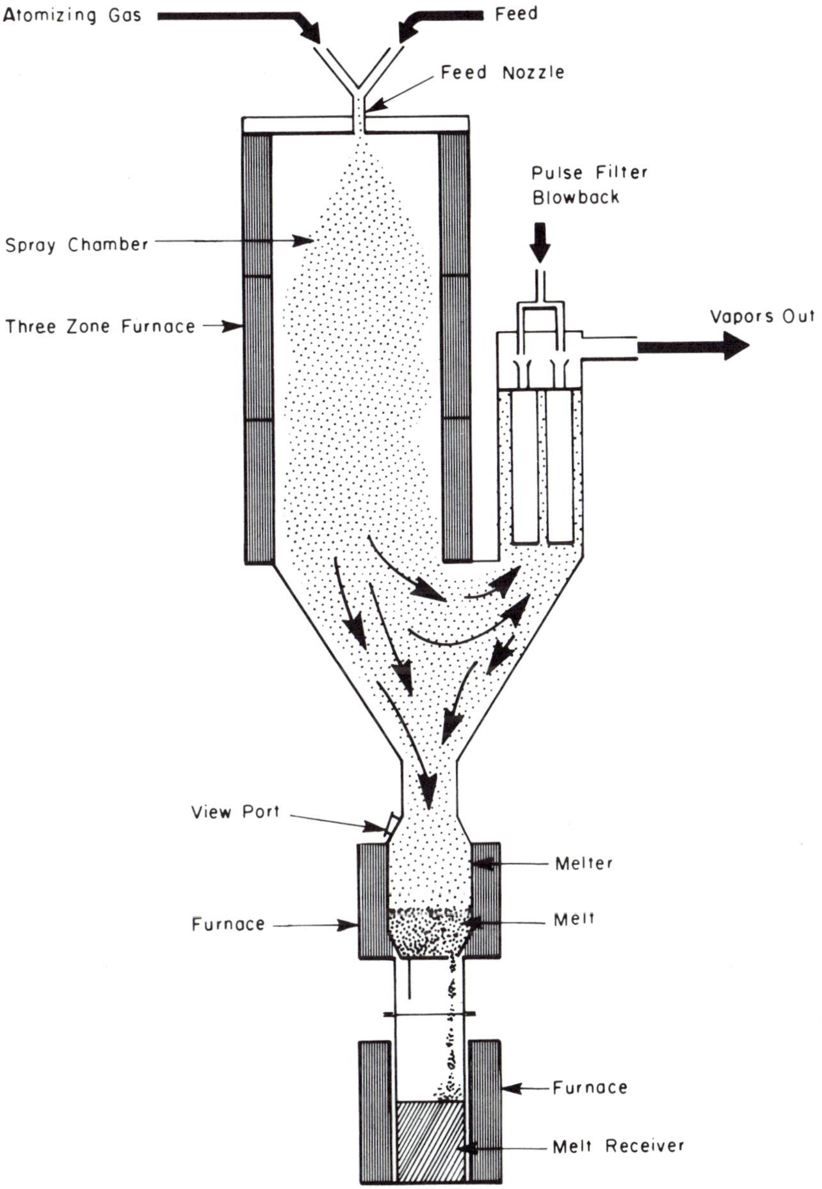

Figure 9. Phosphate glass solidification.

containing all of the melt-making additives is first fed to an evaporator where it is concentrated and denitrated by factors of from 2 to 10 to a thick, syrupy, aqueous phosphate slurry. The slurry is fed to the continuous melter where final volatilization of the water, nitrates, and other volatile constituents is accomplished; the material is heated in the melter to 1000–1200°C to form a molten glass. The molten glass flows through an overflow weir or a freeze valve into a receiver storage pot below. After the pot is filled, it is colled in the furnace, sealed, and taken to storage.

Titanium has been found to be an acceptable construction material for the denitrator-evaporator. The continuous melter must be made of platinum to withstand the corrosive environment there, while carbon steel pots are acceptable for receivers provided they are not heated above 700°C.

If the waste solutions contain sulfate, it is completely volatilized from the melter and a separate stream of intermediate level waste is formed. Because the sulfate effluent cannot be reasonably reused, it requires separate treatment for final disposal. The chemical adjustments required for the phosphate glass process consist mainly of adding phosphoric acid to the feed to obtain a mixture that will provide a solid with a suitable melting point. The mole percent of oxides of the alkali metals is kept to about half that of the total metal oxides in the melt.

FINGAL Process

Still another vitrification process has been developed in the United Kingdom; this process is called the FINGAL process.[23, 24] The feed material for the FINGAL process is highly radioactive liquid waste that has been concentrated by evaporation and held temporarily in a storage tank. The cycle for this process is shown in Figure 10. The concentrate is first transferred in small batches from the main storage tank through a shielded pipeline to a small receiving tank in the cave of the glass making plant; from there it is pumped into the glass-making cyclinder at a rate of about 125 liters/day. Before entering the cylinder, the concentrate is mixed in the line with 60 liters of silica borax slurry per day.

The glass-making cylinder is heated by an electric furnace, and the slurry mixture entering the cylinder is dried and the solid residue fused into a glass. A temperature of 1050°C is required to form the glass, and the furnace maintains this temperature in the area of rising glass surface as the cylinder is progressively filled. Gases from the glass-making cylinder are drawn successively through two other cylinders, each equipped with filters to remove ruthenium and other fission products

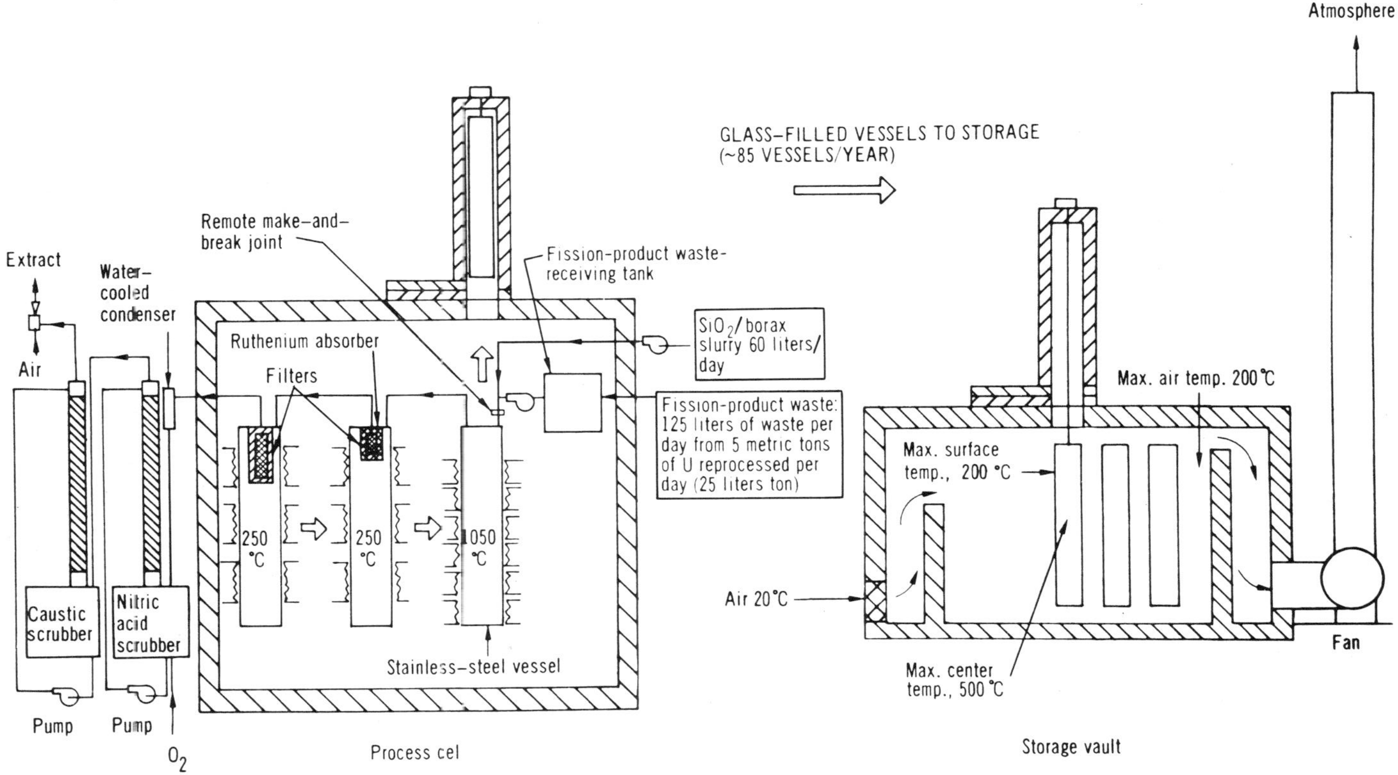

Figure 10. Fingal process and vitrificat:on.

evolved. These cylinders are progressively moved into the glass-making cylinder position, and the filter is then dropped and incorporated into the glass melt. Gases passing through these filters are scrubbed with nitric acid and caustic soda solution before local discharge to the atmosphere. Adequate decontamination factors have been achieved in the off-gas stream during extensive pilot plant testing. Generally, cylinders about 1 ft in diameter and 8 ft high are contemplated for this process.

The FINGAL process has been developed through the pilot-plant stage with radioactive material. Enough information is available for further use of the process. No significant amount of additional work is under way or planned at present on this process.

Other Solidification Processes

In France, two waste solidification processes are being developed, a pot vitrification process and a continuous vitrification process.[25] The pot process is operated at a pilot plant at Marcoule. In this process, waste solutions are mixed with glass frit suspended in a dilute nitric acid solution. The mixture is heated to 1100°C by induction heating in an Inconel pot. The molten glass product is discharged to a crucible receiver by melting a solidified glass plug at the bottom of the pot. The disposal crucible is fitted with a cap and transferred to underground storage pits in a concrete vault. Each pit is large enough to contain 20 crucibles, which are cooled in position by cooling air passing over them at a velocity of 12 ft/sec. The gas released during the process is stripped of volatile fission products by a ceramic trap and directed into a condenser where it is sprayed with condensate by an air lift. The gas in then filtered through a second ceramic trap and sent to a Venturi scrubbing system. Noncondensible components go through a silica gel adsorber before final filtering followed by passage through a spray column on which all purges converge. Condensates contaminated with radioactive ruthenium are concentrated and reacted with formaldehyde in an evaporator. They are periodically recycled through the feed tank of the vitrifying process.

The continuous French process is in an early stage of development with a nonradioactive system. The waste and additives are calcined in a rotary kiln, and the calcine is transferred to a melter for vitrification.

The Soviet Union has also been developing waste solidification processes. Brezhneva[11] described a system for calcining alkaline iron hydroxide sludges using a spray drier with a centrifugal spray nozzle, a cyclone separator, and a melter to form a glass. Bogdanov[12] also described a system for calcining a waste slurry in a fluidized bed of silica at 700°C. The product in this system is collected in a cyclone at 400–500°C and mixed with a glass frit; the resulting powder is melted at temperatures up to 1200°C.

In Germany, a pilot plant for solidifying high-level wastes is being constructed at Karlsruhe.[26] Process steps in this unit include denitration by reaction with formic acid, calcination in a steam-heated spray calciner, and melting the calcine with glass former such as borosilicates in an induction crucible with continuous overflow of the melt into a receiver.

STORAGE OF SOLIDIFIED RADIOACTIVE WASTES

Solidification of radioactive wastes accomplishes a needed reduction in their volume, and generally reduces their mobility and often (e.g., when vitrified) the susceptibility to leaching by water of soluble radioisotopes. However, it does not change the radioactivity or the danger from radiation emanating from it. The solidified wastes still must be separated from man and his environment for as long as the radiation may be harmful. This period of time varies widely. Solidified wastes containing very small quantities of radioisotopes or containing only radioisotopes with short half-lives need be stored for only a few decades. On the other hand, wastes containing significant quantities of transuranic elements or of long-lived fission products such as strontium-90 and cesium-137 must be stored so they cannot contact man and his environment for many centuries. Thus, there is no single storage method that is proper for all solidified radioactive wastes.

Burial Grounds for Low-Level Solid Radioactive Wastes

A number of burial grounds have been established in the United States at AEC facilities and at AEC-licensed commercial installations for direct burial of solid low-level wastes. For each burial ground a hazards analysis is required to establish that the operation will be safe under all conceivable circumstances. This analysis is highly dependent on the location of the burial ground, and serves to establish a maximum concentration of radioisotopes that can be buried at each location.

Very low-level solid wastes are often buried in containers of short-lived integrity such as iron barrels and wooden crates. Solid wastes having higher specific activities are generally encased in concrete or asphalt before or during burial.

Increasingly, the trend is toward greater compaction and higher integrity of containment of solidified wastes in order to minimize the number and size of burial grounds.

On-Site Storage of High-Level Solid Wastes

Some solidified high-level wastes are temporarily stored at nuclear installations; others are stored in facilities that are currently judged to be permanently safe but with provisions for retrieval of the wastes for shipment to a more suitable location if necessary.

One example of this on-site storage is that of the bin storage of the granular calcined waste from the AEC Waste Calcining Facility in Idaho.[27] The product from the calciner at this facility is transferred pneumatically through underground pipelines to a cyclone which empties into bins such as those shown in the schematic diagram in Figure 11. Each of the storage bins is 12 ft in diameter and 40–50 ft high and is constructed of stainless steel. Seven such bins are located

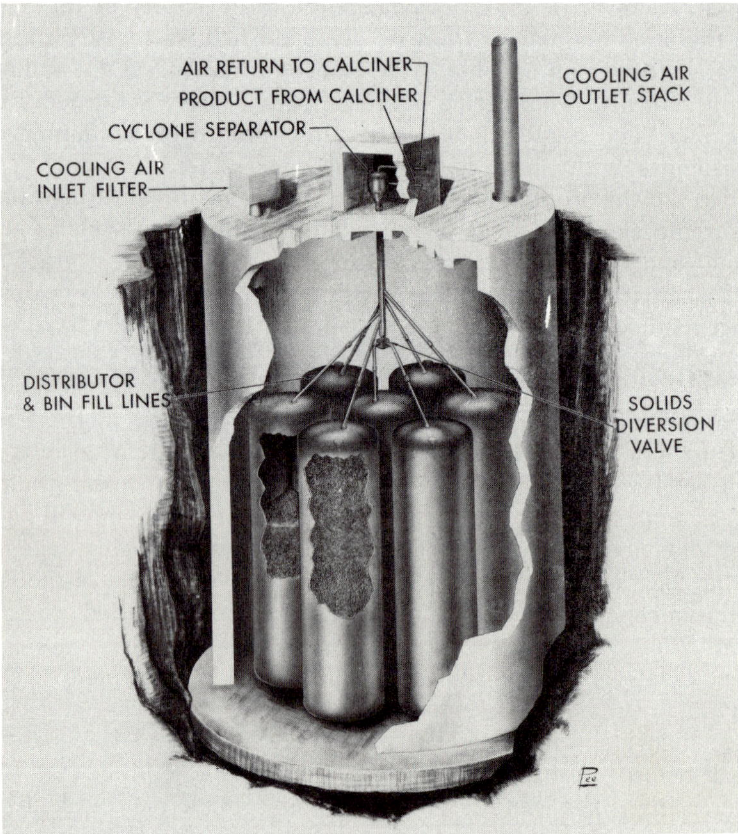

Figure 11. WCF solids storage bin.

inside a concrete vault. Cooling is provided by natural circulation of air inside the vault which keeps the initial centerline temperature of the bins below 700°C. The exit cooling air is monitored for radioactivity, and air circulation can be closed off if any radioactivity is detected; this activity could be present only if a bin developed a leak. Stopping air circulation would increase the centerline temperature of a bin by some 50°C, but the calcined solids have been proven capable of withstanding temperatures of over 800°C without release of any fission products.

The Idaho waste storage bins have been designed to last at least 500 years without leakage of any radioactivity. By that time, the radioactivity level will have decayed to nearly harmless levels. In an extensive safety analysis[28] conducted for this storage facility, it was shown that this facility is immune from the effects of the maximum conceivable earthquake or flood as well as to any chemical corrosion or other damage. While this form of storage appears to be acceptable for the foreseeable future, each bin has been fitted with a pipe to which a vacuum device could be attached to aspirate the granular solids for shipment to a more suitable permanent storage facility if that became desirable.

The General Electric fuel reprocessing plant at Morris, Illinois[4] employs the fluidized bed calcination process for solidification of its high-level wastes. At that plant, however, the granular solids are transferred from the calciner into pots about a foot in diameter and several feet long. These pots will be interim-stored on site for several years in a shielded cell equipped with forced cooling facilities. Within the required 10-year period, they will be shipped in shielded casks to the Federal waste repository in a salt mine near Lyons, Kansas.

Still another example of on site storage of high-level radioactive wastes is the facility being constructed at the AEC Hanford Plant for storing encapsulated solidified compounds of strontium-90 and cesium-137, which are being separated from the stored high-level aqueous wastes there.[29] Each capsule will contain about 10 lb of either strontium fluoride or cesium chloride. A cross-section of the underwater capsule storage facility is shown in Figure 12. The storage area will consist of a series of water pools, each 6 ft × 20 ft × 15 ft deep. Racks of capsules will be moved out of a final packaging cell in a transfer cask, using an overhead crane, and then placed into the water basin. As can be seen in Figure 12, operators working with long-handled tongs will open the cask. Using a hoist located above the storage basins, the operators will remove the rack of capsules from the cask and place it on the bottom of the pool. During these operations, the operators will be continually shielded from radiation by 13 ft of water cover.

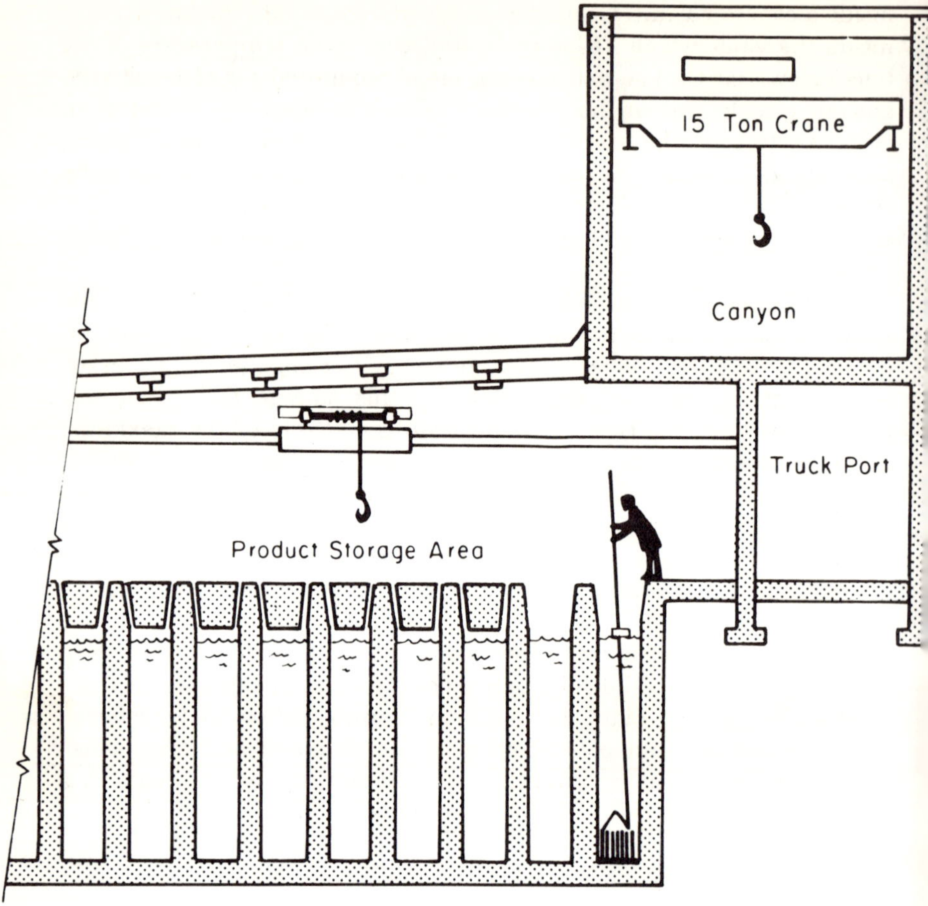

Figure 12. Underwater capsule storage.

In this facility, sintered strontium fluoride is packed into capsules made of Hastelloy C-276, each being 2½ in. o.d. × 13 in. long. This capsule is seal welded and placed in a similar outer container 2⅞ in. o.d. × 15 in. long, which is also welded shut. The molten cesium chloride is transferred by vacuum to stainless steel containers 2 in. o.d. × 24 in. long. After welding, these containers are placed in outer stainless steel capsules 2¼ in. o.d. × 25 in. long, and the ends of these outer capsules are also welded shut. Thus, these long-lived radioisotopes will be stored in double-walled metal containers in isolated underwater pools. Should any leak occur in a container, provisions have been made to reprocess the storage basin water.

Federal Repository for High-Level Wastes

In 1955, a committee of the U.S. National Academy of Sciences surveyed geological structures in the United States and recommended use of natural salt formations for permanent storage of high-level wastes. An experimental program was carried out under the direction of Oak Ridge National Laboratory from 1965 to 1967 in which waste handling equipment and techniques were successfully demonstrated in an abandoned salt mine near Lyons, Kansas.[30, 31] In this program, fuel assemblies in stainless steel canisters were used to simulate waste containers. A shielded carrier was placed over a steel-cased 19-in.-diameter charging shaft, and the canisters were lowered via a cable into a shielded transporter located in the mine some 1000 ft underground. The transporter carried the canisters to the test disposal area and deposited them into 12-in.-diameter × 14-ft-deep holes drilled in the floor of the salt corridor. The canisters were covered with about 8 ft of crushed salt to provide shielding. No detrimental effects were observed at the peak dose to the salt of 10^9 rad, and none are predicted for the 10^{10}-rad doses estimated for future operations.

A Federal repository for high-level wastes is being designed by the Oak Ridge National Laboratory for the Lyons salt mine. This repository will be used for receipt of both high- and low-level solid wastes from commercial nuclear installations in the United States, in line with the recently adopted Federal policy.

FUTURE IMPROVEMENTS

With the development of several processes for solidification of high- and low-level radioactive wastes and the selection of salt deposits as locations for initial Federal repositories, management of radioactive wastes in the United States appears to be well under control. However, as the quantity of these wastes grows and as man becomes more knowledgeable about his environment, it may be that improved radioactive waste management techniques will be desirable.

The most logical basis for improved radioactive waste management would appear to require better segregation of the radioisotopes before, during, or after solidification. In this way, superior disposal techniques would be practical for the small mass of radioisotopes having the greatest biological hazard such as transuranics, strontium-90, and cesium-137. Such separation was first advocated in a pioneering paper by Gleuckauf[32] and possible ultimate disposal schemes which such separation could make practical have been suggested by Slansky and Buckham.[33]

If strontium-90 and cesium-137 were separated by at least a factor of 10^5 from bulk fission products, the bulk residues would be reduced in biological activity (after 50 years decay) by a factor of over 10^6. Clearly, then, present practices would be adequate for the bulk of the wastes. The separated strontium-90 and cesium-137 fraction of the waste could be placed in even safer, but more costly, repositories for ultimate storage with little economic impact on the nuclear industry.

Alternatively, some form of truly ultimate disposal may be developed for the most biologically significant waste fractions. Two such methods have been suggested[33]—shipment by rocket to the sun and transmutation (in a spallation reactor target) to shorter-lived isotopes. Each of these methods of ultimate disposal appears technically and economically feasible but would require a significant amount of technical development.

Meanwhile, presently developed practices for management of solid radioactive wastes are providing the necessary protection for mankind from the possible hazards associated with these wastes.

REFERENCES

1. "U.S. Atomic Energy Commission Status Report to the President on Civilian Nuclear Power," 1967.
2. *U.S. AEC Manual 489,* Chapter 0524, Annex 1.
3. "West Valley Reprocessing Plant, Safety Analysis Report," Nuclear Fuel Services, Inc., Docket No. 50–201, July 1962.
4. "Design and Analysis, Midwest Fuel Recovery Plant," General Electric Company, Nuclear Energy Division, Docket No. 50–268, July 18, 1969.
5. "Barnwell Nuclear Fuel Plant, Safety Analysis Report," Vols. I and II, Allied Chemical Company, Docket No. 50–332, March 21, 1969.
6. Oak Ridge National Laboratory Staff, "Siting of Fuel Reprocessing Plants and Waste Management Facilities," USAEC Report ORNL 4451, 1970.
7. *Federal Register,* **35,** No. 222, 17530–33, November 14, 1970.
8. C. E. Stevenson, A. L. Ayers, C. B. Leek, F. M. Warzel, and F. H. Tingey, "Proceedings of the AEC Symposium for Chemical Processing of Irradiated Fuels from Power, Test, and Research Reactors," TID-7583, Richland, Washington, October 20–21, 1959, pp. 86–154 (January 1960).
9. H. L. Hull and J. W. Langhaar, "Proceedings of the AEC Symposium for Chemical Processing of Irradiated Fuels from Power, Test, and Research Reactors," TID-7583, Richland, Washington, October 20–21, 1959, pp. 172–201 (January 1960).
10. V. R. Cooper, L. L. Zahn, Jr., A. M. Platt, K. M. Harman, and R. A. Schneider, "Proceedings of the AEC Symposium for Chemical Processing

of Irradiated Fuels from Power, Test, and Research Reactors," TID-7583, Richland, Washington, October 20–21, 1959, pp. 17–78 (January 1960).

11. *Treatment and Storage of High-Level Radioactive Wastes, Proceedings of a Symposium, Vienna, October 8–13, 1962,* STI/PUB/63, IAEA, Vienna, 1963.

12. *Proceedings of the Third International Conference on the Peaceful Uses of Atomic Energy, Held in Geneva, August 31–September 9, 1964,* Vol. 14, United Nations, New York, 1965.

13. R. E. Blanco, H. W. Godbee, and E. J. Frederick, *Chem. Eng. Prog.,* **66,** 51–6, 1970.

14. W. Hempelmann and H. Krause, *Chemie-Ingenieur-Tecknik,* Issues 9 and 10, 1970.

15. L. T. Lakey, G. E. Lohse, and W. P. Palica, *Development of Fluidized Bed Calcination of Aluminum Nitrate Wastes in the Waste Calcining Facility,* IDO-14608, May 1965.

16. R. E. Commander, G. E. Lohse, D. E. Black, and E. D. Cooper, *Operation of the Waste Calcining Facility with Highly Radioactive Aqueous Waste,* IDO-14662, June 1966.

17. G. E. Lohse, B. M. Legler, and M. P. Hales, "Calcination of Zirconium Fluoride Wastes in Pilot-Scale and Plant-Scale Fluidized Beds," Proceedings of the Sympoisum on Fluidization, Tripartite Chemical Engineering Conference, Montreal, Quebec, September 22–25, 1968.

18. W. R. Bond et al., *Waste Solidification Program, Vol. 6, Spray Solidification Performance During First Radioactive Tests in Waste Solidification Engineering Prototypes,* USAEC Report BNWL-1391, 1970.

19. K. J. Schneider, ed., *Chem. Eng. Progr.,* **66,** 35–41, 1970.

20. K. J. Schneider, ed., *Waste Solidification Program Vol. 1, Technology for the Pot, Spray, and Phosphate Glass Solidification Processes,* USAEC Report BNWL-1073, 1969.

21. J. L. McElroy et al., *Pot Calcination Performance During First Radioactive Tests in Waste Solidification Engineering Prototype: Waste Solidification Program, Vol. 4,* USAEC Report BNWL-814, 1968.

22. E. J. Tuthill et al., *Ind. and Eng. Chem., Process Design and Development,* **6,** 1967.

23. D. W. Clelland, A. D. W. Corbet, and R. F. Coles, *Chem. Eng. Prog. Symposium Series, Nuclear Engineering—Part XX,* No. 94, Vol. 65, pp. 89–101, 1969.

24. K. D. B. Johnson, J. R. Grover, and W. H. Hardwick, "Work in the UK on Fixation of Highly Radioactive Wastes in Glass," Third United Nations International Conference on the Peaceful Uses of Atomic Energy, Geneva, 1964, Vol. 14, pp. 244–254.

25. F. Laude, G. Rabot, R. Monier, and J. Obernesser, SCCI-153, 1970.

26. H. Krause, KFK-1030 February 1970.

27. C. L. Bendixsen and G. E. Lohse, *Storage Facilities for Radioactive Calcined Waste Solids at the Idaho Chemical Processing Plant,* USAEC Report IN-1155, 1968.

28. L. T. Lakey and J. R. Bower, eds., *ICPP Waste Calcining Facility Safety Analysis Report,* USAEC Report IDO-14620, 1963.

29. J. R. La Riviere, *Chem. Eng. Prog.,* **66,** 42–44, 1970.

30. W. C. McClain and R. L. Bradshaw, *Nucl. Safety,* **11,** 130–41, 1970.

31. R. L. Bradshaw, J. J. Perona, J. O. Blomeke, and W. J. Boegly, Jr., USAEC Report ORNL-3358, 1969.

32. E. Gleuckauf, "The Long-Term Aspects of Fission Product Disposal," *Chem. Process Eng.,* **37,** 23, 1956.

33. C. M. Slansky and J. A. Buckham, "Ultimate Management of Radioactive Liquid Wastes," *Chem. Eng. Prog. Symposium Series, Vol. 65,* No. 97, p. 26, 1969.

CHAPTER VII.4

disposal of chemical plant solid wastes

RUDY G. NOVAK

The solid waste program at the Midland Division of The Dow Chemical Company places stress first on eliminating or reducing the solid wastes leaving the generating area, and second, on disposal of the remaining wastes in a manner that will not create immediate or long-term pollution problems.

The first portion of this program introduces and puts into practice the concept of defining the most economical alternative to achieve zero wastes. This does not mean that processes must run at 100% efficiency, but rather that all wastes generated must be consumed or destroyed within the process. Although this is not practically possible in every case, it appears that if this attitude prevails, it will be contagious and will lead to a climate in which enthusiasm and progress toward zero wastes becomes a part of everyone's job.

Since this first portion of the program is not readily achieved, the program must be continually used and maintained to keep it functional. Top management commitment—exercised straight across the board—is a basic requirement. Outstanding performance must be expected in research, process engineering, and production in order to achieve the desired goal.

This concept is illustrated in the firing of a petrochem boiler. A waste hydrocarbon residue plus a mixture of vent and natural gas is used as fuel to produce superheated steam. This unit is located at and operated by the production facility that produces these waste materials, resulting not only in heat recovery, but also in proper disposal and the release of capacity at the central solid waste disposal incinerator. A flowsheet is shown in Figure 1.

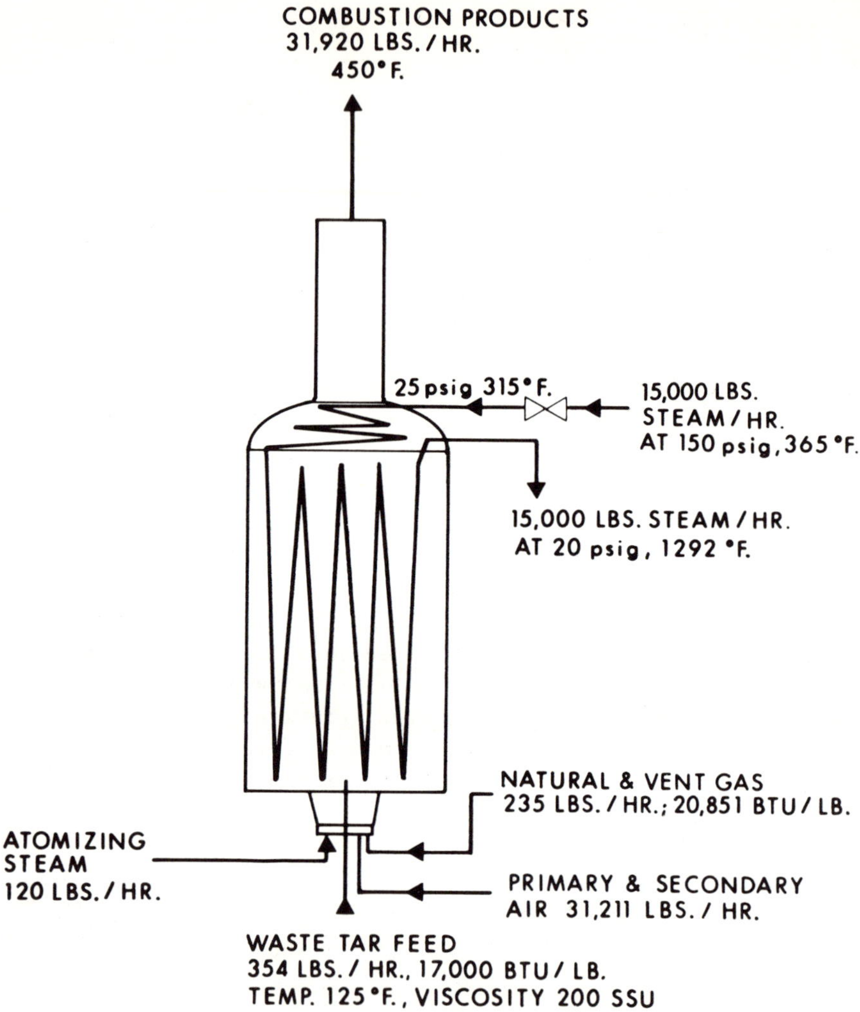

Figure 1. Schematic flow relations in petrochemical superheater utilizing a waste hydrocarbon residue and waste vent gas as fuel.

WASTE DISPOSAL FACILITIES

The second portion of the program deals with the disposal of solid wastes that are generated despite efforts to reduce or to eliminate them at the source.

The primary units for disposal of wastes consist of a 65 million Btu/hr rotary kiln incinerator, which is used for the incineration of chemical refuse and liquid residues, and an 81 million Btu/hr liquid residue incinerator. The rotary kiln incinerator is shown in cross section in Figure 2. This unit is used to dispose of low-ash liquid residues and solid chemical refuse such as plastic trimmings, strands, powders, plastic foam, paper, wood, and other solids of varying Btu content. A pack feed mechanism is used to feed packs and drums of solid waste chemicals into the incinerator.

The incinerator is a waste tar burner used for the disposal of a large variety of liquid residues of varying heat, ash, and chlorine content. Key operating parameters such as temperature and pressure drop are automatically maintained. All water wastes resulting from these opera-

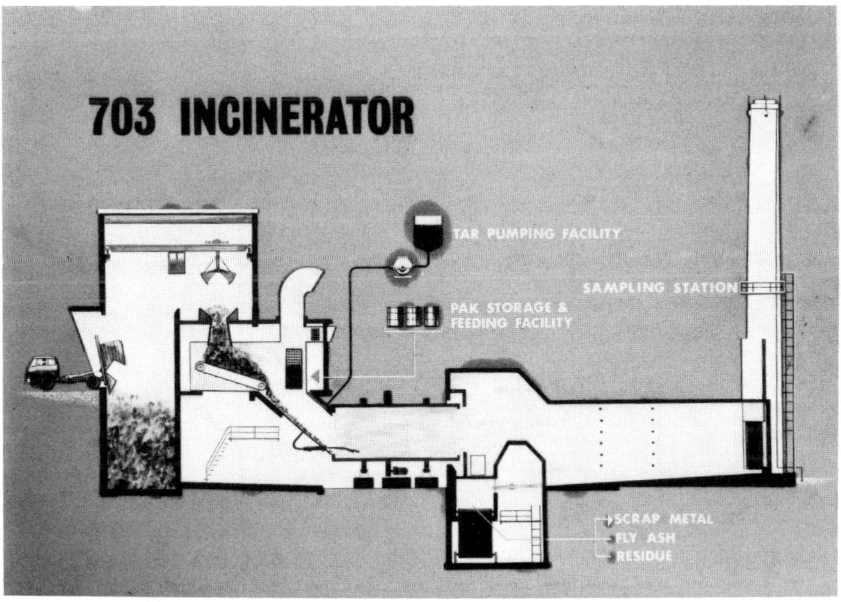

Figure 2. Rotary kiln incinerator for low ash liquid residues and solid chemical refuse.

tions are routed to the waste water treatment facilities. In addition, Dow operates a 200-acre landfill area for selected wastes.

This type of facility is required since many of the solid wastes require controlled high temperatures for complete destruction. Open burning is unacceptable and illegal. Although these units are designed to prevent air pollution, they also prevent water pollution since the solid wastes cannot be treated in waste water treatment plants without

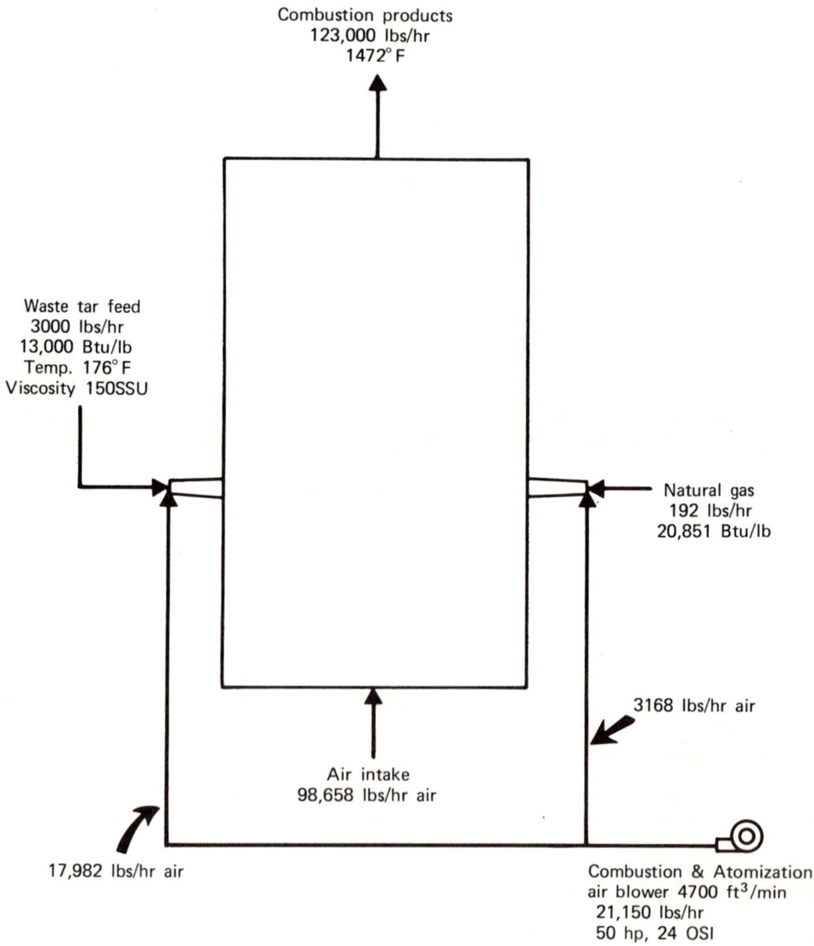

Figure 3. Flow sheet shows flow relations in a vertical liquid-residue incinerator.

causing serious consequences. These units are backed up by one 32 million Btr/hr Hooker liquid-residue incinerator, a 56 million Btu/hr Bigelow Liptak liquid residue incinerator, and a 43 million Btu/hr vertical liquid residue incinerator. A diagrammatic flowsheet of the vertical liquid residue incinerator is shown in Figure 3.

Currently, the facility is receiving by Dempster hauler about 650,000 gal/month of liquid residues, 500 drums (55 gal)/month that contain miscellaneous liquid residues, and about 20,000 yd^3/month of refuse for disposal.

COMPOSITION OF RESIDUES

Liquid residues are by-products, contaminated chemicals, or bad batches that cannot be economically reclaimed or reused. These materials, varying from light solvents to very heavy still residues, must be disposed of without causing safety hazards or air or water pollution problems. Most of these residues are solids at room temperature; they must be kept hot in order to remain liquid. Many residues are chlorinated and can contain as high as 50% chlorine, plus several percent of ash in the form of Fe, Ca, Mg, Na, oxides, and chlorides. Other properties, such as heat combustion, viscosity, ash, are also of interest.

Refuse consists of plastic sheets, trimmings, strands and powders, plastic foam, paper, boxes, wood, plus many nonburnable solids such as cinders, demolition material, dirt, and sludges. The average heat of combustion of the burnable refuse is about 6000 Btu/lb.

Another form of solid wastes is generated in the laboratories. Most of these products are extremely hazardous. Toxic, reactive, or explosive materials—although small in volume—frequently present the most difficult disposal problems. To bring in additional expertise, the operators consult with reactive chemicals committee whenever a potentially hazardous disposal problem arises. A key part of the procedure is to have the individual who created the problem play a significant part in determining the disposal steps to be used.

All requests for hauling wastes are phoned to the incinerator office. Any request not contained in the files is apparently not routine and is turned over to supervision for investigation. All information necessary to back-charge the cost of disposing of this waste to the account of the generating source is entered on a punched card. Pickup service is available 24 hr/day, 7 days a week, although the bulk of the pickups take place on the day shift, Monday through Friday. Costs for hauling refuse run about 60¢/yd^3, either to the burner or to landfill. Costs for hauling liquid wastes in Dempster quantities are about 1¢/gal.

INCINERATOR FOR SOLID WASTES AND LIQUID TARS

Liquid wastes brought to the incinerator in Dempster tanks are transferred to designated receiving tanks that contain compatible wastes. All drums of liquid wastes are also transferred to the receiving tank by way of a drum-dumping dock. The waste is strained as it is pumped from the receiving tank into a burning tank, where it is blended for optimum burning characteristics. All liquid residues are burned in suspension by atomization with steam or air.

Drum quantities of solid tars are destroyed by feeding them into the rotary kiln incinerator via a hydraulically operated drum and pack-feeding mechanism. All refuse, except full drums and packs of material, is dumped into the refuse pit. An overhead crane is used to mix the refuse and raise it to the charging hopper of the rotary kiln (see Figure 2).

While the solid refuse is being fed, liquid tars are fired horizontally into the rotary kiln. As the refuse moves down the kiln, organic matter is destroyed, leaving an inorganic ash, made up primarily of slag and other nonburnables such as drums and other metallic material. The ash discharges from the end of the kiln into a conveyor trough that contains 3 ft of water. After quenching, the material is conveyed into a dumping trailer, and thence to the landfill.

After leaving the kiln, the products of combustion enter the secondary combustion chamber and impinge on refractory surfaces that cause a swirling action. No secondary fuel or afterburners are used. Downstream of the secondary combustion chamber, the gases pass through several banks of water sprays in which the fly ash is knocked down and sluiced onto the ash conveyor floor. Cooled gases pass under a stack damper and then to a 200-ft stack.

INCINERATOR FOR OTHER LIQUID RESIDUES

The rotary kiln does not have the capacity to burn all of the liquid residues. The kiln burns residues that have a very low ash content; other residues are sent to the 81 million Btu/hr liquid residue incinerator, which has a combustion chamber 35 ft long and 10 ft square in cross section. Residues are fed to the unit through a combination of four dual-fired nozzles. Combustion gases are quenched in a spray chamber, followed by a high-pressure-drop Venturi scrubber, and a cooler/mist eliminator. About 3000 gal/min of water is recycled from the primary tanks of the wastewater treatment facilities to furnish scrubbing water. This water flows back to the wastewater plant for

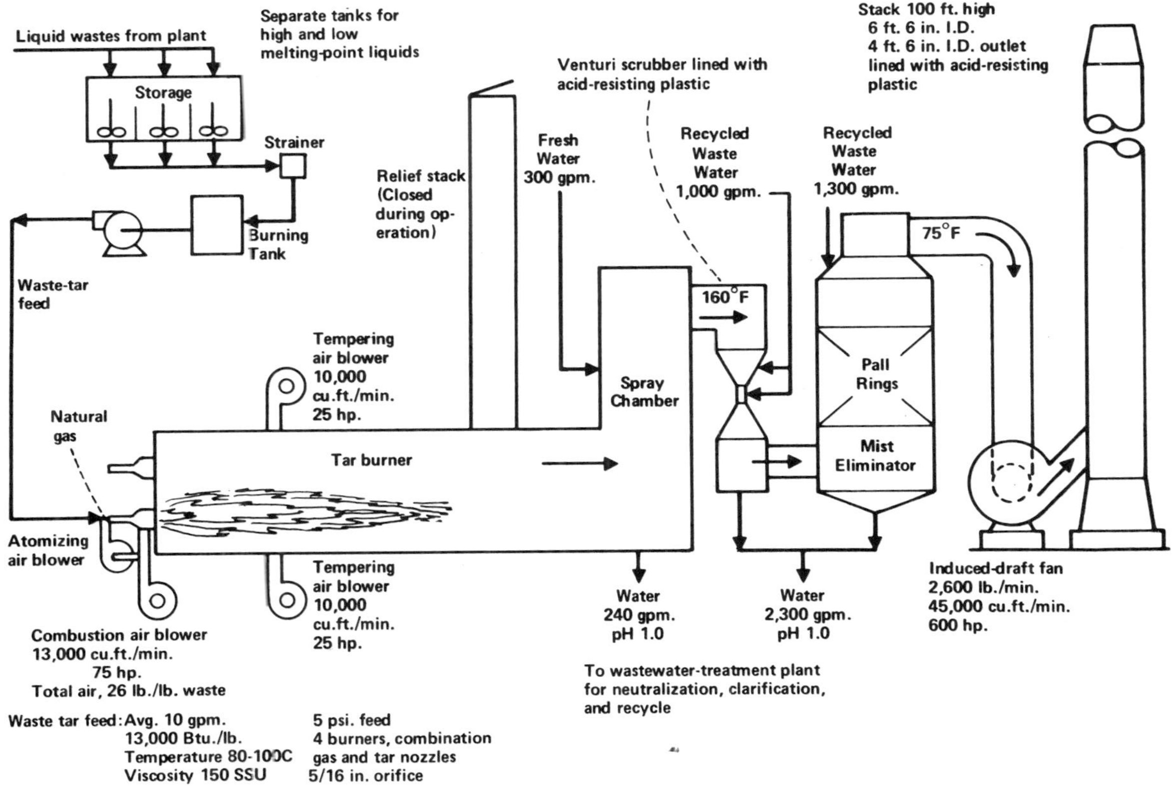

Figure 4. The flow relations in the waste tar burner.

treatment. About 1100 hp is required for this unit. A flowsheet of the total process is shown in Figure 4. Standby units must be kept ready for operation in the event of failure of one of the primary units.

The entire solid and liquid waste operations are maintained by a field service crew, including among other trades a full-time electrician, a machinist, pipe fitters, and a boiler maker. This crew is backed up by instrument men, masons, millwrights, and other personnel located in the central shops. A good deal of supporting equipment is needed for these units. Utilities and utility distribution systems for electrical power, gas, water, steam, nitrogen, and air are required. Storage space for refractories and other special equipment is a must because items such as refractory materials can take up to 9 months for delivery. Wastewater treatment facilities for handling scrubber discharges are necessary considerations, and are key factors for fail-safe operation. Well over $6 million is invested in incineration equipment at the Midland plant.

LANDFILL OPERATIONS

In the landfill operations, about 10,000 yd^3/month of material including rubble and demolition, primary sludge, incinerator ash, certain scrap metal, and special nonburnable solids are handled. For example, attempting to landfill about 200 yd^3/day of primary sludge is a challenging job and requires the best in experience and equipment. These materials are mixed and covered with sand to form a stable fill. A bulldozer operates full time, and a crane and truck part time. Substantial quantities of sand are available in the fill area, and haulings of rubble and demolition. Operating costs at this site run about $1.50/per cubic yard.

In operating the landfill, plant people are impressed that the site is not a junkyard and does not accept contaminated chemical processing equipment. In many cases, the equipment can be sold for scrap or salvage after decontamination.

Size of material going to the fill is also regulated. This eliminates placing large items in the landfill area, which could be extremely heavy, difficult to handle, and take up a large volume of fill. The site is continuously monitored by remote television during daylight and regularly patrolled at night since it might prove an attractive nuisance. Several practice firedrills are conducted each year to review and test established procedures with our fire, service, and plant protection departments.

DEFENSIVE RESEARCH

Incineration performance is defined under specific conditions of operation by actual field tests and measurements. Laboratory and pilot techniques are refined to study the incineration of specific waste materials. Computer simulation studies can also be used to help predict performance of new or existing incinertion designs. With this knowledge, the plant can confidently select the best operating conditions for disposal of a variety of materials that have widely diverse properties.

Another area of concern is construction materials. Accurate operating and maintenance records are vital for evaluating various materials. Selection and incorporation of reliable instruments and controls also play a significant part in the total effort. In some cases, it is advantageous to build up a relationship with equipment vendors who specialize in certain types of equipment.

Forming and maintaining attitudes toward safety at an optimum level include training, preparing meaninful operating instructions, and reviewing these instructions in depth. There is strict adherence to established safety standards concerning raw waste materials coming for disposal. If the standards are not met, Dow refuses to accept the waste and returns it to the source. The standards relate to container types, labeling, and tagging.

Another area of concern is to make certain that equipment meets current FIA (Factory Insurance Association) standards, not only on the drawing board but consistently in the field under actual operating conditions. All shutdown and safety systems are tested at scheduled periods in accordance with written procedures. To operate a unit safely on waste materials having a wide range of physical and chemical properties requires ingenuity and hard work in the design and maintenance of the control system.

Finally, there is the factor of environmental health in the waste processing area. There are frequent consultations with the environmental and medical people who investigate and monitor health problems. These range from noise control to performing health inventories on our employees.

PROBLEMS UNDER STUDY

One of the problems under investigation is the disposal of sludge in various quantities, such as inorganic salts that may be saturated with solvent or organic process sludge. These materials must be hauled through the plant, emptied into storage or feeder equipment, introduced into a destructor unit, and completely reduced to an inorganic ash. These operations are difficult to perform with present equipment, and additional procedures must be developed and refined.

A second problem of concern is disposing of brominated liquid tars and sludges, because large amounts of free bromine are evolved during ordinary combustion. Recovery of the relatively valuable bromine rather than complete disposal appears to be the answer.

A third problem is residue quality. As more of the fuel quality wastes are used within the waste generating processes, other wastes having low Btu values, high ignition temperatures, and poor handling and burning characteristics start to have a sizable effect on the residue mix. In order to destroy this mix safely and efficiently, operating procedures must be modified and, in some cases, additional processing equipment must be installed.

A disposal program for solid wastes in a chemical complex involves much more than the construction and operation of incineration or landfill facilities. It involves first and foremost management. Management in formulating and carrying out a program that involves problem definition, priorities, safety, quality, productivity, reliability, public relations, research, legal questions, monitoring, and economics. The program must be growing and flexible, and requires careful attention to short- and long-range objectives.

CHAPTER VII.5

union carbide chemical landfill, institute, west virginia

E. E. SLOVER and JAMES D. HAINLEY

Union Carbide has been receiving 12,000 yd^3 of solid and semisolid refractory wastes annually into a specially designed chemical landfill near Charleston, West Virginia. The project was licensed by the State in 1965, and it has successfully served three company plants within a 6-mile radius since that time.

Taking advantage of a large supply of low-permeability clay and a relatively undisturbed geologic substructure in a deep, natural basin adjacent to the Institute Plant, Carbide and the soils consulting firm of A. C. Ackenheil and Associates designed and constructed a unique "wet-process" landfill exclusively for chemical wastes. Groundwater protection against chemical seepage is achieved through a clay membrane and a series of peripheral drains. Contaminated water within the fill passes through slag-filled conduits and unique "permeable" retention dike into a collection basin behind an impervious barrier of highly compacted clay.

Approximately 12,000 gal of contaminated water daily are pumped from the basin into the Institute plant's 5,000,000-gal/day industrial wastewater treatment unit.

A wide variety of chemical materials are included in the waste mix fed to the landfill. Pitches, ureas, tars, polymers, inorganic solids, and viscous, insoluble products trapped on filter media make up the major portion.

A rigorous program of segregation, recovery, and pretreatment in the source plants and at the landfill ensures a neutralized, inerted, and stabilized feed that can be worked with standard landfill equipment.

Departure from conventional sanitary landfill procedures came early in the project when it was found necessary to abandon "cell" construction and adopt an earth-blending technique to speed decomposition and control internal gas generation. Water "ponding" on the fill surface, following application of the traditional 6-in. clay seal, caused Carbide to employ its existing wastewater treatment unit as a receiver and led directly to the requirement that such a unit be part of the final chemical fill design.

In December 1969, evolution of the wet-process chemical landfill was completed with the installation of the first phase of a three-step program spanning the 20-year life of the facility. Features of the final design give full control over surface and subsurface drainage, permit both surface and subsurface water pollution monitoring, and defer expenditure of the estimated $250,000 cost over a large portion of the landfill's life.

The average operating and amortization cost for processing 1 yd^3 of chemical waste into this landfill is now $13.80. This is $37 per ton (assuming an average of 750 lb/yd^3) and does not include the cost of segregation, collection, and transportation.

Nationally, the project has drawn attention. It has received an Honor Award for Engineering Excellence from the Consulting Engineers Council of America[1] and the Earth Care Award from *Construction Equipment Magazine.*[2] Discussions have been presented at conferences on solid waste disposal.[3-5]

Carbide's chemical landfill was planned from the earliest stages to meet the air, water, and land pollution restrictions established by the West Virginia Department of Natural Resources–Water Resources Commission (now Division of Water Resources), the West Virginia Air Pollution Control Commission, and the West Virginia State Health Department–Solid Waste Disposal Program.

Since the fill was the first of its type in West Virginia, a decision regarding the prime licensing agency had to be worked out between company and state authorities. In the final analysis the Water Resources Commission was chosen with the stipulation that landfill incidents involving air and solid waste problems would be promptly relayed to the governmental department most directly concerned. All incidents, regardless of type, are immediately reported to the Water Resources Commission and written follow-up reports are required.

License No. 3141, dated July 26, 1965, was issued by the Commission as the state department having jurisdiction in the most likely problem area.[6] The transmittal letter for the license advises Carbide that the permit is of indefinite duration contingent on satisfactory operation of the landfill.[7] A statement of policy specifying that land for industrial fill operations must be owned or leased by the company producing the wastes and that said company must have the means to sample and analyze waste components that might reach underground and surface waters effected by the fill, was issued later.[8]

A July 1, 1967, addition to the Solid Waste Regulations of the West Virginia State Board of Health ensured the continued jurisdiction of the State Department of Natural Resources-Water Resources Division over Carbide's chemical landfill.[9]

SPECIAL REGULATIONS

Table 1 shows the requirements applied daily to all operating, supervisory, engineering, and design activities related to the Goff Mountain Chemical Landfill at Institute, West Virginia. Strict adherence at all levels of management ensures that state codes governing the fill are met.

Table 1 *Goff Mountain Landfill Control*

To maintain our License No. 3141 with the West Virginia State Department of Natural Resources, Water Resources Division we have agreed to the following:

1. NO BURIAL OF "SANITARY" WASTE—Garbage from food and direct human activity must be handled in special landfills licensed by the State Health Department. Ordinary "garbage" attracts rats and flies. We have agreed that there will be none of these vectors at Goff Chemical Landfill.
2. NO POLLUTION OF SURFACE OR UNDERGROUND WATER—Special drains, collection basins, clay seals, and chemical inerting procedures are followed to guarantee that no chemical leaches into any water supply.
3. NO POLLUTION OF THE ATMOSPHERE OVER THE FILL—Fire control and odor elimination are required by the State Air Pollution Control Commission. No breathing hazards or attraction of pests is permitted.
4. NO WASTE OF THE LAND—A definite development program for the landfill is being followed to guarantee that the filled area will be usable when burial is complete in 20 years.
5. NO EMISSION OF TOXIC MATERIALS—No toxic substances may escape from the fill area by any route. Chemical reaction and burial in concrete are used to guarantee that this does not happen.
6. MULTIPLE LAND REUSE IS REQUIRED—Every effort is made to degrade the chemical waste scientifically so that its volume is reduced. Chemical and biological techniques are used. This enables burial of more than one waste in the same place.

MAIN OBJECTIVE: PRODUCTION OF AN INERT, USABLE SOIL OUT OF SOLID CHEMICAL WASTE

DESIGN

Site Selection

Requirements set forth in existing state and federal regulations covering pollution entering all three environmental phases (air, water, land) were carefully reviewed at the outset of the Chemical Landfill Project before any specific commitments were made. Representatives of state regulatory agencies were included in the early planning sessions and their recommendations were incorporated.

By taking these steps at the very beginning Carbide was able to anticipate the tougher landfill laws that came later and design in necessary controls.

Potential sites were then visited and compared with the concept contained in the regulations, and Goff Mountain Hollow next to the Institute Plant was selected. The Goff parcel of land contained 101 acres and was located in a deep, natural hollow typical of this part of the Appalachian region. Five important considerations led to the final choice:

1. Union Carbide owned the land.
2. The land was adjacent to the Institute Plant.
3. An ample supply of a highly-retentive red clay covered a good part of the property.
4. Collection and monitoring of surface and subsurface water would be an easy matter because of the bowl shape of the hollow.
5. Surface and subsurface drainage crossed Carbide property all the way to the nearest river and interceptor-treatment facilities could be installed if needed.

Supervision by the Institute Plant's Environmental Protection Department subsequently proved to be easily accomplished as a result of the fill's proximity and Carbide's ownership of the watershed.

Evolution of Industrial "Wet Process" Landfill

Initially a 20-year life span was arbitrarily selected to provide a basis for cost comparison with incineration. Fill operation was to parallel, as much as possible, that used in standard, well-run sanitary landfills, and any leachate was to be pumped into tank trailers and taken to the Institute Plant's wastewater treatment unit or (up-graded with waste oil) to the process steam boilers as fuel. A series of five small, totally retentive clay dikes was to be built along with any secondary dams needed further downstream to collect any leakage that might develop above or below ground as a result of breaching in one of the clay retention dikes.[10] Clay was spread and rolled on the floor of the fill area behind these dikes, and the dikes themselves were built with clay "keys" down to the top of the first rock layer to prevent the escape of chemicals. Ultimately, the Company expected to recover the filled land as a tank farm and warehouse area.

Severe operating problems (discussed later in the "Operation" section, p. 1017) forced complete abandonment of the sanitary-landfill approach and construction of the unique "wet process" now in use.

A. C. Ackenheil & Associates, Inc., a consulting firm specializing in geological investigation and soil mechanics, was called in to design a barrier that would span the entire hollow and permit "staged" construction over the 20 years of expected landfill life.

A complete geological study of the hollow's substructure, followed by approximately 3 years of design work involving both firms, demolished the "giant-dike" approach as both risky and expensive. In its place grew the wet process idea that was placed in operation during the last week of December 1969.

Figure 1 shows a cross-section sketch of the wet process system. Three construction phases, spanning roughly half the life of the landfill, are planned as a means of deferring capital expense. Each phase involves an additional "leaky" dike with appropriate seals, drains, and "lifts" (terraces). Figure 2 shows Phase I, Figure 3, Phase II, and Figure 4, Phase III.

Site Study and Preparation

Full scale professional investigation of the earth's structure under the landfill site began in 1966. A heavy overlay of clay and site covering relatively unbroken layers of shale and sandstone was found.

Sections of the site contained a clay that could be compacted to an impermeable condition, and these areas were marked as the supply source for the construction of the clay seal, the impervious dikes, and portions of the "pervious" barriers.

Borrow areas for blend-earth to be used in the fill lifts were delineated at this time. Enough was found to last the working life of the fill.

Permeability tests were made of both the clay-seal material and the underlying rock structure, and potential problem areas (such as sheeting joints, sand, and rock fractures) were noted for consideration in the final design.

A back-up plan to grout seal the rock substructure in the event of future clay-seal failure was developed at this stage, and surface and subsurface drainage patterns were mapped for incorporation into the overall plan.

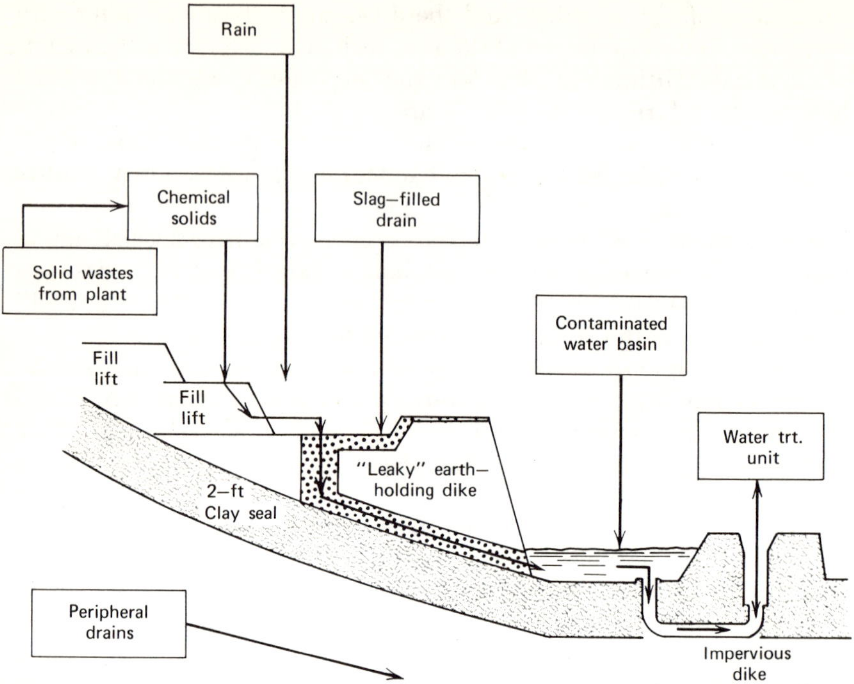

Figure 1. Goff Mountain chemical landfill, Institute plant, Union Carbide Corp. Dynamic wet process biochemical system. (1) No odors allowed; (2) no groundwater pollution; (3) no air pollution allowed (fires, gas); (4) License required.

Site study is an expensive and time-consuming activity, as Table 2 indicates. However, the success of the landfill is totally dependent on its being done well.

Clay Membrane

A 2-ft-thick, selected-clay seal is the key feature that makes this wet process landfill possible. It has to be placed under dry weather conditions and is rolled at 6-in. intervals to a density and moisture content that ensure impermeability.

The seal is used under the entire landfill (fill lifts, dike keys, contaminated basin, and impervious dike). Field and laboratory tests are run constantly during installation, and compaction layers that do not meet the standards are scraped off and relaid.

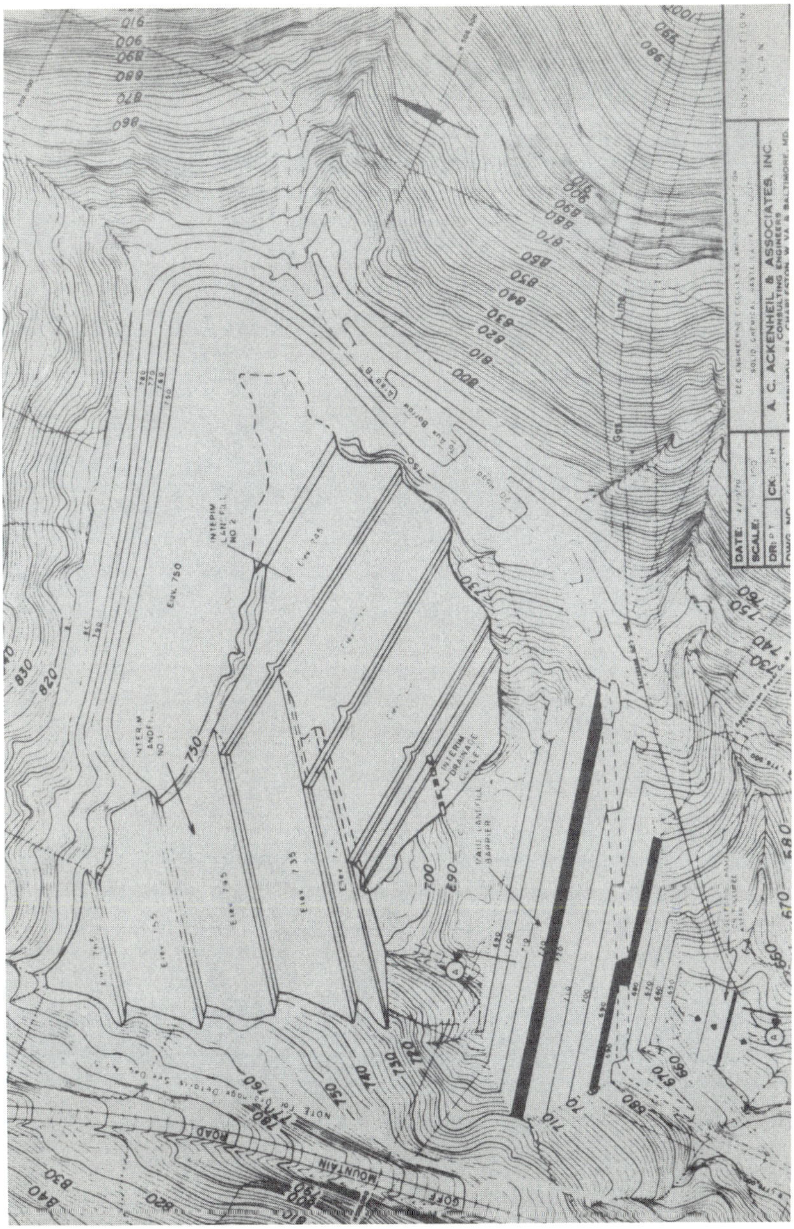

Figure 2. Lifts in plan view, first phase.

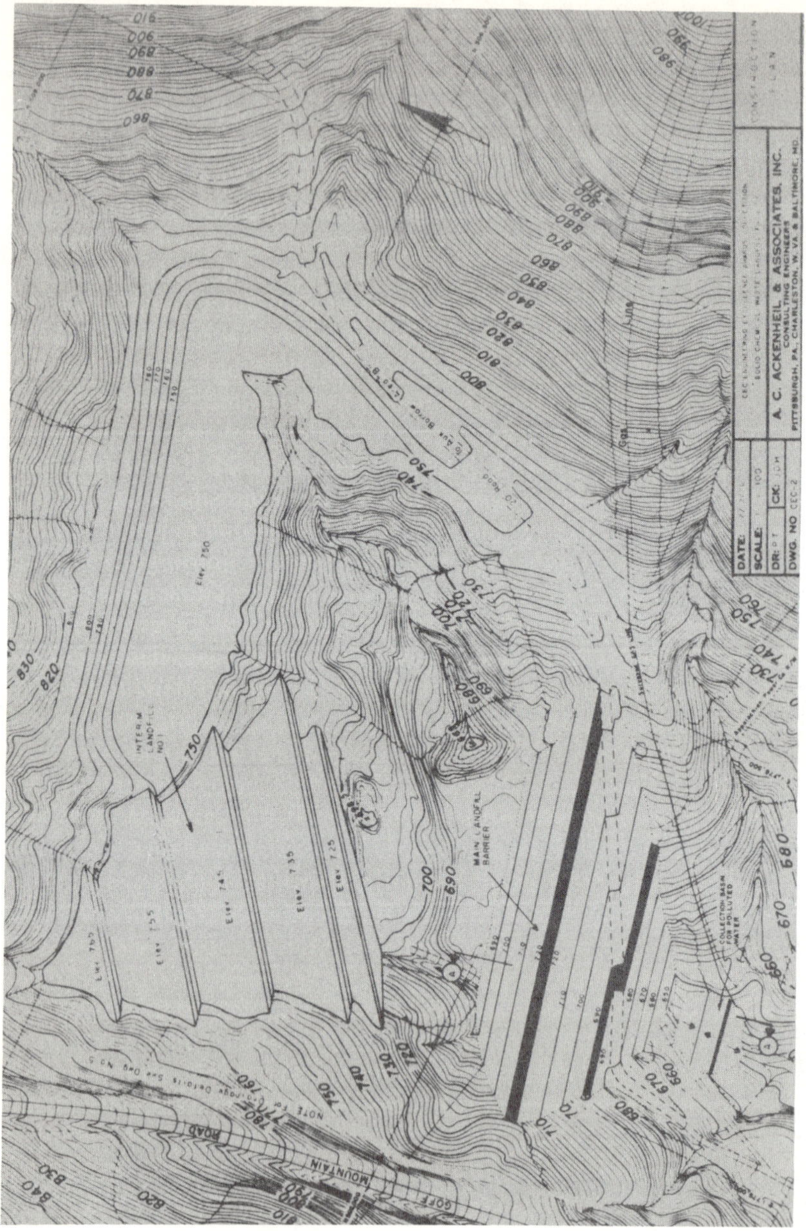

Figure 3. Lifts in plan view, second phase.

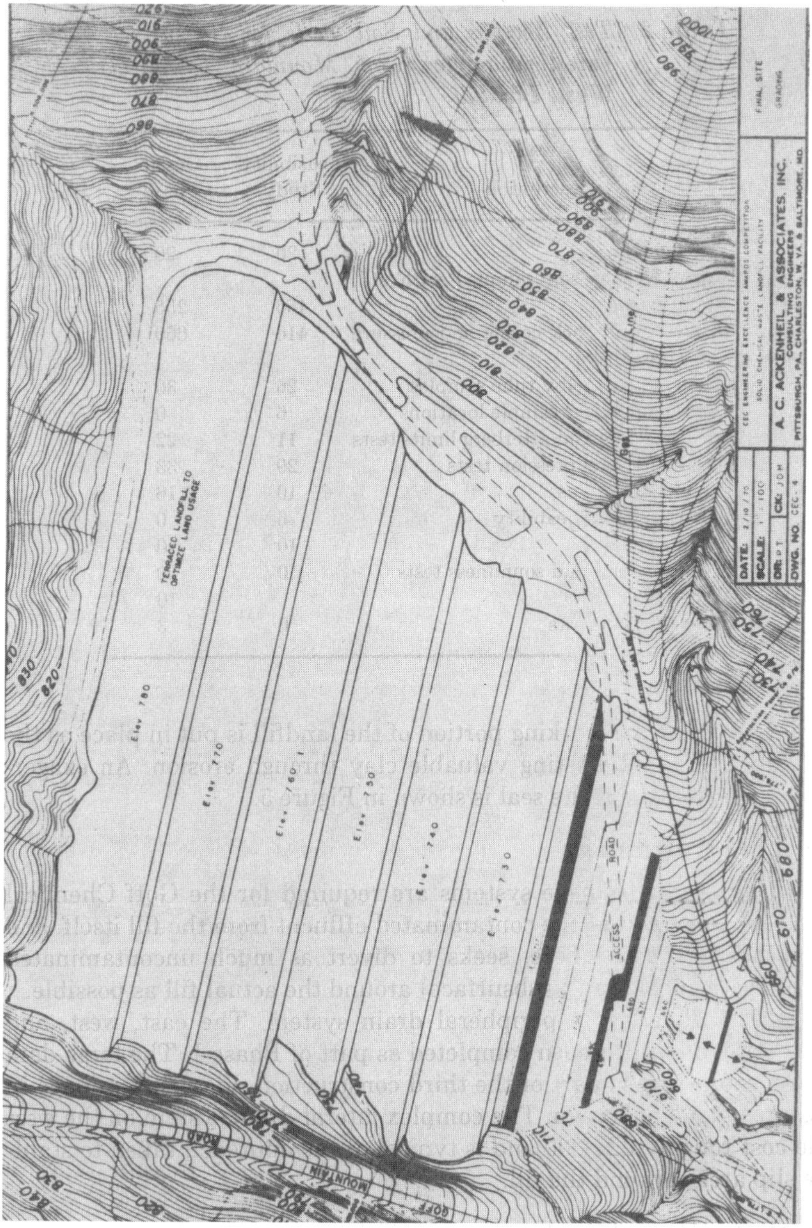

Figure 4. Lifts in plan view, third phase.

Table 2 *Test Borings and Soil Tests for Subsurface Investigation for Goff Mountain Chemical Waste Landfill*

Description	Feasibility Study	Design Phase
Test borings	20	20
Lineal feet of soil sampled with std. split spoon	455	255
Lineal feet of rock coring and pressure testing	416	660
Undisturbed shelby tube samples	26	30
Permanent water sample locations	6	0
Soil gradation and atterberg limits tests	11	22
Unconfined compression tests	29	33
Direct Shear tests	10	16
Laboratory permeability	6	0
Specific gravity	10	0
L. A. abrasion and soundness tests	0	9
Compaction tests		10
Consolidation tests	6	3

Floor seal for the working portion of the landfill is put in place as the fill grows to avoid wasting valuable clay through erosion. An exagerrated cross section of the seal is shown in Figure 5.

Peripheral Drainage

Two types of drainage systems are required for the Goff Chemical Landfill. One handles the contaminated effluent from the fill itself, and the other, discussed here, seeks to divert as much uncontaminated water (both surface and subsurface) around the actual fill as possible.

Figure 6 details the peripheral drain system. The east, west, and north branches have been completed as part of Phase I. The large dam on the south side is part of the third construction phase so drainage in this area is not as shown. The complex lateral drain system on the west side cost roughly $24,000 and is typical of the approach taken to stabilize slip areas around the fill.

Leachate Control System

Early experience during the sanitary-landfill operating period clearly indicated that Carbide was faced with running a wet process because of the heavy rainfall and high groundwater flow indigenous to the site.

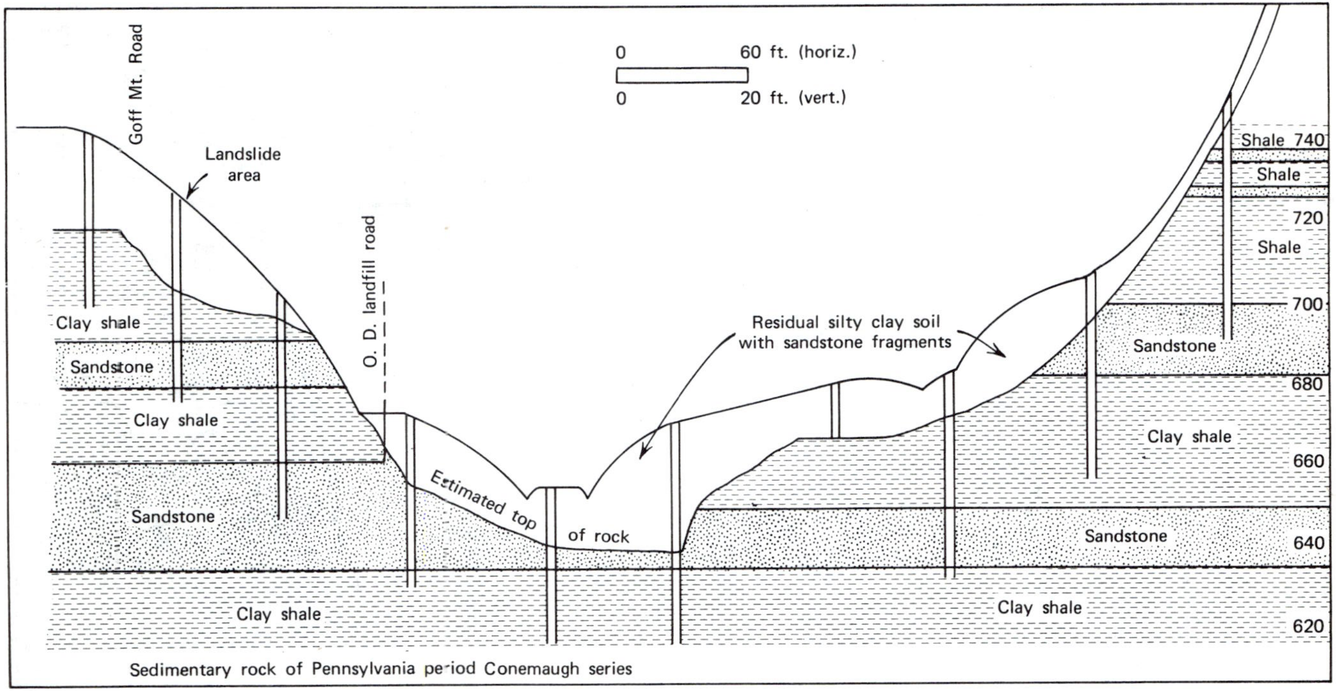

Figure 5. Geologic cross section.

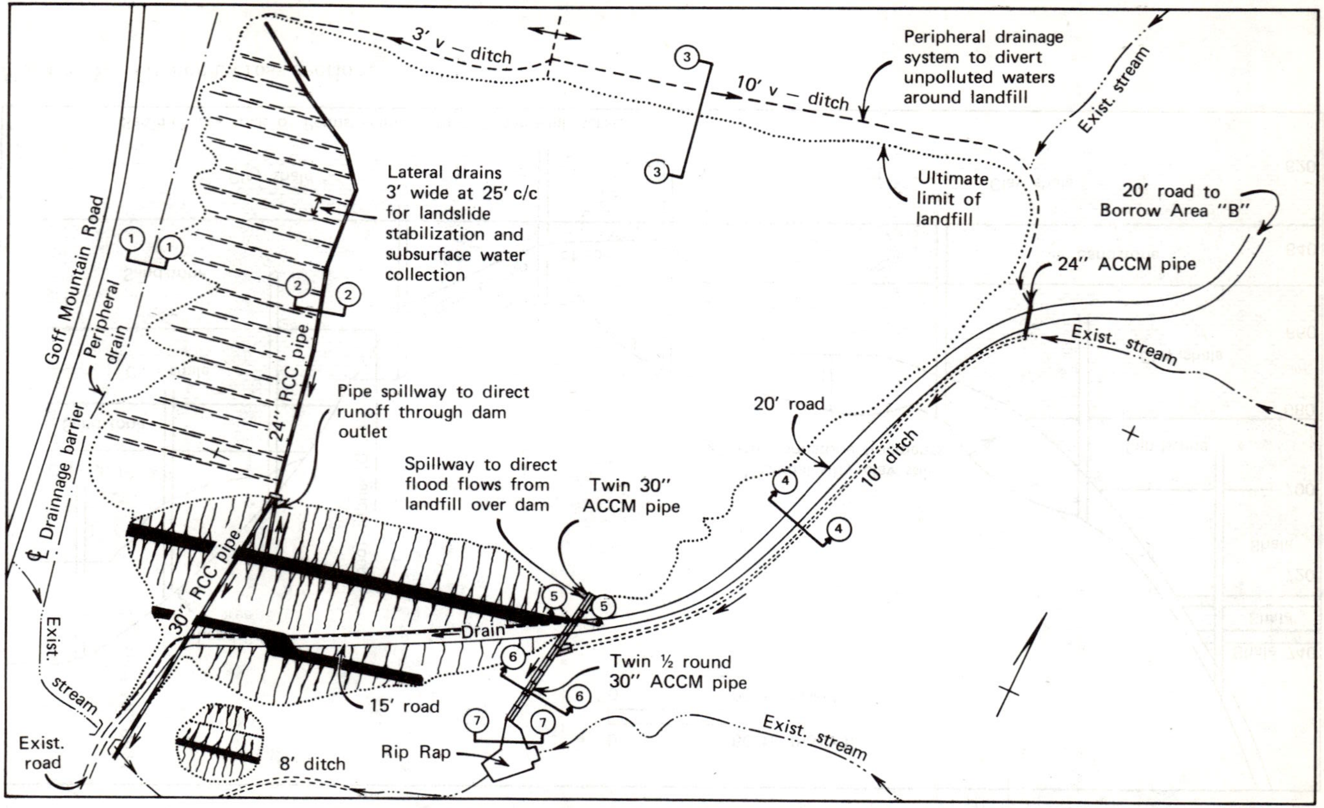

Figure 6. The peripheral drain system.

While searching for a way out of the problem, Carbide and Ackenheil developed the dynamic system shown in Figure 1. The idea represented a departure from the traditional, static, "seal-the-waste-in-soil" philosophy that characterizes modern sanitary landfill practice.

Under the new concept, water is actually encouraged to pass through the fill-waste mass and leach out any soluble components. The leachate is then collected and taken by tank wagon to the Institute Plant's activated-sludge wastewater treatment unit. Sludge from the treatment unit can be returned to the chemical landfill, so the process is essentially a closed cycle from the standpoint of chemical waste.

Tank trailers are presently used to move the contaminated water to the treatment unit, but this procedure will be replaced by a low-cost gravity-flow pipeline system. Table 3 lists the major design features and Figure 7 provides a cross-section view of the two tyes of dike used.

Blend Earth

Overburden not deemed of sufficient quality for use in seal and dike construction is blended, along with soil removed from inplant construction projects, 1 to 1 by volume with the chemical wastes as they are charged to fill lifts. The relatively porous blend stock permits gases that normally form in landfills to vent continuously (in small quantities) from the entire working surface of the fill. No cover layer is applied to inhibit gas flow, so there is little pressure to force gas through seals in fill floors and walls. Internal leachate drains also assist in gas

Table 3 *Design Features*

1. Pervious Earth and Rock Barrier
   Provides drainage for landfill, aids operations, and reduces pollution. (See Figure 7.)
2. Polluted-Water Collection Basin
   Collects polluted surface and subsurface water. (See Figures 2 and 7.)
3. Peripheral-Drainage System
   Channels rainfall runoff and clean groundwater around the project. (See Figure 6.)
4. Subsurface Lateral Drainage
   Intercepts groundwater outside landfill area and stabilizes potential landslide.
5. Terraced Landfill
   Optimizes land usage without requiring deep landfills with high barriers.
6. Groundwater Pollution Monitors
   Provide means for checking groundwater pollution through downstream wells.
7. Surface-Water Pollution Monitors
   Provide means for checking pollution of water draining from landfill.
8. Multiple-Phase Construction
   Defers capital expenditures and utilizes operations for construction of next phase.

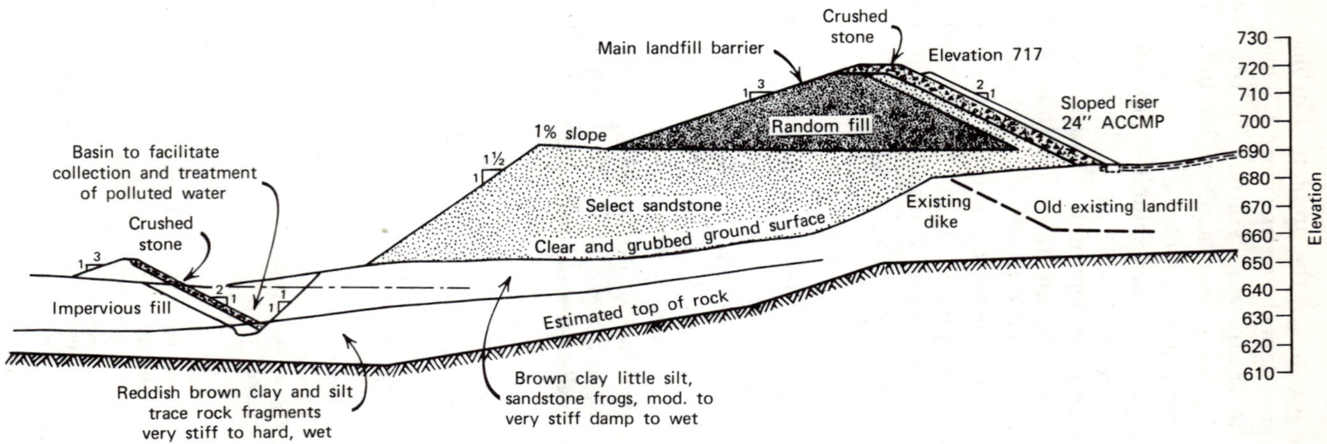

Figure 7. Cross section of two types of dikes used.

control by providing subsurface vents. Odor has not been a problem because bacteria in the soil bring it under control before it emerges.

Fill Lifts

Waste is charged directly onto the fill lifts from dump trucks, portable boxes, and cans. It is immediately mixed with blend earth and not permitted to form pockets of organic material. Cells of the type used in conventional sanitary fills are not constructed because they cause buildup of dangerous levels of methane gas and do not promote rapid degradation. Figures 2–4 show the lifts in plan view as they appear at the completion of each phase.

Access

Vehicle movement on any landfill is a continual problem. Access roads and equipment work surfaces were planned from the start for the Goff operation. The internal-draining feature of the fill itself has gone a long way toward ensuring all-weather operation.

Much of the road construction will be done with overburden removed from selected locations within the working area of the fill, and each phase of construction includes the necessary road alterations (see Figures 2–4).

Monitor Wells

Test borings 3, 9, 11, and 18 were equipped with screened wellpoints and turned into permanent underground water observation holes during the feasibility study (see Figure 8).

Regular weekly sampling and analysis warns of any seal failure and sets the stage for cement grouting if it is needed. None has been required since the holes were bored in 1966.

Analysis

Standard tests used to identify the company products entering the wastewater treatment unit are applied to water samples taken from points within the chemical landfill. Included are standard coliform tests, total carbon analysis, color tests, and specific gas chromatograph analyses. With a few exceptions (iron, sodium, potassium, calcium) metals are recovered and not placed in the fill. The Institute Plant Laboratory is equipped with atomic absorption and x-ray equipment to handle the identification of metals in wastes.

Radioactive materials are strictly excluded and monitoring through the Industrial Hygiene branch of the plant's Environmental Protection Department channels "hot" items to AEC disposal facilities.

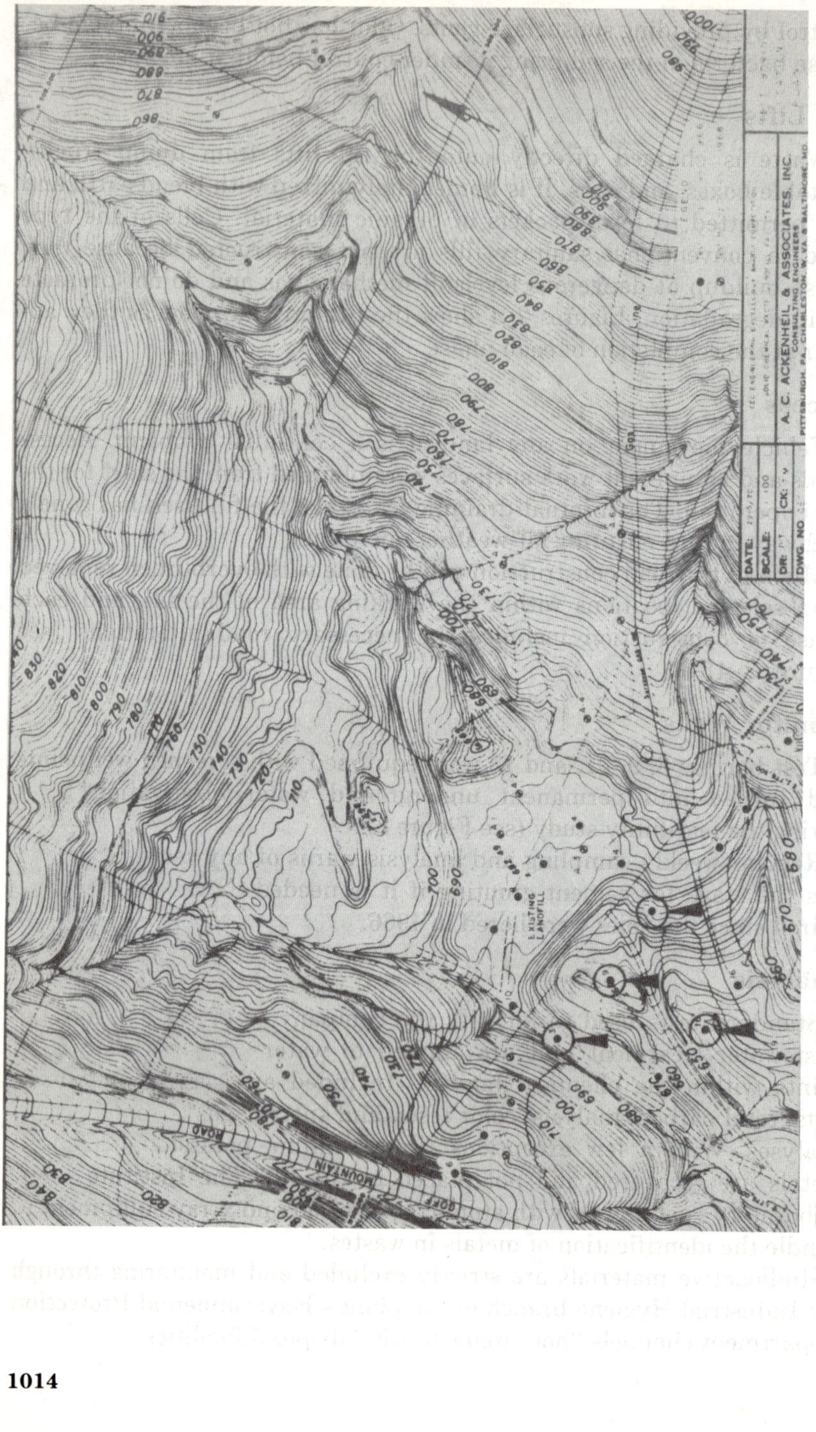

Figure 8. Test borings during feasibility study.

A vital part of the operation of this continually flushing chemical landfill is pH control. Fortunately, a large quality of both acidic and basic solid wastes is available and greater or lesser amounts can be blended into the fill lifts as daily readings dictate. Generally, effort is made to keep the leachate around a pH of 6.8–7.2 to meet the needs of the wastewater treatment unit.

There is some indication that the leachate will tend to go acid (pH around 5.2) if no effort is made to keep the fill at relatively neutral conditions. Since the operation is basically anaerobic, this is not too surprising.

BOD tests are run occasionally, but the real control is the daily total carbon analysis, which gives results in a matter of seconds. When the landfill is working well the total carbon level is generally around 6000 ppm. When waste loading occurs and insufficient blend earth is being used the figure can climb to 14,000 ppm.

Although it is somewhat difficult to relate total carbon and BOD, there is enough correlation to make a rough comparison with the BOD figures on "sanitary" landfill leachates.

Daily samples are taken from the contaminated collection basin and run for total carbon, pH, color, foam, and carbonate-carbon. The latter is necessary because the total carbon instrument picks up carbon from all sources, including ground minerals. To get a true picture of the organic carbon requiring treatment, the carbonate-carbon is subtracted from the total carbon reading.

Uncontaminated water from the wells and peripheral drains is sampled on a cycle that catches each point about once a week. A total carbon level of 130 ppm is indicative of uncontaminated water at this landfill location. (Naturally occurring organics at other landfill sites would make this figure different.)

A complete air quality control laboratory is available through the Institute Plant's Environmental Protection function. Problems with odors and gases are handled through this group.

Treatment

Oils leaving the fill lifts and collecting on the surface of the water in the contaminated basin are skimmed off and taken as fuel to the plant's steam boilers.

The contaminated water remaining is transported to the institute's activated-sludge wastewater treatment unit for bacterial stabilization. Sludge is removed and collected in an anaerobic basin and the treated water is released to the Kanawha River. Excess sludge is sent back to the chemical landfill to complete the process cycle. Figure 9 is a flow diagram of the treatment unit.[11]

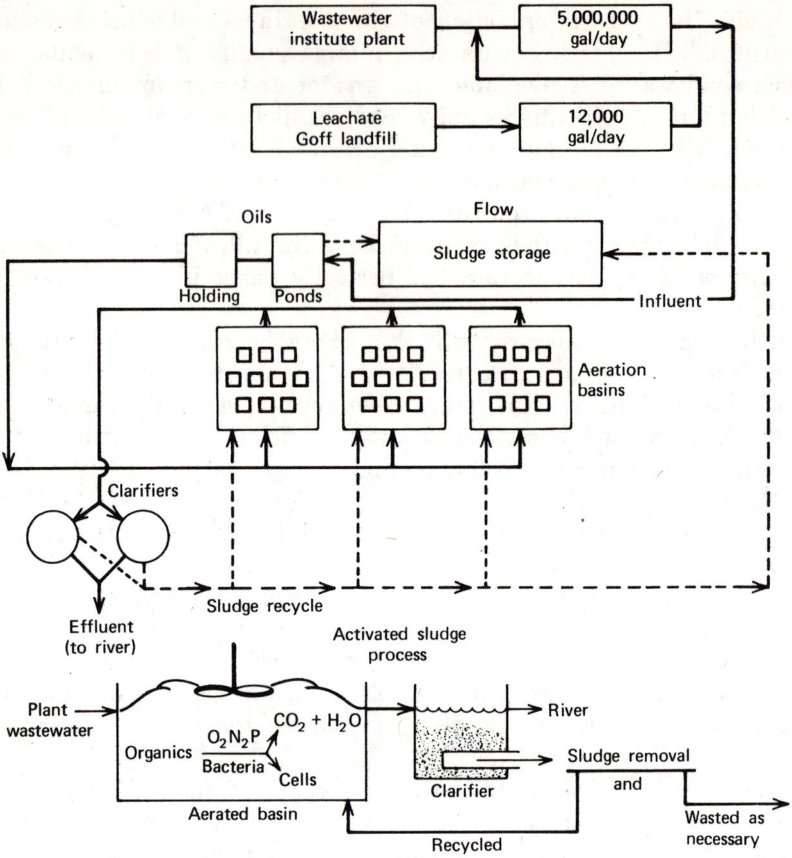

Figure 9. Activated sludge water treatment unit (Union Carbide Institute Plant).

Standards

Basically, the Goff chemical landfill is capable of processing any chemical that is acceptable to the bacteria in the wastewater treatment unit. Standard oxygen-uptake tests using treatment unit bacteria are performed in cases of doubt.

Metals (with the exception of iron, sodium, potassium, calcium, and aluminum) are not generally accepted into this landfill because of their toxicity to bacteria both in the fill and at the treatment unit. Recovery is recommended wherever possible and elaborate encasement in concrete is provided in the few cases where reclamation efforts fail.

Major groups of chemical waste charged to the Goff Chemical Landfill are discussed in Table 4, and the quantities are listed in Table 5.

WASTE CONTROL

Source Responsibility

Goff chemical landfill's flexibility in taking a wide variety of chemical wastes cannot be construed as an open invitation to pass all the materials in each customer plant through this process. Pretreatment, cost, human toxicity, biodegradability, and fill capacity must all be considered. Accurate chemical compositions (lab analyses

Table 4 *Major Categories of Industrial Waste Going to Goff Landfill from the South Charleston and Institute Plants of Union Carbide*

Wastes that cannot be incinerated, returned to process units, or passed through the Institute Wastewater Treatment Plant are scheduled for this landfill. For the most part, they are innocuous, solid materials. However, strict control is maintained over every load placed in the fill. A special disposal sheet is prepared for each one and kept on file. Continuous summaries (prepared weekly) showing the kinds and quantities of waste sent to this facility are maintained and reported with water pollution loadings.

At this writing, the materials fall into the four groups shown below. The State of West Virginia will be notified when changes are necessary in either the major groupings or the materials within a group.

Group I—Spent Filter Paper

Included here are the filter papers themselves, the filter aids, and viscous, insoluble products trapped within the papers and aids. The materials trapped fall into the subgroups: (a) plasticizers; (b) detergents; and, (c) oils.

Group II—Solid Organic Chemicals

Pitch, ureas, and tars make up this group. These are materials which cannot be easily solubilized. They come from the process units as solids, and they are residues from organic chemical production.

Group III—Polymers

Solid plastic waste, unusable organic filler powder, and waste plastic fiber are placed in this dump zone. As the group name implies, the wastes scheduled for this part of the fill are polymeric and totally insoluble.

Group IV—Inorganic Solids

Solid, inorganic materials are placed in this zone. Nickel compounds,[a] phosphates, and alumina are presently going into this part of the landfill.[b]

[a]Reclaimed. Material never sent to landfill.
[b]Daily quantities for each group are described in Ref. 2.

***Table 5** Major Constituents Going to Goff Landfill from Union Carbide Plants 512 and 514*

Constituent	Rate[a] (lb/day)
Group I—Spent Filter Paper (viscous chemicals on filter paper and aids)	
A. Plasticizers	2,519
B. Detergents	1,454
C. Oils	5,510
Group I total	9,483
Group II—Solid Organic Chemicals	
A. Pitch	1,573
B. Urea	251
C. Tar	563
Group II total	2,387
Group III—Polymers	
A. Plastic solids	70
B. Organic fillers	500
C. Fibers	1,210
Group III total	1,780
Group IV—Inorganic Solids	
A. Phosphates (discontinued)	204
B. Alumina	48
C. Nickel compounds[b]	75
Group IV total	327
Total to landfill daily (1965)	13,977[c]

[a] See Ref. 12.
[b] Reclaimed. Material not sent to landfill.
[c] 1971 load is approximately 22,000 lb/day.

are performed if these are not available) and good measurements of the waste quantities are required before any steps are taken toward disposal.

Responsibility for the environmental effects brought about by individual wastes must be placed on the producers, and this is done through the Institute's rigid in-plant control system. Union Carbide cannot allow a gross upset in its biooxidation systems, and this fact is impressed daily on all plant departments. Pressure is constantly exerted by landfill supervision on all users to seek the disposal route combining the optimum in environmental acceptability and process economy.

Disposal methods available are shown in Table 6. Figure 10 illustrates the supervisory organization necessary to maintain the program. Notice the variety of experience and the group's position as another plant production unit. This eliminates the idea that environmental protection is a "service function" and emphasizes the point that pollution control is a regular part of doing business and everyone's responsibility.

Help with any environmental problem is available through the

Table 6 *Institute Plant Disposal Methods*

Method	Advantages	Disadvantages
1. Reprocess the waste	Best return for UCC No environ. problem	Investment cost; technical difficulty; complex waste might result
2. Sell the waste	Return for UCC No environ. problem	Toxicity problems, shipping problems, market problems
3. Burn the waste	Heat recovered as steam	Air, water, land pollution, and transportation problems
4. Biooxidation of waste	Protects the Kanawha River	High cost: (2.5¢/lb. of contained organic chemical) High investment: ($27/lb of contained, continuously fed organic chemical) Strict control required to avoid air, water, and land pollution
5. Landfill the waste	Ensures final disposal of waste	High cost: ($14 per yd^3 or $37 per ton on the average) Strict control required to avoid air, water, and land pollution Requires land often difficult to obtain, reuse, and replace

Note. Steps 1–3 return some value to the company and add little to the environmental pollution problem.

Methods 4 and 5 stabilize the waste but represent a sizeable cost to the company.

In the case of each waste, Institute Environmental Protection considers the possibility of following each course (beginning with item 1) before deciding on the final disposal.

Every possible effort is made to avoid using steps 4 and 5

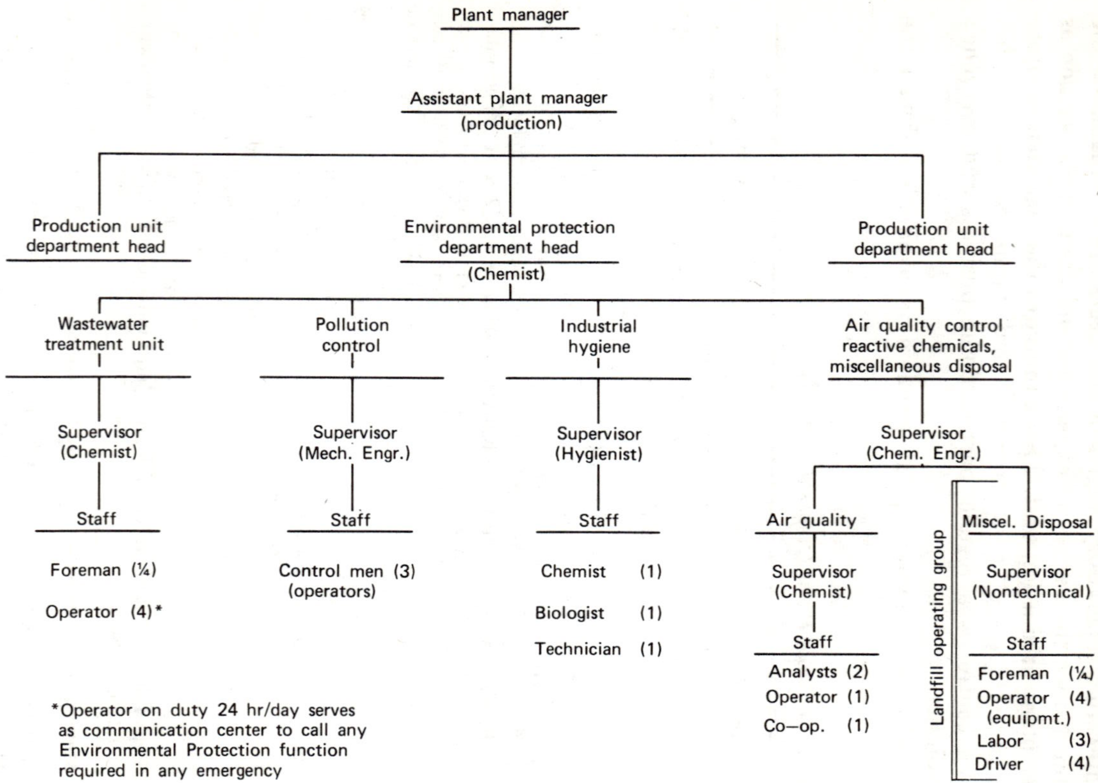

Figure 10. Institute plant environmental protection organization. (Note: A special 11-man environmental lab serves all functions.)

Plant's Environmental Protection Department (Figure 10), but the waste-producing unit bears the cost and responsibility for complete, effective disposal. Round-the-clock availability, variety of talent, and a close working relationship (not limited by particular specialty) make the Institute's environmental team highly effective.

A branch of the plant laboratory is devoted entirely to problems in the environmental area. Although it is not shown in Figure 10, it is an integral part of the team and provides the critical function necessary to insure full environmental control.

Monitor Program

Close monitoring of incoming wastes is the key to successful operation of the Goff Chemical Landfill. Full control of all wastes carried in moving vehicles is retained by the Institute plant's Environmental Protection Department.

The supervisor of the miscellaneous disposal function (Figure 10) permits no vehicle-borne waste (liquid or solid) to enter any Plant disposal facility without a properly completed, signed, and stamped "Order for Waste Removal."

Persons requesting waste disposal are required to submit an order bearing their signature, a complete chemical description of the waste, special instructions, and an account number to the Miscellaneous Disposal Supervisor before waste containers or transportation are supplied.

The disposal supervisor reviews the order with the requester and makes certain the latter has entered proper instructions covering hazards to the disposal personnel (and the environment), method of disposal, chemical composition, protective devices, physical properties, and the account number to receive the costs attending disposal. An important item at this stage in the process is the requester's signature, because each order is treated like a legal contract. Any incident arising from failure to identify the waste or the hazards is made the responsibility of the disposer. All reprimands, costs, investigations, and methods to prevent recurrence become his problems along with the writing of necessary reports. With the full backing of the plant's management in the same line organization, there is no difficulty in getting cooperation of people with waste problems.

When the disposal supervisor is satisfied that an order has been correctly completed, he assigns a special code number, affixes a special red stamp, and notifies the appropriate driver or crew to pick up the order from his office and handle the waste according to the written instructions. Each load must carry an order in transit. No plant driver will accept any waste without a red-stamped order. Through training and experience, the drivers themselves have become excellent monitors. They have become accustomed to the appearance of the more common wastes and refuse to pick up any unfamiliar materials. Changes they observe are immediately reported to the disposal supervisor.

Disposal unit operators must receive a ticket for each shipment of waste or they will not accept it. They, also, require the driver to sign and date the order upon arrival at the facility, and they forward the signed tickets to the disposal supervisor. A log book and a file of returned tickets is retained by the disposal supervisor as a control on the disposal facilites he uses. In the case of the landfill, the returned orders are the final records from which the feed mix and volume loading are calculated.

Instructions are provided with each book of blank forms issued (Table 7) and spaces for all the data needed are provided on the form itself (Figure 11).

Segregation

Solid waste production at the Institute plant is fairly characteristic of all the Goff Chemical Landfill users, so a glance at Table 8 will provide some idea of the general breakdown, how it is handled, and a rough (1969 figures) indication of amounts.

Instructions for maintaining the necessary solids-segregation are provided each production department Table 9.

Recovery

Recovery of any waste either as-is or through reprocessing is the most desirable method of disposal whenever it can be practiced. Environmental problems are generally eliminated and the company nets an economic return. Carbide believes in this approach and applies the philosophy in its plants and in its design and development work on new processes.

Each "Order for Waste Removal" and each request for design criteria involving wastes at the Institute plant are considered in the light of the disposal methods shown in Table 6. Most desirable, of course, is the first (reprocessing), and least is the fifth (landfill).

Selling the waste can often benefit the corporation if a buyer has equipment Carbide does not possess and is willing to follow instructions governing toxicity and reuse. Toward this end the corporation has provided a formal, in-house organization that helps its plants recover value from certain lines of waste. The Institute makes liberal use of this

Table 7 *Order for Waste Removal Instructions (Cover Page for Each Pad of Forms)*

OBVERSE

Use ball point pen. Read all instructions before completing form. This is NCR 3 part form. Remove 3 sheets or put in backing board. This form is to be filled out in its entirety by person requesting disposal, and returned to Waste Disposal Co-ordinator for approval; in duplicate. It is the responsibility of the person filling out this order, to list all personal precautions that must be taken by personnel handling this material in the source production unit. Failure to insert this information, resulting in injury or illness to personnel handling this material, will compel an investigation by the Industrial Hygiene Department. Fill in all components contained in mixtures. Mail completed form to Building 8112.

Material—Fill in name of material completely. Do not write chemical terms. The men who handle this material are not chemically trained. If you have a mixture or compound, write in all components that make this mixture, unless it is a trade named item.

Ordered by—Name of person filling out form and who is requesting the removal. This should be one and the same person. No filling out forms for someone else.

Building—Where material is to be picked up.

Date—Date form is filled out.

REVERSE

Shop order number.—The shop order number issued to your department for disposal of material. This is important. Form will be returned if number is not listed. If the proper number is not known, find out and write it down somewhere on this cover.

Quantity—How many pounds or gallons of material are to be disposed of? This information is important. Four drums will consume one yard of landfill at a cost of $14. For every four drums removed from your unit you spend an additional $36.00 (Container Cost). Total cost: $50/4 drums.

Description—Indicate the consistency of the material. Is it water thin or thick? Is the material solid? Are polymers or solids present in liquid? Mark appropriate box. Be sure and indicate the type of container you have. Check appropriate box.

Components—If a single material, write it in. If a Mixture of several materials or a brand name Plant Mixture, list all materials that make up this material and their percentages. This is important and order will be returned if not filled in.

Hazard rating—To be filled in by Disposal Supervisor.

Destination of material—To be filled in by Disposal Supervisor.

Unit hazard operating instructions—This must be filled in completely. Common phrases will not be accepted, such as "Normal Precautions", etc. Full precautions required by unit personnel must be listed. If there is no hazard, give authority.

FORM: 512–1747–G

O R D E R   F O R   W A S T E   R E M O V A L

Disposal No. 1

Material: ______________________ Phone ______________

Ordered By: ______________ Located at ______________ Date ______ Power House

Shop Order No. Quantity Yds., #, Gal. No.

( ) GOFF LANDFILL	WASTE TREATMENT		No.	No.	
Cobb	( ) Plant	( ) Liq. Thin	( ) Drum 55 Gal.	( ) Dump Pan	M.D.C.
( ) Burn & Store	( ) Sump	( ) Liq. Thick	( ) Drum 30 Gal.	( ) Dump Hopper	FILE
( ) Inert Fill	( ) Sludge Pond	( ) Solid	( ) Can 5 Gal.	( ) Dump Truck	
( ) Store & Crush	( ) Dowtherm Tank	( ) Polymer Present	( ) Fiber Pak 50	Dump Tank No.	
( ) Return	( ) Oil Stg. Tank		( ) Fiber Pak 30	Tank Trailer No.	
( ) Clean & Scrap	( ) Steam Plant				
( ) Atlas					

UNIT HAZARD OPERATING INSTRUCTIONS

ARE AS FOLLOWS: ______________________

COMPONENTS:

Name	%	H	F	S

HAZARD RATING

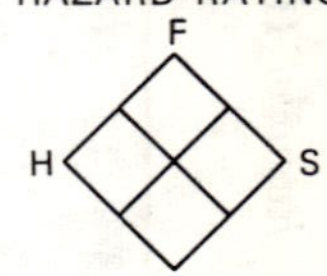

PRECAUTIONS: Chemical gloves—coverall goggles—coveralls required. Do not allow skin contact with material—Do not breath in any vapors. Follow unit hazard instructions listed.

Disposal crew must sign By: ______________

Date: ______________

Figure 11. Form order for waste removal.

Table 8 Solid Waste Production, Institute Plant, March 27, 1969; Time Base: 260-Day Year

Waste Class	Method of Disposal	Amount (yd^3/day)	Amount (tons/day)
Burnables	Compaction and incineration in Charleston city incinerator	66	2.0
Chemical	Treatment and burial in licensed chemical landfill—Goff Mountain (UCC owned)	20	10.0
Inerts	Crushed into inert landfill—Cobb property (UCC owned)	106	39.7
Mixed burnables	Crushed into inert landfill—Cobb property (UCC owned)	60	1.8
Metals	Salvage, Sales	6	3.8
Power plant ash	Ponding and landfill at low points around plant and adjacent property	291	367.2
Power plant lime sludge	Sluiced to flyash landfill	2	1.4
Waste treatment plant sludge	Ponding and further biodegradation	33	28.0
Daily solid waste totals		584	453.9 (907,800 lbs)

service by making it the second most desirable disposal route (Table 6).

Heat recovery is a third, and rather marginal, choice as a waste-handling process. Benefit in terms of free Btus is realized, but it can be offset by environmental problems and costs that go with processing certain wastes.

To date the biooxidation and chemical landfill processes in use at the Institute location have not been operated as recovery units (although some work in this area has been done on a limited number of chemicals). Consequently, every effort is made to avoid using these two systems unless it becomes absolutely necessary.

One might say that the plant's attitude is one of fighting to keep wastes out of the treatment plant and the landfill rather than one of "using them because they are available."

Pretreatment

A cost roughly equal to the annual operating expense for the Goff Chemical Landfill is spent each year in transporting and pretreating miscellaneous wastes at their source units.

Table 9 *Institute Plant Miscellaneous Wastes and How to Handle Them*

Institute plant miscellaneous wastes fall into the categories shown below, and they should be handled in the manner indicated:

	Proper Receptacle	Removal Request	Disposal Method
1. Burnable (Paper, rags, cafeteria waste, food and beverage containers, cardboard under 3 ft in any diagonal dimension)	Closed-top hopper marked "Burnable Waste" Only	Call truck dispatcher at 6264	Charleston City Incinerator (via 512 compactor truck)
2. Chemical (Any chemically contaminated medium, foams and plastics, kitchen grease, spent catalysts, contaminated dowtherm, used oil, urethanes (except standard insulation sections), earth, raschig rings, limestone, unrecoverable pipe and equipment)	Consult Disposal Coordinator at 6475 or 6573	By "Order for Waste Removal (Form—512-1737-E)" Only (Submit to: Disposal Coordinator Bldg. 23 Trailer, Plant 512)	As specified by 512 Disposal Coordinator
3. Inert (Uncontaminated insulation waste, finished urethane insulation sections, building demolition waste)	Open pan marked "Non-Burnable Waste," or dump truck	Call truck dispatcher at 6264	Cobb property inert landfill
4. Mixed burnables (Cardboard over 3 ft in any diagonal dimension, wood, empty fiber paks, crates, wood reels)	Open pan marked "Non-Burnable Waste," or dump truck	Call truck dispatcher at 6264	Cobb property inert landfill
5. Metals All types not chemically contaminated beyond recovery)	Ground at unit stations marked "Materials to Salvage"	Call truck dispatcher at 6264	Salvage

Note. The "Disposal Co-ordinator" mentioned above is the Miscellaneous Disposal Non-Technical Supervisor (Figure 9).

Good waste preparation in advance of landfilling seems to be an important factor in the Goff facility's performance. Overwet slurries modified with more earthlike wastes from other sources are more easily blended into the lifts. Acid sludges are mixed with ash absorbents, then blended with crushed limestone to give an almost neutral mix that can be safely worked by landfill equipment. Plastic dopes are more easily managed if they can be absorbed on waste filter aids taken from presses handling compatible chemicals. Health hazards, flammability, and reactivity govern the type of pretreatment given any one waste, and processing in both the source unit and at the landfill may be necessary.

Again, the fundamental limits on the degree of pretreatment are the hazards of the specific chemicals involved and the effect on the landfill and treatment unit bacterial populations.

An up-to-the-minute working knowledge of the chemicals in the waste source plants is essential to the landfill supervisor (Figure 10). Carbide's Reactive Chemicals Reference Service provides a large part of the needed information.

Dynamic Wet Process

Carbide views its chemical landfill as a dynamic wet process biochemical system and not a vault for the impoundment of solid wastes. Its flow-through design permits the entire fill mass to be treated as a chemical process where many of the principles of solution chemistry and biological stabilization may be applied.

Earth-Waste Blending Technique

Each chemical waste charged to the Goff landfill must be worked into the lifts according to the hazards it presents, its physical state, and its degradability. This calls for a variety of techniques from reagent-filled treatment pits to simple surface mixing with the landfill's blend soil. On the average, some form of surface blending will be employed at some point in each disposal, so it is discussed here as the primary landfill process.

The procedure currently used evolved as experience with both the older impoundment-type landfill and the newer wet process was obtained.

Waste-cell construction, commonly used in sanitary landfill, was abandoned after only a few weeks of operation in 1965. Rain water collecting in the cells before coverage and on the surface after applying the necessary clay seal caused a swampy condition that mired our largest bulldozers and made it impossible to get good contact between the soil and the waste.

A primitive form of the method Carbide uses today was the next evolutionary step. To avoid the miring problem and get the most waste coverage without too much movement, a crawler crane with a clam bucket was employed. Waste was dropped into pits (sometimes 20 ft deep) from the transport trucks and the crane clammed a small amount of earth in on top. Then, a mixing process with the clam bucket was conducted. It was basically a lift-drop-lift procedure that gradually gave a mixture that resembled soil. When the mix appeared much like the soil being added, it was lifted from the pit and scattered over as wide an arc as the crane boom would permit and simply leveled with a bulldozer during the next stretch of dry weather. The crane-blending method worked, but it was extremely slow and expensive. Ponding after every rain finally forced the company to search for a better way to operate the fill.

Work on the design for permanent impoundment dikes was under way at this time, so the problem of overcoming the ponding was included. In 1969 the resulting wet process was installed and the ponding difficulties disappeared.

Today, waste is either spread on the lift surface in depths no greater than 6 in. or it is placed in trenches with depths limited to 1 ft. The width of the spread is limited to that of the machine doing the blending (generally an end-loader with a 1½-yd bucket). Blend earth is stacked at one of the narrow ends of the spread then bulldozed over the waste to a depth of 6-in. This gives a "sandwich" of earth and waste that is approximately half one and half the other.

Up to this point in the process, the operator has been careful to maintain the 6-in. layer of earth between his machine treads and the waste to avoid picking up unblended material and clogging his cleats. Next he begins to walk his machine over the sandwich. Weight of the end-loader forces the track-cleats down into the blend earth and pushes the latter into the waste. By observation, the operator decides when the earth-waste mixture seems most homogeneous. At this point he proceeds to spread the entire mixture over as wide an area as possible on the lift.

Rubber-tired vehicles have been tried in this work, but they are unable to make good blends (particularly on wet days). Metal-track vehicles not only do the blending well, but they also overcome the objection to placing rubber in contact with organic-chemical waste. Both bulldozers and end-loaders with tracks have been used on this landfill. Operators, when given a choice, prefer the end-loader because of its superior ability to place earth exactly where they want it. Given equal sizes, the end-loader seems to do this type of landfill work more rapidly.

FILL-SOIL REUSE AND SUPPLY

Ackenheil's study indicates that there will be enough overburden and "borrow" earth to sustain this landfill's blending requirements for the full 20-year life span. Supplies of this earth are removed as they are needed rather than all at once in order to keep as much natural cover on the ground as possible.

As a hedge against unforeseen problems in obtaining blendearth at the site, the landfill supervisor has taken option on all borrow removed from in-plant construction projects. What is not used at the chemical landfill is taken to the company's nearby inert landfill.

Operation has continued long enough to permit some estimation of the time required for stabilizing the waste mix being fed to the fill. Conversion to the wet process required excavation of some of the older waste-soil mixtures, and it was found that some could be reused as blend soil after as little as 18 months. It may be possible to use less than the predicted 14.7 acres over the 20-year life of the fill.

Gas Generation Control

Gas generation within the landfill has not been a problem since the wet process has been installed. The combination of blending, water flush, and no seal on the lifts seems to minimize formation and vent off dangerous concentrations uniformly over the surface and through the internal drains. Explosimeter readings and checks for hydrogen sulfide have given nil readings.

Ponding Control

Ponding is a surface phenomenon that results when rain water collects on poorly drained, clay-sealed, landfill surfaces. Seals made with semipermeable clay can produce a variation of the same problem even with proper drainage. In the latter case the semipermeable clay absorbs just enough water to become spongy. Generally, the affected area appears to be the same as the rest of the fill surface, and the condition does not come to light until some unsuspecting heavy equipment operator runs his machine into it and finds he cannot get out.

Ponding problems at the Goff property disappeared when the wet process was put into use. Without the surface seal, pools cannot form, and the internal drainage removes water from any semipermeable fill zones. Regular addition of wastes containing limestone and alumina with a particle-size of ½ in. keeps the fill surface open and provides the added benefit of keeping the ground from freezing in the winter time. This is particularly important to the continued operation of the waste-blending operation.

Drainage

Much of the necessary drainage protection has been designed into this landfill (Figure 6). There are day-to-day problems, however, that must be resolved by the operating crew.

Most of the difficulties relate to drainage immediately around the fill-lift area. This is inside the peripheral drain shown in the design plan, and the trenches change position as the size of the fill increases. Minor local slides and siltation cause blockage in the trenches and must be controlled by the landfill equipment operator. Periodic removal of suspected slippage zones and regular scraping of the trench bottoms is a regular part of his job.

Slag drains in the "leaky" dike have shown a tendency to plug when watery sludge wastes are processed. Some removal and replacement of the slag has been necessary as a consequence. (A better pretreatment of sludge wastes seems to be correcting the problem.)

Maintenance of drain planes across the top of the fill lifts has called for close attention. A 2% slope was recommended in the original wet process design, and this seems to be adequate as long as the surface is kept free of slight high spots.

Slippage

Slippage is a problem around the Goff fill site because of the large clay deposits and rather steep slopes on the sides of the hollow. Rolling shear, similar to that experienced in West Virginia highway construction, occurs after long periods of high rainfall.

Landfill supervision has to be ever alert to the first signs (slight bulges on hillsides) of slip so action can be taken to prevent a major slide. Dikes within the fill have been no problem because they are keyed to the top of the bed rock, but this technique is difficult to apply to the hollow's walls.

Most of the time it is necessary to determine the full area of the slip, strip it off, remove soil to the top of underlying rock, then rebuild with interlocking riprap as a base. Concrete is sometimes required. When the repair is complete, grass and trees are planted to hold it in place.

Equipment

Because of its ability to lift and place earth easily, Carbide has made the track-equipped end-loader its basic piece of chemical landfill operating equipment. It is assigned continually to waste-earth blending on an 8-hr/day, 5-day/week basis. Through arrangement with a local equipment rental agency, a backup can be obtained in the event of failure.

State landfill regulations specify that coverage equipment be available at every approved landfill at all times. The end-loader is left at the chemical landfill site to meet this requirement. A 30-ton crawler crane rigged with a special perforating weight for drum crushing also remains at the site. If coverage or fire problems arise, this machine can be quickly rigged with an available clam bucket. Some of the drainage and lift-surface work is done by rental bulldozers on a part-time basis. This keeps the end-loader on the blending operation and speeds disposal generally.

Gradalls, scrapers, and similar personel are hired when necessary to dig some of the special pits needed in pretreatment, maintain the peripheral drains, and repair slips.

Dump trucks and special pan-trucks are provided by the plants that use the landfill service. The Institute keeps three pan-trucks, one dump truck, and two flat-beds busy full time in solid waste transportation.

Waste Containers

The practice of grabbing any old can and jamming solid waste into it for treatment at the chemical landfill cannot be tolerated. Drums do *not* deteriorate readily (one removed from this fill after 5 years' burial still looked new and had active chemical in it), they are costly in themselves, and they take up fill volume without using it efficiently. Closing the spaces between drums placed in landfill cannot be done without considerable expense. The unsealed voids represent landfill volume not devoted to waste degradation, and they provide air passages that promote underground fires.

The Institute follows an active program of drum elimination at the waste source. Lined, slope-sided garbage cans with lids are used wherever possible. The waste-filled liners slip easily out of the cans onto the chemical fill, and the cans are taken back to the source. Plastic liners break easily under the end-loaders's weight so the waste processes satisfactorally.

Dump trucks and dump pans (both lined with plastic and unlined) deliver the bulk of the chemical waste. Problems with these vehicles center around proper loading and cleaning processes to avoid littering roads and emitting odors or material along the transport route. Haulage practices that do not meet ICC regulations give the disposal supervisor the right to refuse landfill use to the disposer. "Orders for Waste Removal" are withheld pending correction of questionable waste-handling plans.

Tank trailer rigs are currently used to move the contaminated water from the landfill. An inexpensive pipeline connecting to the sewer feeding the treatment unit will replace them. Oil skimmed from the contaminated basin is pumped into dump tanks that are handled by the same trucks moving the waste pans. An electric pump at the contaminated basin is required to load out the water, and air-driven and gasoline-powered pumps are available as portable back-up units. High-pressure water cleanup of the many containers used in solid-waste service is often necessary; therefore a cleaning area with drains to the contaminated basin has been installed.

On-Site Inerting and Stabilization of Waste

Certain wastes must be treated outside the confines of the disposing units when systems are not available to prepare the waste properly. If the disposal supervisor is satisfied that proper pretreatment is impossible anywhere inside the plant, he will help the waste producer set up a system at the landfill. All costs for this special work are borne by the producing units, and total personnel and environmental safety must be ensured.

Several unique methods (underground scrubbers, pit reactors, neutralizers, etc.) have been developed to satisfy pretreatment requirements. Liquid can be skimmed off some waste materials into the contaminated basin before the waste is charged to the fill. The Institute has purchased aeration equipment that can be used on this basin to reduce the BOD load returned to the waste treatment unit. Sludge resulting from this aerobic process will be cycled back into the landfill proper and the liquor sent on to waste treatment.

Wet Weather Operation

Many of the difficulties encountered early in the history of this landfill were eliminated with the conversion to the wet process in 1969. Foregoing sections on ponding and earth-waste blending discuss this is some detail.

Care of the access roads and drains is a problem in wet weather. Heavy slagging of the former and silt-dams in the latter have given some relief, but more work is needed in this area.

Chemical-Biological Balance Control

A landfill pH in the range 6.8–7.2 has been arbitrarily set as protection for the wastewater treatment unit. Wastes are charged at specific rates to the fill in order to guarantee an effluent in the desired range. Heavy additions of limestone waste are frequently required to raise the pH.

Fire Control

Every effort is made to prevent "pockets" of organic material in the landfill. Experience dictated that Carbide take this position after two underground fires early in the project nearly destroyed the heavy equipment. Complete excavation of the fire zone was necessary because pressure-injected detergent-water streams had no effect. Blending with fill earth was the only way that successfully extinguished the blaze in each case.

Drums placed in this landfill in the unsmashed condition seem to be an open invitation to fire. Air passes readily into the fill through the voids between drums and aids spontaneous combustion. The totally organic contents of the drums function as centers that feed these spontaneous fires. As a fire deterrent, all drums are now crushed then buried at random intervals across the lifts. Surface fires can be controlled with earth placed by either the crane or the end-loader.

Two drums have ignited since 1969 and set fire to unblended solid wastes. In each case the plant fire department responded and brought the flames under control with detergent-water foam. High-volume solid-chemical extinguishing equipment was tried without success during the more recent of these two incidents.

Odor Control

Odor is not a problem with this landfill. Several features, including the wet process itself, seems to be responsible for the satisfactory record. Natural location of the site near the center of a 101-acre bowl provides peripheral protection in the event that drum processing causes trouble.

The intimate earth-soil blend in the fill lifts keeps odor from emerging in the working area. A more serious concern here is the odor of the incoming wastes before processing. Considerable pressure is put on each disposer to bring any odor problems under control before the material leaves his unit.

Oil on the surface of the contaminated-water basin will produce an objectionable odor unless it is pumped off regularly. Apparently, many of the volatile acids produced during anaerobic decomposition in the landfill find their way into this layer and enter the atmosphere under a warm sun.

Maintenance

This is a continuing effort just as it is in any chemical production unit. Where metal kettles, and the like are the reaction containers in a production process, the seals, lifts, dikes, basins, and drains in the wet chemical landfill are the equivalent. The same maintenance attention and care must be given the landfill system, or it risks a shutdown that can slow or halt production in user units.

The Institute has taken care of this by using plant maintenance personnel under the direction of the Environmental Protection Departments Miscellaneous Disposal Supervisor. It is an unusual marriage of traditional maintenace and operating functions, but it has been successful for the six years the fill has run.

Safety

Constant surveillance over the whole landfill operation must be exercised because chemical landfill is inherently dangerous. Many wastes have been classed as such because they contain compounds that the disposer himself finds difficult to handle and he is looking for a place to "get it off his hands." A large part of the landfill supervisor's time is spent fixing disposer responsibility for his waste until it is safely stabilized.

A major tool in this is the "Order for Waste Removal" (Table 7). Hazards must be clearly spelled out and certified by the disposer himself or the material will not even be picked up. If breathing air, full-suit protection, special handling, and so on are required for the persons in the unit producing the waste, then they must be specified for all disposal personnel who will be involved.

Generally it is necessary to supplement instructions provided by producting units because people in these facilities do not engage in close-contact work as the crew at the landfill does.

Since maintenance workers operate the chemical fill, a way had to be found to simplify chemical information for their use. This problem is met through a hazard signal system and written instructions that are added to the "Order for Waste Removal" by the disposal supervisor himself. The signal appears in the lower right-hand corner of the order and it is based on a relatively simple 1-through-4 grading system for each chemical in the areas of health (left-hand block), flammability (upper center block), and reactivity (right-hand block).[14]

Added to the chemical dangers are the customary landfill mechanical hazards. Dump-blocks and anchor chains are required for trucks doing pit-dumping. Roll-bars and seat belts are necessary on most of the heavy equipment.

Two unusual landfill safety problems are posed by the crane and the tank-trailers that must be used at this site. The Institute has already experienced one near-miss incident involving boom-buckling as a result of unrecognized previous strut damage, and men loading water out of the 500,000-gallon contaminated-water basin must guard against surface oil fires and drowning.

Heavy-equipped operators must be constantly alert for the telltale "bulges" that signal wall-slippage around the hollow and keep their machinery off. Slips can move at speeds sufficiently great to bury or overturn a bulldozer.

The danger of fire or accident makes it imperative that men working on the fill be able to summon help quickly. A telephone is provided in the personnel shelter near the fill lifts, and most of the trucks bringing in waste have radio tied into the plant network. The plant dispensary, fire department, and rescue squad can all be reached through either medium. Help can be on the site in five minutes.

Manpower

Figure 10 shows the manpower structure necessary to operate the Institute's chemical landfill. One or two heavy-equipment operators are always on the site, a three-man disposal crew collects drums from the plant and does the special pretreatment work, one man works the contaminanted basin, and an operating foreman and a maintenance foreman make periodic checks and see that the disposal supervisor's instructions are carried out.

All operations at the fill shut down on weekends and holidays and no night operation is permitted, so all men work an 8-hour day, 5-day week. All waste receiving stops at 2:30 p.m. to allow any necessary blending before the fill closes at 4:00 p.m. All access to the landfill ceases when the operator locks it up for the night.

Recognizing that the materials processed into the Goff Chemical Landfill system require a degree of attention and understanding beyond the capability of ordinary landfill operators, Carbide controls the project through a trained full-time supervisor and a chemical engineer who have had long service in environmental protection work at the institute plant.

COSTS

Site Study and Fill Design

Carbide's costs, given below, for chemical landfill design have been higher than they would be for a comparable fill built today because the concept was under development at the same time.

Geology, testing, and design (Ackenheil)	$56,270
Construction engineering supervision (Carbide)	21,225
Total	$77,495

Capital Costs

The estimated cost of the completed project (all three phases) is $250,000, excluding those design costs that were applied to earlier plans rejected in favor of the wet process.

Capital money expended since the new system was installed in 1969 is given by:

West and north peripheral drains	$32,000
East peripheral drain	12,000
Phase I "leaky" dike	68,500
Contaminated dike and 500,000-gal basin	85,500
Total	$198,000

Annual Operating Costs

Annual costs for running the chemical landfill, maintaining it, hauling water, and providing on-site pretreatment of some of the waste fed average $165,000 for the wet process at the institute.

Land Costs

Land is the least expensive feature of chemical landfill construction. A parcel similar to the Goff Hollow tract would cost some $100,000 in 1974. Over the 20-year expected life, however, the actual fill will use only 15% of the area.

Unit Disposal Costs—Volume Basis

It is important that standards covering the weight and volume of each waste landfilled be established so charges can be made against the disposing units and checks can be made on the fill depletion rate. At the 12,000-yd^3/yr design rate, disposal of 1 yd^3 of waste costs $13.80.

Unit Disposal Costs—Weight Basis

Comparisons between chemical landfill and sanitary operations handling only everyday municipal-type solid wastes are best made on a weight basis. Institute sends its municipal-type burnables to a local incinerator for $6.50 per ton (excluding transportation).

One tone of chemical solids processed through the Goff fill costs $37, or 5.7 times as much as the municipal-type waste.

Many factors cause the high landfill cost, but most prominent is the special treatment that must be given each chemical as opposed to the simple bulk handling that can be applied to the relatively innocuous municipal-type material. A summary of the cost picture for the life span of the fill is given in Table 10.

Table 10 *Total Estimated 20-Year Cost*

Study and design	$ 77,497
Capital costs through Phase I	198,000
Land cost	100,000
Estimated Phase II and Phase III costs	52,000
Operating costs for 20 yr	3,300,000
Total cost for stabilizing 240,000 yds³ of refractory chemical wastes	$3,727,495

REFERENCES

1. *Consulting Engineer,* June, 1970; Ackenheil & Associates, "Pollution Abatement Solid Chemical Waste Landfill—CEC/US 1970 Engineering Excellence Awards Competition," Report, April 1970.
2. *Construction Equipment,* September 1970.
3. E. E. Slover, "A Licensed Chemical Landfill," presented at the 3rd Ohio-Michigan Pollution Conference, Cleveland, Ohio, May 13–14, 1970.
4. E. E. Slover, "Goff Mountain Landfill," presented at the Manufacturing Chemists Association's Solid Wastes Management Committee Meeting, Washington, D.C., December 3, 1970.
5. E. E. Slover, "Guidelines for Operation of Industrial Solid Waste Sites," presented at the First Texas Solid Waste Management Conference, Houston, Texas, May 27–28, 1971.
6. "Water Pollution Control Permit No. 3141," State of West Virginia, Department of Water Resources, July 26, 1965.
7. E. N. Henry (Head, Industrial Waste Section, State of W. Va. Department of Natural Resources, Division of Water Resources), "Transmittal Letter—Water Pollution Control Permit No. 3141," to L. N. Dickinson, Mgr. UCC Institute Plant, July 26, 1965.
8. Bern Wright (Chief, State of W. Va. Dept. of Natural Resources, Division of Water Resources), "Policy Letter—Burial of Industrial Wastes," to L. N. Dickinson, Mgr. UCC Institute Plant, December 3, 1965.
9. "Solid Waste Regulation," West Virginia State Board of Health, West Virginia Administrative Regulations 16-1, Series II, Section 6, Paragraph 6.01-a, July 1, 1967.
10. "Application for Permit No. I-34—Disposal of Industrial Wastes," UCC Institute Plant to State of West Virginia, Department of Natural Resources,—Division of Water Resources, page 5, paragraph, 6, March 24, 1965.

11. J. F. Hall, G. W. Kumke, R. W. Oeben, "Conversion to Activated Sludge of Union Carbide's Institute Plant," *Water Pollution Control Fed.*, 1408–1422, August 1968.

12. See Ref. 9, application for Permit No. I-34, p. 8.

13. See Ref. 9, Application for Permit No. I-34, p. 9.

14. J. J. Duggan, "See Hazards at a Glance," *Chem. Eng.*, 162–166, February 23, 1959.

PART VIII mineral and metallurgical wastes

This part deals with specific areas of the one billion tons of waste from mineral processing, which involves mainly moving material from one place to another, but in the course of treatment, changing its physical character by grinding, sorting, classifying, and extraction. The tonnage of mineral waste is discussed in Chapter VIII.1. Coal as a fuel is not considered in this section since burning culm piles (the waste of anthracite processing) have been extinguished and utilized. The supermarket and shopping area built on top of waste piles at Wilkes Barre, Pennsylvania, in the anthracite region, was untouched by the Susquehanna floods that devastated the city in 1972. Strip mining, a cheaper mining method with far few accident possibilities (USBM) than underground mining of coal, is discussed as a satisfactory method, with efficient and productive land reclamation in Georgia, in connection with kaolin and clay in Chapter VIII.2. Land building in connection with strip mining of phosphate in Florida is treated by Custred in Chapter VIII.3. Hoskins deals with lead and zinc mining waste in Chapter VIII.4.

Mineral wastes, fine sand, and the like cover large areas and need to be stabilized or held in place by the methods and practices discussed in Chapter VIII.5. As the result of processing, slags from electric furnace zinc find utilization as described in Chapter VIII.6 and slags from steel blast furnaces are now important engineering structure materials as told by Bauman in Chapter VIII.7.

Omitted mineral industries are bauxite in Arkansas, but most of our bauxite is imported; mercury in California, now effectively shut down; manganese in the Dakotas, now not operating; molybdenum in Colorado, effectively handling its waste problems; and strip and open pit mines of American copper producers, where, effectively, materials are just moved from place to place in areas where density of habitation is low, and, if the mine were not there, would be zero. Diatomaceous earth, fullers earth, talc, lithium minerals, feldspar, and aplite dust problems have been solved and mineral dressing wastes are low. Land reclamation and development of growth of shrubs and trees on mined land is now regarded necessary.

CHAPTER VIII.1

mineral wastes

Minerals are the foundation stones of civilization. Extracted raw from the earth, minerals are processed to yield such basic substances as metals, chemicals, building materials, fertilizers, and fuels. Our industrialized society could not exist without these essential commodities. They are the basis for nearly three-quarters of our gross national product.

But in extracting and processing minerals, and in using the many products that contain them, wastes inevitably incur. Sometimes they are generated in mining. In a western state a yawning open-pit copper mine each hour delivers 300 tons of copper ore to a mill. To get to this ore, which has a copper content so small that only the most economic extraction method can make the mining operation profitable, thousands of tons of earth and rock overlying the mineral deposit had to be stripped away. This "overburden" dumped near the mine site gradually forms a huge man-made mountain from which the wind can tease air-polluting dust and the rains can carry stream-polluting acids.

Seldom has it been considered economic to reclaim lands disrupted by this kind of mining. Consequently, many parts of our country today remain as marred and unsightly symbols of the continuing quest for the minerals.

Wastes are frequently produced in the processing. In a southern state mammoth machines scoop phosphate-bearing pebbles from the ground—the first stage in a process that supplies fertilizer minerals. Later, separation of phosphorus compounds leaves behind about 700 tons of waste every hour. To avoid polluting surface water supplies, the slimes are dumped into enormous pits where they await the development of an effective disposal method.

Over a billion tons of such mineral-processing wastes are produced in this country each year. The total accumulation of these materials to date covers approximately 2 million acres, Table 1, about the same area as that requiring rehabilitation after surface mining.

Additional millions of tons of waste accumulate as we use minerals and mineral products. In a large northeastern city coal is fed at the

Table 1 Waste in Surface Mining (Thousands of Acres)

State	Land requiring reclamation	State	Land requiring reclamation
Alabama	83.0	Nebraska	16.8
Alaska	6.9	Nevada	20.4
Arizona	4.7	New Hampshire	5.1
Arkansas	16.6	New Jersey	21.0
California	107.9	New Mexico	2.0
Colorado	40.2	New York	50.2
Connecticut	10.1	North Carolina	22.8
Delaware	3.5	North Dakota	22.9
Florida	143.5	Ohio	171.6
Georgia	13.5	Oklahoma	22.2
Hawaii	[a]	Oregon	5.8
Idaho	30.7	Pennsylvania	229.5
Illinois	88.7	Rhode Island	2.2
Indiana	27.6	South Carolina	19.3
Iowa	35.5	South Dakota	25.3
Kansas	50.0	Tennessee	62.5
Kentucky	79.2	Texas	136.4
Louisiana	17.2	Utah	3.4
Maine	21.6	Vermont	4.2
Maryland	18.1	Virginia	37.7
Massachusetts	25.0	Washington	5.5
Michigan	26.6	West Virginia	111.4
Minnesota	71.5	Wisconsin	27.4
Mississippi	23.7	Wyoming	6.4
Missouri	43.7		
Montana	19.6	Total	2040.6

[a] Less than 100 acres.

rate of 400 tons/hr into a giant furnace, where it is burned to generate steam for electric power. Of that 400 tons, roughly 40 tons is converted during combustion into fly ash. Electricity from the plant's generators serves millions of consumers, but the consumers of fly ash are relatively few. Every year millions of tons of this unwanted material must somehow be disposed of.

Clearly these wastes are materials of considerable diversity (see Tables 2 and 3). They are not easily categorized according to geographic origin, chemical content, or physical form. Frequently, it seems that all they have in common is their association with the

Table 2 *Waste in Consuming Certain Minerals and Fuels (tons)*

Junked automobiles	12,000,000
Fly ash from coal-burning powerplants	25,000,000
Metals discarded in refuse	12,000,000
Total	49,000,000

minerals that play so many fundamental roles in our daily lives. But, unlike the minerals that serve us, mineral wastes have given us nothing but problems.

The ugliness of these waste accumulations, and their potential for damaging our environment and our health, have made them the targets of growing public antipathy. But, in the long run, an even better reason for doing something about these mineral wastes is the needless loss of valuable natural resources that they represent. For, almost invariably,

Table 3 *Waste in Processing Minerals and Fuels (Thousands of Short Tons) (USBM)*

Industry	Mine waste (other than surface mine overburden)	Mill tailings	Washing plant rejects	Slag	Processing plant wastes	Total
Alumina	—	—	—	—	5,350[a]	5,350
Anthracite	[b]	—	2,000	—	—	2,000
Bituminous coal	12,800	—	86,800	—	—	99,600
Copper	286,600	170,500	—	5,200	—	462,300
Iron and steel	117,599	100,589	—	14,689	1,000	233,877
Lead and zinc	2,500	17,811	970	—	—	21,281
Phosphate rock	72	—	54,823	4,030	9,383	68,308
Other	[c]	[c]	[c]	[c]	[c]	229,284[d]
Total	419,571	288,900	144,593	23,919	40,233	1,122,000

[a] Of this total 1,700,000 tons are discharged directly into the Mississippi River annually by two alumina production plants in Louisiana. Annual accumulation thus is 3,650,000 tons.

[b] Total not available, but quantities negligible and are included in washing plant wastes.

[c] Not available.

[d] Wastes of remaining mineral mining and processing industries—20% of total wastes generated.

these waste materials contain significant mineral values that an expanding population and a healthy, growing economy require.

The recycling of wastes, of course, implies economic feasibility. To qualify as an acceptable source of minerals, wastes must prove competitive in cost with "natural" sources like ores. Several factors favor wastes in this regard: their abundance; their ready accessibility above ground, which eliminates the necessity for expensive mining operations; and the growing desire of waste producers to find disposal methods more socially acceptable than mere dumping. This desire is an important stimulant to both government and private research on recycling.

Scientists are concerned with the resource potential of mine dumps and mill tailings piles. Samples taken from major dumps across the nation are being analyzed to determine the metallic and nonmetallic constituents offering the best prospects for recovery.

The chemical makeup of one waste material in this category, the "red mud" generated in processing aluminum ore, is already well known. Every year the nation produces over 5 million tons of this red mud. Besides containing substantial quantities of unrecovered aluminum oxide, along with soda and titanium, the mud ranks close to incinerator residue as a source of iron.

The extraction of coal, metals, and nonmetallic minerals often produces still another kind of waste, one for which new uses are scarcely needed. This is the earth and rock stripped from the land during surface mining operations. Such material, called spoil, has only one true use: reclaiming the scarred land from which the desired minerals were extracted.

A project in Pennsylvania involved experiments in reclamation at several surface mine sites to develop comparative cost figures for alternative techniques. The knowledge gained was to supply a valuable reference point in the government's nationwide study of strip and surface mining practices.

Wastes produced by underground mining are used to refill underground voids and support the earth overlying mined-out areas. Thus, large areas of the surface would be saved from the threat of defacement by strip mining, protected from the subsidence problems often associated with abandoned underground mines, and kept ready for use in meeting future needs for living space.

CHAPTER VIII.2
kaolin and clay

Open-pit mining for minerals, despite the fact that it is the cheapest method and employs minimum labor, is deplored by the environmentalists as being unduly destructive. We need clay for our paper industries, and underground mining either cannot be done or is not competitive with imported clays.

Here is a case study where the environmentalist will find that all his objections are met, competition against imported materials is met, the land is rapidly restored, and the operation is profitable. The waste is handled in large volume mechanically, overburden is returned to its place of origin, and the land is green again.

It is estimated that U.S. production of wet-processed kaolin, restricted almost entirely to Georgia, runs 2.5 million tons/yr. Plants using wetprocessing techniques are most numerous in an area within about 50 miles east of Macon in a band approximately 15 miles wide. The open-pit mines of Freeport Sulfur Co. are situated in this area near Gordon, Georgia, where the processing plants are located. Five properties within a 3-mile radius of a centrally located degritting plant are operated. From there the clays are slurried and piped some 6 miles to the plant site. The company's conservation program embraces both the mining tracts and the water courses adjoining the plants.

For several years state and federal officials throughout the United States, as well as various conservation groups, have raised objections to the unsightly and hazardous conditions often associated with areas where open-pit mining has been undertaken. The "roll-over" method as developed by Freeport, is the key to the company's land reclamation program at Gordon, where open-pit mining waste is handled and the landscape restored.

Instead of leaving mined-out areas in a disrupted state where vegetation will not grow immediately, the roll-over method leaves a relatively flat surface where a minimal effort is required for land reclamation and reforestation, and where little or no stream pollution occurs. Such a landscape contrasts sharply with the traditional strip-mined surface where unconsolidated soil conditions, poor water percolation, and steep slopes are conducive to extensive erosion with resulting stream pollution.

Freeport's mining method consists of the removal of the overburden and leveling of the surface with scrapers. The balance of the stripping and the mining itself is by two walking draglines, handling 4½ million yd^3/yr.

Before the mining operation starts, all the marketable timber is sold, while unusable growth and underbrush is removed by bulldozers. Next, 70–90-ft-wide cut is staked out at the edge of the orebody, parallel with the drill patterns.

The dragline is placed at one end of the cut, at a right angle to the longitudinal direction of the pit, and digs a box cut. The overburden is removed to the bare clay and placed 180° away. The dragline is then moved around to the front of the pit, mining the clay and casting it between the edge of the pit and the removed overburden. The dragline is next moved again to the same side from which it dug the first box cut and digs another cut, continuous to the first one, and removes the overburden, again casting 180° away. The machine is then moved again to the front of the cut and the clay is mined and placed between the overburden just removed and the edge of the pit. This procedure is followed for the total length of the cut.

There is a windrow of overburden and a windrow of clay stockpiled next to and parallel to the edge of the cut. The dragline is turned around and starts digging a new cut, of the same width, casting the overburden on the void left on the first cut. The overburden so cast is rough-leveled by the dragline and further leveled by a bulldozer. The clay is then mined and placed on top of the smoothed overburden surface. This procedure is followed to the end of the cut and back and forth across the orebody.

The roll-over method provides flexibility in mining and leaves a relatively flat surface with a windrow of overburden at one end of the pit and an open cut at the opposite end with windrows of clay piles in between.

In 1965, Freeport inaugurated a reforestation program in mined-out areas where the roll-over method had been employed. The first step in the program was experimentation with various types of ground cover to establish the simplest and most economical seeding method. Experiments indicated that native Georgia "slash pine" would be best suited to the soil and climate. The application of 10-10-10 fertilizer to seedlings in 1967 established the value of this fertilization. Later in 1967, 50 more acres were planted with improved 1-year-old slash-pine seedlings, produced by the Georgia State Forestry Commission. The initial results for sycamore seedlings are promising.

Improved seedlings provided by the Georgia State Forestry Commission have now become the standard for all future plantings. By the end of 1970 up to 300 acres had been reclaimed by reforestation.

CHAPTER VIII.3

land building with phosphate wastes*

U. K. CUSTRED

Mining wastes have created problems in many parts of the world as mining companies recover essential minerals from the crust of the earth. Rebuilding industrial waste heaps into useable land is a big and difficult undertaking, requiring new ideas and new techniques.

Dramatic changes are being made in central Florida where phosphate rock is being mined and phosphate wastes have left thousands of acres of scars, many of which may be removed only by nature over many years.

Phosphate is one of the essentials of life. Without it, men and animals die and plants will not grow. In Florida, phosphate is found in huge beds from a few feet to 100 ft below the surface. To reach the deposits, miners use electrically operated draglines, some with booms 285 ft long, buckets handling 75 tons per bite, and 4000 tons/hr. The phosphate ore, also called matrix, consists of approximately equal parts of phosphate, sand, and clay. The matrix has physical properties that make it readily adaptable to hydraulic transportation. It is broken up by powerful water jets to create a slurry and is pumped to a beneficiation plant which may be several miles away. Here, phosphate is separated from the gangue.

Phosphate mining wastes include overburden, sand tailings, and phosphatic slimes. In 1961, eight phosphate mining companies, under the auspices of the Florida Phosphate Council, issued a formal land reclamation policy. The mining companies voluntarily pledged themselves to carry out their activities in such a manner "that the land involved would help meet the esthetic and practical needs of the community." The overburden and sand tailings are used to reclaim the mined-over land. Under this program, council members have achieved

* Originally published by the Society of Mining Engineers, AIME.

a 9-year record of reclaiming land equal to 65% of that mined during the period.

The disposal of phosphatic slimes has presented a problem to the phosphate industry since the mining of phosphate rock first began in the 1860s. As the flotation process was introduced in the late 1920s, the problem of slimes disposal was increased significantly since slime is present in proportionally greater quantities in reserves yielding flotation concentrates. Because of their physical character and the tremendous surface area per unit volume, these clays retain large volumes of water as they are slurried during processing.

The phosphatic clays in the Florida land-pebble phosphate deposit consist primarily of apatite, wavellite, and montmorillonite. In the beneficiation process, the −150 mesh material is removed as waste. However, the particles are essentially smaller than 325 mesh, and more than 70% are smaller than 3μ.

A great deal of work has been done through the years in an effort to find an economic use for the phosphatic slimes. They contain about 30% of the phosphate values in the matrix. They may be used, for example, for direct application on crops as a source of P_2O_5 and K_2O; as a filler for fertilizer; as an excellent light-weight aggregate for building products. However, all these possible uses fail in the face of the vast quantities of water that must be driven off in order to arrive at a useable product. Impoundment in settling areas has been the only feasible manner of water disposal.

The disposal of slimes and the recovery of a portion of the water require settling areas with a capacity ranging up to 2 ft^3 for each cubic foot of phosphate matrix mined. According to a recent study by the USBM, the average cost for disposal of these wastes is 24.5 cents per ton of product (USBM 8404). This is a major cost of phosphate rock production.

Many years after the clear water is drawn off and a settling area is abandoned, the slimes a few inches below the surface have reached a density of only 25–30% solids. Thus, as far as can be foreseen, today's settling areas will never be useable for anything other than limited agriculture.

American Cyanamid Company decided the most profitable course of action would be to devise a method of using the waste slimes, sand tailings, and overburden in such a manner to build useable land. This method is presently being used at the Haynsworth Mine to dispose of approximately 50% of the slimes leaving the plant. Eventual development is foreseen to the point where 100% of the slimes will be disposed of in the original pits with over one-half the volume recovered as firm soil. (See Table 1 for screen analysis of mine wastes.)

On the average, slime constitutes one-third of the matrix mined in the central Florida phosphate field. It is separated from the sand and phosphate by violent agitation with water, by hydrocyclones or hydroseparators, and by decantation.

In the laboratory, it was discovered that the addition of small quantities of sand to thicken the slimes greatly increased the release of clear water, the slimes thickening rapidly to 12–25% solids (see Figure 1). These thickened slimes formed a good soil mixture when blended with additional plant tails. Several test plots were made and seeded with grass. An excellent soil of good bearing strength and high fertility was produced. The laboratory procedure is shown in Table 2.

Based on the laboratory results, a plant-scale test was run in an attempt to thicken slimes to 10–20% solids in the plant thickener. These thickened slimes would be used to repulp dewatered tailings, and the resulting mixture used to build land in mined-out cuts.

Since start-up at the Haynsworth Mine, plant tailings have been dewatered and the treated water returned immediately to the flotation section. The tailings have then been repulped with return water from the circulation system and pumped to disposal sites.

The water clarifier at the Haynsworth Mine has a bowl 550 ft in diameter. By making major flow distribution changes in the plant, primary hydrocyclone slimes were routed directly to the thickener.

Table 1 *Screen Analysis of Mine Wastes, Haynsworth Mine, American Cyanamid Company, Bradley, Florida*

Mesh Size (Tyler)	% Weight of Overburden	% Weight Tailings Sand	Mesh Size (Tyler)	% Weight of Clay-Slimes
+20	2.6	1.2		
−20 + 35	12.1	13.7		
−35 + 48	18.1	32.0		
−48 + 65	22.8	34.1		
−65 + 100	15.9	11.0	+100	0.15
−100 + 150	8.5	4.6	−100 + 150	0.20
−150	20.0	3.4	−150 + 200	0.45
			−200 + 325	0.90
			−325	98.30

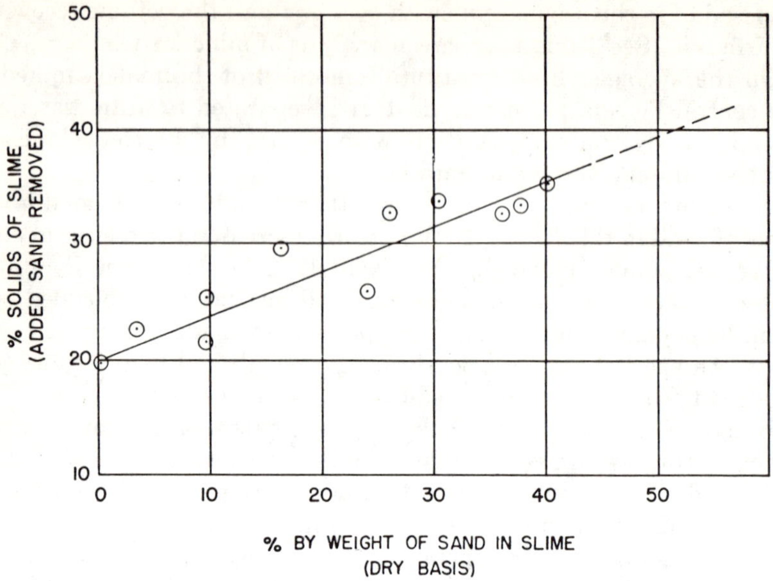

Figure 1. Thickening effect of sand on slime, Haynsworth Mine, American Cyanamid Co., Bradley, Fla.

It was quickly discovered that slimes could not reach the desired 10–20% solids with a single plant thickener. In fact, the very best density achieved has been 7%. The idea of installing several thickeners to raise the solids above the 7% level was studied, and abandoned as being totally unacceptable economically. Therefore, the test program was modified. A schematic flow of the test circuit is shown in Figure 2; this is the process that has been in use at the Haynsworth Mine since September 1968.

At Cyanamid's Haynsworth Mine, a cut was available 5000 ft long with an average depth of 38 ft and an average width of 136 ft. It was filled with water to a depth of some 35 ft. This cut was chosen for the test since it was believed to have sufficient length to allow clarification of the water for reuse.

In the field program, dewatered plant tailings were repulped with thickener underflow. This underflow averaged 3.5% solids, the best density sustained at that time. The mixture was then pumped through a 16 in. pipeline to the cut.

This pipeline from the plant alternately fed two 16-in. lines discharging along opposite sides of the cut. The sand-slimes mixture was pumped into the cut until it reached ground level. In the meantime, the

Table 2 *Procedure to Test Amenability of Wastes to Land Building, Haynsworth Mine, American Cyanamid Company, Bradley, Florida*

1. Secure samples of plant wastes.
2. Thicken slime-clay sample to maximum % solids and viscosity. This may be accomplished by aging and decanting water, and chemical flocculents may be used to improve or speed up the thickening. Aged thickened slimes can sometimes be secured from the plant area. Gentle agitation usually speeds up the thickening process.
3. Dewater the plant tails or wastes that readily thicken to a dense solid.
4. Build a table resembling a mine cut in miniature, a wooden table 5 ft long, open on one end, with a collecting launder for released water. Make 18 or 36 in. wide with a partition down the middle. The sides should be about 8 in. high and the bottom solid and flat. In operation, the table should be sloped enough to allow for water drainage. A cup for measuring load bearing strength will be needed; this is described later.
5. Set limits on weight ratios of waste mixing tests. Good results with phosphate wastes have been demonstrated at 1 part slime (dry basis) and 9 parts sand (dry basis).
6. Weigh clay-slime into buckets. Start with the maximum density slime material. In additional buckets, weigh thick slimes and dilute with water. Repeat this until enough samples to fill all boxes have been obtained.
7. Weigh the proper amount of tails (coarse waste) to get the proper ratio of sand to slime. Stir the mixture and pour into the settling table at the end opposite the drain launder. Observe the results, and at this point the condition of mixing may be observed.
8. Observe the slope of the mixture poured on the table since this is an important clue as to the operable range of the process. The conditions for suspension of the coarse tails in the thickened clays are being sought.
9. Record daily for about 2 weeks the following information:
   a. Angle of slope of material in box.
   b. Amount of water released.
   c. Load bearing strength of the surface.
10. A plastic cup or bottle top is all that is needed for bearing strength. Measure the bearing area, weigh cup and add lead shot until cup shows unquestionable signs of sinking. A simple calculation will give bearing strength in pounds per square inch, or in any desired units.
11. Select best mixing conditions, and seed mixtures with grass or other suitable vegetation for final solidification.

discharge line on the opposite side was being moved ahead 100 ft. Pumping was then switched to the other side while the first line was extended (see Figure 3).

As was expected, the inflowing sand-slimes mixture tended to dissociate, the sand settling rapidly while the slimes migrated and set-

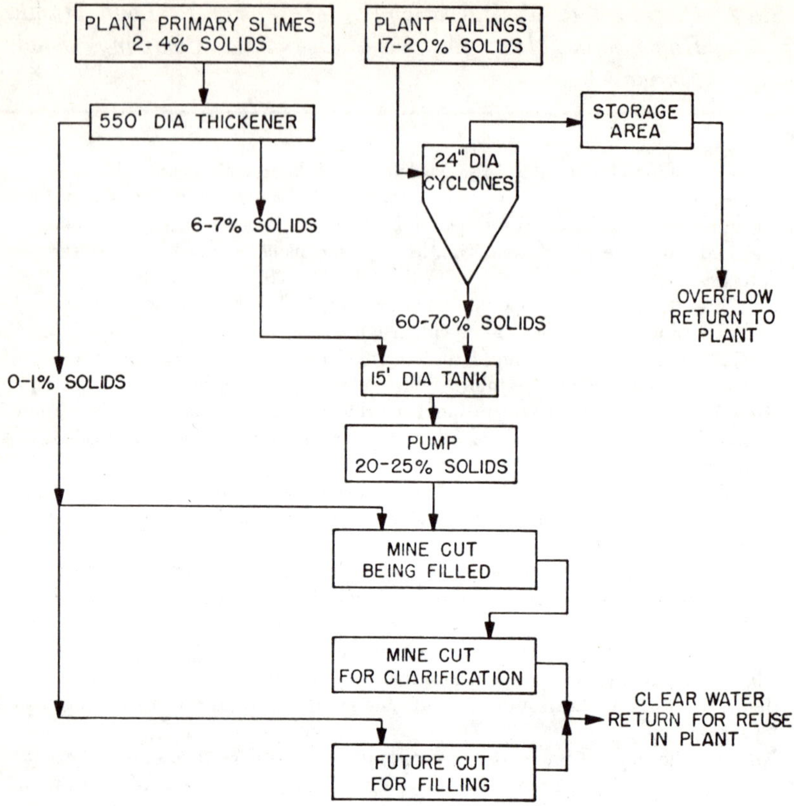

Figure 2. Flow sheet of land building project, Haynsworth Mine, American Cyanamid Co., Bradley, Fla.

tled slowly. However, as the slimes settled, the incoming sand tended to flow over and mix with the thickened slimes. The gentle motion of the sand encroaching on the thickened slimes liberated large volumes of water, eventually creating masses of settled slimes which approached 35% average density. As the sand was added to this slime, the final mixture attained a density of 70–80% solids.

Subsequent test drilling confirmed that to a great extent a mixture was attained, concentrated of course, in the lower levels. The random drilling consisted of six holes in the 5000 ft length. Results of the drilling are shown in Figure 4.

While the total test disposed of only 22% of the −150 mesh material from plant operations, in the latter stages up to 45% of the total slimes

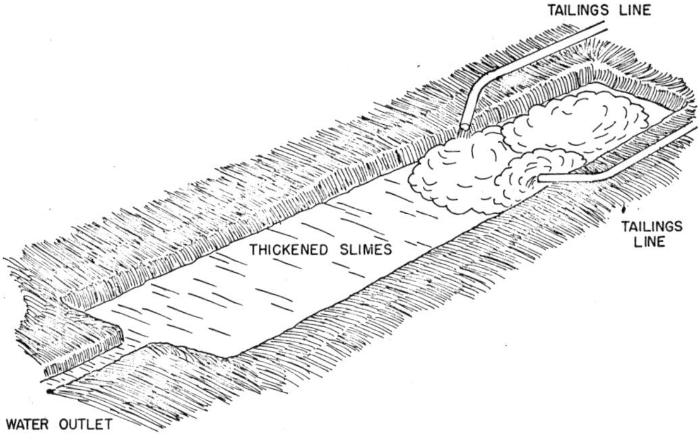

Figure 3. Plan view of cut, Haynsworth Mine, American Cyanamid Co., Bradley, Fla.

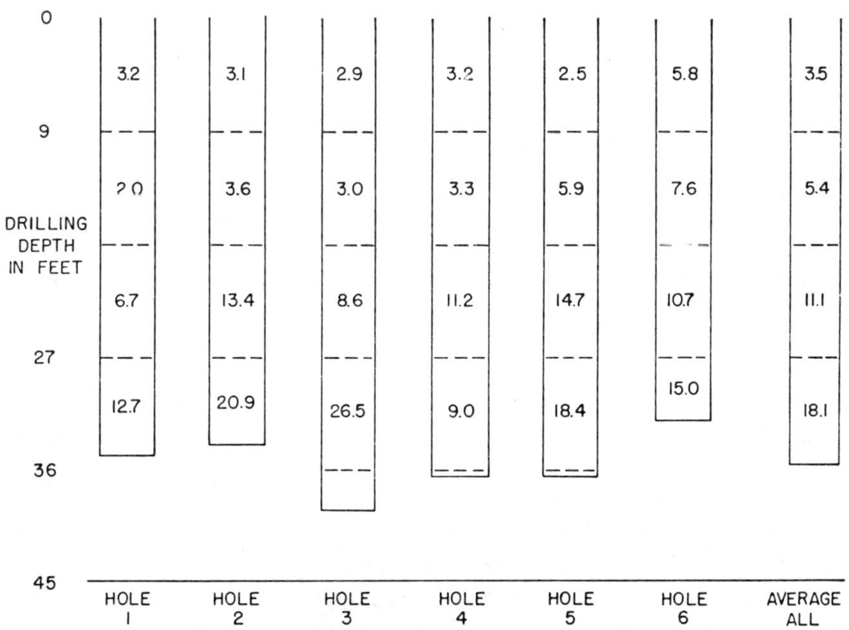

Figure 4. Percent −150 mesh in drilling core sections.

produced was contained and mixed. This is a long way from perfection, but it reduced circulation and settling requirements substantially and extended the life of the existing settling area considerably.

The total volume of the finished fill was 595 acre feet. Using conventional means of waste disposal, the sand would have occupied 560 acre feet of volume, and the slimes, at 20% solids, would have occupied 252 acre feet. Therefore, the effect of mixing the slimes and the tailings was to release 217 acre feet of water over and above that which would be recovered from conventional settling procedures.

The resulting fill is approximately 80% solids—excellent filled land compared to the 20–30% solids that is the maximum normally found in an exhausted, conventional settling area. Furthermore, this fill has good bearing strength. A few days after completion, it was supporting D-8 tractors and other heavy equipment.

The area was fertilized, then seeded with a mixture of Bermuda and Bahia grass. Germination and growth were rapid. An adjacent tailings plot containing no slimes showed almost no response to similar treatment.

At the end of this first plant test, up to one-half of the slimes produced by plant operations were disposed of in such a manner as to compact them to 14% of the volume they would normally occupy, increasing water recovery, substantially extending the life of our existing conventional settling areas, and creating useable land (see Figure 5).

A second field test is now under way. Slimes are being flowed through an existing settling area to an old, water-filled, hydraulicallymined pit. Here the slimes are being allowed to settle for a minimum of 3 months under the surface of the water. This 3-months delay has allowed them to thicken to the 15–25% solids usually found in a conventional settling area.

After the settling period, tailings, dewatered at the plant and repulped with thickener underflow, are being pumped into the settled slimes. In this test as in the prior ones, of course, the clarified water is being returned to the system for reuse. Preliminary indications are that it will utilize up to 75% of the plant slimes, although several months must elapse before we will be able to confirm this.

Sufficient information has been obtained to indicate that the circulation and settling ratio will be reduced at the Haynsworth Mine by one-third. This will reduce circulation and settling costs by one-third. At the same time, it will increase water recovery and extend land reclamation by greatly reducing the conventional settling system.

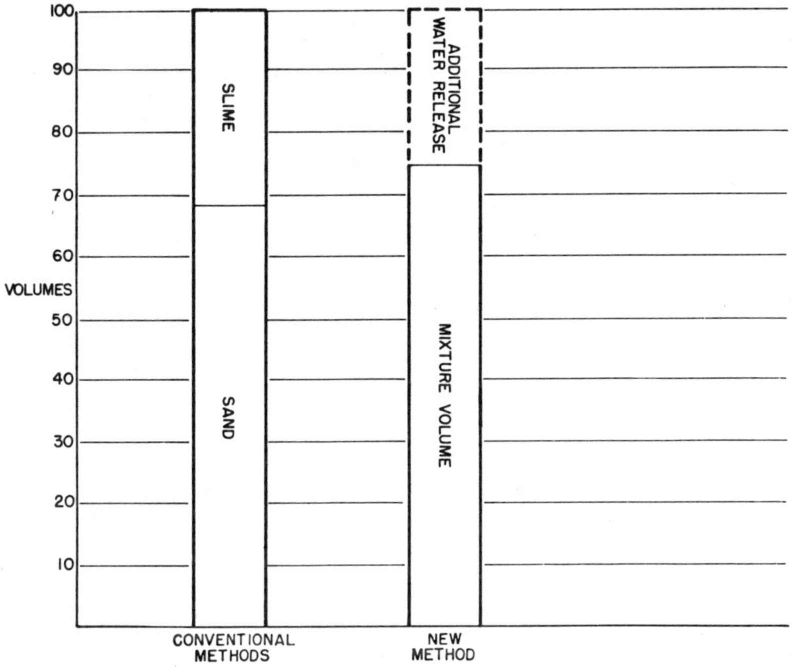

Figure 5. Volumes occupied by plant wastes.

After initial settling is complete, the key to forcing the slimes to give up more of the trapped water is to promote a rolling and squeezing action by the addition of sand. Properly applied, the sand will do this. The thickened slimes need not migrate; in fact, it might be preferable if they did not. It has been found that the highest slimes concentration in the initial cut was just in advance of a submerged retaining wall of sand that previously had been pumped into the cut. In the mining operation itself, such retaining walls can be thrown up with spoils by the dragline during mining (see Figure 6).

Under operating conditions, the procedure at the Haynsworth Mine will be as follows:

First, provide a cut of sufficient length to allow 3 months minimum retention of the slimes before the sand fill is pumped in.

Second, stack the overburden as steeply as possible during mining to provide maximum open cut area. The overburden remaining above the hydraulic fill will be leveled later.

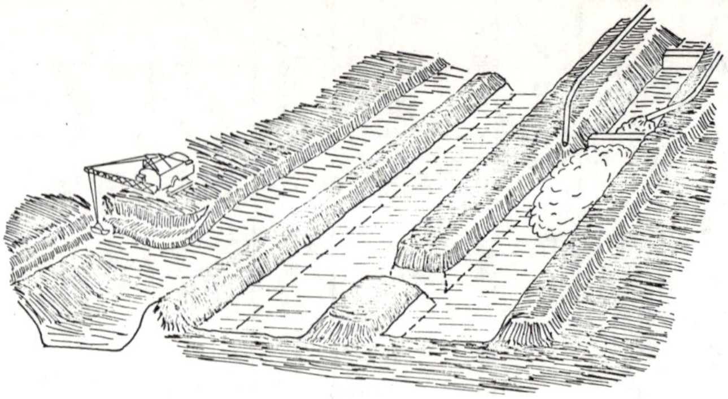

Figure 6. Mining layout for building land with phosphate waste.

Third, fill the cut with slimes to approximately the original land contour and allow to settle for a minimum of 3 months. During this period, maintain a high liquid level to prevent crust formation due to evaporation.

Fourth, after the settling period, pump in dewatered plant tailings, repulped with thickener underflow. To a great extent, the sand will mix with the slimes forcing out substantial quantities of water in addition to that already recovered in the initial settling period.

Fifth, grade the overburden back across the tails, and the land is ready to be seeded.

Certainly some of the savings will be offset by additional pumping costs, but the saving in land reclamation will be substantial.

Any company can implement a similar procedure. All that is required is a suitable cut or series of cuts to allow adequate settling time, and the required pumping equipment.

At Cyanamid's Haynsworth Mine, the ultimate goal is to convert all plant wastes into useable land. Realizing all the answers have not been found, plant scale tests are being conducted utilizing various materials handling techniques. Flocculents have proven to be ineffective with present techniques. Definite progress has been made in cutting waste disposal costs, in reducing the dams built, in increasing water recovery and in extending the possibility of land reclamation at a minimal cost. American Cyanamid Company's Haynsworth Mine Management is confident that with continued effort the final answers and the goal of 100% land reclamation will be attained.

CHAPTER VIII.4

lead and zinc mining wastes

J. R. HOSKINS

Lead is a primary metal essential to modern technology and like most metals cannot be replaced by substitutes for some uses, can be replaced at considerable expense and technological changes for other uses, and in some cases substitution is a question of reasonable economic consideration.[11] The two main sources of supply are underground mining of ore containing lead minerals and reuse of lead wastes. In all cases production of lead in a finished form creates wastes. It is useful to have some concept of the beginning and ending of the lead cycle in order to appreciate the problems involved in accompanying waste production.

NATURAL SOURCE OF LEAD

Lead is most frequently combined in nature with sulfur forming the mineral galena (PbS) with a composition of 13.4% sulfur and 86.6% lead. The mineral frequently is associated with sulfides of silver, cadmium, selenium, zinc, antimony, bismuth, gold, iron, and copper in very finely divided physical mixtures. Sometimes the amounts of these minor elements are so small that economic recovery of galena only is attempted.

Two minerals of lead that have almost disappeared as mining has reached greater depths, are the oxidized ores, anglesite ($PbSO_4$) containing 68.3% lead and cerussite ($PbCO_3$) containing 77.5% lead. There are at least six other lead minerals that are oxides or chlorides combined with phosphorus, vanadium, chromium, molybdenum, tungsten, and other elements. The major lead and zinc (the element most frequently associated with lead) mining areas of North America are shown in Figure 1. The "ALL U.S.A." designation refers to the largest lead district centered around the state of Missouri. Table 1 shows the respective tonnages from each district for a recent year.

An ore is a mechanical mixture of metallic minerals such as galena

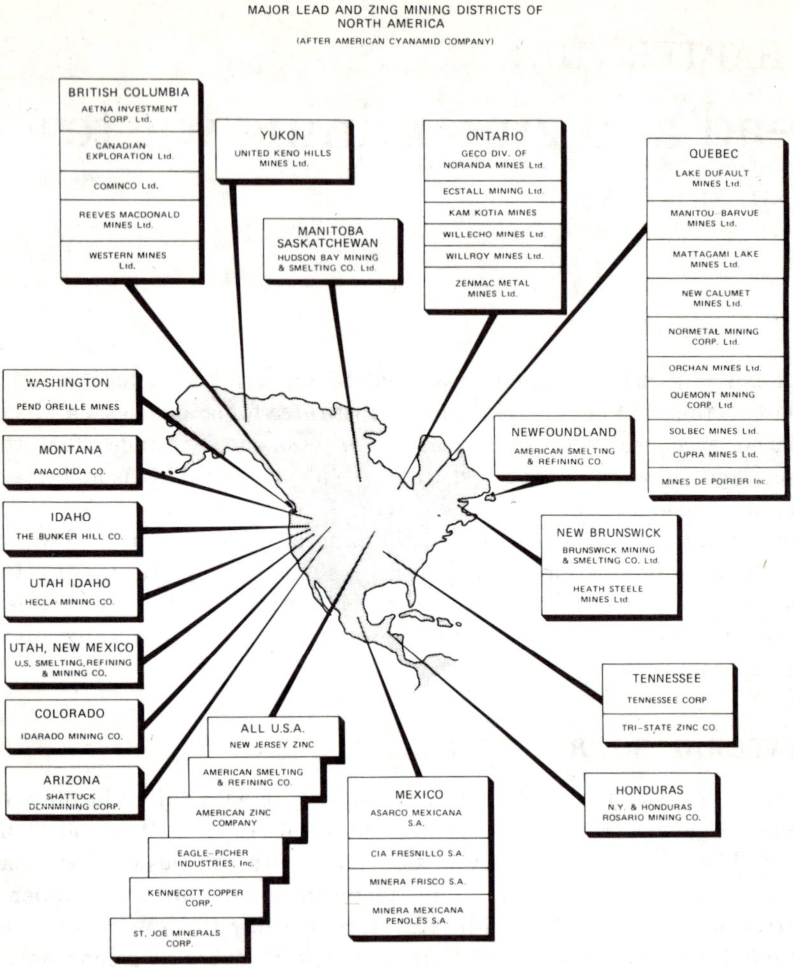

Figure 1. Lead mining areas of North America (Courtesy American Cyanamid Co.).

and nonmetallic or rock-forming minerals of no economic value in which the total mixture can be mined at a profit. Frequently the lead content in the form of galena is as low as 2–4% of the total rock, which means only 40–80 lb of lead per ton of ore removed can be recovered. Volumetrically the ratio is even lower since the specific gravity of galena is about 7.5 and that of the host rock is 2.7. For recovery purposes this requires the removal of approximately 12 ft^3 of waste rock for potentially 40 lb of lead, with less than 1% by volume being retained.

Table 1 *Approximate Yearly Lead-Zinc Production from Major Mining Districts of North America, 1965–1970*

District	Lead (tons)	Zinc (tons)
1. British Columbia	200,000	255,000
2. Yukon, Manitoba, Saskatchewan	8,000	75,000
3. Ontario	7,000	295,000
4. Quebec	5,000	270,000
5. New Brunswick, Newfoundland	70,000	185,000
6. Tennessee		20,000
7. All USA (from companies holding mines in many districts exclusive of others mentioned)	160,000	540,000
8. Washington, Montana, Idaho, Utah	60,000	55,000
9. Colorado, Arizona	15,000	30,000
10. Mexico, Central America	170,000	225,000
Total	695,000	1,950,000

SECONDARY SOURCE OF LEAD

The reuse of lead wastes, primarily discarded storage batteries, provides a large percentage of the lead used. Along with smaller percentages of lead shot, pipe, electrical sheathing, and alloy waste, the 1968 lead waste production accounted for 550,000 tons of the 1.3 million ton U.S. consumption. This exceeded the mine production of about 500,000 tons, the remainder being import balances.

Waste lead products, although generally of high lead content per unit weight, are often alloyed with metals that require considerable processing to remove. In most cases pyrometallurgy is required; this produces gaseous wastes or by-products in which some loss occurs.[17]

USES OF LEAD

The greatest single use of lead in 1968 was for storage batteries (37%) in the form of lead dioxide (PbO_2, 19.8%), as an oxidizing agent, and as lead plates (Pb, 17.3%). Gasoline uses approximately 20% of the 1.3 million tons of lead consumed in the United States yearly in the form of tetraethyl lead [$Pb\ (C_2H_5)_4$] and tetramethyl lead [$Pb\ (CH_3)_4$]. Pigments formed by lead carbonate ($PbCo_3$), white lead [$2PbCO_3 \cdot Pb(OH)_2$] and/or sublimated white lead ($2PbSO_4 \cdot PbO$) formerly used over 8% of the total but now are outlawed for paints and coatings, and lead in ammunition uses more than 6% of the total in a lead-tin alloy. Other industrial and electrical uses account for the rest. Only the pigment and tetraethyl lead (28% of the total or 360,000 tons) is unrecoverable. Lead is fifth in tonnage in metals produced in the world.

All products of lead require exacting specification; hence, production processes are mainly concerned with elimination of impurities and addition of alloying metals. With few exceptions pyrometallurgical processes are required for purification. Table 2 shows the marketable grades of refined lead to be: corroding, chemical, acid, copper, common desilverized A, common desilverized B, and soft undesilverized. It can be seen that alloying agents or impurities are accountable to 0.001%. These specifications combined with original quality from the supply source dictate the number and type of processes required, each with its waste production. Wyllie[35] shows a comparison of today's problems with the economics involved.

MINE AND MILL WASTES

In developing a mine so that ore extraction can take place, it is necessary to sink shafts and drive drifts and crosscuts to get to the ore.[14, 18, 19, 21, 22, 26] These are usually in waste material that must be hauled to the surface and disposed of. This is frequently a continuing process since exploration continues during mining to search for additional ore deposits.

The average 6 × 8 ft drift would produce approximately 4 tons of waste per foot; similar production occurs for shafts. Many mines have several tens of miles of openings that represent approximately 20,000 tons/mile or a waste pile of dimensions 100 × 100 × 25 ft high. Frequently this serves as surface plant foundations, road gravel, or mine fill; it is of minor proportion compared to mill waste. Development waste often is left underground to be used as support of stoped areas.[24]

The mining process normally calls for removal of solid rock by drilling and blasting. The rock containing finely disseminated particles of the lead and other minerals is transported to a mill. With few exceptions, lead mining is from underground mines rather than open pits, owing to geologic relationships. Hand sorting, a process of removing the easily identifiable metallic galena from the white gangue, is seldom practical. Increasingly low grade results from the fine dissemination of the mineral particles (about 0.3% by volume).

Table 2 *Chemical Specifications for Commercial Pig Lead*

		Corroding Lead	Chemical Lead	Acid Lead	Copper Lead	Common Desilverized Lead A	Common Desilverized Lead B	Soft Undesilverized Lead
Silver	max. %	0.0015	0.020	0.002	0.020	0.002	0.002	0.002
Silver	min. %	—	0.002	—	—	—	—	—
Copper	max. %	0.0015	0.080	0.080	0.080	0.0025	0.0025	0.04
Copper	min. %	—	0.040	0.040	0.040	—	—	—
Silver and copper	max. %	0.0025	—	—	—	—	—	—
Arsenic	max. %	0.0015	—	—	—	—	—	—
Antimony and tin	max. %	0.0095	—	—	—	—	—	—
Arsenic, antimony, and tin	max. %	—	0.002	0.002	0.015	0.015	0.015	0.015
Zinc	max. %	0.0015	0.001	0.001	0.002	0.002	0.002	0.002
Iron	max. %	0.002	0.002	0.002	0.002	0.002	0.002	0.002
Bismuth	max. %	0.05	0.025	0.025	0.10	0.15	0.25	0.005
Lead (by difference)	min. %	99.94	99.90	99.90	99.85	99.85	99.73	99.93

Corroding lead, lead refined to a high degree of purity.
Chemical lead, undesilverized lead produced from southeastern Missouri ores.
Acid lead, made by adding copper to fully refined lead.
Copper lead, made by adding copper to fully refined lead.
Common desilverized leads A and B, fully refined desilverized lead.
Soft undesilverized lead, produced from ores of the Joplin, Missouri district.
Source: C. L. Mantell, *Engineering Materials Handbook*, Lead section McGraw-Hill, New York, 1958.

The first step in processing is usually milling on the premises. This entails crushing, grinding, and regrinding until theoretically each galena particle is completely free from every other mineral particle. When finely ground, the particles appear not as isolated minerals but as particles containing both galena and gangue minerals and in some cases three or more minerals.

Table 3 is a screen analysis of the finely ground product. It can be seen that even a reduction of particle size to 200 mesh (0.0029-in. diameter) does not liberate the galena completely; however, many particles of gangue completely or predominantly free of galena are now present.

The milling process attempts to make a complete separation of gangue and ore minerals utilizing mineral characteristics such as specific gravity, chemical reaction, and surface attraction. Present methods generally achieve a recovery of 90–98% of all lead in the feed to the mill. The remaining lead is discharged from the mill as a reagent-laden slurry of sand and slimes contained in the bulk of the gangue. This is called mill tailings. Kidd Creek Ontario is a typical modern operation showing the complexity of a mining and milling where lead makes up 1% of the ore.[3]

The mill tailings can be discharged into ponds, dams, or drainages. Although discharge into streams was common at one time, it is rapidly falling in disrepute. The pond if properly designed[12] will drain the effluent, which should be acceptably pure chemically and free of solids.[30] Most of the solids either settle out and help form additional embankments or are returned underground to be used for stope support by filling the previously mined area.[2, 5, 14, 27, 32] Figure 2 is a schematic dia-

Table 3 *Tyler Screen Analysis of Crushed Ore (Lead)*

Mesh	Weight (%)	Cum. Weight (%)
+65	0.2	0.2
65/100	1.4	1.6
100/150	6.2	7.8
150/200	11.6	19.4
−200	80.6	100.0
Total	100.0	

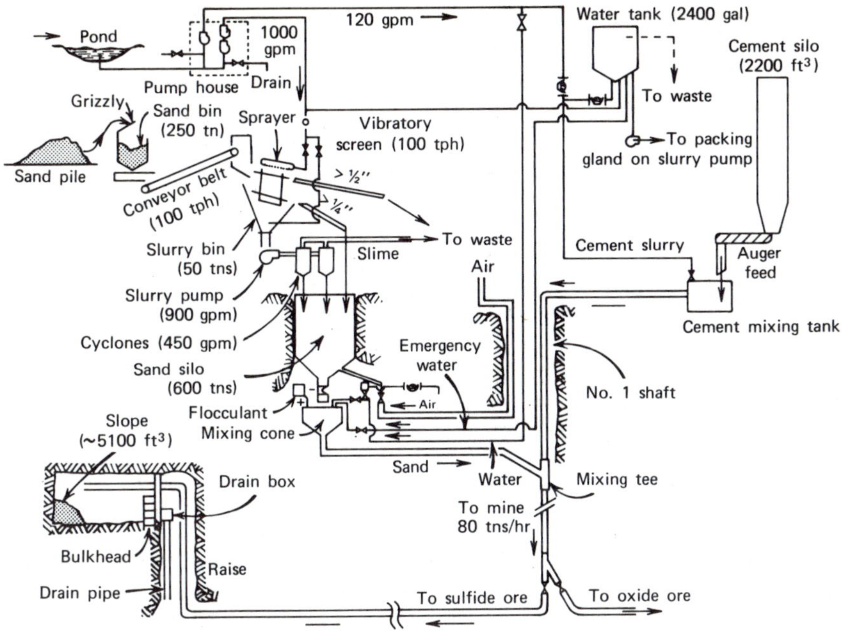

Figure 2. Waste fill of mine stopes with mill tailings (Kennecott Copper Corp.).

gram for one of the Kennecott lead–zinc mines showing the process that must be used to safely and economically place sand fill hydraulically in a mine stope. This can apply to either mill tailing sands or surface sand of other origin.

It is impossible to return all tailings to the mine because of the uncontrollability of slimes (−400 mesh). The fine particles act as colloids and require excessive time for settling; hence, they can easily flood a mine. These fine particles are contained in the surface ponds along with sufficient sands to form the embankment. In some operations where mining methods do not permit filling, the bulk of the mill tailings is stored in ponds.

Table 4 lists several mines and their daily tonnages of waste. Tonnages are given for sand, slimes, and effluent as well as mine waste, and value of mineral.

During milling, the concentrates containing 90% or more of the original mineral still contain some of the waste material. Concentration ratios of 5:1 to 100:1 can be expected. For the former this indicates that for every 5 tons of the original ore containing 100% of the lead (ap-

Table 4 *Mine and Mill Waste Products During Lead Concentrate Production*

Mine or Mill	Daily Tonnage to Mill	Daily Tonnage of Waste
Restricted[a]	750	400
Pandora	1,200	1,000 (approx.)
St. Joe Lead[b]	23,700	22,140
Restricted[a]	2,500	2,360
Shattuck Denn	200–1,200	4,500,000/29 yr. (35% slimes, 65% sand)

[a] Information gained was confidential.

[b] This production is from 12 mines processed in four mills (5 days/wk).

proximately 40 lb/ton in a 2% ore) only 1 ton of concentrate containing 36 × 5 = 180 lb of lead (at 90% recovery) remains. The other 4 tons containing 20 lb of lead is mill tailings assaying 0.25% lead. This process is practiced for economic purposes both to save freight charges on valueless material and to reduce the tonnage to be smelted. No process has yet been able to get 100% recovery economically. Frequently the tailings are saved as at the Eagle, Burgin, and Pend Oreille mines rather than returned. Sometimes it is anticipated that new metallurgical developments will enable them to be reworked at a profit for the remaining values.

In a typical mill process of a lead–zinc mine, in a recent year, production was about 270,000 tons of lead–zinc ore; 10,000 tons of development waste; and 85,000 tons of back cave waste. The above waste stayed in the mine for back fill. A production of 200,000 tons of unseparated sand and slimes were stockpiled in a decant pond 3½ miles away. The stockpiled tailings contain 35% sulfur. Feed to the mill was about 10% zinc and 1% lead with 95% recovery. High sulfide tailings can be beneficial for cementing of back fill but can cause heating.

The Pandora Mill operation for 1969 produced 403,500 tons of ore containing 2.52% lead, 3.75% zinc, with some copper, gold, and silver. Although no specific figures were given, the mill tailings would normally amount to approximately 320,000 tons, which in this case were not separated and went directly to a surface pond. Mining wastes were negligible and stayed in the mine.

In a generalized mill circuit in which a process known as the heavy media or sink-float separation is made, coarse ore in which ore mineral particles are not so finely divided is separated by artificially producing a fluid of high specific gravity. The fluid will float rock at a density of 2.75 but will not float lead or zinc ore at a greater density. The result is a floated waste product that is either returned to the mine or used as surface gravel. It also contains some unrecovered values. Tonnages are small in comparison to flotation tailings; however, during World War II a major lead producer reworked an entire valley containing old mine and mill waste rock at 1500 tons/day. What was considered waste became profitable with improved beneficiation practices.

A typical midwest operation on ore containing lead, zinc, and some copper, in a typical year, produced 335,135 tons of development and stope waste of which 179,452 tons were returned to the mine. Ore produced was 6,050,839 tons, of which 5,645,507 tons were placed in ponds as slimes at assays of 0.08% lead, 0.07% zinc, and 0.02% copper. The 405,000 tons of concentrate containing metal, sulfur, and some waste rock were shipped to the smelter.

Another operation in the area just starting operation reports that little mine waste is produced after initial startup. Generally after foundations and road graveling this presents no surface problems and later production of waste remains underground to fill old stopes. Ore production of 900,000–1,000,000 tons/yr of 5.5% lead plus less than 1% zinc and copper results in about 850,000 tons of mill tailings at 0.18% lead and 0.1% or less of zinc and copper. Tailings are slimes that are ponded on the surface. Research is under way to separate +325 mesh material to be sold as agricultural limestone.

A western producer of lead-zinc concentrates has collected 3 million tons of mine waste and mill sands separated from slimes with assays of about 0.2% combined lead and zinc. This is considered a future potential source of metal rather than waste.

A mining operation in Arizona has had the following history. In 1942 the concentrator started as a 200-ton/day mill and developed gradually to 1200 tons/day in 1968. Over 29 yr of operations, 4,500,000 tons of tailings were stockpiled having iron and sulfur contents shown in Table 5. These are approximately 35% slimes and 65% sands and have a surface area of about 45 acres. The pond is being covered to prevent wind erosion and tailing loss as well as area pollution. Annual ore production gradually increased from a 1939 production of 65,000 tons with 1,750, 000 lb of lead, 5,000,000 lb of zinc plus gold, silver, and copper, to a 1966 production of 330,000 tons with 12,000,000 lb of lead, and 38,000, 000 lb of zinc and other metals. Total mine value was $120,000,000 at the mill and $68,000,000 after freight and smelter costs.

Table 5 *Twelve-Year History of Mill Tailing Accumulation (Waste) for a Southwestern Mine*

	Heads		Tails	
Year	%Fe	%S	%Fe	%S
1956	15.77	16.33	18.80	13.90
1957	15.19	15.77	18.57	14.55
1958	15.20	15.67	18.95	14.99
1959	15.14	16.10	18.81	15.60
1960	14.74	15.42	18.27	14.40
1961	15.18	16.19	17.65	14.39
1962	14.18	15.10	17.15	13.30
1963	14.40	14.37	15.28	12.07
1964	13.13	13.96	14.79	11.96
1965	14.38	14.78	15.81	13.41
1966	13.07	13.71	14.49	12.27
1967[a]	12.83	12.53	13.58	11.04

[a] First six months.

Rather than close the operation completely, a program of research to utilize the tailings for other purposes has been started. This is shown in the following quotation:

> Investigations over the years by our staff and by consulting firms in our behalf have confirmed the amenability of these tailings, and a concentrate of the tailings, to conventional processes such as SO_2 gas generation by roasting for the making of sulfuric acid, liquid SO_2, or elemental sulfur. Lack of adequate markets for such products at that time turned our attention again to the agricultural market potential for a product to help solve the prevalent iron chlorosis, micronutrient deficiencies, and high alkalinity problems of vast farm areas. Basic research in this field was initiated in 1960. Iron King Superferrite was developed and proven a valuable soil supplement. . . . We now offer Micromix, a similarly processed product containing 25% iron, 30% sulfur and 3% zinc, to the fertilizer blending and formulating industry. Both products are granular and closely sized to trade standards. Research in the farm market has been a continuing function as lawn and garden business was being developed.

Although the production of these soil supplements and the possibility of expansion into a series of similar products has occupied our major attention since 1960, we have not lost sight of the broader potential of our raw material as a base for more-integrated chemical operations, starting with sulfuric acid making. The data presented herein indicates a substantial reserve for sulfur recoverable by roasting, which would make feasible the chemical recoveries of several other elements and the base for fertilizer making, in addition to a large range of micro element recoveries.

SMELTER WASTES

Lead concentrates from mills, and lead ore directly from the mine if of high enough grade, are further processed to free the galena from other minerals and to free the lead in the galena from the sulfur chemically attached to it. The process is called smelting. Particular processes may differ but all are principally a system of oxidizing the sulfur in the galena (PbS) by blowing air through the heated material containing galena to produce either a lead oxide for direct use or a product that is later reduced to metallic lead. Commercial lead oxide can be produced in this process only from ores free from other metals.

The roasting process for blast furnace feed produces a sinter that is combined with coke, silica, and lime. Again air blown through this heated charge produces carbon monoxide and carbon dioxide with flue dusts and molten lead. Some other metal sulfides convert to metal in the slag. Precious metals and some of the copper, nickel, etc., metals collect in the lead. The slag produced to extract 2000 lb of lead is about 80 lb containing 8–15 lb of lead.

The slag is treated in a zinc fuming furnace, and then in other processes to remove the remaining metal from the slag or matte. After approximately 98% recovery of the lead in the smelter feed, the slag is discarded. The lead bullion then either is pure enough for some commercial purposes or follows several refining processes to remove copper, tin, antimony, arsenic, gold, silver, and zinc. In some cases electrolytic refining to remove bismuth is required. Little waste except for sulfur dioxide fumes results from the refining. Standard smelter designs are available.[34]

In 1967 The Bunker Hill Company in Idaho had produced over 3.3 million tons of lead and 368 million ounces of silver at its lead smelter. The accumulated slag pile is about 50 acres; however, the SO_2 fumes have been troublesome. Both hydrometallurgical processes and SO_2 precipitation are the objects of a multimillion dollar research program to eliminate smoke and fumes.[16] It has been estimated by Rohrman[28] that lead smelting in the United States through reduction of galena produces 146,000 tons of SO_2. Most smelters have reduced the concentration of SO_2 to economically acceptable concentrations. Additional research is being carried out to reduce or eliminate SO_2 waste or convert it to commercial products.[10] New developments in smelting techniques show promise for reducing the SO_2 gas production.[6, 9, 31]

Scrap lead as smelter feed requires a heating operation with provision for removal of impurities as done in lead ore refining. Technically the processes are similar; physically they differ somewhat and the production of waste products is much less than in refining.

WASTE PRODUCTS

Review of the processes required to produce commercial lead products shows that waste products are (1) mine waste rock, (2) mill tailings, (3) smelter slag and gases, and (4) refinery slag and gases. Tonnages of waste are in approximately this order also, with exceptions for various mines and districts.

MINE WASTE

Annual mine production of lead in the United States is approximately 500,000 tons. Each mine is distinctive; however, mining operations in any one district are similar. The largest lead-mining districts in eastern and southwestern Missouri are within a few thousand feet of the ground surface and the ore bodies themselves are confined to predictable horizons. Development shafts and drifts to reach the ore plus stope preparation, all in waste rock, make up approximately 1% of the total mine rock removal.

Since the mine cannot operate without these openings, the development material must be classed as waste produced. Generally this waste is hauled to the surface at the beginning of operations but later most waste is left underground. On occasion this material may be used for road or other purposes, but in most cases even when left as waste dumps it is a small percentage.

Removal of the ore, which averages 2–6% lead, therefore entails removing 40 lb of lead per ton, approximately 7 lb of sulfur, and about 3 lb of other metallics for a total of 50 lb to go to the smelter for every 2000 lb of rock. Waste production in terms of concentrator sand and slimes averages 1950 lb per 40 lb of lead or about 48–49 lb of waste to 1 lb of lead. This material is concentrator tailings composed of carbonate or silica rock with some sulfides. It is generally stored in surface ponds where water is decanted and drained. The waste pile can be treated and rehabitated to grasses and growth only with difficulty.

The second largest producing district is the Coeur d'Alene district in northern Idaho. Here production comes from deposits distributed vertically downward along regional faults. They have extended from surface to a maximum present depth of 7300 ft, or 2000 ft below sea level. Indications are that they will continue in depth. Considerable exploration and development work is required for ore discovery and removal. The waste-to-ore ratio may be as much as 4 to 1. In these deposits only the presence of silver makes removal profitable. Again the tonnage and waste is the same as the previous example except that the presence of zinc, copper, antimony, cadmium, and iron increases the amount of metal per ton of ore to about 100 lb. The waste production is relatively the same.

In these mines, tailings are returned to the stopes in as large quantities as possible. Only sandy material of +400 mesh is used, owing to containment problems underground. The −400 mesh or slimes are stored in tailings ponds on the surface.

Both accumulating tailings and tailings from past mine production are examined for reworking; most tailings are uneconomical for reprocessing. Rehabilitation of active tailings ponds is not possible, and inactive ponds are rehabilitated only with great difficulty.

Some variation exists in other mining districts, but essentially they fall into these two catagories. Tailing discharge into streams or lakes is increasingly curtailed or treated to acceptable limits and the practice of stream pollution will eventually be eliminated.

SMELTER WASTE

The mine products to the smelter today are almost exclusively sulfides in silicate or carbonate rocks. As such in the smelter, 1 lb of sulfur is produced for every 6½ lb of lead. The sulfur as SO_2 is collected as flue gas or exhausted into the atmosphere. Generally 98% of the sulfur dioxide is collected and converted to sulfuric acid. Market demand for the acid controls the amount converted. If public demand requires maximum collection, excess production of sulfuric acid may result and to alleviate the acid storage problem, conversion to native sulfur will become necessary; then the sulfur will either be discarded or stockpiled.

During smelting the furnace matte containing impurities and metals must be retreated for recovery of residual lead. The slag from flux, coke, and gangue from the ore after removal of zinc and lead by blowing is either dumped directly in a molten condition or granulated. In either case it is a waste product with little economic use. There still remain up to 0.5% lead and traces of silver. Slags may be siliceous or basic with high iron content, but in all cases the specific gravity must approach 3.5 to maintain a matte and slag separation.

The bullion with little remaining waste is either oxidized or refined. The matte is treated for copper recovery and other by-products with some slag resulting. Often it returns through the smelter circuit as a recirculating load.

Flue dusts from dry, wet, or electrostatic processes contain lead, sulfur, iron, copper, silica, zinc, magnesia, and some gold and silver. This also is usually returned as a blast furnace charge. Most sulfur is oxidized in the roasting process; flue gases are usually not scrubbed for sulfur dioxide removal.

Waste products from the Bunker Hill smelting and refining operation are shown in Table 6. The figures are approximate and will not balance metallurgically.

The Bunker Hill Company constructed a sintering facility and an

Table 6 *Typical Analysis of Bunker Hill Wastes*

	Mill Tailings	Smelter Slag
Silver	0.3 oz/ton	0.04 oz/ton
Gold	Trace	None
Copper	0.01%	0.04%
Lead	0.04%	0.4%
Iron	11.1%	—
Iron oxide	—	34.0%
Cadmium	0.01%	0.003%
Insoluble SiO_2	70.1%	33.0%
Lime	1.06%	18.5%
Sulfur	1.9%	0.82%
Zinc	0.3%	—
Zinc oxide	—	2.50%
Antimony	0.03%	0.02%
Arsenic	—	0.03%
Magnesium oxide	—	1.33%
Manganese oxide	—	1.96%
Aluminum oxide	—	7.74%

acid plant at the smelter to recover SO_2 and H_2SO_4 at a combined cost of $6 million.

REFINING WASTES

Many processes are required for refining and production of the many lead products, accessory metals, and precious metals. The wastes are almost exclusively recycled to the smelter with the only effluent escaping being gases with some particulate matter. Other waste is wash water from slimes in the electrolyte tanks.

SCRAP RECOVERY

Lead scrap, except scrap batteries, usually is charged in blast furnaces to follow the circuits previously mentioned. Such smelters produce a flue dust equal to 3% of the charge. Slags and bullion with little matte are produced with the slag being discarded in its usual manner.

SLAG CALCULATIONS

Slag production in a blast furnace can be calculated. Knowing the composition of the ore in detail and the type of slag it will produce, the chemical composition of the slag is predictable. From this the quantity of charge is calculated with the amount of fuel, ore, slag, and flux that must be added and their compositions. Any of the slag forming components such as iron oxide, metallic iron, silica, and lime are known from the charge analysis. A calculation of this type on silicate ore containing SiO_2, FeO, CaO, MgO, Al_2O_3, PbS, plus other metals and S would require balancing the total of these constituents with those added from the fuel and fluxes since they also contain most of these elements. A 1000-lb charge would contain approximately 15 lb of coke ash, 100 lb of slag, 500 lb of ore, 270 lb of iron, and 115 lb of limestone. Lead bullion would vary with the charge but could be 60–400 lb, the remainder being slag.

If we assume that 100 lb of lead is produced for 675 lb of slag, then the domestic production of lead from mined ore (500,000 tons) would account for 33 million tons of slag annually. Actual practice differs somewhat from the theoretical calculations, but the latter are usually accurate enough for most purposes.[23] At a specific gravity of 3.75 the slag would have a volume of 8.5 ft^3/ton or a total volume of less than 4 million ft^3—less than 10 acres 10 ft deep.

REFERENCES

1. W. E. Crandall, "Underground Mining," *Mining Cong. J.*, 170–175, February 1968.
2. W. E. Crandall, "Sand Fill Methods at the Star and Lucky Friday Mines," 1960.
3. Finn B. Domaas, "Kidd Creek—A Huge Mining Success in Ontario," *Eng. Mining J.*, 87–108, April 1969.
4. "Two Missouri Lead Deposits To Start Up In Early 1968," *Eng. Mining J.*, 102, July 1967.
5. "Cemented Sand Fill Gains New Popularity," *Eng. Mining J.*, 90–94, August 1967.
6. "AIME Holds Its 97th Annual Meeting," *Eng. Mining J.*, 94–95, April 1968.
7. "Mining Fattens Capacity at Record Rates," 91–93, February 1969.
8. "World's Biggest in Lead-Zinc Diversifying (Cominco Canada)," *Eng. Mining J.*, 194–202, September 1969.
9. F. W. Gibson, "New Buick Lead Smelter Incorporates 40 Years of Technical Advances," *Eng. Mining J.*, 62–67, July 1968.
10. Frank P. Haver, K. Uchida, and M. M. Wong, "Recovery of Lead and Sulfur From Galena Concentrate, Using a Ferric Sulfate Leach," US R.I. 7360, 1970.
11. International Lead and Zinc Study Group, "Lead and Zinc—Factors Affecting Consumption," United Nations, New York, pp. 17–33.
12. C. D. Kealy, "Seepage Patterns in Mill-Tailings Dams as Determined by Finite-Element Models."
13. Ilmar Mikiver, "Developments and Outlook for the Lead and Zinc Industry," *Mining Cong. J.*, 148, February 1969.
14. Gordon M. Miner, "Use of Cement in Hydraulic Fill," *Mining Cong. J.*, 46–47, November 1966.
15. *Mining Cong. J.*, "Lead and Zinc," 65, February 1967.
16. *Mining Cong. J.*, "Bunker Hill's Lead Smelter Observed 50th Anniversary (Idaho)," 22, September 1967.
17. *Mining Cong. J.*, "Significant Amounts of Lead and Copper Have Been Recovered Through Experiments," 11, January 1968.
18. *Mining Cong. J.*, "Underground Development Work Finds High Grade of Lead (Colorado)," 17, April 1968.
19. *Mining Cong. J.*, "Lead-Zinc Industry," 86, February 1970.
20. *Mining Cong. J.*, "American Smelting & Refining Company is Sinking a Shaft to a Lead-Zinc Orebody (Colorado)," 9, December 1969.

21. *Mining Cong. J.*, "Discovery of Commercial Lead-Zinc Ore Reserve at Camp Bird Mine (Colorado)," 19, February 1970.

22. *Mining Eng.* "Bunker Hill Postpones Opening of Missouri Lead-Zinc Mine," 16, July 1967.

23. Joseph Newton, *Extractive Metallurgy,* Wiley, New York, 1963.

24. Robert Peele, *Mining Engineers' Handbook,* 3rd ed., Wiley, New York, 1941, pp. 10-247–10-274.

25. Jan H. Reimers, "What's New in Lead Zinc Extraction Metallurgy," *World Mining,* 20–25, July 1970.

26. R. F. Pettit, Jr., "Engineering Work at the Coeur Project," Northwest Metals and Minerals Symposium, AIME, Coeur D'Alene, Idaho, April 1959.

27. Gary R. Rice, "Sand Fill Stoping at the Star Mine," Pacific Northwest Regional Conference, AIME, Seattle, Washington, April 1959.

28. F. A. Rohrman and J. H. Ludwig. "Sulfur Oxide Emissions by Smelters," *J. Metals,* **20,** No. 12, 46, December 1968.

29. John Songstad, "Bunker Designs New Stoping Method," *Eng. Mining J.*, 101–105, January 1967.

30. *Standard Methods for the Examination of Water and Wastewater,* 11th ed., APHA, Inc., New York, 1962.

31. D. A. Temple, "Is the U.S. the Next Target for an ISP Plant?," *Eng. Mining J.*, 93, January 1967.

32. W. A. Vine, "Proceedings of the Symposium on Hydraulic Fill," Montana School of Mines, Butte, May 1958.

33. *Water Quality Criteria,* 2nd ed., McKee & Wolf, State Water Control Board, Sacramento, California, Publication #3-A.

34. O. H. Woodward, *A Review of the Broken Hill Lead-Silver-Zinc Industry,* 2nd ed., West Publishing Corp., Pty Ltd, Sydney, 1965, pp.78–192.

35. *World Mining,* "Lead-Zinc of Touisset-Bov, Becker-Alzi," 7, April 1968.

36. *World Mining,* "Base Metal Mines and Concentrators Throughout Finland," 26, November 1968.

37. R. J. Wyllie, "Here's What's New Today in Lead and Zinc Smelting and Refining," *World Mining,* 24–27, July 1969.

CHAPTER VIII.5
stabilizing mineral wastes

Mine, mill, and smelter wastes are scattered across the country as unattractive barren piles that mar the beauty of the land.

Congress, through the Solid Waste Disposal Act of 1965, delegated to USBM responsibility to conduct research designed to prevent air and water pollution from mineral accumulations. Waste from the processing of minerals and fuels amounted to 1122 million tons covering approximately 2 million acres. Of this, approximately 40% is fine-sized material. Stabilization methods (physical, chemical, and vegetative, and combinations thereof) are used or have been tested and proposed.

Massive coverings of coarse slag, concrete, and soil prevent erosion of waste piles. Chemicals can bond particles of fine waste into a relatively inert mass. The most promising method is vegetative stabilization. Numerous species of plants can germinate, grow, and reseed in waste materials that have been adequately prepared to support plant life if environmental conditions are not overly severe.

Preplanning of waste disposal is now common for initiating new operations. Preplanning was frequently ignored in the past, and waste accumulations are often haphazardly deposited in piles with dangerously steep banks. Thus, although stabilization by vegetation might be preferred because of esthetic considerations, it is often not applicable.

Many materials have been tried for stabilization of fine tailings. The material most used is rock and soil obtained from areas adjacent to the wastes to be covered. Soil often has an advantage in that it provides an effective cover and a habitat for local vegetation, but it is not always available in contiguous areas.

Crushed or granulated smelter slag has stabilized inactive tailings ponds. The slag-covered portions are subject to burial from shifting sands or from new tailings. Slag cannot provide a favorable habitat for vegetation.

Other methods of stabilization include bark covering, the harrowing of straw into the top inches of tailings, and the placement of windbreaks. Bark and straw was beneficial at the Anaconda, Montana, copper operations, and reed windbreaks were successfully employed in South African plantings on gold and uranium mill tailings.

Chemical stabilization involves reacting chemical with the waste to form an air- and water-resistant crust or layer. The heterogeneous tailings vary in permeability, pH, and salt content. Chemicals have been tested on samples of copper, lead–zinc, clay, uranium, and vanadium process wastes. The coherency of the tailings surfaces is determined by using airjets and waterjets to simulate wind and water erosion. Chemicals include sodium and potassium silicates, lignosulfonates, pyrite treated with sulfuric acid, cement lime, cationic neoprene emulsion, elastomeric polymers, resinous materials, bituminous base products, cellulose compounds, tail oil fractions, amines, and wax, tar, and pitch emulsions. The resinous, elastomeric polymer, lignosulfonate, bituminous base, wax, tar and pitch products have proved effective stabilizers for one or more types of fine-sized mineral wastes. Several of the more promising chemicals have subsequently been evaluated by field tests on tailings from different mineral processing plants.

A 34-acre uranium tailing accumulation at Tuba City, Arizona, was stabilized during the summer of 1968 as a field trial. An elastomeric polymer was applied to the dike areas of the tailings pond. Calcium lignosulfonate, a waste product from paper manufacturing, was applied to the beach areas of the ponds.

Mill wastes usually (1) are deficient in plant nutrients; (2) contain excessive salts and heavy metal phytotoxicants; (3) consist of unconsolidated sands that when windblown, destroy young plants by sandblasting or burial, or both; and (4) lack normal microbial populations. The sloping sides of waste piles receive greatly varying amounts of solar radiation, depending on direction of exposure. Photosynthesis of plants is not continuous while the sun is shining; under high temperatures, photosynthesis may almost completely stop. Most accumulations of mill tailings are light in color and may reflect excessive radiation to plant surfaces, thus intensifying physiological stresses. Vegetation that may be effective on northern and eastern exposures may not be suitable for southern or western exposures.

Mining companies have planted old tailings accumulations in efforts to achieve both stabilization and an attractive site. Tailings at the Miami Copper Co., Miami, Arizona, were covered with about 4 in. of country soil obtained from areas adjacent to the tailings. This country soil was of darker color than the tailings and provided some protection from heat reflection. Miami seeded this material with mixtures of seeds. Seedlings or sprouts of trees and shrubs also were placed by hand. Vigorous growth of planted species and the rapid encroachment of native species resulted. Natural succession will produce a cover capable of self-perpetuation without irrigation or fertilization.

Seeds of many species have been tested for germination and growth in samples of tailings. The domestic species were more reliable germinators than were the wild species. Plants that showed considerable promise include sweet clover, alfalfa, winter wheat, rye, barley, wheat grasses (western, crested, intermediate, tall pubescent), sorghum, love, Kentucky blue, orchard and shrubs such as big sagebrush, rabbit brush, and Siberian pea tree.

Salts in the wastes or mill water are deterrents to plant growth. Difficulty was encountered in growing vegetation in milling wastes from Kennecott's Utah Copper Division. Duplicate plantings were made (1) in flotation tailings, containing water-soluble salts but assaying only 0.05% copper, mostly as sulfides; and (2) in a tailing containing little soluble salts but almost 0.25% copper. Seeded test plots were watered with tap water and tailings effluent water. Both germination and growth were adversely affected when effluent water was used. Toxicity was the result of high salt content of the water and tailings.

Plant growth tests on uranium mining wastes, stripped overburden from the Federal Partners Mine at Gas Hills, Wyoming, showed that yellow sweet clover, crested wheat, and tomatoes could be established, but that salts solubilized by watering are brought to the surface by evaporation and after a month inhibit or destroy growth.

Stratification and mounding of tailings improved salinity problems. Growth tests were made in unstratified plots of tailings material, and in "as-deposited" tailings capped with a layer of sands and then with a stratum of slimes. Both plots were permitted to dry throughly, and then watered once with sufficiently salt-free water to force the "wetting front" several feet into the underlying tails. Subsequently, water was added only when plants showed signs of wilting. Appreciably more profuse and healthy plant growth was obtained in the stratified plots.

In an attempt to reduce the rate of erosion, five chemical stabilizers were applied to newly erected and seeded ridges. Stabilizers influence erosion rates and the germination and survival of plants. Chemical stabilizers tested included a resinous adhesive, ammonium lignosulfonate, elastomeric polymer, and a catatonic neoprene emulsion. After 28 wk, seedling survival counts showed that almost 40% more seedlings grew on the treated plots than on the control. The catonic neoprene emulsion treatment supported only 50% as many seedlings as the control plot.

Tests showed that 1000 ppm copper had little effect, and that nickel concentrations above 100 ppm and zinc concentrations above 10 ppm were toxic. When using a composite, toxic effects became evident at 10 ppm and became pronounced at about 100 ppm. The effect achieved with the low concentrations tested indicates toxicity to be independent of osmotic pressure.

Addition of fertilizers enhanced germination and growth on some tailings but had the opposite effect on others.

Tests on several types of mineral wastes indicated that adding 45 lb per acre of urea and calcium triple superphosphate provided there is sufficient fertilizer to stimulate growth of most plants.

Basic and acidic tailings, adverse for plant growth, can be neutralized by adding lime or limestone and sulfuric acid. Impervious crusts are sometimes formed if the neutralizing agent interacts with the acidic or basic causative agent. Fine water spray may be effective for eliminating excess acidity.

Uranium tailings from Wyoming and Colorado had pH values of 2.3–4.5. Additions of 70 and 6.7 lb of lime to the pH 2.3 and 4.5 soil, respectively, to the top 2 in. of tailings per acre, raised the pH sufficiently to sustain plant life. Vitro Minerals and Chemical Co. of Salt Lake City, Utah, had both acidic uranium tailings (pH 3.5) and basic vanadium tailings (pH 8.5). These were mixed in equal portions to produce a material of pH 6.3 in which seeds readily germinated. Fertilization, however, was necessary to sustain adequate plant growth.

The New York Ore Division of Jones & Laughlin Steel Corporation at its Benson iron ore mine in Star Lake, New York, has 10,000 tons/day of solids pumped out to a pair of tailing ponds—one active and the other inactive.

The Jones & Laughlin Corporation began experimental planting to see if anything would grow on the dry sections and hold them in place. Many types of seeds and seedlings were tried, with only partial success. At the bottom of the pile, where the wind was not strong and the sand was fairly moist, some grasses and shrubs did take root. Higher up, any plant, even a weed, was soon sand-blasted out of existence. Nearby, Newton Falls Paper Company, at roughly the same time was looking for a place to dump waste paper pulp from its mill.

Every few days a tank truck from Newton Falls drives up to the Benson tailing pile and spreads a load of gray, slushy material over the sand. It now covers over 100 acres and plants are beginning to sprout where the waste has been bedded on the tailing. At first weeds started to show; then some seedlings, evergreens, and shrubs took root.

Suitable chemicals to stabilize surfaces of mill tailings and encourage growth have several advantages. Sandblasting of the plants is decreased. Moisture is retained in the tailings. Germination is promoted and wilting is decreased by creating a dark heat-absorbing surface. Plant food may be added.

Tests have corroborated the preliminary results, which showed that resinous compounds produce the best plant growth and surface stability. Several types of elastometric polymers also proved effective.

A 10-acre plot of Kennecott's McGill, Nevada, tailings is located in a climatically harsh site for vegetation. The growing season is short, the temperature range is wide, and precipitation is only 8.4 in/yr. Wheatgrasses and various tree seeds were mixed and planted by drill seeder at a rate of about 35 seeds per square foot. Prilled urea and superphosphate fertilizers were added to the seeding mixture at a rate of about 45 lb/acre each of P_2O_5 and nitrogen, planted in furrows 10 in. apart. A week later, a resinous adhesive diluted with four parts water was applied over the entire area at rates of 0.18–0.66 gal of dilute solution per square yard.

A germination count of the plants was made 1 month after seeding and showed that 24%, 12%, and 52% of the legumes, grasses, and winter wheat, respectively, had germinated. The seedling plants were vigorous and healthy, up to 5 in. tall. One week later, a killing frost destroyed most of the legumes.

Another survey of the seeded area, in May and September, proved that the primary objective was attained: the area was stabilized against wind erosion and native plant species were encroaching onto the stabilized plots. The Siberian pea tree was the only one of the 11 species originally planted that either did not germinate or failed to survive to the September survey. Eleven wild species had encroached and were growing.

An accounting was kept of material and application costs associated with treatment of the 10-acre plot. The resinous adhesive was applied in quantities ranging from 0.18 to 0.66 gal/yd.^{2} During a year of observation, no appreciable differences were noted in the stabilization of plots treated with the different amounts. Table 1 presents the costs based on application of 0.18 gal/yd.^{2} The estimated cost of $1355 for the 10 acres could probably be reduced if larger acreages were treated or if proper equipment were available.

One acre of the beach area of the White Pine Copper Company tailing at White Pine, Michigan, was vegetatively and chemically stabilized in September 1969. The test plot was planted with seeds of six grasses, three legumes, and an oat grain. Calcium triple superphosphate, ammonium nitrate, and potassium chloride, were applied as fertilization. After planting, resinous adhesive was applied to prevent blowing of the sand and to eliminate cutting or burying of the plants. A

Table 1 Cost of Stabilization Procedure per 10 Acres

Item	Amount	Cost
Seeds	37 seeds per square foot	$ 56
Calcium treble superphosphate	45 lb P_2O_5 per acre	51
Prilled urea	45 lbs N per acre	62
Resinous adhesive		
At 20¢/gal	0.18 gal/yd²	348
Freight at 8¢/gal		140
Labor, at $3 per hour	3 men for 3 days	216
Equipment expense:		
Seeder and grader		100
Water truck		382
Total		$1355

germination count made by the company 6 weeks after planting showed a germination rate of 20%. The White Pine Copper Company planted much more extensive acreages in July of 1970 and the growth has been deemed encouraging.

Approximately 2 acres of a lead–zinc tailings pond at Flat River, Missouri was stabilized in a similar fashion in September 1969, in cooperation with the St. Joe Lead Company. The plot was inspected in March 1971 and the plants were enjoying healthy, vigorous growth. A flock of about 400 geese were feeding on the plants and had been doing so for over 3 weeks, as this was the most succulent feed in the area.

An acre of lead–zinc tailings pond at Page, Idaho, belonging to American Smelting and Refining Company, was also cooperatively stabilized by the basic vegetative-chemical procedure in September 1969. An inspection of the area in June 1970 showed that most of the fall germination plants had died during the winter and early spring. Plant growth was spotty and only 2–6 in. high. The cause has not been determined.

In the late fall of 1969, the wastes of the Interpace Company at Mica, Washington were prepared and planted. A 10-acre plot of the area was covered with 5–12 in. of overburden from the mining operations, and 500 lb/acre of gypsum was disked into the surface. The area was fertilized with 18, 60, and 30 lb/acre, respectively, of N_2, P_2O_5, and K_2O. Rice hulls were spread over the surface and the area was planted with hard fescue, bluegrass, and crested wheat grass. Another acre adjacent to the 10-acre plot was fertilized with N_2, P_2O_5, and K_2O at 34, 144, and 42 lb/acre, respectively, and then planted. The plant seeds used included eight varieties of grasses, four legumes, and three grains. Portions of the two areas were stabilized.

Inspection of the Interpace property in June 1970 showed remarkable plant growth on the areas. The grains were about 3 ft high and full seed heads had developed. The crested wheatgrass was about 2 ft tall with seed heads developing. Short grasses were growing exceptionally well, and although the legumes were sparse, the growth was healthy.

How much of our land is actually spoiled by mining activities? The USDI states that "Only a small percentage—about 0.14% of the total land area of the United States—has been disturbed by surface mining." (Surface Mining and Our Environment, USGPO, 0-258-263, 1967). The challenge, therefore, is to reclaim as much as possible of this 0.14% of U.S. land. It will also make the land free of such problems as soil erosion or dust storms in dry-tailing ponds and mined areas.

DUST STORMS

The tailing pond of the Utah Copper Division of Kennecott Coper covers a 5100-acre site near Magna, Utah, and receives some 100,000 tons (30% solids) of flotation tailing slurry every day. Spray vehicles treat the dried tailings against wind abrasion.

To prevent the wind from lifting tons of dust in the air, the corporation had developed a soil-stabilization program for its vast tailing pond sites in the west. Essentially, the treatment consists of spraying an emulsion of petroleum oils and resins, which imparts a permanent cohesiveness to soil by coating dust particles and forming adhesive membranes over adjacent particles—but without fouling pond water, which must meet stringent company conservation standards.

Among techniques tried out for containing tailings were water sprays, snow fences, and plowing, but none of these proved satisfactory: the tailings still blew and drifted, aided by fairly strong prevailing winds. Chemical additives were then tried, but had to be discontinued because they had an adverse effect on the water recovered and pumped back to the concentrator plant.

Tests covered various product and application alternatives, which included sprays with water and various compounds, by motorized vehicles and helicopters. Spraying from helicopters was too costly and troublesome because the rotor blades churned up the dust; this method, however, is used in the current program to apply a water-containing polymer on those wet areas that cannot be reached by vehicle.

The perimeter of this site measures 13–16 miles; its diameter from north to south is 3 miles. The depth of the dry tailings varies from 200 to 300 ft according to the contours of the area before filling started, and increases about 3 ft/yr. Water drifts to the northeast corner of the diked pont, and is eventually reclaimed and recycled.

Each week, 6000–8000 gal of emulsion concentrates are delivered by tank car to the pond and pumped into a 1200-gal Terracruiser. Equipped with low-pressure balloon tires, this vehicle travels from the rail siding across a highway to reach the tailing pond.

At the pond, the concentrate is pumped into a 12,000-gal sprinkler wagon and mixed with the water in its tank for 1-to-10 ratio. The emulsion is then pumped back into the Terra-cruiser and also into a 1200-gal Juggernaut equipped with tracks. These two units then apply the agent through a double spray stream at 60 psi, traveling at 4 mph and pumping 100 gal/min. Application rates average 0.6 gal/yd.[2] The emulsion penetrates immediately to a depth of about ⅛ in. With a 1200-gal load, each vehicle covers about half an acre.

The pattern of spraying in the rectangular pond area is rectilinear with treatments made downwind. There is no set pattern since the drying of new fill, and the movement of fresh dust determines which areas are sprayed. Only approximately 30% of the site requires dust treatment at any given time since the remaining acreage is under water for the most part or wet enough to keep the dust from blowing. The application cost, which includes chemical and spraying, is about $178 per acre.

CHAPTER VIII.6

disposal of slag from the electrothermic zinc smelter of st. joe minerals corporation

GARY E. WELCH

The Josephtown Smelter of St. Joe Minerals Corporation is one of the world's largest producers of zinc. Located in western Pennsylvania, adjacent to the Ohio River, about 30 miles downstream from Pittsburgh, its capacity is about 250,000 tons of zinc equivalent per year.

The St. Joe-developed electrothermic process uses resistance-type electric furnaces. The principal feed to the vertical shaft furnaces is sized sinter and coke. Electricity flowing through the furnace charge supplies the energy required for the smelting reaction. Temperature of the furnace charge varies from roughly 900°C near the wall to about 1400°C in the center of the furnace.[1]

Residue discharged from the bottom of the furnaces comprises a semidry heterogeneous mixture of partially smelted sinter, coke, fused slag, and globules of ferrosilicon. Residue handling can be conveniently considered in two stages.

INITIAL RESIDUE PROCESSING

Initial residue processing is summarized in Figure 1. Table 1 shows the approximate chemical analysis and the disposition of the various materials.

Initial residue treatment consists of crushing the residue to a top size of 3 in. The residue is then screened through a 1¼-in. trommel with the oversize going directly to final residue treatment. The undersize from the trommel is fed to a magnetic separator which removes iron particles, which are a low grade of ferrosilicon. The magnetics have a composition of about 82% iron, 13% silicon, 3% copper, and 1% nickel.

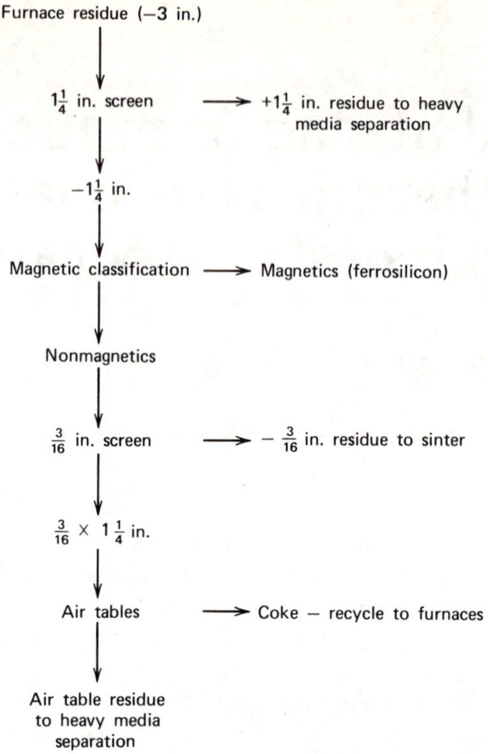

Figure 1. Flowsheet of initial residue processing circuit.

Table 1 *Chemical Composition and Disposition of Materials Handled in Initial Residue Treatment Circuit*

Material	Chemical Analysis			Disposition
	% Zinc	% Carbon	% Ferrosilicon	
+$1\frac{1}{4}$-in. residue	7	15		To heavy media separation
Ferrosilicon	1		93	Sold as substitute for silvery pig iron as high density aggregate and used to produce media for heavy media plant
−$\frac{3}{16}$-in. residue	18	15		To sinter plant
Reclaim coke	3	70		Recycle to furnaces
Air table residues	16	15		To heavy media separation

Following magnetic separation the residues are once again screened. The nonmagnetic undersize (−3/16 in.) from this screening operation goes directly to sintering. This nonmagnetic undersize residue is valuable because of its contained zinc and because it supplies needed carbon and silica to the sintering operation. The remaining residue, now sized 1¼ × 3/16 in., is air tabled to reclaim about 90% of the carbon as coke. This air table product assays about 70% carbon and is suitable for recharging to the electric furnaces. Slag from the air table is sent to final residue treatment.

The only product exiting the circuit in the initial residue treatment is ferrosilicon. Before 1966, this material was used for road ballast. Installation of more sophisticated magnetic classification equipment has upgraded the metal content of the ferrosilicon so that the material can be marketed as a substitute for silvery pig iron and as a high-density concrete aggregate. A modest amount of the ferrosilicon produces media for the heavy media plant. The initial residue processing circuit operates three shifts, 7 days each week. The circuit is operated by one man on each shift.

FINAL RESIDUE PROCESSING

Final residue processing is summarized in Figure 2. Table 2 shows the approximate chemical analysis and the disposition of the various materials.

Before 1966, air table residue was ground and recirculated via the sintering operation. Oversize residue was magnetically classified, screened, and air tabled for coke recovery. The screened slag was sold as road ballast.

Since 1966, final residue processing has consisted of magnetic classification, air tabling, and heavy media separation.[2] Material enters the residue treatment facility across a high-intensity magnetic separator. The air table residue goes directly to the heavy media unit. The +1¼-in. residue, which in handling undergoes some degradation, is again screened across a 1¼-in. screen. The minus 1¼-in. fraction from this screen is diverted across an air table for coke recovery. The nonmagnetic coke-free slag is then fed to the heavy media separation drum. The sink product (containing 75–95% of the zinc in the feed material) is crushed to a nominal size of ⅝ in. and returned to the sintering operation. The float product is sized and sold as road ballast. The metallurgical performance of the heavy media plant is summarized in Table 3.

A feature of the heavy media circuit is the by-product ferrosilicon as

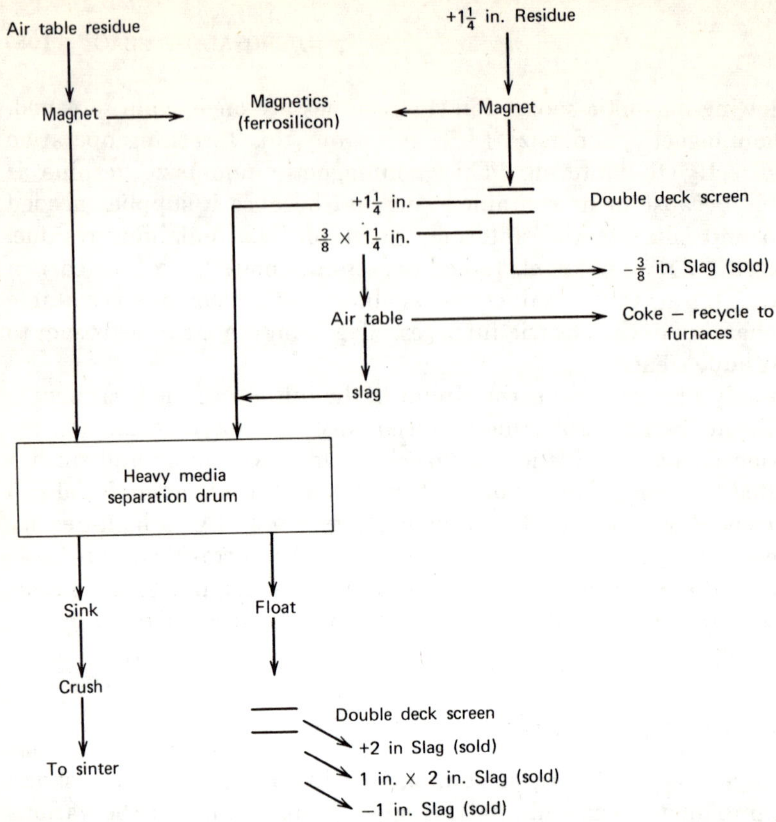

Figure 2. Flowsheet of final residue processing circuit.

Table 2 *Chemical Composition and Disposition of Materials Handled in Final Residue Treatment Circuit*

	Chemical Analysis			
Material	% Zinc	% Carbon	% Ferrosilicon	Disposition
Magnetics	1		93	Sold as substitute for silvery pig iron
Sink product	25–30	2		To sinter plant
Coke	3	70		Recycle to furnaces
Slag −⅜ in. ⅜ × 1 in. 1 × 2 in. +2 in.	2–3	20		Sold for road ballast

Table 3 *Performance of St. Joe Heavy Media Plant*

	Oversize Slag	Air Table Residue
Assay % Zn in sink	24.5	31.5
% Zn in float	2.3	2.0
% Zn in head	6.7	16.5
Zinc recovery (%)	73.0	94.0
Weight distribution, % sink	20.0	49.0

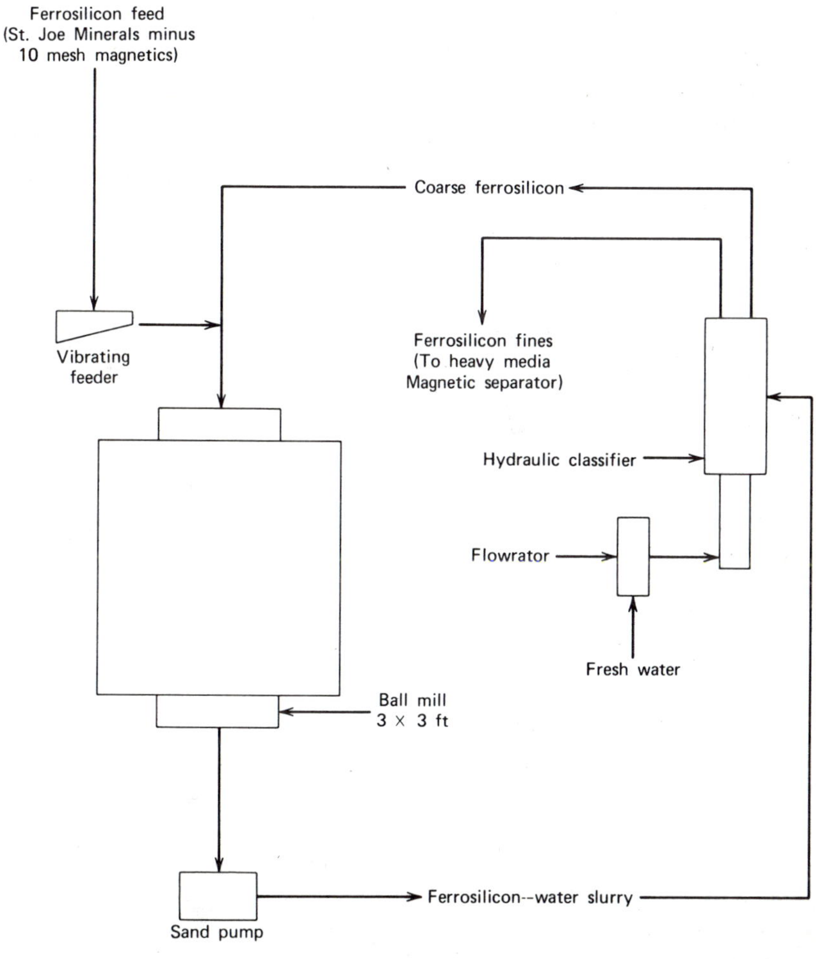

Figure 3. Waste pile prior to recycling were growing day by day.

total medium requirements for the circuit. A flowsheet of the media-producing facilities is shown in Figure 3. The medium is produced from the −10 mesh fraction of the by-product ferrosilicon. About 90% of this size fraction is comprised of 10 mesh × 48 mesh particles. The key to producing good medium is the hydraulic classifier. The ball mill, together with the hydraulic classifier, produce a medium acceptable for in the heavy media plant.

The final residue processing circuit operates one shift each day, 5 days/wk. The circuit is staffed by four operators and one group leader.

CURRENT STATUS OF RESIDUE TREATMENT

Continuous development and refinement of St. Joe's residue treatment facilities have meant substantial economic gain to the smelter. In addition to recovering substantial quantities of previously discarded zinc, residue treatment circuits now produce high-metal-content sized ferrosilicon suitable as a substitute for silvery pig iron and washed, coke-free, sized slag useful as ballast in road building.

REFERENCES

1. R. E. Lund, J. F. Winters, B. E. Hoffacker, T. M. Fusco, and D. E. Warnes, "Josephtown Electrothermic Zinc Smelter," *AIME World Symposium on Mining and Metallurgy of Lead and Zinc, Vol. II, Extractive Metallurgy of Lead and Zinc,* C. H. Cotterill and J. M. Cigan, Eds., pp. 549–580, Port City Press, Inc., (1970), pp. 549–580.
2. R. E. Lund, G. E. Welch, J. Ferrighetto, and B. E. Hoffacker, "St. Joe Reclaims Zinc With Heavy Media Separation," *Mining Eng.,* **20,** 81–83, December 1968.

CHAPTER VIII.7

unprocessed blast furnace and steel slag

E. W. BAUMAN

Slag is the fused residue separated in the reduction of metals from their ores; volcanic scoria. The term "slag" means a waste product to the general public, but that definition is too broad. In public thinking, such products as coal mine waste and boiler cinders are included as slag. For instance, after a disaster in Wales a number of years ago, when a landslide engulfed a school killing many pupils, reporters describing the incident identified the material in the slide as slag, when actually it was coal refuse or red-dog, a term sometimes used around the coal mining areas.

The principal slags resulting from the reduction of ores include copper slags, phosphatic slag, and iron blast furnace slag. Certain slags also result from refining of metals such as steel, or foundry slags and precious metal, i.e., gold and silver, slags. Since the primary function of reducing or refining is to produce elemental metal, the resultant slags are normally considered waste materials, and where quantities produced are small and in areas where disposal is no problem, the slags are generally wasted with little concern as to their disposal. This is the case with most of our precious metal slags in the western United States. However, if slags are produced near metropolitan centers and if the quantity is appreciable—as is the case with iron-producing blast furnaces—the disposal area can take up valuable property and at the same time create unsightly and polluted environments.

BLAST FURNACE SLAG

Of all the metallic slags, those produced in the greatest quantities are the iron and steel slags. Unless these slags can be converted into a useful commodity, the problem of disposal is costly and troublesome. A small group of men have been dedicated to the notion that this by-product of iron-producing furnaces—blast furnace slag—could be converted into a useful commodity. They were so successful that today, some 50 years later, blast furnace slag is recognized by the construction industry as an all-purpose construction aggregate. In some areas, the demand actually exceeds the supply. These men succeeded and they developed the methods to convert a waste and unwanted product into an industrial commodity.

At the turn of the present century, when the U.S. steel industry was coming into its own, production of pig iron was already at a high annual rate and for every ton of pig iron there was an equivalent half-ton, or more, of slag. Thus, the piles of wasted slag began to grow, and additional space for disposal had to be found. In certain areas this was accomplished with the help of the railroads, which argreed to haul much of the slag to outlying areas away from the plants in exchange for the favor of hauling raw materials and finished steel products as revenue-paying freight. However, because of the competitive involvements between freight carriers, this practice was suspected of creating undue advantages for certain carriers, with the result that the Interstate Commerce Commission issued a ruling (ICC Investigation and Suspension Docket No. 516, ICC Reports, Volume 34, 1915) prohibiting the carriers from hauling slag free as a wasting service, if in so doing it served as a means of securing shipments from the steel companies.

EARLY HISTORY OF INDUSTRY

Although there were scattered attempts at this time to process this type slag by individuals working independently of the steel companies, they were only scratching the surface of slag's potential as a commodity, making little or no inroads toward solving the disposal problem. Consequently, with the ICC ruling interfering with the common method of disposal of hauling to outlying areas, the steel companies became seriously concerned for the first time with attempts of their own for converting the slag into a usable or saleable product. Although E. C. Brown[1] predicted, on the basis of his studies, that "Perhaps the most extensive use of commercial hard slag today is in road improvement work," he also suggested other uses where slag could be successfully used, such as railroad ballast, concrete aggregates, filter material, especially in sewage disposal plants, and as roofing materials.

Since slag had been regarded as a waste product, extra effort was needed to convince the construction industry that it could in fact be acceptable and desirable for their purposes. And while sales and promotion of slag remained static, production of pig iron, with its attendant production of slag, increased yearly and slag dumps got bigger and bigger.

At this time two executives of a producer of crushed stone in Ohio decided that blast furnace slag would constitute a good mineral aggregate; since the supply around the steel mills seemed inexhaustible they set up a slag plant near Youngstown, Ohio, and processed and sold slag for this use. Their first efforts were confined to roadway material for macadam base and surfacing, the principal highway construction in the area. Crushing and limited screening of slag were similar to producing crushed stone, and their skill in the latter field led to early success as slag producers. Within a year after the first plant was established they set up operations at other steel plants, and soon several slag plants were under way, with most sales still going for roadway macadam construction. Now that the railroads were hauling much of their product as freight, the railway engineers were soon persuaded to try slag as a railroad ballast; after the success of a few test sections, this use became established (Figure 1).

In spite of strong opposition from competing producers of stone and gravel, sales of slag aggregate increased steadily and others joined the

Figure 1. A well ballasted RR track. Slag aggregate plays an important part in maintaining track alignment and necessary drainage.

ranks of producers. By 1919 there were 38 plants, operated by 14 independent companies, with a daily production capacity of about 25,000 tons.

Although this figure seems impressive, it was not nearly enough to keep up with slag generation, not to mention reducing the accumulated stockpiles. Realization of this fact prompted the producers to form a trade association, the National Slag Association, in 1918. This association established a laboratory, at first on a modest scale; but when the value of research and control became obvious, a modern well-equipped laboratory was built and staffed.

The slag association has conducted research on the properties and uses of slag, has established control procedures, and issued many symposium volumes and special reports (see Bibliography, items 1 and 2).

It was shown (Bibliography, item 3) that the rough vesicular surfaces of slag provide characteristics appropriate to a filter medium, and this use of slag has become established.

Specifications have been developed setting forth the required properties of blast furnace slag in its various uses; many of these specifications have been issued by ASTM.

SLAG DEFINITIONS

In addition to specifications it was found useful also to have a uniform definition for blast furnace slag. The definitions originally proposed by the industry were ultimately adopted as items in ASTM Standards; C-125 as follows:

Blast Furnace Slag.—The nonmetallic product, consisting essentially of silicates and alumino-silicates of lime and other bases, which is developed in a molten condition simultaneously with iron in a blast furnace.

Air-Cooled Blast-Furnace Slag.—The material resulting from solidification of molten blast-furnace slag under atmospheric conditions. Subsequent cooling may be accelerated by application of water to the solidified surface.

Expanded Blast-Furnace Slag—The lightweight cellular material obtained by controlled processing of molten blast-furnace slag with water, or with water and other agents such as steam or compressed air or both.

No mention has been made about the several forms of blast furnace

Table 1 *Types of Slags*

Type	Size	Uses
Air-cooled	$1\frac{1}{2}$–$\frac{3}{8}$ in.	Concrete plain and reinforced
		Bituminous pavements
		Skid proofing
		Macadam surfaces and bases
		Railroad ballast
		Trickling filter media
		Built-up roofing
		Slag wool insulation
Expanded	$\frac{3}{8}$ in. to 4 screen	Concrete, lightweight
		Masonry units, lightweight
		Fireproofing
		Floor fill
Granulated	−4 screen	Pipe backfill
		Concrete floor sills
		Concrete block, lightweight
		Cement manufacture
		Highway base

slag in the solidified form. Slag, molten as it comes from the furnace, is converted into three types of solidified slag products which are air-cooled, expanded, or granulated. The type produced depends on the manner in which the molten slag is treated or cooled when solidification takes place.

The selector guide in Table 1 lists size and uses for the three types.

Table 2 *Average Chemical Make-up*

	Percent
Silica (SiO_2)	33–42
Alumina (Al_2O_3)	10–16
Lime (CaO)	36–45
Magnesia (MgO)	3–12
Sulfur (S)[a]	1–3
Iron oxide (FeO)	3–2
Manganese oxide (MnO)	2–1.5

[a] Sulfur present mainly in the form of calcium sulfide or calcium sulfate.

For a definition and brief description of granulated slag the reader is referred to ASTM C 595-.68 "Blended Hydraulic Cements."

For information regarding characteristics and physical properties of these various types of blast furnace slag see Refs. 2 and 3.

The chemistry of all types of blast furnace slag is quite uniform (Table 2).

PLANT OPERATIONS

While technical data were being developed to prove slag's suitability as a mineral aggregate, work had to be done in the processing plants to properly prepare the product for market. The physical properties, particularily weight, soundness, and toughness, could be controlled in some measure by special handling of molten slag through its early stages of hardening or air-cooling. Pits to permit "curing" of slag before crushing and screening were developed (see Figure 2). Special watering practices of slag in the pits were worked out to produce the best results. Producing expanded slag from the molten materials as it comes from the furnace has been a particularly difficult operation. Here temperature of the slag and its chemical make-up play important roles.

Figure 2. Aerial view of modern slag plant showing pits for curing air cooled slag.

Figure 3. Slag plant illustrating straight flow line production from raw intake to stackers of finished product.

Many patents have been issued on slag and its processing techniques, many concerned with the method for expanding slag. Another problem in production that is primarily associated with slag as contrasted with gravel or crushed stone is the removal of iron. Since most slag, when tapped from the furnace, carries along with it some of the molten iron, special magnets had to be designed to remove the solid iron during processing operations.

This has been accomplished, with the help of the equipment manufacturers, by so-called magnetic pulleys set up in the conveying system for handling both raw and finished material. This operation of iron recovery by the slag processor worked out as an incidental benefit to the steel industry. In addition to handling the disposal problem, they were also recovering heretofore lost iron that had been carried off with the wasted slag. It has been conservatively estimated that the amount of iron recovered annually during slag processing by the industry is equivalent to the production of one additional blast furnace. This service was especially important during World War II, when the iron and steel industry was pressed for ever increasing output of steel.

The crushing and screening operations for slag processing are quite similar to those for crushing and screening in the production of natural aggregates. An up to date crushing and screening slag plant is illustrated in Figure 3. Use of concrete is shown in Figure 4.

Figure 4. Slag aggregate used throughout concrete structure in Buffalo, N.Y.

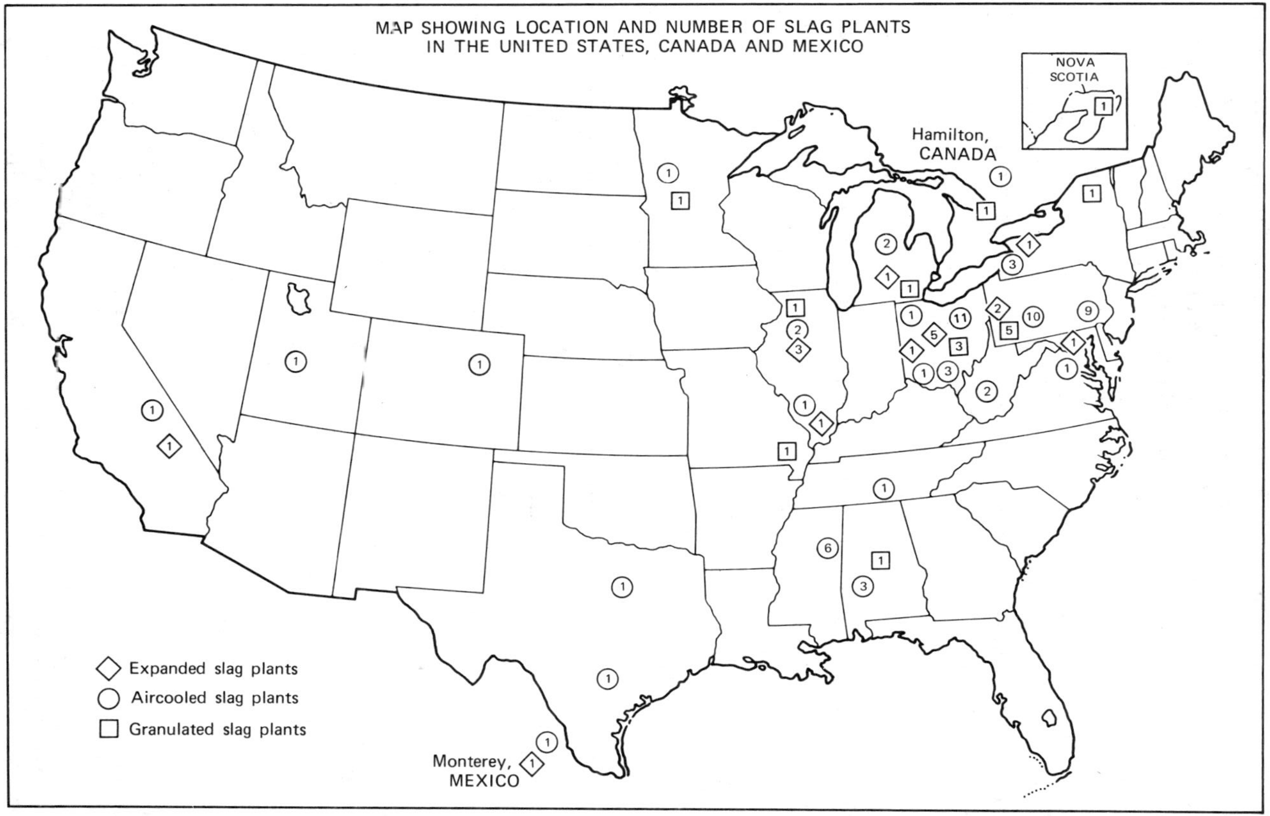

Figure 5. Location and number of slag plants in United States.

As technical data dealing with properties of slag were being developed and as special processing techniques were adopted, more and more slag was being used and its acceptance by the construction industry became a matter of routine. Today slag as a mineral aggregate is listed in all federal, state and local specifications, along with the ASTM Specifications already cited, covering mineral aggregates.

The slag plants situated at or near the steel producing centers of the United States are shown in Figure 5. Statistics on consumption, taken from a recent survey of the slag industry conducted by the National Slag Association, are listed in Tables 3 and 4.

STEEL SLAG

Up to this point, only the utilization of blast furnace slag has been discussed. As previously mentioned, in the refining of iron for the making of steel, another type of slag—differing in chemical makeup

Table 3 *Air-Cooled Iron Blast Furnace Slag Sold or Used by Processors in the United States, by Uses (Thousand Short Tons and Thousand Dollars)*

	Screened		Unscreened	
	Quantity	Value	Quantity	Value
Aggregate in:				
Portland-cement concrete construction				
Structures	1,461	$2,959	—	—
Pavements	933	2,628	—	—
Bituminous construction (all types)	3,462	6,387	—	—
Highway and airport construction	10,013	16,647	275	$337
Manufacture of concrete block	426	739	—	—
Railroad ballast	4,103	5,908	—	—
Mineral wool	389	613	—	—
Roofing slag				
Cover material	425	1,262	—	—
Granules	48	224	—	—
Sewerage trickling filter medium	7	13	—	—
Agricultural slag, liming	2	3	—	—
Other uses	1,057	1,821	777	463
Total	22,326	39,204	1,052	800

Table 4 *Granulated and Expanded Iron Blast Furnace Slag Sold or Used by Processors in the United States, by Uses (Thousand Short Tons and Thousand Dollars)*

	Granulated		Expanded	
	Quantity	Value	Quantity	Value
Highway construction and fill (road, etc.)	1,952	$2,440	—	—
Agricultural slag, liming	69	115	—	—
Manufacture of hydraulic cement	1,210	NA	—	—
Lightweight concrete	—	—	320	$968
Aggregate for concrete block manufacture	367	NA[a]	2,061	6,030
Other uses	261	279	75	264
Total	3,760	2,834[b]	2,456	7,262

[a] NA, not available.

[b] Excludes value for manufacture of hydraulic cement and granulated aggregate for concrete-block manufacture.

and physical properties—is produced in the open-hearth, electric, or BOF furnaces. This is normally referred to as open-hearth or steel slag. Until recent years little attention was devoted to possible commercial utilization of steel slag. This was so primarily because the total annual volume produced in the steel mills was less than half that produced in the iron blast furnaces. Also, much of the production was recycled through the skull-cracking plant for recovery of scrap steel and the crushed slag reused as flux-stone in the blast furnaces. Consequently, for many years, there was not the pressing need for disposal as with the iron blast furnace slag. Now, however, the trend in refining iron is shifting from the open-hearth furnace process to the BOF furnaces where lime is principally used for fluxing, or to electric furnaces where the volume of slag is low. Consequently much of the steel slag currently produced is only cleaned of steel scrap and wasted, so that the problem of disposal becomes important. Because of the success that slag processors had with blast furnace slag, it was natural that the steel industry would turn to them for help in disposing of steel slag as well. Although properties of steel slag are markedly different from those of blast furnace slag, the experiences gained for successfully handling the latter makes the problem much easier for processing steel slag.

Steel slag is much heavier per unit of volume than blast furnace slag and its chemical makeup is different, as reflected in the solidified particles. Because of these differences, steel slag cannot yet be used with as much confidence of performance of mineral aggregates as is the case for iron blast furnace slag. Nonetheless, the slag industry is turning its attention to the problem and some progress has been made toward demonstrating its suitability for use as railroad ballast and as fill material in highway or building construction. Of course, there is still much work to be done—both on research in the laboratory and field investigations involving project applications—before it can be offered for the more sophisticated construction uses in which blast furnace slag and other types of natural aggregate are normally involved. For instance, at present the industry advises it not be used in concrete. One of the large slag processors in the South, where blast furnace slag is in short supply, has been successful in the blending of the two slags to help serve the growing demand for slag aggregates.[4] On the basis of limited experience, there is evidence that the slag might be suitable as an aggregate for use in bituminous mixtures. In all instances it is noted that unusually high stability and excellent skid resistance are provided—both desirable in asphaltic mixtures when used for highway surfacing.

There are marked differences in the chemistry of the slags. One is the presence of some phosphate in steel slag while it is rare in blast furnace slag. The presence of phosphate offers some possibility for its use as fertilizing material. Small tonnages of steel slag, properly prepared, have been used in this country for this purpose during the years. A British firm[73] has developed a process for upgrading basic slag from steel making furnaces to comply with stated minimum requirements of specifications for fertilizer for soluble phosphate content. The process features a high-intensity magnetic separation technique that splits finely powered slag into a phosphate-rich fraction and an iron-rich fraction. In this country the practice of fractionating finely crushed and dried steel slags for reclaiming the steel has been employed, but there has been no work for separating the phosphate concentrations. One of the slag processors in the Detroit area has been successful in developing means for hauling molten steel slag from furnace to plant site in large portable ladles where recovery of steel and processing of slag can be accomplished away from the steel operations.

A recommended specification for steel furnace slags as promulgated by the National Slag Association lists the following under "General Characteristics": "The Aggregate shall consist of hard, strong, durable pieces of slag produced in open hearth, oxygen or electric furnaces as a by-product of steel making and refining processes, or of combinations thereof. It consists essentially of a fused mixture of oxides and silicates." It also carries the precautionary provisions that: "Sources of supply and quality of material shall be approved by the Engineer before delivery is started. Sources consisting of highly contaminated deposits of mill wastes, mixtures of slags of unknown physical and chemical properties, and those containing excessive brickbats and other foreign materials shall not be approved. For material from sources of unknown or doubtful performance, and/or those which are not thoroughly wet by suitable watering systems during production, stockpile aging of processed slag for a period of two months may be required." Also, "NOTE: The quality requirements above are believed adequate for control of materials from established sources of supply. It is recognized, however, that large variations may exist between materials from different sources. Consideration should be given, therefore, to the possible need for additional investigations of new and untried materials from the standpoint of durability and deleterious materials." On the basis of these precautions it is apparent that the industry is proceeding slowly in the promotion of steel slag sales. Nonetheless, for the year 1968, a total of 6,209,000 tons of steel slag were processed and sold. This represents approximately half of the current annual production.

REFERENCES

1. E. C. Brown, Proc. Eng. Soc. W. Pa. **31,** 884–927.
2. C. L. Mantell, *Blast Furnace Slag as Mineral Aggregate in Engineering Materials Handbook,* McGraw-Hill, New York, 1958.
3. G. Josephson, F. Sillers, and D. D. Runner, Iron Blast Furnace Slag, USBM, 1949.
4. Open Hearth Slag, *Rock Products,* **71,** No. 5, May 1968.
5. *Chem. Eng.,* **75,** No. 3, January 1969.
6. C. C. Hommon, Slag in Trickling Filters, *J. ACI,* **2,** No. 2, October 1930; Concrete Aggregates, *J. ACI,* **3,** No. 6, 1931.
7. ASTM C33 Concrete Aggregates, D694 Macadam Base Courses, D1663 Paving Mixtures, C330 Light Weight Aggregates, C 331 Concrete Masonry, D 1863 Mineral Aggregates.

BIBLIOGRAPHY

Slag—General

1. Hardness of Slag as Measured by Mohs Scale, N.S.A. 163–4.
2. "Preparation and Properties of Blast Furnace Slag," N.S.A. 101–18.

3. "Structure and Mineralogy of Blast Furnace Slags," N.S.A. 101-18a.

4. "Slag—The All-Purpose Construction Aggregate," N.S.A. 50th Anniversary Brochure, illustrating the history and uses of slag.

5. E. W. Bauman and D. W. Lewis, "Five-Year Field Corrosion Study of Steel Pipe to Determine Effects of Backfill Materials" N.S.A. 165-1.

6. Albert C. Smith, "How to Work With Slag Aggregates" N.S.A. 156-3.

7. "Blast Furnace Slag as Agricultural Liming Material," N.S.A. 104-42H.

8. G. W. Volk, R. B. Harding, C. E. Evans, "A Comparison of Blast Furnace Slag and Limestone as a Soil Amendment" Ohio Agricultural Experiment Station Research Bulletin 708, N.S.A. 104-42I.

9. "Report of Investigation of Crushed Blast Furnace Slag as a Medium in Sewage Trickling Filters," N.S.A. 64-39.

10. "Slag and Stone, Crushed; Gravel, Crushed and Uncrushed (for Sewage Trickling Filter Media)." Federal Specification SS-S-448, N.S.A. 166-5.

11. George T. Lohmeyer, Supt. Sewage Treatment Plant, Univ. of Florida, Gainesville, Fla. "Trickling Filters and Operation Tips" N.S.A. 155-IA.

12. "Economics in Railway Labor to be Derived from the Use of Various Types of Ballast" Reprinted from Proceedings, A.R.E.A., Chicago, Ill., 1955, N.S.A. 155-6.

13. "Blast Furnace Slag Ideal Backfill Material for Steel Sheet Piling" N.S.A. 166-2.

14. William T. Darden, "Slag Builds Highways in Pennsylvania", Contractors and Engineers, October 1958, N.S.A. 158-6.

15. J. W. Carlson, "Granulated and Air-Cooled Slag for Road Construction," Public Works, December 1958, N.S.A. 159-1.

16. "All Weather Material—Grading an Airfield With Slag," *Roads and Streets,* February 1959, N.S.A. 159-3.

17. "Bethlehem Plant to Use 500,000 Tons of Slag," N.S.A. 168-2.

18. "Granulated Blast Furnace Slag." Tech. Bull., The Standard Slag Company.

19. CSI SPEC-DATA Sheets, covering uses of slag aggregate in concrete, masonry units, bituminous paving, bases and subbases, railroad ballast, and rooting applications.

Slag—Concrete

20. "Selection and Use of Aggregates for Concrete." Report of ACI Committee 621, N.S.A. 162-3.

21. "Materials Required Per Cu. Yd. for Concrete and Cement Mortar," N.S.A. Concrete Proportioning Tables.

22. "Flexure and Compression Tests for High Strength Air-Entrained Slag Concrete," N.S.A. 163-3.

23. "Blast Furnace Slag as Concrete Aggregate," PCA Technical Release No. ST 2, 1937, N.S.A. 94–54.

24. "Slag vs. Stone Concrete Compared in Pushout Tests for Composite Design." N.S.A. 163–1.

25. "Blast Furnace Slag Concrete, NSA Tech. Bull., N.S.A. 168–5.

26. "Heat Transmission of Lightweight Aggregate Concretes," Heat Transmission Coefficients—Expanded Slag included, N.S.A. 105–37D.

27. "Heat Transmission Coefficients for Building Materials," Reprint from A.S.H.V.E. Guide 1951, Air-cooled and Expanded Slag values included, N.S.A. 94–52B.

28. R. H. Bogue, "Blast Furnace Slag in Concrete" Reprinted from *ACI Journal,* October 1955 issue, N.S.A. 155–8.

29. "Lengthened Runway is Prepared for Jets," Slag concrete specified because of its greater resistance to heat transfer, N.S.A. 152–3.

30. "Slag Plays Prominent Role in Construction of Multi-Million Dollar Steel Plant," (N.S.A. 164–2.

31. Wm. R. Quirk, "Slag Aggregate Used in Manufacture of Reinforced Concrete Pipe, 30–120-in. Diameter," November 1953, N.S.A. 154–1.

32. "Slag Block Rated Ne Plus Ultra in Fire Resistance," N.S.A. 165–3.

33. "Slag Concrete Masonry Units,"—N.S.A. Manual (revised edition.)

34. "Blast Furnace Slag Concrete Evaluated by Noted Concrete Consultant," N.S.A. 163–8.

35. "Nailability of Concrete Blocks," N.S.A. 105–44A.

36. "Expanded Slag Concrete—Important Factor in Construction of 22-Story Building," N.S.A. 163–5.

37. "Lightweight Aggregate Concrete for Structural Use" J. J. Shideler, Journal, ACI, Oct., 1957, Reprint. (N.S.A. 105–37G.)

38. D. W. Lewis, "Lightweight Concrete Made With Expanded Blast Furnace Slag" *ACI, Journal,* November 1958, Reprint N.S.A. 105–37F.

39. E. W. Bauman, "Wire Mesh Reinforcement Embedded in Slag Concrete Pavement for 35 years, Entirely Free of Corrosion," N.S.A. 151–1A.

40. D. W. Lewis and Fred Hubbard, "Flexural and Compressive Strength Properties of Air-Entrained Concrete with Air-Cooled Blast-Furnace Slag Aggregate," ASTM Reprint, N.S.A. 156–2A.

41. E. W. Bauman, "No Alkali-Reactivity with Blast Furnace Slag" N.S.A. 159–4.

42. "Lightweight Concrete Frames Terminal." N.S.A. 105-37H.

43. "Housing Authority Finds Concrete Cheaper," Slag specified in Buffalo, New York Housing Project for its fire resistance and good workability, N.S.A. 152–2.

44. "Cost of Sawing Concrete Pavement Joints as Affected by Types of Aggregates," N.S.A. 158–3.

45. "Fire Resistance of Reinforced Concrete Floors," Portland Cement Association, N.S.A. 153–4B.

46. Bryant Mather, "Laboratory Tests of Portland Blast Furnace Slag Cements" Waterways Experiment Station. *ACI Journal,* September 1957, N.S.A. 157-5.

Slag—Bases and Flexible Pavements

47. "Slag for Use in Bitunminous Concrete," NSA Tech. Bull. N.S.A. 167–2.

48. "New Techniques Speed Turnpike Resurfacing" N.S.A. 166–3.

49. "Speedway Faster Than Ever After 60 Years," N.S.A. 168–4.

50. "Old Road Gets New Life With Slag Slurry Seal," N.S.A. 168–7.

51. "Recommended Compositions and Practical Methods for the Design and Control of Hot-Mixed, Hot-Laid Asphatase Concrete Paving Mixtures," N.S.A. Bituminous Concrete Tables.

52. Neal McRae, Street Supt., Birmingham. Ala., "Reduces Street Maintenance Costs and Licks Dust, Mud and Pothole Problem with Paving Most People Can Afford," Reprinted from *American City Magazine,* February 1960, N.S.A. 160–2.

53. "Slag-Macadam Base Topped by Hot Mix," N.S.A. 105–47.

54. "Slag Bituminous Surface Treatments on Road Shoulders," N.S.A. 151–6.

55. "Thin Bituminous Concrete Retreatments Used in Alabama, Florida and Mississippi, N.S.A. 163–7.

56. Charles R. Waters, "Slag Plays Important Role in Highways".

57. "NSA Announces Increased Use of Granulated Slag as Base/Fill Material" N.S.A. 164–3.

58. C. W. Allen, "Granulated Slag in Highway Construction" and F. J. Kimble, Ohio State Highway Department.

59. Charles R. Waters, District Engineer, New York State Highway Department, "Heavy Base Lifts for Resurfacing" N.S.A. 153–1.

60. "Slag Rumbler Strip Warning at Stop Signs," N.S.A. 158–1A.

61. William L. Chilcote, "Resurfacing Roadways with Bituminous Slag," N.S.A. 163–2.

62. W. S. G. Britton, Maintenance Engineer, Virginia Dept. of Highways, "Seal Treatments on Bituminous Concrete Pavements" N.S.A. 158–2.

63. "Building Safety Into Highways by Constructing Rumbler Surface With Slag Aggregate," N.S.A. 158–1.

64. "Building Safer Highways with Slag," ARBA Tech. Bull. No. 264, 1968, N.S.A. 167-3.

65. "Slag, Stone, and Gravel, Crushed (for Bituminous Paving Mixtures)," Federal Specification SS-S-445, N.S.A. 166-4.

66. W. J. Turnbull, "Untreated vs. Stabilized Granular Bases for Flexible Pavements," N.S.A. 168-6.

APPENDIX I

abbreviations

AGENCIES AND ASSOCIATIONS

AAAS	American Association for the Advancement of Science
AAR	Association of American Railroads
AASHO	American Association of State Highway Officials
ACI	American Concrete Institute
ACS	American Chemical Society
AEC	Atomic Energy Commision
AGA	American Gas Association
AIChE	American Institute of Chemical Engineers
AIEE	American Institute of Electrical Engineers
AIME	American Institute of Mining, Metallurgical Engineers, New York, now American Institute of Mining, Metallurgical and Petroleum Engineers
AISI	American Iron and Steel Institute
Alcoa	Aluminum Company of America, Pittsburgh
APCA	Air Pollution Control Association
APHA	American Public Health Association
API	American Petroleum Institute
APWA	American Public Works Association
ASA	American Standards Association
ASAE	American Society of Agricultural Engineers
ASCE	American Society of Civil Engineers
ASHAE	American Society of Heating and Air Conditioning Engineers, formerly American Society of Heating and Ventilating Engineers
ASHVE	American Society of Heating and Ventilating Engineers

ASME	American Society of Mechanical Engineers
AWWA	American Water Works Association
BCF	Bureau of Commercial Fisheries, USDI
BDPC	Bureau of Disease Prevention and Control
BDPEC	Bureau of Disease Prevention and Environmental Control
BSWM	Bureau of Solid Waste Management
CPEHS	Consumer Protection and Environmental Health Service
ECA	Environmental Control Administration
EEI	Edison Electric Institute
EKCo.	Eastman Kodak Co., Rochester, N.Y.
EPA	Environmental Protection Agency
FPL	Forest Products Laboratory
FTC	Federal Trade Commission
FWPCA	Federal Water Pollution Control Agency
FWS	Fish and Wildlife Service
FWQA	Federal Water Quality Agency
GSA	General Service Agencies
HEW	U.S. Department of Health, Education and Welfare
IACS	International Annealed Copper Standard
ICC	Interstate Commerce Commission
I & EC	Industrial and Engineering Chemistry, ACS, Washington, D.C.
Inco	International Nickel Company, New York
ISO	International Standards Organization
JAN	Joint Army-Navy Specifications
NAPCA	National Air Pollution Control Administration
NASMI	National Association of Secondary Metals Institute
NBS	National Bureau of Standards
NCUIH	National Center for Urban and Industrial Health
NCA	National Canners Association
NLMA	National Lumber Manufacturers Association

NSA	National Slag Association
ORNL	Oak Ridge National Laboratory
OSWMP	Office of Solid Waste Management Programs
PCA	Portland Cement Association, Chicago
PHS	Public Health Service
SAE	Society of Automotive Engineers
SO	Standard Oil NJ, now Exxon Corp.
TAPPI	Technical Association of the Pulp and Paper Industry
TVA	Tennessee Valley Authority
UCC	Union Carbide Corporation
USAF	United States Air Force
USBM	United States Bureau of Mines, USDI
USBB	United States Bureau of the Budget
USBR	United States Bureau of Reclamation, USDI
USDA	United States Department of Agriculture
USDA Tech Bul	United States Department of Agriculture Technical Bulletin
USDC	United States Department of Commerce
USDD	United States Department of Defense
USDI	United States Department of the Interior
USGPO	United States Government Printing Office
USGS	United States Geological Survey
USHEW	United States Department of Health, Education and Welfare
USN	United States Navy
USP	United States Patent
USPHS	United States Public Health Service
USSG	United States Standard Gage
VAM	N. N Vailafvoer Maatschappi
WHO	World Health Organization

APPENDIX II

abbreviations

UNITS OF MEASUREMENTS

A	ampere
A/ft^2	ampere per square foot
A/in^2	ampere per square inch
a%	Atomic percent
a	coefficient of linear expansion; surface energy per unit area of material, synthesis to sludge ratio.
abs	absolute
AC	alternating current
a-c	alternating current (as adjective)
actual ft^3/sec	actual cubic feet per second
alk	alkalinity
Ahr	ampere-hour
amu	atomic mass unit
Å	Angstrom unit
approx	approximate
Ar	Argon
atm	atmosphere
at wt	atomic weight
avg	average
B	(Boron; brittle strength; magnetic induction
β	coefficient of expansion
b	horizontal dimension
bal	balance
bd	board
BHN	Brinell hardness number
bf	board foot
bhp	brake horsepower
bhp-hr	brake horsepower-hour
BOD, BOD_5	5 day, 20°C biochemical oxygen demand in mg/liter
BOF	basic oxygen fumace
BOP	basic oxygen process
bp	boiling point
Btu	British thermal unit
Btu/lb	British thermal units per pound, calorific value of a gas
Btu/ft^3	British thermal unit per cubic foot, calorific value of a gas

C	Carbon; capacitance or electric capacity; conductance; corrosion allowance in inches; degree Centigrade; hundred; sludge growth constant
C	heat capacity of fluid; specific heat
C_O	initial dissolved oxygen
C_5	dissolved oxygen saturation
cal	calorie
cm^3	cubic centimeter (liquid, meaning milliliter, ml)
COD_m	National Canners Association; modified COD, mg/liter
coef	coefficient
conc	concentrate; concentration
cond	conductivity
conf	conference
C_p, C_p	specific heat at constant pressure
cp	chemically pure
cps	cycles per second
C_v	specific heat at constant volume
cwt	hundredweight (112 lb)
D	density
D	diffusion coefficient; dissipation factor; tube diameter
D_A	density of ingredients, D_{20} density at 20°C
D_B	density of ingredients
D_C	density of ingredients
D_t	particle diameter
d-c	direct-current
deg or °	degree
diam	diameter
DO	dissolved oxygen, mg/liter
ds	current element of length
dynes/cm	dynes per centimeter
d_1	nominal outside diameter, inches
d_2	nominal inside diameter, inches
E	efficiency; electric intensity or electric field strength; voltage or potential; modulus of elasticity, or Young's modulus; excellent
E	logarithmic-loss factor
E	specific inductance capacity
E_c	compression modulus of elasticity
ed.	edition
E_c	elongation, modulus of elasticity
eff	efficiency

eng	engineering
CE	Civil engineer
cg	centrigram
CI	cast iron
cir	circular; circumference
circ	circumference
cir mil	circular mils
cks	centistrokes
cl	centiliter
cm	centimeter
cmf	circular mil-foot
cm/sec	centimeter per second, measurement of velocity
COD	chemical oxygen demand, mg/liter
COD_s	soluble chemical oxygen demand, mg/liter
el	elevation
elec	electric
elong	elongation
emc	equilibrium moisture
emf	electromotive force
E_n	energy
esu	electrostatic unit
e_t	thermal or total emissivity
Eu	European
eV	electron volts
$e\lambda$	spectral emissivity
ext	extension; external
f	frequency in cycles per second; farad
fbm	feet board measure (board feet)
FC	furnace cooled
Fed	Federal
fl	fluid
fnp	fusion point
fp	freezing point
FS	Federal Specification
FSS	fixed suspended solids
ft	feet or foot
ft^3	cubic feet
ft/min	feet per minute
ft^3/min	cubic feet per minute
ft/sec	feet per second
ft-lb	foot-pound
FW	face width
G	conductance (G), the reciprocal of resistance; good; gravity; shear modulus; specific gravity of fluid; net sludge growth rate
g	gram
gal	gallon
g-cal	gram-calorie
gal/ton	gallons per ton
GP	Georgia Pacific
gal/day	gallons per day
gal/min	gallons per minute
gal/sec	gallons per second
H	(Hydrogen) differential; horizontal shear; magnetic field strength; magnetizing force; viscosity of oil at 100°F or 100 VI

h	heat transfer coefficient; film coefficient of convection, in Btu/hr/sq ft/°F; henry, (electrical)
h	original height in inches
$h_c + h_r$	combined coefficient of convection and radiation
h^2	diffusivity
hex	hexagonal
h-p	high-pressure (adjective)
hp	horsepower
hp-hr	horsepower-hour
hr	hour
I	(Iodine) total moments of inertia
I or *i*	measure of rate of flow or change of charge
ID	inside diameter
ihp	indicated horsepower
ihp-hr	indicated horsepower-hour
in.	inch
in.^{3}	cubic inch
in.-lb	inch-pound
incl	inclusive; including
insol	insoluble
int	internal
i-p	intermediate-pressure (adjective)
IPS	iron pipe size
ipy	inch penetration per year
j	joule
J	Journal
K	(Kalium) potassium; factor; bulk modulus; karat; degree Kelvin; endogenous respiration rate: BOD velocity constant; also hydraulic gradient (common log) cm/sec
κ	proportionality factor, thermal conductivity; also dielectric constant
kc	kilocycles per second
kcal	kilocalorie
kg	kilogram
kg-cal	kilogram-calorie
kg-m	kilogram-meter
kg/m^3	kilograms per cubic meter
kg/sec	kilograms per second
kl	kiloliter
km	kilometer
km/sec	kilometers per second
Kss	coefficient of thermal expansion
kV	kilovolt
kVA	kilovolt-ampere
kW	kilowatt
kwh or kwhr	kilowatt-hour
L	length of cylinder; henry; inductance; viscosity of of oil

	at 100° of zero vicosity; lambert
LP	Liquid petroleum (gas)
l	distance; length; thickness
l and A	Represent length and area, respectively, in consistent units by which resistivity is specified
L	length of cylinder; henry, inductance; viscosity of oil at 100° of zero viscosity index; lambert
LP	Liquid petroleum (gas)
lb	pound
lb/yr	pounds per year
lb/A N	pounds per annum of nitrogen
lb/cap	pounds per capita
lb/cap/day	pounds per capita per day
lb-ft	pound-foot
lb-in.	pound-inch
lb/lb	pounds per pound
lb/cap/day	pounds per person per day
lb bhp-hr	pounds per brake horsepower-hour
lb/ft^3	pounds per cubic foot
lm	logarithmic mean area
l-p	low-pressure (as adjective)
M	modulus of rupture
m	mass; meter(s); thousand
m^3	cubic meter
mA	milliampere
max	maximum
$1000\ ft^3$	thousand cubic feet
mdd	milligrams per square decimeter per day (corrosion rate)
MeV	million electron volts
mg	milligram
mg/kg	milligrams per kilogram
mg/liter	milligram per liter
mh	millihenry
mho	reciprocal ohms
mil	thousanths of an inch
min	minimum; minute
m-kg	meter-kilogram
ml	milliliter
MLSS	suspended solids in aeration unit
MLVSS	volatile suspended solids in aeration unit
mm	millimeter
MW	molecular weight
mp	melting point
mph	miles per hour
mphps	miles per hour per second
mpsi	thousand pounds per sq in.

mscf	thousand standard cubic feet
mt	metric ton, 2250 lb
μ	permeability
μa	microampere
μf	microfarad
μ in.	microinch
μv	microvolt
μw or mu w	microwatt
mV	millivolt
Mw	megawatt or thousand kW
N	(Nitrogen) neutron; appropriate exponent; refractive index; specific gravity
NA-P	nonabrasive-polishing
nat	natural
N:COD	nitrogen to COD ratio
Nd	index of refraction
NEC	National Electric Code
NEMA	National Electrical Manufacturers' Association
NFPA	National Fire Protection Association
ng/m³	nanograms per cubic meter; pptr; parts per trillion
NI	magnetomotive force
No.	number
Nom	nominal
NPK	Fertilizer designation, % nitrogen, % phosphorus, % potassium
NR	not recommended
NRC	noise reduction coefficient
OD	outside diameter
OFHC	Oxygen free high conductivity
ohm-cm	ohm-centimeter
OQ	oil quench
OR	stress
oz	ounce
oz-ft	ounce-foot
oz-in.	ounce-inch
oz/ton	ounce per ton
P	(Phosphorus) breaking load in pounds (rupture strength); poor; pounds per square foot (wind velocity-pressure calculations); power or power factor; internal working pressure
PD	permissible dosage
PE	Population equivalent
%	percent
PFA	puzzolan fly ash
PFC	plow furrow method of waste disposal
pg	gas density
pH	negative log of the hydrogen ion concentration, measure of acidity (1 to 7) or alkalinity (7 to 14)

p,P	particle density
ppb	parts per billion
ppm	parts per million
psi	pounds per square inch
psia	pounds per square inch absolute
psig	pounds per square inch gage
Q	constant; delivery rate in cubic centimeters per hour; quantum of energy; quantity of heat; measure of quantity of electricity; flow rate
R	BOD or COD removal rate, mg BOD or mg COD per mg MLVSS per day; recommended; degree Rankine; thermal resistance; resistance in ohms; resistance electrical; gas constant in pressure and volume relationships
r	roentgen
R	radius; volume resistivity
rbe	relative biological effectiveness
red	reduction
rem	remainder; roentgen equivalent man
rep	roentgen

	equivalent physical
res	resistance or resistivity
RH	relative humidity
RH M	rum (radiation intensity of 1 r per hr at a distance of 1 m from the absorber)
rms	root mean square
rpm	revolutions per minute
rps	revolutions per second
RS	rapid setting
RT	road tar
S	flexure
satd	saturated
SCS	Soil Conservation Service, USDA
scf	standard cubic foot
scf/min	standard cubic foot per minute
scf/min/MW	standard cubic feet per minute per megawatt
SDP	slowing-down power
sec	second
seq	sequence
ser	serial
Set S	setteable solids
SIC	Standard industrial classification
sin	sine
Soften point	softening point
soln	solution

spec	specification
sp gr	specific gravity
sp ht	specific heat
sq	square
SS	suspended solids
SSI	sub soil injector, method of waste disposal
std	standard
STP	Standard temperature and pressure, 32°F or 0°C at 1 atm.
SW	solid waste
SWP	Solid Wastes Program
T	absolute temperature; temperature; torque in pound-inches
t	time
tons/A	tons per annum
tons/cap/day	tons per capita per day
tons/cap/yr	tons per capita per year
tons/day	tons per day
tons/wk	tons per week
tech	technical, technology
temp	temperature
TSF	total fixed solids; tin-free steel
therm	heat content of 1000 ft^3 of 1000 Btu/ft, or 1,000, 000 Btu
t_i	thickness
TKN	total Kjeldahl nitrogen mg/liter
TLV	Threshold limit value
TOC	total organic carbon mg/liter
Tor	unit of low pressure, mm of Hg in height
TP	total phosphorus
T PO_4	total phosphate, mg/liter as P
TS	total solids
TVS	total volatile solids
U	thermal transmittance in Btu per hour per square foot per degree F difference in temperature between the warm and cold sides, (Btu/hr/sq ft/°F); over-all coefficient of heat transfer; viscosity if oil at 100°F
U	average velocity of fluid
g/liter	micrograms per liter, equivalent to ppb
UHF	ultra-high frequency
UTS	ultimate tensile strength
V	velocity in cm/sec; volume; wind velocity in miles per hour (pressure calculation)
V	volts
VA	volt-ampere
VHN	Vickers hardness number

VI	viscosity index, the relative degree of change in the viscosity of oils with respect to temperature changes, using an arbitrary scale	Whr	watthour
		W_s	weight of solvent
		wt	weight
		wt%	weight percent
		yd	yard
		yd^3	cubic yard
		yr	year
vol	volume	Σ(sigma)	summation
VSS	volatile suspended solids, mg/liter	ν(nu)	Kinematic viscosity of a gas
W	particle size; weight	μm	micron
W	watt	Ω,λ (lambda)	ohm
wc	water column (pressure)		

INDEX